New Universe Order:

How the Universe Works with ESS, Macht and Bauen Matter

By Clinton James Terry

Copyright

New Universe Order:
How the Universe Works with ESS, Macht and Bauen Matter

© 2017 By Clinton James Terry. All rights reserved.
Löwenmensch

This book is dedicated to my family, specifically my parents,
sister and grandparents for also being interested
in the world and universe around us.
It is also dedicated to all intelligent life that has ever looked up and
wondered how the universe works and also
to those few in the universe that have solved it.

Contents

Part 2 - New Universe Order

Forward or Why I Wrote This Book and What it is About

You are going to learn how the universe works. That is the quick answer as this book contains the answer to the simple question of how the universe works and it is divided into two parts and aimed at two audiences so is accessible by all. The first part is a necessary history of what humans have done to answer this question from the beginning up until today and why we thought we were right and what we have learned about matter and space along the way. The second part of the book takes off from where modern science is currently stuck with multiple ad hoc theories and nonsense about physics breaking down. The work contained therein is my own and is a simple model explanation of how the universe works without any of the mathematics that is unreadable to most. I will use a physical description of the universe that explains step by step how everything we see or observe around us is built from the smallest bits of space and matter to the largest structures of the universe not just planets and stars but the whole universe itself.

If you are a lay person or someone with nothing more than a healthy curiosity about how the universe works then read on as you will learn everything about the exact progress of human understanding from the concept of elements such as water, fire, earth and air to early Newtonian physics and right on through to string theory and the like. You will then be guided smoothly past the sticking points of all modern theories including quantum mechanics, general relativity and anything new like many worlds, strings and the dark theories. You have nothing to lose and only an understanding of the universe to gain which brings me to the people directly involved in theoretical research. For those of you reading this who are actively working in and intensely familiar with current theories and yet frustrated with their failures then read on as you also have nothing to lose and everything to gain in terms of finally seeing how the universe works. You will be shown the only way the universe is put together that reduces all variables to the lowest number possible and can be used to solve all, not some but all unsolved problems currently facing modern science.

So there you have it this book contains the history of how humans have learned about the two components of the universe which are the space and the matter within it. The path of the history is focused on understanding the universe and nothing more and this leads to the real answer of how the universe works which is my space physics. At the end of the book the journey of human understanding is over as the construction of the universe and all observable phenomena within it as well as all unsolved problems is laid bare. I will show you how this is done using the absolute smallest number of variables that can be tested for in normal laboratories of three dimensions, matter and energy.

A final note to all who do work in cosmology, theoretical physics or anything else aimed at understanding how the universe works, please I ask you to be patient and follow my work and my explanation step by step. You may have to let go of long held beliefs or simply accept what I tell you at face value until I have walked you through the whole universe on paper. I will be using a gentle approach to my explanations such that it is as if I am simply in a room speaking to you and I will avoid the heavily sectioned and rigid structure of a textbook. You will be guided hither and yon through the explanations of the universes most secret machinations and as such stay with me as I cover topics in a general grouping rather than every problem laid out perfectly bullet point by bullet point. This allows me to leisurely bring all of you up to speed with the history of science and of my work.

For those seeking a perfect explanation of the universe you will get it but just remember that in order to cover a topic as big as the entire universe some of the exact details will come naturally last. Also I would like you to keep in mind that I will be introducing you to what may be thought of as new physics from my work and I will need to refer to an entire concept sometimes by a simple name. This means that future uses of these names as needed for descriptions of seemingly unsolved problems and the answers to them cannot be taken lightly and all the rules I outline for a certain description must be thought about carefully and fully when being applied to a specific topic. This is the reason I chose to use a gentle approach to the explanations of my work and a general non-textbook flow to the entire book.

If you can do this and remember each lesson I teach you as I build the universe from a broad explanation of space and matter to the single interactions that underlie anything from black holes, quantum entanglement to the end of the universe itself you will see crystal clear how literally every problem just vanishes before your mind's eye. I know that for those whose minds are open and are not clinging desperately to a world that still believes modern day impossibilities akin to a flat Earth at the center of existence then you will see how we can move forward into a future where the Earth is round and orbits the Sun. This is the answer that science has claimed it has been looking for and wanting to find forever and most especially the last century. It is not time for niceties or holding back but a direct attack on modern theories as science has simply been stuck for far too long and now is the time to see a new universe order.

Too long have people whispered quietly amongst trusted colleagues for fear of being ostracized for their beliefs when perhaps they have known that the current ad hoc theories are doomed. I have no reputation in science, I do not work for an elite institute and I have no fears so I will be the one to say what modern science has is wrong and I what I have is right and it is time for people to look at it. If you are one of the open minded whisperers then throw me to the wolves I will defeat them for you and never stop whispering as those whispers are louder than any ad hoc theory attempting to hold onto old beliefs. Whispers lead to round Earth models orbiting the Sun and today's whispers can lead to a true understanding of the universe through my work.

The ad hoc theories have had over a century and countless people trying to make them work all with no success so to those that love the ad hoc theories the floor is yours go ahead and find the answer. However the rules of the game are you must find an answer that solves all unsolved problems in one fell swoop, one that creates no extra variables that are not even falsifiable and that explains everything currently observed as well as all misunderstood phenomena that create the universe exactly as we see it today. In short you must meet me in scientific combat as you must come up with a model that uses fewer variables than mine, exists within the normal type of 3+1 universe that science is built upon and this single model must cross all boundaries to solve all of the unsolved problems faced by modern physics at once.

Modern science wanted a theory that provided a paradigm shift as overused as that phrase is, it wanted a theory that reduced variables, it wanted a theory that tied gravity and all the four known fundamental forces together, it wanted a theory that did not fail at answering all unsolved problems and it wanted a theory that did all of this without creating new particles, dimensions or other exotics and in a beautiful and simple way. This is it, this is what science has been waiting for. See it, learn it and be one of the first to finally understand how the universe works because this is it.

Introduction

We are smart, we are human. I'm not saying we are the only clever life on Earth but after a while it is always us studying creatures like dolphins and making documentaries about them and how smart they seem to be and broadcasting those around the globe and not the other way around. In other words our little cranial spheres evolved to the point where we study other forms of life and then launch satellites on top of rockets into space so we can use another invention of ours called television to sit back and self deprecate ourselves about how some species or other used a stick to get bugs? We can figure out anything if left alone long enough, so why cannot we answer the oldest and simplest of problems? How the universe works. That is it, nothing fancy to it really and yet that question has been asked for at least 5000 years and probably tens of thousands of years longer since we became sentient. Proof of this can be found on cave walls and in archaeological findings world wide. It may be a scene painted on a rock or a figurine depicting a god or gods that we have used to explain what and how the universe worked as it appeared to us at the time. No matter what it was we have asked this question far longer than 5000 years and still it remains.

The story of humanities interest in the universe is certainly not a new one but one that stretches back thousands of years and as if to echo the future of scientific discovery with institutions such as the Max Planck and the HADRON it begins in Europe. All of the earliest understandings of the universe whether they are constructions or inventions are found in Europe and are so old that many were not found or understood until the twenty first century. The earliest ever recorded evidence that humans were interested in the stars was found in what is today Scotland and dates back to 8000 years BC, it was a calendar. Not a calendar printed on paper or digitally displayed by a computer, instead it was a circular arrangement of stones forming a Mesolithic monument. Located in the Dee River valley of Scotland's Aberdeen shire and understood to be a calendar predicting the motions of celestial objects. Predating all other calendars by 5000 years it showed the summer and winter solstices, the phases of the moon and was constructed out of a series of twelve pits. One of its other amazing features was that the whole site was reshaped hundreds of times over the ages in order for the calendar to stay accurate.

The significance of this is that ancient Europeans were obviously aware that the stars and therefore the universe itself, is a changing system which they needed to correct for. The idea today that such ancient people understood the workings of a changing universe and could correct their calculations for it is simply incredible considering Einstein in the twentieth century believed the universe to be static and immovable. Similar circles such as the Goseck Circle in Germany discovered in 1991 were found and proved that this desire to understand the universe was spread across Europe. The meaning of the circle was not determined until further archaeological specimens had been uncovered. Not only Scotland but Germany, Austria and of course the infamous Stonehenge contained these circles all discovered and interpreted as calendars dating back to 5000 BC or there about.

From large Mesolithic structures humans next sought to refine their inventions and miniaturize them much as we do today turning vacuum tube powered building sized computers into smaller microprocessor powered desk top devices. Found close to the Goseck circle the Nebra Sky Disc is a truly amazing piece of ancient engineering measuring 30cm in diameter, weighing 2.2 kg and being made of blue-green patina bronze metal inlaid with gold. It was used at least as early as 2000 BC and shows a crescent moon at approximately four or five days old, the Pleiades star cluster and contains a 354 day twelve month calendar. What is fascinating is that the calendar contains the necessary markers that tell the user when to insert a leap month every two to three years in order to keep it accurate with the seasons. This mathematical feat was not seen anywhere else in the world until Babylonian calendars were found around 500 BC putting the Nebra Sky disc a full 1500 years ahead of anywhere else in the world. So we have evidence surpassing our traditional understanding of human history that our species was not only fascinated by but driven to understand the universe as a whole. We were making measurements in an attempt not only to use things like the seasons to our agricultural advantage but in an effort to answer the question of why the universe is as it is and what makes it work. More progress would have to be made though to find our answers.

We sure have tried and to some extent we have been on the right line of thinking; the universe is big and filled with stuff. However when you take a closer look at this seemingly simple answer you find that it is anything but simple, for example what constitutes stuff? The quest for stuff has been carried out since long, long before recorded history began and most certainly dates back to when our ancestors used cave paintings as a way of writing and symbolizing their understanding of the world. True there is a limit to what we can achieve by depicting the world in various dyes and charcoals on stone but it demonstrates our species great need to understand the cosmos. Curiosity never killed nor will it ever kill any cats, instead curiosity will help them to achieve a full understanding of the universe and to travel to all of its farthest boundaries.

To me I revel in the fact that humans also share this lust for curiosity with our beloved feline friends as it is the single unstoppable driving mechanism behind everything we have ever done to understand our universe. One of our proudest traits and one we should all embrace daily since the day we are born; curiosity put our cave dwelling ancestors onto a path that would carry on till this day. In this pursuit of a complete understanding of the cosmos and an answer to the biggest question of all time, how does the universe work, we as a species have made great progress to date. However we have not fully answered that biggest and most important of all questions yet so what is it that we our missing in our understanding of the universe? Considering the universe is big and filled with stuff how far have we come in our understanding of stuff?

Chapter 1
The Stuff or Beginning to Understand Matter

Elements to Gods

Let us start where many people start with the ancients Greeks who produced many philosophers that tried their best to logically understand the world around them but of course failed. Many ancients such as the Greek philosopher Aristotle tried to rationalize the universe in ways that involved a naturalistic view of existence and then prove that existence through reasoning. Aristotle believed for example that everything arose from water. A simple view and certainly incorrect but a window into the thought of early science that demonstrates a desire to understand the universe beyond the tired old belief that a god or gods simply made everything the way it was, because…well no one really explains why they desired to build the Earth as they did apparently being a god gets boring and so you make people to grovel at your feet. That missing detail of why the universe is as it is forms the whole basis for wanting to understand the universe in scientific terms. After all a scientific explanation of the universe does not disturb the ancient belief in the gods creating the universe it merely allows for an understanding of why the universe is as it is.

So the stuff in the universe was made up of water, or the elements or any other number of strange and impishly elusive materials that the ancients could create to explain things. This creates a problem however which is this: not everything in the universe is stuff. Some of the universe is the absence of stuff. The ancients had some problems with this because to them the logic flowed as follows: stuff is stuff, the opposite of stuff is not stuff and therefore nothing and nothing is not something and therefore not part of the universe. Only things they could touch or interact with in some way with their five senses was part of the universe, nothingness could not be perceived by them and so was stricken from thought. Leucippus was one such ancient to propose the idea of something other than stuff, he proposed a vacuum. The vacuum he describes was still not a true lack of stuff but rather a strange indefinite element in which the other elements were embedded. While this primitive definition of a vacuum is rooted in the philosopher Anaximander's ideal of the substratum or *arche*, it at least allows for the universe to be constructed of a second ingredient. It is true that the concept of something which was not truly an element to the Greeks was not suggested by anything they could measure and rather by a logical deduction of how not everything could be reduced to an element. The idea that a second ingredient to the universe was possible did however put them on the right track to understanding the true nature of the universe as a whole.

So this is one example where we see humans being on the right line of thinking and where we can examine a universe as we understand it today. We understand a universe with space and matter essentially not really much different from the ancients. No new forms of anything have been proposed; no new stuff and no new absence of stuff. In fact we have gone so far as to say that our matter is more solid than the ancient's elements and that our absence of stuff is more devoid of everything that makes up the universe than theirs. We understand in modern physics really only matter and gloss over space almost entirely, to many people space is simply the place where matter lives.

That being said it is extremely important to look at what matter actually is in the universe and how it orders itself. Matter is mostly solid stuff and stuff is what we need to put into our model of the universe, so far so good as this would certainly please the ancients. What the ancients could not see was that matter is not just large scale things but rather small scale as well, in fact the smallest scales we know so far…so far. Leucippus and Democritus first proposing the idea of an atom, which essentially means indivisible, is actually rather useless in understanding the small scale of matter. The reason it has little use is that their idea of an atom was still just a large scale object. They thought that if you split a boulder you would get a rock, a rock to a pebble, a pebble to sand and sand to dust until you could not split it any further. The last thing you could not split was to them an atom and therefore the smallest object in the universe. The whole problem with this is that their atom never stopped being a smaller form of a boulder. No matter how small a piece of boulder you have it is still a boulder and not a true atom or sub atomic particle.

We now know thanks to numerous experiments using particle colliders that the smallest forms of stuff from say a boulder, is actually made of tiny particles. These particles we have managed to rough up quite a bit and coerced them into revealing smaller particles still. Not content to leave those teeny particles alone we have repeated the process both in the lab and on paper until we are at the limits of smallness with respect to matter. Fantastic, we know that the universe even with its infinite size is made up of almost infinitely small particles of matter and this would not be a problem if matter stayed small. Matter hates to stay small though and anywhere you look in the universe you find matter hanging out with its sub atomic friends making what we observe as humans as large scale objects. Those objects could be as small as a mote of dust but that tiny size is monstrously robust for even the weightiest of sub atomic particles and so confusion reigns in terms of matters behavior. Is matter the small quantumly elusive wave/particle we observe in super colliders or is it the solid unblinking large scale objects we encounter empirically everyday? Making a quantum particle behave almost wisp like so you can pass an object through or between it is easy but try that same almost magical feat with a brick and your head and your on your own. Although both are the same fundamental material they behave in strangely different ways and to some degree modern physics is drawing lines in the cosmic sand and picking sides. Do we fit our models to the large scale universe filled with stars, black holes and covering billions of light years? Maybe we should choose the smallest scales on the quantum level and hold fast to the experiments that we can physically control and measure in labs?

This is the reality of the universe that small creates large and a set of rules and laws must be established to govern both. Some people have argued in the scope of all things that we do not need a set of laws for both and this is of course folly because regardless of whether one theory describes the large well and one the very small with accuracy is wholly irrelevant as they both exist. Large and small things do inhabit the universe together and do interact with each other and with space constantly to form that universe and after at least fourteen billion years do not show any signs of wanting to go away and clear things up for us.

What is needed is a way to incorporate both as one in a seamless manner and with fewer rather than more variables. This problem is what remains unsolved in science and this problem is in fact the question at hand; how does the universe work? To resolve this problem is to answer the question of how the universe works because merging the small forms of matter with space so that these form a complete universe will explain why we observe large scale matter behaving as it does. Once complete all observations we as humans can make whether at the quantum level or at the scale of the universe itself, will be answerable because the method of interaction of everything that makes up the universe is smooth and scalable. In other words the same rules and mathematics that govern matter and space for subatomic particles can be used to explain the largest and most influential objects in the universe like black holes, galaxies and the big bang itself.

One thing nature has shown us is that the simplest answer is always the right answer. Every complicated idea humans have thought about and put forth with at the time the most honest and sincere intentions have inevitably proven to be incorrect. We at one time thought the universe was built upon many different gods, one for the sky, land, sun, moon and so on until we had filled the empirically observable world with a pantheon of explanations. These gods served the purpose of explaining how the universe was created, why it functions as it does and what is responsible for all of the phenomenon we observe everyday as humans. Upon closer examination of the world around us we realized that it did not make sense for gods to actually be the various events or objects, such as the sun, that we observe. What was left for us to do but rationalize the universe into smaller simpler categories whilst leaving aside the debatable question of whom or what created the universe. Leaving out the divine we could concentrate on the mundane like a rock and why a rock is so rockish.

This is where in human history we first simplified the world and the universe to reduce the number of variables responsible for making things that we encounter daily to a smaller more manageable number. The numerous gods necessary to create tides, hurricanes and water itself for example could be reduced at a fundamental level to elements. Elements were what everything was made of according to many ancients and while the gods may have been responsible for how they behaved it was the elements themselves that made up the universe. Water was water and air was air even if a god was angry and turned them into a terrible storm with driving rains that wrought our lands with ruin and brought misery and woe to those who survived the tools of the god's destruction were still basic elements. After all a rock is a rock.

However are all rocks their own unique element or did they all fall under the jurisdiction and guiding forces of just one? To the ancients the answer was a step forward because it was simpler; all rocks belong to the same fundamental element which was the Earth element. This elemental view of the universe rocked (see what I did there?) as it simplified the model needed to understand our world by not over complicating the number of ingredients needed to explain the universe itself. Sadly for the ancients however the idea of elements was wrong as it was still too complicated to explain the whole universe. Disregarding the fact that it was impossible for ancient peoples to truly study the cosmos beyond their own eyes and the use of annual calendars it would have been a massively complicated system of elements needed to define everything we now know that makes up the universe. Take a moment to think about the infinite reaches of space and literally everything in it and pair that up with a unique element for each. Perhaps and of course just to list a few we would need one for space, one for cosmic rays, one for gravity and of course one for Uranus. Absolutely an acceptable joke right there. What you are left with is a list of elements too long and complicated to provide a clean model of the universe and certainly complicated enough to begin tripping over themselves in attempting to do so. Once again a new system simpler in everyway must be devised.

Today we have done away with the notion of gods, elements and the like and settled upon a much simpler view of the universe, one that involves only the two ingredients known as space and matter. It is the various permutations of matter and space that produce the universe we see all around us empirically just as the ancients did but unlike them we can study matter beyond our senses and learn much about its true nature. This unseen world of matter is how we know our simpler view of things like rocks to be of a superior fit when compared to the ancients. Studying the matter of rocks and water allow us to conclude that at the sub atomic level the two seemingly different materials are in fact constructed of identical building blocks. These are not at all the same types of atoms that the Greeks thought of and quite frankly stop crediting the ancient Greeks for discovering the atom which of course they did not nor could they ever do. The arrangement of the blocks is important however as manipulating them can take on a near infinite number of shapes.

The first level of blocks is of course the atom which can range in size and complexity from a simple Hydrogen atom to exotic atoms many of which are produced only experimentally in laboratories and which exist for exceedingly short lengths of time. Take some of the different atoms in the universe and put them together in the correct order and you can create things like crystals and molecules. Building planets or people is really the same thing at an atomic level and yet looking at the two from a distance will make you understand why it is difficult to believe that this can be true. One is massive and seemingly dead while the other is quite alive and studying the former with a brain built of and powered by the same kinds of atoms that make up the planet. Particular arrangements of atoms and their interactions with one another construct everything no matter how large in the cosmos and all through a simpler model of the laws that govern our universe.

So by doing what the ancients could only dream of and study matter in its most minute detail we have once again proven that simpler is better. By extension the next step we take in our understanding of the universe must be one that remains simple and does not add multiple variables or require numerous solutions to numerous problems. The difficulty is that we do face numerous problems when trying to solve the universe and decades of observations everywhere we look in space confirm that we are missing something fundamental in our understanding of how everything works as one. To make matters worse and add more confusion to how we attempt to calculate and rationalize the universe all of our observations done on Earth in our own laboratories refuse to cooperate with our large scale observations. The universe is right in front of us and all around us yet it would seem we are missing a piece of the cosmic puzzle. The truth is no pieces of our puzzle are missing, there simply cannot be because there is nothing other than the universe in the universe.

Put simply we only need to understand how all of the universes pieces fit together to understand how to build and run one in an acceptably simple way. The trick to finding the key to the problem of how the universe works and unlocking it is that there is precisely one and only one way in which all the pieces fit together just like any normal puzzle. Scientific models are what we use to try and fit the puzzle together and over the millennia we have proposed and discarded many of these when they became worn out. Born of the best intentions each model was applauded and simultaneously embraced by the scientists of the day. These models seemed to fit everything we needed at the time but this only lasted until our knowledge grew to the point where we could not explain the universe any further using the same model. A new model was needed to replace the old and very often a whole new way of thinking about the universe. This was also often accompanied by an upsetting of the apple cart with respect to scientific status quo. Long cherished and engrained beliefs about nature, the world and the universe had to be challenged and surpassed in order for the next step of humanities scientific quest to be taken. This process of challenge and change of established ideals is of critical importance in the advancement of our understanding of the universe because without it we would be stuck with things like the elements. Many who do inhabit the comfortable realm of the status quo however are not so receptive to change and renege those who put forth these often times seemingly bizarre proposals for change.

One of the most notorious of incidences regarding scientific censorship came about as a result of religious persecution in the seventeenth century and involved Galileo Galilei. It was the popular belief at the time that the universe was centered around the Earth and in turn the Sun revolved around it. While it is obvious to us now in the modern world that the Earth in fact revolves around the Sun and the Sun around our galactic center, the notion that the Earth was not at the center of the universe was practically unthinkable then. After all empirical observations which could be made by the common person supported the idea of an Earth centered or geocentric universe. When you look out around you the Earth appears to be unmoving and the Sun is the object rising in the east and setting in the west. The Sun moves and we do not therefore the Earth must be at the center of the universe. To further underscore this point numerous religious creation myths put the Earth at the center of the cosmos and this was done by the divine beings who people believed created everything. As far as the followers of religion were concerned this proof alone was not only correct but superseded any other logic and to them the empirical proof that the Sun rose and set was only a secondary reinforcement of their beliefs. To deny that the Earth was at the center of the universe was to deny the very beings that created the cosmos and risk displeasing them and ruining your chances of eternal happiness in the next life.

Suffice to say that when Galileo voiced his support for heliocentrism or a Sun centered solar system, he was met with a less than toasty reception. At first those of faith simply grumbled quietly and then a bit louder that he really should reconsider his attachment to a Sun centered universe and concentrate on less heretical sciences. Remember at this and most periods in history worldwide it was science that was frowned upon by virtually every religious group to ever have existed as science was seen as questionable in its nature due to the fact that science is very good at poking holes in divine ideals. Galileo seemed to find himself at the hands of one of religions least favorite poking sticks as it was capable of disrupting a key concept in the divine teachings of the Old Testament. Putting this in modern context for a moment reveals just how startling this geocentric universe would be today as three world religions share in the teachings of this one book. It is roughly estimated that four billion people worldwide follow the Christian, Islamic and Judaic beliefs going from most to least followers and the world population is estimated at seven to eight billion and growing. If we still believed in a geocentric universe today half the population of the globe would be instantly challenged in their faiths by this new heliocentric cosmos from these three religions alone. Not to mention the numerous other religions that also believed in an Earth centered universe but do not follow the Old Testament. It is easy to see why Galileo met with such quick criticism of his work when it was proposed in 1610 and was denounced to the Roman Inquisition only five years later for his refusal to back down from his own beliefs about the universe. By 1632 after he published his work entitled *Dialogue Concerning the Two Chief World Systems*, the Roman Inquisition had in their eyes tolerated enough and after a laughable one sided trial sentenced him to house arrest for the remainder of his life. Galileo died nine years later as a result of this erroneous persecution and stands historically as one of many interesting figures in the story of humanities understanding of the universe.

One of the things that make the story of Galileo and his challenging of the status quo most interesting is the final outcome of the battle between the geocentric and heliocentric universe and how we view the universe as a result whether it is through scientific or religious eyes. Despite religions vehement rejection of sciences model of the universe ultimately religion had to admit defeat as the Earth does revolve around the Sun. Further more people now understand that it was in fact science that was right all along and a rethinking of ones religious beliefs was necessary in light of it. This shows that people are capable of rethinking their deepest held beliefs and can learn to accept new ones when the evidence is compelling enough and it is this fact that has been repeated to the great benefit of our understanding of the universe time and time again. Today modern science is faced once again with the fact that its most accepted theories are just that, theories, as none of them can explain how the universe works and all of them contain multiple massive flaws.

The prevailing view of the universe throughout much of the world was Ptolemy's view of the universe or one that was close to it, which was such that the universe is constructed of perfect spheres with the Earth at the center. Various spheres housed the stars, Sun and moon but all were perfect circles and obeyed the Earth as their master. This view was held since about 150 AD and is a prime example of how the status quo of thinking has to be challenged and a new model accepted in order for us to make any real scientific progress in understanding the universe. It was in fact Nicolaus Copernicus who first proposed a heliocentric cosmos in his published work *Dē revolutionibus orbium coelestium* or *On the Revolutions of the Celestial Spheres* in 1543 long before Galileo was to be ridiculed. Copernicus himself suspected he would be met with scorn and rejection should he publish his work and did so only reluctantly after urgings from his friends and colleagues. This was a premonition to how the world would view a new and radically different model of the universe and yet Copernicus was correct in guessing how his work would be viewed. The proof he was correct came at the expense of poor and persecuted Galileo.

Out of such ridicule however comes the irrefutable reality that challenging old ideas and accepting new ones no matter how different and at first wrong they may seem is absolutely necessary to advance our understanding of the universe. After all we do not believe the Earth to be at the center of the cosmos, nor is it flat and the motions of all the bodies in the universe certainly do not revolve in perfect circles. The passing of old ideas in favor of new ones helps us to fill in the holes in our understanding of everything around us from the teeniest forms of matter to the almost unimaginable vastness of space. Our ancient ancestors observed the universe and sought answers for it and to them no matter how hard they tried filling in the gaps in their understanding of the universe this was simply impossible. Not because they were not clever, far from it in fact they were very clever constructing logical models of the world around them to explain what they could not measure directly. Some of their best reasoning was making good progress towards the true workings of the universe as is illustrated by the revelation to many ancients that materials of differing nature were not governed by different elements and rather were in fact the same thing. The hot metal of a sword for example if left alone will become cold metal and so a fire and water element are not necessary to describe an object that can have both properties as the object itself remains unchanged despite its temperature.

We have moved on tremendously since then and once again thanks to our beautiful curiosity we have filled in the missing holes possessed by the ancients' understanding of the universe to create a nearly complete picture of matter. Matter is after all the stuff that makes up the universe and one would do well to learn as much about it as possible in order to fully understand the cosmos. So let us take a look at how we have filled the holes of matter and what we found was needed to fill those holes be they observations, experiments or just good ideas.

Chapter 2
The First Modern Ideas Emerge

Newton and Gravity

The first good idea we had leading to today's understanding of the universe came from a man born on December 25th 1642 by the name of Isaac Newton and later known as Sir Isaac Newton following his knighthood. Perhaps the greatest scientist of all time Newton made contributions in the fields of physics and mathematics which we still use today. Just because his work came centuries earlier than other noted scientists does not diminish it in anyway, in fact Newton's work was so influential on science as a whole that many of our latest observations on the universe obey his laws. One of the most important aspects of his genius was his work on gravity, a force that during his time and ours continues to hold center stage in physics. In fact just about all of our large scale and many of our small scale observations of the universe are directly tied to the effects of gravity in one way or another. The expansion of the universe, the motion plus all of the related behaviors of black holes and the simple falling of an apple to the ground are all directly related to gravity. Newton not only gave us the working idea and understanding of gravity but also single handedly invented the entire field of physics and so all who work in modern science owe him much.

Newton's revolutionary work was entitled *Philosophiæ Naturalis Principia Mathematica* or Mathematical Principles of Natural Philosophy and was published on July 5th 1687. The book itself probably never would have seen the light of day without the help of his friend Edmund Halley. Halley provided both encouragement and financial support for Newton's work which ultimately resulted in *Principia* being published. Edmund Halley himself was a noted scientist involved in astronomy, geophysics, meteorology and mathematics who is best remembered today for the comet which bears his name and whose orbit he successfully calculated. In his book Newton established his now famous laws of motion and of course universal gravitation. His laws of motion formed the backbone of classical mechanics and have held up to this day. The first of his three laws simply states that an object will tend to stay at rest or in motion unless it is acted upon by an outside force. What this means is that objects do not spontaneously change their behavior and must require some external force to do so. A ball for example will not start to roll by itself and will only do so if one were to push or kick it. The second law simply states a way of calculating the vector sum of the forces on an object and is summarized in the textbook familiar $F=ma$. Here F is the force, m is the mass of the object and a is the acceleration vector of the object. How much force the ball has is calculated using this formula and changing any of the variables will change the force of the object. Let us say two different weights are dropped from a building and wind resistance and other factors are insignificant to our calculations. The first weight is twice as heavy as the second and both experience the same acceleration force due to gravity. However since force is equal to the mass of the object multiplied by its acceleration we know that the first weight will hit the ground with twice as much force since it has twice as much mass. This formula can be rearranged to solve for any of the three variables you desire which makes it extremely useful and powerful. The third and final law of motion states that any object exerting a force on another object will experience a force equal to but opposite in direction to itself. In other words one object exerting a force on a second will feel a force equal in strength but opposite in direction from the second object on it.

Newton's laws have held up completely or in part to this day and for this reason are held to be some of the most important laws in physics. In quantum mechanics for example things like momentum, position and force operate below the speed of light in the quantum state and consequentially Newton's law used to govern classical objects work perfectly well for these quantum operators. Only at higher velocities and speeds approaching the speed of light do we see Newton's calculations losing accuracy.

Newton's universal gravitation was and is also hugely successful in advancing science, specifically astronomical observations in the cosmos. Choosing the Latin word for weight, gravitas, Newton was the first to precisely describe the effect that would become known as gravity. Simply stated and avoiding the intellectual controversies between Newton and Robert Hooke a fellow scientist of the day, gravity is a measure of two objects' mass and the distance between them. Every mass in the universe has gravity and any two masses will have a line intersecting their respective bodies along which the force of gravity lies. That

$$F = G\frac{m_1 m_2}{r^2}$$

force is thusly calculated in its modern form: . Here F is the force being calculated, G is the gravitational constant at 6.673×10^{-11} N·(m/kg)^2 , m1 is the mass of the first object with m2 being the mass of the second object and r is the distance between the centers of both masses. We say modern form because the gravitational constant was not determined until results from experiments carried out by fellow British scientist Henry Cavendish were performed 111 years after Newton published his *Principia*. Nevertheless this equation represented a massive leap forward in what we could now understand about the universe as we could determine the how as well as the why gravity attracts two bodies together. A positive boom in astronomical science was to follow in the centuries after Newton's publication. Gravity is the one force that binds the universe together and functions of the next wave of discoveries in the quest to answer how the universe works were to come about from advancements made in the nature of matter.

Real Atoms and a Hint of Quantum

Atomic theory as a way of explaining matter as we have discussed was first proposed by Greek philosophers and since based on philosophical thought and not experimental proof was not only incorrect but also abandoned for many centuries. It was not until 1789 when Antoine Lavoisier formulated his law of conservation of mass. The law of conservation of mass states that the total mass of two chemical reactants will always be the same as the product of their reaction. If you have ten grams of one reactant and twenty of a second you will always get a total of thirty grams of product once you have combined them. What is important about this discovery is that it holds true no matter what form the reactants or products take. In other words a liquid and a solid can react to produce a gas and that gas product will have a total mass equal to the liquid and solid added. The state of the matter does not actually matter so to speak. Ten years later in 1799 the noted chemist Joseph Louis Proust proposed his law of definite proportions which states that a compound reduced to its constituent elements will always produce the same proportions of the constituents. The source of the original substance or its quantity will not matter as it will always produce these ratios. This is of course taken as the perfect outcome in both experimental cases as they reflect through the expression of their respective laws the theoretical yield and not the actual yield of any reactions involved.

Building immensely upon the work of Lavoisier and Proust was a man named John Dalton who would perform the ground breaking work necessary to thrust a modern non Greek atomic theory of matter to center of the physics stage. What he did was develop the law of multiple proportions which stated that a ratio of masses was at work in the reaction process, from reactants to product. Most significant was that the ratio would always be made up of small whole numbers. Dalton examined experiments conducted by Proust and noted that when he combined tin with oxygen he got a product that always combined in a ratio explainable by small whole numbers. There was no way of explaining this using the current models available and so Dalton determined that an atomic model would provide the simplicity needed to explain these ratios. Calculation based on Proust's experiments revealed one hundred grams of tin would combine with 13.5 or 27 grams of oxygen. A ratio of 1:2 is made from 13.5 and 27 for the oxygen and so one tin atom would combine with one or two oxygen atoms.

Embracing an atomic theory of matter Dalton further proposed that the ability of water to absorb various gases in different quantities was based off of the atomic shape of the gas, its mass and how its particles behaved when compared to a different gas. He experimented with dissolving similar amounts of different gases into water and measuring how much the water could absorb. Using nitrogen gas and carbon dioxide gas he found that water was far more willing to accept quantities of carbon dioxide gas than nitrogen. Modern physics shows us that nitrogen is a far simpler, smaller and light gas than carbon dioxide, although Dalton had no idea of modern molecular models he was in fact right all along.

Lastly after reviewing his experimental data Dalton concluded that the various laws of proportions, ratios and water absorption results could be elegantly explained by atoms that make up each element. Each element could not have its atoms destroyed or altered by chemical means but chemical processes could be explained by the combination of these different elemental atoms to form various chemical compounds. Chemical compounds are often thought of today as molecules which can be combined with one another to produce many different structures. Crystals for example can be made of one or more elements; diamonds are made from only carbon atoms while rubies are made primarily of aluminum and oxygen. Molecules form the basis of the ruby as the aluminum oxide form is repeated to form the crystal lattice of the gem, where as diamond is a repeated form of carbon that forms the crystal lattice. Regardless of how compounds are constructed and what their final product is be that a red ruby or red blood, diamonds or DNA all things in the universe that we view as matter is made of atoms. Dalton published his full works in 1808 in a book entitled *A New System of Chemical Philosophy*.

Dalton did make a couple of mistakes however that would make his view of atoms, elements and compounds almost but not perfectly correct. One such mistake was that he believed in simplest forms elements would always exist as a single atom and not as a molecule. So to him for example nitrogen would exist as a single atom and not as two nitrogen atoms bound together in molecular form as we find with nitrogen gas in our air. The second mistake he made was in assuming that when elements combined to form compounds of the simplest form they did so only in a one to one ratio of atoms. So to Dalton water would be one hydrogen atom and one oxygen atom and not two hydrogen atoms and one oxygen atom as we know it to be today. Forgiveness must be granted to him however as he was the one developing this atomic model for the first time coupled with crude equipment and for us it is easy to see that hind sight is twenty-twenty.

The next great leap in our understanding of matter occurred in 1897 and was discovered by J.J. Thomson who would show the world that the atom was not the smallest form of matter in the universe. What Thomson would do was to be the discoverer of the first subatomic particle called the electron and usher in an explosion of understanding with respect to the fundamental nature of matter. Such an explosion would not be equaled until humans began abusing particles inside of colliders in an attempt to probe matter down to its limits. With a little help from a Crookes tube an invention derived from a Geissler tube first made in 1857 by a German physicist named Heinrich Geissler, Thomson was able to demonstrate the existence of electrons.

A Geissler tube demonstrates electrical glow discharge, think of neon lighting and you can picture it easily. The invention was ingenious as it forces electrons to dissociate from the gas molecules in the chamber causing ionization of the gas. When the electrons recombine with the gas photons are emitted through a process known as fluorescence. The gases used inside these tubes are rarified gases, ionizable minerals or conductive fluids, essentially anything that the electric current can flow through making the electrons dissociate. What is of extraordinary interest in these Geissler tubes is that we now know it is at the smallest fundamental levels and by quantum interactions of subatomic particles that this simple fluorescing effect takes place. If you want to understand the cultural significance of the Geissler tube beyond neon lighting you need to look no further than the television. Televisions owe their direct lineage to the Geissler tube as it was further developed into the cathode ray tube, the device that powered the first televisions and without which all modern televisions would not exist. While unknown at the time these experiments demonstrate how quantum effects can be observed long before they are fully understood just as today we are no doubt observing effects throughout the universe that contemporary science does not truly grasp.

The Geissler tube was originally marketed as a novelty item partly due to their beautiful effects and mainly for demonstrating the effects of electricity, similar to the sale of plasma balls today in many science curiosity shops. Scientists did however take a far deeper look at Heinrich's scientific achievement and refine and develop these tubes into Crooke's tubes so named for their creator William Crookes. Crookes developed a tube that was made from blown and sealed glass containing a partial vacuum and fitted with two electrodes, one the positive anode plate and the other the negative cathode plate. Applying a current between the two plates causes a stream of electrons to be emitted which at the time were dubbed cathode rays. Initially J.J. Thomson deduced that the rays were made up of negatively charged particles that he called corpuscles, later scientists would rename them the familiar electron that we know today. Thomson began experimenting with these cathode rays and found that the beam emitted from the plates could be deflected by an electric field. Magnetic fields can also deflect electrons and this was known at the time as well but it was the application of electric fields that Thomson used to conclude the existence of his corpuscles. By measuring the mass of a proton and its charge coupled with the charge of an electron in a ratio Thomson worked out that the mass of an electron was eighteen hundred times smaller than that of a hydrogen atom. Since the nucleus of a hydrogen atom is a single proton he now knew the charge and mass ratios of two of the most fundamental particles in the universe and had proven that the indivisible atom was in fact divisible. Poor Leucippus and Democritus, the wall of their indivisible atom had been breeched and as science progressed over the next century or two many more walls in the atomic theory were to fall as well. For now however science had made a great discovery into the true nature of matter as Thomson's work allowed him to construct what became known as the plum pudding model of atoms and consequently matter. In order for Thomson to balance the positive and negative charges of atoms so they resulted in a total neutral charge he proposed that the corpuscles which he knew to carry a negative charge were uniformly distributed amongst the positive charges of the atom. The charges cancel each other to produce not only a neutral charge for the atom but result in a distribution of electron plums in a sea of positive pudding, the plum pudding model was born or baked whichever you prefer.

Thomson's work is groundbreaking as it demonstrated a whole new view of the atom that would shape the future scientific understanding and application of atomic theory by dividing the atom and proving it consisted of multiple parts. His view of the make up of an atom with respect to its overall internal construction though we now know to be flawed and it was Ernest Rutherford a student of his that discovered this flaw and refined the atom into the shape that we use today. While we have learned much more about atomic structure since Rutherford the basic shape of an atom has remained essentially unchanged, something at the center with some other somethings around it. What Rutherford did was discover that the positive charge of an atom must be concentrated at its center rather than spread evenly throughout. This in turn leads to placing the electrons around the central concentration of positive charge to balance the atom and its charges until it once again is neutral. A massive step forward but how did he do this?

Rutherford designed an experiment to test the Thomson model, his colleagues Hans Geiger and Ernest Marsden executed an experiment of his design known as the Geiger-Marsden experiment or the Rutherford gold foil experiment. The goal of which was to study the plum pudding model by shooting a stream of alpha particles at a sheet of gold foil and observing the effects on a detector screen afterwards. Alpha particles consist of two protons and two neutrons bound together and are well known as the nucleus of a helium atom, although at the time the idea of a neutron was unknown in physics. Alpha particles are of course highly positively charged and extremely heavy when compared to the negatively charged electrons. The expected results given the plum pudding model was a low deflection of the alpha particles due to the belief that matter was made of a plum pudding arrangement where positive and negative charges are uniformly spread-out. The uniform spread, low positive charge and high mass of the alpha particle should result in a neat barely deflected pattern on the target screen, what was found was the opposite. A small number of the alpha particles experienced a very high degree of scattering when viewed on the target detector screen. How could this be unless the plum pudding model was inaccurate? Rutherford's wise conclusion was that there must be a concentration of positive charge at the very center of the gold atoms because this would explain how heavy positively charged alpha particles would be deflected by the positive electric charge at the nucleus. The electrons must therefore not be spread uniformly throughout the atom but instead be somewhere else in the atom and the only place that could be is in a cloud surrounding the nucleus. Rutherford dubbed his new model the planetary model of the atom which consisted of a small concentrated positive nucleus surround by a cloud of electrons occupying a much larger volume of the atom. This model explains the results of the gold foil experiment, disproves the plum pudding model, keeps the charges balanced and gives us the general atomic shape we know today.

Like Thomson before him however Rutherford's model was not quite right and advancements had to be made in order to forward atomic theory sufficiently to match experimental observation. This is where we enter a quantum explanation of how the atoms that make up matter work and it was given to us by Niels Bohr in 1913. Bohr replaced Rutherford's model with circular orbits for electrons instead of a cloud. The simplest way to view his model is that of the solar system. Think of a hydrogen nucleus which is just a positive proton like our sun and an orbiting electron like the planet mercury. The electron orbits the hydrogen nucleus just as mercury orbits the sun as seen from a top down view of the solar system, there is a slight difference between the two systems however. Unlike gravity which holds mercury in place around our sun it is the attraction of electrostatic forces that keep the electron orbiting the proton in the hydrogen atom.

So why did we need this new model anyway and what was so special about finding a quantum explanation for atoms? Well it begins with shortcomings of the Rutherford model in explaining two things about electrons, their charge and motion around positive nuclei. First of all electrons were known to emit electromagnetic waves according to the Larmor formula and this would spell disaster for matter in the universe as electrons would go crashing into their parent protons. It takes energy to stay in orbit around the nucleus and as an electron lost energy due to the emitting of electromagnetic waves it should quickly decay its orbit and plunge into the center of the atom. Obviously this was not happening to matter and therefore a new mechanism was needed to explain how electrons kept their distance from the nucleus and maintained stable orbits.

The second problem arose from measurements taken by a new type of experiment called spectral analysis. Spectral analysis is a fantastically wonderful way of learning about atoms and molecules by looking at the absorption lines they produce from light they emit and was originally founded by Sir Isaac Newton. What Newton was the first to do was place a prism in a beam of light and split the incoming white light into its component colors to produce a spectrum of light wavelengths. The rainbow you see on rainy days is created by the same effect but until Newton performed his now famous experiments no one was sure what was causing the effect. The familiar colors generated by a spectrum tell us a great deal about the photons that generate each color of particular interest and is their wavelength and energy. The lowest energy and longest wavelength is found in the red photons and progresses up to the highest energies and shortest wavelengths as you move through the spectrum to orange, yellow, green, blue, indigo and finally violet. Newton did not directly find the dark absorption lines himself but it was his work that influenced German optician Joseph von Fraunhofer to investigate spectrums further. Fraunhofer's work led him to be the first human to observe and record the absorption lines in our own sun and consequently the spectral lines of the sun are named Fraunhofer lines in his honor. He did this by looking at the orange light emitted from the fire burning in the furnaces where he worked producing glass for optical purposes. Using the spectroscope he invented in 1814 Fraunhofer noted dark lines in the light from the fire and he wondered if the same lines appearing in the orange light of his furnace would appear in the orange light of the biggest furnace in our solar system, the sun. What he found was 574 different dark absorption lines in the spectral pattern emitted by the sun.

Today we could not know even a fraction of what we do about the universe without the work conducted by Fraunhofer and his spectroscopes. Spectroscopic analysis is how we know whether an object such as a distant galaxy is moving with respect to us and what its composition is. Some galaxies are moving towards us, as is the case with our nearest galactic neighbor the Andromeda galaxy, or away from us as most galaxies seem to be doing. The atomic compositions of stellar bodies are also determined this way. All of our latest research in the field of extrasolar planets is also being conducted through spectroscopy.

Telescopes such as the now famous Kepler space telescope look for Earth like worlds in systems beyond our own by looking at the spectral absorption lines of light from the parent star of the extrasolar planets. The goal of the Kepler project and others is to find a planet conducive to life such as that found on Earth. The biggest reward of course would be finding the tell tale signs of a habitable planet by the gases present in its atmosphere. What scientists are primarily looking for of course is the presence of oxygen and water vapor. Such a find would be one of the great achievements in modern astronomy giving us our first glimpse of an Earth like world light years from our own home system.

Closer to home when spectroscopy was applied to the hydrogen atom curious peaks were found in its absorption spectra and Bohr had a solution. He reasoned that electrons could only encircle their nucleus in specific orbits, these orbits contained specific energy and angular momentums that the electron must obey. Therefore given these variables the distance from the nucleus was fixed for each orbit, determined by radii and proportional to its energy. What this did was solve the energy loss and consequent decaying of electron orbits due to the realization of Bohr that the electrons could only jump between orbits by instantaneous quantum leaps. When an electron makes a jump between energy levels whether up or down it will absorb or emit a photon proportional to the change in energy of the orbit and therefore reveal the frequency at which the photon did so. The point at which these photons were excited lies along the lines of the absorption spectra and so solved this riddle too.

The Birth of Quantum Mechanics

What is needed now is an introduction into the origins of quantum theory as from this point on in scientific history the world of quantum interactions comes to be synonymous with our quest to understand matter. Thanks to Max Karl Ernst Ludwig Planck a German physicist responsible for the creation of quantum theory we have a way of understanding matter and some of its seemingly peculiar behaviors in ways classical systems could not provide. He was hired to solve a problem known as black-body radiation by electric companies to improve the brightness of light bulbs while keeping energy levels low. A black body is one that is known in physics as a perfect absorber of all incident electromagnetic radiation. In other words light will go into it but no light will be emitted, reflected or refracted by it. The basic problem was that no one could explain adequately the intensity of electromagnetic radiation emitted by a black body and how it was linked to the frequency of the radiation. After studying the problem known as black-body radiation and out of a self described act of despair, Planck came to the conclusion that energy could only be emitted in quantized form. It was 1900 when Planck presented his work to the world in what has now become known as Planck's postulate. The postulate states that energy can only be emitted in quantized form and takes the form of: $E = h\nu$. In this equation E is energy, h is Planck's constant and v is the frequency of the photon. So the energy of the photon is its frequency multiplied by the Planck constant. Originally he used the term quanta to describe what we today call photons but the concept and calculation of quantized energy would forever change the face of physics and our understanding of matter.

Max Planck is undoubtedly the father of quantum physics and he not only shed light on the beyond microscopic world of matter but without his help future scientists would never have been able to make the discoveries necessary to advance atomic research further. One such scientist who enjoyed his help was a fellow German by the name of Albert Einstein who won a Nobel Prize just as Planck did before him however not for his work on relativity as many might expect but instead for his research into the photoelectric effect. An interesting note on the relationship between Max Planck and Einstein occurred following the publication of Einstein's three papers on his special theory of relativity in 1905. At the time Max Planck was a well known and respected scientist not only in Germany but throughout the world whereas Albert Einstein was a complete unknown and as such could not get his work reviewed by the scientific community in general. Planck upon reading the 1905 relativity publication saw its importance and through his name and influence within the science community was able to get Einstein's work widely accepted.

As far as Einstein's work on the photoelectric effect is concerned we need to first take a look at what the photoelectric effect actually is. The photoelectric effect is essentially a property of metals where upon being subjected to light spontaneously emit electrons. The electrons emitted by this process are called photoelectrons. The problem stems from explaining phenomena associated with this effect using the classical electromagnetic theory. Essentially light was viewed as a wave and waves have two properties similar to sound waves or water waves, they have an amplitude and a frequency. According to classical electromagnetism then if you were to change either the amplitude or the frequency of the wave you would alter the rate at which electrons were emitted from the metal upon being exposed to light. Another problem was that a straight forward wave interpretation of light dictates that a delay in the metals ability to emit an electron after it receives light energy should be found.

This is where the problems with the photoelectric effect began, as neither of these two outcomes was observed by subsequent experiments. What was happening instead was electrons were only being emitted if a threshold frequency is met or of course one goes beyond that frequency. This can be thought of as what happens inside the human body every time a nerve impulse fires. Nerve impulses fire because a difference of electric charge carried inside and outside a cell wall occurs and eventually is exceeded. As ions build up on one side or the other of the wall charge builds up as well, but the nerve signal will not be triggered unless a certain charge is met, this is called the action potential. Action potentials inside your body fire only at one hundred percent and not before, essentially they are all or nothing events. The dislodging of an electron from a metallic surface due to the photoelectric effect is remarkable similar, without meeting that threshold frequency you do not move an electron even if you are at ninety nine percent of that frequency and you alter the amplitude or allow the light to shine upon the metal forever. This is where we see a solution provided by Einstein to the photoelectric effect and one that offers up a new view of the humble photon.

What Einstein proposed was that light was not simply a wave as had been thought before; rather it was a packet of energy. Discrete wave packets and the energy of each photon was hf, just as Max Planck had said in 1900. An explanation of the photoelectric effect that satisfactorily answered the experimental observations and new description of the photon which removed the limits of a wave only theory was enough to earn Einstein a Nobel Prize in 1921. The man who confirmed Einstein's work was Robert Andrews Millikan, it was after Millikan's 1914 experiments into Einstein's photoelectric work that Einstein received his award. Robert Millikan himself went on to win his own Nobel Prize in 1923 not only for work on the photo electric effect but for measuring the elementary electronic charge.

The next evolution of quantum theory and of our quantum understanding of matter came from a physicist by the name of Louis de Broglie and would come to be known as the de Broglie hypothesis. The de Broglie hypothesis suggests that matter itself regardless of its type has both wave like and particle like properties. Originally de Broglie stated that electrons possessed these wave and particle properties similar to the way photons did. Written as his PhD thesis in 1924, he also provided the mathematical relationship describing electrons as $p = \dfrac{h}{\lambda}.$ In the de Broglie's equation we see the familiar Planck constant h and wavelength λ as well as p which relates to the electrons momentum. What made his work so important was that it applied Einstein's concept of a photon behaving as discrete wave packets and transferred it to an electron. Photons as we know are energy and electrons are matter, so his concept of bridging the two and showing that they both behave in the same manner was important to advancing quantum mechanics. As our understanding of matter at the quantum level progressed over the years from refinements of theory and experimental results we are now aware of the fact that all matter and not just electrons show both wave and particle properties. It was a significant improvement in our understanding of matter but in 1924 was unproven and in order for Louis' work to be accepted across the scientific community firm evidence had to be found.

It would be three years before proof arrived from two scientists working at Bell Labs by the names of Clinton Davisson and Lester Germer. In order to study de Broglie's theory of the wave and particle nature of electrons Davisson and Germer decided to test diffraction on matter not energy. Diffraction can be seen easily enough on the macroscopic scale, simply go to a body of water and toss in a stone. The stone generates ripples that travel outward from the point of impact with the surface of the water. Since water is a fluid and obviously displays wave properties passing those ripples through a narrow slit, say two cattails, will display a diffraction pattern as they lap up on the shore. The Huygens-Fresnel principle describes this in terms of classical physics because the ripples formed in the water are many times larger than anything at the quantum level. A quantum description however is non-classical yet the principal of diffraction is the same at the quantum level due to the fact that at such microscopic levels energy was known to behave as a wave. The now famous Davisson-Germer experiment set out to test de Broglie's work by passing matter through a slit and looking to see if a diffraction pattern could be observed. Instead of water Davisson and Germer used electrons fired at a target made of crystalline nickel. What they observed was in fact diffraction from the incident electrons as they struck the nickel target; they had proven de Broglie correct, matter did in fact have both wave and particle like properties. This revolutionary experiment earned Clinton Davisson a Nobel Prize in 1937.

The work of Louis de Broglie caught the eye of another physicist, an Austrian by the name of Erwin Rudolf Josef Alexander Schrödinger. The concept that matter such as an electron could behave as a particle and a wave interested Schrodinger sufficiently enough to ask the simple question can we better explain matter as a wave than as a particle? At the time of course what was meant by matter was an electron only but in time as we have seen it will become applicable to all forms of matter not just the electron. Beginning to formulate his ideas in 1925 and publish them in 1926, Schrodinger unveiled his now famous Schrodinger's equation and in it he describes an electron solely as a wavefunction rather than a particle. The equation will take a different form depending on the system you are studying but the most basic form of the equation is provided for a system changing over time and is stated as: $i\hbar\dfrac{\partial}{\partial t}\Psi = \hat{H}\Psi$.

The Greek symbol psi or Ψ is the symbol denoting the wave function itself. Appearing in *Annalen der Physik* the Schrödinger equation describes the wave equation over time for a particle and predicted the correct eigenvalues of the hydrogen atom. The significance of Schrodinger's work was immediately apparent as it provided a quantum solution to motion for subatomic particles just as Newton's second law can be used to describe motion for classical systems. What is more is that because everything in the universe is comprised of matter and matter comprised of subatomic particles, the Schrodinger wave function is theoretically a description of all matter at all scales. A subatomic particle behaves in a quantum nature and matter in a classical nature to us in our everyday lives, but since our everyday lives are formed from quantum particles one can argue that everything we see is purely quantum.

This raises some interesting questions about the world we see everyday and the universe our world inhabits. The question currently receiving much debate is: what is actually real? Are we as humans for example simply quantum wave functions which can be described as nothing more than energy states in mathematical formula? The solid objects that we encounter everyday are made of tiny subatomic particles that appear to exist as nothing more than fleeting waves of energy which as we will see are so elusive that we cannot know everything about them. While Schrodinger's first publication in January of 1926 has been touted as a turning point in physics, chemistry and one of humanities greatest triumphs of the twentieth century it creates as many new quantum possibilities as it solves. Ultimately however Erwin Schrodinger would receive the Nobel Prize in 1933 for his significant contribution to quantum understanding.

What Schrodinger's equation described for the hydrogen atom was the behavior of its electron as a wave and not a particle but did not describe the electron itself properly. A partial interpretation was given by Schrodinger that described the electron in terms of charge but did still not adequately answer the problem. The wave equation did solve some of the spectral line discrepancies of hydrogen atoms that Bohr had encountered previously and so the scientific community knew Schrodinger's work to be correct. What was needed however was a better understanding of the electron and shortly after the last in a series of 1926 papers published by Schrodinger a solution was presented by a man named Max Born.

A German physicist and mathematician Max Born was interested in Schrodinger's concept of a wave function for the electron and set out to determine how best to interpret the electron of the hydrogen atom. What Born calculated was that it was more accurate to describe all the probabilities of the electron with respect to the electron, than simply describing the electron itself. In other words when describing an electron around a hydrogen atom what one is actually doing is calculating all of the possible locations of the electron and the probability of finding it there. The most common interpretation of this in modern physics is removing electrons from their orbits and placing them in atomic orbitals. The orbital is viewed as the space in which you have the greatest probability of finding the electron around the nucleus of the atom and is a dismissal of the neat orbits predicted by Bohr. Schrodinger gave us the orbital which can take various shapes as you move upwards in energy levels for electrons, but regardless of what state you view the electron you are only looking at the possible locations of it in that state.

Now we see through the brilliant work of Max Planck, Louis de Broglie and Erwin Schrodinger that matter is in fact quantum, a particle and a wave and we now have the equations necessary to study them, quantum physics would seem complete but for one small problem. When viewing electrons, indeed all matter as waves and particles it is realized that a mathematical impossibility occurs; predicting the position and the momentum of an electron around the nucleus. This is evident in the answers one gets when calculating electron behavior in an atom and all that can be derived are probabilities.

The Uncertainty Principle and New Quantum Unknowns

Here enters one of quantum mechanics most famous puzzles and most famous scientists. The scientist was one of the worlds most brilliant and influential theoretical physicists, a German scientist called Werner Karl Heisenberg and the realization he made over electron behaviors has become one of the cornerstones of quantum mechanics known as Heisenberg's uncertainty principle. What Heisenberg saw in probability fields for electrons was an inability to know both position and momentum with certainty. If for example you tried to calculate exactly what the position of the electron was you lose sight of its momentum. So your probability of finding the electrons position increases while your probability of determining its momentum decreases. Of course the opposite can also be calculated, increased accuracy in calculating the probability of an electrons momentum is made by sacrificing accuracy for the knowing of its position. This lack of certainty for both variables has rightly been called the uncertainty principle.

First published by Heisenberg in 1927 it took the physics world by storm and to this day permeates all quantum experiments to varying degrees. In fact it is so prevalent that many quantum physicists take the uncertainty principle into account while conducting experiments but treat it almost like a force of the universe which we can ignore; it is simply something that is there. The truth is that quantum mechanics does not want to easily give up its secrets about the wave nature of matter, yet it can be forced to do so by making an observation of a matter wave of interest. When scientists want to know what a particle is doing yet can only guess at the probability of what it is doing they will make an observation which seems to force the particle into one definite possibility. At that moment all the probabilities surrounding the particle disappear and it can be viewed as it appears in that instant of time. Heisenberg when faced with solving the uncertainty principle stated that the act of observation could be a solution to the problem. Over time however it is now known through additional experiments that any object described as waves will show uncertainty. The 1932 Nobel Prize was rightly awarded to Werner Heisenberg for his incredible creation and contributions to quantum mechanics and our overall interpretation of matter.

This uncertainty in our quantum understanding of matter exists in modern physics today and is of extreme importance in many quantum experiments. As we have discussed the uncertainty of wave states is often accepted and somewhat ignored in the lab as something that simply is, but it is sometimes studied directly. Many researchers for example instead of accepting and ignoring it are accepting and not ignoring it, rather they design experiments to directly probe what can be done to resolve this mystery. The problem with modern day physics is that despite valiant efforts and clever experiments aimed at getting rid of the uncertainty of wave systems it simply has not gone away.

For a moment lets leave matter waves behind and look at the discovery of another subatomic particle, the neutron, as this occurred at roughly the same time in scientific history as much of the quantum work. Ernest Rutherford who we have seen was responsible for giving us the planetary model of the atom did so by firing alpha particles at gold foil. Rutherford however shot alpha particles at different substances besides the gold foil, one of which was hydrogen gas. Just as he observed strange effects in the form of deflected alpha particles from the gold, he noted that subjecting nitrogen to alpha particles produced hydrogen atoms. Unaware of the neutron Rutherford believed that he had caused the hydrogen nuclei to be emitted directly from the nitrogen gas, in other words from nitrogen atoms themselves. Knowing that positive charge was associated with hydrogen nuclei as was rough atomic mass, Rutherford called these particles protons and surmised that they must by a fundamental particle in all atoms.

What he also noted however was that as mass of the nucleus increased the number of hydrogen atoms did not increase sufficiently to account for the increase in mass. Something else was giving the atomic nucleus mass yet it was not adding to the positive charge of the atom itself. Since it was contained in the nucleus and obviously not coming from negatively charged electrons around the atom Rutherford speculated a new type of subatomic particle he named the neutron. A neutral particle seemed to be the answer as it provided no additional charge to upset the total charge balance of the electrons and protons in the atom but did provide this missing mass. While Rutherford conducted his nitrogen gas experiments in 1917 it was not until years later that proof of the neutron was discovered.

During 1928 a German scientist with a passion for nuclear physics by the name of Walther Wilhelm Georg Bothe conducted an experiment which seemed to produce a new particle of neutral charge. Walther Bothe would also go on to win the Nobel Prize for his work on wave-particle duality of radiation which he shared with Max Born. What he did was subject a beryllium sample to a stream of alpha particles and observed that a neutral type of particle was radiated from it. As radioactive particles are emitted from matter their weight can be gauged to a degree by how destructive they are. As Newton stated force is equal to mass times acceleration so a light particle such as an electron will do less damage as it has less mass than a heavier particle. An alpha particle which is a helium nucleus consisting of two protons and two neutrons will be enormously more destructive than a beta particle as its weight is many thousands of times heavier than a single electron. The force of radiation from hydrogen nuclei or protons was already understood and this was not it. So when Bothe observed particles with no charge and doing more damage than hydrogen nuclei he was able to ascertain that these particles were also heavier than protons. Rutherford had proposed the idea of a neutral particle through his mass discrepancies of atomic nuclei and Bothe had discovered neutrons through his beryllium experiments. The picture of the neutron was almost finished and it would be an English scientist called James Chadwick that would finalize the picture.

James Chadwick began studying the effect of this beryllium radiation as it was sometimes called in 1932 and decided to compare it to gamma radiation. Gamma radiation is created from extremely energetic photons emitted from the nucleus of atoms. We now know that gamma rays can be created in different ways but in the classic sense it comes about from radioactive decay. Radioactive nuclei will gradually lose energy and decay to a nucleus of a lower energy state, as this occurs energy from the nucleus is lost as high energy gamma ray photons. The term gamma ray was coined by Ernest Rutherford who examined the work of a French physicist called Paul Ulrich Villard. Villard had been conducting studies on radium and in 1900 discovered a form of radiation being emitted from his samples. What he had discovered was gamma radiation which he found by placing a lead shield in between a sample of radium and photographic paper. Lead was found to be capable of stopping alpha particles and therefore would not show up on the photographic paper. Villard noticed that the photographic paper however was still being affected and he concluded that two other types of radiation were responsible. Of the first of these two we find beta radiation which is made of electrons and can be deflected away from the target by a magnetic field. The second type of radiation that still penetrated the lead shield however was not deflected and so was a new type of unknown radiation altogether. Villard at the time did not provide a name for his radiation and so three years later fellow scientist Ernest Rutherford named Villard's newly discovered radiation gamma rays.

James Chadwick was aware of these gamma rays and of the fact that they were an ionizing form of radiation. Gamma rays being projected onto metallic surfaces caused electrons to be emitted just as these new neutral particles were emitted from beryllium after being bombarded with alpha particles. Measuring the amount of ionizing energy from gamma rays and these neutral particles of the beryllium radiation he calculated that there simply was not enough energy in the gamma rays to cause such powerful ionizing effects. So he began to study the energy of these particles more directly and to this end he turned to the tried and true method of subjecting samples to radiation. His radiation was obviously the beryllium radiation but his samples consisted of a wide variety of substances and all produced these new electrically neutral particles. The most important detail he worked out about these particles from his radiation experiments was that all the particles being emitted had significant amounts of mass at least equal to the proton and so therefore could not be made of gamma rays as they are massless. Chadwick had proved the existence of Rutherford's neutron and confirmed the first experimental discoveries of the neutron by Bothe in his 1932 paper titled *The Existence of a Neutron*. James Chadwick received the Nobel Prize in 1935 for his work on the existence of neutrons in atomic nuclei.

Now its time to take a look at what we know physically about matter and its properties in the universe. Our understanding has moved from a purely philosophical albeit rational attempt to understand matter in a natural and non theological way, to an incredibly detailed map of the teensiest pieces of matter we can conceive of. The discovery of atoms led to finding the charges of two particles and therefore the knowledge that matter was made of at least two types of distinct particles rather than one indivisible form. The basic structure of atoms was revealed to by non-homogeneous rather than a uniform mix of these two charge particles. The central positive particle has been confirmed as inhabiting the very center of atoms and is much, much more massive than its negative counterpart. The negative particle was determined to first encircle, then orbit and finally inhabit an area of probability density around the nucleus. This probability density reworked the original orbit model into a more satisfying orbital model and was formulated after it was discovered that matter like energy behaves as both a particle and a wave.

Today we ultimately describe matter not so much as a particle with wave like properties, but rather the reverse and state that matter is more wave like with particle like properties. Finally we have discovered that a third type of particle also resides within the nucleus of most atoms. It is the neutron, neutral in charge and larger than the proton. Other than hydrogen nuclei which are nothing more than a single proton all atomic nuclei contain a mix of protons and neutrons. The next heaviest element up from hydrogen is helium which already contains not one but two protons at its nucleus along with two neutrons. So our final view of matter is that it is composed of atoms which contain nuclei made of protons and neutrons surrounded by a probability cloud of electrons. This description of matter is to all accounts wonderfully complete if matter were the only thing to exist in the universe and never interacted in any way with other matter, but it is not and it does not. In order to make our universe work we will need more than an understanding of just the physical structure of matter and the first thing we shall examine is how matter behaves when it moves.

Brownian Motion, Neutrinos and Relativity

But, you say to me, we have covered the motion of matter already and we established some incredibly powerful and useful laws along the way, three of them in fact. Well yes that is true the work of Sir Isaac Newton is brilliant and some of his laws of motion are the only physics which remains indestructible in our modern understanding of the universe. But matter works, interacts and moves on many different scales throughout the universe and as we have seen the rules can change on the very, very small scale of the quantum world. It would appear that the large scale universe can play by its own rules too; all we have to do is look at the moon and realize it is not where it should be.

This notion of course is silly as the moon really is exactly where it should be, but according to classical Newtonian mechanics the moon is being a bit playful and bending the rules. What one finds when applying Newton's equations to the orbit of the moon is that the moon will be slightly behind in its orbit from where you have calculated it to be. The moon appears to be moving slowly and therefore has not travelled as far as it should given the elapsed amount of time. In order to solve this problem Einstein proposed that space through which the moon travels and time are not separate entities rather they are one continuous material. This so called space-time material is responsible for slowing the moon in its orbit when compared to Newton's equations. We will examine this in more detail later but for now we will examine Einstein's *Annus Mirabilis* papers as this will give us an insight into the moons seemingly tardy behavior and also the body of Albert Einstein's work.

First published in 1905 in the *Annalen der Physik* are four articles written by Einstein pertaining to a number of unsolved problems in the world of physics. Just as it is today so it was at the beginning of the last century that a list of unsolved problems seemed to plague the world of science and these four papers promised the answers. The papers addressed the photoelectric effect, Brownian motion, special relativity and the matter-energy equivalence or $E = mc^2$ as we know the formula today. The first of the articles appeared on the ninth of June and covers the photoelectric effect as we have previously discussed.

The topic of Brownian motion was reviewed in the second paper published on the eighteenth of July. Brownian motion is a phenomena named for the man who first observed it in 1827, Robert Brown. At the time Brown was a botanist who happened to be studying pollen grains suspended in water through a microscope. While attempting to examine the pollen sample he curiously noted that the pollen refused to stay in just one place in the water, preferring instead to move around. Even more curious to Brown was that the pollen was not set in motion in any way, it simply started to and continued to move of its apparent own free will. Not only that but the motion did not seem to obey the Laws of Newton as the motion of the particles was random and ever changing. This is a violation it would seem of Newton's first law of thermodynamics as an object in this case pollen should not alter its course unless acted upon by an outside force. Yet as far as Brown could tell there was nothing in his samples other than pollen and water, so what was pushing on the pollen? In order to satisfy Newton's first law of thermodynamics a suitable force had to be found to produce enough force to change the motion of the pollen grain suspended in water.

What Einstein proposed in his second paper was that it was the molecules of water themselves that were supplying the necessary energy to move the pollen in different directions. If this was in fact what was causing the pollen grains to move then the only driving force behind the water molecules were the component atoms of water. In essence what Einstein was suggesting was that it was the atoms that were supplying the necessary forces required to move the water and therefore move the pollen. This was somewhat controversial at the time because up until this point the idea of atoms was not fully embraced by all scientists. Today we understand that atoms make up matter and we are quite comfortable with that, during the early 1900s however the actual reality of atoms was not wholly accepted within the science community. It was understood that atoms provided an excellent framework and theoretical tool for explaining many observed phenomena such as those seen in chemistry but no empirical evidence had yet been found. By explaining Brownian motion as the product of the movement of atoms the scientific community finally had empirical proof of the existence of atoms. What is more the limited scientific apparatus of the day which hampered scientists because they could not possibly observe atoms directly were in a sense bypassed using Brownian motion. Pollen grains are of course many billions of times larger than any atoms from which they are made and while tiny to human eyes can be observed without special tools. In fact a simple light microscope is all that is needed to make accurate measurements on the movement of the pollen grains and through these measurements scientists could study atoms. This sort of macroscopic observation of the atomic and subatomic world is what enables science to begin bridging the gap between the small unseen world of matter and our own empirically observed world of macroscopic matter.

One modern day example of such a phenomenon is seen in a process known as neutrino oscillation. Neutrinos are known to have three shapes or flavors as they are called and are electron, muon and tau respectively. The idea that neutrinos have this oscillating property was first proposed in 1957 by the Italian physicist Bruno Pontecorvo. Pontecorvo was obsessed with the study of neutrinos and their properties which eventually led to his discovery of neutrino oscillation as well as methods for detecting anti-neutrinos created in nuclear reactors and of fundamental differences in neutrinos associated with muons and electrons. What neutrino oscillation is concerns the unknown mechanism of neutrinos behavior whereby a neutrino generated in one flavor will be measured as having changed to a different flavor at a different time. Each neutrino eigenstate has a different flavor and mass which will result in a specific neutrino oscillation. The three different masses of the neutrinos will all transform into different flavor states at different rates of propagation due to their mass differences. The resultant mixing of flavor and mass states can be viewed over time as the neutrinos propagate through space.

The neutrinos do not travel through just totally empty space but through matter as well and both of these factors affect the rate at which a neutrino will change its flavor. The driving force behind the change in flavor states is quantum mechanical and each phase state of the neutrino is made of a mixture of the electron, muon and tau flavors with only one flavor holding dominance to the others in that phase state. So an electron neutrino flavor is predominantly electron but contains a mixing of the three. As the neutrinos oscillate through their three flavor states the mixing of each component changes in a periodic fashion and as a result if one watches a particular neutrino long enough it will turn back to its original state. An electron neutrino will start out its life at the time of first observation as mainly electron and over enough time and distance will cycle through the other two states being dominant and return to a state where the electron flavor is dominant in roughly the same mixture as when it was created. A neutrino will oscillate as long as it can maintain what is known as coherence in its quantum state. The important detail here is that the coherence length for a neutrino oscillation is incredibly large when compared to the miniscule differences in neutrino masses for each flavor state. This difference results in the extremely small world of quantum fluctuations within the neutrino to be observed over incredibly large distances, in some cases tens to hundreds of kilometers. What this means is that a microscopic effect can be studied very easily at a macroscopic scale. Essentially we are seeing the same thing today with neutrinos at a quantum scale as what Robert Brown observed at the molecular level in 1827 with his pollen grains in water.

The concept of neutrino oscillation helped solve a problem that went unresolved in physics for decades known as the solar neutrino problem. As you have probably guessed by the name solar neutrino problem it was a result of neutrino measurements taken from our Sun that could not be resolved. Inside our Sun are two things we all love dearly even if we are not aware of them. The first is lots and lots of hydrogen atoms specifically in the form of hydrogen nuclei or protons and the second is gravity, more gravity than anywhere else in our solar system. Why we love these two things is simple they create nuclear fusion without which everything on Earth would not be alive. Every second of the Sun's life since it started to shine it has been fusing an almost uncountable number of hydrogen nuclei into helium releasing energy in the process. This energy provides us and all life on Earth with the light and heat we need to survive. Of all the energy and particles emitted by the Sun however light is not the only one.

Another particle emitted is the neutrino and scientists have long been aware that neutrinos are emitted in nuclear reactions such as the fusion taking place in the sun's interior. The proton-proton chain reaction converts hydrogen nuclei into helium and a wide assortment of energy and particle products most of which are slowed in their journey from the core of the Sun to its photosphere and eventually to us. A humble light photon for example is capable of making the journey from the Sun to Earth in as little as 8 minutes and 17 seconds even though the sun is approximately one hundred and fifty million kilometers away. That same photon however will spend the first ten thousand to two hundred thousand years of its life just travelling out of the Sun itself due to the way photons are involved in re-absorption in the solar plasma. On the other hand we have the neutrino which barely interacts with normal matter at all and can escape the radiative zone of the sun and travel all the way to Earth in as short a time as only two point three seconds; that is the time it takes to reach the surface of the photosphere and leave the Sun not the time it takes to travel the distance from the Sun to the Earth. Astronomers and physicists have been well aware of this fact for years and naturally were quite interested in studying neutrinos as much as possible to learn more about how they could achieve this feat and of course their role in the life of the Sun.

A problem was noticed however when predictions about the number of neutrinos generated by our Sun through the process of nuclear fusion were compared to what scientists here on Earth were actually measuring. What was being recorded here on Earth by even our most advanced detectors was a neutrino count only one third of what we predicted should be coming out of the Sun based on our best models. The models were backed up by also measuring neutrino emission in our own nuclear experiments here on Earth. This was the solar neutrino problem. The expected neutrino emitted by the Sun's fusion reaction is the electron flavor and is dependant upon the temperature of the nuclear reaction that creates them. The neutrino spectrum was analyzed and it was found that either the Suns core temperature had shifted to such a degree that neutrino production was compromised or another explanation was necessary. It was long believed that neutrinos were massless particles that would travel at the speed of light and it wasn't until the famous supernova observed in 1987 and consequently named supernova 1987A that it was first thought possible that neutrinos had mass. The time that neutrinos were measured arriving at two different neutrino detectors varied to a degree that scientists thought the neutrinos were travelling slightly slower than the speed of light and therefore had mass.

The first detector to measure neutrino oscillation was the Super-Kamiokande detector in Japan. What it observed was the changing of muon neutrinos into tau neutrinos through cosmic ray interaction with Earth's atmosphere. These observations were the next step in solving the solar neutrino problem. The absolute convincing proof of neutrino oscillation from our Sun that would explain the solar neutrino problem came in 2001 from the Sudbury Neutrino Observatory in Canada by using heavy water in its detectors. The Sudbury detector can measure all incoming solar neutrinos and distinguish between the types to the degree that it can sort which neutrinos are the electron flavor and which are not and therefore muon and tau. Of the neutrinos detected thirty five percent were electron and the remaining sixty five percent were of the other two flavors although the breakdown between them was unknown. By accounting for one hundred percent of the predicted neutrinos from scientific models the problem of the solar neutrinos was solved. Overwhelming proof that microscopic effects can be viewed on the macroscopic level to solve the unsolved problems of the universe.

So what we have seen through Einstein's explanation of Brown's pollen grains is that the very movement of matter is important to how we observe our own universe. We have seen through Brownian motion in 1827 and neutrino oscillation in 2001 that the quantum nature of matter affects how we perceive it on the everyday macroscopic level. The analysis and understanding of how matter moves is therefore fundamental to fully understanding how the universe works. If we leave the microscopic quantum scale for a moment and focus on only macroscopic systems it is soon realized that a different set of rules seems to apply to matter. If one were to try and apply our current understanding of quantum mechanics to large objects the size of planets, stars or galaxies you would not be able to explain the macroscopic observations we see constantly in the universe. The orbits of stars around galaxies, how black holes work or the deceleration and acceleration in the expansion of the universe cannot now be explained through the microscopic quantum world. What is needed is a new way of incorporating everything we do know about matter, space and the universe into a single theory that explains just how the universe does work. Such a theory will have to wait for now, but that does not mean we cannot explore one theory proposed by Albert Einstein that attempts to explain some of what we do observe.

The motion of large objects in the universe and how they interact with each other received a satisfactory explanation on the 26th of September 1905 in the third paper published by Einstein. While it certainly does not provide an explanation to how the universe works or how to incorporate the microscopic quantum world with the large scale macroscopic one, it does explain well the ways in which large objects affect space and each other. The interaction and observation of things like the Earth and our moon plus why the moon does not appear in its orbit where we might expect using classical mechanics is explained using the special theory of relativity. Originally submitted as a solution to some of the current day discrepancies between theory and observation concerning the moving magnet and conductor problems the paper would go on to become the most famous bit of thinking Einstein ever did.

A scientist named James Clerk Maxwell had presented a set of equations designed to answer problems of moving bodies, magnetism and electricity which were known as Maxwell's equations. The equations were initially successful and seemed to provide a mathematical framework for much of the experimental observations made into these systems. The equations work well in everyday environments but at speeds much faster than we normally experience, such as the speed of light they tend to be less useful. At the time it was believed that space and time were independent of each other and that since light was a wave it must have some sort of medium to travel through. Waves on a lake for example conduct their energy through water, sound conducts its energy through the air and it was thought a light medium akin to water or air must exist to transmit light energy as well. These perceived facts were what was impeding any further progress from being made to fix the problems such as those found in solutions to Maxwell's equations. What Einstein proposed was a solution to these problems that combined space and time and removed the aether or light medium from models of the universe. Of the two postulates he proposed the first states that regardless of your frame of reference of motion the laws that govern mechanics, electrodynamics and optics are all governed by the same laws of physics. This is the part of his paper where Einstein introduces the term relativity and from which he will eventually develop his general theory of relativity.

The second proposal of Einstein's pertains to the speed of light which he states is unchanged no matter what the inertial frame of reference of the body that emits it. This is simplified by saying the speed of light is always the speed of light regardless if the object that emits it is stationary or moving. The reason this is of importance is that in classical mechanics this would not be true. In classically described systems velocities are additive and this can be easily demonstrated with a simple analogy. If for example you are standing still with respect to the Earth and you were to throw a stone at fifty kilometers per hour then the stone would be moving at a speed of fifty kilometers per hour relative to the Earth. Now if you yourself were moving at fifty kilometers per hour with respect to the Earth and you threw the same stone at the same speed then the stone would now be moving at one hundred kilometers per hour relative to the Earth. The two velocities are added to find the total speed of the stone relative to the Earth. However this is not true of light as its total velocity is not added. If you are stationary or moving at fifty kilometers per hour and you turn on a light source the photons emitted from that source will always travel at the speed of light and not the speed of light plus fifty kilometers per hour. These are the two main theories proposed by Einstein in his third paper and will be evolved somewhat in 1915 when Einstein publishes his general theory of relativity. We will discuss the meaning of this 1905 paper and the speed of light in more detail later but for now let us look at the last of the four papers to comprise his *Annus Mirabilis*.

Einstein's last paper for 1905 was published on the twenty first of November and presents his famous $E = mc^2$ equation. The title of his paper almost seems to be an obvious give away to his equation and contains a literal dry explanation of that equation but in the form of a question. The question he asks is: Does the Inertia of a Body Depend Upon Its Energy Content? In other words does an object having mass possess a relation to its equivalent energy? The answer to his own question is the energy to mass equivalency equation. Albert Einstein's famous equation however was not made possible without the work of two other brilliant scientists by the names of James Clerk Maxwell and Heinrich Hertz and he mentions them directly in his paper. James Clerk Maxwell was a Scottish mathematician and physicist who gave us the classical theory of electromagnetic radiation. Maxwell gave the world one of the most important discoveries of all time which was that both electric and magnetic fields travel through space in wave form and the velocity at which both travel is equal to the speed of light. Published in 1865 his *A Dynamical Theory of the Electromagnetic Field* would become a cornerstone of modern physics and our understanding of the universe. Ultimately this would prove to be one of the greatest realizations in physics as we consider electromagnetic radiation as one of the basic elements of the universe and something that we utilize constantly in science and technological applications. His work would also lead to the successful prediction of radio waves and a greater understanding overall of many electrical and magnetic phenomena. Considered one of the greatest scientists of all time Maxwell's work was to be built upon by another of histories great minds that of Heinrich Hertz.

Born in 1857 Heinrich Rudolf Hertz was a German physicist who studied the work of James Maxwell and electromagnetic radiation. Hertz came into contact with another famous German scientist by the name of Hermann Ludwig Ferdinand von Helmholtz who took Hertz on as an assistant in Berlin. It was after the two had known each other for a time and Helmholtz saw the potential in Hertz that he suggested his assistant compete for the Berlin Prize problem at the Prussian Academy of Sciences. The prize would go to anyone who could prove part of Maxwell's theory which stated that the depolarizing or polarizing of insulators should produce an electromagnetic effect. Although Helmholtz was confident of Hertz's abilities Hertz himself thought the problem too difficult to solve and instead looked at studying a different aspect of Maxwell's equations. Hertz built a device that would eliminate all outside sources of contamination from wireless phenomena and study radio waves. Working on measuring the distances between an oscillator and zinc plates he timed the radio pulses created by electric currents and determined that they were travelling at the speed of light as predicted by Maxwell's equations. This work he conducted between the years of 1886 and 1889 proves the existence of Maxwell's electromagnetic waves and helped Einstein formulate his theories on energy and mass equivalency.

Chapter 3
Nuclear Physics Emerges and Explodes

Early Nuclear

Now that we have seen where Einstein drew some of his inspiration lets go back to his mass to energy equivalence equation, the famous $E = mc^2$ that inspired his fourth and final paper for 1905. First let us break down the equation into its parts and then we can look at what it means. In it we find E for energy, m for mass and c is the speed of light which he has multiplied with itself. The most important part of the equation is the energy as it is not kinetic or potential energy which was understood and described in classical models beforehand. What Einstein is referring to is the rest energy of the particle which is the energy that any massive particle contains within itself. Not kinetic which would be the energy of motion of the particle and falls under the laws of Newton's F=ma, nor is it potential energy which could become future kinetic or perhaps chemical energy. This is a distinct new type of energy that he proposed was simply contained inside of any massive particle. What this does is imply that on some level matter and energy are one and the same thing. For our purposes here we will not take that too literally as modern science does not quite have a handle on what matter is at the most fundamental level and debate exists as to what matter really is and therefore how it fits into the whole universe in general. After all modern physics is still deciding whether matter is a particle, a wave or some form of other exotic multidimensional material and in fact it might be found that there is no such thing as matter at all. In that case energy would not equal matter, rather energy would equal energy. For Einstein's purposes in 1905 however simply demonstrating that matter possessed a rest energy unique from other forms of energy and that he had calculated an equation necessary to work out its conversion from form to form was enough. So that much matter was equal to that much energy. Now scientists could calculate energy and mass equivalencies for numerous types of observed phenomena. This includes phenomenon that at that time in history were either only in the theoretical stages or in some cases not even realized at all. Most of the initial interest in applying Einstein's equation was of course purely theoretical as it gave scientists a new concept to work with. As with any new idea it met with a range of responses from rejection to acceptance and in many cases simple curiosity. One area of physics that was interested in it was atomic theory as it offered a window into the properties of the atom. Another field for which it found quick acceptance was in nuclear physics. Both types of nuclear reactions, fission and fusion, rely on this equation in order to calculate what is going on with the nuclear fuel at an atomic level. The most sought after aspect of course was in finding the energy outputs from these reactions and harnessing them.

Nuclear physics can be broken into many sub fields and each has its own particular uses and applications. In essence a nuclear reaction involves two or more particles, these can be nuclei from atoms or subatomic particles such as an electron, neutron or proton, which is really still a nuclei of hydrogen. The two particles or nuclides will collide with sufficient force to produce a different set of nuclides from the two put into the nuclear reaction. There are some exceptions to these rules however; the first is if no reaction takes place. A reaction is only said to take place if new nuclides are produced from the interaction of the particles, if no new nuclides are formed then the result is termed nuclear scattering.

The second exception to this rule is that theoretically any number of particles can take place in the reaction process. Since it is such a low probability that three or more particles would meet perfectly at the exact instant necessary and with sufficient force to cause a nuclear reaction it is virtually unheard of to have a three or more nuclide reaction. This does not mean in any sense that it can not happen only that the probability is so low it is rarely considered. Some three nuclide reactions have been theorized and calculated such as the triple alpha process. This is a set of fusion reactions that exists in the cores of older stars that have stopped fusing hydrogen into helium and are now fusing just helium nuclides or alpha particles. The basic reaction runs as follows: two helium fuse to beryllium 8 this element is unstable and decays quickly. As time progresses more beryllium 8 is created before it can decay faster than a helium nuclei can fuse with it and so the beryllium 8 fuses with the helium. The resultant element is stable carbon 12 and represents the end of the triple alpha process.

The third exception to the rule is that a nuclear reaction must be made to happen in other words a natural creation of a new nuclide does not count. Being made to happen does not imply artificial creation of the reaction, for example fusion inside a stars core is made to happen by the natural force of gravity on the nuclear fuel. An example of a natural reaction that was not in any way made to happen would be radioactive decay. Therefore this is a form of atomic decay and not of a nuclear reaction thereby satisfying the third exception. Radioactive decay reactions however are still subject to the energy mass equivalence equation of Einstein as we shall see.

The first type of reaction discovered that involves Einstein's equation was that of radioactive decay. This falls under the category of problems known at the time of Einstein's 1905 publication but had yet to be fully understood. The problem of what radioactivity was begins with a French scientist by the name of Henri Becquerel in 1896. At the time there were no such known phenomena as radioactivity and so Becquerel was investigating phosphorescence in various materials. Most people are familiar with phosphorescence from children's toys that glow in the dark after being subjected to a light source; the brighter the source or the longer the light source is made to shine on the toy or both the more intense and longer lasting the glow in the dark effect. What is happening to these phosphorescent materials is that they are absorbing energy and slowly emitting it later in the form of light photons. Most materials will absorb and reemit the energy almost immediately but a quantum effect of phosphorescent substances allows them to slowly leak the energy over minutes or hours. Henri Becquerel believed that the same phosphorescent activity was present in cathode ray tubes and was responsible for the glow they gave off. He knew the cathode ray tubes produced x-rays and so he set out to test whether various materials that produced phosphorescent effects also produced x-rays. His method of experimentation was simple and involved wrapping photographic plates in black paper to protect them from light and then place the test materials on them. He used numerous phosphorescent salts but found he only got a reaction on the photographic plates from uranium samples and so was found the radiation known as Becquerel rays. What confused this finding however was the realization that non-phosphorescent test samples still produced a reaction of the photographic plates. Since this unexpected result was observed it became obvious that something else was passing through the paper and affecting the photographic plates.

Alchemy turns out to be the answer for this process, not the transforming of base metals into gold but of transmutation. Transmutation in alchemical terms is the changing of one substance into another and realistically this is exactly what is occurring in radioactive decay. In radioactive decay one atom is transmuting literally into another although it is known as nuclear transmutation in science. While the notion of becoming wealthy through the use of alchemy had become extinct by the time of scientists like Ernest Rutherford and Frederick Soddy, a colleague of Rutherford's, it turns out transmutation was the answer to an observation they had previously noted. What they had seen in their laboratory in 1901 was the nuclear transmutation or radioactive decay of radioactive thorium into radium. They also observed that no outside input was made into the sample of radioactive thorium to make it change into radium; it was a seemingly natural property of the radioactive thorium. What they had discovered was alpha decay, a process by which an atom loses an alpha particle thus reducing its total atomic mass and number and decaying into another element. Essentially three types of radioactive decay exist and they are alpha, beta and gamma decay. Alpha decay is the emission of a helium nuclei and typically comes from only the heaviest of elements. Beta decay involves the release of an electron or a positron from the nucleus. Lastly gamma decay occurs when a photon is emitted from the nucleus but usually not directly. Commonly in gamma decay we find a radioactive element that undergoes alpha or beta decay and release what is known as a daughter nucleus. Daughter nuclei are simply the decay products from a radioactive element and are usually unstable themselves. It is from these unstable decaying daughter nuclides that energy is emitted in the form of gamma radiation and therefore is the source of the gamma decay process. The discovery of radioactive decay also led to the discovery of half lives.

Ernest Rutherford is credited with the naming of both alpha rays and beta rays to distinguish the differences in the types of radiation. He noted that alpha rays had far less penetrating power compared to beta rays as he expanded on Becquerel's photographic plate experiments. He noticed that thick sheets of paper could easily stop the alpha rays but not the beta rays. We know now that an alpha particle is made of a helium nucleus and a beta particle of an electron. Since a helium nucleus is made of two protons and two neutrons it is far more massive than a single electron and has a much higher probability of interacting with and consequently being absorbed by the paper used to shield the photographic plates. An electron on the other hand is almost non existent by comparison and can therefore penetrate the paper easily thereby darkening the photographic plate. This is of immense importance because Rutherford began noticing that the element thorium which itself is radioactive seemed to be generating a kind of radioactive gas. While it is known that all forms of thorium which was named for the Norse god of thunder Thor are radioactive the most stable and naturally occurring thorium decays by the process of alpha decay. This common form of thorium is far more plentiful than uranium and it is believed that there is four times as much thorium in the Earth than uranium. Rutherford was undoubtedly experimenting with this naturally occurring type of thorium and witnessing alpha decay. What this gas was is known as a daughter nuclide that is resultant of the decay of the parent nuclide in this case thorium. More proof of alpha decay and plus led to Rutherford making the observation that for any given sample size of this test material he used it always took the same amount of time for half the sample to decay. If Rutherford started with one gram, two grams or three grams of thorium it did not matter, after approximately eleven and a half minutes the sample had reduced to half its original size.

Half Lives, Reactors and Sunna

Rutherford while working at McGill University with his colleague Frederick Soddy began examining this problem in more detail. Soddy eliminated any outside sources of experimental contamination and found that the decay was caused solely by the natural properties of the sample itself and that alpha particles in the form of helium gas were always present. Conducting their research since 1900 it was in 1902 that a full published work of all their experimental findings was released and entitled a Theory of Atomic Disintegration. We take radioactive decay and half lives for granted today but in the late 1800s and early 1900s this was an incredible claim. After all it had not been all that long since the existence of atoms was definitively proved and scientists learned the true structure of matter beyond mere speculation. At the time it was still believed that atoms being the base unit of matter were at least indestructible and therefore all ideas of radioactive decay or nuclear reactions were nothing more than speculation. The fact that Rutherford and Soddy had researched carefully and discovered essentially everything we initially knew about decay processes is remarkable. Rutherford also gave the name half life period to the length of time it will take a sample to decay to half its original size, although today we simply shorten it and say half life.

The decay rates of radioactive elements is traditionally described as a type of exponential process due to the number of half lives passed and the remainder of the starting sample size. To visualize this relationship let us look at a few elapsed half lives and the effect on a starting sample, the arbitrary size of the initial sample does not matter it is calculated from one hundred percent. So what we see after zero half lives is the original sample at one hundred percent, after the first half life the sample is half its original size and therefore at fifty percent. Two half lives will leave one quarter of the sample at twenty five percent, three half lives will leave one eighth the original sample at twelve point five percent. In other words as the number of half lives increases linearly the sample size decreases exponentially but theoretically will never reach zero. In reality one hundred percent of the original sample will eventually decay into another element and so on until a stable element is the end product, one such end product is often times lead.

This illustrates a discrepancy between what actually happens in the universe and what theoretically happens or to put it another way what we observe and what we calculate. Observation is right always and there are zero exceptions to this rule in the universe, not in our laboratories or in the infinite reaches of space. What often times happens is we run into trouble matching our mathematics to observation and this is the case with half lives. According to the math the sample will never truly disappear as we can keep using smaller and smaller fractions of the original, one eighth, one sixteenth, one thirty second and so on. Of course what actually happens is we are left with fewer and fewer atoms in the original sample and eventually the last atoms will decay leaving zero percent of the original sample. This is illustrated in the possibility of having three remaining atoms undergoing radioactive decay; you will not get a product equal to one and a half atoms as this is impossible. You will only be left with a whole number one or two and even that will eventually decay completely. Still the formula for calculating half lives is commonly given as such that you can calculate for an infinite number of half lives passing and therefore reducing the original sample an infinite number of times. Think for a moment of the problems facing modern science today as the mathematics tells people that something should be there in the universe, well maybe. The trouble is trusting the mathematics implicitly can leave you chasing a figment of the mathematical imagination and nothing exists there at all so do not spend too much effort looking for what you will not find.

Nevertheless one of the most important tools in science has been evolved through the use of measuring half lives and is known as radiometric dating. This involves knowing the half life of a particular element and seeing what percentage of that element is left after time has elapsed. The elements involved are obviously radioactive and must be taken up by the test material in some way. The most common example is known as radio carbon dating and is practically a household concept. The way this process works begins with cosmic rays creating an unstable radioactive isotope of common carbon in the atmosphere. The carbon becomes carbon 14 and is present in the air where any living organism can uptake it usually through respiration. Air bubbles in glaciers can also trap air and therefore trap carbon 14 showing in some instances inorganic trapping of carbon 14. The half life of this radioactive form of carbon is 5730 years and so from this known fact we can determine the age of a preserved specimen or sample in the lab. Lets say we wanted to know the age of the Nebra Sky Disc found in Germany, the first thing we need is a sample of carbon 14. The disc is metallic but organic material encrusted on the disc and inside of its various joints can be dated. Extracting a sample of this material we examine how much of the carbon 14 has decayed and once we know the percentage finding the age is as easy as looking it up on a chart. This simple tool of radiometric dating is one of the most universally used tools in virtually all disciplines of science, from biology to physics to archaeology. All of this made possible by the initial work carried out by Rutherford and Soddy.

Rutherford did have one last contribution to make to the world of alchemy and science, as he was the first person to transmute one element into another. While he was a purely scientific man and one who would surely scoff at the idea of alchemy it was in 1919 that he performed his first transmutation of one substance into another. Although Rutherford may have thought alchemy to be nothing more than antiquated science if he had managed to change lead into gold he may have changed his opinion. What he did do was change nitrogen into oxygen and thus carries out the world's first nuclear reaction. When most people hear the words nuclear reaction they immediately think of the splitting of the atom and of either nuclear bombs or nuclear reactors. Yet the splitting of the atom first carried out by Otto Hahn would not occur for years to come and we shall examine those experiments later. Nevertheless Rutherford had carried out a nuclear reaction as he began his experiments with two different nuclides and obtained a product of a different nuclide through the reaction of the two.

Now working in England in the year 1919 he was continuing in the long line of experiments popular amongst physicists of the time where by some form of element or radiation was bombarded onto a test sample. In this particular case the popular alpha particle was being shot into samples of ordinary air. What he found was unexpected as a radioactive gas was being produced in these experiments that behaved very similar to hydrogen. Eventually he began experimenting on nitrogen gas only as he believed this to be the source of the strange radiation. Now firing alpha particles into pure nitrogen gas he was transmuting the nitrogen into an uncommon form of oxygen and a spare proton. Oxygen 16 is by far the most abundant form of naturally occurring oxygen in the universe and what Rutherford had created was the less common oxygen 17 type. The proton however was causing him problems as at the time of his experiments protons were not known to exist. The mysterious radiation he noted was very similar in most respects to the hydrogen nuclei he had encountered in some of his own previous experiments. So nitrogen was producing hydrogen nuclei and since hydrogen was the lightest and smallest of known elements Rutherford proposed the idea that not only was hydrogen nuclei a part of a nitrogen atoms nucleus but a part of all nuclei. He made this conclusion based on the fact that when subjecting hydrogen gas to alpha radiation he still got hydrogen nuclei out of the reaction. Combining this with the fact that hydrogen is the smallest element Rutherford suggested in 1920 that hydrogen nuclei are indeed a new fundamental particle. Rutherford had thus proved the existence of the proton through his experiments with nuclear transmutation.

We have now seen the first type of nuclear reaction and will move on to the type of nuclear reactions known as fission reactions. This is where we see the term splitting the atom first appear and where the tremendous amounts of stored rest energy are released in accordance with Einstein's equation. It begins in 1932 with two of Rutherford's fellow physicists who had been experimenting in their own way with transmutation. The first of his colleagues was John Douglas Cockcroft an English physicist and the second was Ernest Thomas Stinton Walton an Irish physicist. The pair working at Cambridge University built the world's first particle accelerator known by their names as the Cockcroft-Walton generator. It was made from a circuit that resembled a ladder with diagonally connected bars rather than horizontal ones and worked to dramatically increase the voltage you put into it compared to the voltage you could extract from it. The power from this circuit is supplied to the test tube through which they fired electrons, neutrons and protons depending on their experiment at the time. When they first succeeded in splitting the atom however it was the proton or hydrogen nuclei that they used. The target sample was lithium and in 1932 they became the first people to successfully split an atom. The product of the reaction was two helium nuclei proving that they had indeed split the nucleus of an atom. Science was now entering the age of nuclear research properly and would see many new advancements in both pure science and in technological revolutions. Due to their instrumental work in essentially kick starting the nuclear age they would receive and share the Nobel Prize in 1951.

The person responsible for giving us nuclear fission was a man named Otto Hahn and without him we would not have such a detailed understanding of nuclear processes. Otto Hahn was a German chemist who had been fascinated by science and in particular chemistry since his youth. While most atomic research was being conducted by physicists it was Hahn a chemist who was gifted at discovering new elements through his investigations of radiochemistry. Some of the new elements or isotopes he discovered were forms of thorium, lead and ionium. There was great interest in radioactive elements at the time of Otto Hahn and others due to the fact that many were used in medical treatments. The use of radioactive substances as tracers was first conducted in Germany in the 1920s and radio tracers are still used today in certain medical examinations. Therefore the discovery of new elements that could be suitable for medical applications was much desired by researchers at the time just as it is today. Of great importance was in finding radioactive isotopes that were substantially cheaper than those currently used since production of all such medicines was expensive. In order to spread the use of radioactive substances in medicine and to allow it to be available to a wide section of the population news of freshly discovered elements was often made public. Hahn conducted his early work on discovering new elements in Canada and Germany around 1905 and 1906, but it would be three decades before he gave the world his greatest gift.

It was in 1938 that Otto Hahn was experimenting with what were known as transuranium elements. At the time it was believed that uranium was the heaviest of all elements and that in order to reach heavier elements a sample of uranium had to be bombarded by neutrons. This tried and true method of bombarding samples with neutrons as we have seen was widely accepted and popular due to the first nuclear transmutation experiments. So elements higher than uranium were thought to be created in this way and then in 1938 when Hahn and his assistant Fritz Strassmann found that traces of an Earth metal appeared in their sample they knew something wrong or new was happening. Strassmann who was also a German chemist and worked with Hahn for years was familiar with radiochemistry and together they could not find a fault in their experiment. Double checking their samples and equipment they performed the deciding experiment in late 1938 and no matter what they did the experiment invariably produced what they thought was barium. This was a problem as it was strongly believed in the scientific community that bombarding uranium with neutrons would only create heavier elements and not an element one hundred nucleons lighter. Otto Hahn had successfully split the uranium atom to produce nuclides of existing elements and give birth to nuclear fission. Hahn described this splitting of the uranium nuclei as bursting in his first interpretation of the results. Otto Hahn received his Nobel Prize in chemistry in 1944 and left a legacy that stretches far beyond nuclear fission alone.

All of the resulting technology from Hahn's work has undoubtedly changed the world we live in, from nuclear reactors to nuclear bombs to most importantly a greater understanding of how the universe works. Nuclear fission is seen commonly in two types of industries today which are power production in nuclear reactors and the military in both fission and fusion weapons. Although the way the energy is harnessed in each application is quite different from one another. After all our nuclear power plants do not explode in order to produce the necessary heat we need to generate electricity. The fission reaction occurring in a nuclear power plant is designed to be slow and controlled so as to avoid a meltdown. Since the desired product of these fission reactions is heat energy sufficient cooling is required in order keep the nuclear reaction from running away as was famously seen in the debacle of the Chernobyl disaster in 1986.

When running smoothly a nuclear power station draws its power through nuclear pellets stored inside a nuclear fuel rod. The rod is lowered into the reactor core where the radiation being emitted from the rods produces a fission reaction in the core and releases heat as a waste product. The waste product of the reaction as it turns out is precisely what we want to find in great abundance inside a nuclear power plant. The plant itself produces all of its electricity from this waste heat. The heat is used to boil water into steam which is concentrated and fed under pressure into a near perfectly balanced turbine. The turbine spins and the spinning motion is transferred to a generator which creates electricity. All of this of course is far more complicated than it sounds to actually construct and operate a nuclear power plant but the mechanism by which the power is generated is really that simple. Otto Hahn's bursting uranium atoms have given us the modern nuclear power plant and Albert Einstein's mass energy equivalence equation explains how much energy we get out of the reaction. It is estimated that if you burned ten million kilograms of coal you would need only three and a third kilograms of uranium to get the same amount of energy released. This demonstrates the phenomenal power of a nuclear reaction over a chemical reaction in terms of pure energy output. Technical challenges in constructing and operating safe nuclear power facilities as well as total construction and operating costs means that today about eleven percent of the world's power is generated through nuclear power and around forty percent through the burning of coal.

There are other types of nuclear reactors that are used for research and science that also employ the fission reaction mechanism. Both are forms of research reactors that are not concerned with the production of electricity per se. The first type of reactor produces copious amounts of neutrons for medical research such as isotope reactors or nuclear testing of materials. The second is called a breeder reactor and is also interested in power production but at the moment many of the technical limitations still need to be overcome. What a breeder reactor does is consume essentially all of its fuel by continuing to utilize its own waste products as more nuclear fuel. The reactor in a sense breeds new fuel hence its name and while this technology promises to greatly increase power output it also serves two other functions besides. The first is research into nuclear power and reactor design as the numerous technical challenges facing the smooth and successful operating of such plants still needs to be refined before widespread commercial construction can begin. The other aspect of this reactor type is that it can theoretically clean up radioactive waste by burning it as more fuel. The reactor is designed to convert the waste products it creates into more fuel already but it is hoped that waste fuels not produced in a breeder reactor can be consumed by one as well. This would greatly help not only in meeting energy demands of the future but in cleaning up the rather large amounts of nuclear waste that has and is being produced with today's fission reactors. At present however most nuclear reactors are not breeder reactors and due to the abundance of uranium deposits the global power industry does not feel a significant urge to invest wholly in redesigns of existing reactors or construction of new breeder reactors.

Now it is time to examine the last of the major nuclear reactions known as nuclear fusion. In an interesting turn of events we find that nuclear fission is the type of reaction we use to generate power on Earth and we embrace it for its ability to transform so much more stored energy in our fuel to electricity than other forms of fuel. However we are aware of the massive superiority of nuclear fusion when it comes to transforming energy from fuel over nuclear fission and we are trying to build nuclear fusion reactors. Yet despite our best efforts we so far can not generate a self sustaining fusion reaction that extracts more power than is used to maintain the reaction. So all though all of the nuclear reactors on Earth are fission with no fusion ones present at all it is ironic that the process of nuclear fusion itself is by far the most common reaction between the two in the universe. We may not have a fusion reactor here on our world but we do have one in our solar system that we use everyday to harness energy from, our Sun. The modern name Sun that we call our home star is derived from the Old Germanic word Sunna and describes a goddess who each day travels across the sky as she is pursued by wolves which are evil. In the Norse book Prose Edda it is said she moves swiftly across the sky as if being pursued and that she is fearful of what evil follows. It is from these ancient religious works that we are introduced to the idea of a being so bright and beautiful that her radiance shines down upon and lights our world. We understand now that what shines down are photons in the form of visible light and that her brightness and radiance is brought about by a self sustaining nuclear fusion reaction inside her.

This however does not change the fact that we have been using the Sun as a source of nuclear power since before humans even existed, in fact before any life existed on Earth. It is after all the rays from the Sun than were partly responsible for the chemical reactions that created a world in which life could evolve. The sun also provided fuel and opportunities for natural selection to early life as well thus helping to shape the rich and diverse world we see today. One example of this besides plants that are better at photosynthesizing sunlight when compared to their competitors is early animal life that used sunlight to either avoid being eaten or to eat others.

Perhaps most importantly we find the Sun directly supplying power to us through the mechanism of photosynthesis. Photosynthesis is name of the molecular pathway by which plants are able to harness the solar output and transform it into more plants. Plants become the backbone of the food chain for all other organisms on Earth, including ones that do not directly eat plants or animals but live off of the by products of the main food chain. We as humans are omnivores and evolution was pushing us towards a more and more carnivorous path meaning we would become increasingly dependant on plant eaters and the plants they eat. It takes a lot of bio-energy to fuel a large brain capable of the advanced thinking we humans can accomplish today and it would not suffice for us to spend all day grazing to meet our energy requirements. Consuming plant eaters instead provides a concentrated source of energy that allows us to spend less time eating and more time thinking. In this way the mother's milk of our goddess Sunna can be thought of as the life giving photons generated by the core of our nearest star. On an interesting side note it is further written in Poetic Edda another ancient religious text of the Norse that Odin says that if the shield of the Sun, a proto reference to a sun god were to fall from the sky the mountains and sea would burn up. What Odin means of course is that the world our Earth would be destroyed in fire and this is fascinating as ancient peoples were obviously aware of the Sun's great power. Even though they possessed no way of measuring the incredible heat energy generated by the sun and the warmth of the Sun is merely pleasant here on Earth a reference to the Sun's unmatched power in our solar system was still present. In fact in the future the Sun will expand to a red giant star and the photosphere of that star will reach Earth's orbit destroying our world. So yes the ancient Norse were correct in saying the shield of the Sun touching our world will destroy it forever.

The unmatched power that our Sun creates is first made deep inside of the Sun at the very core where the temperature and pressure is of sufficient strength to cause natural fusion to take place. Nuclear fusion is by definition the reaction of two nuclides such that they produce a third new type of nuclide and energy. In any nuclear reaction we see the release of energy which accounts for the difference in mass between the reactant nuclides and the product nuclides. Since mass has a rest energy the difference in mass results in the creation of some energy in accordance with the equation $E = mc^2$. The key to what makes nuclear fusion different to nuclear fission is not just in how much energy is released but the unique mechanism by which it is released. In nuclear fission two or more elements are brought together in order to split the nucleus of one or both and these elements are typically very heavy. Most elements we use in our nuclear reactions are uranium or plutonium and the thorium chain is a possibility too. These elements are very heavy when compared to fusion materials as uranium has an atomic number of ninety two and hydrogen the lightest of fusible elements is atomic number one. Fission and fusion processes are opposite to one another with respect to power output versus power input and the elements they react. Fission for example gives out far more power than put into the reaction when using heavy elements like uranium. Fusion is the opposite and gives out far more power than put into the reaction when using light elements like hydrogen or helium. The lighter the element you use for fission the less power you receive out compared to power you put in. For fusion the heavier the element you use the less power you receive from the reaction. In fact after fusing all elements up to iron a star no matter how large cannot fuse iron or anything heavier than that element. Nuclear fusion will actually absorb more energy in the actual fusion process of iron than can be extracted from the reaction itself. This inverse relationship from one another is just one of the ways nuclear fission and nuclear fusion differ.

This desire of nuclear fusion to only involve light elements was surprisingly found using the same methods that nuclear pioneers around the world were all using and having much success with. Yes once again scientists were namely firing radioactive elements and particles at test samples and observing the effects. Rutherford used this method to perform the first nuclear transmutation in a laboratory, Cockcroft and Walton used to perform the first splitting of an atom and Hahn used it to be the first to perform nuclear fission. The man to first perform a fusion experiment in the laboratory was an Australian physicist named Mark Oliphant. Oliphant was an enthusiastic scientist who was interested in working with atomic physics where he would eventually come to work for Ernest Rutherford in England at Cambridge University. Working in the famous Cavendish laboratory Oliphant spent his first years constructing his own particle accelerator and using it to confirm the findings of fellow Cavendish scientists Cockcroft and Walton. He found another use for the accelerator in 1933 when the laboratory acquired some heavy water and the Cavendish team began experimenting with it. The discovery of the helium 3 nuclei is credited to Oliphant, Rutherford and others after firing the heavy water with numerous test samples and observing the reactions. What Oliphant did himself was to then fire at the heavy water things like helium 3 or heavy hydrogen nuclei such as tritium. The results he recorded showed that the energy released from the reaction contained more energy than the starting particles possessed. This was the first instance of nuclear fusion in a laboratory and a demonstration of the third type of nuclear reaction that powers our Sun and all stars.

The first person to speculate about fusion was Sir Arthur Stanley Eddington a British astronomer and physicist. Eddington was the foremost astronomer in Britain at the time and arguably the world. He is famous for multiple reasons and he is credited with many discoveries on stars and the universe as a whole. One of his most remembered accomplishments was the translation of Einstein's theory of relativity from German into English making the work available to people in multiple countries. He would further go on to help prove part of Einstein's theories of relativity by photographing solar eclipses. According to Einstein space would become bent by the gravitational well of a star and that light travelling through space would likewise be bent. In order to demonstrate this Eddington travelled to both Africa and South America to photograph solar eclipses in the hopes of seeing stars close to the sun that should actually be behind it. The idea being that these stars are really behind the sun and should not be visible at all but that the bending of space allows the light from those stars to be bent around the immense gravity of the sun thus making them visible. These accomplishments made him one of the leading scholars of relativity theory in the world not just in Britain.

His contributions to astronomy ranged from creating a mass to luminosity relationship for stars and writing his 1926 book entitled *The Internal Constitution of Stars* which would become a staple text for decades to come. Eddington's other contribution came in the form of insights into the unseen working of stellar interiors. Eddington had a knack for thinking a problem through and coming to an intuitive conclusion on the problem and this he applied to a problem of the day known as Cepheid variables. Cepheid variables are stars that change in both their diameter and temperature radially to create what are viewed as pulses of intensity. These stars had been puzzling astronomers for over a century since in 1784 astronomers like John Goodricke noted their curious behavior and attempted to find an explanation for it. Eddington solved this problem in 1917 where he proposed differences in the stars heat would affect the luminosity of it and thus cause a change in the stars brightness and size. The stars size would change because as the gas in the star was heated not only did it glow brighter due to increased thermal radiation but the gas would expand as well and cause the size of the star to swell. We now know Eddington to be correct as the opacity of helium in the stars atmosphere changes with its degree of ionization. Ionized helium is more transparent than doubly ionized helium and so as heat increase in the star more helium is doubly ionized and thus the star appears more opaque. The changes in opacity are due to changes in the amount of ionized to doubly ionized helium in the stars atmosphere.

Chapter 4
Looking Inside Stars for the First Time

Stellar Cores

From dissecting these problems Eddington continued his work on the interior workings of stars. It was surmised at the time that stars were prevented from collapsing under gravity due to their own internal thermal temperature and Eddington made the important discovery that it was also radiation that must support the stars as well. He was able to calculate that thermal pressure alone was insufficient to prevent the collapse of the star and that additional pressure from radiation was necessary to support the star against its own immense gravity. Now calculations about a stars interior could be made at any point of the interior and used to determine the pressure, temperature and density at these points. Discoveries such as these led Eddington to suggest that fusing the light elements inside a star might be the source of its power output in 1920.

The first direct mechanisms by which stellar fusion could be accomplished was made by a Dutch-German physicist by the name of Fritz Houtermans and a British astronomer named Robert Atkinson. The two worked in collaboration to show that a star could derive its power from nuclear fusion in its core through the fusion of ionized gas. The incredible temperatures inside a star would be more than enough to strip away the electrons from light elements such as hydrogen and helium and cause them to become ionized. The high temperatures and high pressure at the core would provide the necessary mechanism to cause the nuclei of these light gases to come into contact with one another at sufficient velocity and force them to fuse forming new heavier nuclei in the process. These nuclei could then in turn be made to fuse with one another again and create still heavier elements. In addition they calculated that the fusion process would follow Einstein's equation for mass-energy equivalence and provide the release of energy needed to power the star. Houtermans and Atkinson worked out a chain of fusion events to demonstrate their idea beginning with hydrogen and working up to larger nuclei all the while calculating the energy release from the change in the rest masses of the nuclides involved in the fusion process. Their combined efforts were astounding as it fit perfectly with both the theories of Eddington and Einstein. Arthur Eddington was proven correct that it was a fusion process that powered the interior of stars and that this nuclear radiation was necessary in supporting the star against the inward force of its own gravity and crushing itself to death. Plus it factored in Albert Einstein's mass-energy equivalence equation to show how much power was released for each successive fusion event. When they published in 1929 they could not have known that their own work would be backed up by Mark Oliphant at Cambridge University who proved experimentally that nuclear fusion was possible. As remarkable as their work was they did not have the exact fusion chain worked out for stellar fusion. Though they were able to prove through their calculations of successively heavier elements that Einstein was right in the loss of rest mass and release of energy it would be another physicist who worked out the actual fusion chain of events.

The man who did work out the sequence of successive fusion events was a German physicist named Hans Bethe. Bethe received a worldly education beginning in Germany in Frankfurt and Stuttgart, then travelling to England to study at the famous Cavendish laboratory and finally the United States at the Cornell University. Along the way he met and worked with some of the most brilliant minds in science at the time and was eventually invited to join them at a physics conference in 1938. The conference was on exactly the sort of thing that Houtermans and Atkinson had been working on almost a decade earlier; namely the power generation of the Sun and specifically what the exact process was that ran it. So the question was posed at the conference to offer an answer as to why the Sun had the exact properties it had namely its chemical composition, temperature and density. Explaining what mechanism produced these properties should answer the question and Bethe and a fellow physicist named Charles Critchfield started to work on the problem. Together the two came up with what was to be the proton-proton cycle of stellar fusion from light nuclei up to heavier ones. They began their fusion chain with protons and were able to work up to heavier elements such as beryllium. In this way they were the first two scientists to prove through a series of nuclear reactions what gave the Sun its high energy output.

At the time there was also another scientist, a German physicist named Carl Friedrich Freiherr (Baron) von Weizsacker who in 1938 was also working on the details of stellar fusion chains. Bethe continued his work into stellar nucleosynthesis in his laboratory and would go on with von Weizsacker to work out the series of fusion events up to the heavy element oxygen. Weizsacker developed his theory in 1938 and Bethe in 1939, together the pair published a paper that described this process and was known as the Bethe Weizsacker process. Oxygen is not really a very heavy element in terms of all the known elements in the universe but it is very heavy when compared to a single hydrogen nucleus that was believed to be the only driving force in stellar fusion for years. Bethe had worked out what would come to be known as the carbon nitrogen oxygen cycle for stellar fusion. Bethe's explanation of the carbon nitrogen oxygen cycle as a method for the working of stars earned him the 1967 Nobel Prize.

The CNO cycle which stands for carbon nitrogen oxygen cycle is now a well understood and important fusion chain in the life of many stars. The fusion achieved by this cycle differs from the proton-proton chain reaction in that the fusion involves multiple heavy elements which act as catalysts. The carbon, nitrogen and oxygen isotopes involved in the chain serve as catalysts for the nuclear reaction much the same way any normal catalyst functions for a chemical reaction. A catalyst of course is any substance that facilitates in a reaction to speed up the rate of reaction but does not get used up in the reaction. This is true in stellar fusion involving the atoms carbon, nitrogen and oxygen. Each different atom takes part in a part of the chain that will result in the release of a helium nuclei or alpha particle. The other products of course include energy and smaller subatomic particles which take the form of positrons, gamma rays and neutrinos. There are a few interesting facts about this fusion cycle however that must be noted as not all stars rely on this cycle as their primary power source. Typically the carbon nitrogen oxygen cycle is found dominant in only one of two classifications of stellar masses. Any stars roughly one point five times the mass of our own Sun or smaller will not use this cycle as their primary fusion chain. Stars larger than this will rely much more heavily on it as opposed to the proton-proton chain. The reason for this is in order to achieve a self-sustaining carbon nitrogen oxygen cycle inside a star a core temperature of around 15 million Kelvin is required. Once a self-sustaining temperature has been reached the cycle can begin to provide power to the star and as the temperature of a stars core rises the efficiency of the cycle goes up as well. What this means is that the bigger and hotter the star the higher the stars core temperature and therefore the more power output from this fusion chain. Any large hot stars of course will derive most of their power through this cycle and since our Sun has an internal core temperature of approximately fifteen point seven million Kelvin it relies more on the proton-proton chain. This does not mean that the carbon nitrogen oxygen cycle is absent from our Sun it is not yet it is such a minor player in the Sun's energy output that it only accounts for one to two percent of the Sun's total energy output.

The actual mechanism of the carbon nitrogen oxygen cycle is very similar to what you might expect from a standard proton-proton chain in that it begins with four hydrogen nuclei that will eventually fuse into one helium nuclei. The catalyzing atoms merely aid in this process of fusing a helium nucleus and also produce two neutrinos and two positrons. The solar plasma of course is incredible dense and moving at high rates of speed due to the extreme temperature and as a result the two anti-matter particles created, the positrons will not survive long. The positrons will shortly meet up with two corresponding electrons and participate in an annihilation reaction which will release two high energy photons in the form of gamma rays. The neutrinos which are nearly massless and travel close to the speed of light will take some of the resultant energy from the reaction and carry it out of the star as neutrinos rarely interact with matter. Even the dense solar plasma of the stars core is not dense enough to impede the neutrino on its journey outward from the reaction. The total momentum of both positrons and neutrinos will be conserved in their respective creation however. Of the two particles one may be generated with a higher momentum than the other or it is possible for one to be generated with zero momentum and the other with one hundred percent; this results in the two created neutrinos having a wide range of energies. The daughter nuclide from which these particles are born is in comparison to them so massive that its momentum is essential zero. The neutrinos created are of the electron flavor as opposed to the tau or muon and the positrons are of course always positrons.

Another interesting factor in the carbon nitrogen oxygen cycle is that the reaction can be either cold or hot. This of course is a misnomer and a convention in the distinguishing of the two reaction types as both occur at a temperature of many millions of degrees Kelvin. The first type the cold type is so named because the rate of the reaction is dependant of the rate of proton captures in the cycle. Beta decay or the emission of electrons through the process of nuclear decay is at work but is not the deciding factor in cold type cycles. This colder reaction cycle is very stable and can stay self sustaining for an incredibly long time making these types of stars maintain a healthy equilibrium. Hot carbon nitrogen oxygen cycles differ in that their rate of beta decay is now more important than the rate of proton capture. The name hot comes from the beta decay itself as materials that are radioactive are said to be hot. This cycle also occurs in reactions that have a much higher temperature than cold reactions and so are typically found in stars that tend to live fast and die young so to speak.

The three catalyst atoms carbon, nitrogen and oxygen that make up the fusion cycle do not appear in exact one to one ratios with respect to each other. The amount of each specific isotope of the three atoms is also not equal, for example the amount of carbon-12, carbon-13 or carbon-14 will vary over the life of the star and can be recycled many times over. Since the solar plasma is directly connected to the convective zones over the life of a star using the carbon nitrogen oxygen cycle changes in the ratios amongst the three catalysts can be observed. As various components of the cycle travel through the convection zone some rise to the top while others will sink towards the core. These differences can be viewed in the spectra of the star although the timescales involved in such measurements currently makes them impractical as a tool for learning about stellar fusion. This does not mean however that they do not occur and that they cannot be viewed only that we would need to observe a star for millions of years and perhaps even longer to truly see this mixing of ratios in detail.

Meanwhile von Weizsacker would go on to work out the formation process of solar systems from nebulas made of gas and dust. Using our own solar system as the model and analyzing the apparent masses and elemental compositions of our Sun and the planets von Weizsacker was able to work out what percentages of material would go where during the formation process. To begin with only one percent of the nebula gas and dust cloud was made of heavy elements as the other ninety nine percent was hydrogen and helium. As is expected the Sun got the Lion's share of the elements and left only a tenth around it to form the planets from. As the gas cloud collapsed the matter inside it began to spin in accordance with the conservation of angular momentum. The swirling motes of dust inside the remainder ten percent of the cloud would eventually coalesce into the planets we see today including our home the Earth. The condensations of gas and dust farther from the Sun would be able to gather more of the material for themselves and therefore form larger planets. This we see today in the outer four gas giants out beyond our asteroid belt, while the inner planets are smaller and less massive by comparison.

The breakthrough implication for von Weizsacker's revolutionary discovery was that in order for our solar system to work out as it did a certain set of universal laws governing the formation of all solar systems throughout the universe must apply. In other words von Weizsacker was the first person to offer a clear and logical model explaining the fact that other solar systems in our universe must also have planets. This means that for the first time science could safely assume that virtually all non Earth systems had planets and our solar system was truly not unique. Planets must be one of the most common features of the cosmos and in fact far more common than stars as one star can have a multitude of planets in orbit around it. The hunt for exo-planets at this time was theoretically possible although it would be decades before the necessary technology was available. Carl Friedrich von Weizsacker's work is so widely accepted and supported that to this day no other model has supplanted it only a few adjustments to the numbers and timelines he proposed have been made.

What we have now is a fairly complete history of the discoveries leading up to our understanding of nuclear fusion. While all three types of nuclear reactions are important and help to make up our universe in various ways it is fusion that is by far the most important of these reactions. The main reason being that stellar fusion or the fusion that drives a stars core is responsible for essentially building our modern universe as we see today. I said modern because the mechanisms by which fusion has shaped our universe did not come into play until several hundred million years after the Big Bang. The make up of our universe before stars began to shine was dominated for the most part by quantum interactions and gravity. The early universe is a subject that we will look at later and for now we will concentrate on how the fusion age of our universe built all that we see today in terms of atoms.

The elements from hydrogen to lithium can be explained as products of the force that created and powered the Big Bang. It can be approximated that the elements after those were created by the fusion reactions in stars. Even elements higher than iron for example which are not fused in stars were still indirectly created through a fusion mechanism and this includes some of the heaviest elements we know today such as uranium. Many of the elements were created directly in the actual process of fusion and some were created at the end of the fusion process. To begin with let us look at the elements that stars can fuse such as hydrogen and helium.

If we take the simplest view of stellar fusion we can begin with a very early type star that would have existed in the first generation of stars after the Big Bang. Although gases with nuclei heavier than hydrogen were created in the actual Big Bang itself we will simply begin by taking a star made of hydrogen only. Early stars were not able to utilize the carbon nitrogen oxygen cycle as these elements did not exist yet and it is interesting to think that many of the early stellar mysteries can be solved without the use of elements heavier than hydrogen. It is believed that the early universe was made of many large and hot super giant stars that burned themselves out quickly due to their high temperatures. It is probable that some of the anomalous findings that we record today from the most distant and therefore oldest parts of the universe can be explained by huge stars being forced to rapidly and violently fuse through nothing but hydrogen and then exploding as nova or hyper nova. The early universe was much, much denser than the universe of today and this likely led to strange relationships between gravity and this super dense soup of matter. Conventional fusion models will not suffice to explain what was happening at this time in the early universe with respect to not only the first stars but also the first super novae they created. Of course we would then have to examine the second generation of stars born into this still dense, chaotic and now black hole rich universe. The complexity of these interactions between massive light element only stars, dense space first void of black holes and then filled with the remnants of countless first generation novae is worthy of investigation and we will do so later. For now we will look only at a single peaceful star formed from the collapse of a hydrogen gas cloud to begin our fusion chain.

This simplest stellar fusion chain is known as the proton-proton chain as it involved only hydrogen nuclei to begin with. It begins with four hydrogen nuclei, protons and ends with the creation of a helium nucleus. This being said we will see what other reactants and products come into play throughout the chain of events. Before the star goes into a fusion burning process the hydrogen atoms created by the big bang collect in a super dense and therefore super hot ball. As the temperature inside this ball of super heated gas rises it reaches a point where the hydrogen atoms become ionized and they lose their electrons, the remaining hydrogen nuclei is a proton. As the temperature rises still further the speed at which particles are moving around one another also increases dramatically. Below a certain critical temperature the particles are not moving sufficiently fast enough to overcome the electrostatic forces of repulsion generated by the protons positive charge. Above this temperature and the particles can overcome the electrostatic force until they essentially touch and can be fused. This temperature is approximately ten millions degrees Kelvin and at that temperature the electrostatic force basically loses the war with momentum. The actual temperature needed to start nuclear fusion however varies widely with the pressure of the system and what nuclides are going into the reaction. Trying to fuse different atoms for example will require different temperatures to be met and so no simple answer exists for all nuclear fusion reactions to begin. For a star however due to their massive internal pressures at the core we can say that stellar fusion is turned on at ten million degrees Kelvin.

Once the temperature is high enough and the electrostatic force overcome the two nuclei will meet finally. At this stage in the reaction a second force from within the proton takes over and it is the nuclear force. The nuclear force is what holds the nuclei of atoms and their subatomic particles together. A helium for example contains two protons and two neutrons all held by the nuclear force. The two neutrons by the very definition of their name are neutral with no electrostatic charge. The two protons however exert a repulsive force against one another due to their positive electrostatic charge. The reason that the nucleus of a helium atom does not simply disintegrate from electrostatic repulsion is because the nuclear force which binds all four subatomic particles together is stronger than the electrostatic force at short distances. The distances inside an atoms nucleus are microscopic when compared to the distances two atoms feel for one another from the repulsive electrostatic force. Approximately a grain of sand can be thought of as the nucleus of a hydrogen atom and the electron orbital would be as large as a sports arena in volume around it and so the reach of the electromagnetic force for example is the entire arena while the reach of the nuclear force covers only the grain of sand. Therefore when the temperature is high enough for two nuclei to meet by overcoming their repulsion the nuclear force will hold them together and they have fused successfully. This basic concept of overcoming the electrostatic force with sufficiently high temperatures and pressures which therefore allows the nuclear force to bind nuclides together is how fusion works. The energy needed to fuse nuclides is called the binding energy and changes as the amount of electrostatic repulsion changes for an atom as well as the number of total nucleons present. As far as nuclear fusion is concerned speed is king and protons feel the need, the need for speed.

So what we can see here is the basic mechanism for all fusion and certainly for the fusion of hydrogen nuclei in our example star. What the two protons have made is not yet a helium nucleus but instead a deuterium nucleus along with a positron and neutrino. An atom of deuterium is hydrogen that has acquired a neutron in its nucleus, in this case as a result of the two protons fusing one has become a neutron. What is quite common during this event is that a diproton is created instead which decays rapidly back into regular hydrogen through a proton emission. Once the rare deuterium has been created however another set of fusion events take place and once again another proton is added to the reactants. Now for the first time we see the emergence of a new element in the fusion chain and it is a helium isotope. It is a helium nucleus consisting of two protons and only one neutron, this nuclide will go on to meet another of its kind and fuse with it. The result from this nuclear reaction is the familiar and very stable helium nucleus made of two protons and now two neutrons. This is the very simplified version of how hydrogen is fused inside a stars core to produce helium. The helium is of course ionized as it has lost its electrons but once it cools sufficiently electrons will fill the helium atoms orbitals and regular helium is created.

It must be noted however that this is a very simplified view of the full proton-proton chain and many smaller reactions and decays occur throughout it. Along each step in the chain we see electron neutrinos, positrons and gamma rays produced. Likewise with the carbon nitrogen oxygen cycle the fate of all these products is the same. Neutrinos are expected to flee the stars interior quickly, gamma rays will do much the same and positrons will annihilate with free electrons in the solar plasma. There are other particles created and lost as the probability of stable nuclides like deuterium being created is an uphill battle. Examples of two other atoms created in this process are beryllium and lithium; however the fate of those nuclei is often alpha particles still. Nevertheless this is a good introduction into how all the various nuclear reactions unfold inside stars as they all involve multiple steps to reach a final product. The products are not just new atoms but of course energy in the form of radiation and this radiation is what makes a star shine. All of the radiation produced by a star is in some way or another involved in maintaining the equilibrium of a star through its main sequence phase of its life and in affecting the gradual evolution of the star as well.

Chapter 5
Stellar Nucleosynthesis, Supernovae and More

Star Interiors Expanded

A stars evolution was first proposed in a sense by a man named Sir Fred Hoyle an English astronomer who contributed much to science that was well ahead of its time. The birth of stellar nucleosynthesis can be credited to his 1946 paper outlining the creation of heavy elements in stars. Heavy with respect to the light hydrogen atoms that people of the day well understood powered stars. Hoyle's paper would outline the process by which elements heavier than helium would be made in stellar fusion. Part of his discoveries suggested that a stars interior was far more complex than initially thought. Firstly the temperature of stars will rise to billions of degrees as a star ages and changes what ratios and pathways of elements it uses for fusion. He offered up a mechanism for this by suggesting the interior of a star was not just a simple core surrounded by a mantle of hot gas but numerous shells. Hoyle's idea was that a series of shells existed around the core of a star and that each of these shells would be dominated by a particular synthesis process. All would be a result of high temperatures and high concentrations of particular nuclei that would result in the creation of newer and heavier atomic nuclei.

Another famous paper which he authored was a collaboration with three other scientists by the names of: William Alfred Fowler, Geoffrey Burbidge and Margaret Burbidge. The paper was published in 1957 and the name of which was almost a certainly a comical reference to the subject matter. Called the B2FH paper after the author's initials it resembles a chemical formula with each initial a reference to a different element. The paper of course dealt with stellar nucleosynthesis which involved the process of creating new elements from lighter ones much as Hoyle's 1946 paper did. What was new and expansively covered in this paper were two new processes that would be known as the s process and the r process. Both of these new processes dealt with neutron capture as a way of synthesizing heavy elements in stars besides fusion chains. The s process is a simple mechanism whereby a neutron is added to the nucleus of an atom in a star to create a heavier element. The heavier element is synthesized when a beta decay occurs in the newly created nucleus as one of the neutrons there changes to a proton. Now a new element has been synthesized and the s process is complete. This s process can create a wide range of elements many of them heavier than iron as we will see later, for now it is the starting point for the creation of many new elements and a testament to the work of Fred Hoyle and his colleagues.

As we have learned the radiation released through the conversion of mass into energy is necessary to support the stars own weight against the force of gravity. The light produced by this nuclear process is also what gives all life to our world and has undoubtedly shaped everything we see around us on Earth today. Not only plants and animals but also many of the planetary features of our world are created by the Sun's rays. The rays from our Sun and all the products created by its active core are what are needed to fill a universe with numerous types of atoms and therefore countless arrangements of those atoms to form everything we see now in the cosmos. That being said most of what we see today in terms of atoms was not created by the fusion inside a stars core but rather by the end of fusion inside a star.

So we can observe the universe around us and note that many things are made up of simple atoms repeated in various ways. Looking at the universe around us especially the near universe like our own planet Earth, we can observe many different atoms other than those produced in stellar fusion. While the complexity of stellar nucleosynthesis is very diverse and quite good at producing atoms up to iron it cannot explain where atoms heavier than that come form initially. Remember that an atoms nucleus has a specific binding energy which generally increases as the size of the nucleus increases. This is wonderful for releasing energy from the process of stellar fusion but a limit exists within atomic nuclei and it begins at iron. Around this point the binding energy of an atomic nucleus becomes negative and continues to become more negative for elements even heavier still. What this means is that it will essentially take more energy to power the fusion process than you will get out of it. Inside a star when iron is the only thing left to fuse the heat generating mechanism that supports the collapse of the star is turned off. No fusion means the star will be crushed under its own gravitational weight and this crushing force can only end in so many ways none of which are good for the star.

When you turn off a light or blow out a candle what happens? Well nothing really as you have stopped the reaction responsible for the generation of photons. No energy being input into the system and you will get no light coming out of it nor any heat or other products of the reaction. When a star is turned off however this is exactly the sort of thing that does not happen. Specifically in stars with about ten times the mass of our own Sun or more you will get an even more energetic reaction known as a supernova. Supernovas are the violent explosive ends to many stars in our universe and this is where the extra energy comes from. Supernovae are such extraordinary events with such spectacular energy outputs that for a brief moment they will outshine the entire galaxy in which they are found. This increase in luminosity will last for at least a few weeks but can last as long as a matter of several months. Differences in the size, type and composition of the supernova will determine how long it shines. The massive increase in a stars luminosity means that a previously dim star becomes much, much brighter even visible to the naked eye. It also means that a star that was previously too dim to be seen and therefore a seemingly invisible star can suddenly be seen. The earliest observations of supernova no doubt predate recorded history as mankind has always gazed to the night sky and been awestruck. We can find depictions and representations of the cosmos in various artifacts around the world all of which show a marked interest in astronomy and from a time period far earlier than what we consider recorded history. Prehistoric megalithic structures such as Stonehenge are proof of this as we know today that they were used as astronomical calendars. The knowledge necessary to construct ancient wonders such as these reveal that our ancestors were well versed in the motion of stars and planets. It would be necessary to observe the skies for extended periods of time in order to discern their inner workings and supernova were surely observed as a result. The difference between the ancient times and today is that we now have access to astronomical tools that would make our astronomer ancestors truly proud. Supernova in the past could only be detected if they occurred in our own galaxy or possibly a nearby structure such as the Large Magellanic Cloud or Andromeda. Both are galaxies close to our own Milky Way galaxy and would be capable of producing supernova explosions bright enough to be seen here on Earth. The most important supernova of recent history and certainly one of the most famous is known as supernova 1987A and was named for its discovery on February twenty third 1987. The supernova itself actually occurred some 168 000 years ago as that is the distance the light had to travel to reach us here on Earth. The reason that this supernova is so important is that it was the first to be studied using modern equipment. Discovered twice within one day by two different teams of astronomers it was quickly the object of interest for astronomers worldwide. The two discoverers were Ian Shelton a Canadian astronomer and Oscar Duhalde an assistant working in South America and the second was Albert Jones an amateur astronomer in New Zealand. The supernova was observed not only by hundreds of ground based telescopes but also the Astron space telescope which was capable of viewing the supernovas details in the ultraviolet spectrum. This single supernova and its remnant have been studied by countless amateur and professional astronomers alike as well as with newer instruments such as the Hubble Space Telescope. The amount of raw data collected on supernova explosions has made tremendous contributions to our understanding of supernova and the processes that shape them. It has also allowed us to study what a supernova leaves behind.

One of the most beautiful of cosmic phenomena occurs after a star has exploded as a supernova and are known commonly as either a nebula or supernova remnant. What can be observed in these wondrous objects is the left over gas and dust that made up not only the star that exploded itself but also the rest of the loose materials in its former star system. The debris from the dead star glows with light caused by heating of the outward rushing materials due to an enormously powerful shockwave created from the supernova. The shockwave is one of the largest scale fast moving events we can observe in the universe as it travels at around ten percent the speed of light. While this may not seem that fast compared to the countless numbers of small subatomic particles we observe regularly travelling at much closer to the speed of light it must be noted that this is a massive amount of matter by comparison. A typical supernova will cause the star to explode in such a fashion that most or all of its matter is unbound and blown off into space. Unbinding refers to the amount of kinetic energy a particle obtains from the explosion which is sufficient for the particle to escape the inward pull of gravity and other binding atomic forces. Ironically it is almost as if the star has provided the necessary energy to its own particles to reach the escape velocity that is being created by its own particles. The amount of matter ejected is phenomenal and can be thought of as measureable on scales equal to our entire solar system.

Let us look at this closer and say for example a star of twenty solar masses, or twenty times the mass of our own Sun goes super nova. When a star of this size dies it will collapse inwards on itself forming a dense core which is a neutron star and is maintained by neutron degeneracy pressure. Neutron degeneracy pressure is a specific form of quantum degeneracy pressure and can be most simply thought of as the pressure with which neutrons themselves resist collapsing inwards. Most of the material for the star however will be blown outwards by an explosion and shockwave created by this rapid inward collapse. The amount of material ejected from the star can be ninety percent and in a twenty solar mass star this would equal eighteen of our Sun's give or take a few percent. The Sun accounts for approximately 99.86% of all the mass in the solar system and so a twenty solar mass star would eject roughly eighteen full solar systems worth of matter into space. This is a huge amount of matter now travelling at ten percent the speed of light which is why these events are so incredible and some of the most violent observed phenomenon in the universe.

The dramatic increases in luminosity are also accompanied by dramatic energy outputs. These outputs will be more than the star was able to generate in its entire lifespan. The reason for this is that over the long hundreds of millions to billions of years a star burns fuel in its fusion reaction only a small amount of energy is actually created by the nuclear burning of matter. The nuclear reactions take place in a small region of the star and do so slowly under natural conditions in equilibrium with gravity. When the star goes supernova however the inward pull of gravity is about twenty three percent the speed of light for a core collapse supernova and the outward moving shockwave travels through all of the stars matter at once. This causes chaotic fusion reactions to take place often referred to as run away fusion events. Essentially the majority of the stars matter is super heated, much of it goes into fusion reactions and even more energy is released in the unbinding of stellar materials. The result is a one time energy output from the star that is far greater than the output over the long slow burn of its whole life. This final burst of energy from the star coupled with the ejected material from itself is what creates these beautiful nebulae and supernova remnants. The different colors seen in various nebulas are caused by different elements shining from this explosion. The nebulae themselves are ever changing due to the passage of time, motion of the ejected stellar core and external influences such as the interstellar medium or galactic magnetic fields. Now that we have learned a few of the interesting facts about supernovae lets take a closer look at the two main types of nova commonly observed. The first is a core collapse and the second a type 1a supernova both of which are created in different ways but produce similar results. Time now to go back a bit to our candles and see what is needed to cause the creation of more and heavier elements in the universe.

A candle or light bulb has no source of energy to make more light with and one would be forgiven for thinking the same is true in a now turned off star. After all a star cannot fuse elements heavier than iron and fusion stops seemingly taking away the energy input for the creation of photons. A star does have one more trick in its arsenal of energy producing pathways though and this is gravity. While the star cannot generate any more energy on its own the gravity that its mass has created can. Gravity is able to accomplish this feat of new energy production within the star by essentially crushing the star to death under its own weight. The inward pull from gravity comes instantaneously from all sides and is directed towards the very center of the stars core. Often called gravitational potential energy this simply refers to the power of gravity inside the star. Another way to think of it is the falling inwards that the stars matter would like to do but is prevented from doing by the outward push of radiation and thermal energy from fusion. Now that the driving source behind the thermal energy and radiation is gone the matter is free to fall inwards on itself due to the force of gravity it itself creates. This inward rush of matter creates a number of problems inside the core of the star which by now is a multilayered structure with each structured layer being dominated by a specific element.

Commonly the structure will begin with iron at the core and go through silicon, magnesium, neon, oxygen, carbon, helium and finally hydrogen at the outermost layer. As the stars material rushes inwards the layers or shells of the star are rapidly compressed and can reignite fusion reactions of the various shell elements. Which elements are reignited depends on factors like the size of the star, the temperature inside it and the abundance of elements at each layer. Sometimes only one will begin the fusion process again and other times multiple layers can once again burn. This burning process is very short lived though and the inward collapse continues although the rapid fusion of some of the shells can cause them to explode further disrupting the star. A process of run away fusion can occur as the rapid burning of fusion is reignited in these shells and the result can be a supernova explosion. As a core collapse star dies and matter rushes inwards a dense core of neutrons can be created which will push on additional inward falling material. The matter inside a star can be said to bounce backwards from collapse and ripple towards the surface which can also provide a destabilizing mechanism that will result in an explosion and once again it varies from star to star.

As these processes play out inside the star unimaginable numbers of neutrinos are created which will help to further push the explosion outward from the core. Though nearly devoid of mass neutrinos do produce a significant outward push due to the incredibly high numbers of them that push outward simultaneously. Regardless of which mechanisms are involved the star will go supernova and explode leaving behind the remnants of its core. Often times all three of these events will play out in the stars death and therefore a supernova is the culmination of the explosive re-ignition of fusion materials, the bounce of inward collapsing materials and the generation of copious amounts of neutrinos. Nevertheless a star that undergoes a supernova event will lose most or all of its matter in this process and as a result there are two possible end products to a supernova of the core collapsing type. The first is a neutron star and the second a black hole, both of which are theorized to result from a stars mass exceeding its electron degeneracy pressure and gravity continuing to crush it into one of these two forms. Neutron stars and Black holes will be discussed in much greater detail at a later time.

The core collapse method for triggering a supernova event is the one most prominent in people's minds and yet another mechanism exists for creating supernova in a completely different fashion from core collapse types. Core collapse supernova can occur all by themselves and require no outside influence to initiate the supernova. The second common type of supernovae on the other hand does require additional outside help to initiate an explosion. This type of supernova is referred to as a type 1a supernova and differs from a core collapse in that it involves two stars in order to make the supernova event happen. As one would expect a binary star system is the usual culprit but in theory star systems with more than two stars can result in a type 1a supernova as well. While the probabilities needed to make such an event happen are of course slightly higher the basic requirements do not at all differ. What is needed for this type of super nova are at least two stars who share an orbit close enough to one another to interact with each other later in their lifecycles. The initial condition of the stars can vary quite a bit but the end product must be that at least one of the stars is a white dwarf and usually a carbon-oxygen white dwarf. A white dwarf star is the remnant of a star that has shut off its nuclear fusion reactions and is left to slowly cool over a period of at least a quadrillion years. Even after such long time spans the star will most likely have a temperature of a few thousand degrees Kelvin and slowly, eventually become a black dwarf. A black dwarf also known as a cold black dwarf is thought to be the eventual destiny of a white dwarf star but the time it takes for a white dwarf star to become a cold black dwarf star is incredibly long. Given that the age of the universe as we understand it is around thirteen point eight billion years old and it must take at least a quadrillion years for a white dwarf to cool into a black dwarf it is theoretically impossible for a single black dwarf to exist in our known universe. A fascinating possibility are cold black dwarf stars but at present no more than a theory and while interesting only a side note to the type 1a supernova.

The second star in the binary or trinary system can be of virtually any type star at all. The star could be a supergiant or another small star similar in size to the white dwarf the key is that the second star gets bigger. Also known as a companion or neighboring star the sole job of this second star is to expand in size as it ages. Since all stars do expand in size as they age specifically as they burn through their supplies of key nuclear fuels it is easier and easier for the white dwarf to draw materials off of it. The companion star will burn fuel and decrease it's mass therefore the outer layers of its gaseous atmosphere will grow in size with time. The white dwarf is now no closer to the actual core of its neighboring star but it is much closer to that stars outer atmosphere. Therefore the white dwarf whose mass has not changed is now able to pull in and accrete some of the material from its companion.

Over a long enough period of time the white dwarf will increase its own mass and raise its temperature sufficiently high to reignite the fusion burning reactions that once powered it. The reaction is a result of the white dwarfs temperature rising to the level where carbon can once again be fused. Additionally a white dwarf can rarely collide with another star and this interaction will again raise the temperature and pressure of the white dwarf to the point of carbon fusion. In either of these two possible scenarios the fusion reaction of the carbon is not a slow stable self correcting type as is normal of stellar fusion. Instead a violent form of rapid fusion is experienced that will soon run out of control resulting in a fantastic release of energy. This release of energy will easily destabilize the stars structure and result in an explosion that tears the star apart as a type 1a supernova. The star if not destroyed can continually accumulate more material on its surface from the neighbor star cause this to fuse and explode and then repeat the process again; this is a novae and will be discussed shortly. An interesting and widely exploited feature of type 1a supernova is the brightness with which they can be made to shine through this reignition sequence. A white dwarf can only attain a certain maximum mass which is about one point four solar masses or one point four times the mass of our Sun anything more and it would collapse into a neutron star. The reason is that a white dwarf larger than this would have sufficient gravity to overcome what is known as the electron degeneracy pressure.

Electron degeneracy pressure is a term used in quantum mechanics to describe the lowest common energy state that the electron can exert outwards to keep an atom intact. More precisely it refers to a violation of quantum states where two identical spin particles would attempt to exist in the same space. One can think of a regular atom with multiple electrons in orbitals around it. The electrons are kept separate from one another in their own orbital and moving between orbitals would require a change in energy to the system. When an electron is in a higher energy state it can freely move to a lower one through the loss of energy from the atom commonly emitted as a photon. If however the electrons are already in their lowest energy orbitals there is no where for them to go and so attempting to force two or more electrons together in the same orbital space is resisted. If the pressure is great enough though the electrons cannot resist enough of this pressure to ensure that they are the only particle in a volume of space. What results is a breakdown of the electron degeneracy pressure so to speak and the electron in order to not violate quantum laws is forced from its orbital altogether.

When a white dwarf star has a mass that exceeds approximately one point four solar masses the gravity will be strong enough to force all the electrons around all the atoms of the star to leave their orbitals. When this occurs the white dwarf can be pressed into an even more compact form of star a neutron star which is made from the same atomic nuclei as the white dwarf but does not have the large empty spaces around these atoms normally occupied by electrons. No orbitals means no empty space and therefore a much more compact stellar core than a white dwarf. Since however during normal gravitational collapse of large stars a neutron star is formed directly this mass limit of one point four solar masses does not exist. It primarily exists in the form of a white dwarf and this unique feature is what results in type 1a supernova being almost always uniform. Since the temperature and pressure of a white dwarf as well as its atomic structure is almost always identical to every other white dwarf so is the amount of incoming material needed to reignite fusion. It is easy to calculate the amount of infalling material from a companion star necessary to cause the carbon fusion to be turned on and cause a supernova explosion. The fusion process will reignite as soon as enough material has been acquired by the white dwarf and there will be no delay in this reaction. Therefore the moment enough material is added to a known mass white dwarf runaway fusion begins and the star explodes like clockwork every time. Clockwork explosions like this produce clockwork increases in luminosity for the observed star. Additionally the specific light spectra of type 1a supernovae has been long since catalogued and therefore the telltale signature of these dying stars is unmistakable. This is how astronomers are able to determine which supernova are which and not confuse a type 1a supernova with a core collapse supernova.

While this is a simplified view of things there are other differences in such explosions that astronomers use to distinguish between the various supernovae types. The type 1a supernova spectra however is unique even when ratios of elements differ or when other trace elements appear in the spectra there is no mistaking its spectra for something else. No matter where a type 1a supernova is found the amount of material burned in the explosion is the same and so the luminosity never changes. Observations of type 1a supernova do produce differences in luminosity as we see it from Earth and what this means is that we are seeing differences in distance between ourselves and the supernova. The luminosity produced by the dying star is always the same so a difference in brightness can be used to measure distances in the universe with a high degree of accuracy. If a supernova of this type appears very bright then we know it to be close to us and if it appears very dim then the opposite is true and it is far away from us. This method of measuring distances by type 1a supernova produces what is known as an accurate standard candle.

A standard candle is exactly what it sounds like a standard by which we can measure things based on their respective luminosities. One of the useful tools of these standard candles is that they are bright enough to be seen in distant galaxies. Without the aid of a telescope one is limited to essentially only Milky Way supernova detection as our ancestors were using the naked eye. Large modern telescopes however allow us to see objects in galaxies outside our own and type 1a supernovae are one such object. Being able to judge the distance to these neighboring galaxies using the standard candle method provides us with an accurate measurement of cosmic scales. The ability to accurately judge distances in astronomy has plagued scientists for centuries and many of the key concepts of the universe can only be surmised through knowing very accurately how far apart things are. This works hand in hand with our use of the speed of light as another cosmic marker so to speak. Since we know the speed of light to be a constant three hundred thousand kilometers per second in a vacuum the distance that light travels in a single year is always the same. The distance light does travel in one year is roughly ten trillion kilometers and therefore can be used as another standard measure by astronomers. Measures such as the luminosity of a type 1a supernova or a light year make accurate cosmic maps possible and this is of extreme importance to scientists in determining how the universe works.

So now we have seen the two main different types of supernova and seen that they can be useful in studying the make up of stars and in use as standard candles. There is however one more type of stellar explosion out there and it can almost be thought of as a supernova imposter. These explosions are known as novae not supernovae and can cause confusion due to the fact that they also produce dramatic increases in luminosity of stars. There are two main differences between these explosion and those caused by supernovae. The first is that the star is not destroyed in the explosion itself and the second is that they only involve white dwarfs. The reason that we see an increase in luminosity at all is because the star in fact cannot quite go supernova. The white dwarf star involved in the nova is part of a binary or greater system just like a normal white dwarf in a type 1a supernova system. Likewise the white dwarf star will slowly accrete matter drawn off of its companion star. As with all white dwarfs the star has ceased its own natural energy production through nuclear fusion but it can momentarily reignite with the help of its companion. The gas drawn off of the atmosphere of the companion star is primarily hydrogen as this is the lightest and therefore highest layer of gas available and also the most abundant. The hydrogen accumulates on the surface of the white dwarf star raising its surface temperature to the point of reigniting nuclear fusion. The hydrogen takes part in the familiar carbon nitrogen oxygen chain and can under rare circumstances become stable. Most instances of this reignition reaction however are unstable and the burning of hydrogen is accelerated to the point where huge amounts of hydrogen are burned at once. This rapid depletion of fuel and subsequent conversion into heavier elements which the star cannot under these circumstances fuse is what triggers a runaway nuclear reaction. The end product of which is an explosion of nuclear energy that causes the star to glow very brightly as it loses the accumulated materials to space. The time in which the star brightens varies from star to star but usually is fairly rapid and the gradually fading of peak luminosity can take weeks to months. This rapid rise in luminosity followed by a slow decline and accompanied with a light spectra similar to a regular supernova are what makes them difficult to distinguish from the real thing.

White dwarf novae also differ from supernovae in that they do not destroy the white dwarf star. Only the accumulated matter from the companion star is lost as it is exploded outwards at velocities up to ten thousand kilometers per second. The amount of material that is spewed out into space from the nova is relatively little only about one ten thousandth of a single solar mass as compared to eighteen solar masses worth in our hypotetical core collapse supernova. While the elements ejected from these novae events are wide ranging from hydrogen to magnesium and many in between it does not provide a large source of material with which to enrich the galaxy. The interesting thing about novae systems though is that because the white dwarf star is not destroyed in the actual explosion the event can be repeated over time. The companion star will keep losing materials to the white dwarf and so a build up of accreted gases once again occurs. After a certain length of time another novae occurs and then once again the cycle repeats until the companion star runs out of fuel to donate or some cataclysm befalls the whole system.

Two factors determine the time interval between novae events the first is the rate of accretion of gases from the companion star and the second is the mass of the white dwarf star. The most important factor is the second as the mass of a white dwarf star determines the minimum temperature required for reignition and burning of the newly acquired hydrogen. The more massive the white dwarf star the greater its gravity and consequently the lower the mass needed to reach the point of nuclear fusion. Therefore these large white dwarf stars will have a shorter interval between novae as they need to accrete less material for the novae to take place. Some novae can occur on a time span as short as a decade although this is rare and some are estimated to take one hundred thousand years. These novae are thought to be very common and at least ten are detected each year alone in our own Milky Way galaxy. The nearby Andromeda galaxy can see up to twenty five per year and estimates for both galaxies are actually higher than this.

A common feature of all these different types of supernovae and novae is that they help to produce and also spread out elements heavier than those found in Big Bang nucleosynthesis. The common elements created after the Big Bang are said to be made during Big Bang nucleosynthesis and refer to elements such as hydrogen or helium. These various elements are necessary to make up the universe that we see today and also to make up the life that inhabits it. Now it is time to investigate in more detail the work of Fred Hoyle and his colleagues and how this relates to nucleosynthesis of heavy elements. The two processes described by Hoyle are known as the s process and the r process; the first we will look at is the s process. The s process is a neutron capture process whereby a neutron is added to a nucleus of a light element to create a heavier one. The time scales involved in the s process are on the order of tens to thousands of years with respect to neutron captures and beta decay events. The rate at which neutrons can be captured by an atom compared to the rate of beta decay is slow and thusly the s process is known as a slow neutron capture process of nucleosynthesis. A beta decay event can occur on the order of less than one minute while a neutron capture event could be decades and so there is a near one hundred percent probability that beta decay will occur before another neutron capture.

The result is a steady production chain of stable isotopes inside the star by the creation of a proton through beta decay. The new nucleus has a proton added and has increased by one its atomic number. The process of adding neutrons to atomic nuclei is well understood and explored as we have seen be Earth scientists for a hundred or more years. This process of stellar nucleosynthesis has been going on for billions of years since the Big Bang inside of stars and therefore accounts for some of the heavier elements. What are needed is obviously a free neutron and sufficiently high temperature and pressures to give the particles the kinetic energy to fuse with a free nucleus. The other ingredient is a seed nucleus which is simply the nucleus that will receive the new neutron and this is where the abundance of free elements affects the outcome of new isotopes. A very early star for example let us say one of the first generation stars after the Big Bang will be made mostly of hydrogen and some helium with the slim chance of lithium and beryllium being present in trace amounts. This dictates that the only seeds available for the s process to use are light ones and since beta decay will almost certainly occur before a second neutron capture only other light elements can be created. What we see is that things about as heavy as carbon or neon might be created in these early stars. Of course the proton-proton chain and carbon nitrogen oxygen cycles help in creating newer heavier seed elements as well. Very heavy nuclei such as thallium for example will never be created in these first generation stars through the s process as the chances of extremely rare events all coinciding to make heavy elements is far too improbable using the s process. As we will see in the r process however it is possible to make heavier elements in young metal poor stars. What is more likely though is new seed elements much heavier still will be created in the first generation of supernovae after the Big Bang and this will be the main source of s process production of very heavy nuclei. We will look at this supernova process of nucleosynthesis later for now though we can use the heavier seed elements created by supernovae in the s process.

Since the s process can add a neutron to almost any atomic nuclei and produce a stable isotope through beta decay it is obvious that the second generation stars will utilize these heavy supernova elements. The creation of very heavy stable isotopes using the s process is now simple as a seed of molybdenum 42 for example can have a free neutron added to it to create technetium 43. This was included in the famous B2FH paper as well as a possible explanation for the presence of technetium in red giant stars. The stars themselves are on the order of billions of years old and yet technetium will last at best four point two million years meaning that it must be newly created in the star to appear in the light spectrum of red giants. This helps to prove that the s process was instrumental in creating new elements from heavy seed nuclei following supernovae. Since it is believed that hydrogen and helium were the main elements created through Big Bang nucleosynthesis all other heavier elements must have been made by stars. The s process is one of the mechanisms that stars used to create these heavier elements in abundance. Stellar fusion as we have seen through proton-proton chain reactions and the carbon nitrogen oxygen cycle can create a host of heavy elements. Ultimately we see that direct fusion inside a star will stop at the creation of iron nuclei and the s process is one way to create elements beyond iron but without fusion. In fact it is thought that much of what we see in the universe in terms of different elements can be created with the s process.

Approximately half of the elements beyond iron can be explained using the slow neutron capture mechanism of the s process. In fact very heavy elements such as mercury 80 and thallium 81 can be made using the s process and these are massive when compared to iron 26. The creation of such diverse and heavy elements helps to explain how the universe was filled with all the different elements we see today. Proof of this process can be found today inside meteorites that have been found on Earth. Asteroids forming in the early solar system were made from materials in the accretion disc and trapped the particles of dust inside for billions of years. Some of what they captured was stardust which is the microscopic particles of stars that were lost through various events such as novae. Many different methods for the formation of stardust exist but one simple fact unites them all and that is they hold inside them the purest forms of isotopes created by the stars. The ratio of isotopes in the spectra of stars is physically preserved in these tiny grains that help make up cosmic dust as a whole. The asteroids fall to Earth as meteorites and if they remain intact after impact will keep these grains of stardust locked away and safe for billions of years. Scientists have collected these space rocks and analyzed the spectra of the isotopes found inside and found grains containing silicon-carbide in the exact ratios expected in s process stars. The fact that we can study what occurred in the extremely violent stellar environments of long dead stars billions of years ago and light years away is truly remarkable.

The method for element synthesis inside stars is by now well understood for stars that are for lack of a better word alive. What this really means is they are stable and usually in the main sequence phase of their life cycle where a balance has been reached between the explosive force of stellar fusion and the relentless crush of gravity. We have also seen many of the heavy elements can be created without the need for a cataclysmic event occurring in the universe. Aside from the Big Bang however which produced the subatomic particles in our universe and the elements hydrogen and helium there is another group of explosions that are second only to the Big Bang itself. These of course as we have seen are the supernovae which routinely populate the universe with stellar materials. Infrequent when viewed by human timescales these supernovae events are in a sense almost constantly happening when viewed on a universe long timescale. Couple this with the literally countless number of stars in the universe and it is easy to see how frequent these epic explosions take place. The violence of supernovae and their frequency has been critical for the galactic chemical evolution of our universe. This is due to the fact that a different set of nucleosynthesis reactions occur as a massive star dies and is crushed by its own gravity. It is during the collapse phase of a core collapse supernovae that we find these unique conditions. Stars that are alive create new elements and it also turns out that stars that are dying create them too. These dead stars help to create the elements necessary for life in the universe such as our own species.

This death synthesis of new elements was also proposed by Fred Hoyle and his colleagues in the famous B2FH paper of 1957. In this paper we have seen the s or slow process at work and now we will examine the r or rapid process in detail. The rapid part of the processes name comes from the fact that unlike the slow process neutrons are captured as one would think much faster. Like the s process it also requires seed nuclei in order to synthesize heavier elements and like the s process is also responsible for many of the elements heavier than iron. Once again we see the importance of creating elements heavier than iron as iron is the stopping point for the direct fusion chain of stellar nucleosynthesis. The problem was first detected when in 1956 two scientists published a new list of elemental abundances in the universe. This is simply a way of determining how much of each different element we see in the universe no matter where or how we observe it. Hans Eduard Suess and Harold Clayton Urey were both American scientists who analyzed the elemental composition of meteorite fragments. As we have seen through stardust trapped in meteorites this method is extremely useful in studying the elements of the cosmos without actually leaving the surface of the Earth. What they could not explain was the peaks in abundances of elements like germanium, platinum and xenon. The nuclear shell model predicts additional neutrons could not be added as the addition of new neutrons would push them past the neutron drip line.

The neutron drip line very simply put is a point where a nucleus will not accept a new neutron as it immediately decays away to reduce itself to a stable isotope. If one thinks of making new nuclei by adding a neutron or proton one at a time a certain limit is reached where the nuclear binding forces cannot hold the new subatomic particle and it decays. Add one proton and your probably fine, add two and things get less stable keep adding protons and the electrostatic forces of repulsion amongst so many protons in the nucleus will win. The strong nuclear force cannot produce enough binding energy to hold the additional proton in the nucleus and it will decay away. The strong nuclear force is stronger than that of electrostatic repulsion over short a distance which is what you find in an atomic nucleus but it has only so much strength. This is where you have reached the proton drip line and a newly created element will decay back to a stable one. This use of protons for the illustrative purposes of explaining drip lines with respect to electrostatic forces and nuclear forces provides a simple example for picturing the neutron drip line as well.

An interesting side note worth mentioning here concerns the work of Urey and one of his graduate students Stanley Lloyd Miller which led to the Urey-Miller experiment. Urey had been the first to propose that the early atmosphere of the Earth was comprised of numerous elements three of which were hydrogen, methane and ammonia. Urey and Miller performed what is really a very basic experiment where they took the obvious step to place these three gasses inside a sealed container and subject it to electrical discharge. The electricity was used to simulate lightning strikes in the early atmosphere which would have provided the necessary temperature and energies to alter these gasses. What they hoped to find was the synthesis through natural means of new molecular compounds. What they found was amino acids which are used in the building of larger protein molecules and they therefore demonstrated a possible mechanism for the creation of more complex organic molecules in the early atmosphere. Harold Urey would eventually go on to win a 1934 Nobel Prize for his work in discovering deuterium and would go on to work with the Apollo moon missions and publish nearly half his life's papers on lunar subjects.

Since the nuclear shell model could not adequately explain these abundance peaks the proposed r process of Hoyle and his fellow B2FH cohorts provided the necessary mechanism for the synthesis of these heavier elements. The r process states that as a star collapses at the end of its life into a standard core collapse supernova a massive amount of free neutrons is created which can lead to the formation of new elements. In addition to this abundance of neutrons extremely high pressures and temperatures are needed to catalyze the process. Essentially as a star collapses the inward rushing material will create very high temperatures which will result in a massive increase in the rate of electron captures and subsequent neutronization of matter. The high electron counts in the collapsing core prevent beta decay from destroying the newly created neutrons. In this collapsing moment of the stars life and the following inevitable explosion the rates of neutrons being captured by seed nuclei is much faster than the rate of beta decay. As we saw in the s process beta decay was faster than neutron capture and so would decay into a proton. Here we find that neutron capture is much higher than the beta decay so we are left with nuclei that are quickly loaded with additional neutrons well beyond the neutron drip line. The unique conditions of a dying massive star can in a sense cheat the neutrons past a point where they should no longer be able to stay in the nucleus. Naturally these newly created nuclei are exceptionally unstable and can and will decay into more stable forms, most of these are isotopes heavier than iron and laden with additional neutrons. Therefore most of the stable isotopes produced by this r process are created not when the elements get larger and heavier but when it decays back down to a more stable form. The s process works upwards the r process works upwards and then back downwards so to speak.

The violent nature and insanely chaotic environment of the entire supernova event gives ample opportunity for seed nuclei of all kinds to acquire new neutrons. During the collapse phase new elements are created as well as the explosion phase and in addition to this if an unstable nucleus decays quickly during this process it can itself become a seed nuclei. While unstable these newly created seed nuclei will continue to interact with all free protons and neutrons during their violent egress from the star during the actual explosion of the supernova. Again of course many of these new isotopes will be unstable and very lopsided in their proton to neutron ratios and once again decay to form stable elements. Some of the heaviest elements in the known universe can be created through this process alone and due to the specific conditions that must occur for this to happen and the short time span with which it does explains why many of these elements are so rare in the cosmos. If the rate of neutron capture slows too much the r process brakes down, also as we have seen the bigger the nucleus the less stable it is and so a maximum size is imposed before it decays. The larger the elements and the heavier they are the more likely they will fall prey to spontaneous fission as they split themselves apart.

One last factor in arresting the r process comes from photodisintegration where the massive number of free high energy gamma photons created during the supernova will actually force a neutron, proton or alpha particle form the nucleus. Regardless of how the r process is terminated the resultant elements have now been created and are blown off the star as lost stellar material that will permeate the galaxy. One other interesting note is how the r process can work in very early stars comprised of only hydrogen and helium. Given the fact that many of the earliest stars in the universe where very large and made of nothing but these two elements the r process helped to create some of the first elements heavier than helium. Any large star during its collapsing phase of a supernova will create the necessary conditions for electron capture, increase neutron density and the r process exceeding the neutron drip line. Therefore helium and even hydrogen nuclei can and were used as seed elements in the early universe to create heavy elements during the first generation of supernovae. Both the s process and the r process had a hand in not only creating but quickly increasing the total number of elements in the universe allowing for incredible atomic and molecular diversity to follow.

The number of ways in which different elements can be synthesized is quite extraordinary and we have not yet examined all of them. It is interesting to reflect for a moment on the significance of this fact as it was once completely unclear how any elements came into being. Without an exhausting exploration into the plethora of creation myths that have been invented by our species it is still only very recently that science had any idea how elements were created. The first rational explanation for the elements in classical thought came from the Greek's and their erroneous attempts to explain matter using elements. Not the elements that you find on the periodic table of course but things like water, fire, earth and air. A complete lack of any scientific models or experiments to deduce the nature of matter would plague scientists for centuries. Even as recently as the eighteen hundreds virtually no good explanation for matter as composed of elements as we think of them today even existed.

Then in one brilliant awakening science was able to sort through the nebel or fog and piece a few things together enough to definitively know that matter was made of atoms. Shortly after that of course people still wanted to know was matter in all of its various elemental forms always present in the universe? It would be decades again before the first accurate models of nucleosynthesis would come into the science spotlight. The irony is that our first models of how to build elements so to speak came from objects so distant from our own planet. It was the inner working of stars that shed the first photons onto our understanding of elemental construction. Yet closer to home and in fact right over our heads was one giant science demonstration in nucleosynthesis that had been running since the birth of our planet. A pity our ancestors did not even realize this as we would surely have been quick to investigate it long before stellar synthesis of elements. The driving force hidden right before our eyes comes in the form of cosmic rays smashing into our atmosphere. Since the dawn of time humans have looked up through this atmosphere to understand the farthest reaches of the universe and one of the key components to building all matter in that universe was only a matter of kilometers away.

Cosmic Rays

Time to find out what cosmic rays really are as most people whether they have an interest in science or not have probably heard the term used before. Cosmic rays are in fact not rays at all, but rather particles and it was a simple a misinterpretation of the particles in them. Originally it was believed the experimental results found from cosmic rays were because the rays were part of the electromagnetic spectrum. Naming parts of the electromagnetic spectrum as rays was the standard scientific convention of the time and so it was an obvious step to make in naming particles from the cosmos as cosmic rays. In reality though there is nothing electromagnetic about the cosmic rays and they should probably be called cosmic particles as this is literally what they are. All cosmic rays no matter how light or heavy they are have mass and possess extremely high amounts of energy. The energy that cosmic rays possess is far above anything that we can achieve here on Earth even with our largest particle accelerators. One humbling estimate puts the highest observed cosmic rays at forty million times the energy of our biggest collider the HADRON. Their energy is essentially one hundred percent kinetic as they travel at very high rates of speed and do possess of course mass. The term cosmic obviously denotes their origin as being from space not Earth and yet every type of cosmic ray is commonly found on Earth. The particles in the rays are mainly nuclei of various elements that have been relieved of their electron counterparts. Almost all cosmic rays are made from the nuclei of atoms but some cosmic rays are made of electrons or their anti-particle, positrons.

The rough estimates as we understand them today put cosmic rays in the following percentages: ninety percent are hydrogen nuclei, nine percent are helium nuclei and the remaining one percent is made from the nuclei of various heavy elements. Anti-matter does make up some of the cosmic rays but only in the form of positrons and anti-protons as no complex anti-matter structures have yet been detected. This rules out therefore anti-helium nuclei and anti-heavy element nuclei that would correspond to the normal matter helium nuclei and heavy elements nuclei. No matter what cosmic rays are most definitely dominated by regular matter. The heavy nuclei are commonly called HZE particles which stands for high atomic number and energy. The high is denoted by the capital H, the capital Z for atomic number and the capital E for energy. Additionally HZE ions will have a charge greater than plus two as a helium nucleus has a charge of plus 2 as well. HZE cosmic rays can be made of many different types of atomic nuclei including carbon, oxygen, magnesium, silicon and iron. Cosmic rays because they are the constituent components of atoms or their anti-particles all carry a charge. The smallest of these of course is a hydrogen nucleus which is of course a single proton and has a charge of plus one. Helium nuclei or alpha particles are plus two and HZE are greater than plus two and can have positive charges as high as the atomic number of the nuclei. Some of the heaviest and most dangerous cosmic rays are that of fully ionized iron atoms which can have a positive charge of twenty six as this is the atomic number and therefore total proton count of iron. Most cosmic rays travel at a significant fraction of the speed of light right up to very close to the speed of light giving them incredible kinetic energy. The kinetic energy plus high ionization of these rays make them extremely dangerous to biological life and machines alike. The most dangerous of course are the hydrogen nuclei because although they consist of only a single tiny proton when compared to a massive iron nucleus the sheer number of cosmic ray protons makes them far more destructive overall. Although miniscule in number when compared to the amount of normal matter cosmic rays anti-matter cosmic rays are no less dangerous per particle. This is of course due to the fact that when these anti-matter cosmic rays come into contact with matter they do so much more than just cause ionizing damage like regular cosmic rays. The anti-matter must annihilate itself with a normal matter counterpart and release incredible amounts of energy in the process. The amount of anti-matter cosmic rays however makes them little danger when compared to normal cosmic rays.

Cosmic rays were first discovered in 1909 when a German physicist named Theodor Wulf invented a device called an electrometer. An electrometer is a sealed device that is sensitive to energetic particles carrying a charge. It was thought in the early nineteen hundreds that radiation in the air in the form of ionization was the result of radioactive substances in the Earth. It was understood that various ores and radioactive salts gave off radiation that would cause ionization in test samples. The logical conclusion was that radioactivity was caused from earthbound sources and since many of these came from below the Earth as they were mined, it was assumed that radiation of the atmosphere would decrease with altitude. This was the very idea that Wulf wanted to test with his recently developed invention and set out for a then new landmark to test his theory; the Eiffel tower. If the ionization was coming from ground sources then the amount he detected up the tower should fall to fifty percent of the ground rate by a height of eighty meters. Instead he noted that even at the very top of the tower some three hundred and thirty meters up the rate of ionization had not yet fallen to fifty percent of the ground rate. The only way that this could be explained was if some other source away from the surface of the Earth was causing the ionization. In other words something from space was emitting a source of radiation that could penetrate the atmosphere and ionize the air and be registered in his electrometer. The source had to be space because the only thing above the Earth is space itself and at the time no such human sources of radiation like nuclear reactors or weapons existed. Theodor Wulf had discovered cosmic rays and the search for the sources of cosmic rays would begin although to this day many of the secrets of cosmic rays still elude modern science.

The facts we do know about cosmic rays are however numerous and paints an almost complete picture of what they are and what they do. As we have seen thanks to Theodor Wulf cosmic rays come from space and we also have been able to determine what particles make up these rays. These cosmic rays can be thought of as belonging to two main groups although both share many of the same properties such as high kinetic energies and ionization. One type is referred to as solar energetic particles and as the name implies are generated by our own Sun. The second type is galactic cosmic rays which is slightly misleading as they do come from outside our solar system and from within the Milky Way but they can also come from outside our own galaxy. Indeed it is theoretically possible for a cosmic ray to come from almost anywhere in the universe and still arrive on Earth.

Solar energetic particles are made from mainly protons but also alpha particles and a small amount of HZE ions. The prime sources for these particles is thought to be violent solar flares or as an associated product of coronal mass ejections. Galactic cosmic rays are far more of a mystery as no one knows to date the true sources of these particles. Many different theories have been suggested falling in and out of favor over time and include such sources as supernovae, quasars and active galactic nuclei. The active galactic nucleus is most likely a super massive black hole that is affecting the matter surrounding it to such a high degree that some of it is accelerated away from the galaxy as ionized particles. The mechanism by which this is achieved is referred to as centrifugal acceleration and is thought to be the result of intensely powerful magnetic lines emanating from super massive black holes and pulsars. The concept is simple enough to understand and simply deals with a particles magnetic attraction to the powerful rotating magnetic field of the host object. As the magnetic field rotates around the active galactic nuclei or pulsar the particle is dragged along with it until it has reached speeds that are a significant fraction of the speed of light. Once ejected from the system these ultra high velocity particles become cosmic rays. Supernovae have been proven to be one source of cosmic rays but they cannot be the sole source of these rays. The amount and type of cosmic rays produced by supernovae are simply not enough to account for all the various types of particles observed in the wide spectrum of cosmic rays. The mechanism for any of these methods of producing galactic cosmic rays is not fully understood by contemporary science at this time however.

Cosmic rays are further divided into two different types called primary and secondary rays. Primary rays are particles as they are initially produced which means that they are not modified in any way. They typically originate outside our solar system and are mainly hydrogen and helium nuclei; a smaller fraction of them are made of other heavier nuclei. When primary nuclei enter the Earth's atmosphere they will interact with the atoms and especially molecules that make up the atmosphere. Since the Earth's atmosphere is very dense and molecules are so abundant and also are much larger than atoms many are impacted by cosmic rays with significant effects. The molecule ozone which is simply three oxygen atoms bonded to one another, can be destroyed by incoming cosmic rays. This is probably the most important molecule broken in the atmosphere by cosmic rays as ozone is necessary to stop harmful ultraviolet rays from penetrating the atmosphere. Atoms within the atmosphere are also affected heavily by cosmic rays and one of the most important reactions that takes place is the conversion of stable carbon 12 to an unstable isotope known as carbon 14. The isotope carbon 14 is unstable and will decay to half its original mass in 5730 years. The reason that this reaction is so important is that carbon 14 has been used for decades in a process known as carbon dating or radio carbon dating. Radio carbon dating is one type of many half life dating techniques more generally known as radio isotope dating. The ratio of carbon 14 to carbon 12 in the Earth's atmosphere has been fairly stable for eons and so determining how much carbon 14 has decayed through successive half lives allows archaeologists to accurately date many artifacts. This is only one use of radio isotope dating and a powerful scientific tool that would not be possible without primary cosmic rays impacting the atmosphere.

Many other reactions take place in the atmosphere as a result of primary cosmic rays interacting with matter and the end products are referred to as secondary cosmic rays. A whole host of other particles have since been detected as secondary cosmic rays and these include protons, alpha particles and the usual candidates as well as more exotic products like muons, pions and gamma rays. The general name for this is called an air shower as an incoming cosmic ray will strike a molecule or atom and cause a shower of new particles to be created in the air. It is quite common to see particles like neutrons and muons created in these showers. The neutrons are created directly by knocking a neutron free from an atomic nucleus and the muons are created by a decay process of smaller subatomic particles.

The various particles can be detected in a simple and ingenious device called a cloud chamber. Cloud chambers were invented by Charles Wilson a British physicist in 1911 and consist of a sealed chamber usually filled with alcohol vapor. The alcohol in the chamber is super saturated but cannot condense out of the air unless it has what is known as a seed onto which it can form a droplet. Incoming ionizing radiation will hit particles of alcohol in the air and form the necessary seeds that will cause the vapor to condense into wisp like trails. This allows for a visual examination of incoming radiation and was widely used in the study of particle physics in the early to mid nineteen hundreds. What is even more interesting is that each particle produces its own tell tale signature. A large alpha particle for example will produce a large wide spread mist pattern as the particle causes multiple collisions in a somewhat random manner. A tiny but powerful electron on the other hand is able to travel in virtually a straight line and produces a thread like pattern. The particles entering the chamber are ionized and therefore will have a positive or negative charge. Placing a magnetic field around or through the chamber will allow for the deflection of the particles along their trajectories through the vapor. By combining these two ideas an American physicist named Carl David Anderson was able to discover the positron in a cloud chamber in 1932. The particles he was observing had the same characteristic signature of an electrons vapor trail but carried the opposite charge as was detected by deflection in a magnetic field. This new particle was identical to the electron but opposite in charge and would later become known as the positron or positive electron. Most newly discovered anti-matter particles receive their names by simply adding the word anti in front of them and this makes a reversal of there charge understood. For example we do not refer to an anti-matter proton as a negaton but simply as an anti-proton. Anderson would go on to discover muons in cloud chambers as well and ultimately be awarded the 1936 Nobel Prize for the discovery of the positron.

The results of primary cosmic rays impacting the atmospheres or surfaces of other planets and structures in our own solar system can also be detected from Earth. The secondary cosmic rays produced in these solar system located air showers can be viewed by observing the gamma ray photons that are often produced in these events. Telescopes and detectors on Earth can capture these images giving us valuable data on the composition of these other worlds as only certain types of cosmic ray impacts will produce certain gamma rays with specific wavelengths.

Another aspect of cosmic rays is that the rate and volume with which they bombard the Earth is not constant. The amount of particles is determined by the strength of the event that created them, the strength of the Sun's magnetic field and solar winds and of course the Earth's own magnetic field. The first factor is of course very important as the countless events in the universe creating cosmic rays are not static. Some events will only produce a burst of cosmic rays such as a supernova and as such the incoming particles are finite in number and will eventually be depleted. Other sources of cosmic rays are longer lasting such as a pulsar or active galactic nucleus which it is thought can produce cosmic rays fairly continuously. Sources such as these especially active galactic nuclei will have life spans many trillions of times longer than the age of the universe itself and so can essentially be thought of as never ending; they will certainly outlast the life span of our own solar system.

The second factor in determining the strength of cosmic rays on the Earth is due to the Sun. Both the Sun's magnetic field and solar winds play a part in stopping cosmic rays from reaching Earth and any of the planets within our solar system. The magnetic field generated by the Sun is enormously far more powerful and far reaching than any other source in our solar system and as a result it fills a volume basically the size of our own solar system. The estimates put the limit of the Sun's magnetic field and the influence of its solar winds at a maximum of one hundred astronomical units. An astronomical unit is the distance of the Earth in its orbit from the Sun and is approximately one hundred and fifty million kilometers. This means the end of the Sun's influence on the interstellar medium is around fifteen billion kilometers from its center and thirty billion kilometers in total. This is a rough shape of course as the tail of the Sun's magnetic field and solar winds are elongated in one direction away from the flow of interstellar particles and the motion of the Sun around the galactic center. The end point of the Sun's influence is called the heliopause and denotes a region where the Sun's solar winds are stopped by the galaxies interstellar winds. The two cancel each other out and no one side holds sway over the other.

Long before this region is reached however a boundary to the solar winds exists called the termination shock. The termination shock is simply a way of differentiating the region of space around the Sun where the speed of the solar winds drops from supersonic to subsonic speeds due to the winds from the interstellar medium. The Sun's winds are still stronger however and though the strength of the solar winds is somewhat diminished enough energy still exists to drive them out to the heliopause. The strength of the solar wind therefore determines how many cosmic ray particles can actually reach the inner solar system at all. The stronger the Sun's protective solar winds the less galactic cosmic rays can penetrate the solar systems depths. During times of increased solar activity such as increases in solar flares, temporary increases in the Sun's magnetic field and coronal mass ejections the amount of galactic cosmic rays decreases. The outflow of energetic particles from the Sun pushes back against the galactic cosmic rays and we observe a drop in these rays reaching the Earth in both primary rays and secondary air shower rays.

Lastly the Earth's own magnetic field determines how many cosmic rays reach the atmosphere and the ground. The Earth's magnetic field is our last line of defense against the relentless streams of highly energized, highly kinetic and highly ionized cosmic rays. The stronger the field the more particles are stopped either by deflection into space or by funneling those particles towards the poles. These cascades of particles at the poles are what cause the auroras we see on clear nights and especially during solar maximums when solar winds are increased. Called the aurora borealis in the north and aurora australis in the south these light shows are the result of cosmic rays being directed toward the Earth's magnetic poles and ionizing the air. The drop in energy of the atoms in the atmosphere after these events releases a photon as the electrons of those atoms drop back down one orbital level. The wavelength of the photon determines its color and is based off of the chemistry of the air the cosmic rays impacted. This is why the auroras appear to be liquid, fluid and shimmering in the sky because in actuality the atoms and molecules in the air that are reacting to the cosmic rays is a fluid that ebbs and flows in the atmosphere.

The active mechanism by which these changes in atmospheric chemistry occurs is a process known as cosmic ray spallation. Spallation is a general term that can be used to denote any sort of material that has been sheared off from a parent source so to speak. Think of a rock cut in a highway and a boulder that has broken free from the hill, the boulder has spalled off of the rock face. So in a sense any thing that breaks into smaller pieces goes through a process of spallation. This is exactly what we are seeing in cosmic ray spallation a piece of an atom or molecule is broken off from the original and therefore spalled. The driving force of course is a cosmic ray and the reaction falls into a category of nuclear fission reactions. The process begins with two different nuclides which undergo a fission reaction whereby two new nuclides are created and energy is released. Normally special conditions must exist for nuclear fission to take place and these conditions require extremely high energies in order to actually split the atom. Such conditions readily exist in the atmosphere and especially in the upper atmosphere where cosmic rays enter with velocities equal to a significant fraction of the speed of light. The kinetic energy they carry is easily enough to allow a kind of natural nuclear fission to occur. Thus we see the splitting of molecular bonds very easily into either smaller molecules or their constituent atoms. We also see the direct splitting of atoms to produce various isotopes in the atmosphere as well. In fact a wide array of atomic isotopes is created by the interaction of cosmic rays with the numerous molecular and atomic compounds in Earths atmosphere. The key to identifying these is by analyzing their half-lives as many of the product nuclides in these reactions are unstable and will undergo decay processes rapidly. For example a highly unstable form of chlorine known as chlorine 34 will exist for approximately thirty two minutes before one half-life has elapsed. While on the other hand a very stable isotope beryllium known as beryllium ten will last for as long as one point four million years before a single half life for it has elapsed. In between these extremes are various other atomic isotopes such as those made of argon, sodium and silicon all with different half-lives.

Cosmic ray spallation is thought to be an active process throughout the universe as well and identification of this process is expected on exo-planets. If we were to look at a distant planet and see spectra of many of these various atomic isotopes we would have a good idea what the parent nuclides were. Such information would allow us to determine not only the composition of the atmosphere of other planets but also the type and rate of atmospheric changes present. Using our own planets atmosphere as well as the planets in our solar system as a chemistry set the backwards calculation of reactions on exo-planets would be simple to achieve. The prospect of actively seeing from day to day the changes in an alien world's atmosphere would be a treasure trove of information for planetary scientists that would no doubt explode our knowledge of stellar chemistry in the universe.

This of course can be coupled to the actual chemistry of the universe which we are positive also used the process of cosmic ray spallation to create many elements present today. When the universe was first created after the Big Bang many subatomic particles were brought into existence and gradually cooled enough to form the first atoms. The two prime examples are of course hydrogen and helium however large amounts of lithium, beryllium and boron were also present. Big Bang nucleosyhnthesis as we will see does not account for sufficient production of these three light elements and so another process must have been at work to create them. The answer of course is cosmic ray spallation which is believed to have generated much of the early stable lithium, beryllium and boron in the universe. As we have seen in earlier discussions of nucleosynthesis it is possible to form these light elements in different ways but it is important to show that cosmic ray spallation in the early universe was a major factor in their creation.

Cosmic rays can be further divided into two separate categories to reflect the differences in when they were formed. The two different types are primordial and cosmogenic. It is fairly obvious that the term primordial cosmic ray refers to those rays formed after the Big Bang. The term cosmogenic refers to those that were formed from atoms inside our own solar system, in other words after the formation of the solar system and not before. The reason for these distinctions is to demonstrate the formation processes and to account for some of the distribution of nuclides in the universe. For example we have seen that lithium, beryllium and boron are three types of cosmic rays that were formed after the Big Bang and we see that these three atoms are heavily present in cosmic rays. The amount of each that we can observe in cosmic rays is larger than can be accounted for by cosmic ray spallation events that would occur inside the solar system. This difference helps to explain part of the chemistry of not only primordial cosmic rays but also the early universe. Cosmogenic rays on the other hand are formed in the solar system and obviously therefore many billions of years after the Big Bang occurred. Even though the same production method can be found for both types of cosmic rays by definition they can only belong to one class or the other. No matter what type of cosmic ray is defined all demonstrate a natural process of nuclear fission that helps to create more new types of atomic nuclei in the universe.

Synthetic Elements

So far we have focused on the various pathways that atomic and subatomic particles can interact with one another to form new elements. An examination of the ways in which the universe can create new elements has been followed from a simple proton-proton chain all the way up to things like the r-process and supernovae. Some have been stable elements that can last for almost eternity and some are isotopes of those elements that may only live for a few minutes. The importance of these elements is obvious because all are needed to build the rich universe we see around us. There are however a few elements that need discussion that in many ways do not follow this traditional path and these are known as synthetic elements. What is a synthetic element then and why are they different?

A synthetic element is an element that does not naturally occur in the universe and therefore must be made by artificial or synthetic means. The half-lives of these elements are significantly different from those of naturally occurring elements. A long lived synthetic element will exist for a few months maybe a year while on the other end of the spectrum short lived synthetic elements will blink out of existence in a millionth of a second. This of course leads to one important question which is what use are synthetic elements? The answer is not much as the short life spans of these elements does not allow for them to be harnessed in any practical way beyond simply studying them. This is not to say that in the future a possible application or industrial use for them could not be found it is simply to say that as of today our current level of technology disallows such things. We can however learn much about the nature of atoms, subatomic particles and specifically the forces that govern their nuclei from these rapidly decaying elements.

Another obvious fact about these elements is that even if they were produced in some natural mechanism after the Big Bang we would not be able to find evidence of them since they would have decayed into lighter more stable elements billions of years ago. This is not to say however that these elements cannot in fact occur naturally in the universe. In fact the term synthetic element refers to the fact that the element is not found on the Earth but says nothing of the whole universe as obviously such a cosmic study of the elements and all the processes that create them is temporarily beyond our grasp. It is interesting to note as well that some elements that were thought to be purely synthetic elements have in fact been found in nature. One such example of this is the element technetium which was artificially created from molybdenum but was later found to be a part of red giant stars and therefore created naturally. Still technetium does not occur naturally on the Earth and so can be considered a synthetic element.

Synthetic elements tend to be very heavy in fact usually much heavier than uranium and can have atomic numbers as high as one hundred and eighteen. This super heavy weight element is known as ununoctium and has an atomic mass of two hundred and ninety four. As of today it is only known to be created artificially and has a half life so short that detecting its elemental spectra somewhere in the universe is thought to be practically impossible. Once again however this does not mean that it could not occur in the universe naturally in fact many elements much heavier than this can theoretically exist even if only for a fraction of a fraction of a second. The reason that this is the case is how these elements are created and specifically how we know to create them in laboratories here on Earth.

The production of such elements begins with a pre-existing seed nuclide that is bombarded with other particles such as neutrons or alpha particles. We have learned that many of the transuranium elements were created in this way and making elements heavier than normal uranium is quite simple. Bombarding things with particles has been a favorite experimental technique of scientists for over a hundred years. To create the transuranium element ninety three for example simply bombard uranium with neutrons until one sticks and beta decays to form a proton. The real trick to creating heavier synthetic elements is to bombard the nuclei of an existing heavy atom with the nuclei of another atom. True a proton is the nucleus of a hydrogen atom and an alpha particle is the nucleus of a helium atom but to create synthetic elements you must use much heavier nuclei still. The simplest way to do this is to use light or medium weight elements such as carbon, calcium or lead to name only a few. These medium weight elements are what can be used to bombard the heavy seed nuclide with. The heaviest element known to date, ununoctium was believed to be synthesized by shooting calcium nuclides at californium targets. It is such a rare chance that this super heavy element will be produced that it is thought perhaps only four of these atoms have ever been created in a laboratory.

The rapid decay rate also means that it is difficult to study these elements in depth as a half life is essentially impossible for us to calculate based on our current technology. Since these atoms are produced one at a time the idea of half a known mass of sample decaying over time is nonsensical. One produced atom of ununoctium cannot decay to half its original value instead it simply decays through spontaneous fission or alpha decay into a lighter element. The chain of elements down from ununoctium is how scientists believe they have created it as they can examine the daughter nuclides that result from its own decay. Ununoctium will decay through alpha release into livermorium which has an atomic number of one hundred and sixteen. This will in turn decay again as it is an unstable heavy synthetic element itself. These elements decay so fast that the isotope of each element that is the most stable will essentially last the longest and so is given as the atomic number and atomic mass for that element. The synthetic elements as we can see are created under artificial conditions but through common naturally occurring means.

The collision of two nuclides with one another is something that occurs continuously throughout the universe. These reactions are nothing more than a form of nuclear fusion after all whereby two nuclides react to produce to new nuclides. The key here is that in nature the chances and conditions by which two large nuclides will meet and fuse is low. It is easy to collide any two nuclides we choose to in a controlled laboratory setting because we select the elements to react in a fusion event. In the universe however the chances of a heavy and light or medium weight element encountering one another in conditions extreme enough to cause fusion to occur is remote. Many heavy and light or medium weight atoms meet constantly but not with sufficient force as to overcome the electrostatic repulsion of their nuclei. Perhaps in the collapse of super massive stars trace amounts of heavy elements can be brought together and made to fuse as the star collapses and explodes as a supernova. Yet this is still a rare probability and the super heavy elements created will ultimately share the miniscule half lives of the man made synthetic elements. The only way to detect them would be during the peak spectra of a supernova explosion and even then this holds little practical purpose beyond curiosity as we do not currently possess the technology to harness them in any meaningful way. The most we might be able to hope for is to better understand the mechanisms that exist in the violent solar plasmas of core collapse supernovae and the properties of the super heavy elements themselves. Nevertheless synthetic heavy elements are real and can exist even if for only a short time compared with very stable elements. Perhaps in the future we will be able to extend their life spans to a point where we can harness their power but for now they do provide us with another means of nucleosynthesis and diversity of elements in the universe.

The diversity of elements in the universe is truly astounding and can range from simple hydrogen up to the artificially produced ununoctium. New and heavier elements can be created naturally or artificially by many means from lighter ones. Adding a neutron to an existing atomic nucleus has been done many times in the laboratory as well as in space. A similar method exists for adding helium nuclei or alpha particles and once again we get heavier elements. It is also possible to break large nuclei into smaller or medium sized ones through nuclear fission. Light and medium weight nuclei can be made to form heavier ones through nuclear fusion and once again we get more new types of elements. There is one small problem with our picture of nucleosynthesis up until this point and yet it is a massive problem in itself. Where did all the small light nuclei come from? None of these methods for creating new elements can account for the formation of the lightest of elements. Light elements are usually thought of as hydrogen, helium and lithium which are commonly described in the various forms of nucleosynthesis but where they themselves come from is essentially skipped. What is needed is a method by which these basic building block elements can come into existence. This last process of nucleosynthesis in the universe is by far the most important as it created all the necessary matter from which everything else was made.

Big Bang Nucleosynthesis

It is a process referred to as Big Bang nucleosynthesis and as the name suggests it was active only at the time of the start of the universe. The reason it is so important is that without it none of the other methods of nucleosynthesis would be possible. Natural or artificial all of these methods use the products of Big Bang nucleosynthesis as building blocks. Some of these building blocks are simple subatomic particles such as protons, neutrons or electrons and others are whole atoms themselves such as hydrogen and helium. There is no proton-proton chain, no s or r process, no fission or fusion and no supernova synthesis without the matter created by the Big Bang itself. Not to mention the fact that all these later processes of nucleosynthesis are responsible for a tiny fraction of all the matter in the universe. A single uranium atom for example is massive and extremely heavy when compared to a simple helium atom and yet twenty five percent of the matter in the universe at the time of the Big Bang was made up of nothing but helium. How much matter in the universe is made up of uranium by comparison? Basically zero and that illustrates just how large the gap is in the raw numbers of one type of atom versus just helium in the universe.

As for hydrogen it is the King of the universe by far with approximately seventy four percent of the universe being ordinary hydrogen. Just for fun lets take a look at the entire universe as it spans as far as we know approximately fourteen billion light years on either side of the Earth and factor in all the mass that must be made from helium and hydrogen. Now compare that almost unfathomable number to the total of four ununoctium atoms we created and which by the way no longer exist to realize how much of the universe is simply constructed from two elements. The remaining elements including everything beyond helium make up the rest of the matter in the universe and there are one hundred and sixteen known elements beyond helium. The importance of Big Bang nucleosynthesis is made immediately apparent and without all of the hydrogen and helium it created we simply would not exist.

Yet we do exist and this means we can explain where all this matter did come from and the idea of Big Bang nucleosynthesis as well. The beginnings of the theory began with the work of Fred Hoyle and his collaborators after the publication of their famous B2FH paper in 1957. The paper we have seen was instrumental in explaining the synthesis of elements in stars through both the s process and the r process. The slow neutron capture of living stars and the rapid neutron capture of dying stars so to speak. This model was widely successful but it could not account for the lightest elements namely hydrogen and helium. The slow neutron capture process in particular was used as a method for explaining where the helium in the universe came from. The s process as it is known does produce helium as we have seen and so does the proton-proton chain but two problems arise from these methods for explaining the abundance of hydrogen and helium atoms in the universe.

The first is that they simply do not produce enough helium to account for what we can observe in the universe. Also it produces helium too slowly for the massive amounts of helium we see in the universe today to be created in anything less but a time span of billions or trillions of years. This specific short coming of the models could not be worked around and scientists began looking for alternative explanations for helium production. The helium enigma however is nothing and pales in comparison to the second problem that arises from these models and that is where the hydrogen comes from.

Hydrogen makes up about seventy four percent of the universes matter by mass and therefore any attempt to explain where it comes from should meet two criteria. The first being the actual production of hydrogen and the second being a ridiculously fast method of producing hydrogen so that the universe can come into being in a time span short enough to account for the levels of other elements we see today at their respective ratios. After all a slow production method for hydrogen that does not outstrip the nucleosynthesis of other nuclei by many orders of magnitude will result in a universe where hydrogen is far less abundant. The other heavy elements produced through s process and r process reactions would be allowed to run long enough and catch up to hydrogen and helium such that elements like carbon or iron for example would comprise huge percentages of the universes matter rather.

Today as we observe the universe one percent of the matter in the universe by mass is made of every other element from lithium to ununoctium. The rate of hydrogen production from all known mechanisms of stellar nucleosynthesis for example is actually of no importance however as each of these models has a huge flaw in them. The proton-proton chain, carbon nitrogen oxygen cycle, triple alpha process, s process and r process are all successful in explaining the production of certain elements except for hydrogen. The reason is none of them produce hydrogen but rather they all consume hydrogen in some form or other at one or all of the steps of their processes. In other words the total amount of hydrogen in the universe could never be seventy four percent because all of the known processes of stellar nucleosynthesis would be steadily depleting hydrogen atoms available. Furthermore they still do nothing to explain where the hydrogen atoms came from in the first place and at seventy four percent of the universes total mass this was a huge problem to be solved.

Stellar nucleosynthesis as the sole mechanism for creating atoms in the universe does not explain how the levels of helium and hydrogen in the form of deuterium came to be so abundant in the cosmos. The solution is Big Bang nucleosynthesis which can explain not only where hydrogen, deuterium and helium came from but also why we see the current ratios of these elements in the universe today. An interesting aspect of Big Bang nucleosynthesis is that it can also explain where trace amounts of primordial lithium and beryllium came from as we will see later. For now we can take a look at the early work done on Big Bang nucleosynthesis by the Russian scientist George Gamow working in the united states and his student Ralph Alpher.

A paper was published by them called the Alpher-Bethe-Gamow paper which describes the concept of Big Bang nucleosynthesis by successive neutron captures was published in 1948. In this paper it was argued that the capture of neutrons could be used to explain the creation of new heavier elements. This approach also produced a rough approximation of the ratios of light elements that we see today, however it was flawed as neutron capture cannot explain all heavier elements. This is in part because a nucleus that is stable and containing five or eight nucleons cannot be created in this way. The result is such that these elements would decay leaving nothing for us to observe today and therefore cannot account for the production of these heavier elements. The humorous side note to this paper comes from Gamow who decided to have a bit of harmless fun at the expense of the scientific community. Gamow added that name of his good friend Hans Bethe who as we have seen was a brilliant mind in the field of stellar nucleosynthesis. Gamow decided that it would be believable to add his friends name to the paper and to arrange the names of all three authors such that they appeared in the order of Alpher, Bethe and Gamow. The joke was that this would be a play on the Greek letters, alpha, beta and gamma even though Bethe did not contribute to the paper in any way other than perhaps discussing it with Gamow personally. In answer to many readers question of is this the kind of joke that a scientific mind finds funny? The answer is yes it is funny to scientifically minded people like Gamow and probably to some of the scientifically minded readers of today as well; it's just how we think.

As for the process of Big Bang nucleosynthesis proper we must first consider the initial conditions of the universe in the moments just following the actual Big Bang explosion itself. The temperatures and pressures present at the time of the Big Bang were so extreme that if a proton and neutron were to come in contact with one another and fuse to form deuterium they would immediately be split apart by a photon. This is the process known as photo-disintegration and was everywhere in the early universe playing a major role in shaping the universe. The reason was the photons in the early universe had such tremendous energy and were present in such large numbers in such a small volume of space that collisions of deuterium and photons would almost certainly occur. The photons had enough energy to split the binding forces of the nucleus and therefore destroy the newly formed deuterium atom. Since this was the case how then did matter such as hydrogen, deuterium or helium for example arise in the early universe just after the Big Bang.

The problem lies in the photons as they possessed too much energy to allow matter to grow to larger heavier forms. A process called photodisintegration was at work whereby high energy photons can be absorbed by an atomic nucleus and imparts such energies that the nucleus will shed a particle in response. Essentially the photon would cause early protons and neutrons to become so excited that they could not stay bound despite the strong nuclear force and fly apart from one another. The universe needed to expand in order for temperatures inside of it to cool sufficiently so that the photons lost some of their energy. With the energy of these photons dissipated matter could withstand the bombardment long enough to stay bound in forms such as deuterium. What came next was the synthesis of heavier elements still such as helium and trace amounts of lithium but this only occurred for a small amount of time itself.

In general Big Bang nucleosynthesis could not occur until the universe had cooled to the point where photodisintegration could be ignored and this happened roughly ten seconds after the Big Bang itself. The universe then however needed to remain dense enough for the actual Big Bang nucleosynthesis reactions to occur. If the density dropped too far then matter would not interact often enough or with enough force to allow the creation of heavier elements beyond hydrogen to occur. It still takes some temperature and pressure to make deuterium and the nuclear chain reactions that will produce elements heavier such as tritium or helium. The universe then became too diffuse to continue this process to any great degree at about fifteen or twenty minutes after the Big Bang.

Two types of reactions must be considered when calculating how much matter and of what types the Big Bang will create. Without taking any risks and simply looking at the standard model for Big Bang nucleosynthesis we see that the first variable is the ratio of protons to neutrons. Protons are very stable where as free neutrons are not and will decay after eight hundred and eighty one seconds. In the very early universe just after the Big Bang the density of particles allowed for the creation of protons and neutrons at about a one to one ratio but this would change as the universe aged. Not long after this point the ratio of protons was about seven for every one neutron. This is due to the fact that protons having a lower mass are more easily created and neutrons were decaying back into protons and electrons. The final ratio was stopped as matter could effectively capture the free neutrons and end the possibility of them decaying.

The second and last variable is the ratio of normal matter which is often times referred to as Baryonic matter and photons. As we have seen photons of sufficiently high energies can cause changes to matter to occur and therefore this ratio must also be factored in. Protons, neutrons and deuterium can be converted to and from helium three depending on the energy and number of photons present. The helium three in turn can further interact with more matter such as tritium to produce helium four. Helium four is incredibly stable and as such will go on to make up twenty five percent of the matter in the universe by mass. So looking at the temperature and density of the early universe and how often early particles were able to interact with each other one can calculate the creation of the abundances of early elements in the universe.

From protons and neutrons to deuterium and helium the final result of Big Bang nucleosynthesis produces a universe made mainly of hydrogen, secondarily helium and lastly trace amounts of very light elements like lithium. So this offers an explanation to how the first light elements were created in the observable universe and of course we still need to examine the creation of the very first particles as well. There is however one thing we have not examined yet and that is the Big Bang itself. Up until this point we have mentioned the Big Bang as the starting point for the observable universe that we see and study today. We have also included it in many of our models of how the universe works but we have not looked at how the idea of the Big Bang came about in the first place. After all the Big Bang itself is a model to explain why the universe is as we see it today too. So the problem remains where did we get this universe from?

Chapter 6
The Big Bang and More Changing Ideas

A Large Expanding Universe

The solution starts with the Big Bang and the Big Bang is a name given to the expansion of the universe by Fred Hoyle on BBC radio in 1949 to describe the motion of the universe. Hoyle who was a brilliant scientist in his own right wrongly believed that the universe was static and unchanging, a view that was held by scores of other scientists as well. The static and unchanging view of the universe was first challenged years earlier in 1912 when an American astronomer named Vesto Melvin Slipher began examining the motions of galaxies. Originally Slipher used the common astronomical tool of spectroscopy to examine planets and the chemical make up of their atmospheres. He then decided to apply the same spectroscopic techniques to galaxies and made note of the spectral shift lines they produced. At the time galaxies other than our own were thought to be nothing more than nebulae as the power of many telescopes of the day were not able to discern a great amount of detail on these objects. These nebulae were commonly known as spiral nebulae due to their shape and it was the light from these cosmic objects he measured. What he found was that the light coming from these spiral nebulae had been shifted towards the red end of the visible light spectrum. A common term today we are all familiar with red-shifts and understand that this is due to the object being observed moving away from us. The basic idea goes that as an object moves away from you the light emitted from it which is a wave has its wavelength stretched. Stretching these wavelengths makes them longer which is why they are called red shifted as red light photons have longer wavelengths than normal ones. The opposite is blue-shifting of light and this occurs when an object is coming towards you and the wavelength is compressed and made shorter which is where you find higher energy blue light on the spectrum; a fuller explanation will follow shortly. At the time however Slipher's work was all new and he had inadvertently discovered galactic red-shifts which are a result of galaxies moving away from our own.

The idea of a galaxy's light being shifted towards the red end of the visible spectrum is due to a phenomenon called the Doppler Effect. The Doppler Effect is named after the scientist who first proposed it an Austrian physicist named Christian Doppler. During the eighteen hundreds a great deal of research was put into the study of wave mechanics such as those found on the surface of water. A popular idea was that all waves obeyed these same rules and so could be applied to other wave types such as sound waves. People had long observed the fact that an approaching noise would grow louder and higher in pitch as it moved towards the observer and quieter and lower in pitch as it passed by and moved away from the observer. The mechanism by which this occurred was not well understood by scientists of the day and it was Christian Doppler who offered an explanation for this behavior of waves. In 1842 he proposed that the speed of the wave's source and the relation of it to the observer would dictate the nature of the frequency of the wave.

This can be easily imagined by a pool of water with waves rippling on its surface and a toy boat for example. If the boat is stationary the waves will arrive at it one at a time and the rate will be constant at let us say ten waves per minute and therefore the frequency of waves hitting the boat is ten per minute. If the source of the waves moves towards the toy boat quickly enough then in one minute it can encounter twice as many waves and therefore the frequency of waves hitting the boat is now doubled to twenty. Reversing the process and having the source of the waves move away from the toy boat could drop the frequency to only five waves per minute. So we can see how physically the Doppler Effect can work and this is how sound for example changes pitch based on the motion of the sound source and you. This is of significance to physics in general as all waves obey this law not just water or sound waves. Electromagnetic waves can be affected by their relative motion to one another as well and this fact Doppler applied to the various colors of binary star systems. It had been known that the temperature of a star also affected its color and spectral properties but Doppler was the first to state that the relative motion to the observer was also a factor in a stars color.

This is a groundbreaking realization for not just astronomy but all of science as the Doppler Effect has been used countless times in a myriad of applications. Doppler's idea that the relative motion of sources and observers changes how the universe can be perceived predates Einstein by over sixty years. In other words Doppler had discovered all of Einstein's relativity work concerning light, space and time six decades before Einstein only he was not interested in applying it to gravity. The concept of relativity as it applies to objects on Earth and the universe was discovered by Doppler and has been proven over and over again by the application of his famous Doppler Effect. Most recently his work has led to the discovery of planets outside of our own solar system and this has sparked a fire in the scientific community to find the first true Earth like planet outside of our own system. The topic of exo-planets however will have to wait as it is too monumental to cover quickly here. His idea of relative motions and changes to wave frequency was proven first in 1845 by a scientist from the Netherlands called Christophorus Buys Ballot through his experiments with sound waves. Doppler's theory was again proved correct in 1848 by the French physicist Hippolyte Fizeau who studied spectral shifts in star light.

This is the technique that Slipher applied to the spiral nebula and his observations of the red shifting of their light spectra. The man who would take these observations the next step further was an American astronomer named Edwin Powell Hubble. He also made observations of these strange spiral nebulae and their red-shifted spectra. The key difference with his work was that he discovered that the spiral nebulae had different red-shifts depending on their distance from our own Milky Way galaxy. Slipher had made note of the red shifts and of course found that they were all unique as well but Hubble determined that a red-shift grew more dramatic as the spiral nebulae grew in distance from us. If he mapped out all the objects he observed he found that the closer the spiral nebulae the less red-shifted its spectra was and the slower its velocity away from us. Spiral nebulae with large red-shifts were not only farther away from us but they were also moving faster. This relationship between galactic distances and their speeds with respect to the Milky Way and one another is known as Hubble's Law today. What this means is that the universe is not a static state system in which nothing changes. In reality scientists and proponents of the static view never believed that the entire universe was completely rigid and unchanging merely that it had a definite size that was unchanging. In other words the objects within the universe could move and change such as a planet rotating around its star, an asteroid colliding with another asteroid to create many smaller ones or stars exploding as supernovae and spreading out as nebulae. What could not happen according to the static model was the universe itself could not change its size or presumably its shape. It is as if the entire universe is a tank filled with fluid in which numerous objects can float about, diffuse through the medium of the liquid and interact with one another but the tank itself is an unchanging box.

This view of the universe was the predominant view amongst many astronomers and scientists of the day and from an observational standpoint it did seem to make sense. The scientific equipment at the time with which to observe the universe was not refined enough to see objects clearly enough or in sufficient detail to allow for a change in thinking. Telescopes simply were not big enough to view a spiral nebula in enough detail to make out the fact that they were separate galaxies with stars, arms and galactic centers. A general description of galaxies was made but due to the view of a static universe and one in which the Milky Way was the universe no change was thought to be needed in the model of the universe. It was a bit like the old ideas about the Earth being the center of the solar system. The idea was based off of good empirical evidence and observation for the day but when new ideas and new models were proposed the geocentric universe was doomed.

This is how much in science is learned as old models are replaced with new ones based on good interpretation of current observations. The new model seems different and to many people initially wrong but will ultimately be proven correct even if it takes decades or more for the necessary experiments, mathematics and observations to be made to verify it. After all it took centuries before we were able to directly launch satellites and humans into space to physically turn around and look back at our world to see that it is orbiting our sun and round versus two older models stating it is flat and at the center of our solar system. This change in accepted thinking is exactly what Hubble's work was doing as it meant that the universe itself could not be static and it must instead be expanding. As always with new ideas about the universe his model was initially met with skepticism or down right rejected but of course he was correct and as time went on proven right.

This overturning of the proverbial apple cart was not the only trick up Hubble's sleeve though. It is true that some work had been done mathematically previous to Hubble that suggested the universe might be expanding but it was Hubble who provided the first observational evidence of this. Slipher was the first to observe red-shifts but Hubble was the first to catalogue enough spiral nebulae spectra to work out that the farther they are away from Earth the faster they are receding. It is also true that a few years earlier scientists such as Alexander Friedmann, Georges Lemaitre, Howard Robertson and Arthur Walker had all done work mathematically on solutions to Einstein's equations they did not provide the observational evidence to prove this. The papers are sometimes known as the FLRW metrics and contain mathematical solutions to Einstein's equations that show the universe is either expanding or contracting but not static. The problem with this is that the work was based more on mathematics than observation or experiment which can and does lead scientists down the path of wrong conclusions. What is worse is that often times a purely mathematical theory will gain such support that people will then build new mathematical theories on ones that are untested have no experimental or observation data supporting them.

In this particular instance Einstein himself rejected the work of Friedmann and Lemaitre initially saying essentially that math does not always provide sound theory. In the case of the FLRW papers Einstein is partially correct as some of the solutions seem to provide equally strong evidence that the universe is contracting and not expanding. While the idea that the universe is not static based on these mathematical theories is not under debate and demonstrates how mathematics can be used to advance scientific knowledge the mere fact that a polar opposite outcome can be equally plausible is unacceptable. This is tantamount to saying if you are standing in your kitchen and a fire breaks out on the stove applying a certain fire suppressing agent to it will either put out the flames or incinerate you and your home. Such wild mathematical inaccuracies are what science and Einstein were being guarded about. It was demonstrated however that the universe is expanding and so the mathematics is somewhat vindicated as at least one of the interpretations of Einstein's equations did say that the universe was not static and it was expanding. Interestingly enough Einstein himself thought the universe was static and would be proven wrong quickly. A fact that Einstein was quick to acknowledge after the fact was proven observationally but wisely not before hand.

Of the authors of the FLRW papers one is more vindicated than most as his work predicted another aspect of the universe that was eventually borne out by observation. His name was Georges Lemaitre a Belgium astronomer who also predicted that the universe must have expanded from an initial point, something he referred to as a cosmic egg or primeval atom. Essentially the Big Bang and this was followed by his further prediction that not only is the universe expanding but that it must be accelerating as well. Again it was with skepticism and rejection that Lemaitre's work was greeted but further investigations by future scientists would back up his ideas. I say back up his ideas because as far as the beginning of the universe from a single small point this was impossible for him to observe and therefore exceedingly challenging to prove. However in the 1980s physicists such as Alan Guth would help explain the validity of Lemaitre's primeval atom through his work on cosmic inflation which we will examine in greater detail later. Then a decade later in the 1990s the world famous Hubble Space Telescope named obviously after Edwin Hubble would make a series of observations based on type 1A supernovae that would demonstrate Lemaitre's idea of an accelerating universe. Further theories and work would ultimately yield even further support to the idea that the universe had a beginning from which the so called everything of the universe began and these topics will be examined later.

Hubble meanwhile did have a second important contribution to make to astronomy and science that in some ways has a greater meaning for our understanding of the universe than simply whether or not it is static. His work on the expanding universe is of course the most important contribution he made to astronomy and the reason that we examined it first but years earlier he made another interesting find about the cosmos which increased its size to the limits that we can see today. The Hubble Space Telescope was launched in 1990 and has revolutionized our view of the universe. This has been achieved not only in the inspirational photographs that it has returned to Earth motivating future generations of scientists to explore space but also in the sheer volume of data that can be collected above the atmosphere. The data collected has allowed us to better understand the behaviors even if not the mechanisms of various cosmic phenomena throughout the universe. Even though its mirror is not the largest ever made for a telescope it is arguably the most powerful as it has an unobstructed view of space without the hindrance of the atmosphere.

Back in the time of Edwin Hubble the most powerful telescope was the one hundred inch Hooker observatory telescope completed in 1917. It is somewhat ironic that Edwin Hubble would use the most powerful telescope in the world at the time and then decades later the most powerful telescope off this world would bear his name. Hubble's main area of research upon arriving at the Mount Wilson Observatory where the Hooker telescope resided was investigating Cepheid variable stars. The term standard candle is one we have discussed with respect to type 1A supernovae and Cepheid variable stars are a type of standard candle as well. Cepheid variables produce regular changes in the brightness and temperature that we on Earth can observe. These changes combined with the period of the changes known as pulsation period can be used to accurately determine the distance these stars are from Earth.

We have discussed this previously with our examination of Arthur Eddington's work as he provided a workable theory of Cepheid variable stars.

Hubble was going to take this work and utilize it to measure distances by means of the standard candle effect. He conducted his survey of Cepheid variable stars in various spiral nebulae starting in 1922 and continuing for about one year. It is important to recall that the view of the universe at the time was that the Milky Way galaxy was in fact the entire universe and that it constituted the volume of space in which all astrophysical phenomena existed. This is why when early astronomers viewed distant galaxies outside our own they referred to them as spiral nebulae as much closer nebulae in our own galaxy were well understood. Nebulae were seen as exactly what they are which is a diffuse collection of gas and dust illuminated by radiation from background sources or stars inside the nebulae itself. Astronomers were very familiar with observing the remnants of supernovae that illuminated the gas and dust from the inside of the nebulae. They were also familiar with nebulae that contained new young stars in it that had been newly born and were shining their light again on the gas and dust illuminating it. Therefore when a whole group of strange spiral shaped objects were observed regularly through modern telescopes it was thought that these were in fact just another form of ordinary nebulae. After all they appeared to be made of gas and dust in fuzzy clumps like normal nebulae and they were illuminated from their interiors as well. At the time none of these scientists could have fathomed that a super massive black hole was directly in the center of these galaxies and the hundreds of billions of stars making up the galaxy was what appeared to be the fuzzy gas and dust. Hubble would be the first to propose that a universe consisting entirely of just the Milky Way was not in fact the truth.

Through his meticulous observation of these Cepheid variable stars and their distances from Earth he noticed something strange. All the Cepheid variables located inside these so called spiral nebulae lay at distances far, far away from anything observed inside our own Milky Way. Normal objects within our own Milky Way are found at distances no greater than one hundred and fifty thousand light years from one another as this is the diameter of our galaxy. The maximum distance from Earth would be approximately one hundred thousand light years as our star system lies about twenty five thousand light years from the galactic center and this can be added to the radius of our galaxy. One of the objects Hubble observed was the Andromeda nebula which was believed to be merely one of many nebulae within our own galaxy. The distance from our Milky Way galaxy and the Andromeda galaxy is approximately two point five million light years. One light year is around ten trillion kilometers and so the Cepheid variable star Hubble observed in the Andromeda nebula would have been at a distance of some 25 million trillion kilometers away. This distance is quite inconceivable for a star to exist within our Milky Way and while the exact distance measurements I have just given were not at the time known so precisely it was easily enough for Hubble to make an important discovery. What Hubble discovered was the so called spiral nebulae were actually whole galaxies themselves and that the Cepheid variable stars that were being viewed must have originated from within them.

The idea that objects in the universe could exist outside of our own galaxy had been proposed years earlier by various astronomers but these theories had certain flaws. First was overcoming the obstacle of the Milky Way itself being the whole universe as this was considered to be a scientific fact at the time. The second was that these objects could not be well explained nor defined and were simply dubbed island universes meaning they were separate universes from the Milky Way. Lastly none of the methods for predicting their distances from us could be made with any significant degree of accuracy so that they could once and for all settle the debate. Estimates put forth by those who believed them to be island universes ranged from a few thousand light years to half a million light years away. Furthermore the methods by which these estimates were derived were not exact and instead were more careful scientific approximation. By examining Cepheid variable stars which are an accepted standard candle in astronomy Hubble was the first person to prove once and for all that these spiral nebula were in fact separate from our own galaxy. The universe would now never be the same as the debate as to whether the Milky Way was the entire universe or simple one part of a much larger whole was over. The universe is immeasurably huge to put it bluntly and it is positively overrun by galaxies each containing tens to hundreds of billions of stars. The measurements made by Hubble vindicated all those who came before him who believed in island universes and opened one of the golden ages of astronomy.

Now not only had science made a massive move forward in our understanding of the universe but astronomers would now be busied with the tasks of understanding just what the structure of the universe actually was. In a few short years science had gone from a static universe as believed by many including Einstein to a universe that was both expanding and filled with countless undiscovered galaxies. The composition of the galaxies in terms of not just their spectra but of individual stars was now a source of debate. The spectra of spiral nebula as they had been known for years were well understood but the makeup of the individual stars was not at all well understood. This and a host of other questions were now right in front of astronomers waiting to be answered and all due to the realization that the universe is far larger than we had previously thought. Some of the first insights into these new and wondrous galaxies would come from Hubble himself as he did the most obvious thing that any good scientist would do when confronted with a new subject, he classified them. Classification is one of the first and best tools available for the curious mind to solve problems with as it lets us group things together and then look for patterns and similarities in those groupings. Hubble immediately began applying this careful cataloging technique to all the galaxies he could identify.

Galaxy Types

This first early system of grouping galaxies is known as the Hubble sequence and is still used today. Other systems have been proposed afterwards but all borrow heavily from Hubble's original system and usually only vary slightly. The basic scheme of the Hubble sequence still holds today because galaxies still look the same as they did in Hubble's day. Three main shapes were observed by Hubble and so his arrangement of galaxies follows three branches. The diagram is like a letter y on its side or a tuning fork as some have called it. The first shape of galaxy is the elliptical galaxy which can best be thought of as a ball or egg. These galaxies do not possess the flat spiral structure of our own Milky Way galaxy and appear as large fuzzy blobs of stars with a very bright center. Almost all elliptical galaxies are some form of ovoid shape and not true and perfect spheres.

As such they have a major axis and minor axis that pass through them. The major axis can be thought of as a line drawn through the longest part of the egg say from top to bottom while the minor would run through its middle left to right. Many elliptical galaxies appear relatively featureless and it can be difficult to make out the exact boundaries of the galaxies volume. Most of these galaxies can also be quite large as examples of them typically range from the size of the Milky Way which is about one hundred and fifty thousand light years across to arguably millions of light years across. Due to their large size they are commonly made up of smaller galaxies as well which can aid in observing them as these provide some detail in the otherwise featureless expanses of the ellipticals themselves. These galaxies range in size and shape along the left branch of the Hubble sequence and stop at the fork in the galactic road.

The upper branch of the Hubble sequence is filled with examples of what can be thought of as normal spiral galaxies. All galaxies are normal of course but a convention for classifying them must exist and so we call the upper branch normal spiral galaxies. The lower branch is populated by what are commonly referred to as barred spiral galaxies. This is due to the fact that while they share a similar spiral shape to regular spiral galaxies they also possess a very prominent set of darkened bars that run a long distance across their surfaces and the spiral arms appear to grow out of the ends of these darker bars. Of course variations exist amongst the barred spiral galaxies as well but again this can be thought of as the typical example of a barred spiral galaxy.

Our own Milky Way galaxy is a type of spiral galaxy but due to the fact that we cannot view the Milky Way from afar means that we cannot say for certain what type of spiral we are. There are many examples of spiral galaxies along the Hubble sequence all with slight variations from each other so trying to determine exactly which one matches the Milky Way galaxy best is extremely difficult from inside our own galaxy. This is a bit like saying you must classify what your house looks like from the outside in terms of its size, shape and color but are only able to make observations from within one room of your house. You would not be allowed to travel to other rooms just as we today cannot travel to other parts of the Milky Way galaxy and take measurements from these new perspectives. Therefore it is simplest for us to say that we live in the Milky Way galaxy which is a normal spiral galaxy. As with elliptical galaxies spiral galaxies can have various other structures associated with them such as smaller satellite galaxies or compact dwarf galaxies. The Large Magellanic Cloud is one such galaxy that is a satellite galaxy of the Milky Way galaxy. Galaxies such as these that are smaller than the galaxy around which they orbit are typically thought to be slowly consumed by the larger galaxy thus creating one galaxy that is larger than the two originals.

The next type of galaxy on the Hubble sequence is a cross between elliptical galaxies and spiral galaxies known as lenticular galaxies. Lenticular galaxies appear at the very middle of the Hubble sequence where the three branches of the fork meet. These types of galaxies seem to be an almost perfect fifty-fifty mixing of both elliptical galaxies and spiral galaxies. They appear to have a somewhat bright bulge at their centers with a relatively flattened disk shape surrounding them. These disks are relatively featureless and smooth like one would expect from an elliptical galaxy but are definitely a flattened disk like a spiral galaxy. The disks here in a lenticular galaxy though do not appear to have any well defined structure such as spiral arms or bands as are common in spiral galaxies. For these reasons lenticular galaxies can not truly be considered as either purely elliptical or spiral and so sit in the middle of the branches of the Hubble sequence. Hubble himself had not directly observed these galaxies for himself and instead decided to include them because he believed that a sort of intermediate galaxy class must exist between the two different and well defined classes of galaxies. Years later however observations made by various astronomers including Hubble did provide evidence of the lenticular galaxy and so Hubble's initial hunch was correct. Some of the signature features belonging to lenticular galaxies are a very bright central bulge where most of the galaxies light emanates from and a lack of star forming regions throughout the galaxy itself.

There is one other general class of galaxies that must be included but does not really fit on the Hubble sequence and these are known as irregular galaxies. Irregular galaxies as their name implies do not really follow any set rules that can allow them to be neatly categorized other than they seem to follow no rules. They do not possess well defined shapes and distinguishing their boundaries can be difficult. They do not have any obvious major axis to which they can be oriented nor do they seem to have any central bulge or structure. Most of the time irregular galaxies are smaller than elliptical, spiral or lenticular galaxies and can instead often be found near a larger companion. A perfect example is the Large Magellanic Cloud which as we have seen is a small satellite galaxy around our own Milky Way galaxy. It has an irregular shape without well defined borders and no obvious central structure.

As such irregular galaxies tend to be made up of many dense star clusters held tightly by a common force of gravity. Gravity however will always draw objects together and the law of angular momentum will dictate that this collapsing collection of stars must rotate. The result is a technical subgroup of irregular galaxies that appear to have the faintest hints of a spiral structure. When irregular galaxies are included in a Hubble sequence they tend to be added on to the end of the spiral arm branches. The center of rotation could simply be a common center of gravity or perhaps a small black hole that is growing as the galaxy coalesces into an ever denser space. Given time the black hole could grow into a super massive black hole similar to the one at the Heart of the Milky Way galaxy. It is very tempting to think of these as baby galaxies that were formed from collections of stars and are only now beginning to spiral together forming a very small spiral galaxy of their own. It is entirely possible that this is how many galaxies formed in the early universe as large established galaxies such as the Milky Way or Andromeda would not have existed. Rather these large galaxies would have started out as small irregulars that gradually grew as they attracted more materials to themselves through gravity and time. The small spirals would grow larger with more pronounced arms and eventually collide with one another just as the Milky Way and Andromeda will do one day. The result of these collisions would be what we observe as large elliptical galaxies. The idea that ellipticals are created from the mergers of spiral galaxies would offer a decent explanation for the large size of most ellipticals, the lack of a definite shape and the abundance of older stars in them.

This is a very interesting topic and one that receives much study and debate to this day as no one can conclusively prove just how galaxies do form. Mainly due to the fact that like many theories about the observable universe it is impossible to actually watch these events unfold in a laboratory. We did not witness the Big Bang nor did we see the formation of everything afterwards and even if a similar event were to unfold now for us to observe the timescales involved would be on the order of billions of years hardly a convenient situation. We know that the universe appears to have started from some centralized point and spread outward rapidly in what we call the Big Bang. We know that after this period matter cooled out of the super hot, dense and energetic medium of the Big Bang. After this stars began to shine and today we have a universe populated by large structures called galaxies. How did these galaxies precisely take shape and what were the forces responsible for shaping their evolution is what modern science does not fully understand. A full discussion of how galaxies formed out of stars and other forms of matter can be discussed later but here we can see a good use of Hubble's classification system in action.

If for example we do start with small clusters of stars we can get our smaller irregular shaped galaxies. Some of these due to the law of conservation of angular momentum will start to gently spin and form baby spiral galaxies. These baby spiral galaxies will slowly grow as they collect more stars, gas and dust until they become regular spiral galaxies. Spiral galaxies are quite large and well established with numerous star forming regions spread throughout their arms. The stars in these galaxies can range from just born to very old golden type stars but on the whole they are filled with young stars and at the very least certainly not too many old stars. The small number of old to very old stars probably reflects the original stars from the baby galaxy now spread out through the many new young stars as the baby galaxy's original young stars have matured over billions of years to be the old stars observed today. Eventually these spiral galaxies may come in contact with others of their kind and through a common center of gravity begin to merge. These mergers are quite random and chaotic as most of these galaxies will impact each other at an angle. The result will be a huge collection of stars still orbiting a common center but without a neat spiral shape and the overall galaxy will now be elliptical with a major and minor axis. By the time the collision is fully complete many billions of years will have passed as the galaxy settles into its basic final shape and long gone will be the spiral arms which drive much of the new star formation in galaxies. The final product will therefore be a very large galaxy populated by a higher percentage of older golden stars that have settled into a featureless elliptical shape.

At each step in the evolution of galaxies from small to large we can chart the progress along the Hubble sequence. The sequence almost seems to be reversed in time as we can imagine a spiral galaxy from the top branch and a barred spiral from the lower branch moving from right to left on a collision course. The merger produces a possible lenticular galaxy and as this process evolves over time we are left with a single elliptical galaxy. The only change to the sequence might be to put the oldest flattened elliptical galaxies to the left of the more spherical ones. The reason being that the law of conservation of angular momentum will still apply to even the largest and oldest elliptical galaxies causing them to eventually flatten out once more. The final result of such a flattening would be one single enormous spiral galaxy and perhaps one where the spiral arms will once again drive new star formation.

The possibility for new stars to be born from some of the oldest in a very old and newly flattened elliptical is exciting. Identifying these will be a very interesting task however and will require the greatest attention to detail in every minute aspect of these galaxies. What might be the ratio of young to old stars and the spectral abundance of elements created from different types of nucleosynthesis in these galaxies. Finding the fingerprint of such an ancient and evolved galaxy should be possible and would certainly help to answer the question of what comes after an elliptical galaxy? It is worth noting here that a final somber outcome could simply be that the super massive black holes at the center of such large elliptical galaxies simply consume all the surrounding matter effectively killing the galaxy from the inside. Since elliptical galaxies have typically pulled most of the surrounding materials for many light years into there centers already no new fuel would be left to form stars and the black hole would ultimately win. In either case astronomers are still debating the evolution of galaxies not just at their deaths but also at their births.

What we have seen thus far is an explanation for the evolution of galaxies as brought about by the Big Bang. The abundances of light elements created in the Big Bang filled those early galaxies with the ratios we see today and therefore provide the basis for all reactions that can take place after the Big Bang. The importance of the Big Bang as a model of the beginnings of the observable universe is important because it can be used to explain a great many things. The fact that the universe is expanding can be explained by a Big Bang as one would obviously understand matter being flung out in all directions and expanding as in any ordinary explosion. The Big Bang seems to provide an analogy that is similar on a universal scale except that the matter is unbelievably large such as whole galaxies and the explosion many trillions of times more powerful. Yet the idea that the universe did spread outward from a single point at very high speeds does seem to provide a decent model for why the universe appears as we observe it today. After all any ordinary explosion can be traced backwards to a single point of ignition from which all the outward force, matter and energy of the explosion came.

In order to prove this is true for a normal explosion all one has to do is film the event and then watch the film running backwards to see all the matter, hot gases and energy contract to a single point. Where we are today is at a point well after the initial explosion of the universe somewhere in the expanding debris field of the explosion so it is difficult for us to simply watch a film of the universe running backwards. Without a universal VCR and a tape of the universe's life to view and therefore determine exactly what did happen we need to come up with some other clever roundabout ways of seeing the universe back in time. The Big Bang model is currently sciences favored model for explaining how the universe began and it may not be perfect but it does seem to make a lot of sense. So how can we try and find proof that this model which scientists keep using and falling back on as a way of explaining things as they are today actually happened?

Chapter 7
A Storm of Beautiful Violence or the Big Bang

First Light

One way is to look for evidence that a large explosion originating from a single point or at least a very small volume did occur and one of the first proposed ideas for this comes from the temperature of the universe. After all any explosion is hottest immediately after it begins and will slowly cool over time. So it was surmised the early universe was very hot and as it expanded it cooled. It stands to reason that if we know some facts about the beginning of the universe or vice versa if we can measure certain things about the universe now we should be able to back calculate a point, time and temperature of the early universe.

The temperature of the universe has been predicted through calculations many times over the course of the last two centuries and some estimates have been very close to the actual temperature. One of the earliest accurate calculations of the background temperature of the universe was made by German physicist Erich Rudolph Regener sometime in the 1930s. He predicted the temperature to be approximately two point eight degrees Kelvin and based his predictions on the effective temperature of cosmic rays. Regener was an expert in cosmic rays and in developing accurate techniques to measure their degree of ionization at different altitudes and in different mediums. Some of his early experiments in 1939 may have been the first to be considered space experiments as his instruments were the first to be launched specifically into the high atmosphere.

The device he constructed was known as the Regener-Tonne in German or the Regener-barrel and was a device that could measure the degree of cosmic ray ionization autonomously of the user. It was encased in a steel shell that could protect the payload from outside sources of contamination and the high atmosphere as well as from any dangers of reentry depending on the height of the launch. The whole instrument package was mounted on a V-2 rocket and represented the first scientific steps into space anywhere in the world. In fact the V2 rocket was the first object ever to reach space by humanity and far predates Sputnik as a vertical launch was made in 1944 and all U.S. and U.S.S.R space programs were based completely off of captured V2 rockets. Sadly this era of early space exploration was cut short due to the need to emphasize the military uses of the rocket and to save resources to use against the advancing soviet army.

Nevertheless the accuracy of Regener's measurements of cosmic rays both in the air and deep below the surface of the ocean provided not only the necessary data to construct a near perfect absorption curve for cosmic rays but also provided Regener with a sufficient understanding of the early universe such that he was able to calculate a background temperature for it as well. The key that Regener was on to was the use of cosmic rays as a means of calculating universe temperature and therefore indirectly the proof of the Big Bang. As we have seen the vast majority of energy in the universe was created in the first moments after the Big Bang and cosmic rays were one such form of energy. The same process that produced these cosmic rays also produced other forms of energy such as high energy photons. It is these high energy photons that would ultimately provide another clue to the Big Bang as being a viable model for the beginning of the universe and also that these photons would ultimately lead again to an estimate of the temperature of the universe.

So what are these high energy photons and where did they come from is the next question. Well they came from a time after the Big Bang when the universe had expanded many times from its original teensy weensy size. A whole different set of physics is used to describe what went on in the early universe in the very critical period of about time equals zero until one second and arguably one minute but we will look at that later. For now we are going to move to a time some three hundred and eighty thousand years after the initial explosion. The reason we have to move forward to this point in time is because before this the universe was too hot and dense for matter to coalesce into simple atoms like hydrogen. Remember we discussed photon disintegration? The same sort of thing was happening right after the Big Bang and up until this period three hundred and eighty thousand years after the Big Bang. The photodisintegration was not splitting nuclei but it was splitting atoms in much the same way as regular photodisintegration overpowers the nuclear binding forces.

The period is known by many names and one of the most common is the era of recombination although some refer to it as the period of photon decoupling as well. The photons possessed too much energy for a hydrogen atom to survive being hit by one. The number of photons was also incredibly high and dense due to the small volume of the universe and as such the number of free photons which could strike an atom was unusually high compared to today. The net effect is electrons being kept away from protons and forming stable hydrogen atoms. After a time however the universe cooled and expanded enough that the photons lost some of their energy and the density of photons in the volume of the universe decreased as well. Now conditions were right for hydrogen atoms to be created in obscenely large quantities due to the natural binding attraction of electrons and protons to one another.

Now it is interesting to mention a name often given to primordial photons here which is the first light. The first light of the universe is a slight misnomer in the sense that three distinct periods can claim what is known as the first light. Sometimes the first light refers to the first photons that condensed out of the Big Bang itself and you must remember just before the Big Bang there was nothing that modern science recognizes. A true unknown to modern science even the familiar and seemingly immortal photon was bound up in the thing that was the single point of the Big Bang. Therefore the first photons to be created from this explosion can be called the first light and these photons we know to have existed because they were responsible for the kind of photodisintegration that was afflicting the early hydrogen atoms.

The second candidate for first light was the period of recombination or more precisely when referring to first light, the era of photon decoupling. The reason this is sometimes called first light is because it is the first light emitted by the universe after interactions between photons and matter ceased and therefore represents the first bright flash of light in the universe although photons existed before this. Also this is the phase of the universe we are in now as it is the period when complex matter forms can exist such as atoms and molecules so this light is significant to this phase of the universe's life.

The third type of first light is one that occurs still millions of years later when gravity has had time to work its powerful magical forces on the early gas and create the first stars. The very first stars in the universe igniting their fusion cores cast what some believe to be the first light in the universe lighting up the dark gas for the first time. In some ways this is also correct as the universe though filled with photons did not nearly possess the number and wavelength of photons necessary to make it brighter than pitch blackness. In many ways it was the longest dark period of the universe and so ending this was at least for humans and other sentient creatures the first time the universe was lit up.

So here we have three versions of first light and while they can all be argued validly only one is correct. Is it the last one the starlight shining through the darkness and literally lighting up the universe for the first time? Well no it cannot be because we know that photons did exist prior to this and therefore whether human eyes can perceive this light or not does not really matter as the light was still there. Besides the inward collapse of cold gas by the force of gravity leads to a heating of that gas to a temperature of millions and billions of degrees in some cases before the nuclear fusion reaction that powers the star occurs. This means that the stars in their pre-nuclear fusion phases were actually glowing hot enough to easily shed visible light although that light would pale in comparison to the fusion outputs they would achieve later. One can argue that it was therefore still the act of star formation which led to the first visible light in the universe because it came from the collapse of stellar pre-fusion materials and does not necessarily have to come directly from nuclear fusion itself.

The second answer can not really be correct either for at least one of the same reasons as the third. This reason of course is that pre-recombination photons existed and so therefore simply put the photons produced during the actually recombination event were not the first. How do we know that the photons that we see today from the recombination event are not the only photons from the early universe? Probability is how we know that not all the photons we can see from the primordial universe were produced during the recombination event. The density of the universe was such that while it is true in the period of time before three hundred and eighty thousand years after the Big Bang many of the photons that existed were being in a sense traded back and forth between particles. The probability that one of the very first photons ever created after the Big Bang would have some how avoided all interactions with matter and therefore never have been re-emitted is very low. Remember however that space was expanding during this period as well increasing the distance between objects within the universe therefore increasing the chance that photons would miss other objects and survive as space expanded between them everywhere. So yes it does make sense that many of the photons we see after the first three hundred and eighty thousand years are the result of photon de-coupling.

However the probability that all the photons we see left over from the early universe came from photon de-coupling following the recombination period is equally low. The universe was expanding rapidly and in many cases was expanding faster than the speed of light. The initial faster than light expansion phase was short lived and had ended before the recombination time period but it still allowed for the universe to grow many, many times in volume before the first particles of matter even existed. We will look at this critical event in much more detail later in terms of the very, very early universe as it behaved only a few minutes after its birth. For now though we can be certain that space had decreased its density to the point where photons created shortly after the Big Bang but before the recombination would be able to escape interaction with matter and therefore be primordial yet non-recombination photons. This is a certain probability due to the simple fact that the universe expanded in a nice smooth fashion and did not expand for example in a series of jumps. If the universe had expanded in set amounts than yes the volume and temperature of space would have experienced a sudden increase and decrease respectively in these two variables. This would allow for all photons to be interacting with matter so that when the final jump was made to a larger and cooler universe the photons could simultaneously de-couple and all be born as it were at the exact same Planck time.

The universe of course did not nor could it ever expand in a series of jumps which results in the realization that the universe was increasing its volume and decreasing its temperature in a slow smooth and steady rate. The probability then states that as space got less dense the chances of a primordial pre-recombination photon actually meeting a piece of matter decreases rapidly. This means that a percentage of photons will survive unscathed from any possible encounters with matter. The percentage is lowest the closer you move to the initial point of the Big Bang and increases as you move toward the time of recombination. The probability approaches but never actually reaches zero even for the absolute first photon created in our observable universe by the Big Bang itself. The chances of literally the original photon surviving are beyond small but such a chance will in fact always be possible. Given the near infinite number of reactions in this early dense universe and the quantum speed at which they take place it would be very interesting to work out exactly how small a number it would be. The net result is that while a large portion of the photons we can detect from the early universe did result from the process of photon de-coupling a non-insignificant percentage will have been made before the recombination time period. Therefore the existence of these truly primordial photons should be considered when examining the early Big Bang phase of the universe even if their cumulative effects are slight.

This of course leads to the rather obvious conclusion that first light came from the actual explosion of the Big Bang itself. Not exactly the explosion but from the first photons created after the explosion and which a small number of will have survived until today. The explosion of the Big Bang is unlike anything science is familiar with and should not literally be thought of as one might think of a so called ordinary explosion. Absent are the burst of light, huge fireball and deafening sound of a typical chemical based explosion and we will look at what the explosion that was the Big Bang actually was later.

They have survived because the universe was expanding very rapidly as the first photons were being created and this gave them a better chance of making it through the pre-recombination period unaltered. It should be possible to work out the age of the universe and prove the existence of the Big Bang by examining these first light primordial Big Bang photons but they are only a small percentage of the early photons. Being such a small percentage it is difficult to separate them from the louder background noise of photons that were created during the process of universe wide photon de-coupling. It is these more numerous second generation photons that can give us a clue as to not only the age of the universe but to the fact that the universe most likely began in a legendarily large explosion we have dubbed the Big Bang.

This is the prime importance of the recombination photons really as it comes as close as possible to proving that a large mostly uniform explosion must have created the universe. This is in part because of what we can detect of them today in the form of microwave radiation. Photons of course is simply the generic name of all electromagnetic radiation whether it be high energy short wavelength gamma rays or low energy long wavelength radio waves. It was work by two American scientists by the names of Robert Herman and Ralph Alpher in 1948 who stated there should have been a large number of photons created during the nucleosynthesis phase of the Big Bang. This is of course the radiation left over from the recombination period and the second generations of photons in the universe. The photons before this period were almost all interacting with matter through a process known as scattering. Essentially the photons would only be allowed to move a short distance and then be scattered as they encountered a free electron or proton. The universe did expand and cool until the photons were free of interactions with matter and it is these photons that they postulated must exist. However they had no way to prove this theory and it fell into relative obscurity. The photons that would have been created were of course born into a very highly energetic universe around thirteen and a half billion years ago. These photons began their life most likely as gamma ray photons and slowly lost some of their energy over time. It is this lost energy that would ultimately and somewhat accidentally prove their existence and by extension the fact that the universe was born from a single large explosion.

The proof for primordial photons and for the Big Bang begins with an American physicist by the name of Karl Guthe Jansky. Originally in 1928 Jansky was hired by Bell Telephone Laboratories to find and eliminate sources of static in the atmosphere. The reasoning was simple enough; find the sources of static interference to radio wave signals so clear transmissions could be made across the Atlantic Ocean. Jansky would go on to build what can be in a sense thought of as the first radio telescope although in reality it was more of an oversized antenna. Mounting a large one hundred foot radio receiver on a rotating turntable type base Jansky could listen to radio waves in any direction and therefore locate the direction of interference. After finding where the interference came from he could try to determine what the source was. Initially he found three sources of static and quickly located the origins of two of these sources.

The first source was obvious and these were nearby thunderstorms. The second was almost as obvious and they were distant thunderstorms and of course both of these generate intense ionization in the atmosphere which interferes with radio waves. The third type of static he could not immediately identify though as it was nothing more than a faint background hiss. This of course was quite puzzling as the other two types of static interference had logical and explainable origins. The third type however was not associated with anything obvious like thunderstorms which could be seen locally for the close interference or through weather reports which would explain the second static even though the thunderstorm was far away. So with clear skies above him and no known sources of interference anywhere close to him Jansky was left to simply study the third hiss and hope for the best. After all his employers were interested in identifying all sources of potential interference even mysterious ones. This led Jansky to simply make detailed observations of the direction and timing of the hiss. After one full year of careful observation no known pattern could be found other than the hiss would rise in intensity once a day and then fall again significantly decreasing its intensity once a day as well. The most obvious conclusion Jansky could draw from this was that the Sun must be the obvious source as the cycle repeated once a day.

The case seemed to be a simple one that was neatly solved until observations he made over the next few months suggested it was something else. The cycle of intensity for the third type of hiss was not centered around a solar day but instead around a sidereal day. A sidereal day is the length of time it takes the Earth to rotate with respect to the stars and not the Sun. A sidereal day is not twenty four neat hours but instead is twenty three hours fifty six minutes and four seconds. This time scale is of absolute importance to astronomers because it allows them to track the true position of the stars in the sky and therefore allow them to point their telescopes at objects of interest. The Sun may be the largest and most influential astronomical body in our own solar system but it is not the true master of the Milky Way Galaxy and so sidereal time supersedes it. A solar day has a different time length than a sidereal day and eventually the Sun will move out of position as the strongest source of radio interference. This is exactly what Jansky observed and concluded that the Sun could not be the source of the background hiss. Since he knew it was now a source that was governed in a sense by the stars he decided to compare his notes on when the signal was strongest to photographic maps of the night sky. What Jansky noticed was that the signal was coming from the bright band of stars in the sky that represents the galactic disc of the Milky Way and he further noted that inside the disc the strongest source of these radio emanations was from the very center of the galaxy; the galactic bulge. He made his observations starting in 1931 and by 1933 published a paper on his findings. By accident he found radio waves coming from outside the Earth and in fact outside our solar system a truly remarkable finding which would, unbeknownst to Jansky be the first steps forward in the field of radio astronomy.

Following Jansky's unexpected discovery of galactic radio waves, radio astronomy would be fully brought to life by two men who simultaneously learned and were curious about Jansky's work. The two men were Grote Reber and John Kraus who were both American scientists; Reber in electrical engineering and Kraus in physics. Reber was working in radio when he heard of Jansky's work and would eventually go on to build a radio telescope in his own backyard in 1937. Reber was able to reproduce Jansky's results and was then motivated to begin producing a radio map in 1941 of the sky just as people commonly made optical maps. Reber also discovered an abundance of low energy radio signals coming from space which flew in the face of accepted thought at the time as it was believed that high energy radio sources should be common in the universe. Black Body radiation was thought to be the source of these high energy radio signals and yet it was low energy radio waves that were detected the most. Another of Reber's mysterious discoveries was of the first radio galaxy now known as Cygnus A. These are galaxies similar to our own but which produce massive amounts of radio waves. These are just two of the first discoveries that radio astronomy would bear and over the years countless more have been added.

The other scientist who began studying these newly discovered radio waves from space simultaneously to Reber was John Kraus. Kraus had been interested in radio of all kinds for his entire career and his biggest contribution to astronomy came with his Ohio Sky Survey. Seeing the potential for radio telescopes Kraus set about constructing a large radio telescope while working for the Ohio State University. During this survey Kraus and his students went on to catalogue almost twenty thousand radio sources from space. The significance of this is that unlike Reber who initially set out to simply capture a radio wave photograph so to speak of the sky, Kraus sought to make a very accurate catalogue of as many individual objects as he could. Most of what he did record were unknown sources at the time and therefore yielded a wealth of knowledge for not only the field of radio astronomy but cosmology in general. In addition to this Kraus would also go on to be a member of the first team of scientists to work on the now famous SETI project. SETI which stands for the Search for Extraterrestrial Intelligence is most popularly known for eavesdropping on the conversations or transmissions made by aliens in our own galaxy; at least that's the idea. What it does demonstrate however is another use for radio astronomy that allows for tackling certain problems from another angle and with a tool that was unavailable just decades earlier. Kraus had been one of the first, though not the first to suggest radio telescopes could be used to look for intelligent life beyond Earth. In 1955 he published a paper stating this was so and his article was persuasive enough plus his reputation solid enough that he received the money necessary to build his telescope only two years later. Interestingly enough it was this telescope he used for his Ohio Sky Survey which demonstrates in a way how many people not only wanted to but were serious about finding aliens. Together the work that Kraus and Reber contributed to astronomy in the very early days of using radio telescopes helped to broaden our understanding of the universe in general; there was far more to it than what we could see with just our naked eyes.

CMB

Part of what was far more than we could see without radio telescopes was one of the most significant pieces of evidence supporting the Big Bang. It is known as the Cosmic Microwave Background Radiation, often simply called the CMB and was detected almost solely through the operations and observations made using radio astronomy. Perhaps the biggest feather in the cap of radio astronomers to date the finding of the cosmic microwave background radiation ties together numerous ideas in the field of science from theory all the way through to the most important of all which is always physical observations. Now we can join the Big Bang theory itself, radio astronomy, Big Bang nucleosynthesis, recombination or photon de-coupling, the Doppler Effect and the idea of an expanding universe with multiple galaxies together fairly smoothly to name just a few ideas. So how was this first sown together? Like many things in science of course, by accident. The same way Karl Jansky set out to work on a commercial application of radio technology and accidentally found radio waves from the Milky Way Galaxy as a source of static background noise.

Now in 1964 two American radio astronomers set out to begin work studying the sky and they also had a source of background noise that they needed to identify and eliminate; another moment of serendipity was just around the scientific corner. The two astronomers in question were Arno Penzias and Robert Wilson who had just finished construction of a new and sensitive radio telescope set to receive signals in the microwave range of the electromagnetic spectrum. The energy that they expected to process using their new telescope would be essentially so low that it would therefore be very cold; only a few tens of degrees Kelvin. They could not eliminate four point two degrees Kelvin of static. What this means is the energy coming into their receiver was not in fact zero as most people would have liked and the reason is the cosmic microwave background radiation.

All radiation has energy and therefore in a very real sense heat. If the universe were free of such radiation then the temperature and therefore radiation measured by the microwave radio telescope would have been zero. Just as Jansky had gone to great lengths to eliminate sources of static interference in order to produce the best reception possible for his antenna so too did Penzias and Wilson attempt to eliminate all sources of static interference. The experiments they conducted took place close to urban centers but these were quickly ruled out as possible sources of radio interference. Now comes the famous story, which is famous if you have spent time studying the history of modern science or of the last possible solution to their static problems.

The receiver they had constructed was in the shape of a large metal horn around fifteen meters. The end of the horn was designed around flat edges and possessed the obvious large opening needed to capture long wavelength microwaves. This size of opening being large and flat was apparently also the perfect roosting or at least resting spot for numerous pigeons. Where you have pigeons you will invariably have too many pigeon droppings and this is apparently some law of the universe which physicists must sometimes work around. This was one of those such times and in a last ditch act of sheer desperation it was the copious amounts of pigeon poop that received the blame for the receivers static. The pigeon's offenses were removed from the telescope and to the dismay of Penzias and Wilson the static refused to go away. What started out as dismay soon turned to triumph for the pair as it turns out they had finally discovered the long theorized cosmic microwave background radiation which had just enough residual energy to produce the four point two Kelvin that they continuously detected. Since this time the actual temperature of the cosmic microwave background radiation has been accurately measured at two point seven degrees Kelvin.

The actual nailing down of the anomalous radiation as the cosmic microwave background goes to an American scientist by the name of Robert Henry Dicke and a Canadian scientist known as Jim Peebles. Dicke was a physicist who was also searching for the cosmic microwave background radiation with a new Dicke radiometer, a radio telescope of his own design. Together Dicke and Peebles hoped to conduct their own measurements of the universe and once and for all prove the existence of a cosmic microwave background. The pair had predicted that a cosmic microwave background radiation should be evident in the universe today as a leftover remnant of the Big Bang from billions of years ago with much diminished energy. Upon hearing the news of Penzias and Wilson finding a constant background radiation which they could not explain away, Dicke and Peebles knew they had lost the race to an accidental finding but were quick to point out that accident or not the findings were indeed caused by the cosmic microwave background radiation. An interesting side note is that the temperature of the universe was given by Erich Regener at two point eight degrees as we have seen in the 1930s and the actual temperature of the universe was found to be two point seven degrees some thirty years later. This is remarkable considering Regener did not have access to radio telescopes at the time and made his calculations purely on his ideas of cosmic rays.

The discovery of the cosmic microwave background radiation helps to prove the theory of the Big Bang by demonstrating the process known as photon de-coupling on a universal scale. The high energy photons were born as gamma rays and lost their energy over time but not simply because they just got older. This is where we will be revisiting the work of Christian Doppler and his now famous Doppler Effect. The motion of the source of waves relative to the observer affects whether or not the frequency of the waves increases or decreases. In the case of the Big Bang the newly formed photons from the era of recombination have not been travelling at the same speed relative to us here on Earth. This may sound funny because the speed of light is supposed to be constant and well it is but a slight trick far more powerful than holding the speed of light as a yardstick is at work here.

The space within the universe itself is changing its shape and no matter what space is the undisputed king of everything in the universe. Even the mighty and seemingly untouchable photon is loyal to space and the Big Bang proves this. The original photons created from the recombination were born as high energy gamma rays meaning they had very short wavelengths transmitting their quantized packets of energy. The frequency of these photons would not change if they were left alone and indeed photons generated today can travel for billions of light years without being altered in any way. The only caveat to this is that the space in which they propagate themselves must also not be changing. The thing about the Big Bang as we have seen though is that it is moving from an area of low volume in the distant past and modern science believes this to be an infinitely small point called a singularity. Singularities are interesting concepts in theoretical physics and a topic we will save for later but for now we can be satisfied that the universe started out small. This smallness did not last long however and this is dealt with by another phenomena that modern science believes to have happened called inflation theory and again we will deal with this later. I do need to mention both of these two theories now however as they tie directly into the very birth of the Big Bang and deal directly with the spreading out and increasing volume of space in the universe itself.

The space as it increased and in fact is still increasing in size, means that two points within space cannot be as close together as they were one second, one hour or one billion years ago. The end result is any observer of the original de-coupled photons will be moving away from the point at which they first de-coupled from matter in space billions of years ago. Since the source of the waves is moving away from us the Doppler Effect still applies. Here is where the mighty photon must bow to the ultimate power of space and we see the wavelength of the photons increasing. The relative speed of the photon to us is faster than the speed of light since the photon is moving at the speed of light and we are moving away from the photon. The end result is a photon that still itself travels at the speed of light but whose wavelength is stretched and red-shifted to such a degree that the way we perceive it today is different than how it was generated.

The high energy short wavelength gamma ray has long since transformed itself into a low energy long wavelength microwave. The microwaves have been red-shifted by the expansion of space and constitute the cosmic microwave background radiation. The energy lost by these early photons is exactly the amount theory predicted should be lost if the universe began as a large explosion that underwent recombination some three hundred and eighty thousand years after it was begun. In addition to this the cosmic microwave background appears roughly equal no matter where you look in the sky. Any direction going outwards from the Earth will show you these tell tale leftover photons from the Big Bang and this further helps to prove that the universe should have begun as a single large explosion. Since the space in the universe continues to expand today the space in which these cosmic microwaves travel is still red-shifting them even more. While it would take billions of years to probably notice a significant change in the energy levels of these cosmic microwaves a change will and is still occurring.

So now our picture of the birth and growth of the universe is mostly complete and at least as far as large non-quantum scale events are concerned we can see a straight line from the beginning until today. The work of many scientists up until this point has pointed towards a single event that created the universe as we can observe it now and tracing this path accounts for most of the phenomena we encounter both empirically and experimentally. Empirically because we see a world around us and we know our world is contained within a universe all around it. This is the type of observation that led to our earliest ancestors asking the first, most important and hardest to answer problem any species can ever ask; what is the universe and how does it work?

Experimentally because not only do the observations of the universe beyond our five main senses tell us that there is far more than just what we see in our own cosmic backyard but because our models, instruments and mathematics tell us this must be so as well. The more we probe the cosmos the more we reinforce the need for us to find an answer to that most difficult of problems we posed to ourselves so many tens of thousands of years ago. What we do know has whetted our appetite and inspired ever newer generations of scientists to take on this challenge with as much enthusiasm and curiosity as those that came before us. The universe that we inhabit both empirically and experimentally begins billions of years ago and in a storm of beautiful violence. The universe begins in one single almost unimaginable explosion that brings both matter and space into existence simultaneously immediately spreading out in every known direction and dimension at once. This led to a swirling mix of space, matter and energy in various forms which slowly cooled as it expanded and allowed the first particles to take shape.

Big Bang nucleosynthesis states that this early super hot and super dense broth contained only electrons and protons at first. Shortly after the creation of these two particles neutrons began appearing as electrons and protons interacted and the first atoms were soon to be born. Not long after the Big Bang in cosmic terms of course the first hydrogen atoms were finally able to form during the era of recombination. This period began around three hundred and eighty thousand years ago when photons decoupled from matter to produce the massive wave of light that still pervades our universe today. This light slowly lost energy as the universe expanded to become low energy long wavelength microwaves. These were in time detected using radio telescopes as the cosmic microwave background radiation and helped to solidify the Big Bang theory.

The early primordial atoms and cosmic rays would go on to slowly synthesize newer and heavier elements up to about trace amounts of lithium and beryllium. As gravity took over the gas of the early universe coalesces into the pre fusion stellar masses that will begin lighting up our universe with the first visible starlight. As stellar fusion begins various forms of stellar nucleosynthesis take over as the most dominant forms of atom building systems in the universe. The proton-proton chain, the carbon nitrogen oxygen cycle, triple alpha process, s process and r process just to name a few will give rise to elements as heavy and complex as thorium and uranium. These elements are radioactive as we have seen but are still very stable naturally occurring elements that can last for hundreds of thousands of years before decaying. This is ample time for the universe to create under more exacting conditions rare isotopes that are very unstable with short half lives sometimes lasting only a few seconds or minutes.

The diversity of elements in the universe is at its highest point that can be attained naturally and this is how the universe's periodic table will remain for many billions of years. Eventually some clever little species somewhere in the galaxy such as humans will come along and figure out the secrets of atomic nuclei sufficient to begin building our own elements. Known as synthetic or man made elements we can artificially expand the universe's periodic table once again to the point where the most unstable and heaviest of all atoms can exist for a brief fraction of a second. Along the way a particularly curious species might even begin smashing things apart to see how even those subatomic pieces work. This is where we are today as humans and we can see by looking at all the things we have learned so far that the Big Bang does appear to be the most likely beginning for our universe.

Chapter 8
Beyond the Comfortable, Radio Astronomy and Bremsstrahlung

Problems and Inflation

This is the straight line connection that we can draw from the beginning of our observable universe until today in terms of matter and energy. Matter being anything that takes the form of an atom and goes on to build more complex structures like molecules. Energy in the form of light which we can see and the various other wavelengths of the electromagnetic spectrum that interact with matter. We have seen on at least a simplistic level how these things slowly grew from a large mass of simple particles and photons to a very complex universe. We have looked at the various mechanisms that build atoms and how energy interacts with it to some degree to produce even more complex interactions of the two. The various people and methods by which they arrived at the conclusions and discoveries they made to uncover these interactions has also been examined. So far so good as we seem to know how to explain everything that happened after the Big Bang in the universe up until this point. The beginnings and eventual fates of huge galaxies have even been touched upon and if this were all the universe had to offer we would be quite pleased with ourselves. This however is not all the universe has to offer and what extra it does have to offer is by far the most mind bending we as humans have ever encountered yet. What we have learned so far can be considered as the safe sort of nuts and bolts mechanics of how a universe works.

Taking basic building blocks like subatomic particles and making them do strange things to each other is, well rather basic after all. The next realm of science has to do with the little snags and bumps along this straight line path we have followed. The universe started out as one immense explosion as far as we can tell. The very first stuff in the universe was very hot; it was flung in all directions and eventually cooled to produce familiar shapes and structures as we know the universe today. Except for the fact that when you start to look at some of these models more closely you notice that the puzzle pieces do not really fit so smooth and snug after all. At least they do not if we refuse to dig a little deeper and if we refuse to dig a little deeper we will never get to the truth about how the universe works. The universe blowing itself up and into existence to keep using this example creates some problems such as why is the cosmic microwave background radiation so uniform everywhere we look? This is but one of the little wrinkles in the cloth of understanding that we needed to smooth and we shall use it as an introduction to the small problems of the universe we have encountered to date. Many more problems exist along these lines and many of them stem from solutions we found going back two hundred or more years. Just because we understand how protons, neutrons and electrons merge to form atoms is only a first step. The way in which these particles can 'talk' to each other is quite different. Have you ever heard of a fermion or a boson? To many of the scientifically employed people reading this that question is answered with a comfortable 'yes'. Others however will say no and begin to understand that knowing things like this helps to smooth out some of those little wrinkles. After all knowing your fermions from your bosons and how they play with each other allows you to understand the simple atom far better than just knowing how it is put together.

This is the second phase of scientific understanding of our universe and as such it is somewhat smaller than the first nuts and bolts stage. It is however of extreme importance because it will help set the tone for the third phase which is trying to fill in the holes in our knowledge. This is the simplest and best way to state this fact. There are definitely and unfortunately holes in our understanding of the universe and as much as we hate to admit our limitations we must. This is not something to be afraid of however as our ancestors have faced holes before and we as humans love to fill in holes. Just look at the beliefs we have abandoned such as the fundamental elements like water, fire, air, earth and so on as a means of explaining how our universe is built. We should have no shame in this and should not look back at those who came before us as though they were mad or willingly trying to lure us into a land of error. These ancestral scientists and philosophers tried their best with the instruments and knowledge available to them to make sense of the world and universe in which they lived. Without them we would not be here today able to ask questions they could never dream of concerning the tiniest pieces of matter or the largest and farthest structures in the universe.

Therefore this third phase of science we must focus on harder than anything else we can study today. The trouble is these holes are the deepest and darkest we have ever encountered. It is somewhat ironic that by sheer happenstance one of our greatest scientific holes is actually called a black hole. Not merely due to the fact that they were first erroneously described as holes but because the very inner workings of these objects represents a true hole in the understanding of the universe by modern science. These holes I need to mention now as the only way you get to scientific holes is by not knowing you actually were standing in one in the first place. Time for us to look around and see where our first phase has taken us. To go from the first phase to the third phase of scientific understanding you must ask yourself how can we make our models of the first phase perfect? Or at least as near to perfect as possible for the current knowledge of the day. Since we have been talking about the cosmic microwave background let us use that as a way to enter the second phase and start filling in the holes of our scientific understanding of how the universe works.

The theory of a Big Bang starting everything off for our universe is generally accepted today as a good one and seeks out the holes in our previous understanding of an island universe which was static and fills them in nicely. The problem with the Big Bang theory however is that it creates a few holes of its own that could not be explained through a simple explosion mechanism. Astronomers had known for years that things were moving away from one another more or less anywhere you look in the cosmos and so a large centralized explosion was thought to be a probable solution to this set of observations. Further observations however led to the realization that things were a little too smooth every where we looked. In other words if the universe began in such a violent but normal type of explosion then the space that we observe all the various structures of the universe in should be far more curved than it is. Here is where I will have to give a brief explanation of how three dimensional space can appear flat to an astronomer.

Space we know exists in three movable directions which are up and down, left and right, forwards and backwards. So far so good as this is the normal universe that we perceive every day ourselves and this is exactly what objects larger than us such as the Earth or Sun encounter everyday of their lives. Now however imagine the horizon of the Earth stretching out before you as you stand on a huge grassland or savannah as our early ancestors would have. How does the Earth look to you now; does it look like a large three dimensional object just as round and bulbous as any normal object might or does it look totally flat no matter the direction you shift your gaze too? The answer is obviously the latter and to a single human the Earth is not truly round and therefore three dimensional but it is flat with a two dimensional landscape. Our early ancestors would have been unfortunately easily fooled by such empirical observations and might be forgiven for believing in the flat Earth theory. After all no matter how far you walk on the surface of the Earth it always appears flat because you can infinitely traverse the surface of it and due to its roughly spherical shape never come to an edge to fall off of.

This is how a three dimensional space even one as large as the entire universe can appear flat and two dimensional. Astronomers are aware of such empirical observations and could not help but notice much of the same type of problem when they looked out into the universe too. A regular explosion would produce a highly curved space whereby we could observe this curvature in the pattern of large scale objects embedded in it. Think about a series of circles one inside the other and the galaxies as dots on those circles and at each shell you would see galaxies in a curving pattern. There are hundreds of billions of the large scale markers that can easily be viewed from Earth and these are called galaxies. Yet upon observing these galaxies, as well as other objects, we noticed no such obvious curvature. In fact we found the complete opposite as astronomers noticed that space had a lot more flatness than curvature and this was true in every direction that we look.

All you need is a mechanism to explain why no curvature of space was seen and you have solved this problem. The trouble was no one had come up with an explanation to solve this problem or the other problem with the Big Bang theory. This second problem was the uniform distribution of energy within the universe as found by observations of the cosmic microwave background radiation. The cosmic microwave background was found to be very nearly smooth in the distribution of energy no matter which direction it was measured in the sky. Once again a single large Big Bang explosion would produce or at the very least science thought it should produce large scale discrepancies in the distribution of energy in the universe. The cosmic microwave background was not showing any major fluctuations of energy and consequently this posed another problem for the Big Bang model.

One last problem for the Big Bang was that according to the mathematics at least there was a suspicious lack of large quantities of magnetic monopoles. Magnetic monopoles are exactly what they sound like; magnets with only one pole say north or south but not both. The differences between regular magnets and monopoles are straightforward enough and contain two key factors. The first being that a dipole has a magnetic charge of zero in total as the north and south poles of the magnet are equal but opposite in strength and simply cancel each other out producing a total net charge of nothing. A magnetic monopole consisting of only one pole however lacks the opposing pole that would cancel out its net charge and the result is that a magnetic monopole does possess a total net charge. The second difference between dipoles or regular magnets and monopoles or so called exotic relics is that unlike dipoles which are common monopoles are not. The truth of the mater is simple dipoles are everywhere you look even if you can not see or feel the magnetic field they generate, if you do want to see one though go to your fridge as you probably have something stuck to it with a trusty dipole.

Magnetic monopoles are so rare though that they have never been found or detected in any way ever anywhere the end. That is right magnetic monopoles cannot be observed in the universe no matter how hard or far we look and they cannot be created in the lab. We cannot even create them in the lab for a few millionths of a second like we can super heavy synthetic elements like Ununoctium. Magnetic monopoles exist only in the theory and mathematics of exotic and unproven theories of the universe known as GUTS or Grand Unified Theories. These theories attempt to put everything together and unite the four fundamental forces into one and as such we will definitely be looking at them later. The basic problem with magnetic monopoles and the Big Bang theory is that the theories that predict monopoles from the Big Bang predict that lots of them should have been found. The observations however are always king and through them we have found none; not good news for magnetic monopoles.

So the universe is too flat, too uniform and seems to have misplaced all of its magnetic monopoles. Well of course this is the reality scientists face everyday and that reality is that the universe is never wrong. That is correct no matter how much we think something should be a certain way or something should exist in the universe if we do not see it or cannot measure it then there is a near one hundred percent chance that it is not there. At one point in our history we were very convinced that the Earth was flat, made up of a series of fundamental elements and was the center of not only our solar system but in fact the entire cosmos. We tried to prove this over and over again with our best minds using our best models but no matter what we did we could not make the pieces fit with observation. Wandering stars refused to stay put in the immutable sphere of the non-existent heavens, the orbits of the various celestial objects similarly refused to follow perfect circular orbits and so on. The simple fact is that observation is always correct and theories are not always correct. This does not mean by any stretch of the imagination that they are wrong necessarily, only that they might need a friendly nudge back onto the correct path.

This brings us to one of the twentieth centuries Big Bang nudgers; a man named Alan Guth. Guth is an American physicist who in 1978 began to take interest in the various problems of the Big Bang theory after hearing a lecture on the topic. The Big Bang is thought to have gone through essentially one stage in its evolution which is a massive explosion that brought forth all space, matter and energy into the universe. The resultant explosion using no special rules would then simply spread out to fill the universe and cool off in the process thus leaving a highly developed cosmos today. This was the model of the Big Bang that was producing problems based on observations and needed some nudging. The lecture he attended was given by Robert Dicke the same man who had helped postulate and solve the cosmic microwave background radiation. Dicke was discussing the flatness problem and mentioned that the energy density of the universe was critical or there would be no universe for us to exist in.

Too high a density and the universe would collapse to a singularity before a modern universe as we see it today could evolve. Too little energy density and the universe would drift apart forever with matter refusing to clump together enough to form a universe just as we observe it today. In other words the Big Bang model depended on a precise balancing of energies in order for a universe to work. It was this flatness problem that intrigued Guth to begin working on the very first moments of the universe after the Big Bang and he did so by examining the energies and temperatures that existed at those moments. The newer versions of the Big Bang theory at the time had suggested that the early energy and temperatures of the Big Bang should have produced some specific effects and particles to come into existence just after the actual explosion. The four fundamental forces could be combined into one given enough temperature and pressure creating what some call a Grand Unified Theory. The only force that could not be combined in such a way and still cannot today according to the models available to modern physics is gravity. The others however can be and this early fusing of the three remaining forces coupled with the explosive nature of the Big Bang predicted strange particles like magnetic monopoles to exist. Plus the pesky problem of where all the anti-matter in the universe went cropped up here too. The Grand

Unified Theories state that at the very first moments of matter creation in the universe equal parts of matter and anti-matter should have been created in order to balance some of these energy densities.

The problem arising from this mathematical model of the early universe is that today we find no anti-matter in the universe at all. Anti-matter can be created from events in the universe like cosmic ray spallation or nuclear decay events but anti-matter does not exist in the universe to the degree that normal matter does. In other words why are there not large quantities of primordial anti-matter simply floating in the void of the cosmos? The idea that matter and anti-matter were created in equal parts in the early super dense universe suggests that all the opposing pairs of particles should have met up with their respective doppelgangers and undergone annihilation events. This would leave a universe devoid of matter and awash in a never ending bath of energy in the form of photons. Since we are very much alive and asking the question of how the universe works at all today means that we do not live in an energy only universe. Therefore the anti-matter versus matter wars never took place and something about the theories does not fit the observations. The same sort of problem is encountered when looking at magnetic monopoles, the uniformity of the universe or the flatness problem. All of these rested on the behavior of energy and space at the very first few instants after the Big Bang exploded. Guth realized this and decided to introduce a period of super-cooling to the early hot dense and rapidly expanding universe. This is the critical time when changes to the energies and temperatures of the universe were thought to occur according to the Grand Unified Theory and making a change here was how Guth hoped to solve these problems.

What Guth suggested was that a very short time after the Big Bang initially occurred the universe was extremely hot and dense but this did not last linearly through time. Originally put forth by Guth in 1980 he sought to solve the problem of magnetic monopoles but his work would ultimately go on to offer a solution to the other two Big Bang problems as well. The first thing Guth suggested was to drop the temperature of the universe rapidly. In other words the universe did not simply start out hot and then gradually cool as space expanded. Instead Guth suggested that the universe experienced a brief phase of super cooling where the temperature of the universe dropped by about one hundred thousand times. What this did was allow the universe to momentarily create a false vacuum which acted as a type of cosmological constant. This false vacuum allowed space to expand very quickly at an exponential rate without violating any of the rules of general relativity. A cosmological constant is often factored into equations of the universe to provide a mathematical measure and explanation for the expansion rate of the universe. The false vacuum acted in much the same way allowing space to expand faster than the speed of light. This is perfectly acceptable to physicists because it is believed that only matter and energy cannot travel faster than the speed of light. Space on the other hand is exempt from these rules and so space itself was able to 'inflate' rapidly doubling its own size many times in an almost unfathomably short span of time.

This is what Guth called the inflationary period of the Big Bang and is the key to solving the three main problems with the Big Bang theory. Grand Unified Theories run into a problem when dealing with the early hot and dense universe as they all predict the creation and abundance of many exotic-relic particles. One of the exotic-relics predicted are of course magnetic monopoles which Guth set out to explain why we cannot see them today. The first obvious explanation is that they do not in fact exist but inflation theory offers a second reason why we cannot find them. Grand Unified Theories suggest that magnetic monopoles would have been created in sufficiently high quantities that they should account for a large number of particles today and therefore be exceedingly easy to find. The fact that we find absolutely none today can be explained by the rapid expansion of the universe due to inflation. The time at which the monopoles would have been created coincides with the time of inflation in the early universe. This means that even if large amounts of magnetic monopoles were created by the Big Bang they would have been created in space that was to undergo inflation. The exponential growth of the universe due to inflation would naturally spread out the existing magnetic monopoles to such a high degree that today we would be looking for them in a universe which had nearly totally diluted them. Inflation in a sense spread them out so thin that it is simply too difficult to find them; if of course they exist at all. Still Guth's idea about inflation did seem to solve the apparent problem of Grand Unified Theories and magnetic monopoles. Once the universe had stopped its inflationary period the monopoles were destined to be spread uniformly and thinly throughout the still expanding universe.

The universe itself of course at this post inflationary period would return to its normal hot, dense and still expanding self. The period following the super cooling inflationary period is known as reheating and occurs because the energy stored in the false vacuum is released. The false vacuum or inflation field loses its energy in the form of Standard Model particles and therefore forms many new particles and free photons which can interact freely heating the universe back up many orders of magnitude. So in a nutshell inflation theory states that the universe first blew up in the cosmically large Big Bang explosion. Then it entered a period of super cooling due to energy being stored in the false vacuum generated by the inflation field. The universe doubles in size many times very quickly faster than the speed of light and as the inflation period ends the universe can no longer support the false vacuum and releases the stored energy from the inflation field. The released energy creates Standard Model particles and photons which reheat the universe back to its pre inflation period temperature. This disperses the theoretical magnetic monopoles and seems to tie up all the loose ends nicely. This would be the end of the inflation theory story if it were not for the fact that it appeared as though Guth had also solved the other two major problems with the Big Bang theory and observations of the visible universe.

The problems with the general and early models of the Big Bang were not confined to simply the missing magnetic monopoles. The Big Bang model also produced a universe that was seen as having two other observable and therefore correct problems. The first of these two problems is the Horizon problem and stems from the fact that the universe appears to be very uniform no matter where we look. This is essentially where the cosmic microwave background comes back into the picture. Remember that no matter which direction one looks into space from the Earth it appears that the cosmic microwave background radiation temperature is roughly equal. Not only does the direction not seem to matter but the depth of observation is basically irrelevant as well. In other words you can not only look in any spherical direction from the surface of the Earth but extend that observation back as many light years as you like.

A distance of one million light years produces the same measurements as does a distance of one billion light years. So the volume of space surrounding the Earth no matter where we look appears to be too homogenous in terms of the left over radiation from the Big Bang and the era of recombination. Now I know what you are saying to yourself and it goes something like the universe underwent inflation in a time space far less than one second after the Big Bang and the era of recombination occurred three hundred and eighty thousand years after the Big Bang. Well yes that's true and from that perspective you would be right in thinking that there should be a discrepancy in the age of the universe and the light we see today as the cosmic microwave background. This is exactly the problem that inflationary theory fixes as the old model of a hot only Big Bang cannot account for this. According to that model the universe should have patches of radiation or at the very least energy gradients that differ in the direction and depth you look into the sky. A hot only, in other words a Big Bang without a period of super cooling and inflation, explosion would not allow the matter and radiation in the universe produced during recombination to smooth itself out in the universe.

The reason is time and gravity prevent this from happening as gravity would cause too great of a force on the radiation to allow it to become uniform in appearance in as short a time as only thirteen point eight billion years which is the age of the universe. The expanding space would distort the radiation to the point where it could never appear uniform and thus we would not observe a uniform cosmic microwave background radiation. Since the early universe was expanding faster than the speed of light no two areas of the universe could ever come into contact with one another to equilibrate the energy and temperature between them. In other words any two places in the universe could not interact to the point where they smoothed out their distinct differences in terms of energy and temperature so that when we observe them today we detect them as uniform. We do observe a cosmic microwave background radiation that is consistent everywhere and everywhen we look and so another mechanism must have aided the Big Bang to spread out the universe so evenly.

Inflation aids the Big Bang model by expanding space incredibly fast in an incredibly short length of time as we have seen. The rapid expansion of an early, young and dense universe through inflation means that the space that we are able to detect today was spread out evenly before recombination. Let me explain this in a small amount of detail as it can be hard to grasp for the first time. The universe that we can see is not actually the whole universe but only a part of the entire Big Bang and the universe we cannot see is the part of space that lies beyond our limits of detection. These parts of the universe are moving away from us faster than the speed of light and so no information about them can reach us. Therefore we can only see our patch of the universe and the uniform spread of radiation caused by the recombination era. The spread of matter in our universe in such a smooth pattern was made possible by the inflationary period of the universe after the Big Bang. The space of the entire universe and not just our own small patch underwent rapid expansion doubling in size countless times. The result is that any small quantum fluctuations of space, matter or energy at these tiny scales have been blown up and spread out to the size of our universe as we observe it today. A small difference in the universe back then is now the size of an entire observable universe today and so we see only one version of all the space inflated by the Big Bang. Our observable universe therefore will appear homogenous to us and anything beyond our detection is for now irrelevant. When our tiny part of the universe which was smoothed out by inflation underwent recombination all the matter that was to be involved in the event was therefore already spread out uniformly.

This uniform spread which occurred during inflation happened so fast when the universe was so small that the effects of gravity and time from the old hot only Big Bang model do not matter. All radiation produced by the recombination was produced uniformly spread out all over our observable universe and so we see uniform cosmic microwave background radiation no matter where or when we look. This is how the cosmic microwave background radiation helps to prove the Big Bang theory as a model for how the universe began. Inflation helps to prove how the early universe expanded by using this observed cosmic microwave background radiation and explain how it is so uniform today. Therefore the inflation theory helps to further predict the Big Bang model as well. Maps of the cosmic microwave background radiation made from very accurate observations taken by spacecraft have backed up the inflation theory. The predictions made for the distribution of energy and temperature in the universe by inflation theory has been detected remarkably well by these craft to within a few percent. Without going into the specific numbers it can be fairly safely stated that a mechanism such as inflation must have played out in the early universe to produce measurements of the cosmic microwave background that so closely match those predicted by the theory. This is the second victory of Guth's inflationary theory and not the last; the third problem with the old Big Bang model and current observations of the universe can also be solved through the use of the inflationary theory.

This last problem is known as the flatness problem and deals with the critical energy density of the universe. The critical energy density of the universe states that all the matter and energy in the universe must have a very fined tuned value in order to produce the universe we see today. The reason being that the amount of matter and energy in the universe will have an effect on the very curvature of space as these objects affect gravity. A critical energy density greater than one will give you a curved and spherical shaped universe as opposed to the flat universe we see. A critical energy density less than one will produce a hyperbolic shaped universe that instead of curing inwards like the sphere will actually curve outwards in a sense away from itself. A critical energy density of value equal to one will produce a flat universe and this is what we see today. All of these versions of the shape of our universe are constructed from two dimensional representations of space and we know that space itself is really three dimensional.

The reason why this value is so important is that once again the universe as we observe it today is the correct value. In other words what we see is right and our models have to fit what really exists and not what they think should exist. The universe no matter where we look appears to be flat today and therefore it would appear that the critical energy density has to be equal to one and not less than one or greater than one. This problem was one of the original problems that Guth tackled after attending Robert Dicke's lecture on the universe. Dicke first noted this problem in 1969 and cosmologists had been searching for an answer every since that still fit into the Big Bang model. In other words how could the Big Bang still have occurred to create the universe in a single large outward expanding explosion and yet produce a flat universe as we observe it today?

Here is where inflation theory once again can offer a solution to the old Big Bang models problems. The critical energy density values can be met and yet not violate today's observations because the driving force behind the inflation phase of the universe remains constant during inflation. The density of matter and energy will fall over time as the volume of the universe expands and so a high density early universe becomes a low density universe as far as matter and energy are concerned. During the inflation phase however the value of inflationary energy remains constant as it expands space and remains constant even after the inflation phase stops. In this way the universe can expand to a flat uniform size that is filled with a lower density of matter and energy as we see it today. So a small seemingly spherical early universe just mere moments after the Big Bang can grow to a size where the space has increased in volume to the point where any observations we make today will show a flat universe. Remember the example of the Earth and looking out at the horizon as the Earth is so large even though it is round the appearance it presents to us is that of a flat world. Three dimensionally the universe exists in much the same way and inflation is thought to be responsible for expanding it to such a high degree that any curved three dimensional spaces we can imagine today are flattened out like the Earth appears flat. All the while the critical energy density necessary to create a Big Bang universe that expands and exists just as it appears today is achieved. This explanation by the inflation theory satisfies the requirement of having a very finely tuned critical energy density at the very beginning of the universe. All in all the inflation theory is not fully accepted by all mainstream scientists and some point out that the main reason it should not be fully trusted is that it lacks a certain amount of experimental validation as well as it appears to create additional problems. The verdict is still out today on whether inflation theory is the single solution to the problems of the old hot Big Bang model. Still many in the scientific community point out the fact that inflation theory does solve three of the major problems with the Big Bang model and this cannot be ignored. The overall conclusion is that until mainstream science is presented with a better, more problem free solution to the Big Bang discrepancies, inflation theory provides the best solution available and is backed up by at least some observational evidence from cosmic microwave background surveys.

So for now we have seen the best explanation modern science has yet for the filling in of the holes in the Big Bang model. This is good because it offers us a chance to fill in one of the holes in our understanding of the universe and see things in a more complete way. What inflation theory helps to demonstrate if nothing else is that the observations we can make of the universe today can be answered. These little holes and gaps in our knowledge are for the time being real as modern science does not have an explanation for why everything in the universe appears as it does today. Using newer theories can help to guide us in the right direction for finding the true answers necessary to fill these holes.

Radio Astronomy

The next hole in our understanding of the cosmos comes from things we cannot see as humans and yet can still be seen. What are these mysterious things? Well for the most part they are ordinary objects but they are emitting radiation in wavelengths that we as humans have not evolved to see in. Once again the importance of radio astronomy comes into play as over the decades many of the familiar objects in the sky have done some not so familiar things. The specific hole in our understanding of the universe here comes from the fact that scientists have long suspected many different forms of radiation to have origins in space. These can be wavelengths of the electromagnetic spectrum that are longer or shorter than what the human eye can perceive as visible light. Radio and microwave length radiation has been well understood for decades as we have seen but many of the wavelengths that are more energetic like gamma rays have been poorly understood until recently. Moreover the observation of familiar objects in these new wavelengths has led to some unfamiliar observations. It turns out that many of the most ordinary of phenomena such as stars are capable of doing much more than we could ever hope to see with the naked eye. This has led to many holes in our understanding of the universe that once again must be filled by the observations we make which are of course always right as they are always real. The list here is exhaustive but there are a few that absolutely must be examined as they represent some of the greatest mysteries in modern astronomy and physics today. First we will look at radio astronomy across different wavelengths of the electromagnetic spectrum so as to familiarize ourselves with what can actually be seen in the sky. Secondly we will examine some of the distinct holes and mysteries that arise from these observations and lastly we will examine the theories that have been put forth to fill in these holes.

The first band of the electromagnetic spectrum we will look at is the infrared wavelength of light. This band of light is commonly associated with heat as it falls below the visible spectrum of light and was first associated with increasing temperature as we will see. The discovery of infrared radiation was made by the German astronomer Frederick William Herschel in the year 1800. Herschel is attributed with an incredible list of astronomical discoveries and observations not the least of which was the discovery of infrared radiation. At the time it was believed that the study of binary star systems could produce an understanding of the motion of stars in the heavens as they moved to or from one another. In the mid and late 1700s when Herschel and his contemporaries were conducting their observations technologies such as measuring red shifts was not yet possible. This resulted in great attention to detail being spent on observing the distances of binary stars from one another in the hopes explanations could be found for why they moved as they did. In addition to this binary stars were observed also to measure their parallax shifts from one another as this was hoped to provide the necessary information to calculate the distances between these star systems and the Earth. All of these techniques would greatly increase astronomer's understanding of how objects outside of our own solar system behaved with not only one another as was the case of binary systems but with the universe in general. To this end Herschel began one of the longest and most accurate cataloguing efforts of binary star systems. In fact he attempted to catalogue every star in the sky which led to the discovery that the number of binary and multiple star systems in the universe far exceeded the expectations of the day.

It was during one such cataloguing session of binary stars that Herschel first observed a large disk star in the sky in 1781. His initially belief was that this could not be a star and therefore probably just another random comet in the solar system or the like. His continued observations however would rule out such a phenomena and he provided enough accurate data that the orbit of the object was worked out by a fellow astronomer named Anders Lexell. Lexell concluded that the orbit was nearly circular and therefore greatly resembled a planet rather than a comet. This calculation of the orbit agreed with Herschel's conclusion that the object could not be a star or comet and Herschel placed the newly discovered planet beyond the orbit of Saturn. Herschel had done something no other astronomer had ever done which was discovering a planet without the naked eye. The planet Uranus can in fact be seen with the naked eye but no astronomers before and this includes ancient astronomers had ever identified Uranus as a planet. The reason is from Earth the light reflected off of Uranus by the Sun is too dim to make it appear as a bright inner planet like the ancient astronomers were used too. Anything that dim in the sky to them was thought to simply be another star. In addition to this it takes Uranus eighty four years to complete a single orbit and this fact coupled with its distance from the Earth means it does not appear to wander like familiar Mars does. The net result being Herschel is the true discoverer of Uranus as a planet in our solar system beyond the orbit of Saturn. This discovery is in itself of profound importance and one for which Herschel is still recognized today and yet it was probably not his most important.

The continued interest of Herschel's in stars and their behaviors led to his most important discovery in 1800 which is of course the discovery of infrared radiation. The closest star that astronomers could observe then and today is still of course our own Sun. It had been long known that strange dark spots would appear and disappear from time to time on the surface of the Sun. These Sun spots were of interest to Herschel who was making observations of them using different filters. The mystery of what Sun spots were and how they formed was what Herschel was trying to resolve by examining them in different light gathered through different filters. Upon testing a red colored filter he curiously noted that the light from the telescope was warmer than the light from other filters. Building upon Sir Isaac Newton's previous work of splitting white light into its component wavelengths, Herschel split the sunlight using a prism and attempted to measure the different temperatures of each part of the spectrum. He was keen to uncover what could be causing this change in heat at the red wavelengths and so simply sticking a thermometer into the split beam of light might provide the answer.

The startling thing Herschel found was that even beyond the red spectrum of light on a part of his test surface that essentially contained no spectrum the temperature of the thermometer still rose and rose more than anywhere else in the actual spectrum itself. This part of the test surface onto which he was shining light was close to the spectrum but outside of it and therefore should have represented a control zone for the experiment. The fact that a significant change in temperature was noted meant to Herschel that a new type of unseen radiation must be causing the rise in temperature. This reflects what we have been discussing in the unseen world of the universe and the holes in our scientific understanding of it. Herschel had discovered a hole in our knowledge which was there must be other forms of radiation coming from the Sun and contained if not in visible light at least associated with it and at the same time he neatly filled that hole by discovering infrared radiation. Though he did not know it at the time Herschel had in fact invented the field of radio astronomy in its simplest form. Calling his newly discovered radiation calorific rays Herschel set out to demonstrate that they behaved just like ordinary light. Calorific is of course related to heat as seen in a bomb calorimeter which is essentially a chamber where you place a test sample burn it and see how much heat has been generated by it which reveals the energy inside of the test sample. Calories of food are measured the exact same way through heat release and since the infrared rays he had discovered created great increases in heat beyond those of visible light wavelengths he named them heat rays or calorific rays. The rays could obviously be absorbed as the energy from them was first seen to be observed as absorbed into the thermometer and therefore first measured scientifically by noting the increase in temperature. Herschel soon discovered that these calorific rays could also be reflected and transmitted no different than ordinary light and as such a great interest was placed on them in the astronomical community. The race was on to find other objects in the sky other than the Sun capable of producing these infrared rays. Astronomy had gained a powerful tool to observe the universe in a way never before conceived and it had also gained a new way to find holes in our understanding of that universe.

The early days of infrared astronomy was limited essentially to optical telescopes and the use of prisms in various forms. An interesting and serendipitous side note of Herschel's discovery of infrared rays using prisms is that the infrared light lies right next to visible light on the spectrum. Telescopes are designed to maximize the gathering of visible spectrum light in order to produce magnified images of cosmic objects. The materials used to build a mirror and lens system in a telescope are also suited to collect and magnify light close to the visible spectrum and these will include of course the infrared light as well. So even though Herschel was using a prism to split the light and had no knowledge of infrared radiation his telescopes would still be able to collect large amounts of infrared light. By a happy coincidence all the telescopes available to him and other astronomers of the day were already equipped to be infrared telescopes. The difficulty was in detecting and measuring the infrared light as this band of the electromagnetic spectrum is invisible to the human eye.

Today we can easily employ numerous digital and computer devices to simply be our eyes for us in looking at wavelengths of light beyond our human range. In the days of Herschel however no such devices existed on Earth and so other slightly more indirect and often cleverer ways had to be devised. Again we have a moment of serendipity in the very discovery of infrared light using a prism as a prism can split the light captured by the telescope. This splitting of the light allows us to isolate the infrared spectrum only so that it can be studied even if limited methods of study were all that were available at the time of Herschel. This did not stop Herschel from going on to become the first astronomer to make numerous measurements and observations of astronomical objects and phenomena in the infrared spectrum. Others would follow in his footsteps over the decades and go on to take infrared readings of the moon as well as certain very bright stars from which enough light could be gathered.

This is how the field of radio astronomy began and grew slowly until the early twentieth century as we have seen when it experienced a boom after the work of Jansky, Kraus and Reber. The field of infrared radio astronomy was not to experience its boom until after this and more towards the middle and later half of the twentieth century. Advancements in the ability to detect the infrared photons themselves helped to quicken interest in this field of astronomy. Seth Barnes Nicholson and Edison Pettit were two American scientists who began taking the first modern observations of cosmic objects using infrared telescopes. The vacuum thermocouple they used in their work was what allowed them to take accurate scientific measurements of the moon, planets and nearby large stars. Demonstrating the usefulness of infrared astronomy more and more scientists began to take notice. The effectiveness of infrared astronomy has been widely enjoyed ever since as it can do many things optical astronomy cannot.

For example stars not luminous enough to be detected by visible light telescopes can be detected using infrared telescopes. The discovery of countless new stars and star clusters has been demonstrated in this way. In addition to this information on all known objects in the universe can now be gathered a second time using infrared photons. A photograph of the Milky Way galaxy in visible light and infrared light for example will bring a whole range of unseen phenomena to life. Besides the striking and beautiful visual differences of the two photographs, astronomers can learn much about what lies beyond the spectrum of visible light. The temperature of objects in the universe can be viewed in fantastic detail which in turn can help to teach astronomers more about these objects and the theories about how they work.

The basic concept involves visibly dim objects absorbing energy from the galaxy or universe and heating up which allows them to emit some of this heat as infrared photons. So a black object can be invisible to the visible light spectrum but glow in the infrared revealing something from the blackness and seemingly empty nothingness of space. The latter of the two is by far the more interesting to astronomers as often times it is the theories of how things work in the universe that can most benefit from these new perspectives on familiar things. The observations of these objects are of course correct no matter how they might make us question our theories. The wonderful thing about this is that often times we can use these new measurements to go back and rework our theories to better explain what we are looking at in the sky. The use of infrared astronomy has done this remarkably well by looking at some of the tiniest things in space; cosmic dust.

Cosmic dust is the term for any matter existing in space that consists of small particles usually molecules or small bits of granular material. Some of it can come from comets or asteroids which deposit this dust fresh into the solar system every time they orbit the Sun. Other sources of cosmic dust are stars which produce stardust a kind of clumping together of matter that was pushed away from the star by its stellar winds and condensed. The stars, Sun, comets, asteroids, Kuiper belt and Oort cloud are all sources of locally occurring cosmic dust. A little less locally cosmic dust can come from the Milky Way galaxy itself and from in between galaxies as well. Thus cosmic dust has many different names and classifications some of which are intergalactic dust or circumplanetary dust. Cosmic dust is of extreme importance to astronomers and cosmologists in that it plays a very active role in shaping the universe. After all our entire solar system containing our own Sun and all the planets, comets and asteroids we see today was formed from a huge cloud of cosmic dust and gas. So obviously understanding these cosmic dust grains is of huge importance as this dust permeates the entire galaxy and therefore will have a hand in shaping the evolution of the entire galaxy as well.

Not only this but numerous organic compounds, amino acids and DNA base pairs have been found inside this dust as well as in comets and asteroids. All these various organic molecules are necessary for generating, evolving and sustaining all life on Earth. Many have even suggested that life would not have begun on Earth as quickly as it did or indeed at all had it not been for this abundance of organic compounds floating in space. Some interesting food for thought is the fact that long before our entire solar system was even born, say five billion years ago all the building blocks of life were already present in the Milky Way galaxy. If life on Earth put them all together into self replicating DNA based off of carbon and water chemistry then any and all planets in the Milky Way that are Earth like should theoretically put them together exactly the same way. In other words life as it exists on our own planet should mathematically exist in a very close way to our own. Even if this is only from a bio-chemistry stand point you must remember that all DNA has a certain right handed tendency to its spiral helix and since the base pairs found in space are our base pairs all life built from them should use the same helical twist. This fact alone should massively underscore the importance of studying the dust and debris floating around our own galaxy.

On a more technical astronomy note cosmic dust represents a source of refractory material floating in space that can affect observations of objects. Too much dust and optically you cannot see what is behind it. This is very much akin to looking out your window in the morning and seeing heavy fog in the air. The water droplets suspended in the air of the fog bank act like cosmic dust suspended in intergalactic space. Too much fog and you cannot see across the street which you can easily do on a clear day. Too much dust and you cannot see what is behind it, however unlike fog which will be burnt off by the rising Sun as it heats the ground and air, cosmic dust is not really going anywhere. It is in fact always moving and we know that all space everywhere in the universe is a fantastically dynamic place that never sits still but rather continues to interact with every single other piece of itself countless times per second or per billion years. The cosmic dust therefore is moving but so slowly and at such distance that it will not warrant waiting around for it to clear and on top of this more than likely new dust will simply float into view taking the old dusts place. So how can we see what is behind it and clear all this fog of cosmic dust from in front of our eyes and telescopes?

With of course infrared astronomy and the use of infrared telescopes instead of optical ones. The reason we are able to look behind all the dust is because of the radiative properties of the cosmic dust itself. Cosmic dust like all matter can absorb and emit energy based on its own particular make up. We have seen this demonstrated before with the absorption and emission spectra for things like hydrogen atoms. While there are slight differences in the comparisons of atomic spectra and energy emitted from cosmic dust the principal is essentially the same. Cosmic dust is what we can call cold dark matter. The name is fantastically descriptive as are many names and terms given in science. Cold dark matter is matter that is very cold having in this case a low temperature approaching that of space itself. The matter is also deemed to be dark because it does not make itself seen in the visible part of the electromagnetic spectrum. If we slide down the electromagnetic spectrum just outside of the range of visible wavelengths we enter the realm of the infrared spectrum. This is where the infrared radiation really starts to shine and the cosmic dust ceases to be dark matter. The reason is somewhat similar to the absorption and emission spectra we mentioned a moment ago. Just as hydrogen atoms for example will absorb and emit certain wavelengths of light so too do the cosmic dust. The cosmic dust will absorb the infrared radiation and in fact other radiation from interstellar space as well. Once it has absorbed all it can it will release some of that radiation back into space and in this case it does so in the form of infrared light. This light is invisible to us so we still call the cosmic dust cold dark matter, but to infrared detectors it stands out as bright as any other source of light. In this way cosmic dust demonstrates the importance of infrared astronomy and indeed all forms of radio astronomy from radio waves to gamma rays as it helps us see what we cannot normally see with regular light photons.

Some of what we can see with infrared astronomy and infrared telescopes is all that is hidden by cosmic dust for one. The Milky Way galaxy and all galaxies are filled with cosmic dust including the space between galaxies. In other words the whole universe is filled with cosmic dust and therefore no matter where you look something somewhere will be hidden from view of visible light telescopes. Infrared light allows us to see past this dust to objects of interest behind them and one of the most striking examples of this is our own Milky Way when a comparison of images is placed side by side, one from visible light and one from infrared light. The detail and differences produced by each wavelength is unique and not only beautiful but valuable to astronomers to probe deeper into the galactic center of the Milky Way. Since the galactic center of our galaxy is shrouded by thick lanes of cosmic dust and we are of course looking through all of it as we gaze towards the very center seeing what lies behind is fascinating. In this way we can also see many hidden things that visible light cannot and one example of this is protostars.

A protostar is the name used to denote a star that is not yet born or turned on. It refers to the collapsing, coalescing and accreting disk of pre-stellar material that is heating up due to the force of gravity but has not yet reached the temperature and pressure necessary for stellar fusion to begin. The star has not yet ignited the nuclear furnace of its core and so does not emit much if any visible light. The collapse process does however emit huge amounts of infrared radiation which can be seen in this swirling cosmic dust around the protostar. This is quite obviously a huge benefit to astronomers as they can now look all over the galaxy and the universe to see stars in various stages of formation. Such a tool has helped turn observation into far more concrete theory than ever before. This new information about heat energy in space is far more reaching than just protostars and includes things like the very center of galaxies. The hearts of most galaxies and indeed it is believed of all large galaxies regardless of their shape contains a supermassive black hole. Black holes are by their very nature hard to observe and lying at the center of a galaxy hidden by dense cosmic dust makes observation of them even more troublesome. The reason why infrared astronomy and other forms of radio astronomy are so important to black hole research is due to the simple fact that black holes are black. Visible light does not emanate from them nor can it be reflected off of their surfaces like sunlight off of the moons surface. In order to learn anything about black holes we must look at them using other wavelengths of light and infrared can look through the cosmic dust that usually surrounds them.

What we have learned from techniques such as this is that black holes seem to be prodigious creators of infrared radiation. The production of large amounts of infrared radiation can help us to determine at least in part what black holes must be doing and how they must work despite the fact that the exhibit the curious trait of being somewhat secretive. Often times galaxies are said to have active galactic nuclei which are of considerable interest to astronomers as many of these nuclei contain black holes. The importance of understanding how black holes work and what they really are even just in terms of what they are made of is of extreme significance to science. Black holes are such a mystery to modern science and yet we at least know that they hold the very key to unlocking the secrets of how the universe works. So these active galactic nuclei are hunted with extreme prejudice by astronomers and theoretical physicists world wide. The famous cosmologist Stephen Hawking was made a scientific household name for his work on black holes and his theory of the now named Hawking radiation they must emit.

Another type of active galactic nuclei are those called starburst galaxies and of course they owe their name to the behavior of the stars that inhabit them. One can take the name to mean one of two things and in our modern shallow society it can be immediately construed as the bursting and therefore death of brilliant long lived beautiful stars. The second meaning is far less of a morbid fascination and instead refers to the birth of these stellar wonders by the multitudes as the nuclei of these galaxies undergo a pleasant yet unusually high amount of star formation. The rate of star formation is not at all unusual in the context of the universe but rather is unusual for the average life cycle of a typical galaxy. Most galaxies can be considered to exist in the same part of their extremely long lives as stars are when they are in their main sequence phase. In other words they might experience short lived periods of stellar formation when stellar clouds collapse and protostars ignite their nuclear fusion hearts for the first time but on the whole they settle into the normal cycle of regular star formation and star death. This is the phase that our own Milky Way galaxy is in now and will likely continue to be in for billions, trillions if not quadrillions of years.

Typically starburst galaxies and the active nuclei they possess can fall into two categories: the first being a new and young collapsing galaxy and the second an older merging set of galaxies. New and relatively young galaxies such as those that would have existed in the very early universe will fit into the first category as they are condensing for the first time and forming their eventual much larger spiral disks. The reason that they will undergo starburst galaxy qualities is because the random collections of stars that they will eventually possess are in the process of collapsing under a mutual center of gravity as space around and in between them compresses. As they collapse into a central dense mass of swirling stars they will also drag all of the local gas and cosmic dust down with them. This process will cause the clouds of nebulous gas and dust around them to shrink due to gravity and create new clouds for protostars to form in. As a result of this the end product will be a huge number of new stars being born amongst the original stars of the galaxy. Just as peering through the fog of the cosmic dust helps scientists understand what is happening at the center of protostar clouds so does infrared astronomy allow astronomers a window in to the hearts of these active galactic nuclei in starburst galaxies.

The second type of starburst galaxy is the result of two galaxies either in very close proximity to each other or two galaxies actively colliding together. So in a sense a merger type of starburst galaxy really should be called instead by its plural name starburst galaxies. Even if only one of the two galaxies involved in the actual merger is undergoing rapid star formation the overall process leading to that star formation would not occur if the starburst galaxy was alone. The reason for this is that the process that drives new star formation in starburst galaxies comes about from the force of gravity between the two individual galaxies. As the attraction between the two galaxies escalates the resultant effect is that they move towards one another faster and faster while the space between them rapidly diminishes. Surrounding every galaxy in the universe is an immensely large blanket of cold dark matter in the form of gas and dust that is left over from billions of years ago and most probably some of it comes from the Big Bang itself. After all the only matter produced after the Big Bang was gas and very small clumps of matter constituting cosmic dust perhaps only a few molecules large. These little bits of matter are what started to band together to form the smallest motes of cosmic dust imaginable. These motes of dust went on to form little cosmic specks of granular material that continued to swell in number as the first stars and galaxies formed. Not all of this however would have been completely consumed in the star forming process and in fact a massive percentage of it most assuredly permeated every cubic centimeter of space for billions of years right up until today.

This blanket of dust has been added too by the newly formed cosmic dust and gas produced by violent events such as supernovae until all galaxies have a rich collection of materials surrounding them. This gas and dust also penetrates the galaxies themselves and so any change in motion or gravity of two galaxies about to be involved in an imminent merger will cause a change in the density of this material. This is the same process that occurred with the primordial gas and dust from the Big Bang. The material condenses into hot swirling clouds of protostars which of course leads to the eventual ignition of stellar nuclear fusion and the birth of new stars. The fact that the gas and dust is spread out outside and inside the two colliding galaxies means that there is ample chances for new stars to form as there are numerous places where the gas and dust will condense.

The Milky Way our own home galaxy and its closest galactic neighbor the Andromeda galaxy are in fact set to collide in the distant future as well. Approximately four billion years from now the Milky Way galaxy and the Andromeda galaxy will collide producing a new super galaxy that will result in one or most likely both galaxies becoming separate starburst galaxies before settling into one single large starburst galaxy. Our two galaxies are approaching each other at a velocity of around one hundred and ten kilometers per second or approximately three and a half billion kilometers per year. The merger is thought to take over two hundred and fifty million years to complete and will not be the last merger of its kind for our local group. The Andromeda galaxy is thought to already have undergone one such merger billions of years ago and the Milky Way galaxy and the Triangulum galaxy are thought to both be on collision courses with it as well. The Large Magellanic Cloud is a type of small pre-spiral galaxy that the Milky Way will absorb just as Andromeda absorbed other galaxies years ago. All of these mergers will result in starburst galaxies and all will result in massive amounts of infrared radiation being emitted into the universe. This type of starburst infrared radiation is arriving to Earth from all over the universe allowing astronomers to learn far more about stellar and galactic evolution and the behaviors of these objects over time.

As one might expect there is a telescope named after Frederick William Herschel which studies our solar system, the Milky Way galaxy and the universe in the infrared spectrum. It will be discussed in some small detail as it was not only one of the most significant infrared telescopes ever built but also allows us to examine the challenges faced in making any type of observation along the electromagnetic spectrum outside visible light. Named the Herschel Space Observatory it was constructed through ESA, the European Space Agency and was the largest ever space based infrared telescope. The Herschel Space Observatory had a mirror diameter of three and a half meters and was one and a half million kilometers from Earth which was to last about three to three and a half years but was able to be usefully operated to almost four years. It observed the gas and dust clouds condensing to form new stars to look for evidence of the molecular formation of water vapor in space. The instrument was also sensitive enough to detect water vapor on objects within our own solar system.

The mission illustrates some important factors in the use of these electromagnetic wavelength telescopes. One thing of course being a steady stable environment from which to operate as most of the observations carried out in different wavelengths like x-ray, gamma ray and of course infrared require long exposure times no different than an ordinary optical telescope. Placing telescopes in space offers a very smooth area of operation for these instruments and the Herschel's chosen point in space was picked for just this reason. The craft was parked in orbit in what is known as the L2, or second Lagrangian point between the Earth and the Sun. A Lagrangian point is a position in space somewhere around two large objects where the gravitational pull of the two bodies almost cancels out. If a small object such as a spacecraft is placed in one of these points it will remain suspended and virtually motionless with respect to the two large bodies. In other words it should not start to fall towards either of them and plummet back to the surface. The other interesting thing about Lagrangian points is that the orbital centripetal force of the small object is supplied by the two larger bodies. This is of course another obvious bonus to any spacecraft wishing to conduct lengthy observations. Any two large bodies will produce five Lagrange points within their orbital plane and the L2 point is the point directly behind the Earth. In other words the Sun will always be hidden behind the Earth compared to the space telescope giving it a darker sky to look at and also a colder environment to work in for most of its mission.

Any space telescope regardless of the wavelength it might be studying from gamma rays and down the spectrum requires long exposure times to capture a single image. How long is this really when compared to say your standard point and shoot camera? Well think of taking a regular snap with your handheld camera; let us say a single lens reflex using average shutter speed and film. Chemical or digital film cameras do not matter as it is the time it takes for the light to enter the camera and be imprinted on the recording medium that determines most of the shutter speed. A single picture you take involves the shutter opening and closing as fast or faster than the blink of a human eye. While the opening of the shutter so to speak and exposure of the film on a telescope measuring these different wavelengths can take minutes for a single snapshot. This underscores the need for and the difficulty in achieving an absolutely motionless exposure time for these ultra sensitive detectors.

Now we come to the second and in some ways equally important but in many ways more important factor to these telescopes which is environment. A steady hand is required to take a clear picture but even if that hand shakes you will still get an image no matter how blurry it might be. What if you cannot even take a picture though? The number one problem facing telescopes and detectors that are attempting to take images of these other wavelengths is the environment in which they operate. The environment is so critical to these devices that they can result in no image being recorded at all due to too much background noise, no photons of the desired wavelength reaching the detectors at all or a combination of both these two factors. There are a couple of ways in which the environment can ruin the detection capabilities of these tools and the first is temperature. Taking infrared telescopes as an example, whether they are a purpose built infrared telescope or an optical telescope both must be kept very cold to record infrared photons.

The reason for this is that any object or body above absolute zero has heat and therefore by definition will emit infrared wavelength photons. The temperature ranges vary widely between materials as to how much thermal radiation they will emit but in general some coolant such as liquid nitrogen must be used to chill the telescopes. Any object that has a temperature over one or two hundred degrees Kelvin will emit so much infrared radiation themselves that they would be useless or nearly so as infrared detectors of any appreciable significance. In a sense the telescope itself would drown out the incoming infrared photons with its own infrared photons which is obviously counterproductive. The solution is to place these telescopes in cold environments, to use coolants or both. The Herschel Space Observatory once again exemplifies the practice of both at once as it was placed not only in a very cold environment but also carried on board coolant. Even still the telescopes useful operating lifespan was projected to be three point five years not because the instrument was expected to fail after that much time had elapsed but because the supply of onboard coolant it carried would be depleted and it would no longer be able to record infrared data perfectly. Ground based infrared telescopes face the same problem of battling temperature and for that reason are mainly constructed on top of high mountains, dormant volcanoes or plateaus.

The difference between the space based observatories and the ground based ones helps to illustrate the second problem that the environment in which the telescope operates occurs. This is the atmosphere and more specifically the types and densities of the particles and molecules that are in it. Just as all the various wavelengths telescopes must battle temperature to achieve the optimum in image capturing abilities they all must battle the atmosphere. Astronomers have known of this for centuries due to the simple fact that optical telescopes cannot work effectively through dense atmosphere. Clouds do not make very good companions to late night astronomers. Similarly in the modern age lights from any city in the form of office buildings, streetlights and any other form of urban illumination contribute to what is known as light pollution. While not a real threat as far as environmental sources of pollution go they are horrendous for obscuring astronomical observations. If you have ever seen both a city and country sky you immediately are aware of the difference. In the city on an average clear cloudless night you can see a handful of scattered stars and one or two constellations in the night sky; maybe the Big Dipper or a zodiacal sign such as Leo. The light from all the various sources in the city is travelling upwards and reflecting and refracting off of moisture and other particles in the air to haze out the night sky. Travel to the country and you can see literally tens of thousands of stars, the bright band of our own Milky Way galaxy and if you know where to look even see the nearby future spouse of the Milky Way the Andromeda galaxy all with the naked eye. This form of light pollution is the same background light pollution to optical telescopes that infrared telescopes would see if not cooled down immensely.

The problem is that light going up away from the Earth and being scattered by the atmosphere is not the only problem telescopes face. Light in the form of various wavelength photons coming down towards the Earth from space must also get through the atmosphere and most of the time they do not. I for one am very glad that this is the case because if it were not we would not be here today; we would be dead. The atmosphere shields us from all sorts of high energy photons beyond the wavelength of the visible part of the spectrum that could harm us. You are probably familiar with ultraviolet rays already as they are often vilified by dermatologists globally as harmful UV rays. While from a biological perspective dermatologists are correct and UV A, UV B and to a lesser extent UV C rays are a form of mutagen harmful to life, ultraviolet rays are themselves just a part of the electromagnetic spectrum devoid of evil intentions. The atmosphere and specifically the ozone molecules in it shield us from much of these rays and protect us from harm. The extremely deadly X-rays and definitively deadly gamma rays from space are all shielded by our atmosphere keeping us in various states of being alive and hopefully well. This is the good part of the atmosphere but the bad part for astronomers is that it does exactly that; stop most or all of the infrared, ultraviolet, X-ray and gamma ray photons from reaching ground based detectors. Water vapor is a good absorber of incoming infrared radiation so the higher the elevation of these detectors the less atmosphere they must contend with. Building telescopes on mountains can work for some parts of the electromagnetic spectrum like infrared rays and some ultraviolet but it solves nothing for the X-ray and gamma ray telescopes. This is where space based observations not only shine but are a must for other forms of electromagnetic spectrum telescopes to operate.

Werner von Braun and the V-2s

The very nature of putting craft into space for us today is a simple and fairly day to day sort of thing and indeed we have taken advantage of our advancements in technology to place many satellites, probes and other detectors into space. However this was not always the case as things such as infrared radiation and X-rays were known to have been discovered decades to centuries before we had the capability of launching anything into space. The very first man made object to ever travel into space was the V-2 rocket designed and built by the single greatest rocket scientist of all time Werner Von Braun. Von Braun deserves a detailed history of his own but for now we will concentrate on how his work solved the problems of higher energy electromagnetic telescopes. The first V-2s were produced in 1942 and were fired to heights sufficient to enter space by 1944 and though a first however they were soon under regular production and launching routinely reaching space altitude.

The boundary of space is often considered to be one hundred kilometers as the atmosphere is so thin that there is virtually nothing left of it. In addition to this travelling near and certainly above this one hundred kilometer altitude you will experience the feeling of weightlessness. The typical peak altitudes for a V-2 rocket were between one hundred and one hundred and ten kilometers firmly placing them into the region of space around the Earth. So these rockets of Von Braun's were truly the first man made objects ever sent into space over a decade and possibly a decade and a half before Sputnik; Sputnik's claim to fame is it was the first man made object to orbit the Earth but it was not the first object in space. During war times however the V-2 was designated as a weapon not a research rocket and so carried a one thousand kilogram high explosive warhead rather than a scientific payload. It is a pity that more funds were not available to space research using the V2 because many of space's secrets would have been learned decades earlier than they were.

Nevertheless it was this rocket that marked the turning point in human history where for the first time since we asked that at least 5000 year old question: how does the universe work we as a species were able to leave our planet and enter space. Even if it was only one of our little mechanical friends entering space the V-2 would go on to be used for scientific research into space following the end of the war. After the end of the Second World War the German scientists who had been forced to work on weapons were quickly sought out by the allies. The fear between the west and the east was that the other side would gain all of the scientists leaving none for them. Admittedly both the west and the east were lagging far behind the Germans in developing rocket technology so getting a hold of key scientists was desired. Werner Von Braun and many of his colleagues were offered escape to the west and by extension freedom from the east. Known as Operation Paperclip thousands of Germany's brightest scientists were brought to the west to continue their work as they had always wanted it to be which was simply to build rockets for scientific purposes and the exploration of space. So the V-2 was brought with them and finally allowed to carry out science missions into the high atmosphere and space itself.

This is how we can finally overcome the difficulties of environment facing X-ray and gamma ray telescopes. Finally after decades or centuries of being aware of their existence all the various wavelengths of the electromagnetic spectrum could finally start to be explored. Like anything else small steps were taken at first and some space based observatories would have to wait a few years or decades to be possible, but the proverbial ball was set into motion. The first ever photo taken from space was taken in 1946 by Von Braun's team of German scientists and their new American colleagues which shows the curvature of the Earth and the faint band of our protective atmosphere. The United States military was of course keenly interested in Von Braun for his V-2 rocket and his genius and desired him to build their military and space programs. Astronomers and scientists were also very much interested in Von Braun for all the same reasons but were given a second billing to the military for use of rockets for experiments. Nevertheless many scientists found a clever loophole in the system so to speak and used the military itself as a place to conduct their research. In a sense if you can not beat them, join them and this is what many scientists did in the postwar years enabling them to conduct experiments in space just as Von Braun had dreamed of doing before the war even began.

The first photograph taken from space in 1946 was shortly followed thereafter by the launch of a V-2 in 1949 that carried a scientific payload designed to study X-rays. These research rockets are often called sounding rockets which is a carry over from nautical terminology. The nautical premise was to gauge the depth of the water using weighted lines which was the term given to sound the waters depth. So sounding rockets are used to measure the extreme upper atmosphere and into space by being fired in a parabolic trajectory where the rocket climbs to its maximum altitude and then falls back down to Earth. The nose cone of these rockets, including the 1949 X-ray V-2 carry an instrument package that conducts measurements and observations for part of the flight. Since the V2 was designed to carry a warhead of approximately one tonne the science payload could also be very large and complex. After the mission is complete the sounding rocket's nose cone falls back to Earth where it is retrieved and the data collected. The sounding rockets have a tremendous advantage in that they are much quicker and cheaper to use than a traditional long term space craft. The entire mission can be conceived and executed within months using a surplus military rocket or just the motor. The payloads are small and so they are inexpensive to construct and therefore they can either collect data cheaply or they can be used to test systems that will be employed on much costlier launches in the future. Flying to heights between 50 kilometers and 1,500 kilometers rockets such as these are employed commonly to gather scientific data.

The V-2 launch in 1949 achieved a height of approximately 110 kilometers and carried an X-ray detector that for the first time was able to penetrate the Earth's atmosphere and reach space where X-rays are plentiful. A similar concept in efficient and inexpensive space research was to come from two military officers and two civilians and involves the combination of balloons and rockets. James A. Van Allen, for whom the Van Allen radiation belts were named, was part of this team and he is another example of a scientist who used the military as a way to conduct scientific studies. Joined by colleague S.F. Singer and naval officers Commander Lee Lewis and Commander G. Halvorson the quartet came up with the first rockoons.

A rockoon is a rocket suspended from a balloon filled with helium gas and carrying an instrument payload. The idea behind these systems was simple; use the balloon to lift the rocket higher into the atmosphere so that less propellant had to be expended to reach space as the atmosphere was much thinner and a small portion of the trip had been already covered in terms of altitude. Once the balloon was at its maximum height the rocket engines would fire and the rocket detach from the balloon to carry out the rest of its mission as a normal sounding rocket. Not only is this idea inexpensive but it can be performed almost anywhere and it is no surprise that naval launches of rockoons from military vessels became a norm. This represents the early days of ultraviolet, X-ray and to some extent infrared astronomy in space and all were extremely necessary in overcoming that last obstacle which is the atmosphere. The protective effects of the atmosphere are no longer a problem and the importance of what these new telescopes operating all across the electromagnetic spectrum can teach us about the universe is finally made a reality.

Ultraviolet radiation is the next form of radiation we will look at and like infrared exists very close to visible light on the electromagnetic spectrum. Infrared is just to the left on the spectrum with lower energies and longer wavelengths than visible light and ultraviolet is the opposite. Sitting to the right of visible light on the spectrum it possesses higher energies and shorter wavelengths making it more difficult to detect on Earth but much easier to detect than X-ray or gamma rays. There is no need to mention how ultraviolet astronomy works in terms of its challenges as we have seen that exhaustively with infrared radiation and indeed most or all of the information concerning infrared astronomy can be applied to all other forms of electromagnetic observatories which is why I spent some time discussing it and its history. The benefits of these various types of electromagnetic astronomy are what we shall focus on now and how they help us to understand better the universe and how things in it work.

We shall start with ultraviolet radiation as mentioned and how the universe appears vastly different when viewed only in that light as opposed to visible light. Once again we see the opposite effects of infrared radiation as it can help us see objects that are too faint to be viewed with normal light but glow with heat. Infrared astronomy can show us stars and indeed whole star clusters that are invisible to visible light telescopes because they glow warmly but dimly. Ultraviolet can show us things that we do not see with visible light but only things about these objects that we could not see before. In other words we could always see the object in visible light but some interesting characteristics of it were not readily obvious to us without viewing them in the ultraviolet spectrum. Ultraviolet telescopes see objects that are hot in fact it sees objects that are very hot to the point where most of the universe would dim or fade away from view completely as the objects being observed tend to emit most of their energy as light or infrared radiation. The hot objects that we can see are those that are typically young or old such as stars depending on the stage of stellar evolution they happen to be in at the moment. Other exotic objects which are not stars also tend to give off ultraviolet radiation and these of course are of particular interest to astronomers and physicists because we ask the question what makes them so hot? Ultraviolet light is hampered severely by interstellar dust and this also explains why some of the universe appears to grow dark when viewed by ultraviolet telescopes. What is not obscured by dust however can be identified through the use of ultraviolet spectroscopy and this is one of the most important facets of ultraviolet research in space as we can learn much about the interstellar medium and its composition. The interstellar medium is very important to modern science as we are still learning much about it and many of the ideas concerning exactly how it is composed is debated.

Now we shall move up the spectrum again to X-rays which have an even shorter wavelength than ultraviolet rays and still higher energy levels. Unlike some infrared radiation and some ultraviolet radiation which can penetrate the Earth's atmosphere X-rays cannot and so the only option for making observations in this region of the electromagnetic spectrum is high in the extreme upper atmosphere or in space proper. X-rays have long been known to exist and X-ray emissions from the Sun were seen as early as the 1940s where balloons and the V-2 rockets we mentioned were launched to carry X-ray detectors high into the atmosphere and space respectively. The study of X-rays for simply their physical properties when coming from outside the Earth and insights into how and why the Sun generated them were of prime interest. However scientists wondered whether or not any other objects in the universe might emit X-rays as many had theorized this as possible. After all our star the Sun is not that much different from any other star in the universe and if it produced X-rays surely they did too and scientists were keen on learning the specifics about other stars' X-ray production. In addition to this it was also believed that certain other cosmic phenomena should also be a source of X-rays and so the universe should in theory be at least somewhat awash in them.

It would not be until 1962 however that the first detection of an X-ray source from outside our solar system was observed during a rocket flight. While on the subject of rockets it is interesting to note here that the experiments were conducted by the United States Navy as part of their research programs. Early space exploration was often shared between purely military institutions and civilian scientific ones such as NASA. The reason for this was reflective of the political sphere at the time between the west and the east as both were racing to get into space. With the reality that humans and other life forms would be launched into space, the very first living creature launched into space was a number of worms that were sent up by the Soviets. The scientists involved new that X-rays were not just harmful but downright dangerous to life; the need to study cosmic X-rays in detail was a must for the furtherment of the live exploration of space. Today astronauts regularly must contend with dangerous radiation while working in space and even in low Earth orbit which is still within the Earth's protective magnetosphere. Various forms of radiation are ever present which the astronauts' ships, stations and suits cannot fully shield against. In fact it is a known fact that X-rays, gamma rays and cosmic rays will occasionally hit cells in the astronaut's eyes and cause a miniscule short lived flash of light if they hold their eyes closed. This can be thought of as a sort of biological detector similar to the kinds that you might find if you were looking for neutrinos say in the underground detector like SNO in Sudbury Canada. Most of the time the detector is perfectly dark but every once and a while a stray neutrino will interact with the sensors inside the tank and cause a small brief flash of light. While this is harmless to the detector a flash of light represents a real health hazard to an astronaut as the cells and molecules in their body, not just in their eyes but everywhere are being hit with high energy radiation which is a mutagen.

Mutagens no matter what source they come from or what form they take are the root cause of every cancer and harmful mutation in the world. This is because a mutagen affects a mutation in the DNA of the organism it is exposed to. The DNA upon being hit by say an X-ray has a possibility of being altered slightly even by one base pair or atom which will cause an error in replication of the gene sequence at the particular site of X-ray absorption. The result is something that is not normal to the workings of the host cell and the host organism the cell resides in. In the case of cancer this means that the proteins responsible for terminating cell replication can be damaged and never shut off. This results in what all cancers are which is unchecked cell growth leading to a tumor. The mutation case results in a non-cancerous cell that will simply express a new gene which has been created by the mutagen. Most of the time about 90% the mutations are harmful, about 9% of the time they are neutral and probably less than 1% they are beneficial so no matter what you may have heard about mutations from movies they are not what you want. In fact you must remember that the mutation works as a probability which is independent of every other mutation probability so if you go through all 99% of the harmful or neutral mutations does not mean you will get a 1% good mutation. In fact what it really means is that each time the mutation event occurs it is like you are buying a lottery ticket with only a 1% chance of winning and that is almost a certain guarantee you threw your money away.

The need to understand these X-rays and other forms of space based radiation are of critical importance to our future as humans in space. The 1962 detection of cosmic X-rays confirms the belief that the universe is filled with harmful high energy radiation capable of damaging cells and tissues. The focus of the specific mission however was to simply conduct more detailed analysis of X-rays in space as well as potential sources of them. Originally besides looking at the general soup of X-rays in the solar system the mission was also looking for soft X-ray sources such as X-rays generated by energy captured on the moon's surface and re-radiated as X-rays. The mission was successful in recording more detailed information about X-rays in general and also is noted for discovering the first X-ray source outside of our own solar system as a patch of soft X-rays were detected coming from something that was not our moon. Located in the constellation Scorpius the object was given a beautifully typical scientific name of great mystery and wonderment and it was known as Scorpius X-1. Science as you may have noticed by now and will certainly see with the discovery and naming of new objects likes to be very factual and categorical.

The interesting thing that X-ray astronomy was able to tell us about this new object was that not only did it emit X-rays as astronomers expected but it emitted most of its energy in the X-ray portion of the electromagnetic spectrum. Sources of X-rays outside of our solar system were thought to exist and should be plentiful as we know our Sun is a star and a source of X-rays and the universe is made up of countless stars. What was not expected was that the X-ray output of this newly discovered and named Scorpius X-1 object should be so copious. Despite lying some 9000 light years away it would be practically impossible to miss seeing this object in the X-ray wavelengths because the output of X-rays by Scorpius X-1 is some orders of magnitude higher than the out put of the stars visual wavelengths, in fact about five orders of magnitude more. By contrast the Sun emits only a tiny fraction of its total energy as X-rays and a million times more energy as photons of visible light. This important difference allows us to learn some things about objects in the universe and in this case a neutron star as it is the suspected culprit of all these X-ray emissions.

If ultraviolet rays are emitted by hot objects then X-rays are emitted by molten hot objects that can be as cool as only a few million degrees Kelvin and all the way up to things that are hundreds of millions of degrees Kelvin. The problem here is that stars do not typically burn at such high temperatures and at the same time release most of their energy as X-rays and so an alternative to a traditional star must be responsible. As it turns out Scorpius X-1 is part of a binary star system where it is the neutron star of the pair. As with many binary star systems one star will inevitably be larger or hotter than another when the two begin burning. Typically the larger and hotter stars will run through their supply of nuclear fuel first and therefore enter into their final phase of life first. Sometimes this means they simply burn down to become white dwarf stars as will our Sun one day, but sometimes they will become a supernova. In the case of Scorpius X-1 this is obviously what happened as it would have run out of fuel first, reached critical density for a neutron star during its supernova phase and remained part of the binary system. As a neutron star it has the curious property of incredible gravity for its size and mass and with a still burning companion so close interesting things were bound to happen. Specifically the outer, thinner less dense layers of the companion star would be stripped away by the neutron stars incredible gravity and slowly form an accretion disk. The hot gasses swirling into the neutron star are destined to undergo some unusual conditions before they finally reach its surface.

For starters the gas is going to be sped up as the size of the accretion disk decreases closer to the surface of the neutron star. The second is the density of the gas increases many fold as it is squeezed ever tighter with its neighbor's right above the surface of the neutron star. Lastly the temperature of the gas rises incredibly as the combined effects of gravity and motion causes the gas to begin emitting X-rays before it plummets into Scorpius X-1. The final result is proof of a neutron star slowly accreting its companion stars gasses and releasing them as an intensely bright light made of X-ray radiation. This emission is found wherever a star exceeds what is known as its Eddington Limit, named after Sir Arthur Stanley Eddington. The system is in balance when the amount of force a stars gravity can generate pulling inwards is balance by the amount of force that radiation can generate pushing outwards. In systems such as Scorpius X-1 the inward falling gasses cause the neutrons stars Eddington limit to be exceeded and therefore thrown out of balance. This is what results in the excess energy being vented off as X-rays. So the discovery of Scorpius X-1 proved not only interesting but also a valuable tool in proving scientific theory about neutron stars. This further proves the importance of what we can learn by viewing the universe in as many different wavelengths of the electromagnetic spectrum as possible. Observations taken in radio wave and through to visible light and ultraviolet would not have allowed us to see these important and almost hidden details.

Now that we know that Scorpius X-1 is a neutron star and regular star in a binary star system we can begin to broaden our search for new and interesting objects in the universe. Many of which are in fact X-ray sources suspected to be or known to be binary systems already. Part of the X-ray emissions can come from two regular stars orbiting one another close enough to create a pool of superheated magnetically charged plasma around one or both of them. This plasma is a source of soft X-rays and the study of star systems such as this has broadened our understanding of binary systems. Given the fact that in the early universe many of the stars were huge super giants that burned very hot and very fast, sometimes lasting only a few hundred million years or less, it is simple to deduce the fact that they will have all ended their lives as stars with supernovae. The variation in size of the star before going supernova will determine whether or not it becomes a neutron star or a black hole and also the size of each of these objects. The stars going supernova transforms them into the longest lasting phases of their lives as neutron stars and black holes are thought to last for in theory at least a few quadrillion years to almost forever. So even though these stars will have exploded billions of years ago for us their second phase of life lives on for us to observe now. What has become of these early primordial neutron stars and black holes can along with younger neutron stars and black holes teach us a great deal about stellar evolution. It can also tell us about how these massive stars affect the formation of new stars and galaxies and the whole universe itself. Part of this is already being undertaken through the use of X-ray astronomy and it does in fact involve galaxies and new young stars.

The particular area of interest for X-ray astronomers and cosmologists here is not so much the stars themselves but the space between them and what they are made of and how they behave in the galaxy. Known as the Interstellar Medium this is the space and matter that exists between star systems and between star systems and the entire galaxy. The Interstellar Medium is accompanied by the Interstellar Radiation which constitutes the electromagnetic radiation that occupies the same space and areas within the galaxy as the Interstellar Medium. The various particles and gasses that make up this mixture of materials and energy can be viewed using X-ray spectrums to determine all sorts of things about it. What for example is making up the matter predominantly in one particular area of the Interstellar Medium within the galaxy. How much energy does it possess and what is its temperature. Also the general motion of some of this material can be judged if a large enough chunk of it is moving in one direction.

All of this is invaluable data that can help astronomers paint a better picture of how galaxies evolve not just in broad sweeping strokes from the outside as we might view other galaxies outside the Milky Way but also right here at home. The inner workings of the Milky Way and by extrapolation of other galaxies as well, can be worked out on a far more small scale. After all when we view a galaxy outside our own we see mainly broad sweeping arms, maybe a bar or two and a galactic bulge as there is little chance of even our best telescopes getting a good hard close up look at the space between individual stars at that distance. Right here in our own galactic backyard however we are in the middle of one of the spiral arms. Sitting about 25 000 light years from the Milky Way's center we have a perfect view of space in denser hotter more radiated regions towards the Milky Way's galactic bulge. Looking away from the bulge and outwards to the edges of our galaxy we can see the more dilute parts of our spiral arm and note the changes in the Interstellar Medium there. All of this accompanied by detailed information from multiple wavelengths of observations, including the aforementioned X-ray spectrums, to tell us the subtle variations in a healthy galaxy.

One of the most interesting features of such study is the role of the births and deaths of stars and how this stirs up the Medium. A star will be born from a cloud of collapsing gas and dust comprised from the Interstellar medium approximately ten light years in size, sometimes smaller sometimes larger. The collapsing gas and dust will eventually create a protostar that given enough time will ignite its nuclear fusion core and begin to shine. The very birth of a star however causes a massive amount of energy and material to be pushed away from the newly ignited stellar core and the only place for all this material to go is back into the Interstellar Medium. Here we see one way that the life cycles of stars can stir up the medium and affect the evolution of the galaxy and the local nebulae and star systems around it. These events are easy to detect in the X-ray spectrum as the hot young stars will release quantities of X-rays as they light up and burn. With the star happily burning away in its main sequence phase of its life another stellar phenomena can again alter the flow of the Interstellar Medium.

Stars that are burning in their main sequences will spend the vast majority of their bright burning lives in this phase and a common occurrence for such stars is to release a Coronal Mass Ejection. Coronal Mass Ejections are bubbles of hot ionized plasma trapped inside looped magnetic field lines of the star. The magnetic field lines eventually become twisted enough that they will break and the hot gas plasma is freely ejected in one super heated lump into space at great speed. These ejections occur well within the X-ray temperature range of one million to ten million degrees and can be easily detected as bright sources of X-rays. The gas travels at between one and one and a half million kilometers per hour outwards from the star system and into the Interstellar Medium. Very diffuse, cold and slow compared to when it was born the Coronal Mass Ejection by the time it reaches the Interstellar Medium is however much denser than the medium itself. This will cause turbulence and numerous particle interactions as the two materials meet and once again this alters the behavior of the local Interstellar Medium.

Interestingly as well two stars with similar masses, colors and other characteristics will emit X-rays at different rates depending on their rate of rotation. The faster the rotation of the star the more energy it puts out as X-rays and therefore a link between the dynamo effect of the star and X-rays must be concluded. This is not only another way in which a stars output will affect the Interstellar Medium but is another prime example of how we are learning about the behavior of stars through X-ray astronomy. Lastly as a large star transforms from a burning star to a neutron star or black hole via a supernova explosion we see one more massively potent interaction with the Interstellar Medium. The shockwaves generated by an exploding star are almost unfathomable when compared to the energy output of the star during its main phase sequence. The star will briefly outshine an entire galaxy during this one explosion and emit more energy than our own Sun will over the course of its entire life. Quite obviously not much is going to stop this powerful shockwave from travelling outwards in all directions through the local space it inhabits. The result is a huge shift in the motion and energies of the local Interstellar Medium which will result in gas and particles of the medium glowing with absorbed and reemitted energy some of which is X-ray. This will create a lot of turbulence within the medium that can in turn cause pockets of the gas and dust which make the medium up to be pushed together and increase in density. This increase in density leads to a collapsing of another part of the Interstellar Medium which will cause an area of space lets say 10 light years wide to once again coalesce into another protostar. This cycle of star birth, life and death is obviously very influential on the behavior of the Interstellar Medium and therefore of the entire galaxy itself. Since most of these events occur at high temperatures and X-rays are emitted from high temperature matter studying the X-ray spectrum of these materials is key to understanding the life of the galaxy.

Closer to home for a moment advancements in X-ray astronomy have led to us better understanding our own star and how it works. As we have discussed briefly Coronal Mass Ejections are created by stars and are a source of X-rays and our Sun produces these ejections on a fairly regular basis. The reason why astronomers and scientists are so deeply interested in better understanding how the Sun produces Coronal Mass Ejections is more linked to their desire to understand when the Sun will create one. The reason for this is simply because a Coronal Mass Ejection carries enough energy to destroy most or all of the electronic and computer devices on Earth and therefore plunge us back into the steam age. Current apocalyptic theories aside there is a real danger that a loss of our electronics and computers would lead to a breakdown of society. The simple reason of course is that most of the Earth's population lives in urban centers as opposed to rural ones. All urban centers cannot support their own needs in terms of food, resources or energy production and so shortages would be immediately apparent. Humans sometimes have difficulties with being patient and sharing and the inevitable breakdown of society into less than civilized behavior would be short coming. Try going a week without food, water, electricity and the rest of our modern day amenities and imagine that everyone in the city is in the same situation.

The ability to restore the power grid for example and make a normal country run smoothly once again would take years to decades; far more time than is necessary for society to cannibalize itself into technological reversal. Therefore learning how the Sun generates these deadly events will tell us when the next will occur and allow us time to better prepare for it. Currently technology and models of the Sun's inner workings allow us only to watch around the clock for an ejection and judge its polarity and time of impact. The Coronal Mass Ejection has a North Pole and South Pole in a sense and of course so does the Earth. If the two similar poles meet than they will repel each other magnetically and the harmful and dangerous energy will by bounced off of the Earth nicely. If the opposite poles meet then they will end up attracting one another and cause the maximum amount of deadly particles and energy to be pulled into the Earth's atmosphere and to a certain extent to the ground. The ejection contains super heated, highly energetic and ionized plasma which has enough energy to destroy electronics and computers by melting or shorting out sensitive components within them. Since we believe that the X-ray emissions of a star are linked to the magneto hydrodynamics generated by the star better understanding of what its magnetic field is doing and why will lead us to better understanding of how to predict Coronal Mass Ejections.

The second major source of interest surrounding our Sun and X-ray astronomy comes from once again the corona of the Sun. This time it is less deadly though and simply a curiosity to physicists. The problem is known as the Coronal Heating Problem and is so named because if one examines the Sun at three key areas a head scratching problem arises. The interior of the Sun is the first area and as one would expect due to the nuclear fusion reactor that dwells there it is the hottest region of the Sun at around fifteen million degrees Kelvin. Moving away from a hot object and the center of energy generation and you would expect the temperature to drop. At the surface the temperature has dropped to a mere six thousand degrees Kelvin; so far so good the farther you move away from the center the cooler things get. The surface of the Sun is the second area of interest and now we will examine the third which is even farther out from the core again. This third area is farther away and should be cooler than the six thousand degrees temperature at the surface, but instead it is not and soars back up to five million degrees Kelvin. So here of course is the problem that defies conventional logic as this area of the Sun is the furthest from the core and yet it is significantly hotter than the surface.

In order to heat things up you need to inject more energy into a system and so the gas that exists in the corona must be receiving obscene amounts of energy from somewhere but no one is precisely sure where. Many theories have been proposed and many have been discarded so far the only one that seems at least a little plausible is the idea that magnetic energy from within the Sun is being pumped into the corona and heating the gasses found there. Since magnetic fields are suspected of providing the energy and since X-rays are associated with both magnetic fields and hot gasses, using X-ray astronomy to solve or help analyze this problem is a must. Indeed just viewing the sun in the visible light spectrum and comparing that to the X-ray and all other wavelengths of light will dramatically illustrate not only the differences in appearance of the Sun and its atmosphere but also what information appears and disappears based on the spectrum chosen. So now we have seen the importance of X-ray astronomy in understanding numerous objects in the universe as well as in our own solar system. It is now time to view the universe in the highest energy wavelengths available to modern science and to see structures and to learn secrets about the universe that are the most hidden and most difficult to see.

The last stop on our tour of different wavelength astronomy is at the gamma ray end of the electromagnetic spectrum. Gamma rays have the highest energy and the shortest wavelengths of all the different types of electromagnetic radiation. They suffer almost all of the same problems in detection of the other wavelengths with their biggest challenge being the atmosphere. The Earth's atmosphere is simply too strong for them to penetrate and so any hopes of conducting significant gamma ray astronomy of the universe from ground bases telescopes is remote. The simple detection of gamma rays on the surface is all we can hope to achieve and many of these are the result of cascade showers from cosmic rays interacting with atoms and molecules in the Earth's atmosphere. A gamma ray from across the universe has little chance of surviving the trip to our surface. As a result the need for space based observations is essential and once again all the same challenges facing an orbital observatory or one that travels out of orbit are the same. Staying away from the glare of the Sun, unless of course it is designed to study the Sun in the gamma wavelength, cooling the instruments and maintaining a very stable position from which to take minutes long exposures are all a must.

Gamma rays have long been known to exist and have long been theorized to exist in space as well but due to the stringent limitations of observation and the need for advanced rockets and detectors as compared to when they were discovered, much of the useful gamma ray astronomy has only been conducted within the last few decades. Discovered as we have seen in the year 1900 by Paul Villard it did not take physicists and astronomers long to postulate the existence of these rays originating in space as well as in radioactive materials in the laboratory. One man who was at the forefront of the development of gamma ray studies in space was an American named Philip Morrison who had worked extensively on nuclear research and atomic projects for the United States military. As such he was well equipped to investigate all manners of radiation and in 1940 along with fellow colleague Leonard Schiff wrote a paper on the topic of gamma rays and nuclear reactions. The paper discussed a process known as K electron capture or L electron capture; the K and the L denoting the particular shell from which the electron is used.

The process is simple and involves an electron being absorbed by the nucleus of its own atom and turning a proton into a neutron. The atom is in a highly excited state and must lose some of the absorbed energy in order for it to relax back to its neutral ground state and in so doing will emit an X-ray as well as a neutrino and absorbing a free electron from the local environment. This is the basic electron capture mechanism by which the matter of a normal star upon undergoing a supernova will convert its mass into neutrons to form a neutron star. The loss of the electron in its orbital is the great loss in volume from the matter which results in the neutron only star core being so dense and small. What Morrison had noted in his work was that sometimes a gamma ray would be released in place of the usual X-ray. Since this process involves high energies, free electrons and not much else it did not take Morrison long to write another paper on the subject once his wartime research had been concluded. By 1958 he had written a paper that is considered to be the first work on gamma ray astronomy and had borrowed from his previous paper in 1940. A few more years still would have to pass before the first gamma rays could be detected coming from space since the technology to launch sensitive instruments high enough to avoid most of the Earth's atmosphere was still in its infancy. The tried and tested balloon, rocket and rockoon launches of today are so standard we hardly give them a thought anymore. However back in the 1940s, 1950s and 1960s these technologies were present but not so widespread and so it was not until 1961 that the first gamma ray telescope could be launched into space.

The results were a mixed blessing as while gamma rays were detected coming from space it was impossible to tell where they were coming from in any great significance. Furthermore only around two to three dozen or so gamma rays were actually detected coming from space, the rest were caused by cosmic ray spallation in the Earth's atmosphere and so the overall picture was of a general gamma ray haze everywhere in the universe. Follow up missions would definitively show the general locations and existence of gamma rays originating in space. The Sun is of course one source of these and the main process by which gamma rays were detected came from solar flares. The galactic center was also deemed a source of gamma radiation and a few other bright sources outside of our own Milky Way galaxy. However the resolution of these captured gamma ray photons was still very low and it would be still more years before great leaps in detector and telescope technology within gamma wavelengths was made possible.

One very interesting anomaly was detected during this early period of gamma ray astronomy and like many in science it was made somewhat serendipitously. The organization that made the discovery was the United States military and their defense programs aimed at watching incoming nuclear threats during the height of the Cold War. The west had desired to maintain a close watch on any nuclear activities that might be undertaken by the east during the cold war and since no one on either side was going to freely tell the other what they were up to a method for spying was sorely needed. Spy planes flying over the Soviet Union were not able to watch round the clock what the east was up to nor were they equipped with the instruments to detect nuclear activities beyond photographing the ground for installations. Instead the United States military decided to launch a series of satellites into orbit around the Earth to essentially listen to what the enemy was doing. A whole group of satellites functioning as one is called a constellation and this constellation was dubbed Vela. The way in which the Vela satellite constellation worked was simple and one that was destined to play a far greater science role than military one. The satellites were designed to be sensitive gamma ray detectors that would record the short lived but tell tale signature of a hydrogen bomb being detonated. In this way the direction, energy levels and amount of gamma rays hitting the constellation could be calculated. This would provide valuable information on what type of tests the enemy was conducting revealing how advanced or powerful their weapons capability was.

The fascinating thing about these satellites is that for the entire time they were operational they would detect randomly intermittent flashes of gamma rays. These flashes had two curious characteristics the first of which was they were not coming from Earth and instead originated deep in space. The second interesting point of note is that the flashes though random never stopped. This is not to say they recorded a flash every second or so but rather regular flashes were detected over the years meaning what ever was creating these intense and very short lived gamma ray emissions was a normal part of the universe that was probably not going to stop anytime soon. Now named Gamma Ray Bursts, these flashes of radiation are a puzzling and welcome mystery found totally by accident. The main problem with gamma ray bursts is not a simple one by any standards and is as yet unsolved by modern science. Despite being discovered over 50 years ago no satisfactory explanation of what is causing these phenomena exists anywhere in contemporary, astronomy, physics or cosmology.

So what do we actually know about these strange events? We do know a lot about the actual gamma rays themselves and since the 1980s with the advent of far more advanced technology in the form of detectors and computers a wealth of new information has flooded in. First off we have detected longer duration bursts that can last for multiple hours and this is in contrast with the early detections. The simple answer for this of course is that the Vela constellation was never designed to conduct studies of this nature and being built in the 1960s carried much less sensitive equipment. So the first piece of the puzzle presents itself as millisecond to multiple hour gamma ray bursts are possible.

The second thing we know about these strange events is that they all occur outside of the Milky Way galaxy and not just a little bit outside of it but rather billions and billions of light years away. This helps us to at least focus on a particular area and importantly time of space to study but it raises another nearly insurmountable obstacle. The trouble once again of the universe being large and whoppingly great bloody uber large at that. It would be difficult to make accurate observations of these events if they were to occur to our nearby galactic friend Andromeda but at such immense distances it is currently impossible with our level of technology to determine what object is even responsible for these bursts. On top of that even if we knew which particular object made it we could not discern it from the neighboring objects around it and filter out the background noise for study. So we do have a second puzzle piece and it does provide some information but it also slams the door shut in our face in terms of detailed observations given our current technology.

The explosions while numerous in total are infrequent on the whole and by this I mean multiple gamma ray bursts are detected no problem every year but the frequency and direction from which each originates makes them rare for their parent source. If we look at one burst we can say that it comes from a particular galaxy this many billions of light years away, but that galaxy if it is of average size will produce probably only a handful every million years or so. This puzzle piece does help us a little more actually because despite still being too far away to study in depth it does tell us that there might be a correlation to the composition of a galaxy. The strength, duration and frequency of bursts coming from one or more galaxies will allow us to compare what classification of galaxy is making it. Also we can then examine what stars populate that galaxy as differences in size, temperature and age will all affect the bursts. If the galaxies are all populated by hot young blue super giant stars as opposed to cool red more mature stars we can start to make inferences about what might be triggering these massive explosions of such high energy photons.

The last thing we know about these gamma ray bursts is that they only start out as gamma rays. First detected in the gamma ray range of the electromagnetic spectrum and indeed depositing most of their energy in the form of gamma rays, the bursts actually contain other wavelengths of photons as well. The gamma ray burst starts out with of course copious amounts of high energy gamma rays but will transition over time. This transitioning is a slow fade out of high to low energy photons that will pass downwards from X-ray, ultraviolet, visible light, infrared, microwave and radio. In other words the entire spectrum can be detected and measured from these gamma ray bursts which do lend us an extraordinary tool for increased study of the events. Now we can examine the phenomena in every wavelength and look for hidden correlations in this data to help explain what they are.

So if we put all the pieces of this enigma together we still do not have a complete picture nor do we have definitive proof as to what is causing them, but we do have one fairly good theory at least. One thing we do know about gamma rays outside of gamma ray bursts and the study of them is that all gamma rays are in possession of the highest energy levels of the electromagnetic spectrum we have ever measured. The only way to get these high energies and high temperatures is through very violent events and we know this to be true with our study of other wavelength astronomy. Remember how for example infrared radiation was made by cool low energy objects and as we moved up through ultraviolet and X-ray the objects or events that created them contained higher temperatures and energies. This may seem obvious but therefore gamma ray bursts must come from very hot and very energetic objects or events, the simple reason why this needs to be underlined is because we can eliminate almost all potential candidates as a plausible explanation of these occurrences.

Nothing we have seen and very few things we have theorized can produce as much heat or energy to explain a gamma ray burst and here in lies the problem. Stars are hot and their cores are immensely hot, so hot they can force matter to do what it really would rather not and fuse together and yet this is not enough heat or energy to make a gamma ray burst. What about a star going supernova? Well that will be violent for sure and will spew forth unparalleled amounts of energy compared to the stars lifetime but it still is not enough. Both stars and supernovae do produce gamma rays and frequently at that but never in the amounts or at such high energies as this. So now we can only guess at what are causing these based on the puzzle pieces and what we know about the production methods of gamma rays in the universe.

The so far best candidate modern science has postulated is of an explosion far more powerful than anything we have conclusively observed. If you travel up the scale from supernovae you come to what are called hypernovae and they are essentially the same thing as a regular supernova only far more powerful. The hypernova will involve a star whose mass is far heavier than a regular supernova and might be able to provide the environment and energy necessary to create gamma ray bursts. For starters there is no well defined boundary between novae of different types in terms of non Type 1a supernovae. Remember Type 1a involve binary systems and white dwarfs collecting and accreting matter on their surfaces until a runaway fusion event is triggered that blows off the excess material in order to not exceed the electron degeneracy pressure. Every other type of supernova on the whole follows the model of the star exploding and blowing off most of its outer materials into space, the evidence of these are seen everywhere you look in the universe as nebulae or supernova remnants. The rough values put normal supernovae at approximately 10 solar masses or ten times the mass of our Sun and hypernovae at approximately 100 solar masses or more.

What seems to be the key in creating a gamma ray burst is overcoming the electron degeneracy pressure of matter or more likely the neutron degeneracy pressure of matter. Since a hypernova is so much more violent than a regular supernova the incredible shockwave that drives it can theoretically be responsible for crushing the matter at its core to incredible densities. If the explosion is powerful enough the electron degeneracy pressure which is the force the electrons exert against their neighbors is exceeded. The net result means the electrons are lost from the matter or absorbed by it in the electron capture process and now we have a core made of only nuclei and neutrons touching one another. This we of course have named a neutron star and is one of the possible fates of a star that has undergone a hypernova. The second possible fate science believes can occur if the neutrons themselves are squeezed even further and harder. The only thing supporting the nuclei and neutrons in a neutron star from collapsing any further is the neutron degeneracy pressure they exert. This neutron degeneracy pressure works broadly in the exact same way as electron degeneracy pressure. It keeps the neutrons from being crushed into their nearest neighbors and exceeding this limit will cause another collapse and decrease in the volume of matter just as we saw with the breakdown of the electron degeneracy pressure. If the hypernova is violent enough even the neutrons cannot withstand the incredible forces of the explosion and will be crushed down to a smaller core yet. The most commonly known of these configurations is the black hole and contemporary science has found it to be of the most uncommonly behaving objects in the universe.

Black holes will be given their own section later but for now we just need a hypernova to create one for us. There are other theorized stars as well such as quarkstars, blazars and magnetars the formation of which most likely resembles the formation of a neutron star or black hole. Any of these stars is thought to be born from a hypernova and the energy required to make one is what many scientists believe to be the proof of where and how gamma ray bursts are made. The evidence does seem to bear them out as a hypernova should theoretically possess the energy required to make such great amounts of high energy gamma photons. In addition to this the way the gamma rays fade out over time to X-ray and down through to radio waves is very reminiscent of a normal explosion.

Conventional chemical based explosions rapidly releases much of its energy quickly in the form of intense light and heat and then over time the light and heat of the explosion drops as the matter in it cools and expands. As the glow of the initial explosion fades so does its corresponding, heat, light and energy. While a chemical explosion is everything but the same from a hypernova the pattern we can observe of fading energies is analogous. The other interesting thing about these explosions is that they tend to send matter and energy out in all directions but focus the gamma ray bursts into narrow beams. The gamma ray bursts we observe are apparently narrow beams of high energy photons produced and ejected at the poles of these dying stars. This is an interesting feature of hypernovae as the remainder of the energy is sent outwards roughly spherically as would be expected by a typical explosion pattern. This tends to suggest that another mechanism is working to create these relativistic jets of gamma rays as opposed to a random event.

The most probable mechanism is the formation of a neutron star or black hole because of the incredible gravity they both possess. The escape velocity for a neutron star can be as fast as half the speed of light or 150 000 kilometers per second and the escape velocity for a black hole is the speed of light or greater at 300 000 kilometers per second. A black holes escape velocity in theory could be many times higher than the speed of light simply due to the fact that we know light cannot escape its surface. What this means is that the minimum escape velocity is the speed of light, since modern science does not understand what is really going on inside a black hole the necessary escape velocity could exceed the speed of light. The fact that modern science believes that the speed of light in a vacuum is the fastest velocity possible does not mean that the black hole cannot generate a gravitational field stronger than a 300 000 kilometer per second escape velocity.

How this extreme gravitational force from both black holes and neutron stars can cause a gamma ray burst is related to the in falling material from the hypernova star itself. As the star enters a hypernova phase the core forms first, as far as current models predict and the neutron star and black hole are born and exerting their extreme gravitational forces right away. The remaining material that is destined to explode and spread out as what we see today being a supernova remnant first experiences the intense pull of gravity from the newly formed neutron star or black hole core. The net result is a massive amount of material falling inwards towards the core at speeds a fraction of or very close to the speed of light. The neutron stars and black holes at the center of these hypernova stars begin consuming material as fast as possible and though their appetites are voracious there is a limit to how quickly they can consume materials. As soon as the rate of material falling into the core exceeds the rate at which the core can absorb the material a build up of super heated gas occurs on the surface of the core itself. The high velocity of rotation at the equators of these objects and the large surface area there allows the equator to absorb more material than the poles and so the built up energy from the hot gasses which cannot be consumed by the poles is released as an intense burst of high energy photons. This burst is related to the Eddington limit and of course is our gamma ray burst; if that theory is correct.

The remainder of the stars gases are blown off the surface by the force of the hypernova and so without a supply of excess material falling inwards the gamma ray burst mechanism is shut off. This at least is the theory, it does not fully explain what is going on inside the core of the star and does not fully fit the observations either. In addition we have never been able to observe directly a star before during and after it goes hypernova to get a better picture of what is going on. While we would like to study a star close up within our own galaxy we have not been able to directly observe one in a distant galaxy either with high precision. That being said this theory is the best modern science has for now concerning gamma ray bursts and how they originate even though officially the mystery remains unsolved and in fact could be produced by a different but similar mechanism or one altogether unrelated. One interesting side note on supernovae and hypernovae in general is that although many models predict a remnant core of some sort this is only part of the story as some stars simply blow themselves to pieces and leave nothing behind.

The power of gamma ray astronomy is readily apparent when viewing our own Milky Way galaxy in this portion of the electromagnetic spectrum. The very high energies of these gamma photons is the result of the same sort of mechanisms that create X-ray photons and come from violent high temperature events. The trouble with gamma ray detection is in making accurate measurements of the incoming photons. The particles are very high energy and must therefore be observed from space as they will be absorbed by the Earth's atmosphere. Gamma rays from space can be detected on Earth using ground based instruments but not directly instead the interactions of these gamma rays are seen. The process is one we have already examined and is known as cosmic ray spallation where the after affects of atmospheric interactions with cosmic rays are recorded. The back tracking of these interactions is the method by which ground based observations are made of how the gamma ray hit the particles in the atmosphere. The best information from gamma ray sources is obtained outside the atmosphere and so expensive spacecraft with highly sensitive equipment must be used and not surprisingly this means there are not many of them around. Difficulties in acquiring high resolution and long exposure times are further confounded by the need to filter out the background noise of gamma rays that simply permeate the cosmos.

When all these issues have been dealt with the images and data found using gamma rays in the Milky Way is astounding. The entire band of the galactic disk glows with both X-ray and gamma ray sources despite the huge amounts of dust present. Most strikingly is a feature of the Milky Way's heart that was not visible to us before gamma ray astronomy. The discovery is of two massive plumes of material that are invisible to optical telescopes but glows brightly in the gamma ray spectrum. Extending outwards from the center of the galaxy and perpendicular to the galactic plane are two lobes of roughly spheroid gamma ray emissions. Each one is 25 000 light years in height if you will and extend one from the top of the galaxy and one from the bottom for a total of 50 000 light years. There is a small amount of lower energy X-rays skirting the lower portions of the plumes but the vast majority of the energy is made from gamma rays. The exact mechanism or mechanisms by which these are created is as of yet unknown but most speculate that it has something to do with the super massive black hole that resides at the center of the Milky Way.

As an active galactic nuclei caused by a super massive black hole anywhere in the universe might do, the Milky Way seems to be emitting intense amounts of high energy particles from the poles of the black hole. The possible candidates for what is causing the emission of gamma rays include matter and anti-matter annihilation events from streams of positrons being produced from the black hole poles and hitting electrons further out. The resultant annihilation will produce two high energy gamma rays one from the positron and one from the electron. Another possibility is simply that the excess material being pulled into the black hole as we have seen previously is heating up and some of it being ejected at high velocities probably near light speed. The matter will then travel upwards or downwards away from the poles and smash into the slower moving particles surrounding the galaxy releasing gamma radiation. No matter what is causing it it simply illustrates another fascinating secret revealed through the use of multiple wavelength astronomy and in this case gamma ray astronomy.

Bremsstrahlung

Now it is time to talk of something that is one of the most important aspects of astronomy and physics and is something you've probably never heard of but should understand to appreciate the universe better. It is called Bremsstrahlung and comes from two German words the first being bremsen or brake and the second being strahlung or radiation. Therefore putting the two together and you get breaking radiation which is one of the most important physical processes at work in the universe and in some ways is second only to gravity. This breaking radiation is responsible for many forms of electromagnetic astronomy, activation of the Interstellar Medium and evolution of the universe as a whole since the Big Bang up until today and into the future. The way Bremsstrahlung works is simple; one charged particle which is travelling at speed impacts another particle and the loss of motion from the first particle is transformed into a photon. In other words we have a transfer of kinetic energy into electromagnetic energy hence the perfectly apt name braking radiation.

This Bremsstrahlung is associated with motion and any charged particle and can be found in multiple places in the universe and the laboratory. The Large Hadron Collider in or should I say under Switzerland, does very much use Bremsstrahlung everyday it is operational and collecting data. The accelerator fires two beams of charged particles at one another say protons and lead nuclei both of which are obviously ionized to near the speed of light. The charged particles will hopefully meet and annihilate directly in the center of one of the detectors. As they collide all of or most of their kinetic energy is lost due to Bremsstrahlung and a shower of high exotic particles and high energy radiation is the result. This general particle accelerator mechanism and the use of Bremsstrahlung are overwhelmingly responsible for much of our experimental data concerning matter. The importance of this phenomenon is of course greatly enjoyed by astronomers too and in fact predates particle accelerators by tens of thousands of years. The reason is that

Bremsstrahlung is present everywhere you look in the universe from the Big Bang through to the far distant future. Humans have for tens of thousands of years at least gazed up to the night sky and wondered and in that time we have seen the odd supernova even if we never wrote it down; or maybe we did on a cave wall we are not sure today.

Most assuredly especially for humans living in the North we have witnessed it in the form of the Northern Lights. Think of these multicolored rainbows dancing in bands and ribbons all over the sky and the Vikings and Bifrost the rainbow bridge extended from the home of the gods in Asgard to the realm of men here on Midgard. Part of what we today call the auroras and what the Vikings most likely found inspiration for and called Bifrost is due to Bremsstrahlung. Since these epic times astronomers have explored the cosmos and written their own sagas concerning how the universe works.

The epic prose begins with simple optical observations of the universe for what draws in many amateur astronomers and many professional astronomers and cosmologists as well and this is the ohh and ahh factor. The ohh and ahh factor of course being the sounds people make when they look at something beautiful and I am sure has happened to everyone reading this book at some point in your life you saw a picture of the universe and were awed as this is precisely why you are interested in reading a book like this in the first place. The beautiful nebulae, supernova remnants and galaxies which we all enjoy looking at owe much of their indescribable beauty to Bremsstrahlung. Many famous supernova remnants exhibit the astronomical benefits of the braking radiation phenomena and can tell us a myriad amount about what happened to create the remnant.

The source of particles comes from a supernova or hypernova star and since it originates from the stellar plasma of the stars atmosphere is already partially or fully ionized. This supplies us with all the charged particles we need and the obvious resultant explosion from the novae will move these particles at extreme velocities. Now we have two key factors covered in the form of charged particles and high acceleration of them. The third factor is a set of particles for them to collide with so that they can activate the Bremsstrahlung process. In the case of a supernova or hypernova the first candidate is part of the stellar plasma that was first ejected at slower speeds than the remainder of the plasma. In other words the stellar plasma creates a Bremsstrahlung with itself as different velocity particles impact with one another during the first moments of the explosion.

The second source of materials for the high velocity particles to impact comes from the local star systems own environment. This is made of cosmic dust in the system from comets, asteroids and the left over materials from the formation of the star system and whatever planets formed there as well. In addition to this are incoming intergalactic cosmic rays and the stars own solar wind which it has been steadily producing for millions or billions of years. All of these combine to form the next layer of particles that will sap the kinetic energy from the nova and once again release braking radiation. Lastly as the blown off stellar plasma reaches the terminus shock of the stars system the high velocity charged particles will finally meet the Interstellar Medium which permeates the galaxy in which the star exploded. The Interstellar Medium also extends in between galaxies and fills the entire universe but the potential for particles from a supernova or even a hypernova to reach that far is low. Most likely they will be caught in the gravitational pull of another galactic object and halted for the most part from journeying out through the universe though some do. The Bremsstrahlung caused by interactions with the Interstellar Medium is what we typically see today as a supernova remnant as they stretch for many light years in all directions and therefore exist far outside the terminus shock of the star system that created it.

Each of these three sources of slower moving impact particles have of course different compositions and will release energy in different amounts and at different wavelengths. This data is invaluable for astronomers and cosmologists alike as it can help to determine what the star is exploding into. In other words what is the precise composition and density of the star system that used to exist and of that portion of the Interstellar Medium that it has entered. Studying this sort of emitted braking radiation helps to back track what different parts of the galaxy are made of in terms of free floating, atoms, molecules and dust. The patterns created by each supernova and hypernova are unique as well and would be invisible to us if not for the Bremsstrahlung. A viewing of various different supernova and hypernova remnants will illustrate not only the beautiful visuals of these events but also the precise detonation pattern and impact patterns with surrounding materials. Both of which can answer a lot of questions concerning stellar life cycles and the spread of matter and energy throughout a galaxy and the universe. Looking backward at some of the particular shapes of the remnants will also reveal some of the inner workings of the stars that created them as the explosion patterns when viewed on a parsec sized scale illustrate what was happening internally in the last few seconds of the star before it exploded.

The next way in which Bremsstrahlung can help us answer questions concerning the workings of the universe also helps to underline the importance of the Bremsstrahlung effect as being perhaps only second to gravity. What it involves is the mixing of the Interstellar Medium with respect to the motion and energies of all particles in it. The Interstellar Medium is ever present throughout the universe and contains dilute by Earth standards but dense by universe standards gas and dust. The gas and dust ranges from simple primordial atoms of hydrogen left over from the Big Bang and more complex atoms from various processes of nucleosynthesis. Heavy atoms well beyond the iron formed in stellar fusion can be found here such as thorium and uranium albeit in vary small quantities. This complex series of gases also contains miniscule particles of dust which can range in size from a few molecules to things as large as fine powder. This dust is itself made of various elements and since some of it comes from the recycling of old star systems and their planets can contain forms of these elements already pre-processed so to speak into various allotropes.

The way Bremsstrahlung affects these gas and dust particles of the Interstellar Medium is by altering the energy and motion they possess. When a star is born and it begins to burn its fuel supply in stellar fusion a tremendous amount of energy is released in the form of stellar winds. These winds travel very fast and very far away from the star into the surrounding Interstellar Medium from which that star had collapsed and formed. The energy from this newly born star will push against the gas and dust in the surrounding galactic space and also heat it up both of which is caused by Bremsstrahlung. The gas and dust like any other fluid will begin to move around from the impacts of the high energy charged particles and the radiation released from these collisions. This churning up of the Interstellar Medium will result in it transitioning from a stable cloud of gas and dust to one that has different densities and therefore different gravities. A newly formed pocket of gravity amongst a light year wide swath of gas and dust will begin to collapse under its own force of mutual attraction and create a protostar. The protostar will then ignite and burn once again creating more stellar winds to continue the process of star birth and thus continue to drive the evolution of the universe due to Bremsstrahlung. Some of these stars will end their main sequence phases as supernovae or hypernovae and the resulting explosions will further drive out charged particles to interact with the Interstellar Medium and once again create more stars. In this way it is the Bremsstrahlung and gravity that are responsible for most of the evolution of the universe and the continual changes within familiar things like our own Milky Way galaxy.

Perhaps the single most important aspect of this Bremsstrahlung interaction with gas and dust comes from the build up of progressively heavy atoms after the Big Bang. The simplest hydrogen and helium formed after the Big Bang took hundreds of millions if not billions of years to undergo stellar nucleosynthesis in any of the various pathways we have discussed. The build up of heavier elements were then spread throughout the cosmos as stars ended their main sequence phases in various forms whether a violent explosion or the slow release of gas from a red giant as will happen to our own Sun. The diversity of atoms in the universe increases rapidly and so does the complexity of the molecules they form. Bremsstrahlung can stir up the clouds of gas and dust and impart energy to them which can cause various atoms to meet up and form simple molecules. This coupled with the molecules naturally formed from planets orbiting these early stars and to lesser extent simple molecules from the stars themselves will build up fairly quickly a huge list of chemical compounds. The cumulative result is clouds of gas and dust that are rich in preformed complex molecules that will be incorporated into the next generation of stars and planets around them. The total mixing of gas and dust into new stars which then spread new elements and molecules over and over again is what led to the cloud of gas and dust that formed our solar system.

We know of at least one planet in the universe that supports life and it is the Earth. The necessary building blocks of life came from the gas and dust of the Interstellar Medium from which our protostar Sun formed. What scientists and astronomers have found by examining these cosmic dust specks is the presence of complex molecules formed before our solar system even existed. The molecules are diverse and seemingly very advanced for production by natural mechanisms yet they are there. The fascinating thing about these molecules is that many of them are what are known as an organic compound which is a molecule containing the key elements found in most life forms. Organic compounds are those that contain carbon, hydrogen, oxygen or nitrogen in some form but almost always with carbon involved. These atoms and the molecules they can create are key to life as it has evolved here on Earth and by extension the universe. We have yet to land on another planet containing life outside our solar system and analyze what makes it tick. Therefore we can only use the Earth as our standard frame of reference; however using the Earth is actually a very good idea.

The reason being that scientists used to and some still do, make outlandish claims that life outside the Earth could take on fantastical forms made from fantastical chemistry sets. Their reasoning was based on the simple idea that since the universe is large and we do not yet fully understand it then anything was possible. This is of course not the case as there are certain limits as to what the universe can do given the tools and building blocks at its disposal as well as the starting conditions available to these tools and materials. This limitation on what the universe can do given its available options means that life such as what we find on Earth is a very good analogy for how life probably looks everywhere in the universe from biochemistry to overall appearance and behaviors. To help prove this point let us look at where the Earth came from in the first place and work around this.

The Earth came from the gas and dust of the solar system and this material existed before the solar system was formed or the Sun was burning. Therefore the materials we find in it did not come from some geological or evolutionary process within the solar system but from the surrounding gas and dust that existed before it. The gas and dust was part of the Interstellar Medium and this permeates the entire galaxy and the universe. We know it is everywhere in our Milky Way galaxy because we can observe the Interstellar medium close to home and have found it to be relatively homogenous in terms of composition with slight variances in its density. Since the Milky Way is an ordinary galaxy that formed like any other after the Big Bang it is a safe bet that all other galaxies in our observable part of the universe formed in the same ways.

The Milky Way is probably the result of one central galaxy absorbing other smaller galaxies during merger events over time meaning that we would already have incorporated, if they existed, different types of Interstellar Mediums from different sources. Yet as we have discussed everywhere we look in the Milky Way we see the same gas and dust mixture present meaning that the merger galaxies contained the same building blocks we did and of course the laws that govern the universe are the same everywhere too. This means that these building blocks cannot have been put together in different ways anywhere else in the universe either and the end result is the Interstellar Medium which is ever present throughout the universe must be homogenous as well. So finally we see that what went in to the construction of our solar system and the Earth is what went into the construction of every other solar system and planet in the universe.

So the complex pre-existing organic compounds that went into the formation of our planet and are now used in every biological organism and mechanism here on Earth would be available throughout the cosmos. Entire amino acids which are the building blocks of proteins and DNA bases have even been found in the dust that made up our solar system. Preserved and unharmed by the harsh environment of space since before our solar system existed. The possibilities for life using the same chemistry set as we use on Earth and under the same conditions is for most astrobiologists a given. I will avoid a more vivisected discussion of the topic of life in the universe and its probable forms here as it could easily fill a chapter. However the undeniable fact here is that the Interstellar Medium containing a homogenous mixture of gas and dust in the universe obeying the same set of physical laws and interactions is what goes into the formation of every star system in the cosmos. Life at least one time has arisen out of this arrangement and it is here on Earth the possibility that life in some form arose from this cosmic soup elsewhere is in my opinion one hundred percent likely. The actions by which all the compounds were created and spread throughout the universe and then coalesced back into new star systems is due to both Bremsstrahlung and gravity even if only indirectly.

The universe as we see it today relies heavily on gravity and Bremsstrahlung to churn up its contents and set about the evolution of stars, galaxies and all the structures we can see in the cosmos. This will be the future of the universe too as we find galaxies merging with one another, forming filaments of super clusters or any number of seemingly random events. The two mechanisms were also heavily at work in the distant past however and shaped much of the universe so that it could evolve into the one we see today. In fact in the very distant past including the first few moments after the Big Bang these two mechanisms were at work churning up the cosmic brew. Once the Big Bang had occurred and the universe expanded in size to approximately ten centimeters across, which if inflation theory is correct is also the end of the inflationary period the forces of gravity and Bremsstrahlung begin to shine. In this period of the universes history we find a very hot and very dense mix of energy that is in a sense solidifying out into matter.

The Big Bang did not produce matter in its familiar state today but was so hot that it produced the energy that would become things like protons and electrons and of course any smaller particles that make these up as well. What this means is as soon as matter was forming in this dense universe it was colliding with other particles of matter. The universe was so hot the particles were in a constant state of high speed motion much like the particles making up a very hot liquid will move faster via Brownian motion than similar particles in a cool liquid. The motions will be very much the same overall due to Brownian motion but the rate at which the particles will progress through these motions is very different. This is what was happening to the first primordial particles after the Big Bang and this coupled with the fact that the extreme density of photons prevented simple atoms like hydrogen from forming means that the particles involved all possessed a charge.

So high speed and charged particles will result in Bremsstrahlung constantly occurring in the early universe. The fact that Bremsstrahlung involves changes in speed viewed as kinetic energy and the resultant angular deflections of the two particles involved any dense mix of these particles in the early universe would also help to make it homogenous. The neat somewhat ordered flow of particles from the initial Big Bang explosion would become randomly disordered as particles with higher velocities shared their energy with particles of lower velocity. The end result is an even sharing of the momentums and energies between these particles and is surely one of the mechanisms that led to a smooth evenly distributed universe in terms of matter and energy. This series of primordial interactions is lost to us mainly because the radiation released during the recombination era would outshine it and make it difficult to detect today.

Nevertheless the Bremsstrahlung did contribute to the shape of the universe after the first moments of the Big Bang and gravity would then help as well to hold these changes in place. Once the matter and energy was more evenly distributed and mixed gravity would be able to further stabilize the universe by keeping things held together so to speak. The matter in the early universe under the influence of gravity would be attracted to one another and start to form small diffuse patches evenly throughout the expanding sea of particles. The particles though still in motion would still pull on each other as they moved thus helping to even out the distribution of matter further. This continues until all the little patches had because of their close proximity to one another due to the extremely small size of the universe shared their gravity enough to remove any large imbalances in mass from the initial explosion. The minute quantum imbalances present in this otherwise smooth mixture would as the universe expanded further become the places where matter started to clump up and form the first stars and then the first galaxies. Both of these mechanisms, the sharing of kinetic energy and therefore an evening out of momentums between particles and the mutual sharing of gravity during the mixing process would help to smooth out the matter in the universe leading to a uniform and homogenous cosmos today.

Here we end the journey of wavelength astronomy as we have investigated the overall mechanisms and obstacles behind making accurate observations across the spectrum. In addition we have looked at some of the history of the people, technologies and discoveries made using it as well as some of the profound effects it has on not only our understanding of the universe but how the universe shapes itself. There are of course many more observations to mention, many more technical details about things like peak absorption spectrums and in fact too many to list here. What we can say is that without wavelength astronomy we would not know nearly as much about the universe as we do today and about what makes the universe work. This form of astronomy has helped us to learn more about the various particles and energies that make up our universe and to prove some of the theorized interactions between them no matter if this involves electrons, protons, neutrons or photons. Some of the holes in our knowledge have been filled through the use of this type of astronomy but as with many things in science filling in one hole sometimes just leads to another and another and another. The newer holes we have found in our theories of the universe are starting to in some respects get smaller, after all there is a certain way the universe works and eventually filling in the holes will answer how it works. Some of the holes that we have not discussed yet involve more exotic particles and much smaller ones than we have encountered so far. The next thing we must explore is what these particles are and how they can interact with the larger particles we have seen as this will further our understanding of how the universe works.

Chapter 9
Neutrinos, Particle Accelerators and The Standard Model

Neutrinos

The universe is full of surprises and there are many more particles for us to explore that will further breakdown how we understand matter and energy to work. An understanding of things much smaller than protons and neutrons is needed to fully grasp how everything in the universe interacts with not only others of its own kind but every other type of particle and energy as well. The first stop on our list is to revisit an old friend we have discussed before and one that makes up non-trivial amounts of particles in our universe. I am speaking of course about the neutrino and this little particle is not only necessary to make up our universe it has some very interesting properties that are helping to teach us about matter in general. We have touched on neutrinos a bit before and we have seen how they are involved in or are the products of various reactions, the proton-proton chain for one or the product of cosmic ray spallation in the Earth's upper atmosphere producing shower particles. That being said let us go through a brief examination of how they were discovered and what they are to see the importance of what they can teach us about matter in the universe.

The neutrino was first proposed by Wolfgang Ernst Pauli an Austrian physicist in 1930 in a paper he published concerning his thoughts on the problem of beta decay. One of the interesting concepts of his newly proposed particle was that it had no charge as most particles of the day contained a charge. The proton was positive and the electron negative no matter what they had a charge associated with them. Even the neutron which later was understood to be a combination of charge carrying particles such as a proton and electron through the electron capture process was made of things with charge. The fact that the neutrino had no net charge meant that it was far freer to travel about the universe and could interact with matter in numerous ways other particles could not.

The other feature of the neutrino Pauli proposed was that it was extremely low mass and not more than one percent of the mass of a single proton. Originally named the neutron by Pauli it would later be named the neutrino to avoid confusion with the particle we now know as the neutron which was discovered in 1932 by James Chadwick. Even though Pauli chose the name neutron first his particle would not be discovered until after the Chadwick experiments found the neutron and so the neutrino name was adopted even though it was not his choice. The first discovery of a neutrino through experimentation was made in 1956 by two American scientists named Clyde Cowan and Frederick Reines. The experiment by which they detected the first neutrinos was directly linked to Pauli's ideas of beta decay and it was in fact beta decay events created in nuclear reactors where they first found the neutrino. The beta decay events would create various particles one of which was an anti-neutrino and it was the reaction of the anti-neutrino that they detected once all other interactions had been accounted for. Eventually after many decades of studying it would be found that neutrinos are point like particles that are nearly massless and travel almost at the speed of light with three flavors each of different masses. These discoveries confirmed Pauli's theory of a new particle and would open up a new door into our understanding of the universe as the neutrino was found to be involved in a plethora of subatomic processes over the years.

Though Wolfgang Pauli would die before he saw much of the contributions of his neutrino to modern particle physics it would not be long after his death in 1958 that even more secrets concerning the neutrino were revealed. The neutrino discovered during Pauli's lifetime was the anti-neutrino to the electron neutrino and eventually it would be learned that three types of neutrinos exist. The term for types of neutrinos is called the flavor of the neutrino and can be electron, muon or tau. The second flavor discovered was the muon flavor in 1962 and the third type was the tau flavor found in 1975. Many of the early detections were of the anti-neutrino flavors and the subsequent detections of the regular so to speak neutrino flavor would be found years later. One of the first applications of Pauli's neutrinos to an observed problem in the universe came from within our own solar system. Once neutrinos were discovered and studied sufficiently scientists began incorporating them into their work everywhere. Whether it was a purely mathematical model or actual experiments a new part of the universes make had been discovered and it was time to look for it in as many places as possible.

One place it was found in large amounts is in our own sun and astronomers began measuring the solar output of neutrinos. This was a fantastic time for those who worked on stellar bodies as it allowed them to not only incorporate the new particle into their equations and models of stellar fusion but also to measure them as well. The nearest star to us and the only one we can directly detect neutrinos from is our Sun and so it became the blueprint stellar nuclear fusion furnace for neutrino study in the cosmos. The problem was that the number of neutrinos being detected as coming from the Sun was only around one third of what theory predicted and this became an area of interest around the mid to late 1960s. Scientists new about the three flavors of neutrinos and that the Sun should be producing the electron flavor neutrino; they were only able to measure one third of the electron neutrinos they desired to find. The reason that they could not find enough neutrinos was because of a curious property of neutrinos called neutrino oscillation.

What this is describes how a single neutrino can change into another flavor neutrino over time. The rate of change between flavors and the order through which the neutrino progresses in terms of its flavor is dependant on the starting mass and flavor of the neutrino. As the neutrinos were created in the Sun they were all starting out as electron neutrinos but as they travelled to Earth experiencing multiple variables along the way they would slowly change into the other two types of neutrino flavors. The neutrinos do not really change from one flavor to another instantly like flicking a switch on or off instead they undergo mixing. The mixing of neutrino flavors corresponds to the amount of each neutrino flavor and its dominance within the neutrino. A simple example can be thought of like this and involves all three flavors: one neutrino is 80 percent electron and 10 percent of muon and tau, the overall flavor favors the electron version. As the neutrino evolves over time through neutrino oscillation it may change to a dominant muon or tau form but will always possess some of the other flavors as well. Eventually the neutrino will recycle itself so that it ends up with roughly the same ratios of flavors as it possessed when it was first measured.

By accounting for this change in neutrino flavors scientists were able to find the missing two thirds of the neutrinos and once they had built detectors to measure the amounts of these two other flavors arriving from the Sun they had the physical observations necessary to back up the theory. Since this time the neutrino oscillation phenomenon has been studied more rigorously and the ways in which neutrinos change their flavors is better understood. The curious thing about the neutrino oscillation is that it does not appear to be perfectly constant and will change depending on what the neutrinos travel through from the point of production to the point of detection. The rate at which neutrinos oscillate is slow and in many cases emitters and detectors must be placed kilometers apart in order to give the neutrino enough time to change its flavor. This is in part due to the fact that neutrinos travel at a significant fraction of the speed of light and so they cover a lot of ground very quickly. The oscillation is believed to be a quantum event and so grants scientists a way of studying a quantum mechanical behavior on the macro scale of a laboratory. It is obviously hoped that these neutrino studies will reveal some of the additional secrets of quantum physics as it applies not only to the neutrino but also to other particles as well.

Neutrinos have also helped to solve some of the problems concerning events like supernovae in the universe as well. The end of a stars fusion burning life is marked by dramatic changes to its core and the surrounding stellar plasma that makes up its atmosphere. The stars with masses close to our own Sun and smaller will not explode as their fuel source runs out but will simply live on for trillions of years or more slowly cooling after they have shed their outer atmospheres. On the other hand stars many times larger than our Sun with masses approximately 10 times greater than the Sun will undergo a supernova or hypernova as they run out of fuel to burn. The models available to scientists concerning supernovae were for decades flawed and did not explain the actual explosion itself properly. At the time it was not known that a serious flaw existed within the models but was made apparent when simulations of supernova events were carried out by supercomputers. The available data was fed into the computer for a hypothetical test star concerning its mass, age and temperature and allowed to proceed through its supernova phase. What was found however was that the stars under the old models for supernova explosions did not produce enough outward force to actually make the stars blow up. Instead what was found was a collapse due to the force of gravity once the outward push from the active stellar fusion was removed. No matter what variables they changed the stars would always collapse and not produce the outward push needed to create a shockwave that would drive off the outer layers of the star producing a violent supernova.

It was the neutrino that was in the 1960s postulated to be produced in large amounts during a supernova that would correct the problem. The idea was that during the collapse of a star at the end of its burning phase the inward pressure would crush all the matter in the core to the point where the electron degeneracy pressure would be overcome. The electrons would be stripped from their nuclei and the protons and electrons would be free to interact with one another and produce new neutrons through the electron capture method. During this process a neutrino is emitted and travels outwards from the stars interior to fly off into space in all directions. The addition of the neutrinos into the simulations corrected the problem of non-supernova generating massive stars. The neutrinos produced were created in such large numbers that they would help to drive the outward force of the explosion to an extent but also they would remove a significant portion of the stars gravity. The gravity of the star being lessened means that the core cannot pull hard enough to keep the exploding materials in place and therefore allows the simulation to produce a nice satisfactory supernova explosion. This was a solution to one problem concerning supernova models but it was untested by observation and therefore not confirmed until in 1987 the famous supernova 1987A occurred.

Supernova 1987A was found in the Large Magellanic Cloud about 168 000 light years from Earth and while first detected on the 23rd of February continued to shine for months with a maximum luminosity seen in May. The elusive neutrinos were detected roughly two to three hours before the visible light from the supernova could actually be seen and this confirmed the idea that neutrinos were produced in large numbers by supernovae. The reason why they were detected first also helps us to better understand the mechanisms at work in the interior of these massive stars as they become supernovae. The burst of neutrinos was detected hours before the visible light and lasted for about 13 seconds and it is these two numbers that help us to explain some of the supernova mysteries. All three flavors of neutrinos were expected to be produced in a supernova explosion and all three were found and recorded by multiple neutrino detectors around the globe. The results from the detectors were the proof needed to show neutrino production in large scale amounts in stellar environments during core collapse events and the two numbers also helped refine the models used. The first being how the neutrinos were generated and when during the supernova as they travel slower than light they should have theoretically arrived here after we saw the supernova. Yet this is not what happened and so we see some of the order of events inside a supernova as neutrinos are produced.

First the core collapse does create pressures and temperatures higher than the electron degeneracy pressure and neutrinos are created in copious amounts. Second the neutrinos are starting to be created before the full effect of the shockwave is felt and so can precede the light created by the explosion and reach us first. Third the neutrinos are notorious for hardly interacting with matter at all due to their attraction to the weak force and no charge. This results in the neutrinos having a basically unimpeded path up and away from the core whereas the photons created will be attracted to and hampered by the hot, dense collapsing plasma and be slowed on their way to the stars surface. This is the same phenomenon that can be seen during stars main sequence phase where neutrinos escape quickly on the order of a few seconds and the light will take tens of thousands of years if not more to reach the surface of a typical star. Once free of the surface the photons can return to travelling at near their maximum speed in a vacuum.

The neutrinos however have long since departed and will reach our detectors first and the same events play out in a core collapse supernova. The time scales for the slow down of photons is obviously slower as the collapsing core will explode outward quickly freeing the photons in seconds or minutes versus tens of thousands of years. Regardless the neutrinos travel at almost the speed of light and the fact that they have a head start means that as long as the distance between the supernova and us is sufficiently small the neutrinos will arrive first. A star very far away from us will allow the photons ample room to catch up to the neutrinos, overtake and pass them so that we would detect the light first and neutrino emissions second. The neutrinos were detected first and the time delays agree with the predictions made concerning how a supernova core collapses and when the neutrinos would be produced during that process.

Fourth and lastly the neutrino emissions were detected for over 13 seconds which means that the neutrinos themselves were created over time and not all at once. It also means that the star still had some effect on a percentage of the neutrinos slowing them down more than their neighbors and thus spread out the burst as it reached us here on Earth. This slow down effect is caused by the extreme densities present at the very core of the star which is destined to become either a neutron star or a black hole. The neutrinos are produced throughout the collapsing stellar plasma and so form in areas of different densities all of which can exceed the electron degeneracy pressure. The neutrinos produced at the bare minimum threshold for exceeding the electron degeneracy pressure are in the least dense regions of the collapsing core and can therefore exit quickly. The stars core however is reaching the neutron degeneracy pressure and the surrounding plasma is under far more intense pressure and therefore is far denser. The density here will scatter some of the neutrinos about as they try to exit the core of the star and head for the upper atmosphere.

This delay near the stars near neutron density core is what causes some of the neutrinos to leave the star slightly slower than those produced higher in the stellar plasma and will spread out the burst we record here on Earth. In this way we can see even more clearly the interior workings of a core collapse supernova and all of this was learned from observations made on the neutrino. If the star had been collapsing into a black hole the pressure at the core would have exceeded the neutron degeneracy pressure and the likely hood of any neutrinos created under these conditions escaping the core collapse would fall to zero. The duration of the burst and the type of neutrinos emitted should therefore be able to tell us the differences in physical conditions present just before a neutron star or black hole is formed. Observing the supernova remnant as either a neutron star or black hole will further help to verify what is happening in the core at the moment when the neutron degeneracy pressure fails and a black hole is created. Any of these measurements might allow us a glimpse into the processes at work in forming black holes and help to fill some of the holes in our knowledge concerning them. One way neutrinos can help with this is by alerting us to impending supernovae.

The ability to spot a supernova or hypernova and study it from the very first moments through until it fades to invisibility has been one of the most difficult challenges faced by astronomers. The benefits of such an event are enormous and as such it has been long desired to observe one fully but no reliable method for predicting when a supernova will occur exists. The trouble is the stars that we look at in the night are not as they appear today but rather as they were in the past as it takes light so long to reach us. Supernova 1987A for example did not explode in 1987 but 168 000 years before that at a time when early hominids were still evolving into humans. If you went back in time to the very spot of the Large Hadron Collider or the Max Planck Institute you would not find any brilliant scientists chipping away at the mysteries of the universe you would find early Europeans existing as hunter gatherers and while they may have been brilliant they were not looking at the cosmos with space telescopes and the like. While they undoubtedly did look up at the night sky and wondered what was going on up there they could not see supernova 1987A explode and even though if they could have viewed it with today's technology it would appear as a mature star at the end of its fusion burning life and ready to go supernova.

Yet it would not appear to us to explode as a supernova until 1987 and here lies the problem faced by today's modern astronomers and it is the same faced by our early ancestors. The problem is when you look at a star no matter how much it might appear to be ready to explode at any moment you do not really know if it is. Stars have exceedingly long lives when compared to not just a single human but to the human species as a whole. A star that looks fully primed to explode to us today with our most advanced technology might leisurely wait another couple hundred thousand years to explode, maybe a few million. In other words we cannot in any way simply sit and look at stars and hope they blow up for us. We also do not possess the sheer number of telescopes and observatories necessary to look at every star in the sky that seems to be a supernova candidate and wait around the clock for it to explode. This is where neutrinos can be one of the most useful tools to stellar astronomers who wish to view the full sequence of events surrounding a single supernova or hypernova explosion.

As we learned in the case of supernova 1987A the neutrinos will arrive at our Earth based detectors minutes to hours before the actual light from the exploding star does. This means that our passive neutrino detectors which do not have to be pointed at every star in the sky can simply listen to the arrival of a burst of neutrinos from a supernova. By the time we have interpreted the data as a supernova event we can be ready to view the star as it explodes. We still will not know the precise star that will explode but by comparing the arrival times of neutrinos from multiple detectors around the Earth we can determine which detectors spotted it first. Three detectors sharing information of neutrino counts and arrival times would be enough to at least figure out where roughly in the sky to look. This would give astronomers time to train as many telescopes of all wavelengths in that direction of the sky and spot the supernova as soon as it began.

At this point telescopes would be directly trained on the single star in question giving astronomers the best view so far of a supernova from start to finish. This early warning system of neutrinos would not only alert us to an impending explosion but also allow us to take precise measurements of the event across all wavelengths possible to better understand the inner workings of such explosions. Just as the neutrino data allows us to peer into the heart of a star to fine tune our models of stellar physics so to would the data we collected by looking at a supernova from start to finish. The neutrinos themselves from these events are of course still important to understanding the core collapse type of supernovae due to the fact that it is believed a star will release close to one hundred percent of its radiant energy in the short neutrino burst that precedes the visual spectacle. For all of these reasons neutrinos are one of the best tools we have for studying and understanding supernovae events.

The other application of neutrinos to cosmology is in studying objects that normally could not be studied by conventional means. Conventional means are of course things like optical telescopes as these have been used since we have been able to use them to replace our naked eyes in viewing the universe. Expounding on the optical telescope are other conventional tools such as infrared, ultraviolet or gamma ray telescopes and in fact any telescope along the electromagnetic spectrum. The reason these are seen as conventional means is that no matter what wavelength they employ to study the cosmos they all involve the use of photons. The drawback to using photons for certain types of observations comes from the fact that they are easily influenced by dense matter and other electromagnetic influences. The core of the Sun for example is dense enough and permeated by strong electromagnetic fields that a simple photon will not take a straight line path outwards but will meander for tens of thousands of years before it can escape. A neutrino on the other hand is not hampered by such obstacles and will exit the Sun in a matter of a few seconds.

The reason for this is due to the specific properties of the neutrino as first theorized by Wolfgang Pauli which are very small mass and no charge. By having a small mass the neutrino can easily travel to near light speed and with no charge it is not influenced by electromagnetic fields. This presents scientists with some obvious advantages in using neutrinos for observational instruments as opposed to photons or what we might call normal matter; think of protons, electrons and neutrons. In terms of the neutrinos benefits over photons it is the fact that they can pass through dense objects like the core of a star without being stopped. If we wanted to study stellar interiors directly both normal matter and energy would be stopped by the core and therefore yield no usable data. Neutrinos on the other hand will not be affected by the core and pass through easily allowing us to detect them on the other side and will be carrying useful data.

Neutrinos also possess another major advantage over energy and normal matter in probing the universe for secrets and this is due to the neutrinos weak interaction with just about everything in the cosmos. Neutrinos interact weakly at best with matter or energy and as such allows them to pass through most obstructions with ease. In fact the weakness of interactions of neutrinos and matter is precisely what makes them so difficult to detect and why even the most sensitive underground neutrino observatories record only a handful of neutrinos at a time. This means that if we want to study something that lies hidden behind some sort of obstruction we cannot use energy or matter to view it. Let us for a moment suppose we wished to examine an object of interest hidden by cosmic dust.

Any form of electromagnetic radiation would be easily absorbed or scattered by the dust and the dust would physically block the normal matter through impacts. Neutrinos can pass straight through the Earth without hitting any matter or being influenced by the dense core or magnetic fields and so diffuse cosmic dust would pose no real challenge to it. In the case of normal matter electrons would be easily influenced by magnetic fields due to their low mass and high charge and with low mass comes low penetrating power to pass through the dust. Protons are heavier making them less susceptible to magnetic deflection even though they carry a charge but their larger size also makes them less likely to sneak past any dust either. Free neutrons have no charge to be used in magnetic deflection but they also possess a half life of only 880 seconds and so would not last long enough to reach us anyway. The neutrino on the other hand is so small and fast with no charge that it can pass through most matter easily and allow us to peer to the other side of the cosmic dust.

An interesting fact about matter that people do not often know is that it is almost all empty space and here are a couple of ways to illustrate this fact. The first is one we have touched on briefly and involves the before and after photographs of a star and the neutron star it becomes. A massive star might have a diameter of hundreds of millions of kilometers but once it has undergone a core collapse supernova event the remaining neutron star will be only a few tens of kilometers across. In other words a star that could be as big as the orbit of mars or the asteroid belt will shrink to the size of a city; there is of course wide variability in these numbers as stars vary widely in the universe but the idea is sound. A second way to think about how empty matter is requires picturing a large stadium and trying to imagine a speck of dust at the very center. This is a picture of a single atom with the speck of dust being the nucleus and the entire stadium being the volume of the electrons orbital and therefore representing the physical boundary of the atom. The only reason why matter seems so solid is that particles react strongly to the forces that make up atoms and subatomic particles and prevent them from passing through one another.

This is of course the exception when dealing with neutrinos as they hardly interact with matter at all. If the speck of dust in the stadium is the nucleus of an atom even a simple one such as deuterium with one proton and one neutron we must remember that the neutrino is only one percent the size of the proton and therefore only half a percent of the total deuterium nucleus. Given the fact that the neutrino has no charge to be attracted to or repelled by other subatomic particles it will travel in a straight line as it passes into the electron orbital which is our hypothetical stadium. Now imagine for a moment an object only half a percent as large as a speck of dust travelling into this empty stadium and the chances it would strike anything at all considering the largest thing in this stadium is a single speck of dust at its center. The chances are close to but not zero, however they are so close to zero that experimentally we can consider them to be zero. The only reason why we can detect a single neutrino at all is because we use huge detectors and countless neutrinos pass through them every second and this still allows for only one or two neutrinos being detected at a time.

These are precisely the sorts of facts that have scientists interested in using neutrinos as a way of probing the universe as they can travel far longer distances than normal matter and will pass unhindered through most objects in their paths. While neutrinos from other galaxies might be lost on their trip to Earth it is possible that neutrinos from the center of the

Milky Way can reach us here on Earth. The high radiation and density of the galactic core as well as the incredible amounts of dust that lie in between us and the center of the galaxy might be overcome and allow us to get a better understanding of how the galactic core operates. Understanding how our own galaxy behaves and operates will allow us a better idea of how all galaxies function just as we study our own Sun in order to understand stars everywhere in the universe. The usefulness of the neutrino is only just becoming fully apparent to us but will no doubt be utilized to its maximum potential in the future.

Particle Accelerators

What the tiny subatomic particle that is the neutrino can tell us about the cosmos and how it interacts with matter is only one of many more subatomic particles known to exist and each one we discover helps to fill in some of the holes in our knowledge of how the universe works. This is where we must examine one of the most important scientific tools ever discovered and how it has and is helping to shape our understanding of matter. Particle colliders are of course what I am talking about and they have been used in various forms and in various ways for approximately a century now. Before we go any further let us get something out of the way which is namely the names that these apparatus might be referred to as. There is of course the descriptive particle collider and particle accelerator, also the mighty moniker of super collider and the most action packed of them all the atom smasher. There are probably more names by which these devices have been called and there are probably some enthusiasts that will dicker and argue over the precise usage of each of the terms I have provided. However it is of little consequence as all the names describe machines that we have built which accelerate very small things to very high speeds and that is all there is to it really.

The thing that is not so trivial is just how much one can learn from accelerating said small things to high speeds. As it turns out most of what we now know about matter and particles in general has come about after experimentation with these machines. The machines themselves come in two different varieties which are described by the mechanism by which they accelerate the particles. The first type is of course one we have seen previously named the Cockcroft-Walton generator and is of the electrostatic type of particle generator. The name describes precisely how it works as an electrostatic charge is used to accelerate particles to high velocities. The basic premise is that a charged particle in a vacuum will be attracted to or repelled by an applied electrostatic charge and the result rapidly accelerates the particle. The limitation to these devices is that the strength of the field must be increased if the velocity of the particle is to be increased and in order to achieve very high speed an unrealistic amount of current must be supplied. Still these generators are easy to manufacture and operate and are used quite frequently in a multitude of applications world wide. Typically suited to low energy applications they are quite common and if you have ever had an X-ray there is a good chance you have been on the receiving end of one of these types of accelerators. The Cockcroft-Walton generator named for its creators was used in the study of atomic nuclei in 1932 by which they succeeded in splitting a lithium atom into two helium atoms by way of smashing a hydrogen nuclei into it. By splitting the atom they had employed the use of an atom smasher and from this we can see the common usage for this name emerge in the early and mid twentieth century. The splitting of the atom had revealed some of the inner workings of atoms in general and therefore matter. This is the first example of a particle accelerator being used to probe the mysteries of matter and energy in the universe and would go on to be the foundation of early research in this field.

A great leap forward would be made in particle accelerator design that would also provide for a great leap forward in our understanding of the building blocks of the universe. This new design involves the second type of accelerator which is an oscillating field accelerator and overcomes the energy limitations of the electrostatic design. The oscillating field accelerator was conceived and eventually built by a Norwegian physicist by the name of Rolf Wideroe. Like many scientists in the early 1900s Wideroe was interested in nuclear physics as the field was expanding rapidly and was an inviting and exciting area of study with many challenges. The need for newer technology with which to probe the mysteries of atoms was what drove Wideroe to propose in 1927 the first designs for a betatron accelerator although the name betatron would be applied later. Wideroe's design would go on to revolutionize particle physics as it could achieve energies far beyond that of an electrostatic accelerator. For comparison an electrostatic generator could be used to create energies around 30 MeV and a betatron could achieve energies up to 300 MeV. An MeV is a mega electron volt and the mega denotes a value of one million; therefore 300 MeV is equal to 300 million electron volts of energy. The betatron was therefore ten times more powerful than an electrostatic accelerator and could increase the raw atom smashing force of the particle by an order of magnitude. Wideroe came up with an idea for a new accelerator that used a magnetic field to accelerate particles namely electrons in a vacuum tube. His genius would go on to be developed into what people later called the betatron and would ultimately lead to the most elaborate machine ever built by humans the Large Hadron Collider.

The person who picked up on and perfected Wideroe's idea of using magnetic fields to move particles was a German scientist by the name of Max Steenbeck. It was in 1934 that Steenbeck invented the world's first betatron at the Siemans-Schuckertwerke in Berlin. The way a betatron works is by placing an electron into a vacuum tube that has a circular shape called a torus which can be thought of as a doughnut or inner tube. The electron is then exposed to alternating magnetic fields that will change polarity as the electron passes by them. In the first instance the electron is attracted to the magnetic source and accelerated. After the electron approaches and passes the magnetic source the polarity is reversed and the magnetic field now pushes the electron away further accelerating it. It will then approach the next source of magnetic energy and the whole process is repeated. The fact that the tube is circular means that the electron can be made to orbit many times speeding up continuously as it moves in between magnets. This is where the advantages of an oscillating field accelerator are evident over an electrostatic one. The first advantage is that the whole apparatus can be smaller in size and achieve the same level of energy. For example a circular machine can be made to orbit the electron numerous times lets say ten times which means the electron has travelled ten times longer than the circumference of the vacuum tube in the betatron. In order to make an electrostatic accelerator of the same power you would need to make a single long tube the same length as the number of orbits multiplied by the circumference of the betatron and in this case a single tube ten times longer than the circle of the oscillating accelerator. These numbers are not by any means exact as there is a huge amount of additional variables and forces at work but in general the idea presents itself as a simple comparison in terms of machine size.

The second advantage that an oscillating accelerator has over an electrostatic one is the power needed to move the particle as the latter needs ever greater amounts of electrostatic force to move the particle faster and faster. An oscillating accelerator can use lower powers coupled with patience to achieve higher particle velocities and therefore energies. The amount of energy put into the betatron can be less because it only needs to pull and push the electron a little more each time it orbits the machine in order to accelerate it faster and faster. The longer you wait the faster the electron will go until the maximum potential for the energy you have input into the machine is reached. The only things you need to do are precisely and rapidly switch the magnetic fields affecting the particle and you can move it much faster than any electrostatic accelerator. This form of particle accelerator with its use of alternating fields is the basis of all modern high energy super colliders and has provided scientists with ever higher amounts of energy with which to smash atoms. The ability of the betatron to use more raw force in colliding particles means that it was able to look ever deeper and consequently ever smaller into the nature of matter. Like the electrostatic accelerator our knowledge of the universe would be undoubtedly less without it and like its predecessor comes with some limitations.

Wideroe once again came up with a solution to the problem in 1928 when he published a paper just one year after his 1927 betatron work and based off of discoveries he made from it. What Wideroe proposed was to use radio frequencies to help control and accelerate the particles inside the machines and thereby reduce the voltage and overall power required to accelerate particles. Just one year later after Wideroe had successfully built, tested and published his findings on the subject an American scientist named Ernest Lawrence would use his ideas and build the first cyclotron accelerator. The cyclotron is a device that can generate more energy for a particle and therefore increase the potential atom smashing power. The overall design of a cyclotron is the tried and tested method for particle acceleration of the type of devices first designed by Wideroe and involves the pulling and pushing of particles. Once again we see higher energies involved and therefore more possibilities for uncovering the secrets of matter. The last step in the evolution of particle accelerators will ultimately include all of Wideroe's ideas working in synchronous fashion to produce the highest possible energies of particles ever by humans.

The most modern of all the accelerators and most powerful as well for helping to figure out how the universe works is the synchrotron particle accelerator. Again the concept was invented by Wideroe in his proposal of using radio frequency to help control and power the particles as they travelled around the machine. Cyclotrons had been a good intermediate step in providing more power with which to smash atoms but ultimately would not be able to reach the incredible energies necessary to move particles to near the speed of light. Scientists view the power of accelerators in electron volts or eV as a way of showing how much speed the particle has gained. A simple way of understanding this comes from the idea that a particles mass is equivalent to its energy.

What this means is that in order to move a particle faster it will require you input more energy and although a simple proton might not have increased in size physically in a sense it appears to have grown more massive because of the greater amounts of energy to move it. When approaching the speed of light in a vacuum it is possible for particles to come close to and in some cases exceed their rest mass and the only way that we currently know of to do this is to use a very large synchrotron accelerator. As a particle travels faster its energy will grow and in order to reach speeds close to the speed of light you will need billions or trillions of electron volts worth of energy to push those ever heavier particles. The power of an electrostatic accelerator might be 30 million eV, a betatron might be 300 million eV, a cyclotron 600 million eV but the power of the world's largest synchrotron outpaces them all. The Large Hadron Collider has achieved a power of over 13 trillion eV in 2015 and has easily become the world's strongest particle accelerator.

One of the advantages to synchrotrons is that they do not require the magnetic field needed to control the particles to be present over the entire accelerator ring but only over the actual beam path. A cyclotron might use a magnet with a diameter of meters where as the Large Hadron Collider has a diameter of ten kilometers but a beam aperture width of only a millimeter. Small orbits in cyclotrons means more magnetic field strength needed to control the particles due to lost energy created by high speed particles travelling in tight circles whereas in synchrotrons gentler curve can be achieved due to high diameters. Cyclotrons also cannot accelerate particles close to the speed of light because the increase in velocity of the particle will push the particle out of phase with the radiofrequency driving it and the beam will be lost. A cyclotron the size of the Large Hadron Collider would be impractical due to the incredible amounts of energy necessary to maintain beam path and power the magnetic field. This is a somewhat simplistic view of the problem that does not cover the exhaustive topic of accelerator design limitations but will suffice for now. What it does accomplish is simplifying the problem so that one can accept the fact that in order to reach velocities close to the speed of light you must use a synchrotron. The use of a synchrotron drops the required magnetic power needed to push electrons, protons and ions in the direction and the speeds you want to create high energy collisions. A synchrotron will use more power than other accelerators overall but only because it is exceeding their capabilities and driving particles to very near light speed.

In the end it really is about achieving these very high velocities and therefore very high energies as measured in electron volts that matter as this is what allows us to probe ever deeper and smaller into the workings of the universe. In my opinion the universe is in fact very fair about how it shares knowledge with curious minds as it keeps no secrets so long as you are smart enough or work hard enough to uncover them. If you are smart enough, work hard enough or usually both then the universe will happily reward you and tell you more about how it all works. This is the basic premise behind all particle colliders as people have had to be smart to build better machines and have had to work hard to smash things together at greater speeds. The reward comes in new information on the nature of particles and how they are both created and destroyed in these collisions. To date the Large Hadron Collider is both our smartest and hardest working collider that we have offered up to the universe in exchange for answers. The universe in turn has been kind to us and given up some of its details on the inner workings of matter. What follows is a summary of the Large Hadron Collider itself, what it is capable of and some of the motivations behind building it.

The Large Hadron Collider is first and foremost a scientific tool used to probe the most elusive structures of matter in an attempt to better understand how the universe works. The tool itself is a particle accelerator that was built under the Swiss Alps close to Geneva Switzerland and was constructed by CERN or the European Organization for Nuclear Research. The facility itself is massive and covers many square kilometers mostly due to the enormous ring shape that dominates the structure. The ring consists of a 27 kilometer long vacuum tube surrounded by superconducting magnets used to accelerate and control the motion of the particle beams inside them and is buried anywhere from 50 to 175 meters below ground. What many people may not know is that the entire collider is not one simple machine like the accelerators of old but instead is a complex of various particle accelerating apparatuses designed to work in concert. The entire complex consists of multiple synchrotrons, linear accelerators and detectors that all act together to take normal matter like protons and lead nuclei and make them perform however is needed.

The ultimate goal of the accelerator is to increase the velocity of matter in it to near the speed of light in order to generate enough force to see the smallest particles inside it. To this end multiple linear accelerators of different types can be used to inject matter into the main rings for collisions. The main ring itself is actually two beams that are used to move particles ever faster in opposite directions until they reach the desired detector where it is hoped they will collide. The particles being studied are so small that not one bunch of particles is fired but multiple particles in the hopes that a few collisions will be seen. There are also a series of boosters that will further accelerate the particles along the way or prior to injection into the main ring. Another interesting fact is that the Large Hadron Collider also contains a proton decelerator the purpose of which is to slow down anti-protons so that the properties of anti-matter can be better studied.

The initial concept of the Large Hadron Collider was first proposed in 1984 where a group of scientists held a conference in Lausanne Switzerland and discussed what type of collider would be built and what sorts of particles and collisions it would study. The collider went live in 2008 and had its first test run which also resulted in its first mishap caused by a magnet quench that damaged a part of the ring magnets, vacuum tube and dumped liquid helium into the facility. A magnet quench occurs when the cooling system around a superconducting magnet fails and the magnet is essentially reheated above its normal operating specifications. The result was a loss of liquid helium coolant some of which was vented into the rings which exploded as the gas expanded. The vacuum inside the tube was also lost as the helium gas contaminated the vacuum tube itself during the incident. The necessary corrections were made and the facility was back online 14 months later.

The accident though unfortunate helps to explain another key feature of the LHC which is its need to be cooled to around -271.3 degrees Kelvin in order to operate properly. This extremely cold temperature is actually colder than the background temperature of space and illustrates just how cold the facility must be kept around the magnets. The entire facility is more or less permeated by a giant cooling system as the entire ring assembly utilizes thousands of magnets ranging in size from a few meters in length to 15 meters in length. Some of the largest magnets bend the beam as it travels around the rings and works in much the same way as we have seen in previous accelerator designs. The smaller magnets are used to focus the beam and lastly to increase the chance of particle collisions by making the particles clump together tightly prior to impact. The accelerator cannot be used at full operating power until the magnets have been sufficiently broken in and are therefore able to handle the greater energy loads required of them. This is almost exactly the sort of thing that your cars engine goes through as you must slowly over thousands of kilometers increase the maximum speed of the engine in order to ensure it will safely handle the highest stresses you can place on it. The magnets once fully broken in allow the facility to operate at its full power of seven trillion electron volts per proton beam or 7 TeV. When the two beams converge and impact each other the total power is pushed to 14 TeV but in fact can go much higher when colliding ions such as lead nuclei.

The second function of the collider is designed to probe the depths of matter by smashing very heavy objects together and very heavy is described as something far more massive than a single proton. In the case of the Large Hadron Collider it was decided to use lead nuclei for their stability and a collision from the two lead beams will produce a total energy of 1150 TeV. The different types of collisions will yield a wealth of information on the make up of matter at the smallest levels and highest energies. The reason this type of information is important to the scientists working with the LHC is that these are the sorts of conditions believed to have been present shortly after the Big Bang. It is commonly thought in the Standard Model as it is elsewhere that the universe began in a single huge explosion and that after the explosion dramatic changes occurred in the very nature of matter in the universe. If the universe explodes at time equal to zero then the Large Hadron Collider is hoping to recreate the time frame approximately one trillionth of a second after the Big Bang occurred. In this time scale it is believed that some of the four fundamental forces were combined into one and that truly fundamental particles existed that would go on to make up the matter we see today. Specifically it is thought that the weak force and the electromagnetic force were one at this time and has been called the electroweak force to encapsulate both.

At this time after the Big Bang it is believed that matter such as protons and neutrons did not exist yet and in fact it would be seconds to minutes after the Big Bang that these particles would come into existence. Instead a hot and dense plasma of quark and gluon particles is thought to have filled the universe and constituted the matter portion of the universe along with electrons and neutrinos. It is specifically this quark-gluon plasma that scientists are interested in exploring using the two lead ion beams. The reason the lead ion beams can produce more total energy per collision than the touted maximum 14 TeV of the machine is because lead ions contain many nucleons. The total number of nucleons in each ion and in the total beam far outweighs a simple proton and so the collision energy is much greater. This huge amount of energy is thought to provide suitable quark gluon plasma for study surrounding the early universe after the Big Bang. The proton beam collisions are hoped to yield more insight into the general nature of matter by vivisecting the hadrons that make it up. So we have a general understanding of the biggest most complex machine ever built by humans and what it does. We also have an understanding of the history of these machines, how they work and the general goal of operating them. Now we come to the scientific good stuff and begin to talk about what they have found out concerning the make up of matter, how matter interacts and the newest findings from the Large Hadron Collider. In order to do this I will now explain the basic understanding of matter as defined by the Standard Model of physics. So let us start out by describing the Standard Model in the first place.

The Standard Model

The Standard Model of particle physics is an attempt by scientists around the world to explain how the universe works by incorporating all of the known particles, their interactions with one another and the four fundamental forces into a single coherent theory. The work has no real starting point in time as scientists have been attempting to understand the universe at its simplest levels since before recorded history and determine the interactions taking place throughout it. Steady progress has been made in modern times and has led to a greater understanding of everything in the universe including atoms, photons and how we see them behaving in space. The desire to link all of these things together into a single understanding has been around for a while, but in the middle of the twentieth century a much bigger push towards specifically solving everything began. The purpose driven desire to reduce everything to a single theory of everything was in a sense done by the 1970s when the form of the Standard Model that we know today took shape. The reason scientists like the Standard Model on the whole is because it has the ability to be tested through experimentation and built upon from those experiments. Tools such as particle accelerators are able to probe matter to a degree that allows for the discovery and identification of various subatomic particles that fit into the Standard Model.

In the 1970s quarks had only been theorized as a particle fitting into the Standard Model but by performing various atom smashing tests teams around the world were able to find strong evidence for the quark. In this way physicists propose new particles and conduct tests to find them in order to incorporate them into the Standard Model. The goal of those who favor the Standard Model is ultimately discover all the particles in the universe, all of their force carriers and how they interact with each other. It is believed that this will allow for a complete picture of the universe and how it works as proponents of the Standard Model think the universe is made of subatomic particles and understanding them will teach us about all the other forces and interactions we see day to day. This ties back into the only thing we know to be real and undeniable in the universe which is of course observations. I will stress this again as no matter what anyone says the observations are simply never wrong and it is the interpretation of the observations that evolves in science. To this end the understanding of all particles and their interactions is hoped to provide all the information needed to explain things viewed anywhere in the cosmos.

This is of course a very tall order and one that although the Standard Model has had much success with has not beaten. In fact the Standard Model falls short on a great many things such as: gravity, the acceleration of the expansion of the universe and many exotic objects like black holes. In fairness to the Standard Model however it has had significant success when explaining matter on the subatomic level and like quantum theory cannot be ignored. I mention quantum theory because much of the Standard Model and its world exist down at the tiny and ultrafast realm of quantum mechanics as well. The two are of course used to explain one another and so adds another layer of credibility to both so rather than focus on the short comings of the Standard Model let us now look at what it has discovered.

When I say discovered I am of course being generous as it is not the Standard Model that discovered all the subatomic particles it is merely a classification scheme of sorts to organize them into one simple system. The structure of matter and energy as we have seen existed long before the mid to late twentieth century and already incorporated everything from, electrons, protons, neutrons, neutrinos and photons just to name a few. The Standard Model has done a good job of arranging everything into a simple system from which further experiments can be done and new theories and particles predicted. As such the subatomic world is grouped in three ways which are the actual particles themselves, the force carriers that are associated with each one of them and the fundamental force that each of them corresponds to. This to many is the beauty of the Standard Model because by placing everything into neat little groups and subgroups scientists such as those at the Large Hadron Collider are able to pin down and study single particles at a time. The LHC has already yielded tremendous amounts of data concerning the newest particle in the Standard Model which is of course the Higgs Boson. By tremendous I mean they have found it which is in and of itself a huge leap forward for particle physics and the Standard Model. We will get to the Higgs Boson and the Higgs mechanism later but for now I will show you the groups of particles, their force carriers and the fundamental force they belong to.

First and foremost if you are new to the world of subatomic particles and the Standard Model you will be exposed to a great number of new names and ideas but don't panic. The nice thing about the Standard Model even if it is revealed to never fully explain the universe is it does a great job of explaining matter. You will soon learn everything the Standard Model knows about matter simply, easily and by the time we are done here none of the new particles will seem hard to understand. For anyone who works in particle physics you can skip this section as the necessary simplification I will use means you could probably spend your time better getting a nice hot cup of tea. The first thing we must do is divide what the Standard Model knows about the universe into three groups. These are particles which make up things in the universe, the fundamental force to which each of these particles is associated and the mechanism by which that force acts also known as the force carrier. The second thing we will do is group each of these three things together so that we can see how all of them act in concert as one.

First the particles are divided into two groups consisting of quarks which we have already heard of a little bit and leptons which I have not discussed. People were aware of quarks first even if they were not aware of it as quarks are what make up solid matter like protons and neutrons. Water which was postulated to be one of the elemental forces in ancient times is two thirds protons by atomic nuclei and although leptons were present in that matter no one knew it. Neutrons were also present along with more protons and certainly by weight and mass the quarks were the heavy weights in matter to our ancient relations. The word hadron which appears in the name of the Large Hadron Collider has its root in the Greek word for bulky and leptons was taken from the root word for thin. In modern terms quarks make up hadrons which make up the bulk of an atom at least by mass. Quarks come in a range of types but the first and best known are the up quark and down quark which we can use to make protons and neutrons if we have three of each. A proton for example has two up quarks and one down quark while a neutron has two down quarks and one up quark.

The other key feature of quarks is that they all have an associated charge divided into thirds of a single unit of charge. An up quark has two thirds of a positive electric charge and a down quark has one third of a negative electric charge. So with a little simple math we can see that if you add the two up quarks in a proton you will end up with four thirds positive charge. We know a proton has a positive charge of plus one and so we must take some charge away to balance this to one positive charge and not four thirds positive charge. This is where the down quark comes in with its negative one third electric charge where adding it to the four thirds reduces the charge fraction to three thirds or three over three which is one. There we now have a nice neat plus one positive electric charge on our protons just the way we are used to.

So let us look at the neutrons we are used to as well and add up their quarks. A neutron contains one up quark so that is two thirds positive balanced by two down quarks each adding a negative electric charge of one third each. We get two thirds positive plus two thirds negative equals a net electric charge of zero thirds or neutral. This is how up quarks and down quarks make up both protons and neutrons in ratios of three quarks per hadron and with positive one and zero electric charges respectively. These are the standard if you will two quarks although four more kinds are known to exist but do not worry they can be summed up fairly quickly. The next set of two are the charm and strange quarks and it is simplest to think of them as heavier versions of the up and down quark. The last set of two known quarks are the top and bottom quarks which are even heavier still than the charm and strange quarks and correspond respectively again to the up and down quarks. That is it; no more discussion is needed on these four other types of quarks here as they can be thought of as bigger heavier versions of the up and down quarks. In addition to this the charm, strange, top and bottom quarks are typically not found naturally occurring in nature instead we see them almost exclusively in particle collisions and the like. On top of this even inside a super colliders detectors they tend to exist for a mere fraction of a fraction of a second in many cases and therefore are not understood nearly as well as their lighter brethren. We can obviously talk more about all the quarks but for now we are simply cataloguing the known parts of the Standard Model.

This brings us to an even lighter type of particle the lepton which also helps to make up matter and is probably a more abundant form of particle in the universe over quarks. Leptons were also known to our ancient ancestors but not as matter but rather as energy or the gods. The most common lepton of which they were aware was by far the electron and this tiny particle would have been seen by them in the form of electricity. Since electrons make up lightning our ancestors saw them frequently enough but did not think of them as matter possessing mass. To them it was a form of energy, destructive and certainly one of the fundamental elements such as water or fire. Often times these little leptons were thought to be the handy work of the gods and of course everyone is or should be familiar with at least one thunder god, Thor. After all we named Thursday in his honor as it is known as Thor's day; Wednesday was also named for his father Odin who is known as Woden an ancient god of English and Germanic religion and so today we have Woden's day. People living in even more ancient times let us say cave people from that same English or Germanic area would have undoubtedly worn fur to stay warm. Fur when rubbed with certain substances, perhaps rock or glass will build up a static charge and is demonstrated in numerous science classes world wide. Static charges are of course a discharge of stored electrons which are leptons so our ancient ancestors still knew of leptons.

The electron is one form of lepton that like the proton carries an electric charge however with the electron it is not subdivided by smaller particles like the proton is with quarks. A proton uses as we have seen three quarks to make up its total charge while the electron on the other hand contains an entire one negative electric charge all by itself. The electron is incredibly light compared to the proton and weighs only $1/1836^{th}$ of a single proton yet carries an equal electric charge in terms of strength. Electrons also have heavier relatives just as up and down quarks do and like their relatives also exist for only a fraction of a fraction of a second. The first of the two heavier electron relations is the muon and the second is the tau which is even heavier than the muon, both have revealed less of their secrets to us than the electron.

The second type of lepton was in no way known to our ancestors as it was only just made known to us in the mid twentieth century more or less. First proposed to exist by Wolfgang Pauli in 1930 it would later be detected approximately a quarter of a century later in 1956. These little leptons have been present since the Big Bang but there is no direct way that they manifest themselves visually to us in day to day life and so our ancestors could not incorporate them into their understanding of the universe. We have discovered them and have found that they exist in the multitudes almost beyond rational thought. Trillions are passing through you every second and you are totally unaware of it and this is due to the fact that they hardly interact with normal matter and they are of course neutrinos. We have already discussed them at length and so a quick summary of them is all we need to place them in the lepton category of particles. They are almost but not quite massless, they travel almost but not at the speed of light, are tiny and have a neutral charge, plus they come in three varieties.

With that we have come to the end of our discussion of particles in the Standard Model as everything else in it is either a force carrier or a fundamental force itself. That being said we do have some grey areas to discuss concerning force carriers as they themselves can be though of as little teeny particles too. Why do we call them different things then? Well to answer that we would have had to talk to Paul Dirac who was an English physicist that spent a great deal of time developing our ideas of particles and quantum mechanics. Dirac is a very important figure in the field of theoretical physics as he developed many theories which we use today including string theory and so we will talk more about him later. Particles can be described in some ways by mathematical means as opposed to physical means. In other words when we can not directly observe and measure something and particles are a prime example of this as they are so small and move so quickly, we use mathematical terms to try and understand what they are up to. I would like to point out here that I said 'try' to figure out what they are up to as the mathematics does not always as they say add up. Observations are king and are simply put never wrong while mathematics on the other hand can be wrong easily and often. I am of course not talking about adding two plus two and getting four but rather am discussing the use of abstract mathematics to describe objects barely understood like subatomic and fundamental particles. I am also not implying that any self respecting mathematician would deliberately create false mathematics to solve a problem but rather they make mistakes based on the limited amount of information they have to work with.

This is of course a central problem in science as many times throughout history we see a particular mathematics being modified in some way to fit new observations. Contemporary science today is facing this problem daily as mathematical models still do not accurately reflect what we physically observe in the universe whether in a laboratory experiment or in the cosmos itself. This truth aside some particles tend to behave according to certain mathematical descriptions and one such set of descriptions known as Fermi-Dirac statistics applies to some particles in particular called fermions. Named for both Paul Dirac and Enrico Fermi they are things such as quarks and leptons, the particles they make atoms and are sometimes known as baryonic matter.

The other form of particles are the force carriers again named by Dirac and again being described by a set of mathematical relations. They are still particles but are smaller than a proton or neutron and are sometimes thought of as components of other forces in the universe because they carry the interactions between fermions and the fundamental forces or at least how contemporary science believes them to work. They were named Bosons by Dirac because they are again described by a set of mathematical equations called the Bose-Einstein statistics. The general name bosons can be given to force carriers although there are others. In general force carriers are very small and act as the intermediaries between things like quarks and leptons which means they are sometimes considered particles and sometimes not. This is due to the fact that some force carriers have mass while others do not and yet they seem to perform the same function in between particles.

A perfect example is the photon which is a massless quanta of energy and yet it is the force carrier of the electromagnetic force between say an electron and the proton it orbits. The reason the Standard model accepts this fogginess between particles and non-particles stems from the fact that the Standard Model also attempts to interpret things in the universe as fields. An example of this is the description of a particle as well a particle in some instances but as the excitation of an associated field in another. With respect to a single particle a description of matter as a single particle is useful but it can also be useful to describe the particle as a field excitation in order to describe how it interacts with other particles. When dealing with large numbers of particles it is often more useful if not always more useful to view particles as particles and avoid the hideously complicated description of large numbers of interactions as fields. The fields described by the Standard Model are called quantum field theories and they see the universe and everything in it as a set of fields that interact with one another through quantizations of energy from a particular field. It is this discrete quantization of a field that physicists call a particle; an electron for example is a quantized particle of a particular field whose interactions are carried by the photon which is also quantized. In this way the quantum field theory attempts to explain everything in the universe as a set of interactions of different fields with one another through quanta.

The Standard Model does not incorporate gravity as it has never found nor can it accurately create a particle to carry gravity and as such scientists of the Standard Model are looking for a theoretical particle they have dubbed the graviton that would be the force carrier for gravity and matter. The force carriers and their associated particles are known for many particles and so give support to the theory of the Standard Model and is what drives projects like the Large Hadron Collider forward. So let us look at what the Standard Model describes as the various force carriers for particles many of which they have been able to study at least partially in some detail.

So we will begin with the quarks as they are the most massive particles commonly referred to in the Standard Model and also the first we examined previously. The associated force carrier for the quark is the gluon and it is responsible for the interactions between quarks, protons and neutrons in the nuclei. The gluon is responsible for constructing the three systems of quarks that make up protons and neutrons and gluons possess mass as a result. In the case of the neutron you might think that a gluon would not be needed to hold the three quarks together as they have a net zero electric charge. Two down and one up quark have no electric charge and so do not tend to repel one another but the proton has a plus one electric charge which stems from two up and one down quark. There exists a net positive charge of one which means that the quarks are on some level repelling one another due to there two thirds positive charge on the up quarks and the only thing stopping them from flying apart is the force carrying gluon. More specifically it is the carrier that binds them together in order to make nuclei strong and resist flying apart. If you think about it this carrier must be strong as the force between two or more protons in the nucleus will make them want to fly apart due to the fact that they both possess similar electric charges. The fact that they do not fly apart means that the gluon is doing its job well enough to create very complex and heavy nuclei from light helium to heavy uranium.

The next force carrier is the photon and it is as we have seen different in the fact that unlike other particles has no mass and can travel at the speed of light. The photon is the force carrier between both quarks and leptons, such as two electrons which are leptons scattering off each other can produce a virtual force carrier photon. The force carrier can be felt by quarks and leptons together as we have mentioned already in the interaction of electrons with the atomic nuclei they surround. In addition to this the effect of the photon is felt between atoms as well when they bind into structures like molecules as it is the sharing of leptons or electrons between quark masses or nuclei that creates a molecule. The holding together of many atoms is mediated by the transfer of energy by the photon as the force carrier. The way in which this transfer occurs and at what level of interaction between atoms will also determine the nature of the substance it is carrying the force between.

For example is the matter you are studying a solid, liquid, gas or plasma and what are the properties of each? Basic science tells you that the behavior of different states of matter means it has in some ways different properties. You usually cannot compress a solid and solids make great conductors for energy for example tapping a hard substance and hearing the loud clear transmission of sound waves. A gas made of the same matter however can easily be compressed and is by comparison a terrible conductor of sound and these differences are mediated by the force carrier and the interactions between atoms.

The last known force carrier is the intermediate vector boson and it works with the weak force also known sometimes as the weak nuclear force. The weak nuclear force is so named because it is responsible for interactions between particles in the nucleus but is not nearly as powerful as the strong force which is also known sometimes as the strong nuclear force. It can be seen interacting with both quarks and leptons in a variety of ways in nature and the laboratory. The force carrier here is divided into two groups the W and Z intermediate vector bosons with the W further subdivided into W- and W+ versions. W bosons have either a positive or negative electric charge and magnetic momentum making them each others anti-particle and hence differentiating them by the positive and negative nomenclature. The Z boson is electrically neutral, has no magnetic moment and is its own anti-particle.

The overall effect of intermediate vector bosons is in radioactive decay and the energies associated with it. This takes the form of natural radioactive decay that has been commonly observed for decades in elements like potassium, thorium or uranium where by a sample of known size will decay to half of its original mass over a set amount of time. This idea of half lifetime we have seen before and are necessary for many radio isotope dating techniques is just one of the uses of the weak force. Besides allowing scientists to probe the nature of matter even a hundred years ago the intermediate vector bosons are involved in the nuclear reactions that take place in the cores of stars. Of particular importance to us are the reactions taking place inside our own star the Sun as it is what is responsible for releasing the energy we need to live. This can be simply pictured in the simplest of reactions where hydrogen is converted into helium and the lost mass is converted to energy which we feel as heat. This of course makes us aware at least at some level of the intermediate gauge boson in the universe as we have been aware of the Sun since before we were sentient. What took some time to understand about these intermediate vector bosons is how they do what they do and the jobs of the W and Z types are both involved in nuclear decay but in slightly different parts of the process.

In a typical nuclear transmutation for example it is the job of the W boson to either absorb or emit a neutrino which is associated with either an electron or positron. Again we see the obvious relation between the positive and negative naming system in the W bosons and the positron and electron involved in nuclear transmutations. The Z boson on the other hand is involved in a different aspect of nuclear transmutation which is the behavior of the neutrino involved in the reaction. The mathematics behind these nuclear transmutations dictates that various properties of the particles involved must be conserved in some way and it is through the neutrino that this is done. Let us look at a simple nuclear transmutation whereby a single neutrino is lost from the system and realize that the entire system has a certain amount of energy and other properties associated with it. The loss of the neutrino will take some of the energy from the system and in order to conserve energy without creating or destroying anything in the interaction the differences must be made up in the neutrino. Therefore when a neutrino is emitted from a system it will be the job of the Z boson to dictate what the energy, momentum and spin of the neutrino will be as it leaves. All of these variables can be different from neutrino to neutrino as the chances of any two neutrino emissions being identical are extraordinarily low. This process is sometimes known as scattering and can be observed simply by watching a single electron when viewed in a cloud chamber or similar device.

The cloud chamber was the invention of a Scottish physicist named Charles Wilson and as we have seen was instrumental in early particle physics. If a neutrino beam is shone through an instrument such as this it will appear as though an electron has suddenly been created and travels in a certain direction. Since we have not just witnessed the spontaneous creation of new matter which is of course impossible the electron must have existed previously in some state with a certain momentum. What the neutrino has done is scatter off of the electron and imparted some of its momentum to the electron causing it to move in the direction of the neutrino beam path. What this means is simple since a neutrino has no charge it will not have imparted any electromagnetic energy to the electron. Also these two particles are not affected by the strong force meaning no gluon interactions. Lastly the particles are low in mass as to almost possess none it can be deduced that no gravitational affects are involved in the scattering process. This leaves only the weak force to be responsible for the results and that must be mediated by an electrically neutral vector boson which is of course the Z boson. With the Z boson the W- and W+ bosons make up the force carriers for the weak nuclear force and explain some of the processes of radioactive decay.

As of now we have covered the strong nuclear force, the electromagnetic force and the weak force in terms of force carrier particles the only one left is gravity. Today's weather forecast is foggy with visibility down to zero meters. That is how the Standard Model sees the force of gravity as simply put it cannot. The force of gravity has been since the time of Newton and probably even before a bit of a tricky wicket when it comes to saying exactly what it is for contemporary science. Take Newton's time for example as he stated that gravity is the force of attraction between two objects famously and most probably romanticized in his sitting under an apple tree and having inspiration fall on his head. Newton quite brilliantly figured out that it was gravity that held the universe together whether it was an apple to the Earth or the moon to the Earth the force behind these machinations was the same. Centuries later Newton is still spot on right but Newton also never said what gravity actually was and today contemporary science is still in the same situation. There is no current model in modern science that can explain what gravity is other than the thing that pulls on everything. The Standard Model likes to use particles and the excitation of fields in order to explain which particle is responsible for which force. We have been discussing particles, force carriers and forces a little bit and now it is time to apply these rules to gravity and see where the Standard Model stands.

What the Standard Model tells us about gravity is that it is a force that affects all particles that have mass, such as electrons, neutrinos, protons, neutrons or whatever. The force that is gravity is apparently the longest reaching of any of the forces by far and then some. An electrical force for example is felt over short distances and so is a nuclear force but the force of gravity can reach trillions of kilometers in all directions and act upon all of the particles it finds. There is no way that the electric or nuclear forces from these particles can be felt trillions of kilometers away even from behemoth objects such as black holes. A black holes gravity however is a true force to be reckoned with and not just figuratively but an actual force, the force of gravity. So what makes gravity work in terms of the Standard Model?

The Standard Model says there must also be a force carrier for the force of gravity and has so created the graviton. The graviton would have its own properties just as other force carriers are felt by quarks or quarks and leptons then gravitons would only be felt by massive particles. Just as the other force carriers are responsible for certain behaviors of particles on both the micro and macro scales gravity would also have its boundaries clearly defined. The boundaries for gravity seem to be the entire universe as gravity is the force responsible for shaping everything from our own solar system, our own Milky Way galaxy and all the way up through the local super-cluster and finally into the filaments that make up the largest structures in the universe. The problem for the Standard Model is that while the effects of gravity and what particles it must act on is well defined the graviton responsible for these interactions has never been found. Moreover it would seem that the graviton may not even exist. The Standard Model has no answer save for a force carrier particle named the graviton and this is more than fine for the Standard Model as that is how it is constructed.

The fact remains however that no graviton has been seen and many believe that it may be impossible to ever see it. Even the mathematical models used to predict the nature of the graviton are suspect as they can only guess at what properties it must have. The best the Standard Model has come up with is a plus 2 spin massless particle smaller than anything we have ever seen. The method by which it is generated is foggy, the energy of the particle is foggy and where they come from and go to in order to make the force of gravity behave as we observe it is foggy. Nevertheless despite these obstacles the scientists behind the Standard Model are seeking answers through experiments conducted using the Large Hadron Collider. What they are hoping to find is ultimately the graviton but in the meantime they have found something else and it is something that the LHC was built to find. The thing they have found is the Higgs particle and while it is not responsible for gravity it does seem to be responsible for giving some particles mass.

The Higgs particle has been considered by many a triumph of the Standard Model as it appears to address some of the problems faced by the Standard Model in attempting to explain how the universe works. Namely in how particles have mass when according to the Standard Model they should not. The problem was evident in the early 1960s when it appeared that some of the basic building blocks of the theory were inaccurate. The Standard Model as we have seen is made of fields that become excited and give rise to things like particles. The problem was that according to the theory some of the particles were thought to be massless and yet they did possess mass as was verified through experimentation. The particles in question were the W and Z bosons which seemed to be violating their law of symmetry.

In the Standard Model symmetries are the forces that give rise to the fields that give rise to the particles. This is of course a simplified view but it at least lets us learn where we get our particles from according to the Standard Model. Symmetry a bit more specifically is a physical property or mathematical relation of a system that is preserved even if the system undergoes some sort of change. The simplest way to think of this is to picture a sphere and how it looks and then picture a cylinder and how it looks. If we view the two along one axis the sphere appears to be a circle head on and the cylinder does too. Now if we rotate both along one axis the sphere continues to look like a circle but the cylinder now appears to be a rectangle if we view it from the side. The sphere will look the same from any angle giving it spherical symmetry where a change in the angle of the cylinder will change how it looks meaning it does not possess the same sort of symmetry. The fields created in the Standard Model act in a similar way in that a particular symmetry can be used to describe the different forces. The properties of each force should be uniform everywhere and thus preserve its symmetry, but the Standard Model theory was predicting massless W and Z bosons instead.

The reason is a particles mass is related to its energy and therefore the distance that it can travel or how far it can influence things around it through the force it is associated with. So a massive particle should only act over a short range while a very light particle can act over a much larger range. If you study the force you can learn something of the particle that is associated with it. This is what led to the trouble with the W and Z bosons as it was known from experimentation that they only interacted over a very short range. Since they interacted over such a short range according to the theory they must be very massive particles indeed. The general idea being that the electromagnetic forces and weak forces become one at a temperature of a million billion degrees Kelvin as this would possess the unification energy required to fuse the two. The resulting force is the electroweak force and it is thought to have existed for a short time after the Big Bang. At high energies the electroweak force is strong and should have massless bosons. The electroweak force however does not exist in the everyday universe at present and it was postulated that something broke this force into two separate ones namely the electromagnetic force and the weak force. If something could break the symmetry of this field it would explain why the two forces are now separate and why the bosons associated with the weak force behave as they do.

This idea is not new to physics as certain laws only work under certain conditions and this is illustrated by looking at the work of Newton and Einstein. According to Newton's laws of motion anything in the universe can be calculated with accuracy unless relativistic effects are taken into account. At this time you need to use Einstein's work on relativity to explain the behavior of objects until you encounter extremely massive things like black holes. At this point Einstein's work fails and you cannot use relativity to explain the object and if you view the extremely small such as the quantum realm Einstein and Newton have trouble as well. In fairness to Newton more of his work does hold at the very small scale too. In this way only a narrow slice of the conditions are ever met to describe the universe using our current models. Going too small or too large seems to push beyond our known physics and the laws that govern certain aspects of the universe do not hold any longer. For example you can never use relativity to explain the quantum world and yet you can test the quantum world to realize that quantum mechanics are definitely on to something. Look at the large scale structures of the universe say a star or planet and now the quantum world cannot cope with what we observe. So this idea that laws do not hold under every condition and at every point in the universe is familiar to scientists and it was postulated something was breaking the symmetry of the electroweak force too.

The Higgs mechanism is the answer given by physicists for how this happens and how it makes some of the bosons behave as we observe them today. I say today because it is believed at different times in the history of the universe specifically right after the Big Bang bosons followed different sorts of behavioral patterns. Remember the electroweak force and remember that according to the Standard Model particles are excitations of fields. What this means is that a boson we observe today is merely an excitation of a particular field or set of fields together and therefore during the time of the electroweak force the fields would have manifested themselves in a potentially different way than they do today even though the fields are responsible for the same forces then and now. What the Higgs mechanism does is break the electroweak symmetry so that the two forces appear as we see them today and this affects the nature of the bosons associated with each force.

The theory was proposed by multiple parties in the year 1964 and one of the scientists to propose such a theory was a man named Peter Higgs. Higgs is a British physicist who became interested in the problem of mass and specifically what gave mass to particles early in his career. What Higgs proposed was a field present in the universe that could give rise to mass in particles when certain fields interacted with his new theoretical field. This theoretical field would of course become known as the Higgs field and would remain unproven for decades. Higgs believed that at the beginning of the universe a mere fraction of a second after the theoretical Big Bang started everything off particles had no mass. According to his theory a fraction of a second after that the particles would acquire mass through an interaction with this new field and the stronger the interaction with the field the greater their corresponding mass would be as a result of it. The way this worked was by giving mass to the particles that made up large forms of mass and that of course is atoms. By theorizing that the Higgs field would give mass to quarks and leptons in varying amounts mass could in turn be given to things like atoms and molecules that would go on to build the universe.

This may come as a surprise to many people and certainly to no one in the world of particle physics but most of a proton mass for example does not come from the three quarks that make it up and so the Higgs mechanism is responsible for only a small portion of its total mass. The rest comes from a series of interactions that we will examine later involving force carriers such as gluons. Even so this was a good step forward in explaining where some of the mass in the universe comes from and to a large degree the mass that people think of most when they think of matter. Higgs further postulated that the field necessary to perform these actions must be present everywhere in the universe and always possess a nonzero strength so that the field can be active when needed and thus generate mass.

The problem with his theory was that it was at the time un-testable and his initial paper was met with rejection by CERN. What was needed was a much larger and more powerful particle accelerator to create the energy levels necessary to look for signs of the Higgs field and ironically it would be CERN that ended up building the machine that found it. This machine of course is the Large Hadron Collider in Switzerland that would find the particle and announce those findings in July of 2012. The announcement was made not because scientists detected the Higgs field as this is impossible for our current level of technology. At present humans do not possess the ability to test any force or any field directly but can only view the interactions of these forces and fields with matter. Think of Newton sitting under his apple tree on that English summer afternoon and after you have finished soaking up the pastoral beauty of the scene you realize that Newton never saw the force of gravity or the field it creates. What Newton did observe is the affect that gravity has on matter namely the apple that fell on his head. If you study enough falling apples you can learn all sorts of things about gravity.

This is precisely what scientists at CERN were doing as they searched for proof of Higgs' theories. What they were looking for was identical to Newton's apple albeit on a much, much smaller scale and instead of fruit it was a boson particle they were hoping to find. Just as every other field has some sort of force carrier particle associated with it the Higgs field was suspected of having a similar force carrier particle that would provide evidence of the field. The hope was that a suitable particle candidate could be found that would provide evidence for the existence of the still theoretical Higgs field and therefore proof that the Higgs mechanism was in fact what gave some particles mass.

The particle was announced in July of 2012 based on previous experiments conducted at the Large Hadron Collider and included data on a potential candidate particle that appeared to have most of the characteristics predicted by the Higgs model. What was interesting about this find is that it not only resolved several previous problems with the Standard Model but it perhaps finds proof of the existence of the first known scalar particle. A scalar particle is one that has a spin of zero and other than the Higgs boson no particle is predicted to have or measured to have a zero spin. All particles have various properties one of which is the spin number associated with it which dictates how it and its corresponding field should behave. Our familiar quarks and leptons have a spin number of one half regardless of mass or charge whereas things like gluons, photons and bosons have a spin value of one. What this means is that they are not scalar particles and therefore the fields associated with them are not scalar either. The Higgs boson particle on the other hand is believed to be possessing of a spin value of zero making it a scalar particle therefore associated with a scalar field. Why is this so special and why does it matter if it is part of a scalar field?

The reason is based on the fact that in science and particularly things like the Standard Model and quantum field theory scalar fields have the same value no matter the position of the observer. This is known as coordinate independence and simply states that a value measured for a specific field will not change even if the position of the observers changes in space. No matter what they will always get the same value and this is illustrated by things like the background temperature of the universe. Here on Earth when we measure the background temperature of space we are able to look in any direction we want spherically around us. Also we are able to look individually at different distances from us which can be thought of as different depths in the sphere of our observations. What we observe through our measurements is that the temperature of space is uniform, or extremely close to it everywhere in the cosmos. We can then conclude that the temperature of the universe is scalar and that even if we took measurements billions of light years from Earth we would find the same background temperature. In things like quantum field theory and the Standard Model what this means is that the value of the Higgs field would be uniform everywhere we look too. If the field is uniform and scalar everywhere then the force carrier for it must also be scalar and that means a spin value of zero.

The scientists hope to definitively prove that the spin value of the Higgs is zero because that would mean the Higgs field is indeed everywhere in space and can therefore generate mass in the same way everywhere in space. This would mean that the matter we see here at our point in space and at our time in the evolution of the universe is identical to matter at any other time or point in space. Therefore the laws of our Standard Model hold wherever we look and we can further develop our theories of matter with confidence. This should underline the importance to the Standard Model of the Higgs mechanism and in trying to study it as much as possible. In a sense if the Higgs mechanism and all of its predications are proven correct then we have at least one truly unchanging building block with which to further study matter. In order for the Higgs mechanism to work with a scalar field of uniform value it by its own definition cannot create particles of the same type but different masses at different points in the universe. In other words a boson is a boson and matter is matter and that is that. If the theory holds true then the interactions of other particles and forces can be studied with more confidence as a sort of unchanging benchmark will have been found for mass in the universe.

The Higgs field is believed to break up into four parts that can be used to explain many of the particles we see in today's modern particle accelerator experiments. There are according to the theory two charged parts, one neutral and a fourth separate part. The two charged parts are believed to help explain the W- and W+ bosons as would be expected to these two bosons having charge. The neutral part of course would therefore correspond to the Z boson as it has no charge. The fourth part is what is thought to constitute the massive Higgs boson particle which was the subject of much particle hunting for decades. Each of these particles helps to expand on the Standard Model and how it describes matter and of course the interactions of different fields and particles with each other. As we have discussed particles have various associated properties to them such as energy which can be thought of as mass as well as things like charge. Now we will look at these properties in more detail in order to understand the various ways that particles are described by the Standard Model and how some of these properties interact with the universe around them.

The first thing we will examine is the mass of these particles and in order to do this we must first discuss mass as it relates to particles and the structures they represent on larger scales. So what is in this case a particle can be thought of as anything smaller than an atom and in fact smaller than a proton or neutron for that matter too. Some particles like leptons are already smaller than atoms and according to modern science are thought to be indivisible and this is most easily viewed with a simple electron. Smaller than an atom are also things like protons and neutrons but they are not the smallest particles in the nucleus as we have learned over the decades that these subatomic hadrons are made of smaller particles still. It is interesting to note here that there was a time in history when all of science was convinced that particles such as the atom were the absolute smallest things in existence.

Later we believed that things like protons, electrons and neutrons were the smallest indivisible things in existence. Now most main stream scientists believe that things like, bosons, quarks, leptons, even the Higgs and the like are the smallest things possible based on our most recent and most powerful experiments. Of course some of you are following my line of thinking on this topic and are probably wondering the same thing as me right about now which is are there smaller particles than quarks, leptons and bosons? After all we were one hundred percent convinced in the past that it was nigh on impossible to have anything smaller than an atom and that proved incorrect. So what is to say in the future with far more powerful and exotic experiments we will not uncover something smaller or simply altogether different than what modern science understands as 'everything' today?

That being said it may be that we have reached the very bottom scale of particles in the universe because eventually it has to stop somewhere. The concept of cutting things in half forever is a fun little thought experiment but is of course nonsense when viewed absolutely and one quickly realizes that there is at some point an end to the search for smallness. Back to the particles smaller than protons and neutrons we find things like our familiar quarks and bosons which we think are the smallest you can get too. The interesting thing here is that not all of these particles have mass for example a quark does and is responsible for making up matter like an atom. On the other hand a photon is the force carrier of the electromagnetic force that helps to make atoms and yet it has no mass. The strange thing here is that no matter what goes into making up atoms, molecules and anything that is matter really it all appears to have mass and that simply is not true. If you take an apple and try to squeeze it you meet with resistance and the whole piece of fruit seems to be very solid. This is only an illusion though as most of the apple is empty space and if you were to compress it down to the density of a neutron star or black hole you would be left with something probably no bigger than a single hydrogen atom in volume or maybe even less. What is probably even more surprising to people is that the mass left over in your compressed apple is not even all mass although with the empty space gone you would be forgiven for thinking it was a solid dense ball of mass and nothing else.

This is where the first property of particles comes into play in the Standard Model and ties directly to the Higgs boson which is the particles apparent mass. If we look at a proton as an example we find that approximately one percent of the mass of the proton is actually what you might think of as solid matter. This solid part of the proton and when I say solid I only mean to convey the general macro level understanding of matter as a solid heavy material. This solid one percent can be thought of as coming from the Higgs mechanism and yet there is a large 99 percent left unaccounted for. This being the case how can we be getting excited over a particle that only gives matter one percent of its mass? The reason it is interesting is that the one percent is often necessary to explain why matter stays together the way it does and is brought about by a series of interactions between particles and forces. You simply need a way to generate mass in any amount for the Standard Model to not fall apart. That one percent is not created from nothing as this would violate several basic laws of physics namely the law of conservation of energy but instead it is the manifestation of mass from the Higgs field. The Higgs boson therefore is the carrier for the one percent mass that we observe in matter as it was contained in the Higgs field in the form of energy.

The other 99 percent of the matter that we know to exist comes about from the interplay of particles and forces between things like the one percent mass particles. This interplay of particles and forces all carries energy in some form or another and therefore explains the remaining mass as energy and mass are technically one and the same thing. This harkens back to the way particle physicists describe the mass of things they accelerate in colliders where the weight or velocity of the particle is not measured and instead it is converted to the energy it carries. So really Newton was right all along and way ahead of Einstein as what we are seeing in the particle accelerator is simply his F=ma and all Einstein did was once again prove Newton right. This is exactly the same thing that is happening inside baryons like the proton and neutron where one percent of their mass is derived from things like the Higgs mechanism and the rest is derived from kinetic energy. The kinetic energy inside of themselves comes from the quarks that make them up as they move around at quantum speeds and from the gluons that mediate the strong nuclear interactions throughout them. The combination of these two different forms of energy input when calculated as mass give scientists the remaining 99 percent and explain how the Higgs boson does not have to be responsible for one hundred percent of the mass. If you push on an atom and it appears to have 100% weight by mass it only means you have to supply 100% of the force necessary to move it and that you perceive as being mass which resists your push.

This helps to explain the first and most basic property of the subatomic particles which is their mass and why it is measured the way it is. Mass can be calculated as energy and vice versa so particles are listed in energies and not units of weight like kilograms. The second thing this does is help us to integrate things that should not otherwise be compatible with each other in terms of making calculations. The things that have mass are familiar to us and W and Z bosons as explained by the Higgs mechanism allow us to work with something tangible when discussing mass. Many particles however do not have mass but need to be incorporated into our models and calculations. The conversion of particles mass to energy or their energy which is usually kinetic to being thought of as mass can allow us to integrate everything into a single seamless whole in terms of the mathematics involved. Massless particles can be given energy values that are easily interchangeable with the energy values of massive particles.

Thirdly the mass of particles often times do not have definite values but vary up or down from moment to moment. This is not strictly speaking true and I will explain what I mean by it and why it is important. It is not true for all particles as some have been measured so thoroughly that little doubt can be left as to their true mass. It is true because some particles have not been measured sufficiently and so we do not know their exact values yet. The reason why this is important is because without knowing exactly what the values for every single particle in existence is we cannot fine tune our models to make perfect predictions and calculations for any desired outcome we wish to study. To date we say that values are calculated with this or that level of certainty and aim for what seems closest. This does not however mean that we have examined enough particles of all types in enough detail to nail down once and for all the true values for each. To make matters worse different experiments and different conditions often means that a value can change slightly. While this is usually due to equipment limitations or to human error it does mean that very large discrepancies between theories can arise and make our quest to solve everything just that much more cumbersome than it ought to be.

Remember a very small difference in mass at the quantum level may not seem like much until you scale up that level of particles to real world sized objects that are comprised of tens of quadrillions of them all at once. The small difference now becomes a very tangible problem in applying the quantum to the everyday macro scale. This has led to some of the biggest theoretical blunders in science and can create two opposing camps for how to solve certain fundamental problems in the universe. This is of course an aside for now but a very important one that anyone who works in the field of theoretical science will tell you is not only important but sadly not just confined to things like little mass discrepancies.

What we do know of particles is not confined to just their mass but also various other properties such as the spin of a particle. When thinking of spin and in fact when thinking of many words related to science in general it is best to not be too literal. This can cause some confusion as science is often times very literal and this is easily seen with something like the Large Hadron Collider. It derives its name from three incredibly obvious facts the first being that it is very big so it is called large. The second is that it works with Hadrons and so that was stuck into the name. Lastly it is designed to converge two beam packets of Hadrons into a series of repeated collisions and hence the collider aspect of the name. This of course is where science is very literal but when dealing with things like the quantum realm such as the behavior of subatomic particles with each other and often times on their own particle names words can be used that are not literal translations of what is happening. This is seen in the spin property of particles as it refers to angular momentum and the fact that particles have two types of angular momentum.

The first is seen as the spinning of the particle around something else such as an electron orbiting an atom and the second is the spin of the particle around its own axis. Both are forms of angular momentum and yet only one is classical and in a sense real. The easiest to think about is the orbital angular momentum which as we have explained can be thought of as any old electron moving around any old nucleus. This is a classical interpretation of a quantum effect as all the particles involved behave in quantum manners and the ways they move are quantum related such as making discrete jumps in energy levels from orbital to orbital. However the basic laws of physics that describe the motions and paths these particles are following can be solved classically. After all even a tiny atom still has a lepton moving in circular orbits around a central nucleus and the way you calculate what is going on when is the same.

The second type of angular momentum is harder to imagine as classically it should not really exist and quantumly it should not exist either. Let me explain this simply with a comparison of the two systems. First anything quantum and especially things like leptons which are thought of as the smallest type of particle you can get and is therefore an elementary particle are best described as a quanta. Remember the true nature of anything that is a quanta for a moment as it gets easily forgotten as people hear the word so often and begin to gloss over its real meaning. A quanta is simply the smallest unit package of a thing you can have. That may seem like rather a trite way of getting the meaning across but it works and remember even today science does not really know what is going on at that tiny level. All of the models modern science uses to describe matter which is real are based off of mathematics that are not necessarily right but were selected because they provide a decent way to describe what is happening. If you think I am speaking silly now remember that science is still divided as to what particles really are because at the quantum level they can behave as waves or particles depending on what we are doing to them. A debate still exists despite what either side says as to whether particles really are particles, waves or both. In summation the descriptions we have today of the very small quantum world are only descriptions and not one hundred percent fact as we have never seen or held a single quanta of anything.

The scoffing at such a notion is tenably warranted as of course you cannot directly see or hold such things as far as we know but the truth persists that since we cannot do exactly that we do not know exactly what we are dealing with. So if we translate this to quantum spin and realize that it is described as a kind of angular momentum we are faced with a kind of paradox. Angular momentum is reserved for things that move classically in circles because they are made up of objects large enough to orbit something or complete some similar motion. A quantized particle is essentially a point and therefore one dimensional meaning it cannot possess any size of any significance whatsoever. Remember that any atom is only occupying the volume that it does because of the repulsive forces its orbiting electrons exert on neighboring atoms and this is due to electron degeneracy pressure. Everything in an atom is technically a one dimensional point particle such as the electrons, quarks and gluons and everything about the atom only appears to have three dimensional volume because these particles fill a certain set volume of space. So how can a single point move circularly and therefore have angular momentum? The answer is simple it cannot. This is the paradox as a single particle is not even orbiting anything it is rotating about its own axis and therefore cannot have angular momentum of the classical kind either. While we are on the subject it is further impossible for any point like particle which exists so small as a one dimensional object to have any axis as it would require three dimensions to therefore have a center through which an axis could be drawn. Now that we have covered some of the little niggling problems let us look at why the term spin and the description angular momentum was applied to these tiny particles in the first place and why it is somewhat mathematically validated.

The term spin is also known as quantum spin and the number used to denote its value is referred to as the spin quantum number but many scientists simply call it the spin number as it is well understood to what they are of course referring. We have seen what the definition of orbital spin and classical angular momentum are and now we will look at how quantum spin was found. The effects of quantum angular momentum as related to the quantum spin of a particle were first found in 1922 by two German physicists called Walter Gerlach and Otto Stern. The two scientists devised an experiment to test quantum properties of silver atoms in magnetic fields. What they did was fire atoms through an inhomogeneous magnetic field and observed the patterns that were made when striking a target plate. If things like atoms have no quantum magnetic moment related to their spin then they should produce a relatively uniform pattern upon striking the target. This will be demonstrated through a vertical line on the test target as the random motion of the particles will allow them to be affected by the north and south poles of the magnets and create a nondescript pattern. What is observed however is a pattern of two definite and distinct groupings on the target sheet separated by a path of empty space with few to no marks on it.

The reason this happens is because of the quantum spin inherent to the particles in the atoms and the atoms themselves overall. Since the particles possess a spin factor which is quantum in nature the magnetic field will interact with it and the magnetic moment of each particle will be displayed on the target. Depending on the magnetic moment it will either end up in the top grouping on the target or the bottom one. The particles can be said to have either an up spin or a down spin as seen physically on the target. The reason a magnetic field is used that is inhomogeneous is due to the fact that if the particles were passed through a neutral magnetic field any spin in the dipole of the particle would be cancelled out. If the particles are viewed as classical dipole magnets they possess both a north and south pole and therefore any spin in the particle will feel equal effects of the magnetic field through which it is passing. The net result is that the push and pull effects of the magnetic interactions cancel and the beam of atoms will pass through the magnetic field in a straight line producing a single patch on the target completely unaffected by the field. The angular momentum for a particular axis seems to cancel the information of the other two and so for example knowing what the X axis of a particle is doing means you lose the information about what the Y and Z axis are doing. What is important here is that not only do particles have a kind of angular momentum associated with them but that it produces a discrete spectrum.

We now know what the quantized spin value for every subatomic particle we have discovered is and they take one of two forms. The first of the two forms is that of a half spin number and the second is that of a whole spin number. We know from a previous discussion that we can group things neatly into a few categories and the first are the quarks which have a spin number of one half. The second group is the leptons which possess a spin of one half as well. The third group is the force carriers which include things like gluons, photons and bosons and they all have a whole number spin. The gluons, photons, W and Z bosons all have a spin value of one while the Higgs boson also has a spin number that is whole it is set at a value of zero. The Higgs boson is the only known particle that has zero spin although it is still early on in the study of this particle and more than likely as we learn more about it some of the values associated to it will be fine tuned even if its spin remains zero. The separation of the patches on the target screen is determined to be a set amount and therefore has a finite size difference. The size difference being finite and discrete reflects the quantum nature of the particles and can be used to study quantum effects on the macro scale.

This was the first type of experiment that allowed scientists to study such quantum effects on the large laboratory scale just as modern day scientists are studying neutrino oscillation on the macro scale. This helps to demonstrate the two types of spin and differentiate between classical orbital angular momentum and quantum spin angular momentum. The importance of this work into quantum spin number is reflected in various aspects of today's modern technology such as atomic clocks and Magnetic Resonance Imaging equipment which all owe their existence to the idea that a particles spin is quantized and measureable. The direct observation of a particles spin is still of great importance all on its own as it remains one of the best pieces of proof for the quantization described by quantum mechanics.

Of even greater importance than atomic clocks or Magnetic Resonance Imaging machines is the work of Wolfgang Pauli that came after it. Pauli gave the world his famous Pauli Exclusion Principle in 1925 which would become one of the most important aspects of quantum mechanics. The Pauli Exclusion Principle states that no two identical fermions can occupy the same state at the same time. Fermions have one half spin integers and the example Pauli used was of the electron. In his explanation two electrons shared three of the four quantum numbers and occupied the same orbital around an atom. Due to the fact that the electrons could not exist in the same quantum state if they possessed all four numbers equally then they must have differing and opposite spins in order for only three of the four numbers to be the same.

The importance of this fact cannot be overstated as it is the basis of what holds matter up so to speak. When I say up I mean that matter is solid and does not simply fall apart on its own. The Pauli Exclusion Principle is the underlying reason why every day objects appear and are solid to us. Even on the atomic level the objects although filled with copious amounts of empty space per atom are in fact solid. The reason is that since the electrons in atoms cannot occupy the same quantum state they are kept separate from one another. Furthermore in highly poly electron atoms we find that not even all electrons can occupy one orbital as electrons approaching the nucleus would violate the Pauli Exclusion Principle. The net result is that electrons must occupy various orbitals of differing energy levels and thus give the atom a large overall volume. This was proven rather solidly in 1967 by English physicist Freeman Dyson and included both electron-electron interactions as well as nuclear-nuclear interactions. The interactions are characterized by the repulsive forces things like electrons feel for one another due to a similarity in charge.

These observations of the Pauli Exclusion Principle in action can be applied to the theory of white dwarf star, neutron star and black hole formation. In a white dwarf star the core is held together by the electron degeneracy pressure which prevents the core from further collapsing into a denser matter state. The Pauli Exclusion Principle is what is responsible for the white dwarf star core from collapsing any further and becoming a neutron star. It also helps to explain the Type 1A supernova as the resistance of the core to overcome its electron degeneracy pressure is what results in the outer accumulated layers of materials being blown off the star and the core remaining intact. In the case of neutron star formation we see a point where the electron degeneracy pressure fails and the Pauli Exclusion Principle seems as if it will also fail. In fact it does not and rather than let loose electrons violate the law it refuses to permit this and the end result is the formation of numerous new neutrons in the electron capture process. The final result is a neutron star not violating the Pauli Exclusion Principle and now held together with neutron degeneracy pressure. As Dyson demonstrated it is still the Pauli Exclusion Principle that is responsible for the solidity of matter even at this strange state and only by exceeding the neutron degeneracy pressure once again can we get an even denser form of matter.

The densest form of matter known to modern science is that of the black hole which is the result of a stellar core collapsing under more pressure and gravity than the neutron degeneracy pressure can withstand. As far as modern science knows the black hole represents the last step in the descent down the staircase of matter and density. Modern science also does not know what physics work inside black holes and so this might be a place where the Pauli Exclusion Principle breaks down or maybe it does not and a newer form of particle exists inside the black hole but still obeying the law set out by Pauli. The density of matter no matter whether it is a white dwarf star or a neutron star is still made possible by the Pauli Exclusion Principle being active. Only the immense force of gravity can turn matter into these exotic objects and in our normal everyday lives we thankfully are no where near gravitational fields strong enough to overcome the electron degeneracy pressure. This keeps our matter solid and stable creating the world we see around us not only in atomic form but in all forms. Getting larger than an atom will create things like molecules and crystals which involve multiple atoms sharing electrons and sharing space.

The sharing of space does have a limit however as atoms will not compromise their own personal space which of course is the volume of space that atom makes for itself with its electron orbitals. So even in large structures the Pauli Exclusion Principle is still at work demonstrating the fact that two identical electrons cannot exist within the same quantum space. To see this in action for yourself simply take two solid objects for example two wooden blocks and try to push them together so that they occupy the volume of just one block. There certainly is enough space inside of both wooden blocks as most of their volume is simply void of matter. But there will be no way for you as an ordinary human to get both wooden blocks to exist simultaneously in the same space at the same time and this is due to the electrons making matter solid.

These are all facets of the quantum spin number and what they mean to matter and the everyday world and while they are very influential on how the universe behaves they are not the only factors affecting particles. Another of the factors that determine how the Standard Model arranges particles is the charge of the particle as well. We have seen that some particles have positive charges as is the case with the up quark and some have negative charges as seen in the down quark. These charges are what we would call positive and negative but they are not entirely equal as the charge of an electron is a true electric charge of negative one. Whereas the charge associated with a quark is the result of something called color charge. We will look at color charge shortly but before we do that let us examine basic electric charge.

Electric charge is simply the force generated or felt by matter when placed in an electric or electromagnetic field. Although it is commonly said to be in effect in an electromagnetic field a pure electric field is possible too. Electric fields can be imagined in the case of two point particle charges the same way we would view a large multi-atom bar magnet. In the case of the bar magnet we see magnetic force flowing outwards from the north pole end of the magnet and drawn inwards to the south pole end of the magnet. In this same way a positive point charge will flow electric force outwards and a negative point charge will attract it inward. Since we know that the electromagnetic force can utilize both electric and magnetic fields at the same time we can observe these phenomena on the macroscopic scale. Any time you pass a magnet over a piece of conducting wire you will get a flow of electrons in the wire. This is the result of the electric charge of the matter interacting with the electromagnetic field from the magnet and shifting electrons around inside the wire. It is the charge of the elementary particle that is being seen here. The effect of this property of matter is utilized almost everywhere in the universe all the time and creates the cosmos we see today.

The reason is all matter is comprised of elementary particles which have a charge and the interactions of these particles with one another determine what large structures you get. The smallest particles will form things like protons and neutrons which in turn with other particles like electrons form atoms. The arrangement of these particles to form larger structures is directly related to their charges and must be balanced for matter to exist as it does. The proton for example is made of three quarks as we have seen and each has its own charge that must be summed in order to get the total charge of the proton. We know that a proton is made of two up quarks and one down quark and that each up quark has a two thirds positive charge and the down quark a one thirds negative charge. The proton must contain quarks in this arrangement for it to have a single plus one positive charge as an end product. If the proton had no down quark for example it would have a total of four thirds positive charge which would not balance with the one negative charge of the electron. Protons must be arranged in this one positive charge way for things like atoms to exist at all since electrons have a neat one negative charge. This arrangement of balanced neat single integer charges for protons and electrons is directly related to any large structure that can be created by combining atoms together. The molecular structure that makes up all life in the universe for example is only possible because of the ways that atoms share their total positive or negative electric charges with one another when bonding. The idea of charge is thus carried over directly to atoms as well and not just elementary particles and this is seen everywhere in the form of ions.

The term ion simply means a charged atom or molecule and the simplest one of course is a hydrogen atom which is naturally a proton. Any larger atoms however can also be ions if they gain an electron or lose one. If an atom loses one or more of its electrons it will have fewer total electrons in it than protons meaning the protons win out and the atom is a positive ion. If the atom gains one or more electrons it means that the electrons are the winning side and the atom is a negative ion. When atoms of differing charges meet up they behave like members of the opposite sex in biology and are attracted to one another. This attraction will lead to bonding which will form molecules and the number of atoms, the total charge per atom and the energy of the system will determine what configuration the atoms take. This is how you can make an almost limitless number of molecules from a finite number of atoms and this is evident in the complexity of chemistry in the universe considering really only 92 building blocks are available. All this is due to the charge of molecules, atoms, subatomic particles and finally elementary particles. This sums up charge overall and demonstrates its importance in constructing the cosmos but there is one more aspect of charge that we have not fully explored. This is of course color charge and it is a fundamental part of the Standard Model and explains why things like quarks behave as they do in terms of charge.

So what is color charge? Well first of all it has nothing whatsoever to do with color as that is simply a naming scheme that was devised to divide the three types of charge and so scientists picked the three elementary colors. If one is discussing the three elementary colors with respect to the human eye then we use red, green and blue. If one chooses to use a physical medium such as an artist would then red, yellow and blue are the primary colors. Almost always scientists use the red, green, blue color naming system for color charge in elementary particles. The idea of color charge is related to quarks and explains how they create the total charge of a hadron when combined in groups of three and also how they maintain that charge. The basic premise is in many ways theoretical as it is not fully proven but there is a good amount of explanatory data that can be used to interpret color charge as real.

The main difference between color charge and traditional electric or electromagnetic charge is that it has three values. Normal electromagnetic charge has two which is of course either positive or negative whereas color charge can have three which are denoted as the red, green and blue. This does not mean that there is really some kind of third charge like super positive, super negative or sideways charge for example it is simply a way of distinguishing the charges inside things like hadrons so that they balance out. Balancing out means existing as a single proton or neutron and having a conserved charge of positive one or zero respectively. What it does mean in terms of charge and quarks is that the quarks must always balance the total charge between them in a single hadron and in order to do this they utilize gluons. Just as quarks use gluons as the force carriers of the strong interaction they use gluons as the force carrier of color charge as well. Remember that it is the gluon that acts as the go between quarks inside a proton or neutron holding them together and generating the strong nuclear force. In this case instead of a field generating the strong force there is a color field being created as quarks exchange gluons. This is the key to the color charge theory and it involves the exchange of gluons freely from one quark to another.

To think about it simply imagine you have a proton or neutron and it of course is made of three quarks. Each quark will have at any one time its own specific color charge of red, green or blue. However the colors associated with each quark are not permanent and will change as the system evolves and in order to evolve gluons must be exchanged between quarks. A quark of one specific color will emit a gluon that travels to another neighboring quark and the end result is that both the donor and recipient quark have changed their color. Quarks can have either a red, green or blue color charge and the force carriers the gluons can have any of three anti-colors. The gluons carry anti-red, anti-green or anti-blue charges. Combinations of color charges result in a neutral or white color charge just as mixing red, green and blue light gives you white light. If all three colors are present then you have a neutral charge and this is extended to anti-particles as well. Similarly any single color and its corresponding anti-color will also be white or neutral colored.

The gluons which carry the different color charges between quarks are a little bit different as they always contain two colors. The first color is a regular charge color and the second is an anti-color such as red and anti-green. It is this dual color charge of the gluons that allow quarks to change color charge and still conserve the total color charge of the particle they make. The seemingly random way in which quarks exchange gluons actually leads to a balanced system as one quark emits a gluon of its color the recipient quark becomes that color. Let us say that a red quark emits a gluon and sends it to a green quark what will happen is the red quark turns the green quark red. In order to balance the system which would now have one blue quark and two red quarks the original red quark turns green. Therefore the color charge of the entire system is conserved as we are back to having one quark of each color charge. The gluon possessing a color and anti-color are what makes this possible as the colors and anti-colors it carries are carried to or taken from quarks. Here we can stop looking at different pairings of colors and anti-colors in relation to the color charge changes they produce in quarks. This is be cause there are so many different combinations and if you are someone who works with this sort of physics then your doing fine and understand these already. If you are someone who does not do this sort of thing then you do not need to worry about it as the information is at this point extraneous.

To sum this up quickly there are nine possible gluon charges since there are nine possible color and anti-color combinations yet the math works out to only eight. Regardless of this reality we now understand how quarks possess color charges and how gluons possess both a color and anti-color charge. Also we know how these interact with one another in order to conserve color charge for the entire system that they make up. The color charge fields between gluons that are created can be thought of as exactly the same as any electric field. The only difference is that the field lines do not deviate outward as much as in an electric field as the strong interaction keeps them more tightly pulled in-between quarks. It is here we see an interesting feature of the field that is generated by the quarks and the gluons that bind them as the field behaves somewhat strangely.

The field between quarks is a little different than other fields that you might be familiar with such as a magnetic field or gravitational field. In a magnetic or gravitational field the farther two objects get from one another the weaker the field strength becomes and this is evident when looking at two bar magnets or two planets. When the objects are close together there is a lot of magnetic field strength or gravitational field strength present and when they are separated farther and farther apart the field energy drops. In the case of quarks and the field that binds them we find the opposite to be true as the field apparently gets stronger as predicted by Quantum Chromodynamics. The simplest explanation of quantum chromodynamics is the theory of how things like hadrons are held together. Essentially quarks and the strong forces interacting inside things like protons or neutrons and how these interactions determine what happens to what in terms of the strong fundamental force. This is what we have been talking about with respect to color charge as the charge sharing that goes on inside things like protons involve the strong force and the name is derived from it. Quantum means quantum, chromo means color and dynamic means energetic or changing depending on the vernacularity of how it is used.

The theory of quantum chromodynamics was first constructed with its roots in the 1950s when numerous experiments and models observed or predicted a large number of new particles. The particles were hadrons which are considered to be fairly large as particles go and therefore it was argued that not all of them could in fact be fundamental particles. Fundamental particles are also known as elementary particles in common usage amongst physicists and denote things like electrons, quarks, neutrinos and the like. What makes them elementary or fundamental is that it is believed that these particles are one of two distinct things. The first being that they are in fact the smallest type of particle or excitation of a particular quantum field that you can have and thusly are named fundamental and elementary. The second possible interpretation of these particles is that they are simply the smallest that we as humans have been able to tease out of the universe.

What this means is that they may in fact be the smallest and therefore elementary particles of the field they are associated with or it might be that we have not been able to break them down still further. This would mean that there are in fact particles which are fundamentally smaller than the ones we understand still and we have not yet discovered them. The possibility that this second idea is true is very much validated by the very notion that we simply have not built a big enough and strong enough machine to detect them yet. Remember that until the construction and operation of the Large Hadron Collider the Higgs boson was only a pure theory and now it has some serious backing evidence to support it. So the notion that we can possibly crack up the little bits of matter that we know are there in some form or other a little smaller is definitely a reality and this includes all fundamental particles and the new Higgs boson as well. This does not mean that there is something smaller than say an electron that makes it what it is a very light and negatively charged particle. An electron might be an electron all on its own and doing quite well at it to boot.

That being said the first thing any self respecting physicist will tell you is that the list of what is not at all or at least partially understood about the universe is uncomfortably long for humanity. To this end the concept of new and smaller particles would be welcomed with open theoretical arms as the properties of these new particles and the interactions between themselves and other particles of all types would more than likely fill in some of the holes in our understanding of how the universe works. All of this aside the need of mid twentieth century scientists to figure out what was really going on with all these new particles led a number of researchers from around the world to culminate in the idea and mathematical validation of new particles called quarks. No one person can be said to have created the quark as it took over a decade of investigating the problem and many people along the way making contributions to the classification of these particles until finally agreement was reached and the quark was theorized. With the quark now part of the theoretical models of physics it was soon realized that interactions between them must also be scrutinized in order to move forward. Part of this scrutiny led to two ideas in quantum chromodynamics which persist to the present day and help explain the strange field behavior mentioned between quarks.

The first is the relation of quarks and the field between them to something called confinement. Confinement is what states that the field between quarks acts in the opposite manner to the familiar fields such as the magnetic field or the field of gravity. The strength of the field between any two quarks will remain the same even though the distance between the quarks is changing. So for example if you were to move two quarks farther and farther apart from one another the field strength will remain unchanged and this is seemingly in direct violation of the law of conservation of energy. Since a quark should produce finite field strength as it has a finite mass it seems impossible for the field to be equal over ever increasing distances and you would be spot on for thinking this. The secret behind this confinement of the quark field lies in the gluons which are the force carriers between them and the energy they possess. Remember how most of the mass in a proton is made up not of matter such as the W and Z boson that get their mass from the Higgs mechanism but instead from the energy stored in the movement of the gluons?

Well that is what is happening here again as the separation of two quarks will lead to the field strength remaining intact and eventually if you move them far enough apart from one another you will find something strange happening. The strange little thing that happens is that another quark pair will pop into existence from almost out of nowhere. Since the Standard Model forbids the creation of something from nothing it is obvious that the new quark pair came from somewhere. The answer to where the quark pair came from should be obvious by now as it comes from the energy stored in the gluons that travel between the quarks as force carriers. Remember also that gluons are not really mass particles according to the Standard Model instead they appear to have mass due to the energy mass equivalency relationship they possess. Since the Standard Model works on the idea that fields are excited to produce particles we see that the field energy being perturbed by the gluons is what causes another pair of quarks to be allowed to come into existence without violating any of the known laws of physics. This is a fairly important thing when it comes to how matter is constructed and behaves because what it means is a stable large particle such as a proton or neutron will always be there.

The reason lies in the spontaneous creation of new quark pairs inside the hadron itself as this means that a hadron will never run out of quarks. Since a proton or neutron is made of three quarks, two up one down and two down one up respectively the triplet of quarks must be preserved in order for the hadron to not simply fly or fall apart. By creating new quark pairs immediately upon initial quark pair separation the hadron will always maintain its quark triplet structure and therefore stay in a nice stable state. This idea of confinement helps to build the Standard Models picture of matter and how it exists in our universe in a nice neat way. The problem with the idea is that like many in the realm of the ultra theoretical and ultra small modern science simply does not know if it is right. Modern science does not necessarily believe that it is wrong or that at least conceptually it could not happen only that until proven one hundred percent correct it cannot be accepted with one hundred percent satisfaction. One of the biggest strikes against this idea is that energy from somewhere must still be input into the system to cause the field to create a new quark and this is not fully resolved.

Needless to say people have been trying to prove this correct for a while in order to validate the theory and while no direct proof has been found to show the mechanism is actually working other clever observations have been employed. The basic problem with studying this concept directly is that no current human technology can come anywhere close to accurately isolating a quark pair and then pulling them apart to see what happens. We can barely smash the little particles free in expensive and mind buggeringly short lived experiments thereby allowing us to detect and measure them a bit and that is all. So how do scientists who subscribe to this notion try to prove that it is correct?

They do so by not looking at what they can see directly with quarks but what they cannot see at all. To explain this simply let us look at other particles that we know exist for a fact, in other words we can see them and isolate them and play with them at will not just theoretically or in a mathematical model. Think for a moment of a helium atom and you find the familiar protons, neutrons and electrons all arranged neatly in one stable structure. Now those three particles we have definitely examined in great detail for between one hundred and two hundred years depending on the particle. We find them bound up in atoms all the time and we also find them floating merrily along all by themselves from time to time. In fact the universe is crammed with free electrons, protons and neutrons and all you have to do to prove this is look up at the myriad of cosmic rays that arrive at our atmosphere every second. This is where the quark hunters run into some problems though and their work remains in the theoretical realm.

The reason is things like electrons, protons and neutrons are found in atoms all the time and this can be thought of as their bound form where they create one larger superstructure much bigger than any one of them alone. OK so far so good because we find quarks in their bound form too and we call those things protons and neutrons which we are ridiculously certain exist. Since we find bound hadrons and leptons and bound quarks surely if we find free hadrons and leptons we should also find free quarks too. The trouble is quark hunters do not find them and in fact no one has ever found a free quark anywhere. We have determined the basic properties of quarks many times in particle collider experiments so we know exactly what we are looking for only it is not there. This would suggest that the quark cannot exist in a free form in the universe as it is bound in hadrons and we know this to be true based on observations proving that quarks do not roam freely. Since there are no free quarks this means they all remain in their bound state which is snuggled safe and sound inside a hadron. This in turn suggests that confinement theory is in fact correct as it also states that quarks will remain in their bound state.

So this is the indirect proof confinement theorists use to try and prove their ideas correct. Since separating two quarks according to them only strengthens the gluon field to the point where a new quark pair is generated what you are left with is quarks that are continually bound up into hadrons. This means that no matter how many times or how hard you try to separate quarks in order to generate a free quark what you are left with is the creation of more quarks and therefore stable hadrons. No free quarks because they are locked inside the hadrons and bound by the strong force which holds them tightly together. This is of course a neat little piece of thinking and the absence of free quarks is a possible interpretation of how confinement works but the fact remains that there are other explanations for how this might occur. It could also be that all confinement theory is dead wrong and the simple reason you do not find free quarks is because they are immediately destroyed or decayed upon being freed from a hadron. In addition since we still do not have the technology to directly manipulate quarks on the individual level it is possible and highly likely that there are more secrets these little quarks are hiding.

Quark-Gluon Plasma and Superfluids

This is only one of the interesting aspects of quantum chromodynamics however and the other is currently being studied at the Large Hadron Collider. You might have heard of something called the quark-gluon plasma which is thought to have existed moments after the Big Bang. In fact it is difficult to discuss the Big Bang these days without mentioning this quark-gluon soup as much of the basis of the Standard Model can be built around it. The name is as direct as it gets really as you have two particles suspended together in plasma and they are of course the quarks and gluons that bind them. Known as asymptotic freedom it postulates that at extremely high temperatures a weak interaction between quarks and gluons might exist and take the form of quark-gluon plasma. Plasmas are of course the fourth known type of matter after solids, liquids and gasses and are characterized by the ability to conduct electricity and be influenced by magnetic fields. That is the definition of conventional plasma of any of the major elements and you see this constantly in the universe inside the cores of stars as the hydrogen there exists in a plasma state. A quark-gluon plasma might be a little different though as the particles involved are somewhat different from a normal plasma. Firstly normal ordinary everyday plasma is made of atoms or ions which are simply atoms that have lost or gained a few electrons. In the case of plasmas however the former is more likely and we need look no further than or own sun for verification of this fact.

The dense collection of swirling and roiling hydrogen atoms at the core of the sun or at least just above the surface of the core is made up almost entirely of ionized hydrogen. In other words free protons and we know that this must be the case because if you recall the proton-proton chain we discussed before it requires hydrogen nuclei to work and not hydrogen atoms. This stellar plasma is affected in all the ways we would expect plasma too and both magnetic fields and electrical conductance are present. A quark-gluon plasma however is somewhat uncharted territory though as the particles involved are far smaller than a proton and might react completely differently. They might in fact behave identical to large hadrons like a proton but they might not react to magnetic fields or conduct electricity at all. The quark should at least be affected by a magnetic field as we know that quarks carry charge this is how up quarks make a proton positive electrically and down quarks make things more negative.

However what about color charge and the way they might be affected if at all by magnetic fields? Color charge is as we have seen not exactly a true type of charge and it is certainly not the familiar electric charge which is easily influenced by electromagnetic fields. The ability to understand what effect a magnetic field might have on a quark alone is not well understood beyond the theoretical stage as after all confinement itself says we cannot isolate quarks for study without generating more therefore making them not isolated. This is the rule set out by quantum chromodynamics itself and therefore following its own rules asymptotic freedom yielding quark-gluon plasma is tricky to study at best.

Also what of the gluons themselves as they possess no elements of standard electric charge to speak of and only have color charge properties. In regular plasma charges are subject to a process called screening whereby the presence of other charges will interfere with the distance of the electric charge created by the first. In quark-gluon plasma it is the color charge that is screened and in this respect we do have one similarity with normal plasma. Screening itself is the work of Peter Debye who was a Dutch physicist that conducted experiments concerning charge and the length at which electric fields could be of significance in solutions. Once the effect was lost it was said that the electric charge was screened and the length of this effect is called the Debye length and represents the radius of the Debye sphere which of course is the volume of the electric field created by the charge before screening. The idea of both interacting together even bound weakly means that a mix of charge screening is occurring at once evenly distributed through the plasmas.

Adding to the confusion is the behavior of these particles and fields at such high temperatures as they far exceed the normal day to day ranges of conventional plasma. Everyday plasmas can range widely in temperature from a few hundred degrees in the case of a simple fire to many millions of degrees in the center of a star. At the quark-gluon temperature particles will not be bound as we know them now as hadrons cannot exist since the quarks are moving about and the gluons which are the force carriers of the quarks are free to move about as well. Add to this the fact that gluons are not really particles as quarks are but more or less excitations of a field and that field is operating well beyond normal temperatures and what happens is more of a guessing game. The only place where the theoretical quark-gluon plasma has been momentarily created is in the most powerful particle colliders such as the Large Hadron Collider. It was not until 2000 that such plasma was possible to create in laboratory conditions when it was announced by CERN that they might have found a new phase of matter. Various colliders have experimented with this since then and in 2015 the Large Hadron Collider was able to produce definite quark-gluon plasma through collisions of protons with lead nuclei.

The temperatures that have been recorder for such events vary quite a bit just as do the temperatures of a normal plasma and range from approximately four to five and a half trillion degrees Kelvin. This is quite outside the range of normal stellar plasma that might get up to hundreds of millions of degrees on a good day so to speak. There exists a threshold range which has been theorized but not proven that divides the dominance of confinement and asymptotic freedom of matter at high temperatures. This temperature is thought to be around two trillion degrees Kelvin according to theoretical calculations and the experiments to investigate this further will no doubt be quickly pursued. Above a certain value and the asymptotic part of quantum chromodynamics appears to win out creating quark-gluon plasma. Below this threshold value and the confinement part of quantum chromodynamics wins out and matter behaves normally as we would expect it to in everyday circumstances. This temperature is known as the transition temperature and is a keen area of study.

Another keen area of study is how this new form of matter behaves and this of course takes us back to our discussion about the properties of normal plasma and quark-gluon plasma. It was debated as to whether the plasma would behave like a gas or a liquid and in the end the LHC experiments have confirmed that it does in fact behave like a liquid. This is important in nailing down some of the fundamental properties of something that is really still very theoretical. In addition to this it must be realized that quark-gluon plasma does not just require high temperatures to exist but that it also requires high densities as well. Changing either of these two variables will inevitably change how much you need of the other in order to get the creation of quark-gluon plasma. The threshold temperature of two trillion degrees is the agreed heat needed to transition matter from ordinary everyday matter to this new phase of matter but all of these experiments have been conducted using particle colliders. This is good for studying the quark-gluon plasma in the laboratory setting but it involves very fast ultra-relativistic speeds and violent head on collisions in order to produce the desired results. Such high speed and violent conditions might not be needed to create quark-gluon plasma in the universe itself and specifically at the time just a few millionths of a second after the Big Bang.

The time frame just after the Big Bang that is thought to involve quark-gluon plasma is central to the concepts behind the Standard Model and one of the focus areas of many large particle collider experiments. It is specifically the behavior of matter in these ultra high temperature conditions that is most interesting to particle physicists for the understanding of particle interactions. This separate branch of the Standard Model is called Thermal Quantum Field Theory and simply involves the calculations of various quantum field properties at set temperatures. The nature and behavior of such high energy and small particle based plasmas is not well understood and we must remember that some similarities exist between them and normal plasmas.

The key is what differences exist as they certainly do based on what we know of the other states of matter such as solids, liquids, gasses and plasmas. In each different state matter behaves uniquely after all what you can do to a solid you cannot do to a gas and vice versa. The extremely high temperatures of such quark-gluon plasmas means that the constituent parts of things like hadrons are moving also at incredibly high velocities. There is no understanding of what such fast moving particles and fields are capable of under these conditions and this is key to understanding the early universe. All of this boils down to one thing which is an attempt by the Standard Model to unify all three fundamental forces except for gravity into one.

This brings us to another interesting aspect of such high temperature physics which is the concept of something called quark matter. Quark matter is as the name implies matter that is made entirely of quarks and is directly related to the time right after the Big Bang which is known as the quark epoch.

We currently live in the time after the quark epoch which is known as the hadron epoch due to the fact that hadrons are the dominant form of matter in the universe today. This means the quark epoch is so named because quarks are the dominant form of matter in the universe during its time. The quark epoch is characterized as the time in the life history of the universe where the Big Bang has occurred and we are now about a trillionth of a second into the life of the universe. At this time it is believed that the four forces were present and operating as we would expect them to and the total quark epoch would end at a time of about a millionth of a second old. The quark epoch is of particular interest to scientists as it represents a time when matter did not exist as we understand it in everyday life and the conditions under which it existed were also far different than that of today.

In our everyday life the strong nuclear force has a maximum range of about one femtometer which is a millionth of a billionth of a single meter and is the approximate size of any normal hadron. This is because the strong force is dominant in holding hadrons together and their constituent parts which are the quarks together also. If the temperature and density are high enough as they were in the quark epoch just after the Big Bang then hadrons literally melt. The physicality of matter is not lost however and there are still things which would be familiar to us as solid matter just very small as they are solely things like quarks and a few other particles like electrons and neutrinos. In this form of the super-dense and super-hot quark-gluon plasma however the strong force is able to exert a much larger sphere of influence over the universe than before. This means that the close proximity of unbound quarks to one another permits a universe wide felt effect from the strong force and the quark-gluon plasma is thought to behave not just as a fluid but most likely some sort of superfluid.

Superfluids are fluids that are typically seen as very cold atomic fluids such as Helium superfluid and they have some very unusual properties. Typically superfluids have no viscosity and do not obey surface tension rules or the law of gravity. This in no way means that they are some sort of anti-gravity matter only that without input from external sources they can do things like spontaneously climb the walls of containers and flow out. A classic demonstration of a superfluid involves helium placed in a container that is kept super cold and allowed to move at will. If you have a sealed vessel in which you place superfluid helium liquid and float an empty container in it the liquid helium will climb the walls of the container without external input. As it moves up the outside of the container it will flow into the empty open end and begin to fill it up until it has created its own equilibrium with the liquid level outside the container. This is only one of the interesting properties of superfluids and the other is zero viscosity.

All liquids have some amount of viscosity and viscosity is best thought of as how easily a liquid can flow. Think for a moment about water and the metaphorical molasses in January and you have an idea of which fluid moves more easily and therefore is less viscose than the other. Normally however no fluid should have zero viscosity as this would seem to break the laws of physics and yet superfluids do exactly that. One interesting aspect of ultra-cold atomic materials is that it is not just confined to single atoms. Wolfgang Ketterle is a German physicist and probably the world's authority on studying very cold things as he has dedicated his work to researching near absolute zero atoms and their subsequent behavior with themselves and others. He was the first to produce condensates of numerous gasses, the first to create an atomic laser and most interestingly the first to create molecular condensates. The creation of molecular condensates opens a whole new line of questioning and study for the realm of the ultra-cold as structures much larger and far more complicated than single atoms are now known to be capable of similar feats. The fact that structures much larger than an atom can produce superfluid like states is exciting as it means there should be no qualms about structures much smaller than atoms doing exactly the same thing. Quarks and the like are about as small as you can get compared to a molecule and now we have strong evidence that all three: molecules, atoms and subatomic particles like quarks can be described using the same methods. The quantum dominated realm of ultra-tiny things like quarks can be scaled up to the macroscopic level of things like molecules allowing scientists an easier time of deciphering the world of quantum mechanics much like neutrino oscillations are helping us to do. Ketterle was awarded the Nobel Prize in 2001 for his numerous contributions to the study of ultra-cold systems.

Back to the zero viscosity of superfluids we see this is a very interesting aspect of matter as it is physically akin to superconductors. Superconductors are materials that allow for the flow of electrons or electric current with zero resistance and this again should be breaking the laws of physics. The reason is the same for the superfluids as there should always be some resistance to flow of a liquid or flow of electrons in a material due to the fact that some loss of energy occurs at the subatomic level. In a wire for example even a wire made from a very good conductor such as gold or platinum you find that the electrons will still be slightly attracted to the atoms they are passing through even if only briefly. This means that you will not achieve one hundred percent efficiency in transferring current along the wire. Superconductors have been studied and fairly well understood for years and so the realization that superfluids are real and not just theoretical is also important. The quark-gluon plasma is also expected to have some of these strange properties and this of course begs the question how can this be if superfluids are ultra-cold and quark-gluon plasma is ultra-hot? The exact method is not one hundred percent known but it is believed that a phenomenon known as Cooper pairing might be at work.

The idea behind Cooper pairs came from a man named Leon Cooper who is an American physicist and it helps explain some of the properties of superconductors. The basic concept as set forth by Cooper in 1956 suggested that electrons would form pairs and be bound together under certain conditions and low temperatures. The binding of the electrons in lowered energy states allows for gaps in the energy spectrum of the continuous binding states for the electron. When considered for not just one pair of electrons but for many as you would find in any real world material such as a superconductor the effect is widespread amongst all electrons. The net result is a pairing of numerous electrons all throughout the superconductor and an ability of the electron pairs to move freely as they are more attracted to each other through their bound state than in some ways they are to their own nuclei. The full quantum explanation of what is happening is known as electron-phonon interactions and is too long to include here. However what Cooper had shown with his Cooper pairs is a workable model for how superconductors operate and we have seen that the similarities between superconductors and quark-gluon plasma are real. The superconductors have no resistance to current as they transfer electrons to and fro while quark-gluon plasma is a superfluid that has no friction associated with the moving of its particles and hence no viscosity.

The belief that quark-gluon plasma can also behave as a type of superfluid is thought to occur inside the very heart of a neutron star where the density and temperature are possibly high enough for quarks to exist in a free state. Many astrophysicists have long believed in such strange phenomena occurring inside the cores of these exotic dense objects and have even given rise to their own type of star dubbed the quark star which we will examine a bit later. Inside the hot neutron star core however the nucleons would behave somewhat similar to electrons in Cooper pairs as they could share their strong nuclear force freely. Just as in a superconductor where the effects of the pairing is felt not just locally but all throughout the material so to in a neutron stars theoretical quark center the nuclear force would be felt throughout the quark core.

This would mean that a pairing of sorts can exist even though the temperatures are anything but cold as in a conventional superconductor. The outcome would yield a superconductive and superfluid like core that would behave just as cold superfluids due to a type of Cooper pairing. The extension of this would be to scale up the amount of quark matter in something as small as the very inner core of a neutron star and apply it to something the size of the entire universe. This brings us back to the period just after the Big Bang known as the quark epoch and the existence of a universe filled with nothing but quarks, gluons, leptons and what not. The idea that a superfluid quark matter core can exist in the hearts of neutron stars due to a type of nuclear force Cooper pairing now seems plausible in the very early universe too.

The early universe is much larger than a neutron stars core but at the same time was essentially tiny compared to what it is today and therefore had a small volume. This small volume meant that all the matter in the universe as well as everything else was crammed into a tiny space and existed at fantastically high densities. This would provide the necessary conditions for a similar type of Cooper pairing nuclear force to be felt throughout the universe and therefore dictate how matter behaved. The reason that this is so widely studied in modern science today is two fold and the first is obviously humans are curious creatures. The second reason is how the Standard Model is built and used to try and explain our universe as it appears today. In fact one of the major tenants of the Standard Model is to fit the contemporary theories and models and to fine tune them so that they can explain how we have a universe exactly as we observe it now. This is once again where I have to mention that observation is King and the ruler of all in both science and the universe. It might seem a bit over the top but in reality it is not as what we can observe empirically and scientifically is never wrong not even for a Planck time. The reason is simple: the universe is actually as we observe it and that is that. The theories and models of modern science are wrong or at least not that wrong but they are not one hundred percent correct when compared to observation.

I am of course being a little bit harsh here but I do it for good reason as I want people to think clearly about the observations and draw new conclusions if that is what we need to move forward. Many people out there might disagree with that and I would happily lump them in with all the incorrect theories as being wrong. Any good scientist or thinker is going to say 'hang on a minute and let us try to rethink what we have to fit the observations'. The idea of rethinking old theories or discarding them completely has been utterly necessary for us to make progress in understanding the universe. When was the last time we left an offering or sacrifice at an alter just to try to influence events beyond our control? If you say that I did it just last week I will ask you very nicely not to peak into my windows anymore as I live a very strange existence. The Standard Model is anything but perfect and leaves much too much unanswered to offer a complete explanation of the universe but it does do a fantastic job to its credit in trying to fit what we do know to what we can observe.

This is one of the key reasons that the Standard Model has enjoyed so much success and why hordes of scientists world wide embrace it. Love it or hate it the Standard Model is pretty useful. How the Standard Model has come into being is a direct attempt to link what we do know through experimentation to the observable universe as we see it now. The difficulty is that in order to fit what we can see today to what happened just after the Big Bang is tricky as so much has happened in the intervening time that we cannot observe. All we can do is theorize and try to look at the leftovers from these events. Think of the cosmic microwave background and you will remember what had to be done to try and reverse construct the events of the Big Bang. Think also of inflation theory before that and you see we are still theorizing how to make the universe we see today fit into a 13.8 billion year old explosion. It is like trying to figure out how a gun works by examining the dissipating smoke wafting from the barrel. The Standard Model is looking at the leftovers from the Big Bang today to see how it all started and the understanding of the quark epoch and just exactly how that phase of matter behaves is key to understanding what happened after the quark matter disappeared. The disappearance of quark matter means the beginning of the hadron epoch and our familiar form of matter that we can study easily today. That is why the basic properties of this superfluid quark filled universe are being debated by Standard Model enthusiasts everywhere.

The need to study quark matter is paramount for the Standard Model and indeed the construction of the Large Hadron Collider is proof of this as one of its main jobs besides finding the Higgs boson is too study the universe as it was during the quark epoch. With the discovery of the Higgs boson neatly and satisfyingly checked off of the LHC's to do list more and more effort is being placed on the study of quarks. The obvious place to study quark matter is therefore in very powerful particle colliders because the conditions can be closely controlled in the laboratory. Also the matter can be scrutinized under multiple sets of detectors as various experiments are dreamt up to test for quark matter. Lastly the experiments are reproducible and can be repeated over and over in the controlled conditions of the facility to allow scientists to narrow down the ranges of such matter. The practical limits of things necessary to create quark matter and quark-gluon plasma are not fully understood and variables such as temperature and density can vary in terms of the threshold limit. The threshold limit here is of course the conditions where normal hadrons breakdown and form quark-gluon plasma and this limit is not yet well defined. Currently theory has given us two variables to measure and play with in determining the properties of quark matter and they are temperature and quark chemical potential.

The quark chemical potential is a kind of ratio of quarks to anti-quarks and so increasing the chemical potential means more quarks and fewer anti-quarks. A decrease in quark chemical potential will yield a more balanced ratio of the two quark types. Temperature obviously needs no explanation as the higher the temperature the more energy is input into the system and the faster particles move; this is familiar to any level of physics from classical to quantum. Moving up the temperature scale can create quark matter and moving up the quark chemical potential scale can create quark matter but in different ways. The number of combinations of temperature and quark chemical potential to be considered here is large enough to allow for multiple answers to be extrapolated. Keeping the quark chemical potential low and increasing the temperature is thought to produce quark matter but in short lived phases such as cosmic ray events. In cosmic rays it is believed that very heavy nuclei such as lead ions that the energy imparted by these rays striking particles in the Earth's atmosphere will be sufficient to create momentary quark matter. The problem is the mathematics of thermodynamics works best with large volumes of long lasting matter when trying to work out the properties of something like quark matter. This means that while cosmic ray events are a possible candidate for the natural production of quark matter they are very short lived and very small this means that some of the calculations can become tricky.

The color charge of quarks is of particular interest when talking about the specific properties of quark matter and in these very fast very small collisions you will find multiple threshold values. The angle of the impact of the two particles will impart a different amount of energy to the system, the size of the nuclei will also affect the outcome and of course the flavor and type of all six quarks have different chemical potentials. This leads to a number of interesting collisions but ones that are difficult to study due to the fact that the impacts are so small and quick. This however is the road that is most often associated with the conditions of the early universe just after the Big Bang and as such we see a slight bias towards quarks over anti-quarks. This is interesting because at the start of the road we had no conditions stating that quarks should be favored and anti-quarks neglected yet it would seem that the universe tends to favor the regular kind of quark. This is again one of the key areas of study in quark matter and the behavior of quark-gluon plasma as well.

If we move up the quark chemical potential scale however we tend to find conditions that favor larger volume systems that exist for longer periods of time and this allows for the thermodynamics to be used far more efficiently. These conditions are of particular interest to scientists as it is now believed that a relatively constant temperature with higher and higher quark chemical potentials will lead to a threshold point where hadrons break down and quark matter is formed. The location of this type of threshold is believed to be in the heart of neutron stars that would have a small quark core. The properties of this quark matter core are believed to be superconducting and superfluid and this is because the type of matter present is known as color flavor locked quark matter. This basically means that the colors are paired up as we have seen in a type of color Cooper pairing and it is theorized that of the three color flavors this is the type of matter that will form at the highest densities for three flavor quark matter. This is where we must discuss another form of quark matter that is only theorized but has many scientists quite excited and also comes with a fun name and it is known as strange matter.

Strange matter is best described first by its use of the word strange in its name and the reason is simple it is not made of matter as we know it. To this end there is a second definition which is a bit more in depth but first let us talk briefly about matter that is not what we would call normal and definitely what we would call strange. Normal matter as we think of it can be thought of as atoms which are made of things like electrons, protons and neutrons. Other than the electron which is a lepton the other two subatomic particles are hadrons which are made up of quark triplets as we have seen. In each of these hadrons is the familiar up and down quark and no other types are present. In other words the universe while it is capable of producing things like top, bottom, strange and charm quarks seems to have no plans for popping them into regular day to day matter. We know that the other four quark types exist as we have detected them in experiments and worked out most of their properties from the data collected and it seems to match theory close enough.

So we are fairly happy or at least content with the idea that the universe can make other types of quarks and we also understand that matter or more specifically stable matter is made of only the up and down kind. This is what we observe by examining any type of matter that we have ever come across whether it is a simple atom or subatomic particle we find this triplet configuration of quarks. The universe as we know tends to favor certain things over others for whatever reason it fancies after all we are made of matter and not anti-matter. So there is a tendency for some things in the universe to be considered normal and stable as the matter that we see today is very old, at least 13.8 billions years old. This begs a question however into the nature of why the universe would go to all the trouble of making other types of quarks if it never wanted them to be included in regular matter?

From here we need the second description of strange matter which derives its meaning from the fact that a new and essentially non-normal type of matter has been tentatively theorized. It is of course called strange matter and what makes it strange is the inclusion of a strange quark to the familiar up and down quarks. What sets strange matter apart from other forms of matter is its theorized stability. Normal matter as we see it everyday is legendarily stable and the estimated age of a single proton is immeasurably longer than the lifespan of the universe itself. Quark matter is thought to be very unstable due to the fact that it requires very high temperature, density or quark chemical potential to achieve. The last time scientists think that quark matter existed to any appreciable degree was shortly after the Big Bang in the quark epoch. This time span of course lasted for a few mere microseconds and then the proverbial party was over so no point in being fashionably late if you wanted to study it in depth. This leaves us with a lot of theory and as far as quark matter is concerned theory says that due to very high Fermi energy quark matter tends to be unstable and fall apart rather spontaneously.

Fermi energy is simply the difference in energy states between the highest and lowest values a single quantum particle can occupy at absolute zero. Without crunching the numbers this means that the difference in energy states is so great between the up and down quarks of normal quark matter that they tend to not be stable unless they are more or less forced to behave. Behavior can be thought of as gravity created by a neutron star forcing the quark core to remain safely snuggled under a layer of neutron star crust which thereby keeps the quarks under very high temperature and pressure which overcomes this high Fermi energy. Strange matter is thought to be stable at conditions that are far below the extremes inside a neutron star core and be relatively resistant to lower temperatures and pressures.

The necessary trick is getting a large number of up and down quarks to become strange quarks all at once and even better having the strange quarks pair up afterwards. The more strange quarks there are the more stable the matter would tend to be as the effects of outside conditions can be better resisted by a larger amount of strange matter.

Strange matter is thought to be made of little pieces of things called strangelets which comprise equal numbers of up, down and strange quarks bound together. Normally the newly formed quark containing a strange quark would decay through the weak interaction to a regular arrangement of up and down quarks. The heavy original particle which contained a strange quark is referred to as a Lambda particle and will decay rapidly into lighter particles. The interesting thing here is the disappearance of the strange quark after it is lost from the original hadron and the tendency of the free quarks to be rebound into stable hadrons. Presumably the free quarks would generate new quark pairs through the confinement process due to the fact that temperatures are not high enough to maintain the asymptotic freedom of quark-gluon plasma. This is after all why the quarks refused to stay together in the first place as the required high energy was not present in the system. It is if the universe is forbidding the existence of these strange matter particles yet there does seem to be a way to create stable forms of strange matter. This is where the volume comes into play more and more of the system and it seems an increase in the volume of particles means an increase in stability.

The reason is the lowest ground energy state that does not violate the Fermi energy of the particles seems to be for the existence of an up, down and strange quark arrangement. This is plausible when the only form of matter is that of free quarks and therefore we are outside of the confinement process and hadron zones. Now that we have a large supply of free quarks and a way of making them bind together we must look at how this form of strange matter interacts with regular matter. The basic thought on this subject is that because the strangelet configuration of quarks exists at a lower ground energy state than normal or nuclear matter it seems logical that given enough time nuclear matter will decay into strange matter. In this case nuclear matter is anything conventional like an atom with protons and neutrons made of up and down quarks. The nuclear matter is considered meta-stable and in theory will not last forever so given a long time nuclear matter might decay naturally to strange matter. The time span necessary for such slow decay however is likely to be many times longer than the age of the universe for a couple of obvious reasons.

The first is we have never found free strange matter in the universe so it is unlikely to be decaying in a time span slower than the age of the universe. Secondly we know the stability of normal nuclear matter to a very high degree of certainty and all current data suggests that it is almost infinitely stable, meaning if it does indeed decay at all it happens far too slowly to be measured at the present time. The other possible interaction of strange matter with normal matter, again if all theories of strange matter are correct, is that strange matter has the potential to convert regular matter into more and more strange matter. The process would be simple as first a large piece of stable strange matter would touch regular matter and convert it to more strange matter. The higher energy state normal matter would have to lose its excess energy in order for the process to work and so free nuclear energy is released into the system. This means that the total strange matter in the system has increased so therefore it is larger and even more stable plus it is free to once again convert more matter. The process continues and accelerates as it goes along until whatever the strange matter touched is entirely converted into strange matter and a lot of free energy. This is of course a fairly unfortunate thing to happen to the ordinary nuclear matter and can be considered a theoretical hazard. The problem with this scenario is that you have to have a good sized amount of stable strange matter to start with and under normal conditions strange matter is anything but stable. Cosmic rays are a favorite candidate for many exotic physical phenomena in the universe and it has been hypothesized that the extremely high energy collisions from ultra high energy cosmic rays might have enough momentary energy to create strangelets. The conditions of the high atmosphere however are once again totally unsuitable for strange matter to exist for any length of time past a few nanoseconds. Proof of this again comes from the fact that we have not spied any despite our scientific reconnoitering. This leads us back to space and neutron stars as a possible candidate for stable strange matter.

Since we know that strange matter does seem to be a possibility and that it does require some specific conditions in which to exist we must look for the most likely place to find them. Neutron stars which form after a supernova are thought to be ideal homes for quark matter and strange matter for two reasons. The first is the way that a neutron star is formed and this we know is from a regular core collapse type supernova. It is possible that other types of supernovae can produce them as well but a normal core collapse allows for two favorable outcomes. The first is the survival of the star to the second phase of its life which of course is the post fusion phase.

To be honest I have never truly liked the description of a star as it transforms itself from main sequence fusion to post fusion. Most of the time you will here various terms rooted firmly in morbid fascination that sound like the death of a star or a star at the end of its life. In reality this is flatly wrong as the star is simply changing how it lives from an active to less active form of life. After all a planet once formed never really changes if outside factors can be eliminated and so it is seen as immortal but people never extend this truth to stars. Is a planet only alive while it is being formed and is full of energy, molten and dynamic meaning that once cooled as the Earth is now it is dead? In reality the star is just as immortal because once it has gone through its main sequence of stellar fusion it will enter into the second phase of its life where it more or less becomes a very large, hot and dense planet. Just look at our own star the Sun to realize that once it has shed its outer atmosphere at the end of its red giant sequence it will become a white dwarf star which is one of the longest lasting types of stars known. Many argue that because a star will lose its heat slowly over time that it is dead but in reality this is the same thing as any other planet which we deem not dead. This is just my personal feelings on the subject but since most stars will leave some remnant behind I find it to be more fact than opinion.

Returning to our very much alive neutron stars we find that the two factors they have in the ready for forming quark and strange stars is they fact that they do not die and they do not miss becoming neutron stars. The first factor is simply the rare chance that a star when going supernova will actually destroy itself completely. Most of the time the critical mass for exceeding the electron degeneracy pressure is not balanced on a razors edge and has a wide margin for collapsing straight into a neutron star or beyond. In other words the chances of a star actually destroying itself one hundred percent in a single large explosion is the least likely outcome as this would require just enough but not too much force to achieve. Most stars do not follow strict rules while they are forming and so masses vary above or below this critical limit due to the size of the initial clouds from which the stars formed. So the first factor means that we have something left over once the star has gone supernova and this is of course a necessary step in the process.

The second factor is that neutron stars do not miss becoming actual neutron stars by having so much mass and undergoing such a forceful core collapse that the neutron degeneracy pressure is exceeded and a black hole is created instead. Again stars do not abide by stringent guidelines while forming from their early accretion disks and can easily become large enough for this to happen. It is far more likely than not that in the early universe following the Big Bang the number of first stars that formed were all super giants that burned hot and fast do to the over abundance of gas for them to form out of. This led to the possible formation of numerous black holes as well as filling the early universe with many new heavy elements. Neutron stars however have not exceeded the neutron degeneracy pressure and so do not become black holes which would not form quark or strange stars. This makes neutron stars ideally suited as homes for quark matter and strange matter.

The quark type of star would be what we described previously which is a neutron star with a hot quark-gluon plasma like center. The immense gravity of the neutron star would provide the suitable vice necessary to keep the inner ultra dense core in a quark state. The outer neutron mantle would provide protection for the core from the cold vacuum of space that would disperse the quark matter. This is of course linked to the Fermi energy we mentioned earlier that can only be satisfied if the core is kept very hot and dense and this is exactly what the neutron mantle does. During a core collapse supernova of precise conditions it is theoretically possible for the entire core to resist becoming a black hole and transform completely into quark matter. If this is the case then the star would spontaneously dissolve because the neutron mantle is absent. This is essentially the only way a quark star can exist and indeed it is believed that the first detection of a quark star will be one that appears to be a normal neutron star at first but has slightly different properties belying its core of quark matter. I feel the need to state of course that at this time no quark or strange star has yet been found and we are dealing with theoretical objects.

That being said there is a second type of object that also may or may not exist and it is of course made of strange matter. Sometimes called strange stars or strange quark stars these are made up partly or entirely of strange matter as opposed to quark-gluon matter at their cores. The formation of such stars would proceed in much the same manner as that of a normal, if such a thing can be said quark star. The initial star would be a large one that will inevitably undergo core collapse and transform itself into a neutron star. Just as in the case of the quark star which has a theoretical quark center and by passed the usual full neutron star stage, so too would the strange star. The difference here is that unlike a quark star that must have a neutron shell as a protective layer around it the strange star might not need one at all. Of course it is possible to produce a hybrid star of neutron material and strange matter just as it is possible to form a hybrid star of neutron and quark matter. The key difference is that there is no theoretical way even for a quark star to exist without the protective neutron star shell. A strange star might be able to be a complete mass of strange matter all on its own that has no need for a neutron crust layer.

The reason goes back to the unique theoretical properties of strange matter that suggests it might be more stable than ordinary nuclear matter. This stability means that it can theoretically exist on its own without the necessary pressure and thermal protection that neutron star material would offer any matter acting as its core structure. The near vacuum of space is after all nearly devoid of temperature so that will not be adding any energy to the system. It is also at essentially zero pressure and therefore provides none of the raw force needed to keep matter from flying apart as would be the case with quark matter. The low ground state energy of the strange matter is what keeps it theoretically in its most stable configuration so even with basically no external heat or pressure that entire star mass could be made of strange matter. The only real force acting here is of course gravity which should provide enough strength to hold the mass of strange matter together.

The problem for astronomers here is that it is difficult to detect any real differences in how neutron stars, quark stars and strange stars appear to us as observers. It is difficult enough to tell what is happening tens of trillions of kilometers away at the closest neighboring system let alone exotic ones many times farther away still. This means that while the stars might exist if conditions are right Earth bound scientists are hard pressed to spot major differences between the three types. To this end all the theoretical evidence for the existence of these objects remains just that, theoretical. Further to the point just the possibility that these exotic forms of matter can exist at all is still largely debated as no one can make them in the laboratory. Even if someone thinks they have a glimpse of one it immediately destroys itself and so we are left with nothing.

This is where I must remind everyone of things like man-made elements and the lessons that can be learned from them when applied to theoretical matter. Take for example ununoctium, the heaviest of the man-made or synthetic elements and one that we know can exist as we have made it ourselves. The problem here is that of an experimentally over-enthusiastic cautionary tale. Just because we can make something in the laboratory does not mean that it exists in the universe. True the same conditions for smashing atoms together to get newer and heavier elements takes place in the cosmos all the time but any naturally produced ununoctium destroys itself just as fast as we see in our own experiments. In other words the universe does not tend to favor that kind of a heavy element and while it can truly be made the basic physics of existence forbid its long term survival.

The Big Bang, inflation period, quark epoch and the most violent events we can think of such as hypernovae and the incredible interiors of black holes do not allow for this element to exist for any length of time. So from the beginning of the universe to its present day state where we know that the conditions of the cosmos have changed dramatically nothing in physics can make an atom as large and heavy as ununoctium survive. This can be extended to things like strange matter whether produced as a single hadron in the laboratory or en masse in the end product of a core collapse supernova with all the necessary conditions just right we might find that the physical laws of the universe prohibit strange matter from really existing. If you still do not believe me then I challenge you to look around the cosmos and show me where the conditions exist naturally to turn normal matter made of quarks and atoms and the like into a rocket. I mean this in all sincerity as a rocket that we use to explore the universe is made from stuff in the universe and created by the universe.

It has nothing odd in it at all just regular matter so conceivably can it not be naturally occurring too? The answer of course is no the universe left to its own devices and excluding things like intelligent life will never naturally assemble the building blocks of matter into a rocket. The same might be said for things like synthetic elements and exotic matter like strangelets. While they too are made from ordinary building blocks of matter such as quarks and hadrons the universe has no natural method for producing them in stable forms.

One last word on this subject here is that we have not found any of these things existing around us anywhere. The universe as we all know is uniform throughout and you will not find bizarre sections of it where time flows sideways and gravity works inside out. In other words there will be no oddball circumstances that permit the existence of these objects elsewhere in the universe. If you read or study enough about the universe I am sure you have no doubt heard over and over the idea that everything that is unique or scientifically fun can exist but for some weird reason it is never close to home. No one ever says the most interesting things will be found in the Milky Way or even our local super-cluster and instead they tend to favor conditions which might be ten billion light years away. Poppy-cock. If this were true then we cannot trust any of our observations about the universe because even the simplest things like light would behave in novel ways everywhere in the universe making any data we have collected from it useless. This means we can throw out everything we know as it is based off of incorrect information and since we know this to not be the case we can safely conclude that the conditions we find right here at home are the same everywhere. So that being said where is all of our strange matter and other forms of exotic particles?

The answer of course is in our laboratories which make them interesting to study but we must resist the temptation to visualize it ever present throughout the cosmos. That being said and certainly not to suggest that these things do not exist as we know they do, I only wanted to point out the potential limitations of applying them to the universe beyond experimentation. To wrap up our discussion of these topics it must be said that this is the limit to where our modern day science can even shakily set foot on the subject of matter. Quarks we are fairly familiar with as well as the gluons that connect them and we would be remiss in our duties if we did not at least examine these other forms of potential matter.

So now we know all about quarks and quark matter as well as quark-gluon plasma and strange matter should it exist. All of this collectively came about from a description of what happens at very high temperatures, quark chemical potentials or both. The necessary reason for this was to delve as deep as we can with the Standard Model in an attempt to fully understand matter in the universe. Where modern science is right now is looking squarely in the face of the quark epoch to see what makes it tick and to try and figure out exactly what this means for the state of the universe today. The Standard Model is trying to learn as much as possible about the quark epoch in an attempt to learn about how its properties affected the next epoch to come. The following epoch is the Hadron epoch and we live in a universe fully dominated by hadrons and so anything we learn about how they were born can hopefully help us to better understand what they are up to now. The beginning of the hadron epoch came about after the quark epoch ended and we do know one thing about how this happened.

We know that the energy levels present in the very early universe during the quark epoch were so high that a continual state of quark-gluon plasma was present and essentially not much else. There was in fact a bit more to it as things like certain leptons and photons were also part of the mixture as well as various anti-particles. The high levels of energy meant that the interactions between individual particles particularly the quarks were too great to allow them to join up. It was necessary for the universe to cool sufficiently in order for the total energy per volume of space to drop enough that quarks could bind with one another. As space expanded a few microseconds longer the volume of the universe grew distributing the energy and therefore dropping the temperature. With the energy levels now below the threshold energy of quark separation we could start to enter the hadron epoch. The binding energy levels of hadrons could now be met and we find the quark-gluon plasma disappear in a very short space of time probably not much longer than a few microseconds itself.

The free quarks would now obey the laws of the hadron as the nuclear force became dominant over its familiar femtometer range. The gluons as well would take up their familiar roles of intermediating interactions between quarks. We know that the favored combination of quarks is to form triplets of various up and down compositions to form things like protons and neutrons. So far so good as this is the kind of matter that we are more than well acquainted with and from here we only needed to wait for the universe to cool a bit more for the leptons to join the party. This of course is the era of recombination that we have previously discussed and involves free electrons being allowed to bind to nuclei and form light elements. The photon de-coupling occurs and what many consider the first light of the universe begins to shine. From here nothing has changed in terms of the structure of matter in our universe for some great lengths of time.

To put this in perspective we believe the universe to be about 13.8 billion years old and the quark epoch itself might have lasted only a few tens of microseconds. The photon de-coupling occurred after only 380,000 years and we have the earliest elements ready to go and form things like stars. So this leaves us with a number of 13.799620 billion years for which the universe has had normal matter as we encounter everyday. To put this another way 99.997246% of the universes age has been in the normal matter phase which occurs after the era of recombination. Therefore only 0.002753% of the universe existed in some state that included matter not as we know it and existing in some exotic form. Remember also that I included the 380,000 years after the birth of the universe that included the hadron epoch before photon de-coupling if I had removed this we would find only a few tens of microseconds where matter existed in a truly different form than it does today. The hadron epoch still contained protons, electrons and the like which are all forms of matter that are quite normal they were simply not bound into things like atoms and larger structures. Looking at the universe from the point of view of the hadron epoch until today and you find that the length of time that matter existed in a non-normal fashion falls to almost zero. Why is this so important to understanding how the universe works? The answer is simple matter in its present form is of the utmost importance to studying the universe from an observational standpoint.

Chapter 10
The Universe at Large...Quick Catch It!

Quantum Problems and Schrodinger

This form of matter has weathered everything the universe can throw at it and the worst being time. All of the high energy and exotic states of matter are not how the universe really is meant to be as far into the future the state of matter will exist largely as it does today. Every second that passes even right now as you read this book and the percentage of time matter existed in the cosmos in exotic forms falls ever closer to zero. It will however never truly fall to zero and for this reason we must do our level best to study what did happen in that tiny slice of the universes life that was before the hadron epoch. Most of our efforts however should still be based on what happened afterwards in the time span that covers almost the entire life of the universe. This is because as I have stated the universe likes things behaving in this way. This is a very simple truth and one that must be accepted before more progress can be made. The state of things now is just as they should be and not because they just happen to fit all the precise parameters necessary for a universe or some such other coincidental nonsense. Things only appear to be perfectly fit because modern science does not yet understand how everything is put together and so in a sense there exists still some kind of mathematical mystery and wonder to the universe. In reality there is none and we simply need to find out how all the little pieces fit together in order to make sense of it all.

So here we are at the very end of the road in terms of what modern science knows about matter and how the universe works. Literally there is nothing more that we can say definitively as to how things work and as I have shown even some of the things that we think we know we only guess that we know. Confusing? Well it should be because even some of our most endeared little particle friends are still fairly misunderstood. Again I must remind the reader that despite our best multi-decade and multi-billion dollar efforts most of what we say is 'fact' is only a collection of fairly sound hunches. This collection of decent hunches is coupled with mathematics which in many cases can only hope to do its best when actually describing real physical things. This is something that has been going on for a hundred years or more and can be seen happily continuing up until the present day.

The problem is that this accepted notion of fogginess in our scientific thought processes has been so prevalent throughout the years that it has become taken almost as fact. Many people with nothing but the best of intentions used shaky equations based on unclear models of the universe to take another stride forward in the unlocking of the universes workings. Sometimes this has been not much of a problem and little bits of new information can help to smooth out previous wrinkles but often times the scientific community is rolling so to speak and it is not going to be side tracked by something as bothersome as shaky ideas. This of course leads to massive problems in our understanding of the universe and we are grappling with many of them still. After all modern science has a laundry list of unsolved problems that are rooted in the fact that the ideas used to figure them out are themselves not factual. There are probabilities that all things being equal should produce some safe results. This is of course a gamble as are all probabilities. The lottery is a probability game and it is well understood the chances you have of winning. Well the same is true for working on how the universe functions as a whole from matter and energy to space and time. If you begin calculations with probabilities instead of one hundred percent strong facts you leave yourself open to inevitable failure.

Think of this as shooting a gun at a very distant target if you are perfectly aimed you will of course hit the target dead center. If you are slightly off in your aim or the guns sights have not been properly adjusted then a tiny imperfection in your aiming can wind up missing the target by a great deal many meters away. As we go further from the gun we notice the deviation from the target gets greater and greater. The same can be said for work in science once you leave the facts behind and as you move farther ahead in history you are in a sense moving farther from the target. The little innocent assumptions of a hundred years ago have never been corrected for and so the same misaligned gun sights are with us today. Only now we keep building equation and theory upon older more antiquated equations and theories leaving us with best guesses that are no where near the target anymore. It might not be as dire as I make it sound here but I assure you it is a significant problem as we have many glaring paradoxes between theories and literal dead ends in contemporary science. If we did not then no one would still be asking how the universe works because we would have already answered that question years ago. I feel that a famous physics puzzle from decades ago can sum this up fairly neatly and bring us to the bridge between the known or at least relatively well guessed facts of the universe to the unknown and purely hypothetical. I am of course now going to tell you of the story of Schrodinger's cat.

Schrodinger's cat is a simple and unfortunately cruel thought experiment designed to study the Copenhagen Interpretation of quantum mechanics. As I will tell you later it involves contemplating the life or death of an innocent cat and hence it's odd nature yet it does shed light on a very serious problem of quantum mechanics. First we must look at why the problem even exists in the first place and then we can move on to the Copenhagen interpretation itself. The problem stems from the fact that quantum physics is not really well understood by modern day science and by the scientists of yesteryear who invented it. If we think about matter and our understanding of it since the times of the ancients we have gone from an era of partial understanding to full understanding and then back to partial understanding. The ancients only half understood what was going on with matter because they lacked the scientific data necessary to really dissect it. In the history books far too much attention has been placed on the Greeks and the idea that they had a rudimentary understanding of matter.

This is of course folly as they had no idea whatsoever they were looking at and no mention is given to their great blunders when considering the primal elements like water, fire or air. Similarly the concept of the atom is quite erroneously attributed to their achievements as thinkers like Democritus and Leucippus. All they did was look at things and imagine breaking them down ever smaller until you could not split the last piece. This is hardly a description of an atom or matter and rather a lack of tools capable of cutting the smallest dust motes in half; besides did no one stop to realize that they are in fact dead wrong? The atom they argued was the smallest form of matter and we now know that to be widely incorrect as we have electrons, photons, quarks, gluons, bosons and the list goes on and on. What they did understand however is the basic everyday properties of matter like its weight and the inability to compress things like solids. From here thinkers of the past not just the Greeks could begin to work out how matter interacted with us and our surroundings.

This period of human understanding would last until the eighteenth and nineteenth centuries when new experimental observations of the properties of matter would allow us to fully understand said matter. When I say partially or fully I mean based on the level of technology and theory available at the time. The reason we could fully understand matter is because not only did we still grasp the way matter behaved physically as the ancients did but we also began to figure out why this was so. The first steps of course involve proving beyond random thought experiments that matter was made of tiny particles which we named atoms. The next step we took was in examining these and working out that matter in the form of atoms had smaller little bits making it up as well. Once we had accomplished this feat it was time to work out the ratios and rules of how these little sub atomic particles interacted with each other. This is the last time that we fully understood matter as the next revolution in science would come from the quantum physicists and their attempts to learn even more about the little sub atomic particles that we thought we had neatly squared away.

What scientists of the late 1800s had noticed was that many of the effects we can observe concerning atoms and photons always occur in set amounts. These set amounts are finite values that never seem to change and so they were dubbed as quantized. The father of essentially all quantum mechanics was Max Planck who as we have seen studied problems of black-body radiation in 1900. This led him to the creation of the Planck constant which is an almost infinitely small physical quantity that can be used to describe these quantized events. Any quantum event we can measure is a multiple of the Planck constant and there appears to be no variation amongst recorded observations. This flies in the face of classical mechanics as in the classical interpretation of the world you get many differing amounts of whatever you might be measuring along a gradient. In other words you can measure things to many decimal places rather than just single integer differences as in quantum mechanics. Yet at the smallest scales of matter conceived at the beginning of the twentieth century observations were made not along a gradient but at quantized values. So here lies the problem the smallest measurable forms of matter exist in definite quantized amounts and yet we know the classical world to be real and solid. No matter what is happening at the smallest scales of matter or how well we do or do not understand them large scale classical observations are real and right. However, the large scale is made of the small scale so how can this be and how can we get the two sets of observations to play nice together?

This is the biggest challenge and the last challenge in all of physics because answering this answers everything about how the universe works. The largest structures in the universe and how we observe them are simply not wrong and the tiniest experiments carried out in our labs seem to be right so how can we take the lab and use it to explain say an entire galaxy? The answer is that modern science cannot do this right now based on the nature of calculations used to work with quantum effects. The descriptions of what are happening to particles at the quantum level are far less physical in nature and much more mathematical. This is a huge problem as the physical reality of particles and their behavior is always correct and the mathematical interpretations of what they are doing are only a best guess. To understand this better we must first examine a few of the mathematical ways in which the properties of particles are calculated in quantum mechanics.

The very first thing that must be understood is that in quantum physics a paradox occurs between theory and mathematics. The nature of quantum mechanics is that particles and energy behave in exact precise and never changing ways which is why we say it is quantized in the first place. Another way of thinking of this is that the physical reality of something is always known with hundred percent certainties as the value it possesses is quantized and so exact. The paradox arises from the inability of the mathematics to keep up with physical reality and so we find that all of the calculations for quantum exact states are themselves not exact. In fact they are the opposite of exact and exist only within the realm of probability which makes them only estimations of what a particle is doing. To make matters worse the estimations themselves are massively vague as can be illustrated with the orbital cloud of an electron around a hydrogen nucleus.

When one examines the first shell orbital of a hydrogen's electron you find that the exact position of the electron is not known at all but exists within a probability cloud around the nucleus. Making this matter worse is the fact that the calculations for the orbital do not pin down its location anywhere but only suggest that it inhabits some random part of the space. So the very quantized and exact particle the electron is only described in ways that are completely vague and very much not exact. This is of course a problem as the electron is not really in all places of the orbital at once as this would violate many laws of physics. In order to be in every place at once the particle would have to be moving not at but faster than the speed of light to be in every place at once. To truly be in everyplace at once the particle would have to move at an infinite speed in order to cover every position in ever smaller units of time and in fact could never do so as the graph of this action would approach but never touch the axis.

There also exists the problem of two particles occupying the exact same quantum state in the same space at the same time. This is a direct violation of how particles work and so you would not be able to construct structures larger than a single atom as the overlapping orbitals of say a molecule would allow for electrons to share the same space at the same time. Lastly we know a single electron to have very definite properties such as mass and size and so we know it simply does not fill the whole orbital as it is far too small an object to do so. So to sum up the mathematical calculations for the probability field of an electron around a hydrogen nucleus are wrong. To the lay person reading this you are quite happy with what I have just wrote but to the scientist you are probably fuming and hopefully to a few of the brightest scientists out there you are in agreement with me. The reason why some people might be angry with what I said is that the calculations for electron probabilities is the best we have for describing the actions in the system and they would be right. It is the best we have but it is so far from perfect that it is almost laughable that the mathematics of physics is taken so seriously when the observations are not. People have been for decades trying to fit the real physical world that we know to be correct through observation to the mathematical models of quantum effects that were devised over a hundred years ago. In truth the age of the equations is not the problem the lack of progress in truly understanding the nature of particles over the last hundred years is the problem.

If we think again about the humble electron orbiting the hydrogen nucleus we realize that modern science has not progressed much in the most basic understandings of how the universe works over the last hundred years. Obviously the electron exists in a space around the nucleus we have known this long before quantum mechanics tried to describe it mathematically. If the electron did not live close to its nucleus than it would simply fly away and ridiculous as this might sound this is the basic thinking that must be employed to overcome the need for probability calculations. The idea of calculations using probability has been useful in trying to get many of the aspects of the small quantum world to be somewhat workable in the laboratory. This is where they are most useful but again we know them to be incorrect because if they were not we would not be working on a list of unsolved problems in physics now would we?

So we are left with a set of equations that do not describe the properties of particles in exact quantized amounts but rather only in vague probabilities. That being said we do have to employ those same equations if we want to make any sort of reliable calculation of a particle and its properties. So a few examples of these properties and calculations will follow in a very brief description of how quantum mathematics tries to work. As we have already learned no math in the quantum world is exact and exists as a series of probabilities and this arises from the need to try and nail down certain states of a system.

Let us look at the electron orbiting its nucleus for a moment and realize that it has only one true position and momentum. This is not open to debate it is fact the electron occupies only one precise set of variables at any given time. The difficulty arises in trying to find both at once and so quantum mathematics has come up with some clever and roundabout ways of narrowing down these variables. I will show the step by step approach that quantum physicists use to build up a system and how they do a very good job of isolating only the necessary factors to arrive at what is hopefully a solid probability. First it can be said that any quantum mechanical system exists in a set space and therefore only calculations relevant to this space are considered. By setting up a finite space for the system we now have a somewhat classically familiar frame of reference to help us picture what happens next.

The second step is to treat any of the numerous conditions that can exist with this space as unit vectors which are often referred to as state vectors as well. This allows the calculations to eliminate some of the unnecessary variables that could possibly affect the system and instead pick out only the ones needed. This is one of the whole goals that need to be emphasized here when dealing with quantum calculations. The scientists know that what they are dealing with is by its very nature poorly understood by humans and difficult to visualize. In even more confusing terms it is difficult to reconcile the various quantum mechanics with the classical mechanics which govern all large scale observations. As we know the observations that we make everyday of the universe all around us no matter what the scale, big or small are always correct. So the quantum physicists are left with the unenviable job of trying to reduce everything to the simplest terms in order to avoid extraneous information. This reduction of extra data is one of the clever and fundamental tricks of quantum mathematics.

This leads us to the third step which is to compile all of the related variables into a single equation for the system from which calculations can now be made. This final step is where we get our complex wave function from and is what most people are familiar with in terms of quantum mechanics. The use of wave functions or collapsing the wave function is by now ridiculously almost a household term. This generous use of words however is well warranted as it is the formalized method for calculating quantum systems in a way that gives a simple and hopefully accurate answer. The mathematics are complex indeed but simplified to the point where they can be useful in applying their solutions to real world experiments. Real world experiments are another way of saying observation and so you can clearly see the importance of this type of mathematics.

Now we must mention two types of particles the fermions and the bosons which are descriptions of matter under certain mathematical conditions. The fermions were a result of Fermi-Dirac mathematics and were called fermions by Dirac in honor of Fermi, hence we get fermions. These particles are described in such a way that they obey the laws of the Pauli Exclusion Principle and possess half spin integers. This includes the six quark types and the six lepton types which can be said to make up regular matter so to speak. The second set of particles is the bosons which are given their name from the Bose-Einstein mathematics that are used to describe them and here we see the name bosons used. The difference here is that bosons mathematically do not have to follow the Pauli Exclusion Principle as according to the calculations they can occupy the same quantum state at the same time. This is the class of particles known as the force carriers which seem to bind other particles together and so they have whole number spin integers. Fermions all have mass and are considered to be real matter whereas not all bosons have mass and are seen as having a type of virtual mass. Remember this from our discussion of gluons in a hadron where they have no real mass but their kinetic energy can be interpreted as mass.

The reason we mention these two different types of particles or particles and force carriers is that we see in each case a different set of mathematics used to describe them. This creates small but real differences between the two and yet quantum mathematics must find a way to incorporate them as one into a given system, state or space. Further confounding these calculations is the realization that none of the mathematics used in any quantum calculation is fact and rather a set of probabilities. One look at an average quantum equation and you will see the difficulty in arriving at even a remotely accurate answer. This is of course partially resolved in the use of the complex wave function as a method of reigning in all these variables and unknowns. This has long been illustrated as an imperfect method since the introduction of the Heisenberg Uncertainty Principle.

The uncertainty principle was introduced by Werner Heisenberg in 1927 and discusses the relationship between the two classical examples of quantum mechanics namely the position and momentum of a particle. It can be applied to more than just those two however and in fact there are multiple conditions where two differing variables called complimentary variables are subject to the uncertainty principle. We have already discussed this previously and the familiar concept of increasing the precision through probability for one variable while losing the precision by a decrease in probability of the other is well understood. What needs to be underlined here once again is the fact that this type of mathematical relationship between two complimentary quantum variables underlies most of quantum mechanics. The idea that a certain amount of vagueness about a particular system exists is in many cases ignored and simply taken at face value. Quantum systems are imperfect and so let us now get back to calculating or at least that seems to be the unfortunate mantra of many quantum physicists. This is not to imply in the least that quantum physicists are in any way sloppy and in fact it is quite the opposite as they are trying their level best to reduce these unknowns to as small as possible. No matter how hard modern science tries however a gap in the knowledge of how the universe works and the nature of quantum effects remains and is keeping this nebel with us.

This is in part due to the fact that the scientific community is still divided on whether matter is a particle, wave or both. The generally accepted line of thinking is that matter is more wave-like than particle-like however this cannot be proven definitively. If it were one hundred percent correct than people would not still be discussing the problems of uncertainty, wave particle duality or observer effect in physics anymore and yet of course they are.

The Heisenberg uncertainty principle was introduced in 1927 and was a result of work previously done by both Werner Heisenberg and Niels Bohr. Between 1925 and 1927 the pair had been collaborating together over quantum experiments and ideas for years and had come to some conclusions concerning the nature of how quantum systems behaved. The collection of these thoughts is known as the Copenhagen Interpretation and was not formally coined until some years later. There are two main foci of this interpretation which to this day are difficult to shake fully. The first is that quantum mechanics itself is not a complete visualization of how the universe and everything in it works but rather only a probability of how things will work. This is what we have been discussing and is still being accepted today with things like complex wave functions and the need to use them in order to get data from experiments in a meaningful way.

The second focus is that all of the different probabilities of the calculations in a sense exist at once and only by making an observation of the system itself will the final outcome materialize into reality. In other words the many different possible outcomes are narrowed down into the final measured outcome which is only known to be the correct one after the measurement itself is made. If this made you scratch your head then do not fret as you are not alone in fact every physicist since the dawn of quantum mechanics has had to come to grips with this strange behavior of quantum systems. To begin with let us take a brief look at some of the steps needed to fulfill the Copenhagen interpretation of a quantum system.

The first goes back to what we saw with how quantum calculations are made namely reducing variables until only the necessary are included in a given space and then devising the wave function for that system. Initially one might think that a forgotten variable is present and not part of the wave function allowing for new information to be present at the time of measurement and thus explain the observation effect. This is not the case as scientists have for decades made quite detailed analyses of all their experiments as well as those of their colleagues to remove everything other than the variables in the system itself. Modern science has been unable to locate any hidden variables and so the calculations they make are as air tight as possible and therefore so are the wave functions.

Next it is accepted and this is where the nebel or mist truly appears, it is understood that the uncertainty principle is real and you cannot define each variable of complimentary variables in a system with perfection. You must accept some vagueness amongst variables in order to continue. This point will be delved into again as the very nature of quantum uncertainty could in some way be formalized into a kind of hidden variable the likes of which were omitted previously. The next step involves the actual act of measurement and can take different forms. The traditional is that the observer who makes the measurement is a scientist and so a human but it can also simply mean an instrument that takes the measurement autonomously of the scientist. It is at this point that an observation of the system is made and the wave function collapses to reveal the one true state of the system. All the various other potential probabilities disappear and you are left with the reality of what the wave function was hinting at.

The end result is one where the true state of the system is made readily available to the scientists and thusly the terms by which the results are described will be in classical ways and not quantum ones. The reason for this is that any device be it instrument or human observation that examines the system is a classical device and so a description of what was measured will be in classical terms. This makes further sense as everything in the classical world is a real and definite thing which is in contrast to the quantum realm where things appear to be many different probabilities at once. Therefore you cannot interpret the results in a quantum way at the time of measurement the system is not in multiple states at once it is in only one state hence the collapse of the wave function.

This is where the line between quantum and classical descriptions of the universe get blurred as everything in the universe is both at once. To put this more simply everything in the universe appears to be made of tiny quantum objects so a quantum description at some point must be used. When grouped in large amounts everything quantum naturally behaves in a classical fashion and so a classical description must also be used. This apparent paradox of physics is real and both ways of describing a system are correct as you cannot describe a galaxy in quantum terms nor can you describe the behavior of a photon in classical terms. If you are thinking that a bridge of sorts is needed to make the two different descriptions agree with one another on any scale in the universe then you are doing some good thinking. The contemporary understanding of science does not understand how to bridge the two in order to explain how the universe works and so ever more intricate explanations are required to try and explain what is taking place on micro and macro scales. There are times when a quantum or classical description best fit the observations of what is going on in the universe and a similar phenomenon occurs in quantum mechanics.

This is of course the accepted notion that the wave function describes both a wave and particle nature to matter and therefore we once again see this reference to wave-particle duality. The collapse of the wave function will give you one definite state for the system but the explanation of what has happened in the system can be explained by treating the results as waves or particles. This is a worldwide truth that further adds to the problems of working out what is happening on the tiny quantum scales of the universe. Vagueness in mathematical probabilities is now coupled with an unclear choice of particle or wave descriptions of how matter is behaving and leads once again to a best guess scenario of how the universe works. Many very clever physicists have posed the question of what would happen to a large scale object if it behaved like a quantum system. Those same people would say that in theory at least you could simply wink out of existence only to reappear perfectly intact a short time later. All large scale objects are after all made from quantum particles and disappearing and reappearing is a favorite way for quantum things to spend their day. We realistically do not pop out of and into existence however and so this disparity between the quantum and classical realms continues to confound.

I am of course having a bit of fun with all this as the chances of all the quantum particles in your body deciding to disappear suddenly all at once is an unimaginably large number. Yet physicists take these sort of thought experiments very seriously as no matter how rare an occurrence that might actually be real solid classical macro scale matter does have a little quantum layer underneath it all. So now we see how quantum systems through the Copenhagen interpretation are manifested from start to finish as well as some of the real areas of contention around them. No matter what however the Copenhagen interpretation is still the most widely accepted and useful tool for calculations involving quantum systems. It is the method by which real experiments involving unknown quantum effects can be fairly well tested and is currently used daily.

This however is not the end to the quantum story as there are other interpretations of the collapse of the wave function as well as other ideas concerning wave-particle duality. The first thing we must do however is finally take a look at Schrodinger's cat. Upon examining the general ideas behind the Copenhagen interpretation Erwin Schrodinger came up with a serious problem in the results of the interpretation and in 1935 devised a strange way of examining this with real life objects. The thing he chose was not an object but an animal and from here we get the hypothetical cat that Schrodinger used in his thought experiment. The problem he was addressing was the fact that according to the Copenhagen interpretation it is possible for quantum objects to behave as though they were doing two things at once. From the previous discussion it is helpful to think about the many different probabilities that according to the mathematics exist simultaneously at the same time and yet when observed only one materializes into the correct answer. Before the observation you do not know which probability is correct and therefore the particle can be thought of as doing many things at once.

To prove this point and how it did not seem to hold up to everyday objects Schrodinger designed an imaginary experiment where a live cat was placed into a box along with some other objects and the lid closed preventing anyone from knowing what was happening inside the box. Specifically what was happening to the poor cat inside the box as the other objects placed into the box were a flask of poison being hydrocyanic acid, a Geiger counter and a small amount of radioactive material. The idea being that until the lid is opened no one knows if the cat is alive or dead and until the box is opened the cat exists in two quantum states simultaneously. The reason being that inside the box exists a quantum object that is tied to the fate of the cat while it is in the box and it is the radioactive material. If the radioactive material decays through one of its half lives or even a single atom of it then the Geiger counter will register this quantum event. The act of radioactive decay of unstable isotopes is of course an event governed by quantum mechanics and calculated to happen by quantum mathematics. The decay would release radiation that the Geiger counter would detect which would cause a device to smash the glass flask of hydrocyanic acid and if left alone for a time will unfortunately kill the cat. So his thought experiment begins truly here as the cat which is not at all a quantum object is governed at its core by quantum events and so if the radioactive decay is governed by two possible states existing simultaneously then so is the fate of the poor cat. So the cat is not alive or dead in the box but rather alive and dead in the box at the same time. This is of course quite impossible no matter how convincing the mathematics behind it might be and therefore exists a paradox between quantum states and reality.

In other words when does a quantum state stop being both and become one of the possible states. To delve a little deeper into the interesting point of the experiment the cat itself is of infinite importance not just because it's fate is governed by a quantum event either. The reason is simple the act of observation according to the Copenhagen interpretation will affect the outcome of any quantum system. So far this seems normal as you might be saying yes I understand because I have not observed the state of the cat in the box and therefore I have not affected the system and violated one of the Copenhagen's interpretation rules. However we have not asked the most important person involved in the experiment yet and that is of course our fearless feline scientist.

The cat is also a classical object that is capable of making observations on quantum systems and is part of the experiment at the same time. So now we have the observer outside the box seeing the cat exist in two states at the same time while the other observer inside the box is existing in only one state at a time. If the time elapses and the cat emerges alive then the only existence the cat has known is that of life. Normally the observer should not be able to exist in multiple states at once as we are classical and only quantum objects should be able to exist in more than one state at a time. The cat however is able to exist in both the alive and dead state at once and therefore the paradox arises. This has led to a virtual rift in thoughts concerning quantum mechanics as people struggle to find a way to explain these findings in such a way that satisfies the experiment, observer, classical systems, quantum systems and of course the always correct observations.

Not only has Schrodinger's cat taught us much about what is or is not going on in a quantum system it has also helped us to keep this fact in mind in everyday thinking of physics and how they affect the universe. Many things in physics are often accepted which leads to them being treated as a fact and the end result is people look beyond them to the next step. This is what has happened to a certain extent in theoretical physics and quantum physics in terms of probabilities not being exact values yet being so common they are often ignored and treated as fact so long as the experimenter can move in a forward direction. Schrodinger's cat is such a well known thought experiment that it is useful in cracking through this barrier of accepting probabilities as fact and reminds us that we must consider all potential alternatives. This has been especially helpful as the thought experiment has taken numerous joke forms which can be very useful in reminding people of the paradox and the reality that it is not solved by contemporary physics. For example: "Schrodinger's cat is thirsty and so decides to get a drink and by extension Schrodinger's cat walks into a bar and doesn't." Some of you are smiling right now and others are left scratching their heads but to those of you who see the humor in it you are given a friendly reminder of what it signifies. This kind of reminder is powerful and still echoes strongly today as the Copenhagen interpretation is challenged by some with alternative viewpoints on quantum systems.

To start off with let us take a look at some of Schrodinger's work as it relates to quantum mechanics outside of his thought experiment. What Schrodinger gave us is his famous Schrodinger's wave equation. From this we see the world of quantum mechanics in terms of waves and the probabilities of these waves. Most of the quantum world and how it is calculated is done so through a mathematical theorem known as superposition. Quantum superposition is used to understand how waves behave at very small scales. Superposition is the idea that waves can be represented by linear equations and the laws governing waves therefore also govern the linear equations. For example we all know that if we have two waves in synchronicity with each other and we add them together we will find that their peak amplitude of X is now equal to the sum of the waves, in other words X+X or 2X. If we had three, four or five waves we would be left with 3X, 4X or 5X for the amplitude value. Now if we take two waves whose amplitudes are 180 degrees out of synchronicity with each other the amplitudes are still added but they will now cancel each other out as one waves peak is the other waves trough. So here we add +X to –X and we get zero. Any combination of waves placed upon one another no matter how much they differ from one another will be calculated as a sum of all their respective amplitudes.

This is superposition with waves being used as a simple analogy and they can be thought of as waves on water or sound waves or whatever you wish. In the quantum realm matter is often viewed as a wave as well and so the same type of linear equation superposition can be used to describe the waves in a quantum system. What Schrodinger did was to figure out how to describe a quantum system in simple terms using waves, probability and superposition. The nature of everything in the system at any one time is as we have seen the wave function and it is made up of many different waves all with their own probabilities. By treating the probability of each wave as a separate thing Schrodinger was able to simplify his equations. Each individual wave is not especially difficult to calculate and so can be known with a fairly high degree of certainty. This is because each wave is stationary in its time frame when viewed alone and so what Schrodinger did next was combine all of the possible waves into one long superposition or linear equation. This is what we call quantum superposition and creates a long sequence which theoretically could be infinite and we call the complex wave function. We now have a single equation that describes simply all the possible states of probability for a wave in a quantum system. The equation can then be solved and the final behavior of the particle revealed. There is of course a lot more to the mathematics than just this when working on quantum systems but the general idea of how waves and their respective probabilities are handled should now be understood.

So now it is understood that according to quantum mechanics particles can appear to exist in more than one state at a time and by the act of observing them they are fixed into only one of the states. If we think of superposition and particles we can get an understanding of why there is more than one theory on how to interpret the Copenhagen interpretation itself or offer up alternatives to it. Let us think of a single atom floating by itself in space and it is the only thing occupying a particular quantum system. There are no other variables present in the system at all just the atom and whatever strange quantum effects might be present. If we leave the atom alone it will happily exist in two states simultaneously according to quantum theory. It can be in a relaxed state or in an excited state; we will treat the relaxed state as occupying a smaller volume of space and the excited state as occupying a larger one. One would expect the atom to be in either one state or the other but not both at once which seems to be impossible. If we try to check on the state of the atom by observing it we will see it in only one state which can randomly be the normal state or the excited one. In theory we should get the same results with multiple observations of the system as we would tossing random coins about and averaging out how many times we got one side of the coin versus the other side. This is perfectly normal according to quantum theory and yet there is something we can do to affect the system.

If we introduce an electromagnetic wave to the atom and continually bombard it with energy something curious happens. When averaged out over many observations of the atom as it is bombarded with energy we find a gradual wave pattern that reflects the state of the atom over time. In other words the energy hitting the atom will slowly cause it to be either normal or excited at such intervals that they resemble a sine wave of sorts. This is because as time progresses the amount that the atom will spend in one of the two states shifts slowly. At first it might be fully in the normal state and then become mathematically 75 percent normal and 25 percent excited. Then 50 percent of each, 25 percent normal and 75 percent excited. Finally it will be fully excited and these changes in ratios of time spent in each state when graphed give you the wave shape. This is mainly an analogy to illustrate simply what is happening at the quantum level but this classical description helps us to better understand and visualize what is going on. Remember from the rules of the Copenhagen interpretation of quantum effects and the results of experiments on quantum systems must be described in ordinary classical language and terms and that is what is happening here with the wave description of quantum superposition. It might also be that the wave shape is not a reflection of the wave nature of matter but rather the wave shape of the photons being absorbed by the atom and causing certain levels of electron excitation to be made manifest in a way that when measured seems to make the physical mass carrying electron seem as though it is a wave.

So the quantum effect of the atom is that if left alone it really does seem to be able to exist in both states at once and this is key to all quantum systems. Not just the excitation through change in energy levels of an atom but everything else can be thought of in similar ways. You might be familiar with the idea of spin on a particle and that a particle can possess both up and down spin values at once. What quantum mechanics must face though is reality where atoms do not exist all by themselves in a set amount of space with no outside influence. In reality atoms are constantly being interacted with by other atoms, photons, neutrinos, electromagnetic waves and the list goes on and on. What this means is that the atom cannot maintain its quantum superposition in multiple states at once. The outside influences to the atom will inevitably force it into one state or the other and it will remain there until it is acted upon by another external force. These forces are all quantum objects themselves and yet the way they interact with one another is classic Newtonian mechanics. The object will stay at rest unless acted upon by an outside force in classical mechanics or the quantum state will stay at rest until acted upon by an outside quantum force. In this lies a tantalizing clue of the bridge between the classical and fully observable universe and the quantum experimental universe. Now if we plot an atom as it exhibits superposition and then plot when it loses superposition due to outside forces we get a flattening of the graph. The wave disappears slowly and we are left with a straight line as the atom settles into one of the possible states. The amount of time it takes for the atom to lose superposition and adopt only one state is called decoherence.

Double Slit Experiment

The most well known experiment in all of quantum mechanics must be the double-slit experiment and it is as famous a physical experiment as Schrodinger's cat is a thought experiment. The double slit experiment is the classic example of wave-particle duality in action and has been performed for over two hundred years now. The very first versions of this experiment were made by Thomas Young an English scientist who in 1801 discovered evidence that light was a form of wave. Since the ancient times people have describe light as rays or beams but never waves although it had been suspected that light might not be particle like after all. Young was interested in many aspects of physics and greatly in the study of light and optics. Like Newton before him who had examined the spectrum of light using prisms Young also examined the spectrum of light but instead of crystals he used cards. Young found that if he placed a narrow card in a beam of sunlight he was able to observe that colored light was visible in parts of the shadow cast by the card. By placing a second card before or after the first so that light could not directly hit both cards head on the patterns of color at the edges of the shadows disappeared. This seemed to suggest that light was made from a substance that behaved like waves as the light rippling out from the first card would interfere with light rippling out from the second card. Since light was a wave the patterns of colored light normally seen were also made from waves. Since waves can cancel each other out do to interference the disappearance of the colored light meant that light was a wave that was cancelling out the colored component normally observed at the edges of the shadows.

Young used this experiment coupled with his demonstrations of ripples on the surface of water demonstrating the same principle to establish that light was indeed a wavelike structure. The importance of this experiment has been seen ever since with others following like James Clerk Maxwell who built upon this to propose that light and electromagnetism were one and the same thing. The wave behavior of light has obviously been copied many times in numerous quantum experiments and the double slit experiment that was built upon the original Young's interference experiment is still being performed today. The modern version of this experiment is to use a laser instead of a window and two narrow slits instead of thin cards. The idea however is very much the same which is to demonstrate the wave properties of light through a series of interference patterns. In the laser versions of the experiments one does not see colored light but only the color of the laser itself in most cases a red laser is used. The laser light shines upon both slits at the same time and so only a small portion of the beam actually passes through to the detector screen beyond. The detector is nothing out of the ordinary and is used simply to show the reflected light from the laser beams. When the experiment is run the laser light entering the two slits will produce the familiar interference pattern on the back detector similar to how water waves would create interference patterns on the shore.

The striking thing is that the pattern does not just look similar to how a series of water waves would create an interference pattern it looks almost identical. Water is a purely classical object and the waves produced by ripples on its surface are purely classical in their behavior as well. Light however is known to be quantized into discrete packets of energy which behaves in quantum ways. Yet here we see light behaving in a purely classical way and from this the wave nature of light can be simply demonstrated. What Young initially set out to prove was the wave-like nature of light as opposed to a particle-like nature. His own narrow cards or double slit experiment do suggest he is correct however and it led to the realization that light is not fully a wave. Further investigations into the properties of light would begin a chain of events that led to the realization that light exists in quanta which are therefore discrete and not continuous like a classical wave. At the time of Young there was no such thing as quantum mechanics in contemporary science and so to him the wave properties of light were purely classical. It is now well known to us that light is not a true wave as it exists in discrete amounts and further refinement of the double slit experiment proves this. These new refinements however further prove the strange quantum wave-like nature of light at the same time. The experiments of Young and the modern versions using a laser produce a clear wave interference pattern but they do so by using a continuous beam of light that is made up of many photons all shining at once. This alone would seem to back up Young that light is a wave and not a particle however variations of the double slit experiment have proven otherwise.

Once the concept of quantum mechanics was born and the quantum nature of light revealed the idea to study everything about these new discoveries quickly took off. The classic double-slit experiment allowed scientists the opportunity to do just this and in order to learn more about the strange quantum world many alterations were made to the experiment that showed conclusively that light is a particle that displays wave-like properties. When I say conclusively I of course mean as accurately as anything can be in the realm of quantum mechanics according to modern science. More advanced experiments involve such things as covering one of the slits and then the other, trying to measure which slit a particular photon passed through and most interestingly allowing only one particle through at a time. All of these different methods have created much useful information about the quantum nature of particles and also much debate about them too. Firstly photons are not the only particle to exhibit this wave interference pattern as scientists have performed the double-slit experiment with things like electrons as well and in some cases atoms and even molecules. The end result of these experiments is always a pattern that resembles the classic photon wave interference pattern.

The first of the variations that must be mentioned is the passage of a single particle at a time through the openings of the slits instead of a continuous beam as this allows for more careful measurements of the quantum nature of the particle. In large amounts subtle properties of the system may be lost as the large numbers of quantum objects will inevitably behave in a classical fashion. However isolating a single particle and allowing it to enter the experiment by itself removes this and allows for only quantum effects to be present at least it is hoped. The wondrous thing about the double-slit experiment run with single particles passing through a slit at a time is that you will see a single point on the detector screen for every particle sent. This allows for point by point data about the experiment that can be viewed over time rather than all at once which allows for a few facts about quantum particles to emerge. The first is that matter really is a particle at least in part as a wave will not create a discrete point on the detector screen. In other words Young's view of a wave only nature of light is not correct as light and other particles do have at their core a particle-like structure as it were. It does show however that Young is also at least partially correct as light or any other particle when sent through one at a time will still behave as if it travelled like a wave.

These are the facts that are lost when running a double-slit experiment with large beams of particles. The fact that a single particle removed from interactions with any other particles as it passes through the slit will remove large scale classical observations and allow for purely quantum ones to take over. What is found upon sending a single particle through at a time is that the strike on the detector screen is seemingly random. When the experiment is allowed to continue for many, many strikes upon the detector screen a pattern begins to emerge that is not at all random. The pattern that emerges is once again the familiar interference pattern that we see using large continuous quantities of particles. So each individual particle seems to know how to move like a wave even though it is isolated and unaware of the behavior or unable to feel the effects of future strikes.

The strike is definitive proof that particles do have a particle property to them as they make a single impact on the detector screen. The probability of where the particle will strike the detector screen reflects the wave property of the particle and therefore combining the two will result in a classic wave-particle duality description of matter. The wave nature of the particle is not denied and it is calculated using probabilities which are clearly a wave favorite tool, but there is still further proof that matter has a particle component as well. Experiments have been devised that allow for detection of which slit the particle passes through at a single time and it has been found that particles go through only one of the two slits and not through both at once. A particle cannot go through two slits at once but a wave can and this is demonstrated using a simple wave table such as the one Young used for the water ripple experiments some two hundred years ago.

A single water wave upon reaching two slits will pass through both slits as the wave is a continuous function of the motion carried in water. The wave can then travel through the slits where it will eventually contact itself in the form of two smaller waves from each slit. The waves will interfere with one another and this pattern we see on the shore or detector. A particle however cannot pass through two slits at once and careful measurement have demonstrated that the particles do pass through only one of the two slits at a time and not both meaning that they are not purely waves at all. Once again the particle nature of matter is demonstrated and so matter cannot be purely a wave and more to the point we once again see the wave-particle nature of matter.

The difficulty in using this experiment to work with purely quantum effects is that at some point the experimenter must examine the outcome of whatever has happened in the system. This raises a very serious problem for quantum mechanics as the very nature of the Copenhagen interpretation states that any sort of measurement constitutes an observation of the system whether that be human or mechanical apparatus. Observations according to the Copenhagen interpretation will collapse the wave function destroying all the possible probabilities that the particle could have had and solidify it into the reality that is observed. This means that any act of looking at the double-slit experiment should in fact affect some sort of outcome that is not purely quantum as in quantum worlds everything is possible at once; at least according to theory. Here we see many more problems arise with simply trying to figure out what is going on according to the contemporary view of quantum mechanics. The final result of all these problems is the question of what are quantum particles up to when they are left alone.

This is the way that particles will therefore behave at the most fundamental level and so can possibly tell us a great deal about how the universe works as all larger structures and interactions then are built upon these fundamental behaviors. There is of course the understanding amongst many scientists that the probabilities and wave equation so long favored by quantum mechanics is only a mathematical construct used to describe possible outcomes of experiments involving particles. What this means is that the mathematics used to work on quantum problems is not at all a description of what is really happening in the system. The mathematics are only describing how one can calculate with an approximate degree of certainty where a particle might be or what some of its properties might be at a given time in the quantum system. The mathematics makes no attempt at all to explain what is really happening at the quantum level or why these behaviors are present in the first place. This is a solid fact and is refutable by no one in modern science. This might sound silly but how do we know that quantum effects are not the product of magic or some wondrous work of fantasy?

The answer is we do not know that based on the mathematics we use to describe quantum systems. We only know what a particular variable of the system might be at a certain time based on probability equations. What this oversight in quantum mechanics does is force some theories about quantum effects to be created to fit the math which is absurd. What is needed more than anything is a description of what is really happening down at that tiny quantum level. This will provide the necessary insight into the nature of matter and how the universe works so that we can revise or possibly create entirely new mathematics to describe what we are observing. Remember the observation is the only important thing in all of science as it is simply never wrong. The idea that a particle can be in two states at once or indeed in many states at once is not a full reality but one that exists fully in the mathematics of quantum calculations.

If that statement seems a bit harsh it is only because much of what we know about particles existing in multiple states at once is always based on experiments that require observations at some point and therefore violate their own quantum rules. What never violates any rule is the observed outcome as it is a definite reality and so it is the correct result to any experiment. If the universe wanted a particular particle to exist in a particular state then it decides that without help from us as we have no impact on any quantum force. The act of observation has been challenged by many as not clearly defined with many hazy areas surrounding what constitutes an observation at all.

To put this another way if a quantum tree fell in a quantum forest would it make a sound? The answer is yes it would as whether we are there to observe it or not makes no difference the underlying quantum forces still exist. To suggest otherwise would be to suggest that the universe only exists because we are aware of it and that if no one is looking at the quantum tree it does not exist. This can be proven wrong simply by breathing as the oxygen in the air created by the tree that you just inhaled came from a tree outside your line of sight and so the universe there still exists. It possesses no motion or physical properties whatsoever nor does it move forward, backward or even exist at all in time. This is of course nonsense and a simple experiment can be performed right while you read this book to prove it wrong.

If the universe exists only because you observe it then you do not exist at all either. How? Simple you came from a set of parents who inhabit this world and universe and so it must have existed before you observed it. If you exist now and are observing the universe then no one before or after you can exist so where did you come from and where is the universe going? The universe is going to stay put for many infinities to come, morbid as this might sound go check the obituaries for anyone who died yesterday and notice how you are here reading it today. The universe exists whether or not the deceased person observes it or not and so quantum effects behave exactly as they should whether we measure them or not. In other words every possible state existing all at once all the time unless we observe it is less than a sound theory.

This fact seems to be lost as it was formed from the original set of quantum equations from things like the Copenhagen interpretation and Schrodinger who said that the addition of all wave probabilities in linear equations is only to describe the possible outcomes of any quantum variable. They never said that all quantum probabilities are real only that all possibilities exist at one given time for the purposes of calculation. The reality is that an electron is a finite object and so it cannot exist in every location around a nucleus at the same time as this would involve more energy and velocity than it could achieve. Existing in every possibility at once means its mass is multiplied by the number of possibilities and that in order to be in every location at once it has infinite speed. Since both of these are beyond normal electrons capabilities the idea that all possibilities are real is incorrect.

The idea therefore that we are affecting a quantum system in a physical way is also incorrect and this is not to take anything away from the various experiments that have been conducted which seem to yield strange results. Instead it is a friendly reminder that the mathematics used to describe quantum systems and the actual driving forces behind quantum mechanics is not fully understood by modern science. If it were then contemporary science would not be examining these effects or wondering about why they happen because they would simply know the truth. In addition to this a quick note here is that it should not be at all scary to anyone to acknowledge that we need new mathematics to understand the universe. Afterall if we only had access to the mathematics of 1000 years ago we could not do what we do today in science as the mathematical language simply would be too primitive to work. The same is obviously true of today as our current mathematics might not be advanced enough to calculate the universe and new mathematics must be invented just as we have done many times in the past.

The truth is that the observations are always correct and some of these have been observed in large scale classical systems removed from the quantum realm. This seems at the outset to be a preposterous statement that flies in the face of quantum mechanics; a classical analog for a quantum system? Someone somewhere must be stark raving mad and yet no not really in fact the very wave-particle duality that has been at the center of our recent discussions has been observed with large scale non-quantum objects. Just as it appears that a single photon or electron being allowed to pass through one or both of the double slits in the famous double-slit experiment behaves as both a particle and a wave, we find the same in silicone oil droplets. A very simple little experiment has proven that when silicone oil droplets are placed on the surface of a test liquid and a vibration is applied to the system the droplet starts to bounce up and down with the vibrations. This takes a moment but eventually the droplet will start to bounce according to its and the liquids mutual resonance frequency which will start something quantum to happen. The droplet will begin to feel apparent quantum effects similar to the particles in the double-slit experiment.

The way this happens is interesting but simple; the droplet will rise above the surface of the liquid for a brief instant and then fall back down. As the droplet strikes the surface of the liquid it transfers some of its kinetic energy to the liquid which will cause small ripples to form around the point where the droplet hit the surface. These ripples will spread out as circular waves from the droplet over and over as the droplet continues to vibrate very rapidly upon the surface of the liquid. Eventual past ripples will interact with the droplet to gently guide it in a random direction and away from its point of origin at time equals zero in the experiment. If left alone for long enough the droplet will in a sense self-propel its way along the surface of the liquid and make its way to the double slits in the barrier. The droplet will choose either one slit or the other but it obviously cannot choose both at once. Once it passes through the slit it will continue to follow its own self made pilot wave until it eventually travels all the way to the detector screen which is really just the edge of the tank holding the liquid in this experiment. The point at which it strikes the edge of the tank is random and cannot be predicted before hand, in other words it cannot be predicted before it is observed. The observation comes from the droplet hitting the tank and knowing the probabilities before hand will still not tell you the final position the droplet will take until it actually takes it.

This is a perfect macro scale definition of the quantum double-slit experiment as a particle, in this case a silicone oil droplet, will follow a pilot wave along the surface of the liquid just as particles follow along in their own wave-like paths. The silicone oil droplet will then eventually strike the target in a random place just as it is impossible to predict with certainty where a particle will strike the detector according to modern science. The droplet is of course a single point with finite size just as the particle is quantized and will strike the detector in one place. So we have the wave nature of matter shown here in macro scale as well as the particle nature too.

The exciting thing about such an experiment is that one can now begin to think about how all quantum interactions might play out on the large scale world that makes up the universe. While the absolute underlying mechanisms of the silicone oil droplet experiment are not a perfect match for the quantum mechanics of the double-slit experiment the results of both are identical. This is what has got many people thinking about how much the quantum world can be thought to affect all matter. It is not just the recreation of the double-slit experiment at work here either it is possible to construct other large scale analogies for quantum effects using this system. The way in which one would calculate the final position of the oil droplet is reminiscent of the uncertainty principle for example. This experiment is very interesting as it shows on the macro scale the micro world of quantum effects however it is not a perfect match for all quantum mechanics. One of the most curious aspects of quantum mechanics known as quantum entanglement is still out of the grasp of the oil droplet system.

The interesting thing also about this experiment is not only does it hint that the macro world is behaving according to micro quantum laws but the opposite must be true as well. If quantum effects can be observed on the macro scale like this oil droplet experiment or of course Brownian motion two centuries earlier it also means that what we see as classical must be true in quantum. In other words what works in the classical world is a reflection of the quantum as well and so the idea that large objects behave classically means also that whether modern science understands it yet or not quantum objects must also therefore behave classically more than previously thought.

So modern science will tell you that particles are a combination of wave and particle, it will tell you that you can calculate many details of any quantum system using wave functions which are governed by probabilities and that when you observe a system all other possibilities disappear leaving only the one true reality behind. This is one of the interpretations of the quantum world and it is probably the most widely accepted but it does seem to leave some holes in the total explanation of how the universe works. So naturally people have thought of other ways in which the quantum world can be calculated in an attempt to resolve these holes. One such person was a man named Hugh Everett who was an American scientist that proposed a different way of looking at quantum systems.

Universal Wavefunction

The theory he proposed was called the Universal Wavefunction which sought to offer a different explanation than the Copenhagen interpretation as to the nature of any outcome concerning a quantum effect. Whereas in the Copenhagen interpretation the act of observation causes the wave function to collapse and result in only the one true reality the Universal Wavefunction theory states that the wave function need not collapse at all. First proposed in 1957 Everett's work was widely criticized from the beginning as it is mutually incompatible with the accepted Copenhagen view of quantum systems. To better understand the Universal Wavefunction theory it was necessary to explain Schrodinger's cat thought experiment as it is central to both the Copenhagen interpretation and Everett's work. Which ever theory is looked at they must all answer to Schrodinger's cat as it holds the key to final quantum results as dictated by both theories.

In the Copenhagen interpretation the cat is in both states at once and when one opens the lid of the box only the one reality materializes into existence leaving all other probabilities and possibilities to vanish. In the Universal Wavefunction theory when one opens the box up and looks inside again you will see only one outcome as the cat can only occupy one of the two states at a time. However this is where the theories part ways as Everett's work theorizes that we only see one reality but in fact both have been created and exist at the same time. So you start off with one universe that contains one box and one observer. Then the universe moves to the point right before the lid is removed and the observation made as the scientists peers into the box the universe splits into two separate universes. The first universe contains the same out come as the Copenhagen interpretation and for this argument we will say that includes the cat being alive. Once again a cruel use of the poor little kitty. The second reality is also realized in a separate universe where the lid is opened and the observer sees a cat in the second state which is deceased. Both outcomes are made real and yet each observer is only aware of the one that they see themselves. The two universes are now separate in this respect and so satisfy both quantum states but cannot transmit any information about their particular outcomes to one another.

How does this help solve quantum effects? Simple, the very nature of quantum systems seems to be that they exhibit quantum superposition pretty much all of the time and through experiment we can determine which state is actually the real one. Since quantum superposition says that any quantum object has technically two realities if both are realized then we can satisfy the paradox created in Schrodinger's cat experiment. The observer that opens the lid will cause no paradox with the observer inside the box which is of course our feline scientist. In this way the wave function for any linear system or equation need not collapse at all and all probabilities are made real. This leads to a near infinite if not truly infinite number of universes where different quantum outcomes are made real and therefore any possible outcome has or will happen. So for example if an event did not happen in the past some other universe exists where that event did in fact occur and of course the same goes for any future event.

This is very reminiscent of the statistical tool of flipping coins to see which comes up more often head or tails. Since a coin toss is random and you do not know the outcome until the coin lands you can say this is the same as making the observer effect. Toss a coin but do not look at it on the floor and you can only guess what state it is in heads or tails. In addition the coin has only two states it can be in which is the same as many quantum systems for example an atom in an excited or non-excited state, spin up or spin down. Looking at the floor will show you what state the coin is in and in an alternate universe the coin takes the other side up approach to reality. The idea of reality was central to Everett and his thinking on quantum systems as he argued that what we see in the observation phase of any quantum experiment was always real. The way he explains this is simple before the scientist makes any measurement of the system they are already entangled with the system itself and so the outcome will always be in a sense predetermined. The observer and the observed object which can be any particle we choose are already linked before the experiment is conducted and so can be thought of as already having made an observation. The observation was linking the two to one another and so they will exist in a relative state to each other that always appears real to us when we think we are making the actual measurement.

In this way any observer conducting Schrodinger's cat experiment will see exactly what they are supposed to see because the outcome was made earlier than simply opening the box. The two different scientists in two different universes are both convinced that their outcome is correct because to them it is real, only in this case there are two realities which according to the universal wavefunction equations are both allowed for. Another way of thinking about this is that the two scientists proceed in their respective experiments as if the wave function has collapsed and superposition is lost due to the states becoming decoherent. In Everett's view of quantum mechanics the wave function does not collapse but only appears to do so and so you would end up with multiple outcomes and multiple universes. This is perhaps the meaning of his original naming of his theory as the universal wavefunction. It might be that the wave function creates many new universes or supports them at least. The other possible meaning is that the wave function itself is universal as all quantum states it encompasses always apply anywhere and anywhen in the universe.

Remember that this is all linked to the idea of decoherence which was invented by German scientist H. Dieter Zeh in 1970. The basic idea behind decoherence is that by interacting with something outside of its own quantum space an object will lose coherence in its wave function and adapt to its environment. The process is thermodynamically irreversible meaning that the system cannot be restored to its original state without the addition of energy from an outside source. This means that the energy must come from the outside world and therefore causes a change in it as well meaning the whole system of the object and its environment cannot be changed back to its original form and entropy has increased. The properties of the local environment will dictate what final state the quantum object takes as the wave function describing it collapses into a single reality. Changes in state of anything be it the object or the environment in which it dwells can be thought of as exchanges of information between them. The act of decoherence is a loss therefore of information to the environment around the object although the object and the environment now share this information if viewed in total.

This means that decoherence gives a good explanation of what happens during the act of wave function collapse through observation but does not create it per se. The interesting thing about the theory of decoherence is that the information is not lost to the total system and so in a way superposition is not lost either. This means that the information exists still about the object in both quantum states and so it is as if the wave function did not collapse. This retaining of information is a kind of universal wavefunction in itself; not necessarily a true superposition any more but the information is still there. This is useful in explaining the universal wavefunction as theorized by Everett. If the information of the superposition of the system is not lost by observer measurement then Schrodinger's cat really is in two states at once in two universes. This type of thinking in universal wavefunction theory is what pleases so many scientists as it allows for the many paradoxes of quantum mechanics to be explained in a way that seemingly ties up all the loose ends. One of the loose ends is the famous double-slit experiment where according to the universal wavefunction theory the outcome of any single particle is already determined and all other paths are realized in separate universes.

The universal wavefunction of Everett's 1957 work was largely ignored or dismissed and it was not until some years later that the theory was revisited and popularized. Here is where it gained its new name of many-worlds theory which is unfortunate as it was not the name given by its originator Everett. However the many-worlds name stuck and is the familiar one by which it is now known as well as other versions such as the multiverse theory. The multiverse moniker has perhaps surpassed them all as it is the current favorite amongst the lay person and therefore responsible for much of the theories mainstream notoriety. The theory has not just gained popularity in public circles but also in scientific circles where it is regarded as a very strong contender for an explanation of quantum effects along with the classic Copenhagen interpretation. That being said most physicists still side with the Copenhagen interpretation but the number who subscribe to the universal wavefunction or multiverse theory is not insignificant.

One of the key areas of interest in research regarding the universal wavefunction theory is to see if we can detect the separate universes. The reality is that under no circumstances should we be able to and according to Everett himself as well as many other proponents of the theory once the two universes diverge from one another they evolve separately and so no information can be shared between them. In addition because the universal wavefunction theory takes into account both future events and past events not only do we not know what happened to the other universe after we make a measurement we also do not know anything about the previous divergent universes that existed before we even tried to perform the experiment. In fact before we discovered quantum physics at all as this is how the multiverse theory works. According to it there are really an infinite number of universes that exist close to each other but cannot share information. So how can one even begin to propose a test for alternate universes? The answer is gravity and it should be made immediately clear here that just because gravity is not understood by modern science does not mean that it can be thought of as the cure all for paradoxes or discrepancies between theories.

Nevertheless supporters of the multiverse theory say that because gravity is the only universal force that permeates the entire universe and is usually considered to be not part of a quantum system it might be possible that it is the common link between all universes. Gravity would not be a part of the observation or relative states proposed by Everett in choosing an observed quantum superposition and therefore should still be found throughout all universes whether they have diverged from our own or not. One of the very central tenants of multiverse theory is that the universes created by it are separated by incredible small quantum scale distances and so even though we cannot perceive them they are technically right next door to us. The assumption then is that we should be able to detect any gravitational differences between what we experience in our own universe here and what an alternate universe might be feeling. This of course would be true for any hypothetical scientists in the alternate universe as well as they should be able to feel our gravity also.

Most likely this is just speculation and no such experiments exist to test for these differences if they should exist at all.

One of the difficulties with this approach is that an alternate universe from ours existing in close proximity would not be that much different from ours save for a few small changes. For example the two universes created according to the universal wavefunction theory following the Schrodinger's cat experiment will have zero differences in gravity as nothing has gained or lost any mass and so it is impossible to use gravity to measure these two universes as being real. Remember that the multiverse theory itself states that the separate universes are very close to each other separated by only a Planck length or two in other words very similar to ours which should have little to no difference in gravitational strength. The second problem with this approach to proving the multiverse theory is that gravity is still not understood by modern science and so trying to use it as a measuring tool between universes cannot be done with accuracy on the quantum level.

No matter how many people believe or dismiss the universal wavefunction theory it has left its mark on the history of quantum physics even if not as strongly as the Copenhagen interpretation. While it does solve some of the problems of quantum mechanics and mathematical limitations regarding them it also creates some new ones along the way. What is really needed by modern science is a new system that solves these problems once and for all. In the meantime however one problem still remains at the core of physics and science that has not been resolved by any contemporary science and that is gravity. It has received much discussion and is every present in the universe more so than any other force it would seem and yet modern science struggles with it and loses at best. So now let us take a look at this force and see what the current understanding of it is according to the leading theories.

Chapter 11
Gravity, Time and Entropy

Gravity

Simply put gravity is the most well known and yet the least understood force in all of modern science. Before we go poking at modern science however let us think back to the ancients around the globe and what they thought of gravity. The answer is nothing at all. That is basically correct from the standpoint of any ancient culture around the globe and throughout time. Gravity is just such an everyday phenomena that most of the time it was completely overlooked by our ancestors and the reason for this is simple it is evolution. This is fundamental to our understanding of gravity as no other force is as pervasive throughout the universe as gravity. To our ancestors and to any small child just exploring the world around them for the first time gravity is nothing more than something that simply is. Think about it for a second when you were growing up or even when you were fully grown you probably never thought any more about gravity than you would about breathing. There are of course some of us out there who have been thinking about gravity for let us just say a goodly amount of time though and yet our own frame of reference always started out by not thinking about it. To our ancestors gravity was just there and since you could not measure any changes in it or interact with it in any experimental way other than using it to maybe drop things on your enemies in times of war it was matter that got the center stage when it came to trying to solve the universe.

This entire work that I have written up until pretty much this point has been a testament to the resolve with which humans have tried to understand matter from how it behaves to how it is made. Yet in all those decades, centuries and millennia of study has anyone really poked at gravity a good deal? Well of course not because gravity is treated as one of those fixed constants that you just use to keep quiet and be a good little scientist and get back to calculating things. Somewhere along the way however we left our ancestors and their empirical way of examining the world and we got down to some serious experimenting. This was the golden age of poking at stuff to see how it worked and things were going brilliantly until someone said that we really ought to work out gravity as it is one of the fundamental forces. The trouble is no one in modern science has been able to do that and so it is to new theories that we turn for a possible answer.

The main difficulty arises from the fact that two of the three realms that modern science has examined do not want to share all their little secrets and act in unison. The three realms are what I call the small scale, the medium scale and the large scale. Small scale is the world of particles whether they are molecules, atoms of sub-atomic particles they are very small and seem to be governed by the quantum world. It is here that we see the three forces play somewhat amiably together and they are of course the electromagnetic force, weak force and strong force. We have seen that the gradual road to understanding that humans have placed on matter over the thousands or tens of thousands of years that we have been studying it have paid off. From it we get numerous particles and a good idea of how the forces hold them together and get them to interact with each other and the world around them. From this understanding we see that many of these interactions appear to take place at set levels which are quantized and so we use quantum mechanics to look at and work with the teensy weensy.

The medium scale is where quantum mechanics cannot help us at all because none of the theories, models or mathematics used to describe quantum systems work on large scales.

Anything the size of you or I or bigger like the Sun or the Earth simply do not obey any quantum rules and instead behave purely as classical objects. So the medium realm is made of things roughly that scale and probably a bit larger maybe the size of a galaxy. When I say galaxy I do not include the black holes that are inside many of them as the medium scale fails at explaining them just as quantum fails at explaining things the size of a planet. The medium scale is of course described by general relativity which is a classical system and so does a decent job of describing classical objects like those mentioned for medium scales. Here we see things like gravity come into play a bit more although relativity makes no attempt to say what gravity actually is only that it can have certain relativistic effects on matter for example. The medium scale relativity fails utterly at describing the small scale just as the small scale quantum fails at describing the medium scale. Where the medium scale ends is where relativity once again fails to explain the universe and the large scale is the boundary where relativity fails to explain observation.

The large scale is anything that is the size of a galaxy or larger such as the filaments that make up the universe or even the entire universe itself. In addition any object that possesses incredible levels of power is beyond the comprehension of the medium scale relativity as well. This is why I mentioned black holes a moment ago as they do things to the laws of relativity that essentially prove them wrong. Since the universe is made of everything that there is and this includes things like black holes we can all agree that relativity and the medium scale theories are not correct. Here is where it gets even trickier though as the quantum small scale cannot hope to possibly explain these large scale objects and effects either. So we are left looking for a bridging of all three scales such that one model can be used to make everything work everywhere in the universe and at any time as well. First however gravity must be more fully understood in order to get all the three realms explained at once and so a closer look at the exact workings of gravity is needed even if modern science can only offer up theories at this point.

What modern science does know about gravity is more about what it affects than how it affects it. For example if we look at the other three forces we can see that they all work over very short distances and are not felt by everything in the universe. The strong nuclear force acts over a very short range and is consequently the most powerful of the forces it covers a distance no greater than the diameter of a nuclei. Even though we have looked at the nuclear force being a pervasive force capable of reaching much larger distances it does not do this without outside help. Quark-gluon plasma thought to have existed in the early universe and through extreme pressure created by a small volume of the universes space it was the most important force shaping how matter behaved. The strong force however was not able to achieve this on its own as it took the extremely violent conditions of the entire universe itself to realize this and left to its own devices the strong force cannot possibly accomplish these feats. The next place we might find the strong force dominating is inside quark stars or strange stars but once again they are only able to exist over scales larger than an atomic nuclei because of external forces. Or rather because of external force, singular, this is the force of gravity.

As for the weak force we find much the same story as the strong force it is still stronger than gravity but acts over a massively shorter distance. Just as the strong force needs external influences such as temperature, pressure and gravity to extend its reach so does the weak force. The most likely candidate for a greater area of effect generated by the weak force is the electroweak force which is thought to exist at extremely high temperatures and densities. These were believed to again have existed at the beginning of the universe but only because the universe was so small that the necessary pressure and temperature existed to allow the electroweak force to exist.

The electromagnetic force is stronger than gravity but is the only force to be felt beyond the range of a single atom. This is obvious to anyone who has ever conducted or at least seen the famous science experiments involving magnetism. Practically no student of science has missed the familiar bar magnet with iron filings surrounding it revealing the invisible magnetic field lines. The lines as any child can see are extending well away from the magnet that created them and therefore extend a great distance farther than the nuclei where they are generated. This increased distance of the electromagnetic effect is also ever present through the universe around planets, stars and whole galaxies. Just look at the Sun's own magnetic field and see that it extends a good fraction of a light year out around our solar system to boundaries like the heliopause. The Sun's gravity however extends much, much farther to the distance of a full light year at least as this can be seen with objects well beyond the Kuiper belt. The Oort cloud is held firmly in place by our Sun so much so that every now and again the balance of objects in it shifts and the Sun's gravity will pull one into the solar system proper. There is no way that the Sun's magnetic field stretches out one light year or if it does it must be fantastically weak to the point of non-existence. The force carrier for all of these three forces also has a limited sphere of influence such that they do not affect all matter or energy in the universe. Even the photon which can travel freely throughout the universe cannot interact will all matter and energy and even so is only the carrier not the actual force itself.

Gravity on the other hand is the dominant albeit weakest of the forces that does influence everything everywhere at all times and on all scales. The force of gravity is the weakest of the fundamental forces and this can be simply illustrated by picking up any object you find lying about. You are small and do not comprise much mass or possess much energy when compared to the Earth. The Earth generates a gravitational field many trillions of times more powerful than you and so you are stuck to it and not the other way around no matter how people view this scenario. Yet you can easily pick up something from the Earth's surface and therefore overcome the gravity of an entire planet and in reality not just any planet the largest of all the rocky planets in the entire solar system. Good for you as you are now stronger than Earth and the Earth's gravity and yet if you try to jump away from Earth you will not make it past a meter or less. There is no way for a human to leave the Earth's gravitational pull under their own power and here we see our first strange and weird taste of gravity. You are stronger than gravity you proved this by picking up something and yet you are weaker than gravity because you cannot leave the Earth and head out into space. So which is it to be gravity is strong or gravity is weak? The answer is both at once and this is the wondrous fact about gravity as it is weaker than the other three forces and yet so much stronger than all of them combined.

While gravity might be weaker than say the strong nuclear force the nuclear force inside an atom cannot influence anything outside of its own nuclei. The weak force of gravity stretches across light years to grab all those little atoms and pull them together where they become a star. The force of gravity now continues to grow and eventually it is strong enough to over come the electrostatic repulsion atoms feel for one another and ionize them ultimately leading to nuclear fusion which is made possible by temporarily beating the other forces too. If we go to something bigger like a black hole the gravity there is almost beyond measure and the weakest force gravity easily beats the other three and bends all matter and energy in the universe to its will. This is the unmatched power of gravity and it has been ever present since the universe was only a few moments old where since then it has been the single driving force behind sculpting the universe as we see it today. Any observations we make about the universe are due to the effect of gravity on a small or large scale.

Gravity also influences all particles and energy in the universe a feat that no other force can match. Even light which is thought to have no mass and therefore should not be influenced by gravity is still bound to it. A simple realization of this comes from the warping effect that light undergoes as it is lensed around massive objects like galaxies and black holes. Or perhaps a black hole itself where light is obviously obeying gravity as it enters and becomes trapped inside. Here is the second conundrum we face in modern science as light is a quantum object and yet it obeys classical laws. Gravity and the medium and large scale it influences is purely classical with nothing quantum about it and yet light which has been measured over and over in the quantum world obeys the classical. So what are we left with and how can we reconcile the two?

We are left with merging the two until they can be used to explain all and this means getting as I have said all three realms to agree with one another under one model. The classical world is providing us with undeniable facts about the universe through observation which must be continually stated as never being wrong as it simply cannot be incorrect. The quantum world on the other hand provides us with a means of testing the very small building blocks of the universe to learn a few things about them. Lately there has been too much leaning towards quantum and away from classical in terms of explaining the universe when in truth both must be embraced to find the one true answer. Quantum is a world filled with uncertainties and as I have shown unknowns and probabilities as it is here that our understanding of matter and the universe leaves the realm of fact and enters that of speculation. Perhaps accurate speculation but speculation none the less as nothing in the quantum world is truly a solid fact. If this were so you would not have raging debates between warring factions of physicists on the nature of quantum mechanics. The fact that all quantum mechanics is based off of things we cannot see and measure directly but only infer to through mathematics is perhaps the biggest clue that those who eschew the virtues of a purely quantum universe are misguided. On the other hand those that attempt to explain the universe using only classical methods while ignoring the light shed on matter by the quantum world are equally doomed to not find the truth. So many in the scientific community have tried to get past the shortcomings of both and figure out how to explain the large scale classical force of gravity with the small scale quantum world of particles and as you might have guessed many of these ideas are called quantum gravity.

Quantum gravity is not in any way a complete theory in fact it is made of several competing theories none of which are themselves complete. The basic idea behind all the theories of quantum gravity is to try and assign quantum like properties to gravity so that it can fit into the ideas surrounding the other three forces which quantum mechanics can describe. This is the hope of many physicists as they believe it will allow for relativity and quantum mechanics to be united along with the four forces. The mathematics involved in all theories of quantum gravity is quite involved and I will not bore the reader with a long set of equations for any of them. What I do need to stress about the mathematics is that each theory can be made to seem somewhat valid mathematically if certain conditions are allowed. The condition necessary for one theory to work will cause another theory to fail and vice versa. This is because no one in modern science is quite sure of all the important variables in the universe which could significantly effect their calculations. In addition to this as we have seen there is a considerable amount of disagreement within quantum mechanics itself that serves to further complicate any assumptions made in the equations.

Nevertheless a few theories have attempted to ignore as much as is allowable in the need for a complete understanding of all variables so that they can attempt to explain gravity in quantum terms and they have put forth some interesting ideas. Before we get to these theories it is essential that I list the hurdles in the road to discovering quantum gravity if it does in fact exist. These hurdles I will include in place of the mathematics that describe them and so the concepts will be outlined providing the understanding for the potential theories. The first are the fields present in the universe as theorized to exist by quantum field theory. In quantum field theory the forces that govern interactions of matter in the universe arise from excitations of a particular field, for example the strong force in the nucleus of atoms. In a field view of quantum gravity one of the underlying fields of the universe would be responsible for gravity and the graviton as the force carrier. Mathematically though gravity is nonrenormalizable under quantum conditions and so results in an infinite number of parameters to be accounted for.

Renormalization is a tool used in quantum mathematics to deal with infinite values that will inevitably arise when calculating small and large scales together. For example trying to apply the tiny energies of quantum forces to an entire galaxy will result in small scales being used an almost infinite number of times. If this occurs too often in an equation without a way to reconcile these massive discrepancies then the parameter is said to be nonrenormalizable. This means that under no set of realistic conditions or assumptions can the infinities be dealt with and so the equation is useless. Applying gravity to quantum equations results in this phenomenon over and over again. Here we see something crop up that irks quantum physicists to no end which is at low energies it is possible to work out a decent model for quantum gravity that basically mimics relativity. In other words gravity appears to obey relativity more than quantum mechanics. At high energies quantum mechanics fails again as an infinite number of parameters need to be addressed and so once again you cannot describe gravity in quantum terms.

It would appear that gravity really does behave in a more relative way than it does a quantum one. This however has not deterred quantum scientists form trying to plow through these obstacles and perhaps find a clever loophole around them. The whole idea behind quantum field theory is that all fields and their corresponding energy levels must be explainable in a single framework. There are various ways in which it has been proposed the fields will be excited and interact with matter that have been used by a number of quantum gravity theories. So we can see that the idea of fields which is central to quantum mechanics and the Standard Model in particular must be fully resolved before a complete theory of quantum gravity can be obtained. The full understanding of fields is not possessed by modern science and so best guesses or other criteria are suggested to make the mathematics balance out nicely in the end. There is a second parameter that also affects the nature of various quantum gravity theories and it pertains directly to space itself and relativity.

Throughout the ages as people have talked about space they have tried to figure out exactly how it behaves. This is evident as far back as recorded history goes where ancient thinkers believed that space was actually a sphere of sorts that was set high in the sky and contained the stars. The belief was that the stars were fixed in the sphere and could not move around freely. The only things that could move were the planets closest to Earth and the idea that these were planets was still unknown to the ancients who sometimes referred to them as wandering stars. The first proof that space was not a rigid solid object should have been deduced by those same ancients that saw the planets move as if the sphere were a solid nothing could move throughout it. If the sphere were a fluid of sorts then the planets could move across its surface at least like ships on the ocean. After this notion of a rigid sphere many variation of space took over and for a long time they all involved spheres and even spheres within spheres to produce a neat orderly clockwork universe. Removing ourselves for the need historically of people to make things behave according to perfect circles we find that space eventually was freed from its spherical confines and embraced as a true three dimensional structure that has no clear boundary and through which objects can move at will.

At will that is according to certain laws of the universe and it was here that we truly see the birth of space as a subject of debate in science. Was space constructed of an actual substance that filled the universe or was it devoid of everything and literally a vacuum? In addition to this space was solid and rigid where objects could move through it according to said laws but the actual background of space was fixed in place. If space was not a rigid structure to which everything else was bound then could space itself move about or at least a little? This is where we have a great question still popping up in modern science as space sometimes appears to be better explained as a solid unmoving structure and sometimes as a more malleable one.

The idea of malleability in space is actually quite old as many different scholars throughout the ages and in recent centuries have toyed with the concept of a liquid medium if you will for the void. The first real proof that this was true in the modern era came from the theory of general relativity which states that space can be bent by gravity. Essentially the gravity has an effect on space where by the geometrical shape of space becomes curved due to its presence. This is what is often referred to as the curvature of spacetime and many people will be familiar with the old demonstration of a heavy object suspended in a sheet of cloth whereby a smaller object is thrown into a temporary orbit around it. This is the classical demonstration of relativity as it has been taught for decades. The sheet is the surface of space according to relativity and the heavy central object can be thought of as something with great mass like the Earth. The Earth's mass bends the fabric of space into a gravity well and the smaller object is falling into it all the time but is spared thanks to its orbital velocity. The smaller object in this example can be anything such as the moon or a satellite of human design which orbits around the Earth but technically is always falling towards the Earth. This has been demonstrated as a property of space by examining how light is bent around objects as it passes by very gravitationally strong objects like a star or galaxy. The first direct proof of this came from a photograph of an eclipse taken by astronomers which showed stars that should normally be hidden by the Sun and appeared just at the outer rim of the Sun's edge during the totality moment of the eclipse. The proof that space was not rigid at large scales was proved once and for all and since that time no one has debated this; at large scales however. So if space is said to bend and be somewhat dynamic and proof is at hand how can it be open to question whether or not space really is bendy?

Quantum Gravity Theories

The answer comes from quantum mechanics where at very small scales the effects are not noticeable. In fact it is assumed in all quantum systems that the space that the objects occupy is in fact rigid. The idea that space is malleable is left out of quantum equations all the time and for a fairly good reason too. The scales that quantum effects are measured over are so small that any change in space would not be significant and as we have seen working in quantum mathematics is so difficult anyways anytime a variable can be removed to simplify the calculations involved it is usually done so. The problem is that to explain gravity in a quantum way you must eventually face reality which is of course the universe is much larger than a single quantum system. We know that the universe can include great swaths of space that are bent and therefore the scaling up of quantum gravity usually includes those nonrenormalizable parameters that consequently cause the quantum gravity theory tested to fail. Yet we have observed the universe and know that space is not rigid and so we are left with a challenge for any quantum gravity theory to overcome.

To this end quantum gravity theorists have decided to look at the problem from various angles and multiple theories have emerged. Some involve the idea of rigid space and some assume a more flexible kind. One thing that must be made massively clear at this point and that is the equations used to try and explain quantum gravity by modern science are just that, equations. They exist only within the framework of hypothetical mathematical models and do not describe the real universe. The real universe describes the real universe and all the theories are attempting to do is tweak the variables in case they are lucky enough to get an equation that matches observation. To date none have been able to do so but to date two stand out as the front runners in potential quantum gravity theories.

The two candidates are similar and different at the same time. The first you probably have heard of and it involves little strings. It is called shockingly enough string theory and offers one possible way of explaining quantum gravity using some interesting mathematical trickery. It also proposes that the universe is a little different than one might see in everyday life as you look around you. Most people happily exist in a three plus one dimensional world. The three dimensions of height, width and depth you are all familiar with and move about in freely. These are also referred to as the X, Y and Z axis in mathematical terms which you have probably studied in school. The plus one dimension is not really a true dimension as it is time, but in many casual discussions of physics it is referred to as the fourth dimension. Sometimes it is called the dimension through which the other three move and of course this is no different than you moving about throughout the day.

The reason the fourth dimension of time is not always referred to as a dimension is because at any static moment the other three still exist meaning you do not truly need the fourth in order to create an object. The passage of time is merely a construct used to show what happens to that object from static moment to static moment. On the other hand there are those who treat the passage of time as a very real and necessary thing. In fact they argue that the universe needs time in order to explain a few of the phenomena existing within it like entropy. So whether you believe time is real and actually woven into the fabric of space as Einstein did when he coined the term spacetime or whether you believe it to be a creation of the human mind does not really matter as you can use it or not in various calculations of the universe.

What you cannot go without in string theory is extra dimensions that do not involve time. That is right, string theory works well with more dimensions than just three and in fact in can incorporate up to six extra dimensions in addition to time. This is the only theory to incorporate extra dimensions into its conceptual framework in any serious way and it does so in an attempt to try and work gravity into quantum mechanics. Specifically it is trying to incorporate the graviton as it would then become the force carrier for gravity and all known force carriers are governed by quantum effects as far as modern science knows. This would therefore be the unification of gravity with the other three forces in a quantum framework that in theory should explain how the universe for the most part works. At least how the universe works on a small and large scale as the quantum graviton would be responsible for large scale gravitational effects like keeping a planet in orbit around its star. In order to do this string theory needs a few specific tools in order to have any hope of explaining gravity on the quantum level.

The first is the need to do away with the nonrenormalizable parameters in the equations that make gravity misbehave. At least misbehave mathematically for quantum theory as in the real universe gravity behaves just fine. So the way that string theorists go about this is to find a new way to describe the universe that is not based on the usually point particle physics of quantum field theory. This is where the second tool comes into play and where the theory gets its name derived from. This is of course the introduction of the quantum string which takes the place of the point particle and consists of a small, extremely small piece of quantum string that exists only in one dimension. The way string theory works is each string is fairly similar to one another but they can vibrate, oscillate or move in different ways to each other. These different types of movement will cause the string to behave as different particles, for example a particular vibration might denote spin or electric charge. Two strings that vibrate in different ways from each other will create particles with varying amounts of charge, spin and all the other properties of the known particles we can study today. One of the particular vibrations of these little strings is of course the long sought after graviton which would allow for a particle with all the right vibrations in it to exist in quantum mechanics. The addition of this graviton would be a unique particle that obeys all quantum laws shared by known particles that represent the known three other forces. This would seem like a complete picture of quantum gravity were it not for the rather massive obstacle of only three physical dimensions.

Simply put the mathematics for string theory in three dimensional space will not work out ever and that is that. So now we add the third tool of the string theorist and this is the six extra dimensions we touched on earlier. In order to get the equations to produce any potentially meaningful results six extra dimensions must be added to accommodate all the various string features and behaviors. The major difficulty encountered here is that you cannot add any more dimensions to the normal three dimensional space no matter how you try but string theorists argue that you can. The way they do this is to say that you cannot add more dimensions to the original three but you can add them inside the original three. This is one of the hardest concepts for anyone not familiar with physics to understand and in truth it is impossible for trained physicists to comprehend as well because the human mind is geared to three dimensions. So how do trained physicists tackle the problem of extra dimensions?

Through a process called compactifying. Compactifying is a way of folding extra dimensions into one dimensional mathematical spaces that can then fit into our three dimensional universe. Another way of looking at it is to make the mathematical predictions of various theories that use extra dimensions fit into a normal observable universe. The extra dimensions would normally get in the way and make all calculations literally meaningless and absurd, but if some of the dimensions are compactified then many extra dimensions are folded up neatly into something that can be worked with. This aspect of string theory is one of the biggest points of contention the theory faces as no one knows whether extra dimension exist at all. At the same time it is definitely one of the most interesting aspects of string theory mathematics to date. One of the reasons that it is interesting is that mathematically it can be made to create the graviton which is after all the whole point of any theory of quantum gravity.

For a moment however let us use one more example of extra dimensions to help illustrate how additional dimensions beyond our familiar three might exist in the universe. This example is much easier to visualize and should help the reader get acquainted with the concept if they are not already so. If we think of an ordinary string, wire or any fine filament we can see an object that exists in three dimensions as it has an X, Y and Z axis which we can measure. If we pull that object back from our eyes far enough we will not see three dimensions but rather only one as it now appears to us as a one dimensional line.

Now to take this even further if we were to shrink down to the size of a tiny speck of dust and walk on the surface of the object we would be aware of only two dimensions as we could move left or right, forward or back on it. Since the object is curved the whole expanse before us would always be flat and two dimensional. Now one last thought on the subject suppose we know that the object has three dimensions from afar but cannot see the surface of it. The surface still exists and the three dimensional physical object can be thought of as having a fourth dimension wrapped around the outside of it that we would only experience if we were the size of a dust mote. So depending on how we examine the object it can have one, two , three or even four dimensions despite us only being aware of a certain number given one of the scenarios. String theorists argue this point all the time as they state that we simply cannot see or sense the other dimensions which are compactified right in front of us.

If string theorists are correct that more dimensions exist than the three we are normally acquainted with then gravity stops being nonrenormalizable and a mathematical graviton can exist. This is the basis of all string theory to which there are of course a few different versions. The most prominent of the versions is called M theory which at its core is of course a string theory. The addition of relativity to M theory attempts to make a more complete version of string theory that will in total have an extra dimension over string theory. String theory normally has ten dimensions and M theory has eleven but this does help it try to account for the relativity seen in the universe. The problem with normal string theory is one of the same problems with all quantum mechanics which is it takes place on a static and rigid space background. As we have seen there are two types of backgrounds used in various theories. The problem with most quantum theories is that they treat space as solid and in the case of the hypothetical graviton the effects are dire.

Because gravity is the weakest of the four known forces and because all quantum systems are so incredibly small the addition of any gravitational effects basically do not register in quantum equations. To think of this another way recall your early lessons in mathematics where you learned about the concept of zero and how it is applied to regular numbers. When you first learned about numbers and how you could play around with them everything was simple, after all any child knows if you have one apple then get a second you have two apples and that can keep twice as many needle wielding doctors away. So we learned that one plus one is two and from there all forms of addition and subtraction were a simple concept to master. Even multiplying numbers is fairly straightforward as you simply add apples up very quickly in this method and soon you have a bushel basket of them. Division we all had a bit more trouble with and no one can deny that but at the end of the lesson it really was just multiplication in reverse so not hard to understand. Then we learned about zeroes and how they worked in with our regular numbers.

If you show a child one apple and then multiply that by zero you should have zero apples and yet the apple remains. Congratulations you have yourself now just learned how to confuse a child and taught them about multiplication of numbers and zeroes at the same time. Now if you want the child to quit mathematics straightaway tell them that when you divide the very real and solid one apple by zero it becomes undefined. This will surely confuse the child as the apple is still very much indeed well defined as an apple. It has all the apple properties you would expect: it is probably red, tastes sweet with some seeds inside and keeps the doctor away. So check, check and check it is still an apple despite zeroes best attempts at destroying it. This is a very real problem in all mathematics and one that many scientists and thinkers over the years have underlined quite rightfully. The real physical world is King and simply never wrong and thus our observations of it are never wrong either. Yet mathematics can sometimes fail to make sense of the real world and so if we check over the equations we find that there are no mistakes in any of the actual calculations. So this means that the mathematics has been performed correctly but the equations used to describe real world physical observations are wrong.

To take this a step further one should never use mathematics and the conclusions they can reach to formulate theories and laws about how the universe should behave. Mathematics can state up and down all day until it is red in the face that the apple should be undefined and the observation of the universe is never wrong; we still have an apple sitting on the table in front of us. Quantum mathematics are bound to the same failures and limitations as this and so caution must be used when trying to treat quantum mathematics as a basis for models of how the universe actually works. To throw gravity and zero into this fray we find that the incredibly small systems that quantum objects occupy feel very little effects of gravity at all and especially any distortion of spacetime. So if we add a value that is either zero for gravity or is close to but never actually reaches zero to different aspects of any quantum equation we are most likely going to end up with a value or parameter that falls to zero, becomes undefined or explodes up to infinity. Proof of this is not limited to quantum mathematics either as Einstein clearly has something wrong with his theory of general relativity which we can demonstrate with a black hole.

Specifically the heart of the black hole is thought to be a singularity which is a region of Einstein's spacetime that is so bent gravity becomes infinite and time stops dead. Both of these are of course dead wrong and a singularity is simply the end result of mathematics gone haywire when describing the real universe. The number of paradoxes stemming from such a singularity is practically infinite itself and would have had irrevocable and dire consequences for the entire cosmos long before humans could even evolve to ponder what a singularity might be. So back to quantum mechanics, gravity and spacetime for the given object in a quantum system we find that quantum mechanics traditionally treats it as if it has no real effect. Quantum mechanics in a sense argues that the amount that space would be affected at the quantum level is essentially zero and so it is left out of the equations.

On large scales however gravity does affect space and modern science simply does not know if gravity can have profound effects on space at quantum scales so leaving space as a rigid structure in quantum calculations can be disastrous. This is the strength of M theory however as it tries to incorporate general relativity into its equations and thereby deal with the idea of non-static space. The end result for M theory is not much different than regular string theory as it would create a quantum theory of gravity which would be able to mesh with the three known forces. The only real difference is that unlike regular string theory large scale gravitational effects in the universe would make more sense and it is for this reason a great number of scientists favor the M theory version of string theories.

There are a number of specific points of interest in M theory that deserve mention as they help to explain why M theory is the string theory of choice so to speak. The first is that M theory acts as a kind of umbrella to many string theories all at once. This is a bit of a dodge on the part of M theorists as it seems they have there sticky little fingers in the cookie jars of a great many theories. However it is not unwarranted as many of the different string theories can be useful at different times in the framework of M theory to try and describe gravity.

This brings up a topic known as S-duality and T-duality when discussing these various theories. To start with let us talk a little about the strings in string theory as it is unclear from the name what they really are. The obvious answer is that they are little bits of string and in a quantum sense that is actually quite apt. On a more theoretical level they are small quantum objects, possibly the smallest that string theory can permit and they resemble little pieces of string, hence the name. These strings can come in two different varieties which are open or closed. Open strings look just like a little string with two ends that are not attached to anything. Closed strings are made of the exact same material but their ends have been joined up onto themselves to form a circle. As we have stated the way in which each string type vibrates will denote what type of particle it creates. The actual strings are too small to see and we only observe them as a particle leaving their open or closed states a little bit hidden. No matter what type of string you talk about all the theories aim to give an answer for quantum gravity. Where the interesting thing begins for these theories is that they do not all use the same types of strings. Some use open and closed while others use closed only.

This is where a phenomenon known as S-duality occurs between various theories and therefore helps to place them under the umbrella of the broader M theory. Essentially S-duality is a mathematical way of describing the same thing but in two different ways. So for instance in one particular theory we might find that a set of particles that interact strongly together. A strong interaction is one where the particles readily join up and decay away from one another. A weak interaction on the other hand is just the opposite with particles only occasionally joining up and decaying again. In the example of the strongly interacting particles we find that in a different theory these same particles can be described as multiple weakly interacting particles in another theory. So one theory uses smaller numbers of strong particles and another uses many more weakly interacting particles but in the end the two different theories are not really different; both are describing the same thing only in two different ways. Think of it as 100 grains of sand to get a weight of say X, or one pebble to get a weight of X as both are made of stone you are really just trying to get that weight of X.

The second duality is called T-duality amongst string theories and it introduces something called a winding number. This is an idea that can best be described as looking at a single string and how it moves around various objects and in particular circular ones. For example if the string were to move around a hypothetical extra dimension that was circular you would be able to count the number of times the string was able to wind itself around it. This would of course give you the winding number which can be coupled with the momentum of the string as quantum objects tend to have a momentum as we have seen. So you have a winding number and a momentum which in one theory might be for simplicities sake A and B. Yet in another theory the numbers will be reversed so that you have B and A. This is the T-duality which can be used to link two theories together just as the S-duality can as well. The fact that theories can be linked through these dualities allows them to fall under the general jurisdiction of M theory and consequently helps M theory be far more robust than any one theory on its own.

There is one unusual aspect of all string theories and of course especially M theory that is not found in normal science and exists solely in the theoretical realm of string theories and it is the Brane. A brane is pronounced the same as the word and organ brain yet shares nothing in common with its intellectual counterpart. A brain such as a human brain is responsible for the transmission of thoughts from place to place. A brane in M theory is a solid structure that exists in extra dimensions in string theories and is responsible for projecting extra dimensions onto a point particle in some fashion. Branes are all about extra dimensions and nothing else but it is the way in which things interact with the branes that helps to explain some of the ideas behind string theory. According to normal physics which is seen as classical or relativistic objects exist within three spatial dimensions and no more. In string theory point particles are hypothesized to be constructed of one dimensional strings that vibrate in different ways to produce the familiar particles we see everyday. In this sense a normal particle will look as if it has three dimensions but really only be constructed out of one which it gets from the string. The strings themselves will get their dimensions from branes or at least the way in which they interact with a brane. The brane in turn is the last step in the ladder for string theorists as nothing creates them according to string theory they are just the background parts of the universe.

Branes are thought to inhabit space pretty much everywhere and have certain characteristics that describe them. One characteristic is their world-volume which is the area of space that they fill and it has both a volume and surface area. The word brane comes from the two dimensional surface of a membrane. The second characteristics they possess are various physical qualities such as mass or charge and these help denote what interactions they will produce with strings. The way branes work is that they translate the vibrational information of a string from one dimensional to higher dimensions of a point particle. This is not the only type of interaction permitted as strings and branes can be viewed separately creating various types of dimensions which may or may not always involve a point particle. The combinations are diverse but for simplicities sake I am illustrating the straightforward interaction of the two that give rise to the point particles with which we are familiar. The reason this is so done is because there are no ways in which to test for branes, strings or extra dimensions in the laboratory as all of our senses and equipment operate on only three dimensions. This may of course be for good reason as it is not known whether anything of extra dimensions, strings or branes is even real.

Yet if string theorists want to test their ideas ways of translating string and brane theories need to be found such that normal point particle systems, both machine and human can test for their existence. This is the perfect time to remind the reader of the idea of compactification which is the string theorist's tool for condensing all these theoretical extra dimensions into the normal three that we perceive so that tests can be performed. Each testable particle in string theory would correspond to a specific combination of strings, branes or both. The mathematics of which would dictate what specific properties should correspond to each known particle. To this end the string theorists have devised a system of different branes and string interactions and one such example is the D brane. Not useful to all string theories as some do not use open strings, M theory does use open strings and is the umbrella theory of string theories and is therefore worth mentioning D branes here.

A D brane arises from the arrangement of open strings onto a specific brane called a D brane where the D stands for a mathematical property known as a Dirichlet boundary. The way that the open strings interact with the brane is by placing their open ends so that they lie attached to the surface of the brane. The usefulness of this particular arrangement of strings is such that the behavior of the string is a gauge theory which makes it similar to the gauge theories used by the standard model. Just as with S-duality where one aspect of a theory describes the same thing in another's proponents of string theory are hoping to use this kind of link to bring the naturally occurring string theory graviton into focus with the standard models other force carriers. This would of course be the ultimate goal of any quantum theory of gravity and the various numbers of dimensions brought about by branes and strings is thought to help explain this. A string can have dimension one which we know and a brane with dimension zero can be used to describe a point particle. Currently M theory can mathematically describe hypothetical branes of up to five dimensions but in the future should be able to go higher if such possibilities exist. Combinations of strings and their vibrational states along with different brane types will give rise to multiple dimensions, multiple particles and a complete description of all four forces according to string theorists.

Another interesting aspect of string theories which falls under the dominion of M theory is something called Supersymmetry. Supersymmetry is an aspect of more than one type of string theory which involves particles that can be thought of as mirrors of known particles. Well known are things like fermions and bosons which constitute normal parts of all Standard Model theories and are included in string theories as well. The idea of supersymmetry introduces a new group of particles that are related to the known ones but differ in one of their characteristics. Fermions are the rough building blocks of matter and have half integer spin values. While bosons are the force carriers between these that have whole integer spin values. Supersymmetry gives each pair a mathematical counterpart such that a fermion will have a theoretical boson super partner and each boson will have a theoretical fermion super partner. The difference between these two counterpart particles is a half spin value while they possess the same mass. If we look at a real world example of this we find known particles but the idea is based on the discovery of unknown particles which we know now do exist.

The most basic one is the positron which is the anti-matter version of the electron and at first was only theorized to exist. It is a particle that shares all the same characteristics with an electron but has in this case the opposite charge. So the electrons negative charge becomes a positive one in the positron even though its mass is equivalent. Supporters of super symmetric particles are hoping to find the same sort of real world particle with the electron which would be called the selectron which obviously stands for super electron. Each known particle would then have its own super partner which can be used to help explain gravity on the quantum level. This is due to the fact that when including supersymmetric particles into the Standard Model you will naturally get a graviton particle in the various newly calculated particles.

The inclusion of a graviton would be the final piece in the superstring puzzle to unifying quantum theories with relativity. The name supergravity theory is given to the combination of supersymmetry and relativity which give certain solutions to the quantum gravity equations depending on whether the system is calculated for low or high energies. The interesting thing about supersymmetric particles is that all of the quantum values associated with them can be calculated before any experiments are run. Then knowing these values scientists can begin searching for them with experiments such as those conducted at the Large Hadron Collider. Knowing beforehand what they are looking for is key to proving the supersymmetry theory correct and experiments using particle colliders are well understood and fairly run of the mill to particle physicists. To date however no super partners have been detected using any colliders or other experiments and this is a serious obstacle to the theory as the exact values were predicted and this is akin to putting your theoretical money where your mouth is.

Nevertheless the idea behind supersymmetry has fueled much research as it would fit neatly into the M theory framework and solve some other problems along the way. One of these problems would be the hierarchy problem which to date has not been solved by modern science. The hierarchy problem is the energy discrepancy between the weak force and gravity which does not fit with the theories of the Standard Model. It is often said that gravity is the weakest of the four known forces but it is not often stated just how weak it truly is. The difference between the strong, weak and electromagnetic forces is of course significant but pales in comparison with the difference between the weak force and gravity. Gravity as a force is approximately 10 000 billion, billion, billion times weaker than the weak force. So this is of course a huge problem for modern science and the Standard Model as they both have no way of explaining this huge gap in strengths. What this stems from is a mathematical problem that we have touched on before called renormalization. When looking at very small quantum systems and the larger scale experiments massive differences in values are realized.

This is again the single largest problem in quantum mechanics as any attempt to make a theory of quantum gravity must be reconciled with relativity. While some argue that quantum will one day replace relativity as an explanation of the universe what quantum can never do away with is the very real and observable relativistic effects in the cosmos. Even if someone explains things in a quantum framework they must take into account the observed relativity going on all around us like light being bent by large gravitational objects such as stars, black holes or galaxies. Those effects are relative and real even if one day they are explained in terms of quantum mechanics. This is the problem quantum gravity faces daily as a theory and the huge gaps in values between calculations and observations. The idea of supersymmetric particles hopes to close this gap or explain it fully by providing the missing variables from the equations which consistently do not match observation. If supersymmetry is proven correct then not only would quantum gravity be explained but other problems such as the hierarchy would be solved too. For these reasons many scientists are backing the M theory horse in the race to find a workable theory of quantum gravity that can be used to explain relativistic effects.

The M theory as a candidate for a possible merging of gravity with the quantum world is not the only horse in the race however and the second favored contender by modern science is known as Loop Quantum Gravity. Loop quantum gravity goes about solving the problem of combining the force of gravity with the quantum world in a completely different and similar manner to string theories. The similarity stems from the need to explain at least on the quantum scale interactions of any theoretical gravitons with the rest of the standard model. Of course like all the other string theories it also needs to include the theoretical particle known as the graviton in the first place. This is where the similarities stop as string theory attempts to solve the gravity problem from a purely quantum stand point and only includes relativity as an after thought so long as it does not interfere with the quantum world in any way. Loop quantum gravity tries to solve the gravity problem from the opposite end of the physics spectrum by including relativity straight out of the gate and trying to formulate it into quantum mechanics afterwards.

In order to do this it starts off by acknowledging the fact that space is subject to relativistic effects and that these must be addressed on the quantum scale in order to explain gravity which is large scale and classical in a quantum framework. Therefore according to loop quantum gravity space must be dynamic and it introduces little quantum loops in order to make this happen. The loops are similar to the spacefoam that Einstein rather mockingly threw at the quantum physicists of yesteryears. Einstein was backing the purely classical framework to the universe as his own work is merely a tweaking of Newton's which of course was purely classical in nature. What could not be ignored however was the very real quantum nature of particles and since particles make up matter which is part of the universe at least one part of the universe must be explainable through quantum mechanics. To this end Einstein simply incorporated quantum into his relativity by saying that the underlying structure of space was a very small foam like system. It is about as descriptive as Einstein's supposed spacetime which he never actually describes what it really is only that it exists. No more thought or work was done afterwards in any serious way by Einstein or his followers into the subject and this is what makes it such a glib and backhanded invention towards quantum physicists.

Loop quantum gravity however attempts to heal the wounds of the decades old quantum versus relativity debate by explaining that the spacefoam can be thought of as little loops. These loops would be on the order of a Planck length beyond which anything smaller makes no sense to modern science, quantum physics and the Standard Model. Is there a length smaller than this or a time scale shorter than a Planck time? Possibly and no one can say that there is not, only that based on the current understanding modern science has of particles and the universe anything smaller does not fit with their mathematics. Not fitting with the current accepted mathematics does not mean that it does not or cannot exist only that the mathematics themselves do not allow for it. It is entirely possible that smaller spaces and times exist as the theories favored by modern science have many paradoxes, inconsistencies and holes in the knowledge so simply saying that everything dictated by those theories is correct is folly. After all there was a time when modern science was convinced that only two particles existed in the universe the electron and the proton and that they were arranged in a homogenous fashion rather than in little bundles called atoms.

This quandary aside the loops in loop quantum gravity would be further interwoven with each other to create the background of spacetime that can produce the observable effects seen in the cosmos at very large scales. The quantum nature of the theory comes in two forms the first of which is the inclusion of a graviton by which the field of gravity interacts with particles. The second is that the size of each of these little granules has a discrete value that is quantized and which will change in discrete rigid jumps and these granules can be many sizes right next to one another. The way in which this is explained is by a mathematical system called a spin network. Originally the spin networks can be used to explain spacetime in a purely quantum sense and when taken farther can be used to merge gravity into the Standard Model. This is because according to relativity gravity results from the mathematical geometry of spacetime and so the shape of space in a sense creates the gravity. This means that relativistic effects are directly linked to space and for loop quantum gravity this can be explained at the quantum level using loops and the graviton.

The loops are calculated in two different ways for loop quantum gravity and as such there are now two versions of the theory. The first is a canonical quantization of the theory so as to more accurately focus on relativity and its classical description of spacetime. A canonical quantization is essentially trying to explain a classical theory in quantum terms yet at the same time maintaining as much of the original as possible. In this way one hopes to preserve the classical descriptions for real world observations as much as possible yet through a quantum framework. If successfully done then a workable quantum theory for a macro system can be obtained. Obviously the relativistic observations made throughout the cosmos are macro in scale when compared to quantum so if the loop quantum theorists are successful in explaining relativity in quantum terms then they are one step closer to a quantum theory of gravity. The general mathematical idea behind canonical quantization of a classical system begins with a single particle. In classical mechanics particles have properties like momentum which can be described using a coordinate system. In a static space which is of course non-relativistic it is possible to describe a particle in both classical mechanics and quantum mechanics therefore the quantum description does not disturb the original classical description in any way and so it is allowed. The description in quantum terms uses what are known as operators acting on particles to describe the quantum state of the system.

The second hurdle in describing classical mechanics in quantum terms come when one tries to make a quantum translation of fields. This is a necessary step because not all quantum systems are statics with a static number of particles there are systems where particles can be either created or destroyed which would therefore affect the description of the system overall. Look at the humble photon for example which is part of the electromagnetic field and consider what happens to it if it strikes a particle. The system in which the particle and photon exist will change from its initial conditions as the photon might be absorbed by the particle or perhaps absorbed and reemitted later. In order to address this very real classical system in quantum terms fields need to be addressed head on. To do this it is sufficient to replace the classical variables describing the field with quantum operators. So things like amplitude of a wave are now quantized so that changes are made in step by step jumps like all quantum systems. This is applicable to the excitations of a particle given a particular energy state and so the particle can now be described as a wavefunction. If we take this a step further one finds that the equations of a single particle in a field are very similar to the equations of the field itself so in a sense a wavefunction for a particle describes a wavefunction for a field as well. This led to the naming of this canonical quantization of classical fields to be named second quantizations as they appear to be a duplicate of the first single particle versions. In any event the idea of quantizing a classical system was mathematically achieved and can be used in theories such as loop quantum gravity which involve fields and particles such as the hypothetical graviton.

A second version of loop quantum gravity exists and it is called the spin foam version which shares many similarities to the first version. It too considers relativity a part of its quantum equations, will use a graviton and divides space into small sections but it is how it divides up the space that is a little different. Whereas the spin networks of normal loop quantum gravity are used to describe space the spin foam versions attempts to address spacetime in a geometric way. The spin foams are made up of two dimensional faces which can be used to build up a geometric representation of spacetime. In addition the theory attempts to preserve the relativity of all spacetime such that it does not vary from place to place or from coordinate to coordinate. In a similar way to the normal loops in loop quantum gravity the spin foams attempt to describe space in a quantum manner that can then be used to hypothetically provide a workable theory of gravity by explaining the relativistic effects of space. This is due to the starting premise of the two loop quantum theories which adopts Einstein's idea that gravity is created from the geometrical shape of spacetime. Whether these theories are correct and this includes Einstein's idea that gravity is the created product of the shape of spacetime, then the search will be on for the graviton to help fill in the Standard Model and unite the four known fundamental forces.

There is one more candidate for a theory of quantum gravity which deserves mention and it too emphasizes relativity. We have seen in string and M theories that quantum effects are given center stage in the unification of all four known fundamental forces. We have also seen in loop quantum theories the application of relativity as a way of incorporating quantum mechanics again. However there is another third approach that would not focus on quantum effects that much at all and when it does it is incorporated as almost an after thought. Not that this theory denies the usefulness of quantum mechanics or quantum interpretations of particles but it so much more strongly highlights relativity instead. The theory is called scale theory and the most interesting thing about it is that it suggests something special about spacetime beyond the normal relativity interpretation that the shape it has dictates gravity. In scale theory the very shape of spacetime can also give rise to quantum effects themselves and so explain both gravity and the origins of quantum mechanics in one theory.

The approach for such a feat begins with no particles at all but rather spacetime itself and the geometric shape it takes. Just as loop quantum gravities define spacetime as loops or spin foams scale relativity also has its own unique geometry for spacetime. This geometry is perhaps the most mathematical of the all as it is derived purely from equations about space and how different scales can be described within it. To start with one must remember that in relativity there is no such thing as an absolute value that can be given for a system or anywhere in the universe. An absolute value can be thought of as anything which cannot be changed and therefore represents a sort of benchmark measurement. To think about this simply think of money and for a moment think of a dollar which of course is not an absolute monetary amount as you can break up a dollar into numerous smaller coins. Now if we go to the bottom, in most countries scale of money we find the penny or one cent piece which is absolute as you cannot get change for a penny so to speak. So in this fashion a penny is an absolute value as it is the absolute smallest currency you can possess physically.

What relativity states is that no such values exist in the universe as they can only be calculated relative to one another. So for instance the motion of a planet or even a single particle cannot be referenced to any absolute motion in space but only against another planet or particle for example. The motion is relative from the first object to the second and this idea is used in many aspects of physics by relativity such as orientation of an object as there is no up in space per se or the acceleration of an object as there is no stand still in space either. So in this way any given scale of an object or system can be made to obey relativity as well and as in quantum mechanics you have a quantum state in scale relativity you have a scale state.

The next step in the theory begins with an American scientist named Richard Feynman who was studying quantum objects and noticed something interesting about how they moved. While looking for the paths that quantum objects travelled on from point to point he determined that the paths they took was more often than not highly irregular. Think of a small child in the backyard who left to their own devices needs to cross the yard to go inside and now imagine the likelihood of that same child travelling in a straight line from the back of the yard to the back door. Probability states that the child will get distracted on the way back by perhaps a sandbox, a grasshopper or the family cat sleeping on the porch all before getting to the door. Feynman noticed that quantum particles could not make up there minds which way to go in any given moment and while traversing space from point to point moved only in unexplainable ways almost like a quantum chaos theory. This is where scale relativity attempts to explain geometrically the nature of such behavior through the shape of spacetime itself. The main idea behind scale relativity is that space time is responsible for the creation of gravity through a geometric explanation of relativity. This is basically the same as normal relativity which uses geometry of spacetime but it does have one big difference which is the shape of spacetime is fractal in nature.

A fractal is a pattern that can be represented mathematically such that the entirety of it is constructed by simply making smaller and smaller copies of the original pattern. Think for a moment of a simple triangle as the starting shape and to it add another identical triangle at the very middle of each of the three sides but make this second triangle smaller. Now add a third smaller still triangle to the very middle of the second triangles. Repeat this for infinity and you have created a very simple fractal triangle. One of the more well known fractals is the familiar Mandlebrot Set which was created by the mathematician of the same name; Benoit Mandelbrot.

So what scale relativity is suggesting is that the very geometric shape of space is fractal in nature meaning that the little pieces that make up the very first parts of spacetime are repeated on top of and within it on a smaller scale to infinity. What this has to do with the seemingly random and erratic motions of particles is that scale relativity further hypothesizes that the motion of the particles is linked to the fractality of spacetime. So in a sense when you are watching the path of a specific particle unfold in a quantum experiment what you are also seeing is the fractal shape that spacetime has in that system as well. The particle is almost passive and simply following the path set out before it by spacetime itself. The fractal shape of spacetime would also adhere to the rules of relativity that there are no absolutes and therefore no absolute scales either. Instead scales would be defined as ratios of one to another. So what constitutes a ratio for scale relativity?

The best way to think of this is to picture something lets say a house from far away at great enough distance you see only a smudge on the horizon. Get closer and you can see a rectangular body with a triangular head on top. Closer still and you can make out a door, chimney and windows which are all details you could not see before. The object is the same but the clarity with which you could discern its individual components changes just as resolution changes for a photograph. This is exactly what is meant in scale relativity and the term resolution is used to describe it accurately. So the ratios exist between scales which we can think of as say the far, medium and close up views of the house. Which is the correct view? Well none really as that would presuppose an absolute resolution with which to look at the house and in reality there is no such exact right distance to view the house. So the same is true in scale relativity as the resolution changes so does the ratio of the scales which are relative to one another or the state of scale. In this way it might be possible to explain physical effects in the universe by examining the resolution of the state of scale which might be down to the particle size.

At the particle size the very, very small fractal shapes of space would in theory explain the motion of the quantum object. At larger scales of the fractal pattern things like gravity could be explained as a relative geometry of spacetime and would account for one object orbiting another. This large scale explanation of gravity would be a way of incorporating relativity into a quantum theory of gravity. This is because according to the scale relativity theory just the shape of spacetime itself creates both gravity and quantum effects. It has been suggested that many of the quantum effects observed in experiments might be caused by a fractal spacetime which is due partly to the infinite scales possible. You can after all continuously blow up a fractal or shrink it down a little more and always get the same shape no matter how much you change scale. The same would be true of a fractal spacetime as you would need a type of fluid mechanics in order to move smoothly from one small scale to a large one or vice versa. This fluid motion to the scales is needed to explain the use of fractals on both micro and macro scale systems. If correct then scale relativity could achieve a meshing of the quantum and relative theories. Scale relativity however creates a host of problems of its own when dealing with other unsolved problems of the universe such as vacuum energy overloads and critical densities to space at the time of the Big Bang and inflation.

Problems with Gravity and Mathematics

The question on every physicists mind today is are these theories even remotely correct? It is possible that all of them are wrong and something new must be introduced to explain the force of gravity. If one is found to be correct even somewhat so we might be looking at the next step in our understanding of gravity as a whole and this is something that has been happening for ages. A quick history of the theory of gravity shows that many old models are either completely replaced or they are slightly modified in order to better fit observations made about gravity in the cosmos. The first generally accepted idea of gravity belongs to prehistory where our ancient ancestors observed things falling, being heavy or keeping them from flying like some animals. In the days of Aristotle and his kin the accepted notion was that gravity worked harder, faster and stronger on heavier objects. This was the empirical explanation for why a rock would fall faster than a feather to the ground. Centuries later Galileo began to challenge this accepted ideal that held sway for so long and insisted that gravity acted equally on all objects. The famous story of Galileo's proof of this fact has seen many versions over the years but the basic accepted tale goes that he climbed to the top of the Tower of Pisa and dropped cannonballs of different weights off the side. What he noticed was that they both impacted the ground at the same time or at least so close together that neither he nor any other observer could discern the difference.

This immediately disproved Aristotle and his idea that gravity acted stronger on heavier objects. A victory for science and proving old theories wrong while adopting new ones. This of course begged a new explanation for why a feather fell more slowly to Earth than a rock. Galileo guessed that the air must be offering a form of resistance to the very light feather and therefore held it aloft longer while the cannonballs had no trouble plummeting through the atmosphere. Galileo himself offered no idea of what gravity was or how it worked but he did succeed in correcting a centuries old mistake and set the stage for future scientists. His idea was proven correct very famously in an experiment conducted by David Scott and the crew of the Apollo 15 lunar mission in late July of 1971. While standing on the surface of the moon which has virtually no atmosphere whatsoever Scott held in one hand a hammer and in the other a feather which after giving a speech about gravitational attraction he dropped both at the same time. Sure enough the hammer and the feather impacted the lunar soil at the same moment proving that gravity did in fact act equally on all objects regardless of mass. So Galileo was the successful in superseding Aristotle in the way gravity worked upon objects of differing mass and again set the stage for the next step in understanding gravity and this time almost completely.

The next scientist to work seriously on gravity is arguably the most important and brilliant scientist that has ever lived and his name is Sir Isaac Newton. The reason that Newton is so highly regarded is that his work covers more areas of science than anyone else's and his work is the most useful and correct in explaining the universe. Even at quantum scales some of Newton's laws hold while Einstein's work falls apart even though it is widely accepted that Einstein's theory of general relativity is the successor to Newton's work. What Newton discovered about gravity was not what it is as no theory in modern science knows that, instead he figured out how it works. He mathematically proved that it was an attractive force between two bodies and the masses of both combined with their relative distance from one another could be used to calculate the force of attraction. Here we see another aspect of his genius at work and this is the first real and workable model of gravity. Up until this point theories were simply guess work which led to them being disposed of in favor of new ones. Newton's theory of gravity would and never will be thrown out but rather modified instead to fine tune the model to the observations of the universe.

The famous story associated with Newton's insight into how gravity worked is of course the classic science tale of him sitting peacefully beneath an apple tree in his yard while he contemplated the workings of the universe. As he sits and thinks an apple falls from the branches and hits Newton upon the head where he is struck with inspiration. The inspiration of course is that the same invisible force that attracts the apple to the Earth is what holds the moon he had been looking at to the Earth as well. Part of his genius also comes from his creation of the gravitational constant which he inserted in his famous

$$F = G\frac{m_1 m_2}{r^2}$$

equation: . The force of attraction in his equation is of course denoted by the capital F and it is what you are calculating for between any two bodies. The m1 and m2 are obviously the masses of the objects and the r is the distance between them. The unknown factor is the G and it was unknown until Newton figured out what it exactly was. The G is the gravitational constant of the universe which should be made clear here is not the same as the free fall velocity used in physics calculations on Earth and that is denoted by a little g. What makes it so special is that the G should be a representation of the strength of the actual force or effect of gravity in the universe.

To think of it another way Newton was calculating a value for the force long before big fancy particle colliders were helping to probe the other three forces with accuracy. In fact long before the three other forces were even hypothesized Newton had already figured out how strong gravity is and that is a huge part of his genius. The gravitational constant is still used today hundreds of years after it was invented by Newton and used in his law of universal gravitation as well as centuries later by Einstein and his work in general relativity. The value has remained unchanged since the formation of the universe and so is one of the most important numbers science has ever encountered as gravity is the key force responsible for shaping the cosmos. Acting on a far larger scale than any of the other forces and effecting changes in all matter and energy in the universe gravity has produced some of the most incredible structures ever imagined. Things like stars, planets and whole galaxy clusters which brings us to Einstein and his work on gravity specifically his theory of general relativity.

Newton's laws are incredible and have stood the test of time for the most part since 1687 when he first published them although he had discovered them some years earlier. At certain scales Newton's laws are holding steadfast but at more moderate ones it became obvious that they required some tweaking. While Newton's work on gravity enjoyed massive success and fame when after studying the motion of Uranus it was theorized that another planet should exist beyond it. This was indeed found to be true and Neptune was soon added to our solar family as the eighth known planet. Like a child with a new toy scientists are quick to apply a new idea, theory or model to every problem or system they can get their brains hands on so to speak. To this end Newton's laws of gravity were used to measure and compare every object in the solar system to one another in order to learn more about the solar system in general. While the discovery of Neptune was a success it was noticed that not all of the planets and their motions could be accounted for perfectly by Newton's laws. The planet giving them the most trouble was Mercury as there were apparent discrepancies in its orbit that did not fit with Newton's laws. This revealed that while Newton's laws on gravity were by far the best available and did work perfectly under some conditions it was not a complete picture of how gravity affects objects and something was missing.

This is where I must interject a point of highest importance on the concepts of theory and observation in the universe. We are looking back hundreds of years now to the days of Newton and can recognize that his theory while it explained much did not explain everything and so when a new theory that tweaked his original a little came along to explain things a bit better we accepted it. This new theory is of course general relativity and now that it is 100 years old as well as I write this we would not bat an eye at the idea of using it instead of Newton's laws. The reason we selected it is because it better explained the observations made of the cosmos and so supplanted the old model. Rather than stick to the old accepted ideas of Newton as groundbreaking as they were science had to move on and utilize a new theory rather than try to invent or force the universe to conform to an idea that existed purely in mathematical form. Can you see where I am going with this yet?

Modern science needs to revamp its thinking on the universe and how everything in it works as the currently accepted models do not explain everything and many discrepancies exist within the various theories just as they existed in Newton's. While it may sound like heresy to many scientists today it should not as those same scientists are the ones who would not bat an eye at the notion that Newton was not fully right. Nor would they bat an eye that we do not live in a geocentric universe standing on a flat Earth. So I say that modern science needs a new way of looking at the universe even if it seems to be completely divergent from many of the currently accepted and pursued theories milling around today. The Standard Model, quantum mechanics and general relativity are all incorrect as none of them can begin to completely explain how the universe works in its entirety.

If you do not believe me simply ask your friendly local theoretical physicists which models are correct and incorrect and you will soon see that they can come to no single agreement. Some like one and hate another, for example when the Higg's boson was revealed and its rough mass announced it popped up at approximately 125 GeV which was about the middle of the two predicted and hoped for values of 115 GeV and 140 GeV. A low end value of 115 GeV was to favor the idea of a super symmetrical universe and fit with all the current mathematical predictions for one. A high end value of 140 GeV was leaning towards the multiverse hypothesis which again was rooted in the mathematics and not observations. So two of the leading theories on how the universe is constructed are wrong based off of their own math as neither liked to find out the Higg's boson weighed in at 125 GeV.

The bottom line here is a real world example of mathematical models being used to try and force the universe to be something it is not in order to simply fit the equations. In reality the universe could care less about our little human ideas and will, is and has been doing exactly as it pleases since it was created. Nothing in the universe is wrong it simply cannot be as it is the universe itself and exists exactly as it does period. So just as we discarded old theories such as the geocentric universe or modified existing ones like Newton's laws of gravity it is time for us to do this once again as none of the models used in contemporary science can come close to explaining the universe at all. If they did we would have no paradoxes, no discrepancies and not be still looking for things like new particles. The first people to tell you the Standard Model is wrong are those who study the Standard Model as they know it contains gaps in the knowledge of the universe and are continually looking for ways to fill them. In the twentieth century it was immediately obvious that Einstein's general theory of relativity had flaws in it and was just as inaccurate as Newton's when it came to describing the cosmos. It has simply been kept in use as it currently is the favored model of physicists today. A search is on for a replacement though as we have seen with things like M theory and the like.

This is exactly what the scientists in the late 1800s had to face as all attempts to reconcile the apparent errors in Mercury's orbit with Newton's laws had failed. Just as scientists today are trying to invent things like new particles in order to explain the errors in their own models scientists hundreds of years ago tried to find new planets to explain Mercury's behavior. The way they did this was to look for a new planet closer to the Sun that could explain the orbit. The general idea was actually a very good one at the time as extra planets had solved problems before and in theory should be able to do so again. So in the same way Neptune was perturbing Uranus' orbit it was believed that another unseen and as yet undiscovered planet must so too be acting upon Mercury. It was in a way the reverse thinking of the Uranus problem where instead of looking for a planet farther out in the solar system they needed to find one closer in and in fact it would have been the tenth planet by modern standards. At the time of course they did not know of the existence of Pluto so to them there were only eight planets in our solar system.

On a side note much debate rages back and forth about whether or not Pluto is actually a planet or a dwarf planet and when I say rages I mean exactly that. The two sides are quite heated about whether or not Pluto deserves the distinction of being called a planet or not. Some argue it must have certain planet like qualities such as being able to conform to a roughly spherical shape under its own gravity something which most bodies around 500 km across or greater can easily manage. Well Pluto checks out there as it is over five times that diameter and it even has its own moons to boot, Charon being the largest at $2/5^{th}$ the size of Pluto. One other possible characteristic of a true planet is the ability to clear its own orbital path. Its orbital path is the trajectory by which it circles the Sun and by clear it what is meant is that given sufficient time the planet will pull in any solar materials in its way. This is done by the planets gravitational field which will eventually remove all debris left over from the formation of the solar system and hence clear its orbital path. Well as far as we can tell Pluto has done that as well.

Some however argue that Pluto is not a true planet because of things like the center of gravity created by it and its largest moon Charon. Since the center of gravity of the two objects in the system lies in the space between them and not inside the sphere of Pluto itself some argue that Pluto is not a true planet. The Earth for example contains the center of gravity for itself and the moon a quality some say is necessary to be a real planet. The truth of the matter is that Pluto in my opinion should be considered a planet for the simple fact that it is difficult for anyone to truly say what being a planet really is.

I will give you a simple idea from biology that can be imparted to Pluto and all planets, stars or any object in the universe. To compare any of the cosmic objects with a biological one is simple we must only look at its life history data. In biological terms this is how a creature is born, grows, eats, mates and does everything in its life before it eventually dies. If we consider a human then we are faced with a problem as a human child is indeed a homo sapien but it is not an adult homo sapiens. So which is really a human is it the child or the adult? Well of course there is no right answer both are humans in different stages of their lives. The same can be said for cosmic objects like a planet and specifically Pluto as it is most likely a planetoid left over from the creation of the solar system. That being said there were literally hundreds of such planetoids flying hither and thither throughout the solar system and some eventually ran into each other to form larger objects like the Earth. That being said Pluto is simply a young planet developmentally compared to the Earth but both are planets in different stages of development.

If you need more proof look at the Earth and any of the gas giants and tell me which a real planet is. The inner rocky planets did at one point have gaseous exteriors like the outer gas giants or at least they tried to but the Sun's heat and solar winds were too intense to allow for gas to remain in large quantities. In the outer solar system beyond the asteroid belt the temperatures are much colder and gas and water ice were able to be collected in great amounts by the gas giants. Therefore Earth and the other inner rocky planets should be considered as not true planets because they lost a vital component of their make up with the stripped away gas and water ice. Only the outer planets have all the components that it takes to be a true planet as they were able to clear their orbital paths of all gas and water ice as well and hang on to it until today. This is of course incorrect as the Earth and other inner rocky planets are true planets just in a different stage of development from the gas giants. The baby is Pluto, the inner rocky planets are children and the outer gas giants are adults yet they are all part of the same species Sunna Planthera. So in my humble opinion Pluto is a planet and deserves to be included in the family of our solar system. This concludes our tangent into planets and the Pluto debate as we really should be getting back to Mercury and Newton I only included this as a chance to hopefully help out in the worldwide discussion of the topic.

So Mercury is the closest planet to the Sun as we now know but at the time of the Newton Mercury problem it was widely speculated that a yet unseen plant was orbiting even closer to the Sun than Mercury and whose gravity was affecting its orbit. This closer faster companion to Mercury was however difficult to spot with the technology of the day and to be honest would not be easy for astronomers today. The reason is that any planet orbiting so close to the Sun would be very small to not have been detected already and therefore its small size coupled with the fact that it would be even further away from the Earth than Mercury is would make it appear even smaller still. This means that its orbital radius would also be much smaller than Mercury's making the mean distance it travels away from the Sun at any time very short indeed. In order to view such a planet you would need to essentially look directly at the Sun and hope to catch a glimpse of it as it skirted the farthest points around its orbit of the Sun. If you are familiar with astronomy then you have probably heard many analogies for how difficult this can be such as trying to see a grain of sand one centimeter to the side of a search light from a kilometer away or something equally fitting. The thing that must be underlined is that the Sun is bright, very bright and trying to see a very small object which is very close to it is next to impossible and so the early attempts at spotting a new planet would have encountered much difficulty straight out of the gate. Another way to spy such a small object is to watch for the moment it transits the Sun.

A transit is an astronomical event where a planet moves in front of its star and you can see it outlined as a very dark shadow in your telescope. This is similar to what astronomers are doing when they search for planets outside of our own star system as they wait for a planet to transit in front of the parent star and monitor the dip in light received from the star. Here of course the analogy is even more absurd and perhaps a speck of dust next to a search light and viewed from the moon is more appropriate when talking about such vast distances as those that exist between us and any of our stellar neighbors. The search for a Mercury like planet closer to the Sun also failed in this regard not just due to technical challenges but also because we know there is no such planet around our Sun. In fact the time in which such a small, fast orbiting and close planet to the Sun would spend in transit also makes looking for it incredibly difficult because without knowing where it was you would simply be guessing at when to look.

Needless to say despite the various planet hunting tricks available to astronomers of the day and the present no such tenth planet has ever been discovered orbiting close to the Sun. With the failure of invented planets to solve the problem it was necessary to stop trying to make the universe fit the theory and make a new theory to fit the universe. The problem was that no one knew where to start as Newton's laws worked perfectly except for a few troublesome observations. The answer in fact would not come from any astronomical observations per se but rather from theoretical physics that later would be applied to the Newton discrepancies. The person who would find an answer to this problem was Albert Einstein and his general theory of relativity which he published in 1915.

To start off with we need a general introduction to general relativity as it is two things: the first being not easy to understand if you have no idea of space and various physical concepts related to space and second it is not a complete picture of gravity at all only the best that modern science has to date just as Newton's work had been previously. To this end we will be looking at the general properties of the theory and how they seem to play out in the world around us after which we can look at space more closely in terms of how the theory came about and lastly how the theory is insufficient to explain gravity. This second point should be fairly obvious due to the twin facts that general relativity cannot cope with the very small or very large scale of the universe well and that scientists are still out to create some sort of quantum version of gravity that relies more on quantum mechanics than relativity. Never the less let us dive right into what general relativity is in a nutshell.

The simplest and often most widely understood concept of relativity is that it produces a curvature of spacetime and that is all. All of what general relativity is can be centered on that statement from the theories inception to the final conclusions it will make about the universe. The basic idea before this was that space was a static sort of fixed structure and that things would move about within it. The idea of general relativity is that space is not static and can be bent or curved in some way. This idea was not at all unique to Einstein as others before had suggested that the background of the universe which is space may be somewhat pliable but they had never been able to experimentally show this to any accepted high degree. The German mathematician Carl Friedrich Gauss was an early proponent of a curved space who argued in the nineteenth century that space possessed a unique type of geometry which was non-Euclidian. The general idea of a curved space would directly influence later thinkers such as Einstein who would use a curved space in his idea of curved spacetime. The idea of spacetime being nothing more than the inclusion of time into the already existent fabric of space such that together the two can be equated at the same time. Where Gauss got the genius for a curved space came from his ideas of mathematical geometries and where Einstein got his idea of a curved spacetime came from the same with one difference. This being that Gauss was purely interested in the geometry of space and simply proving a curved space was possible he did not attempt to figure out why it might be curved. Einstein decided that the curvature was caused by something and this something was any object with mass.

What Einstein thought was objects with mass and especially those with enormous amounts of mass would produce a curvature of spacetime. His theory of relativity can work in many cases without this but the theory of general relativity requires it and it is from this latter theory that he argues space can be bent. The larger the object the more space becomes curved according to general relativity and a product of this bending of spacetime is that gravity can be created from it. This was Einstein's major contribution to the study of gravity and in his view gravity is no longer a force as described by Newton but now is simply the result of a bent spacetime. Now if you have been paying attention throughout this book you will notice one of the major problems stemming from Einstein's idea of gravity being not a force but the product of bent space. That problem is that gravity is considered to be one of the fundamental four forces of the universe and all the other three forces are governed by quantum laws. So in other words why not gravity as well? If the strong, weak and electromagnetic forces can be considered as excitations of a field that contains force carriers for that field then why not gravity as a force too?

This is one of the big stumbling blocks of both quantum mechanics and general relativity and is in a sense where all the fuss comes from when one tries to answer the simple question of how does gravity work. No one has found a graviton force carrier yet and no one has been able to use general relativity to explain the quantum world so it is difficult to say which if either is right. For now we will explain gravity as Einstein sees it right or wrong and in his idea of the universe gravity is a result of spacetime curvature. The curvature comes from matter and the more matter the more it curves so the higher the gravitational effect. The interesting thing about how this idea works is that spacetime becomes curved and when I say that the word time is part of Einstein's idea of space. So not only is space curved according to Einstein but so too is time as it is woven into the fabric of space according to general relativity.

The effect of curving time means that time will not flow at the same rate for everyone everywhere in the universe just as gravity is not as strong for everyone everywhere in the universe. Astronauts far away from the Earth will not feel the gravitational effects of the planets mass and will float freely while we on the surface of the Earth very close to its mass do not float anywhere. We fall fast and usually with unfortunate consequences. So how would time change for an astronaut in different places in the universe? According to general relativity the person who is in the least curved part of spacetime will feel a different passage of time than someone who is in a higher curved part of spacetime. In general the higher the curvature of spacetime the slower time will run and the lower the curvature of spacetime the faster time will run. To think of this another way one persons second is not equal to anyone else's second of time because no matter how small we all exist in slightly different curvatures of spacetime. A larger creature itself has greater mass than a smaller one and so more curvature of space and so on the ridiculously small scale actually has a slower second. An adult male human on average will age more slowly than an infant, a full grown male lion will age more slowly than its cub, a tree more slowly than its sapling and so on and so on. At these scales however and due to the overpowering effect of the Earth's mass on spacetime we do not perceive these micro-scale local events but they become more apparent when we travel into space.

The most noticeable demonstration of this comes from any object orbiting the Earth as not only gravity but also velocity affect a change in the curvature of spacetime. Despite the fact that the Earth itself is moving at a speed of 107, 200 km/hr around the Sun and astronauts in orbit around the Earth aboard something like the International Space Station are travelling at only 27, 600 km/hr because they are travelling together the overall speed of the astronauts compared to the Earth is in fact 27, 600 km/hr. The spacetime curvature of the motion of the two objects around the Sun is the same only the curvature caused by the International Space Station is measured. To be expected what one finds upon examining two perfectly synchronized identical clocks where one is on the ground and one in orbit aboard the space station is that the clock in orbit runs slower than its ground based counterpart. An astronaut in low Earth orbit will typically age 0.014 seconds more slowly than anyone else on Earth over the course of a single year. So not too exciting really but still a small difference in passage of time between the two relative clocks. If you were to orbit the Earth for one million years you would only gain 3.88 hours of life on anyone on Earth and so while everyone else was a doddering one million year old fool you would have the super youthful appearance of someone nearly four hours their junior.

These effects of time dilation do not come into play in any great way until much higher speeds are reached and usually these occur at around the 90 percent or higher mark of the speed of light where the spacetime curvature intensifies. This of course means that gravity and velocity can cause this curvature of spacetime and time dilation to occur. In order to calculate the total time dilation both the effects of gravity and the effects of time need to be addressed. When travelling in low Earth orbit astronauts are indeed in a weaker field of gravity and so time should run a bit faster for them but they are also moving very quickly and therefore time should run slower for them. Combining these two differences in spacetime curvature will produce an overall time dilation that we measure as 0.014 seconds less per year in orbit. These differences do add up however and in order to maintain accurate contact with anything in space its position and speed must be taken into account before attempting to communicate with it.

So it would appear that high gravitational fields and high velocities, especially those approaching the speed of light are the two ways to curve spacetime and produce things like time dilation. The predications made by general relativity seem to fit observation fairly well and so modern science has utilized them as the best model of gravity to date and yet we have still seen that it cannot answer fully what gravity is. A new and complete picture of the universe will have to be accepted by modern science in order to fully answer what gravity is and how everything in the universe fits together from the small scale through the medium and up to the largest scales possible. For now let us continue exploring how we came to understand gravity through the lens of general relativity so we can see the steps along the way. Oh and yes you got it that is why Mercury is not where it should be in orbit too.

The first steps of course came from Newton and his brilliant work on gravitational attraction which amongst other things was the first to predict a bending of light due to gravity. Most people credit Einstein with the idea that gravity can bend light but in fact this idea came as a result of Newton's equations and not Einstein's. Applying Newton's laws of gravitation to massive objects like the Sun and to light rays, as they were known in Newton's day, produces a bending of the light along its straight trajectory. The bending came from the force of gravity as Newton describes gravity as a force similar to the other forces now known to science. The amount of bending of light for a given gravitational source as calculated by Newton's equations is half of the actual value which is still amazing considering his work is over 300 years old.

The reason many people accepted Einstein's work over Newton's was the fact that his calculations which were simply minor tweakings to Newton's own did manage to produce a more accurate predication of the degree of bending light would experience given a particular gravitational force. In fact it was the proof of this in 1919 by Arthur Stanley Eddington and Frank Watson Dyson of a photograph of an eclipse where stars that should normally be hidden behind the Sun were visible at the edge of the Sun's disc during total eclipse that made Einstein practically a household name. To follow this path a little further no one in the science community disliked Newton or any of his ideas they are sound and have largely stood the test of time for centuries and many are still in use today even more accurate at some scales than Einstein's. Yet when a new model comes along such as relativity that can predict with greater accuracy the observations made then it is naturally adopted and studied as the new favorite by modern science. To this end we see that gravity was explained by Einstein as a property of the curvature of spacetime and not as a force so naturally people accepted gravity not as a force but as an artifact of deformations in the geometry of spacetime.

This is a point that must be emphasized here however as Einstein does not know that spacetime is even real at all as he only assumes it is in order to make his mathematics workout. The fact that it does seem to make more accurate predications has far overshadowed the fact that not many people have stopped and truly tried to figure out what gravity really is and whether or not it curves time or if time even exists for that matter. The very notion of time has been debated for millennia that we know of and despite many people wracking their brains over this problem for years modern science still has no clear answer. The very idea that gravity can be a force with a carrier particle is somewhat accepted but the idea that time itself is a force with some sort of carrier is utterly absurd to nearly all. Yet both sides of the time debate rage on saying it does not exist or that it does due to things like time dilations. Time itself will get its own special section and discussion later as it is integral to many of the ideas propagating throughout the science community. For now we are going to examine the concepts that led Einstein to general relativity and some of his own thought experiments that helped shape his idea of curving spacetime. So in order to see how Einstein came to his conclusions we need to look at Newton's geometry of the universe and something called frames of reference.

To begin with we need to understand the frame of reference as it applies to science and how frames of reference work into a theory of gravity. First off a frame of reference is simply science talk for looking at any one instant in time that applies to a given object or system. Think for a minute of a simple act such as blinking your eyes and try to describe what is the real time in which you blink your eyes. The answer is that you cannot do this not because you cannot describe it as existing at only one frame in time. Blinking your eyes begins with your eyes open, closing, fully shut, opening again and finally with you eyes all the way open. There are five time points in which you can roughly describe the act of blinking and so this is how science sees the world and especially when dealing with something that modern science has trouble grasping like gravity. One frame of reference for an eye blink might be seen as the time in which your eyes are beginning to close just as one act of gravity can be described as a ball beginning to fall. The frame of reference for the ball is that of it in motion and being acted upon by the force of gravity if the ball is stopped and lying on the ground then gravity is not acting on it in any meaningful way other than keeping on the ground.

In the time of Newton we find that gravity is described as a force that acts upon any object and its inertial motion. Inertial motion is just that the motion of inertia and refers to anything that moves and according to Newton it will continue to move unless its mass is acted upon by an outside force. Gravity is such an outside force and so the inertial motion of the object can be acted upon by gravity altering the straight line path it is following. The frame of reference in classical mechanics is thusly one that describes an object travelling in straight lines through space and time. Since the days of Newton and certainly during Einstein's days scientists felt the need to be loquacious and change the name of straight lines which is easy to understand to words that mean the exact same thing but are rooted in mathematical mystery henceforth straight lines for any object in the universe have been dubbed world lines or worse geodesics. World lines describe the path any object takes as it travels through time that is all and geodesics are simply the straight lines mapped over curved spaces. It is note worthy that geodesics is a perfectly normal word in geometry merely the introduction of it to physics that is unnecessary.

So the straight line frames of reference used by Newton have been usurped by general relativity into geodesics but in the end they really describe the exact same thing which is the inertial motion of an object and the path it moves along. The difference here is of course that general relativity relies upon Einstein's idea that gravity is a property of spacetime curvature and is not a force but if you examine the world line of any object which is its path through space over time you find Einstein has done nothing new. If Einstein is correct then the curved path an object follows is due to it moving straight but following the curvature of spacetime. If Newton is correct then an object moves through space in a straight line and its straight path deviates according to the direction and strength of any force that acts upon it in obedience of his second law of motion.

Once again we see that modern science does not know where to turn in an answer for how gravity works as general relativity states that the path of an object is due to curvature of space because gravity is a property of said curvature. Quantum mechanics while it is not classical mechanics seeks to prove gravity as a force which would mean that objects do behave as Newton stated travelling in straight lines and being acted upon by the force of gravity. The discovery of a graviton in the Standard Model would easily deep six the idea that gravity is a property of spacetime curvature and yet any such model would still have to account for spacetime phenomena such as time dilation. If you are confused right now about how gravity works and think that an answer will never be found you can relax as there are many out there who agree with you that this is one of the toughest challenges faced by human thought.

What Einstein did was conduct a thought experiment in order to come up with his idea for general relativity and it revisits the idea of free fall and frames of reference. The idea that he had was to equate acceleration of an object to the frame of reference in which it can be described. For example if you are standing on the surface of the Earth and you drop an object let us say for example a round ball you will note that the ball accelerates towards the ground at a rate of 9.8 m/s^2 The ball is accelerating towards not the ground but the center of mass of the Earth and the rate of acceleration is constant at the surface of the Earth but it will vary depending upon altitude. Now if we were to enclose you in a windowless room free from all vibrations and accelerate that room upwards at a rate of 9.8 m/s^2 and you dropped the ball it would also fall towards the bottom of the room at the same rate that surface gravity will pull it towards the center of the planet. In other words the frames of reference for each example of a ball drop are identical to an observer and it does not matter according to Einstein that the falling is caused by gravity in one instance and acceleration in the second. The net effect of what is happening to the ball is the same and therefore its inertial frame of reference is the same so in a sense gravity must now be taken into account as a form of free fall due to gravity.

Einstein states that space and time must be taken into account as one and therefore the Newtonian geometry of a straight line path for an object does not transfer neatly to the new mathematics. The inclusion of gravity and time with space creates a new geodesic geometry of space according to Einstein that will result in a curvature of space. If we take an object and trace its straight line path through this new mathematical landscape we find that the overall frames of reference are no longer all encompassing and instead exist in little slices next to the object in motion. Each frame can then be plotted against spacetime as it moves and the result is a curved line created by each slice in spacetime. Adding them all up reveals a curvature of spacetime and ultimately gives you the theory of general relativity. The theory of special relativity can exist without gravity but only operates on small scales where gravitational effects are negligible. Once the small scale is left and we reach the medium scale we find that gravity comes into play and cannot be ignored rendering the use of general relativity more appropriate. The curvature of spacetime as a whole becomes more important at higher gravities and higher relative velocities approaching the speed of light where according to Einstein spacetime bends far more.

An example of curving spacetime can be seen in something known as gravitational red-shifting. The concept is simple and states that as a photon moves away from a high gravitational source it will lose energy and have its wavelength shifted towards the red end of the spectrum. So a high energy photon generated at the surface of the Sun will have a lot of energy and a short wavelength but as it travels upwards away from the Sun's center of mass the curvature of spacetime decreases according to general relativity. The light wave will be stretched and lose energy meaning the light will shift from the blue end of the spectrum towards the red end of the spectrum all due to changes in gravity. The other effect gravity has on light is that if the gravitational force applied to it is strong enough then the path of the light will change from a straight line to one that is bent. Newton himself was the first to suggest that light would bend in 1704 as part of his famous work *Opticks*. Some 80 years later Henry Cavendish in 1784 and then Johann Georg von Soldner in 1801 also backed up Newton and stated that based upon Newton's work light would indeed bend in the presence of a strong gravitational force. Soldner would calculate the amount of light bending that an object would experience given a set gravitational force and later Einstein would repeat Soldner's calculations and find the same value.

Afterwards once Einstein had formulated his theory of general relativity he was able to apply an updated version of the Newtonian math coupled with general relativity to the problem and find a new value for the light bending solution. It would turn out that light would actually bend twice as much as previously thought and in so doing predictions for light bending around large objects could now be more accurately calculated. Newton's universality of free fall first predicted such changes in a lights straight line path and Einstein expounded on this with general relativity and the overall result is that Newton is correct that gravity does bend light and Einstein has fine tuned the mathematics to provide the correct amount of bend.

Seeing both these effects of light in the presence of gravity indicates that whether it is a force or a curvature of spacetime light like all things in the universe must also obey gravity. So much so that another effect of light bending can be predicted and observed around gravitationally massive objects known as gravitational lensing. Originally suggested in 1924 by Russian physicist Orest Khvolson it was a concept that predicted a large object with sufficient gravity could behave like an optical lens to light behind it. Just as an optical lens placed between a light source and an observer can bend the light before it reaches the observer so too can gravity bend light. Imagine for a moment a very large and gravitationally powerful object like a galaxy and now image a second galaxy hidden from view directly behind it. Normally you would not be able to see the second more distant galaxy as it would be hidden from your line of sight by the closer galaxy. The effect gravity has on light is to act as an optical lens and bend the light around the closer galaxy thus making the second more distant object visible. The catch is the second hidden object will not appear as a simple clean image of a galaxy instead it will appear to form a ring of light around the closer galaxy. That is because both of the galaxies lay in a perfectly straight line with you the observer and so the light from the second galaxy hits the gravity of the closer one from all angles and is bent in a circular or oval fashion by the closer galaxy. If the object were slightly shifted from the straight line of you and the closer galaxy you would see it offset and so it would take the appearance of an arc around the closer galaxy. Many examples of this phenomenon have been photographed with advances in telescopes and especially the Hubble Space Telescope. Although first proposed in 1924 by Khvolson it did not garner much publicity and it was not until Einstein again in 1936 discussed the topic that it became more widespread and physical proof of the concept would not come until 1979.

Is the theory of general relativity complete and able to explain all gravitational properties observed in the universe? No. In fact there are a great deal of problems that general relativity fails to answer and based on what the theory is will never be able to answer. Yet general relativity remains the favored model for gravity in contemporary science and so many have begun to study what it says about gravity in the hopes that it might lead to a complete theory of gravity. This is of course nothing new just as Einstein sought to improve upon Newton's idea of gravity new scientists are attempting to do the same. One thing they all must contend with is the idea of time as time is integral to general relativity and many scientific theories in general. It was time in fact that led Einstein to create a relativistic idea for space based on what was happening to an object from one frame of reference to the next and thus gave him the idea for spacetime. The concept of time as being an important piece of the universe is nothing new and is some centuries older than Einstein and general relativity but what time is is a question that goes back even further than that.

Time

Truth be told no one can really say when time was first thought of as a thing that was part of the world and universe as a whole. It is flatly incorrect to rely on the written accounts of human knowledge as the first time anything was ever done. A perfect example is trotting out the Greek thinkers once again and saying that just because something was recorded in history was the first time it was ever conceived of. It is understandable that archaeologists wish to have a factual piece of evidence showing them precisely when something occurred but it is of course not necessary. A perfect example of this is the concept of mathematics which predates any written records produced by humans by some tens of thousands of years. People were not simply sitting around doing nothing waiting for the classical age to arrive so they could begin counting, speaking and thinking about things larger than themselves. All of these were in full swing long before any such classical civilization was even in the womb so to speak.

So this leaves us with the problem of time and where it all began in terms of human thought. Time was known to humans before they were even humans as animals themselves are aware of time and experience it daily. At least at the most basic level animals are aware of the passage of time as one day ends and another begins. Humans would have been capable of making this deduction hundreds of thousands of years ago in some hominid form or another. We see the first signs of culture and civilization dating back to about 50 000 years ago and at this time humans had stopped evolving and were no different than you or I. Barring some genetic flow from one local population to another the modern human was born and therefore the modern human brain along with it. The fact that we find cultural artifacts and numerous paintings on walls means our ancestors were thinking, communicating and creating. This is inarguable as no society focused purely on surviving and even non-human primitives would waste time and resources or have the luxury of wasting time and resources on things like art as can be seen by cave drawings. So at least at this point we know humans were aware of time and were thinking about it and proof of this is found everywhere in the world as diverse cultures attempted to create calendars to track the time passing throughout the year.

Without a doubt the most famous of these ancient relics and one that dates back thousands to tens of thousands of years is Stonehenge. The trouble with time is that despite our ancestors using it as a year long clock much as we use smaller units of time as a clock today they did not understand what time really was. Surely our ancestors would think that tens of thousands of years later their great, great, great and then some descendants would know what time was would they not? Sadly no, modern science and modern philosophy does not have a definite answer for what time really is.

To make matters worse some debate whether time is real or only created by us and other intelligent beings. Neither side seems to be gaining any ground yet at the end of the day only one can be right. For now let us look at time as if it were real and see what we know about it. Time many argue began at the moment of the Big Bang as it was the point which created the universe and before that there was no universe in which time could exist. In fact this is one of the central proofs many feel is necessary to prove that time is real as nothing can exist without a universe and so time did not exist either. Once the Big Bang occurred then everything was spewed forth from that mighty explosion to create the universe we see today. This is of course space, matter, energy and many feel the time in which they all must flow. So that is all there is to it time is real and nothing can exist without time to move in. Or is it?

Well that is just the focus of the problem after all no one really knows if you need time to have a universe. For example time advocates argue that nothing can exist without time as it is the fourth dimension and you must have three spatial dimensions plus one for them to flow through, i.e. time in order to make a universe. But what if you simply look at a single object in a snapshot way? Take the idea of a single particle sitting all by itself and interacting with nothing else and examine it in the smallest time frame possible which as far as modern science knows is a Planck time. A Planck time is the shortest possible time frame that can make sense in quantum mechanics and since quantum mechanics rules the very, very small it is seen as the smallest and most indivisible time unit possible; a time atom if you will.

A Planck time is so small that there are more Planck times in a single second than there have been seconds since the approximately 13.8 billions years since the Big Bang which makes a single Planck time a mere moment in raw existence. More precisely a Planck time is defined as 5.39×10^{-44} seconds or the time it takes light to travel one Planck length which according to theory only is the shortest possible measurements of both time and scale. At that inconceivably brief moment the particle does not actually flow through anything as it is more or less frozen in place not oscillating between any quantum states or superpositions whatsoever. In other words it exists at that brief moment in only three of the excepted four dimensions or another way of looking at it is it exists in space but not in time. From this kind of reasoning you can demonstrate the lack of a pressing need for time to be a part of the universe, a force or medium through which three dimensional objects must flow. So the bottom line here is time is not real as it is not needed and so time therefore is something else.

To this end there are two standard camps in the time wars the first being the side that favors time as a real and wisp like essence that is part of the universe and the second that believes time is not real and merely a product of the human or any intelligent creatures mind. Before we delve into that let us at least describe what time is mostly like and how it mostly behaves. The first thing you must realize about time is that most of our understanding of it comes from what I call normal physics in the universe and by that I mean conditions where we live are not extreme in any way. The conditions that existed a mere Planck time after the Big Bang were a natural part of our universe but very extreme and also extremely short lived they in no way make up the Lion's share of physics in the universe. The 13.8 billion years of the universes age is near 100 percent spent in normal physics that we observe in our own solar system or galactic backyard. That being said where we live time whatever it is or is not flows in only one direction; forwards. So we can at least say that time is something that keeps events ordered from moment to moment flowing ever onwards from the past into the present and heading off towards the future.

The second thing we know about time is that it is measurable and can be used as a tool by intelligent creatures to keep track of events. Whether they are long term like the passing of a year or day as animals understand time or short term such as the seconds or fractions of seconds counted off in a laboratory experiment. Time can be used as a measuring device to aid intelligent creatures in the observation of the universe around them. Here we enter the first bit of trouble with our clever human brains as we understand real solid things like physical objects and know that they too can be measured in terms of length, width and depth. Since we ascribe numbers to anything we measure it is a natural and not necessarily correct assumption that time must also be a dimension because we measure it with numbers as well. Regardless of time being real or not these are indeed two inarguable properties of time as we encounter it day by day. Time flows forward and we can measure the rate of passage of time in units for later thought if needed. Now that we know the basics of time we can begin to look at the two opposing camps and their specific ideas concerning what time really is.

The first side is that which favors time as a real and partially tangible thing and the reason I say partly is because the effects of time are only felt implicitly by us and the objects that we encounter. This first view of time is indeed the very first view of time as the concept of time as a system of record keeping or measurement is more of an advanced concept that would not be calculated until the invention of numbers. It is possible that time as a simple way of counting things however predates this as animals are aware of time as a way of keeping track of day and night cycles or seasons of the year. Yes all animals at least the more evolved ones are aware of this on some level as instinct for example will tell certain species to seek shelter during night to avoid predators or for herds to follow the seasons in search of food. I only mention time as a real thing first because it is the first concept that humans or most other intelligent creatures would be able to think and rationalize about. The idea of time as a record keeping system most likely does predate this but was so common place as to be forgotten when it came to discussing what time was in more scientific terms.

Therefore the idea that time is real was the first and foremost concept of time throughout history as it permeates the cultures of the globe and can be seen in their various creation mythologies whether polytheistic or monotheistic each different myth uses time as a force. A perfect example of this is how every mythology begins with the creation of the universe and the forward flow of time since then always tracing out the history of whatever people dreamt it up in order to explain their surrounding world up to the present day. This is an immutable fact of these mythologies and time is never mentioned merely as a system of counting created by humans and utilized to keep daily events ordered and measured. Time was always created by the gods or god depicted in the creation myth as a force of the universe along with other typical staples such as the earth and sky or things of such a nature. A perfect example of this comes from the Greek's and their god Chronos who literally is the embodiment of time itself and therefore it would be impossible to refer to him as a system of measurement as he was a god and part of the universe.

The idea that time was a true force of the universe removed from such antiquated superstitions was formalized millennia later when true science beyond rationalization of the surrounding world took over. Here we see the use of the term Newtonian time appear as a way of describing time as a force integral to the workings of the universe and of course it derives its name from Sir Isaac Newton. Multiple thinkers have ascribed time to be one of the fundamental dimensions and properties of the universe but it was Newton who in his *Philosophae Naturalis Principia Mathematica* introduced the idea of something he called absolute time and space. The combination of these two entities allows for the use of Newtonian mechanics and provided an important concept in science for future thinkers to use; there would be no Einstein and general relativity without Newton's idea of time and space as this was needed for Einstein to examine in order to come up with his own theories on time and space.

What Newton argued was that space and time were needed in order to construct the objective reality that we see around us. To put this another way you cannot have an object existing in the universe if it does not inhabit both space for its three dimensions and time for its progression through the universe as well. Here I must make a note as the progression of time can be called many things and is indeed called many things when discussing the exact nature of time. Some examples all with slightly different meanings and usages include: progression, passage, flow, persistence, duration and motion of time. The meaning of each is basically the same but they all attempt to put a slight spin on the meaning in order to better convey what each usage of the word can do when describing a specific aspect of time. To use Newton's description of time we find that he will refer to two different types of time and also two different types of space. For now we will look at time first as he describes it, followed by his description of space and lastly how his ideas of both were necessary for future thinkers to formulate their ideas of time and space.

To Newton the universe consisted of two types of time the first of which was absolute time. Absolute time permeates the entire universe everywhere and is ever present no matter what conditions exist in it and it always flows at one set rate. In other words the absolute time of the universe ticks away at a steady pace which cannot be altered faster or slower and is in a sense ticking away in some sort of universe absolute seconds. The absolute time is also indivisible as it only needs to progress at one rate it has no need to be fractionated in any way or summed to form larger units in any other way as it simply is time itself. This form of time Newton states exists whether we can perceive of it or not and was best understood not through empirical means but mathematical ones. It is in the mathematics alone that Newton argues time can be truly understood and utilized in order to make sense of anything that we perceive as reality around us.

The second type of time according to Newton is what he calls relative time and refers to the kind of time that humans can perceive empirically and understand without the aid of mathematics. Think for a minute about the motion of an object and you can begin to understand what Newton meant as the object moves through space you can track its progress from one moment to another. This is the type of relative time that Newton believes humans use to understand the universe around them but is not the true time of the universe and so is not an absolute value for time. This is also the kind of time that humans use to measure things like seconds, minutes or hours and can be useful in making observations of the universe. From here one might be able to think of timing something in a laboratory in seconds or out in space in years but no matter what according to Newton this is an application of relative time. In addition even though relative time is made of seconds which might be thought of as a kind of base unit for time it is not the base second of absolute time and so not true time. After all who is to say what a unit of time is anyway as it might be 60 seconds to a minute or if we had devised a different time keeping system maybe 120 seconds to one minute.

The length of human time is arbitrary or based off of some measurement made pertaining to our environment which is unique in the universe. For example a year passes in 365 Earth days but a year on Mercury might pass in as little as 88 days or a year on Pluto as long as 248 Earth years. So we see here that the specific orbital period of Earth which is our home world determines for humans what a single year is and from here our ancestors broke time down into months, days, hours, minutes and seconds. All of our human time units are independent of the absolute time which governs actions of the universe itself according to Newton. This is actually a very interesting point that Newton makes and in many ways a huge leap forward in thinking that he most likely could never have fully anticipated. The reason is humans do use an arbitrary unit of time to measure events around us yet the universe might in fact have a set clock and it could be hidden in the quantum world.

For a moment think of the actions of a quantum object existing at microscopic scales and at microscopic time frames and remember the Planck length and the Planck time. If the Planck length and time are indeed the smallest units of size and time possible then Newton was right that there is in fact a set of absolute time and size for the universe as nothing can be smaller than these. We do not know at all if the Planck length and Planck time are in fact the smallest of anything as the quantum world cannot fully explain the universe and so all bets are off. The same is true for relativity which is even less precise and can explain even less about the universe than quantum mechanics so it is not correct either. However if the Planck length and Planck time are real then Newton was centuries ahead of his time and time was not the only thing Newton believed had two versions.

Space was another thing that Newton believed has two distinct components and like time there was an absolute value that did not depend on how humans perceived reality. Absolute time is the true time of the universe and relative time is that which is created by humans used to describe the behavior of an object and its perceived reality. Absolute space operated the exact same way and was a fixed universal value that was ever present throughout the cosmos. Relative space is the space motion associated with an object that exists within the absolute space of the universe. So in other words if we have a planet in orbit around the home star we say that the relative space of the planet is in motion as compared to the somewhat stationary absolute space that makes up the background. Relative space can exist in many forms and is not limited to simply one object in motion whether on its own or around another.

Relative space can also occur for objects at rest provided that the object is still in motion and to understand this you need to think of a planet rotating on its axis in space. If the planet were motionless in space yet rotating around its own axis it would still possess a relative space generated by the rotational motion of its body. This rotational motion is independent of the absolute space which is the background of the universe and is not moving in any dramatic way. In other words this means two things the first being the motion of an object creates its own perceived reality that humans view and can measure which is separate from the background space of the universe. The second is that the universe which has absolute space is occupied by numerous areas of relative space which are created by objects existing within the universe. Therefore you can routinely have and do have countless pockets of relative space which are constructed out of the absolute space by the presence of an object and the forces that are associated with it or exerted upon it.

In Newton's view objects were either at rest and therefore in a state of absolute rest or they were in motion at some absolute speed and this could be as we have seen motion from place to place or motion through something like rotation or vibration while not moving from place to place. In the example of the moon orbiting the Earth or the Earth orbiting the Sun the gravitational tension of the force of gravity was centered at the barycenter between the two objects. A barycenter is simply the center of gravity between any two things with mass. So here we see that Newton was able to demonstrate that gravity was a force existing between two objects in the universe as a form of motion and the relative space created by the objects affected a change in the absolute space around them and this includes the barycenter of the two which was devoid of mass. While the concepts of absolute space and time have since been challenged they were and still are a huge triumph for science and discovery as they stood the test of time for going on three centuries and in some cases have not been abandoned despite most people flocking to the theory of general relativity. In fact no such theory of general relativity or special relativity could ever have been devised without Newton's ideas on space and time as they are the direct result of his models.

The models put forth by Newton opened up a series of debates on the nature of reality, the universe and the behavior of objects within them known as operational definitions. To some degree Newton's ideas are in fact still the most applicable we have to the universe as a whole even if they exist in a more debated form. The reason people are still discussing the properties of space and time reveals the simple fact that modern science does not have an answer to what they really are and so no currently accepted theory is in fact correct but merely approximations of what we observe. This in turn leads us to the second and opposing viewpoint concerning the nature of time as it takes a contrary stance to Newton's ideas and states that instead of being something real and actually part of the universe time is nothing more than a figment of the human or intelligent creature's imagination and measurements.

This second idea concerning time is radically different from the one espoused by both Newton and Einstein who despite their differences in theories agree that time is in fact a real thing. Einstein has even gone so far as to say that time is definitely a real tangible object that is so integral to the universe that it is woven into the fabric of spacetime itself and depending on what you do to it you can actually go forwards or backwards in time. By studying Newton's theories Einstein applied the ideas of acceleration to inertial frames of reference and by including gravity concluded that a curvature of spacetime was needed in order to make general relativity a workable theory. By this very definition time is a part of space and Einstein believes this to be so due to the fact that in his concept of relativity the inertial motions of each reference frame carries a time component which appears to react to the same curvature as space does. Since various astronomical problems can be more accurately solved using general relativity instead of classical mechanics many have ascribed to the idea that time is in fact part of spacetime. Therefore time is actually a real force in the universe that has a physical effect over any objects in it and proponents of this theory claim to have proof of this notion.

The proof they say can be seen when comparing the apparent passage of time of two separate clocks that are in relative motion to one another. As we have seen a clock at Earth rest will be in a different frame of inertial motion than a clock orbiting the Earth which is travelling much faster. As we have seen already with time dilation this will produce a difference in the rate at which the clocks mark the passage of time. General relativity states that a decrease in gravity will speed up a clock and an increase in speed will decrease a clock yet when both are factored in an orbiting clock in weaker gravity travels fast enough to produce a very small change in time. This difference in time is what makes some believe that time is in fact real and part of the spacetime fabric so to them it is inconceivable to argue that time does not in fact exist. The problem is that no one in modern science or philosophy has any real clue as to what time really is and this can be simply illustrated by thinking about the known forces in the universe.

All of the known forces in the Standard Model for example have a particle which is believed to be the excitation of a corresponding field and each force also has a measurable force carrier associated with it. Things such as particles and force carriers excite scientists to no end as they fulfill all the requirements of good experimental science. Namely they can be reproduced at will and measured without much ambiguity and so far gravity is the only force to elude them. The Standard Model and its backers would love to find a graviton in order to complete their understanding of the four forces and perhaps come up with a theory of everything. The trouble is that no one has found a graviton and no one in modern science even understands what gravity really is as its behaviors are so different from any of the other forces. It is immeasurably weaker than the other three forces yet can easily overpower all of them in ways that are hard to imagine if enough gravitational energy is applied to any system; think of the strange effects of a black hole or quark-gluon plasma at the early stages of the Big Bang. So what if gravity does not actually belong in the Standard Model after all and what if gravity is the only known force that plays by its own set of laws and does not require any sort of particle or particle like force carrier to interact with matter?

The answer is it might in fact be possible that this is the complete truth as nothing says everything in the universe has to be explained neatly so that it can fit into the Standard Model of particle physics as dreamt up by humans. To this end time is even less well understood and poses even more seemingly unexplainable challenges to modern science than gravity does. If we can throw out the idea that gravity is an excitation of any sort of field then perhaps we can throw out the idea that time is a real thing as well. Whether spoken mockingly and in jest only some have even suggested the existence of a mysterious particle known as the chronoton. This is of course the hypothetical particle named for the Greek god of time Chronos and would be the force carrier for some sort of time field. This notion has been scoffed at to the point of no return in contemporary science and for the above mentioned reasons of things like gravity not playing by the rules can be done so for a very good reason. There is no workable way in which a time force or time particle would be able to fit into the Standard Model as time has virtually no impact on anything and is simply a measurement tool used by scientists to study the other known forces.

To put this another way if gravitational effects are so incredibly weak compared to the other forces that they can be all but ignored at quantum scales how weak must time be at these same scales? Time has barely any effect on anything even at relativistic speeds or gravities as the apparent effects of it are felt exponentially when approaching the speed of light. At lower speeds it is non existent for all practical purposes and for proof of this remember how long you get to live extra if you orbited the Earth for a million years aboard the International Space Station a paltry 3.88 hours. The fact that we see clocks slow down as one travels faster might be a by product of some other physically tangible effect such as heat is a by product of travelling at great speed through the atmosphere. The truth is modern science has no real idea what time is and the only evidence to support the notion that time is real is something which cannot be explained in the least using any known laws or forces. The fact that time cannot be explained in any satisfactory way suggests that we are witnessing some sort of by product effect caused by other much more well understood forces or mechanisms.

Timeless

These realities are exactly the sort of thing that leads some people to the second idea concerning time which is that time is itself nothing more than a construct of an intelligent mind. In other words time does not in fact exist, is not a tangible force and cannot be explained in any scientific way with regards to things like the Standard Model or general relativity. This leaves us with a gap in our knowledge base that cannot be filled and also leaves us with the fact that time is something that seems real and we use everyday yet cannot be explained. So if time is not real what are the arguments to support the idea that it is nothing more than an invention of the intelligent mind? The first of the arguments lists time as nothing more than a system of measurement which can be coupled along with things like the three dimensions and quantities.

Humans and doubtlessly many other intelligent creatures have used three dimensions to make sense of the physical world around them and specifically any object within it as a single object is much easier to grasp than say the three dimensions of the entire universe. Our ancient ancestors long before our time understood the concept of three dimensions as without such a notion great wonders of the world such as early megalithic calendars would have been impossible to construct. In the modern world today we still use three dimensions all the time in our attempt to understand the universe even to the point of debating the structure of the smallest literally point like or one dimensional sub-atomic particles. So which is the real system of measurement to use in examining three dimensional objects is it a meter, a league, or any other number of scales used to measure?

This now brings us to quantities which are nothing more than numbers used to describe how many of something we have. This is once again an ancient concept that every human civilization has understood and can be seen throughout prehistory in the use of numerical symbols or pictures used to depict world events. Again you cannot build a society without the concept of numbers even if it is only a simple system of counting goods to be traded. In the modern world we count all the time and not just in normal society but in the secretive and shrouded halls of mad scientists smashing the very elements of the universe together in their strange underground circular chambers. In fact counting is probably one of the most important tools available to the modern physicist as we are constantly searching for numbers of things. How many protons did we collide with how many lead nuclei and how many of each particle did we get out of it? We count values all the time and once again numbers are not something inherently necessary to run a universe. The universe will continue working just fine despite the fact that various sentient creatures are trying to figure out why it is running in the first place.

Nothing in the universe naturally counts as the ability to examine objects and determine the number of them in any one place is a uniquely biological trait shared by intelligent creatures and simple animals alike; some animals can even assess numbers and measure things too. Think of a pride of lions on the hunt they are constantly making calculations in their keen minds as to how many prey are in front of them, the size and shape of the prey and the distance and velocity variables needed to bring one down for a meal. So intelligent creatures such as lions and humans like to think and count but the universe does not as no physical process no matter how subatomic or how cosmic in scale ever thinks to count once. The number of hydrogen nuclei needed to fuse into a single helium nucleus is a form of counting by a non biological system you say? Balderdash and hokum, the universe is merely smashing things together trillions of times per second in a stellar core and the fact that they eventually stick together is pure scientific coincidence which we as humans have counted up to meaning something. To prove that the universe is not counting and does not follow any rules all we need to do is look at different types of stellar cores to see how they burn their hydrogen. Since the rate at which hydrogen nuclei meet and fuse to become helium atoms is dependant upon two things all we need to do is see that stars are like people; they like us are all different.

A small cool star will have more trouble fusing hydrogen quickly but will happily burn slowly for many billions if not trillions of years. A large hotter star will not burn as long but can easily power through hydrogen atoms quickly and make lots of helium. The pressure and temperature of a stars core determine how fast they fuse things so a small star that is cool has less pressure and lower temperature meaning it will be less likely to fuse two hydrogen atoms as compared to a hotter larger star in the same time interval. In other words the rate or speed or time it takes to get helium differs and it is only a creature like a human that will record the difference. To the stars involved time is irrelevant as they will fuse whatever they want at the rate they want and not a moment before. So once again we see that time even in the non-biological workings of the universe does not seem to matter and once again we see that it is only intelligent creatures that are counting and thus finding significance in these events.

Like the three dimensions we as humans have come up with numerous systems for counting time as we do in counting sizes. The very definition of Earth based time is once again Earth centric as humans derived it based off of things like the length of a year and the length of a day. The length of a year and day on any other planet in our own solar system is different from Earth and so our neat little system of time would not work to explain those worlds at all. So this leaves us with the realization that time no matter how you count it is a human invention as modern science has not been able to find any sort of universal time unit. We can find no single passage of time at all in contemporary science that would be considered as the base unit of time or some sort of cosmic or quantum starting point from which we can construct larger units or intervals of time. This lack of proof of a fundamental unit of time has led many to believe that time is merely a human invention as is supported by the great thinkers of the modern age.

The German philosophers and scientists Gottfried Wilhelm Leibniz and Immanuel Kant who were both incredibly influential in their day and ours postulated that time was only a construct of the human mind, one of many that was necessary for us to conceive a reality based on the physical world around us. With lives spanning the 1600s to the 1800s Leibniz and Kant respectively formed the basis of much modern science and philosophy concerning human perception of time and space. Leibniz himself was said to have invented calculus independently of Newton and is the father of the first mass produced calculators and the binary number system which forms the base of all modern digital computers. Kant is the creator of the ideas that our own solar system formed from a cloud of interstellar dust rather than the hand of any divine beings and further was the first to propose the concept of the Milky Way galaxy which also must have formed from vast clouds of gas and dust centuries before anyone else dare suggest this. These two intellectual powerhouses therefore bring considerable might to bear upon the time is not real arguments and the notion that humans or other intelligent creatures will use time to make sense of the world around them in the form of a personal reality.

This is also given the fact that Leibniz was the first to suggest that space was relative as opposed to Newton's idea of an absolute space and was based off of his work on motion which used kinetic energy and potential energy. Potential energy starts to sound in some ways a lot like the rest mass of an object as seen in relative space and Leibniz even had a formula for his kinetic energy or what he called the living force it was given as Kinetic energy = mv^2. That looks an awful lot like $E=mc^2$ does it not? Nevertheless it was still argued by both Leibniz and Kant that time was only in the human mind and did not need to be part of the universe as a force or some thing in which objects needed to move or exist in. To this end a lot more sense can be made of time than if we were to take the stance that time was an actual thing. Once again I must remind the reader that just because time seems to slow down when for example an object begins to move very fast does not mean time is real it could be a by product of some other process that modern science is unaware of. To think of this another way we often say a north pole and south pole of two bar magnets are attracted to one another when in reality this is untrue there is no attraction between them as there is attraction between people. This is once again our inability to describe what is really going on with magnetic fields and so we use any old wording so long as it can convey the general idea.

Time is a perfect example of something that you cannot truly define without referencing something else. Try it and you will always find yourself using other descriptive words to explain time. Time is something that a clock counts you might say, well that is true but it requires knowledge of what and how a clock functions. Time is a way of keeping everything from happening at once, once again this puts time in the presumed context of flowing forward and knowledge that one understands this prior to explaining time. The number of examples goes on but the fact remains that time is at best a construct of the human mind that relies upon other understood information about time already or objects associated with time. So if we take the stance of time as being nothing more than a human construct we are siding with a great number of people including Leibniz and Kant. Time will therefore not be real, does not have any sort of force carrier like a chronoton or exist in any sort of field in the universe as spacetime. The natural question to follow this is can science work without time being real? The answer is yes and in fact it works better without time than it does with it.

To illustrate this think of a black hole as described by general relativity and you will instantly see that removing time from the equation will also remove some of the problems of black holes. First a quick look at a black hole as proposed by general relativity and what that is constitutes an object that has such high curvature of spacetime that the gravity associated with it goes to infinity and the time associated with it goes to zero. So it seems like quite a strange object and presents a great number of puzzles for modern science. Well actually it is time for me to reveal a little secret about black holes that some people have known probably for decades and that is this: black holes do not have those two characteristics. Yes that is correct black holes do not behave in any way the way that Einstein or general relativity say they do in fact it is utter nonsense.

At the heart of a black hole exists one of the greatest goofs in modern science called a singularity. A singularity is the point where all the time and space strangeness of a black hole reaches its zenith and lies well below the event horizon. Why does a singularity mean that black holes and the bizarre properties they supposedly posses are incorrect? The answer is simple a singularity exists only in mathematics and not in reality and is a prime example of the dangers of treating equations with too much reverence when one should be standing back from it some distance and remarking how it does not make sense any longer. Things that no longer make sense are replaced with new ideas that do make sense but in the case of black holes contemporary science has no better explanation of them and so the singularity remains.

Now precisely why a singularity and black holes in general cannot exist as described by general relativity comes from the fact that it is impossible to have an area of space with infinite gravity and zero time. A singularity is simply sciences way of saying we do not understand what is happening here. Now getting back to time for a moment which exists in general relativity as a part of spacetime and we find that it is a direct result of Einstein's frames of reference including time that leads to this impossible object at the heart of a black hole. Removing the time component from the black hole will result in strange things like time stopping completely inside it going away. The curvature will also be lessened to the point where gravity no longer explodes to infinity which is a very impossible idea as if it were true a single black hole would destroy the universe.

Think for a moment about a universe with a single black hole in it as described by Einstein and general relativity and you have an object with infinite gravity. Gravity is a force that can traverse the entire universe with infinite range and its effects are felt immediately so any object with infinite gravity would slowly begin to collect all matter in the universe around it. Given enough time a single black hole would slowly start accelerating everything in the universe until all matter everywhere were travelling close to the speed of light and falling into the central black hole. You and I are very much alive some 13.8 billion years after the creation of the universe and so no such black hole object has destroyed the universe. The net result is then that black holes do not and cannot possess infinite gravity and once again we see that a mathematical description of spacetime fails to match up with observation. To follow this line of thought to its natural conclusion time must not be a part of spacetime and so time is not real or at least not a part of space.

So once again we see that a problem in astronomy such as a black hole can be fixed by removing time as it results in an object with incredible gravity but no strange impossible set of conditions surrounding it. Time does slow down in certain cases but a black hole might be the ultimate proof that this is due to something else and not to time being real and certainly not part of spacetime. As for the rest of time where the apparent effects of relativity are not present time is certainly only a construct of the human mind. This is evident not just in the different timescales created worldwide but also in the usage of time. Time is used to count and measure nearly everything we do and is used to run almost every machine we have ever built. This does not of course mean time is at all real as it can be thought of in these instances as simply another measuring tool by which humans construct a reality from the physical world around them. That idea is undoubtedly tested for millennia as humans have always used time to describe the world and in this way is completely explainable using the reasoning, philosophy and science of Leibniz and Kant. We live in a world that exists where time is measured we do not nor have we ever sent an instrument to live or record in an area of high relativity and so we cannot say with proof that time is needed. We can easily say that time is created by intelligent creatures as a way to understand the world around us.

This is where the debate on time will end for now as we have demonstrated that neither side can truly gain any ground over the other using their preferred methods as a way of accurately describing what time is or is not. To sum up each side they are according to modern science equally weighted for now as each can be said to be somewhat true. Take for example the idea that time is real and necessary for the universe to work and we have some weak, indirect evidence for this argument in the effects of time dilation. On the other hand science likes facts and the opposing view saying time is bogus has many, which are no theoretical proof of any time force exists in the universe and humans do in fact measure things so time is just one of them. This leaves us with no new information concerning time and instead moves both of these opposing sides to one I dare say fact about time that they can equally agree upon. This is the property of time known as the arrow of time and can also be debated but does overall seem to be real.

The arrow of time is an observed phenomenon of time whereby events flow in only one direction. This is of course from the past to the present and onwards to the future and derives its name from the fact that an arrow always points one way which is from the base to the tip. The direction the arrow points or flies is of course one dimensional as it always moves from past to future and so we can demonstrate that at least observationally time has only one dimension unlike physical systems which modern science describes as having three and sometimes more dimensions. This is of course another effect that intelligent animals and humans are aware of and have been for ages. A human can observe things growing from seed to tree for example, a lion will observe the oldest or youngest herd animals to hunt and so we see the idea of time and age is ubiquitous for all intelligent creatures. Many gods and creation myths from all the world's cultures from primitive to modern days involve the flow of time in one direction and one direction only.

The human mind can create whatever myths it wants to concerning gods creating the universe and invent all sorts of fanciful stories to accompany them but never in the history of our species has a culture described the arrow of time reversing even when all it would take was a creative mind to dream it up. Even those cultures with a cyclical time system are not travelling backwards in time merely forwards to a point where a physical cycle will repeat but not a temporal one. Time will flow forwards for these cultures endlessly as the physical world is born, grows, dies and is reborn again in the next cycle. The cycles of time are not true time but measured time as represented by a human invention called the calendar, the overall time governing the universe even in myth still flows forwards.

In modern science, cosmology and theoretical physics the same is true for both observation and experimentation. We do not look up and watch black holes spontaneously expand outwards and rearrange their matter such that it produces a star rapidly un-fusing the iron at its core all the while sucking up the photons it has emitted for millions to billions of years. The idea of things being reversible in time is a tantalizing one in the laboratory as you can easily perform an experiment and in many cases run it in reverse. Take for instance a chemical reaction whereby you combine hydrogen atoms with oxygen atoms and you of course get water from two different elements. You now have bonded atoms in the form of molecules which are governed by the same quantum processes that appear to be at work in many places in the universe namely the electromagnetic force that is responsible for the atoms sticking together through the sharing of electrons. A human can overpower the very forces of the universe and get the two different elements to dissociate from one another splitting the molecule back into two separate types of atoms. To do this every science student knows to apply a charge to the water and collect the gasses forming at the anode and cathodes in a couple of inverted test tubes submerged in the water bath. So we can in this sense reverse the flow of time as it pertains to the chemical reaction of bonding two hydrogen and one oxygen atom together and at the end of the experiment we have returned the two elements to their original separate forms. What we are doing is reversing the physical process but not the temporal process and this is true for two reasons.

The first is simply that if we were to set up two clocks in the laboratory, one for the laboratory and one for the different parts of the experiment and record the passage of time on each we would see a flaw in the reversing of the reaction. This flaw is the fact that even if it took exactly the same amount of time to get the two reactants to combine into a product and then get the product to dissociate back into the two reactants this would disagree with the laboratory clock. The reason is simple the time for the closed system can be equal for the reaction going forward and then again in reverse but the total time as measured from outside the system is twice as long or longer as we would need to set up the experiments and actually run them. So what we find is the universe time or the time in the laboratory rules over the experimental time and therefore reversing the reaction is not really reversing the flow of time only the physical system. This is a more indirect way of proving that we have not reversed the arrow of time despite reversing the physical effects of the experiment to a state where they mirrored a time before we ran the experiment.

A much more direct way of proving this fact comes from the fact that everything involved with the experiment did not run in reverse as everything in the universe is tied to the experiment. Remember any system within the universe can be isolated from other systems in order to study something or observe something of interest but the systems can never be isolated from the universe as a whole as the universe contains everything and this includes every isolated system someone might create. The significance of this can once again be illustrated from our very simply laboratory experiment involving hydrogen and oxygen atoms reacting to produce water. The system is isolated from outside influences but the system is not isolated from the laboratory and scientists and experimenters make up part of that laboratory so in order to reverse the time in the system we must also reverse the time in the laboratory. This is where the clock in the laboratory comes in handy as we can film the entire laboratory and the isolated system as one. By watching the tape recording we make from it we can see that inside the system time appears to flow forwards as we react the two elements with one another and then flow in reverse as we return them to their original separate states. Watching the tape and examining what takes place in the laboratory which is for this purpose the entire universe we see that the people conducting the experiment are not running in reverse. The scientists are simply moving forward in the universe or laboratory while the system moves back and forth so overall the time in the universe still flows one way. This is of course from the past to the present and then into the future.

This simple analogy involving the creation and destruction of water in a laboratory illustrates what is happening anywhere we look in the cosmos at every Planck time since the Big Bang some 13.8 billion years ago. Time is always flowing forwards and never under any conditions deviates from this path. The significance of this points to the fact that time may not actually be a real thing, force or anything at all in the universe. The reason I say this is that every single other component that humans and presumably other intelligent creatures have observed throughout the cosmos can be seen to work in multiple directions. Any form of matter or energy can be converted from one form to another given the right conditions, and force or field can be excited or relaxed given the right conditions and any apparent changes to space even at the universe sized scale can be altered given the right conditions. In other words everything that we have ever been able to measure empirically, with instruments or even just theoretically can be made to move in more than one direction or take on more than one form. Time is the only thing that does not seem to fit anywhere in the workings of the universe in such a way that it would obey even theoretically the same rules as matter and space do all around us constantly.

For instance we understand a great deal more about a little bit of matter such as a hadron and what makes it up as well as all the little bits that hold the other bits together. We are pretty good at understanding bits thank you very much and while we may not have all the little bits put neatly into their respective niches we do at least grasp the vagueness that underlines most of these relationships. Using our previous analogy of making and breaking water in the laboratory we can do the same with things like a simple hadron from which matter such as water is built. Gluons, quarks, color charge all of this can be made to roughly run forwards and backwards in a manner that satisfies the understanding modern science has concerning the fundamental forces and elementary particles. Time that is used to measure how long these interactions need in order to take place is the only component that does not appear to be involved in the interactions in any way. We do not see a time particle or field being used in the reactions involving fundamental particles or fields and in fact at quantum scales we have demonstrated that certain aspects of the universe seem to be a little unnecessary.

The most obvious example is gravity which is such a weak force when compared to the other three fundamental forces that its effects are largely ignored in individual quantum reactions. Time which appears to be even less tangible than gravity could behave in the same way to the degree of being unnecessary for individual quantum reactions. What this means is that time might have zero influence on the universe at certain scales and therefore with respect to certain particles and interactions of said particles.

A more logical explanation for this is that time does not really exist and is only something that humans and other intelligent creatures have created for themselves in order to make sense or at least measure the universe around them. I feel it is important to for a moment distance the physical workings of the universe from the philosophical ones at this time and examine briefly how the two sides see reality. In the philosophical sense time is needed for things like humans to construct a reality for themselves about the universe and then they can function within it. This is a somewhat fanciful idea concerning the universe as certain proven aspects of the universe itself prove this to be wrong. A perfect example is the quantum world which is something that animals have largely evolved to not see and therefore not perceive and therefore not need in order to construct a reality of the universe. Quantum interactions continue whether animals can perceive them or not and this is definitive proof that the physical make up of the universe is not needed nor was it tailor made for humans or any other creature. So the idea that we need various forces in the universe to construct a reality of the universe is false as certain forces go largely unnoticed and yet are very real.

This leaves us with the second interpretation of philosophy which is really just science that was first speculated about prior to proof of concept. This is that the universe does behave in certain ways whether or not anyone or anything can notice them and therefore that is the true reality. A pure science reasoning and not at all a philosophical one as hard evidence is needed to support this claim beyond simply postulating it. Science has provided such evidence with experiments such as those responsible for the Standard Model. In this view of the universe which exists whether we need it to or not we find that time is even more unnecessary than before as interactions do not need themselves to be measured in order to be real. Even at the smallest timescales that humans have calculated time does not need to be a real thing and is still used only as something that is a form of measurement. Remember the fact that if a Planck time is the smallest time scale and that time is hypothetically real then at any instance of one Planck time or less you do not need time in order to describe the system. Therefore a three dimensional universe can exist without a fourth time dimension and time can be removed as a fundamental force.

Another aspect of time and the debate of whether or not it is a real tangible force comes from the notion of a Planck measurement in the first place. Simply put the Planck system of length, time and so on may not actually be correct as we will see. Throughout history humans have continually rethought the universe or at the very least the world around themselves. This is evident in how we perceive the old theories concerning the cosmos and how relevant they are today if at all. In fairness to the old religions of the world it is probably far more correct to adopt a polytheistic view of the universe than any sort of monotheistic one. The reason for this is that what ancient religions were describing in their various gods was often times a real physical phenomenon.

Take for example the ancient Norse religion and its numerous gods and you can see what I mean. We have already seen the ancient Germanic peoples viewing the star that Earth orbits as Sunna the goddess embodied and later we see Thor who is the controller of lightning and thunder. In modern science these two gods are actually quite accurate as Sunna is the embodiment of stellar fusion and gives light and life to the universe in the form of photons and new elements through various atomic chain reactions. The goddess Sunna gave light and life to the Earth and universe and since we know stars are the reason almost all elements exist in the universe and hence the ability to form rocky water laden planets with life on them is possible the ancient Germanic peoples were actually one hundred percent correct. The only thing they did not quite have nailed down was the mechanism by which Sunna made light and life as what they called divine power is actually fusion power brought about by a very, very hot and dense core.

The same is true of Thor as he is the god of lightning and thunder we know now he is really the embodiment of the electromagnetic force and again the ancients were spot on in describing a force through the use of a god they only missed how the god created or controlled that force. The same is true throughout scientific history as different humors, ethers, homunculi and what not have given rise to a better understanding of biology, physics, and chemistry and so on and of course astronomy and the cosmology of the universe as well. So the ancients used one set of explanations to deal with the four known fundamental forces and we have slowly started to use new ones in recent times. In fact for the tens of thousands of years that humans have been evolved into our modern forms and thinking about the universe even if only through the use of gods it has only been the last few centuries where people have been thinking of the universe in terms of measurable forces or particles.

So here we see the very definition of what the universe is change over time and ages leaving us with a chronological history of new understandings much like the growth rings of a tree. In a sort of scientific dendrochronology it is possible to look at each different theory and what it brought to our collective understanding of the cosmos. At one time we had no idea of scale in terms of the biggest and the smallest and this is reflected in many religions especially the monotheistic ones. To these people their myths involved finite sizes of worlds and unknown sizes of the universe and are probably the least scientific of all the ancient mythologies. The universe was either just a wee bit bigger than the ball of dust that rolls around the Sun or much larger depending on how much you wanted to mentally prostrate yourself in front of your divine one. So here the absolute scales of the universe are at best a haphazard blur of personal realities and faiths. Moving on from these scientifically void beliefs we find people examining the universe in the last few centuries and starting to question how small can small be?

Revisiting the times of Dalton we see that matter can be made of small units called atoms with different properties thereby combining into different forms familiar to us such as molecules. Again we must remind ourselves of the ludicrousness of crediting the Greeks' with the discovery of the atom as anyone who understands modern physics realizes they were not even close and simply calling theoretical dibs on something does not mean you have discovered anything. We do see an interesting idea of scale and absolutes with the Greek's misguided beliefs in their type of atom as to them size was infinitely small. No matter how many times you cut an object in half according to them you could do so again and again thereby getting ever smaller pieces of that original object. We know this to be wrong of course but we also know that atoms are starting to get down to the level of the very small and we have also learned that splitting one of those in half is no small feat. So at the time of the atomic hey day the smallest possible scale to the universe was the atom and the idea that anything smaller was quite preposterous.

Moving to the next ring of our scientific tree coring we find that a revolution in physics would allow for even smaller sizes to be possible and here of course we are talking about the quantum world. Quantum mechanics is easily the most important contribution to physics that was ever made in the twentieth century as it has directly led to numerous new particles, better understanding of forces and to direct applications of its knowledge base to the real world. So when new scales of the universe are described in terms of quantum mechanics scientists around the world sit up and pay attention. We have of course the Planck system of measurements being created as a result of this new subatomic understanding of matter and the universe in terms of how it all works in general. So now we have a definitive set of measurements from which we can construct frameworks of how the universe should work. Or at least that is what we have in theory. The truth is the Planck measurement is modern sciences best guess as to the nature of certain scales in the universe and right or wrong has been incorporated into most modern theories. Any theory that attempts to discard it is severely scrutinized and often times marginalized for attempting to break some of modern sciences golden rules. In reality modern science has no perfect understanding of how the universe works and these flawed understandings are built upon assumptions such as the Planck scale or the speed of light and so on.

Something is rotten in the state of scientific Denmark as none of the current theories in contemporary science come close to answering everything. So the underlying truth is that a few things have gone amiss at the most basic and fundamental levels when it comes to understanding the cosmos. The Planck measurements are one such assumption that while it is true they do make some amount of mathematical sense they also are units constructed by humans and our current understanding of the universe. This is no different than the previous understandings that humans were firmly convinced explained the universe and later were proven wrong. This being said it is true that we are closer than ever to proposing a complete model of how the universe works and of course the day will come when modern science does accept a new model that is correct. After all the universe really only does work in one way regardless of whether or not a human is there to understand it. So in terms of Planck measurements they are the best we have for now and possibly only need some fine tuning in the future but as for the Planck measurement of time it is just a unit created by humans.

The fact that modern science has built its understanding of the cosmos around things like the Planck time being the smallest unit of time does not really matter as the unit is still one we have constructed out of mathematical thin air and not necessarily fact. To put this another way any unit that we say is the smallest unit of time will always be a human unit and not necessarily the true limit of the universe. Think of a half life for a given element and it makes more sense as it refers to a quantum state that is also uncertain at any given time. Remember Schrodinger's cat and the fact that the whole thought experiment and the subsequent Copenhagen and many worlds interpretation are based off of this one concept. The radioactive decay of the sample inside the box is unknown because it is a quantum state which can either be decaying, not decaying or both according to modern science.

In reality however half life is somewhat of a misnomer as it refers to half the sample decaying in a given time and not the actual time for a single radioactive atom to decay. That second time is not reflected in the half life equation for the stated reason that it is a quantum property. So in terms of time and the Plank time we see that even at the smallest and theoretically shortest durations any system can endure time still means nothing as the number of Planck times a specific particle will exhibit any quantum behavior is not set in quantum stone. If one particle exhibits a behavior faster or slower than another it occurs in different amounts of Planck times and if the event for a given quantum object occurs faster than any other quantum object it is possible for it to occur either in reality or on average in say half a Planck time. So once again time does not necessarily obey scientific laws and fit neatly into any sort of mathematic equation or human understanding of the universe.

If this seems hard to grasp I will give you another and simpler interpretation of the Planck time as it refers to the universe and the three dimensions that particles do exist in and the assumed fourth dimension through which they experience duration. This second interpretation is that the Planck time is real and that nothing in the universe can occur in a time scale shorter than a single Planck time as nothing in the universe needs to exhibit any quantum behaviors faster than that. This means that the Planck time is real, the shortest time frame possible in the universe and of course that this is also meaningless. Why meaningless you say? Because at the end of the day the Planck scale unit of time does nothing to prove whether time is in fact real or not. To be honest nothing anywhere says time has to be real and the human constructed measurement of time called the Planck scale is simply our current smallest time unit in modern science and is yet just a human idea until proven if it is true at all. Time does not have to be real and time is not proven to be a tangible force, particle or object in anyway by the use of a Planck time unit.

The significance of this cannot and should not be understated as all of modern science and its attempts to understand the universe hinge on fundamental concepts such as this. This fact is illustrated in the idea of whether or not time is important at the quantum scale as it is at the relativity scale. For that matter if time is real why cannot the two theories just play nice and reconcile their mathematical differences? The answer is simple something is missing from modern sciences understanding of the universe and how it works and so the link between the very small and the very large is lost. In this way we once again see the divide amongst scientific thinkers as to whether or not time is real or just all in our minds. To end the debate a new model of how the universe works will have to be presented and accepted by the forward thinking members of contemporary science and this model will be proven correct by its ability to blend the small with the large, explain all observable phenomena in the universe and make the previous problems of the old theories vanish in a wisp of smoke.

Entropy

Before I get to all of that and I will much later for now we need to examine one more aspect of time and a problem it has presented modern science for decades. A problem which has come to guide the principals of what is generally accepted as good theory and what is frowned upon and to do this we need to re-examine our creation and destruction of water in the laboratory. In our hypothetical experiment we created water by burning hydrogen and oxygen together and we ended up with released heat energy and water molecules. Later we split the water molecules by using an electrical current which imparts energy to the water molecules causing them to dissociate from one another and once again give us back our original hydrogen and oxygen reactants. The idea that we were reversing time was analyzed from this and from a physical standpoint we end up the same as we started out and so it would seem at least physically like we travelled backwards in time. For various scientific and philosophical reasons we have seen that while the reactants returned to their original forms the rest of the laboratory still moved forwards in time. The universe itself contains everything this includes the reactants, end products, experimenters, laboratory equipment and so on such that time overall flowed forwards despite the final outcome of the hydrogen and oxygen. What was never discussed thus far in our experiment and many have probably forgotten about them are the various things such as energy that were emitted or absorbed during the experiments.

In the first instance we created a great number of photons as the hydrogen and oxygen burned to form the water. Most of these photons were visible light that we could see as a glow of color from the flame and the rest were infrared which we could perceive through the heat we felt as the atoms combined. So as time flowed forwards we got water molecules and at least two types or wavelengths of photons emitted in fantastic numbers as released energy. Anything that burns will undergo what is known as an exothermic reaction meaning one that releases more energy than was put into it in order to begin the reaction itself. A small spark is all that is needed to set a whole forest on fire and obviously the spark contains far less energy than that stored as chemical potential energy in the countless trees that will burn in the ensuing firestorm. So this reaction allows us to see time flow forwards easily enough but what about when we split the water molecules back into their component atoms?

If we reversed the process and truly caused time to flow in reverse then as we split the water molecules energy equal to the number and type of photons emitted will be reabsorbed into the atoms during the dissociation process. This is not what happens at all and even if it were it would need to be the exact same photons of visible and infrared electromagnetic radiation that were absorbed and deposited in exactly the same quantum states as they were before the burning occurred. This is of course impossible and what really happens is we now invest a great deal of energy into the water molecules in order to get them to dissociate from one another. This type of energy absorbing reaction is known as an endothermic one.

A simple example of an endothermic reaction is ice melting as the ice absorbs passively ambient heat in the form of infrared radiation into its interior. As the ice heats up it stores the energy in its bonds and eventual undergoes a phase transition to water the outside environment which powered this process experiences a drop in temperature which is really a drop in the total measureable energy it contains. So we have an endothermic reaction powering the reverse process of splitting water molecules into hydrogen and oxygen gas again. The energy put into the system is not visible or infrared photons at all but instead electrical energy from the two wires we have immersed in the water bath. This physical process demonstrates that while we have in fact reversed the reaction of the hydrogen and oxygen atoms burning with each other we have not truly reversed the reaction itself. In order to do that we would have to do the near impossible of cramming all the free photons emitted back into the atoms from which they came without using any electrical energy. This is something that the universe will never do spontaneously as it involves a transfer of energy forbidden by the laws of physics and the law of entropy. Yes entropy has reared its interesting and enigmatic head to explain why we cannot simply reverse the whole water making process in the first place and why we must always use an ever growing number of particles in order to get anything to do anything in the first place.

To some of you out there reading this you have a nice smile across your face as you know exactly what entropy is and enjoy puzzling over it to no end. No doubt you have been waiting to see it pop up in this description of the universe as entropy seems to behave like the arrow of time itself and flow only one way. For the rest of you I will explain exactly what entropy is and exactly why you will be left wondering what it holds in store for the fate of the universe like the rest of the theoretical physicists out there. Entropy is a measure of the amount of disorder in a system and that is the simplest way of stating it. The more complete way of describing entropy is through the laws of thermodynamics and can be thought of as the total number of microstates in a thermodynamic system that is governed by macroscopic variables. Since most people find it easier to simply picture disorder and the reference of disorder and chaos in the universe is more applicable to theoretical physics and cosmology I will refer frequently to the former, simpler definition. A quick history of where entropy came from goes back to the 1800s when many scientists were working on the problem of heat, work and energy transfer in systems. Prior to that even we see the visions of Newton once again appearing as it was Newton himself that quite correctly proposed that heat and light were a form of indestructible matter that was either attracted to or repelled from other forms of matter. While this may seem somewhat inaccurate to us today as we know heat and light to be forms of energy rather than matter if we dissect the meaning of Newton's idea we see he is perfectly correct.

To start with you must realize that Newton lived in a time before the concept of energy as a physical thing and was seen as another form of what scientists and thinkers had understood for millennia and that was matter. Fast forwarding to the early twentieth century and Einstein proves Newton correct to a degree by showing that through his equivalency principal matter can be thought of as energy and vice versa. This is of course not literally true as you can try all you like you are not going to turn photons into hadrons no matter how many of them you have. The idea however for the time of Newton was sound and in fact sound experimentally too as matter was affected by both other matter and energy so the distinction between the two was something of minor importance in the 1700s. As for the part about heat and light being indestructible this is also somewhat true as many students of science have heard the familiar words stating that energy cannot be destroyed only transformed from one form to another. Which is also known as the law of conservation of energy which simply put says the total energy of any isolated system will remain over time unchanged. A German physician and scientist by the name of Hermann von Helmholtz would go on to make a great leap forward in the understanding of mechanical work and energy or heat as it was often being studied.

Being a physician Helmholtz studied the metabolism of muscles and conservation of energy within them and was able to demonstrate that no energy is lost in muscle movement. He expanded on this by treating biological systems and mechanical or physical systems as one and the same proposing that the energy driving each of them was conserved. This was a great leap forward as it ties all, kinetic, potential, chemical and other forms of energy together in one universal understanding of the conservation of energy which can thus prove the theory of conservation true. Helmholtz would go on to make many contributions to modern day physics and physiology including the basis of much of our understanding on the senses of sight and sound. Potential energy becomes kinetic energy, chemical energy can become heat energy, and energy stored in the bonds of atoms can become nuclear energy. At the end of the day however you have not destroyed any energy only transformed it from one type to another.

Here too Newton was correct and lastly his idea that heat and light are attracted too and repelled by different types of matter is also correct. Again more correct from an eighteenth century experimental perspective but correct nonetheless. A cold object can be thought of as attracting heat as it absorbs infrared radiation from an external source and stores it internally. A mirror can be thought of as repelling light as a beam shone at it is nearly perfectly reflected with little to no loss in the strength of the incident beam. Fire can be said to radiate or repel both heat and light away from it as we know today that fire is a form of weak plasma. In this respect Newton was again ahead of his time and so much so that scientists centuries later would build on his ideas of motion and heat and light to give birth to a modern version of entropy.

Newtonian entropy can be thought of simply as the fact that heat and light and by extension all energy are never destroyed but merely transformed in a given system. How this became a modern version of entropy stems from the study of heat or thermodynamics in mechanical systems and how much work you could get out of them. Remembering of course that the industrial revolution was exploding out of places like Britain and Germany and spreading like wildfire across Europe and you can see why so many people were employed in the thinking of thermodynamic mechanical systems. Some of the initial thinking was done by a French mathematician named Lazare Carnot in 1803 and involved energy loss and motion; two of Newton's favorite topics. The concept of work and specifically how much work you got out of a given mechanical system as opposed to how much energy you put into it was where Carnot drew his ideas. What he reasoned was that any time a mechanical system sustained a sort of shock to it a small or large depending on the action an amount of useful energy was lost.

Let us say a wheel travelling along a road and suddenly it hits a bump what happens is some of the energy being used to move the wheel in a straight line along the road is redirected by the bump into making the wheel jump off the road. As the wheel falls again gravity steals more energy as some of the downward force of the wheels impact is absorbed by the ground. This action will also produce sound and heat due to friction and since these cannot be created from nothing they must have come from somewhere. Where the sound and heat friction came from was stored energy carried by the wheel, the loss of which means that energy can no longer be used to move the wheel smoothly in its original straight line down the road. If the goal was to travel from A to B along a road using as little energy as possible then the bump in the road has now taken some of the usable energy. Carnot's own son furthered this by stating that work can be garnered from a system where there is a drop in heat. In other words any time the heat of a system cools the energy must go somewhere and this was in his idea into useable work and one can think of the heat generated in a combustion engine will drop as the explosive force is transformed into mechanical motion.

Much of the work done in the early to middle 1800s on heat and changes in thermodynamic systems and especially that of the English scientist James Prescott Joule would go on to give us the laws of thermodynamics. Joule was in some ways destined to enter science as his family were brewers and much of the brewing industry relies on science and especially chemistry in order to improve batch quality. This also meant his family had enough money and influence to allow James to be tutored by none other than James Dalton himself the man who gave us the first true atomic theory. James Joule was primarily interested in the study of heat and was able to directly influence scientific development in thermodynamics which would formulate the first of four laws of thermodynamics. We will continue with a description of these laws in a moment but for now we will continue on building the history of entropy through experimentation of thermodynamic systems. While the earlier work by the Carnot's had been widely accepted Joule rejected the caloric theory they proposed and suggested instead that heat was a result of a kind of molecular motion. In Joule's mind a rotational force coming from molecules created heat within a system and caused matter to have heat energy. This is very close to the idea of vibrating quantum bonds between atoms using the electromagnetic force exerted by atoms we know is correct today.

In order to explain why the energy between molecular collisions did not eventually lose strength over time Joule suggested that the collisions were perfectly elastic. By being perfectly elastic no energy loss between molecules can occur and therefore the heat they generate can be maintained within the system over time. In other words Joule was proposing that in a closed system mechanical work could be transformed into heat energy and vice versa. He would go on to give the world the first mechanical description of work through this conversion process from mechanical motion to energy and the unit to describe it is called the Joule in his honor. A joule using the notion of J is the work done by an object that experiences a force of one Newton and is moved through one meter also written sometimes as a N.m or Newton meter as long as torque forces are not involved. This is the mechanical side of the coin to see the energy side a joule also refers to the energy radiated as heat when a single ampere current of electricity travels through a resistance of a single ohm in a single second. Joule would go on to prove and measure the conversion of mechanical force into heat through experiments involving solids, liquids and gasses which would definitively prove him correct and lead to other scientists giving us the first law of thermodynamics as a result.

Thermodynamics as a field was one that developed slowly over decades and drew from many different minds in order to be formulated into the now familiar system we know today. As we have seen from work carried out by the Carnot's and Joule certain insights into the study of heat and its ability to explain work and energy were stepping stones to the final conclusions of thermodynamics. While it is true the calorific theory of heat as proposed by the Carnot's has been proven incorrect when compared to the work and ideas of Joule all of the steps in the study of heat have led to this decades long understanding of heat, heat transfer and work in systems. The true founders of thermodynamics as a modern science is threefold and we will look at each and their contributions in one frame of reference. The first we will examine and the most important is without a doubt Rudolf Clausius who was a German physicist and mathematician. The reason Clausius is the most significant of the three is due to the fact that not only did he make huge contributions to the study and development of thermodynamics as a field but he is also the person who introduced science to the modern concept of entropy. Entropy is of course the end product for modern physics and theoretical physics where the study of order within closed and open systems is concerned. Not only can entropy measure the order within a finite space such as a closed laboratory experiment but it can be applied and routinely is to the entire universe as it holds true throughout the cosmos and at any time within its history.

The second scientist involved in the foundation of thermodynamics is William Thomson an Irishman whose name may not ring any scientific bells at first. However you will undoubtedly recall his name when his title of Lord Kelvin is spoken as he is the creator of the Kelvin scale of temperature and is the result of his work on thermodynamics and the idea of absolute zero. Today scientists commonly refer to the Kelvin scale worldwide when measuring temperatures and especially when dealing with the kinds of heat levels involved in theoretical physics. As you may recall the temperatures of various scientific instruments and this includes the largest, most complex and expensive ever created by humans the Large Hadron Collider commonly cool materials down to within absolute zero.

Finally the third scientist involved in the formulation of thermodynamics is a man named William John Macquorn Rankine from Scotland. Rankine was heavily involved in the study of heat and mechanical systems as his first calling was that of an engineer and so the study of these two disciplines together led him to a greater understanding of thermodynamic systems as a whole. His primary focus in thermodynamics involved the first law and was developed by himself as well as Clausius and Kelvin. In fairness to all three of these founding fathers to thermodynamics while it is true that some were more instrumental than others in developing specific laws all had a hand in the creation of them overall. This also includes a number of other scientists worthy of mention from history as well and in particular Helmholtz who we have already seen and another German scientist named Julius Robert von Mayer. The reason I mention all of these five minds is because the exact times at which each one proposed a particular theory or conclusion on the study of heat or perhaps how they stated it or by what means they derived it has been muddled by history. We can agree that Clausius, Kelvin and Rankine are direct early contributors to the field and demand recognition but Helmholtz and von Mayer were also heavily involved in the development of thermodynamics and either contributed to the development of its finalized laws as we know them today or came to independent conclusions that mirrored those of the other minds.

To state this another way think for a moment of heat in a gas or liquid system and can be thought of as the random Brownian motion of particles which are difficult to keep track of precisely. Well decades of research and study into the study of heat and involving five brilliant minds will lead to a history that is difficult to track of precisely as well. Remember many scientists develop ground breaking ideas or experiments by themselves long before anyone else and keep them secret for years without publication which we view as a sort of time stamp. Meaning exact dates are difficult to pin down and you need to look no further than geniuses such as Newton or Darwin who either kept ideas close to themselves or did not publish their works for years or decades for fear of rejection, ridicule or simply they were not motivated to do so. Nevertheless we now have the names of all who contributed in large amounts to the theories and can refer to them when needed as we discuss the laws of thermodynamics in earnest now.

The first law of thermodynamics is rooted firmly within the historical context of the 1800s where the study of work and the efficiency of machines were being heavily scrutinized. This makes perfect sense as anyone seeking to get the most bang for their buck will tell you it is vital to maximize the gain you get from the work you put into a system. This can be seen in the earliest steam engines that were ridiculously inefficient when compared to modern standards and for people at the time any way of improving the power output of such a device only improved the output of whatever job they might be doing. In this case one of the first applications of steam engines and steam power and the job for which the unit known as the horsepower was derived was extracting waste water from mining operations. This job was before the invention of the steam engine done by human or animal power and in fact a horse was so commonly used to turn a wheel that would draw a bucket of water up from the black depths of the mineshaft that this is where James Watt got his unit name of horsepower from as he had to equate it to something mine owners could understand in order to further his product. James Watt was a British inventor who made significant contributions to the field of steam power and so much so that his then new Watt steam engine would be partially responsible for furthering the industrial revolution worldwide not only in his native Britain.

So to see directly how these historical necessities of invention were the kinds of motivation to understand heat, energy and mechanical work the first law of thermodynamics simply states that: The increase in internal energy of a closed system is equivalent to the difference of the heat supplied to and work done by the system. In other words it does not matter how much heat or how much work you are dealing with inside the system so long as the addition of the two balances to the total energy of the system. Note also here that the term internal energy is used as this is key to understanding not only thermodynamics but also entropy as a closed system can be part of a larger world which can affect the overall state of both. When dealing with a purely closed system however we can ignore the outside factors completely and this is especially true for thermodynamics and not entropy where to fully understand entropy and all its meanings the only internal system that really counts is the entire universe as a whole. The idea of entropy and the universe as a single system will be dealt with later after thermodynamics just keep this point in mind for now with respect to thermodynamic laws. So in order to understand the first law of thermodynamics and the 1800s and the industrial revolution we can look at the exact law closely and see how it applies to efficiency.

The total energy of the system is equal to heat and work essentially so if we have a very inefficient machine and of this example we will say an early steam engine or steam piston as they were sometimes known. An inefficient steam piston will contain more heat and less work in its total energy meaning you get less bang for your buck in terms of work and more waste energy in the form of heat loss or friction. An efficient steam apparatus will reduce the heat factor and therefore since the total internal energy must be conserved will increase the work output you receive and therefore you get more bang for your buck. So what we see here in the first law of thermodynamics is a proof of concept of the law of conservation of energy which states that energy cannot be created or destroyed only changed from form to form. What we also see here is of course the direct influence of Joule and his ideas of heat, energy and mechanical work all rolled into one and so the first law of thermodynamics in a sense neatly contains everything from the early studies of heat all the way to practical applications of the industrial revolution. One last thing that the first law of thermodynamics states is that we see a relationship between what can normally be thought of as regular energy in this case heat and types of unseen energy such as work since it can be thought of as stored energy. I say stored because kinetic energy is stored unless the object is acted upon by an outside force or conversely the object itself interacts with another in accordance with Newton's laws. So the idea that mechanical work is equal to energy is correct as it contains kinetic energy and this along with heat comprises the total energy of a system for something simple as a steam engine.

Now we are going to move on a bit and talk about the second law of thermodynamics where things will definitely move towards the less tangible and begin to clear the way for more modern concepts. That being said nothing here is too scientifically scary for the physics initiate but a basic understanding of certain physical realities of systems needs to be examined. So in order to do this we will use a very simple example of a different physical process that is readily understandable to all to explain the second law of thermodynamics. This second law simply states that heat will not spontaneously move from a cold location to a hot location. The basic notion of this law is such that any closed system will tend towards heat equilibrium if left alone for a long enough time. This can be seen easily in a hot object, let us a say a nice hot cup of tea that is left outside on a table during a winter day. While it might look as though the cold is moving from one area and accumulating in the hot cup of tea resulting in it eventually cooling and freezing solid this is in fact not true. To understand this better I will now explain something of a rather significant nature that is almost totally forgotten by all but a few and this is the plain and simple truth that there is no such thing as cold.

Yes that is correct cool, cold and freezing as humans understand temperature do not and never have or will exist. The only thing that does exist in the universe is heat and the variations in how much heat an object or system possesses determines if it is warm, hot or blistering. Every system in the universe as well as every object in the universe has in it heat and this includes things like liquid nitrogen. Liquid nitrogen is fairly common in non-scientific circles and synonymous with something that feels dangerously cold as you would certainly be fool hardy at best should you ever touch it. The truth is however liquid nitrogen only feels cold to us because we as humans like things that are body or room temperature. The liquid nitrogen still has heat inside it or it would be a frozen block of non-matter and I say non-matter because everything in the universe needs heat in order to exist and behave as we see it around us. The lack of all heat would mean the complete cessation of movement of every constituent particle of the nitrogen meaning the nitrogen would most likely cease to be a coherent atom and therefore not be nitrogen devoid of heat. So you must always have some heat in an object no matter how small and therefore cold does not exist and therefore our cup of tea is not absorbing cold in the winter day but emitting its heat to the less well heated environment.

So the second law of thermodynamics holds as everything must flow from hot to cold. This idea of heat being moved from a hot area to a cold area happens not just to our cup of tea but to every object in the universe and is called equilibrium of a system. Equilibrium of a system can involve many such things and indeed the laws of thermodynamics do not solely pertain to heat or heat energy but have derived this name from the way in which they were first postulated. As we have seen it was heat and work in systems that gave birth to modern thermodynamics and so as a result the name simply stuck. Now we come to that different physical process that will help to explain equilibrium of a system in a way that everyone can immediately grasp.

For this example I will use diffusion to explain the flow of heat from a hot to a cold region and I will use the example of food coloring being dropped into a bowl of motionless room temperature water. The water in the bowl is motionless so as to produce no turbulence with which to stir anything dropped into it and the temperature is simply room temperature or tepid in order to illustrate that no vigorous motion of the water is occurring either such as boiling. So we begin to drop our food coloring into the very center of the bowl and in your mind pick whatever color you like best and you can add as much as you like too. What will happen regardless of all these variables is that the food coloring will stay somewhat clumped in the general area that you allowed the drops to fall in. For the next part of the experiment we will do something that excites scientifically minded people and yet is very much akin to watching grass grow or paint to dry.

We will wait and do absolutely nothing to the bowl of water containing the food coloring mainly because we do not want to disturb the system with any outside influences and also because we do not have to. Given enough time the food coloring will spread perfectly evenly from the center of the bowl to every conceivable area of the volume of water available in the bowl and it will also become less bright in color. This is diffusion which is the spontaneous movement of one substance throughout another in all directions from an area of high concentration to an area of low concentration over time. In other words the food coloring has reached equilibrium with the system it is in and in this case it is the bowl of motionless room temperature water. That being said this is a physical process involving food coloring in water and not heat so how is this possible especially if no form of mixing or any motion of any kind was involved?

The answer is simple as Newton's laws forbid the idea that something like this could happen unless a force was acting on the objects to make them move. The object being the molecules of food coloring and the force being Brownian motion which we have learned is a macroscopic observation of the quantum motion of particles and in this case of water molecules around the food coloring. So while it looks like no motion is occurring in the water it is in fact the opposite that is true as motion exists constantly in the water and at the atomic level is ever so gently pushing and nudging all of the food coloring molecules. This is a beautiful example of the classical mechanical and quantum mechanical world coming together and explaining a real world observation.

The same concept behind diffusion is occurring in a thermodynamic system at both the classical and quantum level as energy in thermodynamics is the some total of the heat and work of the system. Heat can take the direct form of infrared photons or it can be classically described as the work each particle can physically exert on each other within the system. The addition of both of these gives you the total thermodynamic energy and therefore links the quantum heat with the classical work. The fascinating thing about these systems is that we can see a link between many different aspects of physics and the world around us all at once and all being described using thermodynamics. The process that underlies all of them is of course entropy which as we have said is more or less a measure of disorder within a system. Before we go any further I should make perfectly clear that entropy was first and foremost an invention of thermodynamics and secondly used to describe randomness, apparent chaos and disorder within a system as many people are familiar with it today. To begin with let us look at what can be called classical entropy as it relates to classical thermodynamics.

In classical thermodynamics entropy refers to the distribution of heat within a system. You can think of this either as entropy or as heat flow but no matter which you choose they are the same as they describe the degree to which a system has reached thermodynamic equilibrium. Here we can see the direct link in terms of heat and entropy to the second law of thermodynamics and by using our example of food coloring in a bowl of motionless water we can easily picture the process in action. Low entropy would be the state where we first dropped the food coloring into the bowl and as such the brightly colored molecules are bunched close to one another. As the equilibrium of the system progresses with food coloring mixing with water we see the entropy increasing but not yet at a point where the system is in equilibrium. Lastly we have waited long enough for the food coloring to spread evenly with the water molecules throughout the bowl and the system is now in equilibrium and thus the entropy has reached a maximum for the system. It is easy to understand from this example that entropy is a measure of how much this equilibrium has progressed in a system from the initial conditions until the time the system is again measured. This simple food coloring experiment illustrates the classical thermodynamics of entropy as it pertains to heat as time progresses we see heat moves from any hot regions to the colder ones until all thermal energy contained within the system has been shared equally throughout it. This is the point at which the system is in thermodynamic equilibrium and since this is so the entropy must therefore be at a maximum for that system. In this sense the second law of thermodynamics describes not just heat transfer within a system but also dictates the flow of entropy in the universe as a whole. This idea of increasing entropy in a closed or isolated system applies also to the universe and we see one of the core beliefs of modern science stated for the first time.

Entropy appears to increase over time throughout the universe and like time flows only in one direction when equated with the entire cosmos according to modern science. This concept of universal entropy will be dealt with further after we have finished with thermodynamics but it does lead us to the second application of entropy as it applies to the second law of thermodynamics. This second application is not classical but statistical and is referred to as statistical thermodynamics. The effects of statistical thermodynamics are more concerned with motion than heat and so can be seen on the whole as focusing more on matter and particles as opposed to energy. That being said energy is of course inexorably linked with matter and so at some level matter and energy is a part of every description of any thermodynamic process. Nevertheless in statistical thermodynamics the random or seemingly random motion of things like atoms, molecules or other quantum governed objects is what results in the explanation of entropy.

I have already given us a perfect example of this process in action as it is the Brownian motion that drives our food coloring to disperse evenly throughout the bowl of water. While the food coloring, water and bowl can be at the exact same temperature as each other the forces of entropy are still at work. There is no heat to disperse throughout the system in order to reach equilibrium and so something else must be doing this which is of course the minute motions of all the molecules contained with the system as either food coloring or water. Just as heat flows to equilibrium from hot to cold areas and cannot be reversed the concentration of food coloring molecules will flow from dense to dispersed areas and cannot be reversed. The processes that drive each are different but similar and yet they can both be measured using the entropy of the system. It is easy to see that entropy is now quite a powerful tool in measuring changes of systems regardless of the make up of the system anywhere in the universe.

Now that we have introduced entropy to thermodynamics we can examine the third law of thermodynamics in earnest and then explore pure entropy in a bit more detail. The third law of thermodynamics deals entirely with entropy and as such is definitely something that is a bit more intangible and theoretical than the first two laws. The third law of thermodynamics states that the entropy in a system will reach a minimum value as the system approaches absolute zero. We have mentioned previously the concept of absolute zero and the fact that it is the result of the work of Lord Kelvin in 1848 but now it is time for an explanation of absolute zero itself. Absolute zero is the lowest possible theoretical temperature in a thermodynamic system for an ideal gas. Kelvin used the ideal gas law to calculate the theoretical temperature to be -273.15 degrees Celsius and the temperature is therefore also known as zero degrees Kelvin. Absolute zero is also the lowest possible values for both the entropy and enthalpy of a system as well and here things get a smidge more complicated than just simple heat energy as with the first two laws.

To quickly introduce enthalpy we need to realize that it is a thermodynamic function that is the measurement of a system resulting from the internal energy of the system added to the product of the pressure and volume of the system. If this seems complicated it has a very simple equation which is: $H=U+pV$, where H is enthalpy, U is internal energy and p and V stand for pressure and volume respectively. You are probably familiar with calculating enthalpy and do not even realize it as you constantly do so at least empirically in your lifetime. Any time we measure enthalpy we must measure the difference between two systems or states of systems to get the change in enthalpy. You have been introduced to endothermic and exothermic reactions already and a positive change in enthalpy is endothermic while a negative change is exothermic. The heat released from lighting a match is exothermic, the sweat drying on your skin and cooling you down is endothermic so you have been aware of these enthalpy changes even if you did not know it.

Any thermodynamic system has a measure of both enthalpy and entropy and absolute zero therefore can be calculated from them. In absolute zero therefore we see that a zero temperature cannot ever truly be reached as the system will always have some enthalpy remaining as any system contains quantum mechanical energy which does not stop even at absolute zero. Quantum mechanical energy can be thought of as what drives the little motions in our cup of hot tea as driven by Brownian motion. No agitation of the tea is made through macroscopic means rather all motion is the result of this quantum mechanical energy causing the particles to ever so slightly vibrate back and forth or move from superposition to superposition. This tiny amount of movement is significant in a cup of hot tea which is macroscopic and at the microscopic quantum level is supremely significant as the systems are so small.

Potential Wells and Absolute Zero

Therefore the quantum mechanical zero point energy of any system even one extremely close to absolute zero will never truly go away and therefore the system does not and cannot reach zero enthalpy. Zero point energy is also sometimes referred to as the quantum vacuum energy or quantum vacuum zero point energy but really these are just fancy ways of saying the lowest energy a quantum object or system can ever possess. A simpler term for it is the ground state energy of a system or object and since it is believed that all quantum systems or objects have wavelike probabilistic properties the energy can never be zero as this would negate the wave nature of the object. Even in the lowest energy state possible a quantum object should oscillate between super-positions therefore giving it some quantum motion and therefore energy. Zero point energy was given to us by the German physicist Max Planck in 1911 as part of his own work in quantum theory and since has been expanded upon by things like the Standard Model. In the Standard Model it is believed that all forces are the result of fields in space and that these fields can also have a zero point energy the net result is a term called vacuum energy as it deals not with particles but the fields from which the Standard Model believes they are excited. To further explain how a particle even at the theoretical absolute zero can have energy is seen by examining something called a potential well.

A potential well is the theoretical well that represents the lowest possible energy state for an object and when illustrated takes the form of a dip or well in which the lowest point of the curve represents the lowest energy state possible for the object. Anything in the lowest point of the well is unable to transfer its energy from one form to another the simplest way to illustrate this is to think of a solid quantum object at the bottom of a potential well. So if we have a potential well describing energy derived from the force of gravity and the object is at the lowest point in that well it is impossible for the object to spontaneously transfer some of its energy to kinetic energy as this would mean moving away from the bottom of the potential well. Much like a ball at the bottom of a hill will not spontaneously roll back up it which would give it kinetic energy but of course that would be impossible for it to do so without being acted upon by an outside force. In classical mechanics an object if acted upon by an outside force can break free from and escape the potential well but this requires as we have shown additional energy input from somewhere else. There would be no other way for the potential energy possessed in the system to escape so to speak and change to another energy form, but it would seem that the universe might allow for a loop hole under certain situations and this involves invoking the power of the quantum world.

Quantum mechanics however behaves a little bit differently at least according to quantum theory and remember much in the quantum world is theory which is not fact. Nevertheless if all of the strange and seemingly impossible behaviors of quantum systems are true then we can see something odd happening to an object at the bottom of the potential well. Classical mechanics and general relativity forbid many of the mysterious and enigmatic behaviors of quantum mechanics and yet we know that irregular behavior such as that seen in the famous double slit experiment is seemingly rooted firmly in experimental fact. This key feature of quantum objects is their tendency to exhibit probabilistic effects such as those made famous with the uncertainty principle. Since a quantum object is described mathematically by probabilities and some of these involve the objects position in space it is possible for the object to exist at the same energy level but in multiple locations. Think for a moment back to the discussion of superposition and how a quantum object can exist in a state which is both excited and non-excited at the same time theoretically. Even though the excited state seems to possess more energy than the relaxed state overall the quantum object exists in both and so therefore its total energy remains the same at one quantum energy level during superposition. Using this idea of multiple probabilities and applying it to a potential well we can demonstrate that a quantum object can technically break free from the well without the addition of any outside force.

The process by which the potential energy can quantumly escape the well is known as quantum tunneling. Think for a moment of the quantum object literally tunneling through the side of the potential well for a basic understanding of how this works. In a sense if we place a quantum object at the bottom of the potential well and then plot every possible position it could inhabit overtop of that from its probability field we see that in a very small percent of the time the object exists outside the potential well. The significance of this goes back to the fact that the object despite potentially existing in multiple locations contains the total energy of one single object just like our superposition quantum object. In other words the potential energy of the object at the lowest point on the graph of the potential well can be maintained and yet a small percent of the time the object can exist outside of the well without any outside help or change in its own energy levels. This is the sneaky behavior that quantum mechanics can employ in order to allow an object to escape this potential well at its lowest point and defy the laws of classical mechanics and general relativity. This is of course only theoretical and must be treated as so as much of everything in the quantum world is purely mathematical or the derivation of another equation. Still much that is strange and quantum appears to be correct and much of that has been harnessed in some fairly physical ways so to overlook these probabilities would be reckless.

So now we have an understanding of all the various things that can occupy an object in terms of what modern science understands as its most basic properties. Heat, entropy, enthalpy, zero point energy and so on and so forth really only amount to how one can think about what an object possesses in terms of energy. The term energy should really be replaced in science in my opinion or new a new word for energy should be introduced to distinguish the various differences in what modern science currently calls energy. The modern usage of energy is far too reaching and all encompassing, an umbrella type word that does not truly reflect what is going on in the universe at all times. To this end I will be introducing my own word for energy later in the book and I will explain how it is different and how it can be used to distinguish from the other forms that make up much of the current scientific vocabulary.

The reason I feel the need to explain this now and take a small detour from our discussion of thermodynamics and entropy is because the single term energy fails to describe what is really going on at the most fundamental levels and those fundamental levels are key to understanding the true nature of matter in the universe. Take for example our discussion now of potential wells, quantum energy, zero point energy and all the other forms of energy you have ever heard about and start to compare them. Do you really believe that a photon is in any way the same thing as the kinetic energy possessed by an object in motion? How about the chemical energy stored in molecular bonds and zero point energy are they the same thing? The answer of course is no not even in the slightest and yet science has for centuries decided to call everything energy.

The problem was only made worse when Einstein gave us his equivalency principle as the famous conversion of energy into matter makes people believe that the two are one and the same when of course the are not. An electron is a particle with mass and very different properties than a photon and this is true for its doppelganger the positron. Get the two particles to shake hands and after the annihilation smoke clears we find two gamma rays which are most definitely photons. So you might conclude that the two are really one and the same just as Einstein stated and this of course is wrong. Electrons have charge, mass and do things like bind atoms together and molecules together while photons do not do any of these things except mediate the electron to nucleus connection. The equivalency formula only tells us the value that each possesses if we convert both of them to the same units and no where does the universe ever say that just because we can do that they therefore must be the exact same thing. Remember how masses are given for particles in the super collider experiments such as those at the Large Hadron Collider and elsewhere in the world. A particle with definite mass is instead given an equivalent energy in order to help with the mathematics. This does not mean that just because we accelerate a particle to relativistic speeds it suddenly becomes energy.

See how this little problem is causing more headaches for theorists as it is a foolish assumption to treat matter and energy as one and yet it is done so the world over everyday. An electron is no more a photon than a seed is a tree despite the fact that we can calculate the values of energy that each possesses. Are the related to each other? Well of course yes and you get the whopping prize of a dunce cap for saying that with such vehemence as everything in the universe must obviously be connected in some way or nothing would be as it is. If electrons and photons for example were not related in some way then atoms would simply fall apart, so yes everything is somewhat related in some way with everything else and this is exactly what people are trying to do when they create a single theory that explains everything. Yet no where in those theories do people say that all of the universe every part we can and cannot see are related. To this end comparing things like zero point vacuum energy to potential energy of an apple on a tree branch is impossible and can create problems when trying to deal with the most fundamental concepts in the universe.

The potential well is a perfect example as energy is used twice in the description of a quantum object at the bottom of a well and yet the energy types are not related. You have potential energy of the object at the lowest ground state in the well and you have quantum probabilistic energy stored in the particle itself. If energy were equivalent in all ways and the word fit perfectly in descriptions of all systems whether they are open or closed would mean that even a quantum object could not tunnel out of the well. Yet the mathematics states that quantum tunneling is a possibility and it gets this ability from a different type of energy from potential energy; quantum energy. This point I am striving to make now will no doubt be the subject of debate in modern science until a single model for how the universe works can be presented and accepted and the reason for this is that until such time modern science has numerous holes in the theories it currently attempts to use to explain the cosmos. If the world were purely governed by classical mechanics then no holes would exist as classical rules would not have to answer for non-classical phenomena and this is of course true for quantum mechanics and relativity as well. None of the three major theories of the universe and how things within it behave can fully agree with each other or more importantly with observations. To this end I would simply like you to keep in mind the differences in apparent energies, how they are theorized to work and how one should be mindful of these differences when dealing with the most fundamental of things in the cosmos.

For now however we can apply these ideas to thermodynamics and specifically the third law of thermodynamics and how this works in with entropy. The idea of energies and particles of differing types will become all the more important as we look at entropy on the largest scale possible which is of course the universe itself. So in the third law of thermodynamics we were discussing absolute zero and its temperature as a sort of bottom end to any heat values for a system to possess. The question becomes then can we actually reach absolute zero and if so what does that mean for matter in the universe? So once again in order to answer this question we are faced with the third law of thermodynamics and what it states that at absolute zero entropy is at a minimum. Since entropy in classical thermodynamics refers to the heat available to the given system we must get that internal heat to zero degrees Kelvin.

The idea sounds simple enough as all one would have to do is cool whatever was in the system that contained heat to the point where the last thermal energy was removed. The super colliders used world wide to probe the very nature of matter routinely cool themselves down to within a fraction of one degree Kelvin. The trouble is that they cannot cool themselves down to zero degrees Kelvin and in fact it is impossible to cool anything down to absolute zero using any thermal methods. Think for a moment of a pitcher of room temperature tea in your kitchen on a summer day and you want to make it into iced tea as this would be far more refreshing. You can put it in the fridge or drop ice cubes into it and both of these approaches requires some patience on your part as the cooling process is not instantaneous. So you wait and sure enough the temperature of your tea drops so you might think that if you had something cold enough you would be able to cool your tea down to absolute zero not that you could drink it then but just for curiosities sake.

The reality however is that your tea is never as cold as what you used to cool it down in the first place. If for example you used ice from your refrigerator freezer you would expect the tea to reach minus twenty degrees Celsius as this is the common setting for most everyday freezers. In reality the tea never reaches this temperature because the tea contains more thermal energy than the ice and so when the system has reached thermodynamic equilibrium we find that the second law of thermodynamics has played out nicely. Heat has flowed from a hot to a cold region driving the system to one equal temperature which is cooler than room temperature but warmer than minus 20 degrees Celsius. So let us add more ice as that should surely solve the problem and cool the tea to the same temperature as the ice. Well no that will not work either as we have simply narrowed the temperature difference between the now cooler tea and still frozen ice meaning the system will reach a new thermodynamic equilibrium. If we repeat this process indefinitely we will make the tea very close to but not equal to the temperature of the ice and the third law of thermodynamics will be upheld as the system approaches absolute zero which in this case is the temperature of the ice cubes, the entropy of the system does decrease. Even at 253.15 degrees Kelvin however we cannot get any substance to reach the temperature of that which is used to cool it as the continual addition of ice to tea produces an endless set of equilibriums that are mathematically asymptotic. This is because the ice is also absorbing some of the heat from the tea which means that at a new equilibrium its own temperature will always be just slightly above minus twenty degrees Celsius.

Therefore the same is true of anything that we try to cool to absolute zero as whatever we use to physically or mechanically cool it down will always result in a total system temperature that is in equilibrium with both the object and the cooling apparatus. Like the tea and the ice we can never reach absolute zero merely a temperature very close to but never reaching it. To picture this mathematically simply draw a graph with axis X and Y and plot any values you like using the formula of 1 divided by X where X is a variable integer. So for example one over one is equal to one of course. If we imagine that the values of X are colder temperatures than one which is our object of interest we can cool that number one down by using greater cold temperatures or in this case large values of X. So we divide one by ten and get of course 0.1, now we divide by one hundred and get 0.01 and lastly we divide by one thousand and get 0.001. We repeat this until we are dividing by trillions upon trillions of zeros and all we get is some incredibly small value made up of near countless zeros but always followed by a single one. In other words the mathematics backs up the physics as no matter how much you try the value of your graph will approach but never touch the X axis which would yield a value of zero on Y which was our goal all along. The reality therefore of the third law of thermodynamics is that while absolute zero represents a total absence of heat energy from a given system and at this point entropy tends towards a minimum we simply cannot reach it given conventional means.

This leads us right back to where we started in viewing what this means to matter at its most fundamental levels. After all the experiments and theories that are currently being batted around and scrutinized in modern science today are only concerned with the most basic properties of matter. If you think about a phase change in matter from one form to another you get a clear understanding of why this is of such profound importance. Take for instance simple ordinary water and how it is truly a unique and amazing substance as compared to most other compounds. To illustrate how understanding the exact and minute details of particles and they way they interact with one another is important we will discuss ice and water. First of all if we had never encountered ice before and assumed that cooling a liquid further and further would only result in a colder fluid we would be blissfully ignorant of the phase change of matter from liquid to solid and of course never mind the fact that we ourselves would be made of various solid materials as well. We might try to cool something to the freezing point just as trying to cool something to absolute zero and never be able to reach zero degrees Celsius or 273.15 degrees Kelvin which is the freezing point of water. Assuming that matter behaved no differently at this temperature would be an incredible scientific blunder as solids behave nothing like liquids and represent a new way for matter to be organized in the universe.

Water stops being a random collection of Brownian motion quantum objects and starts to become a fused quantum crystal that forms intricate patterns which are lost once melted. Our liquid scientists would be unaware of simple things like snowflakes and never once make the discovery that no two are alike no matter how hard children try to find them in pairs. This demonstrates how important it is to understand matter at the level of absolute zero as well as a new form or behavior of ordinary matter might occur. In addition to this if we did know that liquids can turn into solids we might additionally be forgiven for assuming that one of the most common observations of the liquid to solid transition phase of matter does not work for water. Namely when a liquid freezes solid it usually becomes denser and occupies less volume and if we observed hundreds of elements and compounds we might conclude and fairly reasonably too that water also would be denser when frozen. This is of course not the case as water is one of the unique substances that expands as it freezes and you can see this by looking at the ice cubes floating in our iced tea from before since after all they are floating and not sinking.

The curious thing about this expanding water fact is that without it life would most likely not be possible on Earth as a simple change in waters crystalline bonding structure into ice allows it to do something as simple as float on water protecting life from the extreme cold of the atmosphere above. In fact water is at its densest at four degrees Celsius just warm enough for life to survive without freezing solid and perishing. The lack of understanding of such molecular intricacies would leave us in the dark about how life meshes with physics not just here on Earth but elsewhere in the universe and it would also prohibit us from investigating whether or not other materials possessed this interesting trait. You can clearly see that the DNA of not just molecules and atoms but subatomic particles is one of the keys to understanding how the universe works. So the behavior of these at every conceivable level is currently a top priority for modern scientists as how materials exist at absolute zero might give us a clue to how they work at more fundamental levels still.

One aspect of this is as we have learned the possibility of quantum tunneling out of potential wells for objects that might have zero temperature but still possess quantum energy. The very nature of quantum effects on matter is directly linked to the most important aspects of entropy which go beyond simple heat equilibrium and underlay something for more complex and difficult to map in the universe which is disorder. As we have seen two types of entropy exist thermal and statistical the former being of course associated with distribution of heat within systems and the latter being a measure of the increasing disorder within the same system. Now it is time to dive head first into entropy as it pertains to disorder and the universe from not just the Big Bang but also to how it has resulted in the universe we see and observe today.

The man who invented entropy was a German physicist named Rudolf Julius Emanuel Clausius who besides giving science the concept of entropy was also one of the most important contributors to thermodynamics. Initially Clausius had spent a great deal of time studying Carnot's work and found it to be in error as it disagreed with the conservation of energy law well known to be fact in physics. From this he reformulated much of what had been studied with respect to heat and useable work in thermodynamics and began to look more deeply at the transfer of heat within a system. What he deduced was that heat cannot travel from a colder region to a warmer one by itself and of course we have seen this as the second law of thermodynamics for which he was obviously instrumental. While others stopped pondering the meaning of heat transfer within systems as it applied to the second law of thermodynamics Clausius continued to investigate this phenomenon. Instead Clausius sought to mathematically explain what was occurring in such systems as time progressed and heat went from areas of high concentration within the system to being spread evenly throughout it at which point thermal equilibrium was reached. In order to do this he created the term entropy which is based off of the Greek meaning transformative and he used this to explain how heat energy was distributed within a system. His breakthrough idea was published in 1865 and would become one of the most important aspects of physics from then on and still dominates much of what modern science will accept in terms of theory.

Clausius and his breakthrough idea proved that as time went on entropy increased for a closed system and that certain realities of such systems could now be made. The most famous of course is the idea that the entropy of the entire universe increases and will one day reach its maximum. This is not only true for smaller systems such as a simple isolated thermodynamic system but for the cosmos as the universe itself can be treated as a simple isolated system just a very large one. In any isolated system the only place for energy to go is the system itself and therefore an isolated system will always tend to shift the energy it contains from areas of high concentration to areas of low concentration and therefore create equilibrium. The universe exists outside of any closed system and so all closed systems are really just tiny parts of a much larger system which of course is the entire cosmos. The cosmos however exists within the universe as it is the actual universe and so cannot by definition exist outside of itself or be isolated from itself and therefore the universe is one big isolated system as there exists nothing beyond it.

Even for a moment think briefly of all the various theories surrounding the universe and how it might be constructed. If the universe is simply the universe then of course it can easily be an isolated system as nothing exists outside of it. If the universe is infinite in all directions then this still means it has a finite size which is of course the size of infinity as even an infinite size is a fixed value and no matter how far you go you cannot escape the boundaries of an infinite universe once again proving you cannot go outside it and therefore the universe is still an isolated system. The multiverse is the last option for universe sizes and all multiverse theories involve the universes not being able to share information with one another and therefore stating that they are isolated from each other. Whether the multiverse theory states they are separated by a black hole or some extra dimension of space no information is permitted to pass between them and so each universe is still isolated from each other and therefore overall is an isolated system. On top of this all mutliverse theories believe that these universes do in fact talk to each other through the weak force of gravity separating them by a few Planck lengths so no matter what they actually are still sharing energy and thus constitute a single system once again. So no matter how you slice it the universe is just one big isolated system and Clausius has cleverly shown that if given enough time all energy which is a constant in that universe will eventually reach a cosmic equilibrium.

To sum up isolated systems it can be stated that the total entropy for all isolated systems always increases and tends towards a maximum when equilibrium is reached. Isolated systems always increase in entropy and the overall entropy in the universe behaves like the arrow of time pointing in one and only one direction. Just as time will not flow backwards the entropy of the universe will not decrease and so we have here one of the biggest contributions of Clausius and his invention of entropy to modern science. Playing out on a stage of infinite importance to statistical thermodynamics the idea of ever increasing entropy has been one of the most important and seemingly unbreakable rules of theoretical physics for over a century. As all modern theories of the universe and the systems that exist within it must account for entropy at some point or another. Many modern scientists often wonder whether entropy like time can ever be mastered and if so might an insight into how such a mechanism would work offer clues to how the universe is put together. Closer to everyday life many people have heard of something called a perpetual motion machine whereby someone is able to construct some sort of apparatus that once set in motion will run indefinitely without additional power input. The very concept of entropy and how it pertains to isolated systems forbids the construction of any such apparatus as the total energy of the system would seek to eventually reach equilibrium with the entire universe regardless of how efficient the machine ran. So we see that entropy in the universe should be heading for a maximum value even as we speak as any isolated system demands this but there are other types of thermodynamic systems which can at least on the small scale circumvent such an idea.

Now then let us discuss the effects of entropy if we do not have an isolated system and the system is employing some other type of boundary. Boundaries in thermodynamic systems and in discussing entropy come in many different types but are on the whole used simply to describe where a system starts and stops. In the case of an isolated system the boundary is the entire system and this is why hypothetical isolated systems behave similarly to the universe as their boundaries are the entire size of their system plus their surroundings. As we have seen the universe has no surroundings no matter which theory of the universe you ascribe to. The terms, open, closed and adiabatic are common types of thermodynamic systems which are not isolated and permit various levels of matter, energy or work to pass through their boundaries.

Anything that passes through a boundary becomes part of the system and will change the information within the system; thermodynamics and entropy are attempts to account for these changes. Isolated systems might at first seem very similar to a non-isolated system in that in each case we have a finite space but in terms of entropy they could not be more different. Think for a moment of our experiment creating water by burning hydrogen atoms with oxygen atoms and then separating them afterwards using electricity in the standard science class setup. We have demonstrated that the definition of times arrow can be somewhat misleading depending on how you view the system but that overall the arrow of time always runs forwards when looking at the entire laboratory assuming the laboratory constitutes the entire universe. Simply exchange entropy for time in this analogy and we can learn about isolated versus non-isolated systems and entropy easily.

If the chamber containing the reactions was the entire universe and no laboratory existed we would find that entropy does not increase for the universe as the creation and destruction of water molecules would contain all the information about the system. The apparatus however is not the whole universe as the laboratory represents the surroundings to the system and the walls of the apparatus are the boundary to the system. From here we see that during the burning of the hydrogen and oxygen heat and light are released as electromagnetic radiation that will travel to the boundary of the system, interact with and eventually pass through it. This will change the overall entropy of the system and this is even more compounded by the fact that reversing the process will take electricity which will create similar effects that again can penetrate the boundary and alter the total entropy of the system. In this way it is easily demonstrated that a non-isolated system behaves extremely different from an isolated one.

The entropy of a non-isolated system is therefore not fixed and constant and can be changed at will but this comes at a price. The price is that while any change in the system is governed by what we or anyone else affecting it do to it the overall entropy of the surroundings beyond the boundary will always increase. The heat and light for example created by the experiment will possibly leave the system and travel into the laboratory which constitutes the universe in this analogy. The energy, matter and information lost from the system cannot be lost from the universe and so overall the entropy of the universe once again increases as it is an isolated system. The laboratory is an isolated system with ever increasing entropy and the apparatus is a non-isolated system with variable entropy. Regardless of the type of system whether it be open, closed or adiabatic the effects that its boundaries possess over the internal entropy pales in comparison to the entropy of the system plus its surroundings. This is a perfect illustration of a perpetual motion machine and why it cannot work as there is no impenetrable boundary to the machine and so entropy can leak out into the surroundings thereby sapping the machine of some of its useable work energy.

Thermodynamic processes can also be classified as reversible or irreversible and each relates to the initial conditions of the system, the final state of the system and the overall entropy of the system and its surroundings. Reversible processes are as the name implies reversible and can be made to return to an original state of the system with negligible changes in entropy. In this case the steps taken to reverse the process and return to the original state are microscopic and take an incredible amount of time to complete as any rapid changes would result in a large increase in entropy of the system and its surroundings. Proceeding in tiny steps at a very slow pace results in a state whereby the entropy for the system and its surroundings is essentially constant. This also dictates that the surroundings will return to the initial state it existed in beforehand and not just the system. The goal of any reversible process is much the same as any normal thermodynamic system that seeks to maximize work and minimize work loss which takes the form of heat energy and entropy changes.

This is in contrast to irreversible processes where entropy will always increase for the system and the surroundings and usually by a great deal. What this means is that a lot of heat energy or waste is generated as it is not efficiently transformed into work and the penalty for this is increased entropy. Just as the entropy and final states are affected by a reversible process so to are the entropy and final states of an irreversible one with the difference being that entropy is not a constant. In an irreversible process it is both the system and the surroundings that will experience a total increase in entropy as time progresses. Almost any combination of systems can be achieved when viewing thermodynamic systems and can be created from open, closed or adiabatic systems that may be reversible or irreversible given set conditions. It is clear to see here that the scope of what thermodynamics can cover is quite extensive as not only is it highly detailed but also broad in its scope as it can define a small isolated laboratory state as well as the overall state of the universe. To further add to the considerations of thermodynamics and entropy to the universe we see that both classical thermodynamics and statistical thermodynamics must be employed to fully describe the cosmos.

This is due to the fact that despite much of modern science shunning classical mechanics in favor of newer more complicated models classical mechanics still dominates the cores of each of these new theories and in many cases has never been supplanted merely its appeal has been usurped by ever more devious descriptions of physics. To this end we find that classical mechanics which deals with relatively large particles in its systems and you can think of things like atoms, molecules and what not are needed in order to describe large scale observations. Something like the heat and entropy of a star for example would best be describe classically or the tidal friction and resultant heat energy caused by a large planet deforming one of its moons would also best fit a classical description. This is the kind of thing that can be calculated using classical thermodynamics and entropy as the system here can be thought of as the planet and its heat and gravitational energy as well as the heat and gravitational energy of its orbiting moon with the surroundings being the rest of the solar system.

Typically thermodynamics does not deal with such large systems and instead focuses on more down to Earth sized events but the idea is the same for any large scale system. The average values of these particles would take the form of familiar measurements such as temperature, pressure and entropy to name a few. Leaving the large scale and looking at the small scale we find that classical descriptions are not exact enough to account for what each of these particles or even smaller constituents of these particles are doing at the most fundamental levels. To this end statistical thermodynamics was developed as a way to account for variables and motions of these tiny particles that used to be explained in terms familiar to classical mechanics and now familiar to quantum mechanics. The very motion and behavior of atoms, molecules and the like can be described in mathematically statistical ways using measurements such as spin which of course is a purely quantum property.

Here we have entered into the domain of the modern concept of entropy as it is thought of most often when discussing how the universe works. When scientists talk about the universe and talk about entropy they are almost always referring to statistical thermodynamics and the entropic properties it possesses. You may have heard the terms disorder, randomness, chaos or even disarray when people are discussing the entropy of the universe and in reality all are really referring to the same thing. This is of course the ever increasing entropy of the universe as it is comprised of a single mind bogglingly large isolated system. Over time thermodynamics dictates that entropy of an isolated system will increase regardless of all other factors and so scientists have been applying this to the cosmos itself. What they have found is that on the biggest possible experimental scale possible, the universe it does appear that the laws of thermodynamics are true and randomness and entropy of the system is ever increasing. You will no doubt recall the idea that the universe began in some sort of giant explosion called the Big Bang and that from there the cosmos has taken the shape we are familiar with today.

We can examine entropy in the universe using the Big Bang theory and apply the growth of the cosmos over time to see if disorder is indeed occurring. To begin with the Big Bang theory says that everything in the universe was compressed into a single space infinitely smaller than a single atom called a singularity. Right from the get go if the Big Bang theory is correct the universe was in its most highly ordered state with the least amount of entropy as everything was in a single uniform structure known as a singularity. From here the actual explosion occurs for reasons modern science does not know or understand but for this examples sake we will simply allow something to trigger the actual explosion and in a sense unpack the singularity from its neatly ordered state. So the explosion occurs and we have the next evolution in the life of the universe as it has gone from a single infinitesimally small point containing everything to allowing some of the things within it to split apart and grow.

Two of the things are space and energy which are now separate entities according to the Big Bang theory and both are spreading rapidly outwards from the point that was where the singularity resided. This is the earliest possible stage of the universe precisely after the unpacking of the singularity and well before anything machines like the Large Hadron Collider could ever hope to examine and already we see entropy has increased. The initial state of the system involved only one object and therefore a minimum of information was needed to describe it but now the system has evolved over time to include two objects and since two is a larger integer than one we see that the information necessary to now properly describe the universe has increased. In other words the overall entropy of the system has increased and the system is now what modern scientists would say is more disordered.

What I would like to point out here is that the universe itself upon unpacking itself from a singularity will always create some interesting problems for statistical thermodynamics to explain as the universe sort of breaks the normal rules here. What I mean by this is the normal definition of a system in thermodynamics as applied to something in the laboratory which has familiar things such as the volume or space of the system, its constituent parts and the amount of time allowed to progress from when one initially examines the system to the final state when the last observations are made and gives you the change in overall entropy of the system. Why is this so tricky when talking about the universe and looking at it through the lens of statistical thermodynamics?

The answer is simple the universe itself is creating space, time and matter as it grows so the very definition of entropy changes within the system which of course is the entire expanding universe itself gets blurred. Normally we look at a system and see how much volume it has or in other words how much space it occupies. This is necessary of course to see how long heat will take to move from a concentrated area to fill the entire volume evenly at which point we can say the system is in thermodynamic equilibrium. The universe upon growing and expanding itself is changing the rules a little as it is making ever more space and therefore ever more volume with which the energy and heat must now expand to reach equilibrium. In a sense the early universe was at equilibrium immediately with no change in entropy as the incredibly hot energy present in it was uniform throughout the finite space of the Big Bang. Still we must look at even more variables such as time which before the Big Bang is thought not to have existed at least to proponents of that particular theory.

In this case the very tool used to measure changes in entropy of the system from its initial condition to any final state was either absent as in the case of the singularity or being created at the time of measurement. In the instance of the singularity time did not exist and was in a sense frozen or at least the singularity was because without time how can we measure or even theorize how long the singularity existed? Millions, billions or trillions of ages of the entire lifetime of the universe itself, was that how long the singularity existed or was it perhaps only a Planck time of a Planck time? With no time present all physics understood by modern science fails immediately. Then we come to energy which before the Big Bang also did not exist and this of course includes energy that would eventually cool down to form matter.

Going back to the whole purpose of the Large Hadron Collider and its continuing quest to understand what went on in the very first moments after the Big Bang we find the emergence of the quark gluon-plasma that would eventually form matter we see and are familiar with today. Before that time however modern science again does not understand what was truly going on in the universe and one example might be the absence of whatever was needed to make anti-matter. Was that component of matter forgotten somehow by the Big Bang or did it all get wiped out with the normal matter created as well or maybe the normal energy that would go on to create the normal matter annihilated the anti-energy before it could create significant amounts of anti-matter.

This once again illustrates how the universe is breaking the rules when examining a system in terms of entropy by creating what it wants when it wants and not leaving us a trail of breadcrumbs to follow. The formation of matter is critical to understanding the entropic changes in the early universe as the number of particles of matter to anti-matter will determine what entropy we have after annihilation events occur and thus explain some of the total energy we find in the universe today. If we allow these variables to be temporarily bypassed we can now focus on things that modern science does understand much better such as the quark-gluon plasma. While still under investigation and with many new details to be worked out quark-gluon plasma is at least something we can get our hands on and muck about with a bit; only a bit. To this end we see rapid increases in entropy in the early universe as the number of collisions and interactions between matter and energy is megalithic to say the least. In other words the amount of information needed to describe all the collisions, scatterings and other occurrences increases with each passing second and indeed fraction of a second.

The quark-gluon plasma also adds for the first time that we can see one of the four forces in the Standard Model to the system as the nuclear force dominates this time period and must likewise be factored into entropy changes of the system. The quark-gluon plasma phase of the universe is short lived however and we find ourselves finally settling into the hadron epoch. This further adds forces and interactions as we now have all of the four fundamental forces present and contributing significantly to the total information content of the universe. The total level of entropy in the universe continues to rise as particles collide with and bounce off of one another still unable to form atoms. The reason for this is that the universe is still too dense and hot for electrons to stick with protons and create hydrogen. The temperatures are too hot for this to occur and the highly energetic photons present are also knocking any electrons free from forming potential atoms. The rate at which these interactions are taking place is almost unimaginable and so is the increase in entropy.

The universe after further expansion has increased in volume sufficiently that the temperatures inside of it can drop below those necessary for electrons to join with protons. Likewise the energy available to free photons has decreased and they no longer possess the energy needed to disrupt electrons and protons from pairing up. Now we have the first atoms of hydrogen and shortly afterwards other light elements in trace amounts. This is the most familiar phase of the universe that modern science understands as it is the phase in which we now live and has existed for nearly the entire 13.8 billion year age of the cosmos itself.

Truth be told it might be in fact the only phase of the universe as modern science has only ever observed it and no other phases have ever been detected naturally anywhere else we look. It is entirely possible the Big Bang theory and the quark-gluon plasma epoch are just theories and never occurred at all. Things modern science has thought up in terms of theory or experiment are interesting but does not necessarily mean that those ideas ever existed or will exist naturally in the universe. Nevertheless we are very familiar with this phase of the universes growth and we do know that entropy increased steadily once this phase was reached. The volume of space continued to expand and is still expanding today all the while the temperature of the universe is decreasing along with it. The entropy of the universe is increasing as the number and complexity of interactions of particles grows ever larger and adds continually to the information needed to describe the universe. The nice neat order of the singularity, initial space and energy and the relatively homogenous quark-gluon plasma have long since passed leaving a small number of more highly ordered particles to create a huge periodic table of elements which in turn can produce near infinite compounds. If left alone long enough it is predicted the universe will continue to expand and increase in volume which over time will allow the entropy and disorder of the universe to grow along with it.

Information

The entire universe can therefore be theoretically described using entropy and by the information needed to describe it. The information to which modern science commonly refers is a mathematical description of a system or part of a system. To think about this simply picture a single fundamental particle motionless in space and you can describe everything about the particle using certain mathematical information. The particles charge, mass and spin can all be described mathematically and if extensive enough will contain all the information necessary to describe that single quantum object. Now imagine putting that particle in motion and you would need to add information describing how it moves in addition to its initial properties in order to describe it mathematically. The simple act of giving the particle a little friendly scientific push has added to the information needed to describe that particle. Now have that particle in motion strike another particle and think of all the information you would need to describe such a small and simple interaction. You now need all the information from the first particle, the information from the second particle and the information for the system in which this interaction takes place. This will include three spatial dimensions for the system and of course we are measuring the interaction using time. Now we have gone from a simple stationary particle to two colliding particles, all associated forces generated from the collision, the system in which it occurs and any by products created from the collision such as heat friction. A small scale system involving only two particles and one interaction is one of the easiest to describe and yet still requires a plethora of information to do so accurately. The descriptions will only become vastly more complex as the number of particles or interactions or both increases along with increasing size to the system.

The amount of information that would be needed to describe something simple like the motions and interactions along with all associated forces and by products of a simple glass of water sitting very boringly on a table for one second would easily contain more information than the entire human race has ever produced. Every painting on a cave wall or megalith, every scroll or book and every form of digital information added together for the last ten thousand years would still contain less information than a glass of water. So why do scientists even bother to try and calculate entropy in the first place? The reason is simple and it is called quantum information. The thing of it is we know that the interactions on very small scales can have very large repercussions and that understanding these can lead to a greater understanding of the universe as a whole. Think for a minute about inflation theory and what it says for the fact that the universe is on the whole relatively uniform and homogenous. Inflation theory says that a tiny quantum fluctuation in the initial moments following the Big Bang can be blown up trillions of times and become hugely significant in the cosmos we see today. A small quantum fluctuation in other words will produce a uniform visible universe to us such that a section of the universe beyond our range of detection might be very different given its small quantum fluctuations at the time of the Big Bang.

This is of course one example of the small dictating what the large might do or become and it also shows us another important aspect of understanding entropy which is looking backwards. Scientists have long looked backwards in time by looking out into the universe in an attempt to see what the universe was like early on. Science has also looked backwards by examining things like the acceleration velocities of galaxies away from us like Hubble demonstrated or by examining the Cosmic Microwave Background to see where the energy came from. Our hopes are that by looking at these clues and working backwards we can learn the initial conditions of the universe. Understanding the quantum interactions of particles and the concept of entropy means that it is possible to work backwards from an initial state and learn about a previous one with respect to a particular system.

For example if I were to place an ice cube on a patio table in summer and wait the ice of course would melt as expected and the structure of the ice cube would appear to be lost. In other words the information contained in the structure of the ice would be lost from the universe. However understanding entropy and quantum systems it is theoretically possible to reverse calculate every conceivable interaction involving the melting ice and reconstruct the ice cube from a puddle of cool water. Since we know the quantum interactions to be real and we know that we can describe using information the changes in the system as it tends towards increasing entropy it is always possible to reverse any system to an earlier time frame. That being said while this is an incredibly powerful tool for understanding the universe the simple act of un-melting an ice cube mathematically as it were is probably beyond the abilities of human computation for the foreseeable future as the amount of information is staggering and this is coupled with the fact that modern science does not fully understand physics yet. What modern science does understand is that information and entropy in terms of disorder in the universe appears to keep increasing over time and this affects much of the rules used in understanding the universe overall. One rule is that most modern theories that will be accepted by contemporary science must all accept an increasing amount of entropy and theories that do not are somewhat shunned. This is not to say that a theory which proposes entropy to decrease or a mechanism by which entropy in our universe can be balanced elsewhere is invalid only that it has some serious hurdles ahead of it in terms of acceptance.

Chapter 12
Three Views and Unsolved Problems

Classical, Quantum and Relativity

So now we can start to look at these hurdles in earnest as the number facing any new theory or any old theories for that matter are significant. We have been discussing every facet of thermodynamics and entropy and have seen how it fits into the universe and we have seen how important it is in describing what is going on in that universe. Whether it is a small system or the universe itself the ever increasing entropy is something that must be considered when trying to figure out all the secrets of the cosmos. Accounting for entropy is only one of the hurdles and fortunately or unfortunately for modern science it is not the only one. Previously we have looked at some of these hurdles such as the horizon problem, flatness problem or homogeneity problem which ever aspect or name you prefer with respect to the overall bigness and appearance of the universe as a whole. There are others that exist on the very small scale as well such as quantum entanglement which has been proven to be real and yet escapes explanation by all the theories modern science has at its disposal. The universe also appears to have slowed down in its expansion phase sometime after the Big Bang and again modern science has no answer to why this might have happened.

Similarly the reason for the universes apparent acceleration in its expansion flies in the face of the deceleration phase and requires still more explanation. The very nature of matter is still not fully understood and the nature of space even less so than matter or energy. Come to think of it there are many in modern sciences who think we still have not seen all the matter in the universe anyway and this is evident in the flat rotation problem we examined with stars and the galaxies they inhabit. Another hurdle to be sure and one that any new theory must address concerning missing matter, missing energy or missing physics to explain what is really going on. Shall I continue here? I think not as the list can be exhaustive and the list of potential answers even more so from the numerous candidate theories. Instead let us look where modern science stands in terms of the most accepted views of the universe and later we can revisit these hurdles and some of the possible solutions presented for them.

The accepted views are of course the ones that give us the most satisfactory explanation to observations made of the universe or explanations to experimental workings within the universe. The absolute first modern view of the universe is the classical view which was essentially the gift given to us by Newton that truly began the study of physics and the modern examination of the universe that has remained an unbroken chain from his days until ours now. Classical physics or classical mechanics as it is more widely known treats every thing in the universe as an object which can be described according to set laws of motion. These laws of motion are governed by forces that will act upon the objects and by examining all the forces on all the objects at a given time we can understand any system or the universe itself. So what is the success rate of classical mechanics today given that it has been in some ways usurped by quantum mechanics and general relativity?

Classical mechanics success rate is near one hundred percent and easily scores the highest of the three with the other two major views of course being quantum mechanics and relativity. The reason it scores so high is that it is the most applicable of all the views and has enjoyed the longest life. Without a doubt it is the grandfather view of the universe and has dominated for more decades and centuries than quantum and relativity combined. Even though certain aspects of the universe are now described using quantum equations or relativistic equations it is the mathematics of classical mechanics that made these other two possible and in fact is still embedded in them today. Classical mechanics uses objects and forces which can produce large scale effects seen in the universe. Today we use quantum mechanics to describe these objects as particles or force carriers as well as the Standard Model tying together the idea of forces with all energy and matter.

The predictions of classical mechanics on large scale objects in large scale systems is the direct precursor to all relativity and for example the idea of light bending around a gravitationally massive object like a star is a classical idea and not a relativity one. All relativity did was fine tune the mathematics to get a better prediction of the amount of bend and it used Newton's equations to do it but the big jump, the huge new idea of light actually being bent by gravity came from classical mechanics. To think of these things another way imagine the very first car ever built by Herr Karl Benz in 1886 in Germany and a modern car that follows it in the twenty first century. Is the original invention of Benz useless, unimportant and essentially antiquated junk when compared to the new car? No of course not in fact the most important car ever built was that first car which created the most explosive worldwide impact on the human race of the modern era. Nothing in our world would be as it is today without the first Benz motorwagen and the impact it had on cars, trucks, roads, infrastructure, industry, engineering, human culture and history.

This is how one must view the invention of classical mechanics and the ideas it embodies such as treating everything in the universe as an independent object acted upon by forces and producing large scale events. Some of the laws and ideas of classical mechanics have also not been abandoned and in fact are still used today. At the level of microscopic systems the laws of relativity do not apply whereas the laws of classical mechanics do so much more accurately and this is also true for speeds slower than that of light. So the use of classical mechanics is still practical today in its own right but also in the fact that classical mechanics are embedded in the ideas and mathematics of things like quantum mechanics and relativity.

The next view of the universe currently accepted by modern science is what can be seen as the microscopic version of classical mechanics and it is of course quantum mechanics. Quantum mechanics while more precise in its descriptions of objects and systems than classical mechanics is essentially the same thing. It involves forces acting on objects and does not include large scale relativistic effects such as light bending around a star. Remember quantum mechanics and relativity have exactly zero compatibility with each other whereas classical mechanics is essentially compatible with both. It was classical mechanics that first gave us the idea that light would bend around something like a star and not relativity so classical mechanics being able to account for large scale effects while discussing very small systems involving objects and forces is in a way the theory closest to a theory of everything that modern science has. Considering also that it was developed some two centuries prior to the others and did not enjoy the spoils of all the scientific progress made in that time, more accurate instruments and experiments we can easily see how close classical mechanics probably came to finding a single solution to how the universe works.

However in fairness to quantum mechanics and relativity classical mechanics does not have a full understanding of the universe and modern science needed a greater understanding of particles as a whole in order to progress further. Quantum mechanics provided this understanding and allowed for far more accurate descriptions of individual objects such as particles and how they behaved. The most important aspect of quantum mechanics by far is that it provided absolute and definite values to particles whether that is some innate property or interaction with other particles or forces. For this reason quantum mechanics enjoys the second highest success rate of the three views of the universe and how it works accepted today by modern science. The laws of quantum mechanics describe how interactions between particles and forces will play out as well as measurable values for each particle at a fundamental level. These can all be obtained using repeatable experiments which is one of the cornerstones of science and lends great weight and support to the quantum theory overall. Through the study of quantum systems science has learned much about particles and the forces that make up the universe and so much so that new particles such as the Higgs boson could be theorized from existing information. Factual and repeatable experiments such as those carried out routinely at the Large Hadron Collider provide evidence of their existence and allow for measurements of them to be made. All of this of course is fed back into the overall understanding of quantum mechanics and expands the Standard Model of particle physics.

The shortcomings of quantum mechanics stem from its inability to account for gravity and some of its own predicted and bizarre quantum effects. Gravity we have seen is unexplainable by any of the accepted modern theories of science whether that is classical, quantum mechanics or general relativity and so quantum mechanics inability to explain gravity keeps it from being a complete theory of the universe. Another shortcoming of quantum mechanics comes from its lack of understanding of certain quantum effects such as wave-particle duality and quantum entanglement. This is not a flaw of quantum mechanics as quantum mechanics never claimed to be a theory of how everything works only a way of describing observable phenomena in quantum mathematical ways. Nevertheless the lack of any solid explanation of things like the double slit experiment and quantum entanglement mean that something is missing from the overall picture of quantum mechanics. Still quantum mechanics produces measurable values for objects that can be produced through repeatable experiments which are definitely a feather in its cap.

The last of the three main views picks up where quantum mechanics leaves off and is the theory of general relativity. General relativity deals with large scale objects and large scale phenomena that quantum mechanics cannot. Quantum mechanics has no framework for dealing with scales beyond that of the microscopic and once you start to deal with objects even the size of a grain of sand quantum mechanics is already pushing the limits of size it can cope with. No workable concepts for quantum interactions can detail what happens to large collections of quantum objects. So in other words many quantum particles making up the solar system for example cannot have any of their interactions with one another detailed through quantum mathematics; quantum laws do not explain why something as simple as the Moon orbits the Earth. Relativity deals solely with these aspects of the universe and does a good job of explaining mathematically why things like the Moon do orbit the Earth as they do.

This is the true area in which relativity shines as although it does not introduce anything new that classical mechanics did before it such as light bending around a star it does manage to greatly improve the accuracy of such predictions. The effect of light being bent by a star according to classical mechanics is half the value predicted by relativity and when measured the amount of actual light bending agrees with relativity significantly. This success of relativity can also be applied to much heavier and more distant objects such as whole galaxies where astronomers have long observed peculiar light distortions near the rims of large galaxies from objects many light years behind them. Extending such gravitational effects on light relativity also accounts for gravitational red-shifting whereby light leaving a strong gravitational field will be shifted towards the red end of the visible light spectrum.

The theory of general relativity's last contribution to modern science is the effect of time dilation where time can be made to run faster or slower as seen by an observer. A simple demonstration of this is seen in clocks synchronized and measured here on Earth and in orbit around the Earth where it is shown that the clocks in orbit run slightly slower than their counterparts on Earth. Overall however relativity scores third and lowest on the overall scales of usefulness in understanding the universe and the reasons for this are somewhat severe. Without dissecting each one too much we will lump all of them into one general reason which is this: nothing says general relativity is right or actually makes sense. I hope I got a few people's attention there as the idea of relativity being imperfect and greatly imperfect has been around for some time and is starting to gain more momentum. Here I will only state my own feelings on this subject as well as a few general ones shared by many theoretical physicists today.

First off what is a spacetime anyway? Sounds odd to some but if you stop to think about it what exactly is this ether of a spacetime described by Einstein anyway? The answer is even Einstein does not know. A physical description of spacetime is never given anywhere in the theory of general relativity only that it is curved. The whole concept of bent spacetime was only dreamt up by Einstein to better fit classical mathematics and frames of reference to observation. I do approve of this as I approve of any model that relies on observation more than equations to create a new theory and this is partly what was done in creating the theory of general relativity. The problem is that no explanation is given to what spacetime is and therefore how it can be incorporated with the very real and testable quantum mechanical world. Time itself is also a topic of much debate and absolutely zero proof throughout history has been provided to suggest that time is actually a real physical phenomena. Time can only be argued to exist or not exist and so to simply incorporate it into an imaginary and extremely poorly, in fact not explained at all spacetime is reckless at best. Time appears only to slow down at high velocities such as those found in orbit around the Earth for example but what if another mechanism was at work?

Take for a moment a clock based off of heat energy so that the temperature of a particular clock determines how fast time passes for it. If we create two synchronized clocks and place one in a normal laboratory on Earth under standard temperatures and pressures and one in a very cold environment such as the polar caps we will find a discrepancy in the flow of time for the two. The control clock will count time normally and the experimental clock on one of the two poles will be in a colder environment and so run slower resulting in it counting time passing less rapidly than for us in the toasty warm control laboratory. One could reasonably come to the conclusion that time is linked to changes in heat and so create a heat-time in order to account for these differences. Predictions made using heat times and different temperatures would always agree with heat-time theory because the clocks themselves are based off of using heat energy to count the passage of time. A heat clock is analogous to the mechanical clocks we use on Earth and in orbit so if a heat-time may not be real than why should a spacetime be real? Remember that matter is affected by spacetime in such a way that it appears to slow down under certain conditions and using our heat clock and heat-time the matter of the clock would appear to slow down as well. Some might argue that we can explain heat easily and therefore would point out that a clock based on heat-time can run slowly no problem. I agree we can explain exactly what is happening to the matter on a quantum level and using quantum interactions and forces to explain why cold things make our heat clock run slower. We do in fact understand through quantum mechanics a method by which heat can be used to slow the passage of time for matter. So by extension I argue that we can find a quantum explanation for how matter is affected in areas of high gravity or high velocity that does not involve the invention of a spacetime and instead is simply some currently unknown to modern science quantum effect that is making time pass more slowly for a regular clock. In other words spacetime need not be real and for that matter time itself need not be real only an explanation of why certain quantum reactions slow down under certain conditions. Another unseen or unexplained mechanism could be at work by which the passage of time for our clocks is slowed in what Einstein calls relativistic frames of reference. I have found this mechanism and will be explaining it in a later section of the book.

The last major flaw with relativity of course is its inability to connect with quantum mechanics. This is the other major reason that relativity takes the third place spot in our examination of the three main views of how the universe runs currently accepted by modern science. Quantum mechanics while still very foggy in some areas is in fact real and tangible and has been used successfully for decades to create real world solutions. This is expressed both in experiments where predictions are made using quantum laws and then found later by testing those predictions. That is one of the very cores of science and is always accepted over any theory especially when applications of experimental results yield real world technologies or discoveries. So any competing theory of how the universe works even if only how part of the universe works must be compatible with quantum mechanics at least on some level.

General relativity cannot under any circumstances mesh successfully with quantum mechanics as it does not contain any testable small scale values which can be empirically measured using quantum experiments. The large scale observations which the mathematics of relativity can calculate for are simply too large to be reproduced in the laboratory so that one can compare the two theories side by side. In fairness to relativity this does not mean that its calculations of large scale effects such as the light bending around a star are inaccurate. In fact quantum mechanics cannot explain such large scale phenomena at all and this is why relativity is still used when viewing macro phenomena in the cosmos. The main reason why quantum mechanics scores higher than general relativity does for explaining in part how the universe works is we can test it with high accuracy in the lab down to the smallest fundamental levels to understand what is happening in a given system.

General relativity involves observations made on regions of space or objects so vast, so remote or both that for now humans cannot possibly test them with accuracy at the fundamental level. If we view the minute nature of these interactions over light year scales we might find an underlying quantum mechanism at work and this is of course what many theorists are trying to do with quantum theories of gravity. So overall general relativity is useful for measuring certain macro scale phenomena in the universe but scores very poorly at saying what is actually causing these phenomena and for providing a nuts and bolts explanation of the mechanisms involved. All general relativity theory is based solely off of visualized frames of reference for accelerating objects and simply mapping their respective trajectories over time it does not attempt to explain why spacetime behaves this way in the first place. It does state that matter is what is responsible for gravity and that gravity is the result of a curvature of spacetime to which objects with matter are falling into and if that does not sound like a circular argument with no explanation of why matter should make spacetime behave as it does than I do not know what is. So instead we will leave the equations for general relativity relegated to making calculations of observed phenomena but not used for explaining why these happen in the first place.

Whether you agree or disagree with my reasoning for the order of these three views the historical facts speak for themselves and the triumphs and failures of each coupled with their chronological development have led us to where modern science is today. Where that is leaves modern science still scratching its head as to how the universe works and pondering the substantial laundry list of unsolved problems in physics. What science does have on its side is a very good understanding of how much of the universe works and this is of course all thanks to the three main views we have been discussing. For a fact what does science know about the universe in general? The first thing we know is that the universe is made of two main parts which are space and matter.

Space is space and matter is sometimes called energy and vice versa but the bottom line is we have space and the stuff that goes inside space. Space has gone from being a nothingness to a series of immutable spheres of some pristine crystalline nature to a void filled with an ether and immutable at the same time to finally a space that is made of curved geometries and built out of something modern science has no clue about. To be very forward for a moment modern science has precisely zero concrete ideas about what space is and can only say that it appears that non-Euclidean curved geometries better explain observed phenomenon than other geometries do. Does this mean that space truly is made of curved spacetime and that general relativity is correct? Of course not it only means that using curved geometries fits observation a bit better but it does not mean the way space behaves cannot be better understood with a model built on something other than curvatures.

Indeed space must be built on something other than curvatures or it would have long been taken at face value that general relativity is completely correct and can explain everything in the universe. Of course general relativity cannot possibly hope to do this and this is precisely why modern science is still wracking its meta-brain for a better way to model the universe. The one thing modern science can say with confidence is that space is a dynamic and changing force that helps to govern the interactions of matter in the universe. This knowledge is useful in incorporating space into any theories dealing with matter and energy at fundamental levels or conversely at cosmological scales. This brings us to the matter in the universe or matter and energy if you like.

Modern science knows for a fact that the properties of space can greatly affect both matter and energy and use this in turn to better understand the cosmos and matter and energy as well. A perfect example of this is the simple red-shifting of light seen throughout the universe as a physical effect is being wrought upon the photons that are being red-shifted and it is derived from space itself. On Earth we may never have observed red-shifting or blue-shifting of light because the differences in space geometry here on Earth do not vary by any significant amount. Therefore simply observing these effects would be nigh impossible and the vastness of space has been instrumental in teaching us about some of the properties of energy. Matter on the other hand is much more easily experimented here on Earth as it tends to stay put rather than wander off if you turn your scientific back on it like energy. Name me one type of energy other than potential energy that does not disperse and usually rapidly away from where it is generated. If you said magnetic energy then very good you guessed the most useful form of energy that tends to stay where you left it and makes it relatively easy to study apart from being completely invisible. The matter making up something like a bar magnet however is much more highly visible and easy to get your hands on and experiment with. This hands on property of matter is what has led us to such a detailed understanding of how it works and taken us from things like the primeval elements such as water, air or fire to name a few and landed us firmly in the realm of real elements such as hydrogen, oxygen or thorium. To this end we have a detailed understanding of what makes up these elements such as protons, neutrons and electrons.

Further study has revealed the forces that hold these particles together and are responsible for helping to give matter many of its characteristic properties. So modern science also knows a great deal more about matter than ever before in recent centuries and all is explainable except for a few little details. These details are primarily quantum related problems whereby matter behaves in strange ways that do not make sense to modern science. Think for a moment about the apparent quantum fact that particles can spontaneously appear and disappear from time to time. There exists a small probability that at any given moment a quantum object will momentarily vanish before reappearing once more. This kind of behavior flies in the face of rational thought as no large scale objects just blink out of existence for no apparent reason only to pop back into existence again completely unchanged. This is the sort of problem that must be dealt with before a complete understanding of the universe can be reached as macro scale objects are made of micro scale objects. All macro scale objects firmly obey the laws of classical mechanics in everyday life and yet they are comprised of particles which do obey some classical laws but also have their own set of quantum mechanical laws. This is one of the areas where modern science stops knowing about matter and starts guessing about it similar to the way that modern science guesses about space. A greater understanding of one will surely lead to a greater understanding of the other as the two are interconnected.

We will leave space and matter here for now as this broadly sums up what modern science knows about the two, what it does not know is in some cases greater. So what else besides space and matter does modern science not yet know about the universe? For starters the evolution and growth of the universe is not fully understood. What is known for sure is that the universe started out much smaller than it is today as all the galaxies seem to be flying away from each other and the farther out you go the faster they are moving away from one another. So we know that by back tracking the motions of these galaxies the universe shrinks down to a smaller and earlier size in its development. This is the last thing modern science knows for sure about the life of the universe as continuing the shrinking backwards in time results in theoretical scenarios.

Remember the Big Bang theory no matter how compelling has never been fully proven and never been fully accepted. Case in point is the recent rise in thinking that the Big Bang never in fact happened at all and instead a different mechanism was involved in creating and expanding the universe. To say that the universe and everything in it including all of its countless billions of light years of space and every single piece of matter and energy within it started out as a singularity that was a single point in space infinitely smaller than a single atom is fairly outlandish. Outlandish that is if you say that this is how the universe started with one hundred percent certainty but theoretically acceptable if it is only an idea.

The experiments conducted at the Large Hadron Collider aim to study the first few moments after the Big Bang is thought to have occurred and study the quark-gluon plasma found there. Is quark-gluon plasma how the universe started out though? Well not necessarily as the ability to get matter to behave in ways that we do not see normally in the universe is interesting it does not mean that these behaviors are in any way a norm in the universe or even that they have to exist at all naturally. We can make a cheeseburger for example out of ordinary elements and a cheeseburger is certainly commonplace on Earth but naturally occurring arrangements of atoms into cheeseburgers in the universe is probably just a gorgeous fantasy for now. So too the quark-gluon plasma may just be a neat piece of experimenting that will undoubtedly teach us about how matter and energy behave but it does not mean that we can say for certain that the universe started out with a large quark-gluon phase right after the Big Bang. In fairness to these evolutionary theories of the universe it is of course exceedingly difficult to prove any one of them definitively however I am simply stating that we do not know them to be factually true. So here we have another of our little problems cropping up as the complete understanding of how the Big Bang or something like it was able to expand a universe so that it fits the appearance of today eludes modern science.

The last major area that modern science does not understand about the universe is many of the extreme small scale and large scale interactions of matter and energy within the universe and how they relate to the fundamental forces. These interactions are actually the most problematic of all the unsolved issues modern science is facing as they are the greatest in number. Here I will only list two as clear indicators of how important they are to understanding how the universe works. The first I will mention is gravity and black holes and to be even more specific why black holes behave in all the strange ways they do. Everything that modern science understands about black holes comes from one power source and one power source only and it is the weakest of the four fundamental forces which is of course gravity. In fact unit for unit of strength gravity is practically unrelated to the three stronger forces of electromagnetism, the weak force and the strong nuclear force. Yet under the right conditions gravity can overpower all of them at once and with extreme ease.

One such set of conditions is that of a black hole where matter has been compressed to such incredible degrees that the way it creates gravity is beyond anything else observed in the universe. The gravity it creates is nearly infinite in strength and far more powerful than the gravity created by similar amounts of matter that are not formed into a black hole. Think of it this way a star that will one day create a black hole does not gain mass rapidly as it collapses into a black hole yet the gravitational effects it can exert after its metamorphosis into a black hole exceed what it was capable of before hand. Remember just because it is said that a black hole has an escape velocity equal to the speed of light says nothing that it must rip the atoms apart that fall into it. This is clearly something that modern science has no understanding of and yet is a reality that must be solved in order to fully understand how the universe works.

Light will Destroy the Darkness

Another of the interaction problems is seen when examining the galaxies closest to us and look at how they behave internally and with one another. Internally they behave fairly straightforward we see areas of star formation associated with spiral arms as they circle the galactic centers, the centers are bright and hot as we would expect and all their stars orbit the center at seemingly the same rate. Wait a minute that last one seems a bit odd as the stars closest to the center should be moving faster than the stars more distant and yet we observe them moving at very nearly the same speeds. This is known as the flat rotation problem and has baffled astronomers and physicists for decades and is linked to an even older problem still. This older problem is how the galaxies get along with their universal neighbors and especially when grouped together in galaxy clusters. The galaxy clusters are sticking together nicely and show no signs of drifting apart from one another which is odd because as far as astronomers can tell there is simply not enough matter present to provide sufficient gravity to allow this to happen. It is as if more matter would be needed in order to make the current excepted models of gravity used by modern science make sense.

Originally proposed in the 1930s the idea of dark matter was suggested as a solution to this problem. The naming of dark matter came about simply because it was assumed that if we cannot see the matter it must be dark and yet still be normal matter just a kind that does not radiate any electromagnetic radiation that we can detect and so is technically dark in nature. So here we have another set of interaction problems as galaxies do not possess enough gravity it is believed by modern science from normal matter to hold themselves together and they should according to those theories fly apart. On a larger scale still the local galaxy clusters should not have enough gravity to maintain their shapes and the individual galaxies should fly apart destroying the cluster. Neither of these things has or is happening with galaxies in the universe and so we are faced with another unsolved problem that requires new thinking to fix it. The only problem is there are no new proofs available to modern science despite valiant attempts to spot them. Scientists are nothing if not patient, methodical and meticulous and so every effort has been made in the last century to explain this and many other unsolved problems still lurking in the knowledge of modern science yet the answers still elude them.

Where this leaves us now in our understanding of the universe to date is with modern science having found many different pieces of what the universe is made of and even many of the rules that these pieces obey but a large number of holes exist in the total knowledge of how the universe works. I have up until now described what science knows for certain and even some of the most favored new theories of modern science to date but in order to attempt to explain these holes we must now list them and how the three main views of the universe stack up against them so that we know where we stand on each of them in turn.

Where we are now is essentially at the beginning of a new story as far as what modern science understands about the universe and how it works. The first story that we have been experiencing up until now involves what science knows about the universe as divided into two parts. The first part is what we know for an absolute fact and concerns things like proven measurements of particles, forces and observations about the universe around us. It does not necessarily mean that modern science understands why these things are happening perfectly only that they have been measured so as to remove 99 percent of the doubt surrounding their properties. The second part of the story involves trying to fit these known facts into the rest of the universe as best as can be done by modern science. This part includes the theories which are currently being used to try and explain the unsolved problems in physics today. Think of it as the shaky middle ground in our understanding of the universe with the known facts acting as the bottom ground which provides the stable basis from which we build our knowledge of the universe. The next story which we will be starting now is a sort of top down approach to our previous bottom up story.

The unsolved problems in physics and indeed in science in general are at the very top of our understanding of the universe and from there we head down to that shaky area filled with questionable theories that cannot explain observed phenomena in the cosmos. To clear something up right now the top is not an area to be afraid of and in fact should be the most fully embraced of all three areas of our tower of understanding of the universe. The reason is simple the top is never wrong and it actually could never have been wrong from the get go because it was created by the universe and not intelligent creatures such as humans. The top incorporates the observable phenomena of the universe from the smallest quantum systems to the largest unexplained behaviors of the universe itself. These phenomena are real and not theorized or measured by humans and so cannot be wrong it is us and our theories that must bend to the powers of the universe in order to understand how the cosmos works.

This being said the top is the area where we find our unsolved problems because current theories and models used by modern science are incomplete, wrong or simply heading off in the wrong direction when it comes to making the known measurements at the bottom of the tower work smoothly and seamlessly with the unsolved problems at the top of the tower. This shaky middle ground is what must be rectified in order to understand how the universe works and so I propose that the best way to do this is to solve the top area of the tower first. No small task to be sure but by solving the top first we can then clear up the shaky middle ground because all the flaws existing within that shaky middle ground will vanish as the known measurements blend with the observed phenomena as one. To do this it is necessary to simply step back and understand the top as it is observed instead of trying to get the observations to conform to some shaky middle ground theory. So where shall we start in our understanding of these top tower observations which have been called the unsolved problems of physics? Well like any good story we shall start at the very beginning and in the case of the universe and modern sciences understanding of it this is the Big Bang.

Currently modern science believes on the whole that the universe started out with a Big Bang which was a colossal explosion that brought all space, matter and time into existence from a single point infinitely smaller than a single atom known as a singularity. Other less accepted views also circulating in modern science say that the universe existed before this single explosion and that everything was not simply created at this precise moment. While hotly debated amongst theorists the one thing that they can agree upon is that the universe did have some very large and violent explosion in the distant past that closely resembles a Big Bang event. While not fixating on the particulars, details and the why not's too much a single large explosion appears to have spread matter and space evenly as far as our human eyes can see. So if everyone is in agreement with this concept how can a problem exist?

Simple, the rate of expansion of this explosion does not make sense to any known theory or any of the three views used by modern science to explain how the universe works. This is due to the fact that unlike normal explosions that start off fast and over time end up slow with a smooth deceleration to the velocity of the matter being expelled the universe changes speeds at will it seems. The universe started off with a normal high speed explosion that spread matter rapidly outwards from the point of the initial blast. The explosion was so powerful and violent that it moved faster than the speed of light and in fact many, many times faster than the speed of light. This was due to the speed at which space was expanding and space can travel much faster than the speed of light. The matter inside space did not travel faster than the speed of light inside space itself but it was dragged along with the expanding space to spread out into the void at beyond the speed of light.

As we look around the universe today modern science has no idea of how the mechanics of space work and how space can travel faster than light. The closest idea is an unproven theory known as inflation theory which we have discussed before and at its core it still has no idea how space can move faster than light and the physical properties of space itself. What it does do is explain how a period of rapid expansion faster than the speed of light could potentially solve the problems we see with how the universe appears so uniform today for one. This accounts for space expanding quickly and dragging matter with it during the early moments of the universes life but it does not explain what happens next. The universe slowed down in its expansion and then sped up again. If we ignore the initial blast of the Big Bang as modern science needs entirely new physics to deal with space and how it behaves we can overlook any seemingly strange interactions that might occur at such times. Once we get to the post rapid space expansion phase however we cannot put our scientific heads in the space sand and pretend that the rest of the explosion makes perfect sense according to the known laws of modern science. The reason is a normal explosion slows down after the initial blast and the universe does appear to have undergone a period of deceleration in the expansion of the universe. However the universe has now begun an acceleration phase in its expansion which is totally opposite to how a normal explosion would progress over time here on Earth for example. This would be like setting off a bomb and watching the initial blast wave travel outwards rapidly and then gradually slow down but instead of the expanding fireball continuing to slow until it reached the maximum size it could the fireball instead speeds up and continues expanding well beyond the normal volume it should be able to occupy.

This is what the universe is doing and it has baffled modern science since it was discovered around the turn of the last millennium. Modern science is thorough in its measurements and these have been verified over and over again the universe is still expanding and the rate of expansion is not slowing down at all but instead speeding up constantly over time it seems. This is the first of the truly unsolved problems facing modern science and physics today as no known currently accepted theory can explain why this is happening and it is difficult to fit it into any of the three accepted views of how the universe functions. Classical mechanics seems to make some sense in that the large scale objects are behaving just as Newton predicted they would if acted upon by an outside force. Enough force will slow them down and conversely enough force will speed them up. The trouble is no forces have yet been detected to account for the changes in velocity of expansion.

Quantum mechanics would have no problem incorporating a new particle to explain the changes in velocity but to date nothing new has been found to fit into the Standard Model that would explain these results. To make the quantum headaches worse only extremely hypothetical particles using multiple dimensions or with totally undiscovered properties can be used to make any predictions and they exist only in the most derived and mathematical forms not in real physical particles. As for relativity it is at a complete loss as according to relativity the universe should be slowing down based on the curvature of the universe pulling things backwards on themselves again. Einstein went so far as to invent a constant that would keep the universe at a static size until it was shown that he was wrong and the universe was expanding. Relativity is also more of a calculation tool then an explanation of observable events and new particles or energies that could solve this problem cannot be incorporated into its framework. So overall the three main views cannot offer an explanation of what is really transpiring out their in the cosmos and modern science has its first major unsolved problem of the universe.

The second many believe to be linked to the first and it appears to concern missing matter in the universe. We are of course talking about why the universe appears less massive than it should based on our observations of motions of things like galaxies and galaxy clusters. Astronomers have for decades since Hubble have been looking at the velocity of galaxies as they move around us and found that some are coming towards us and most are going away from us. Take for example the Andromeda galaxy which is definitely heading towards us and in fact it and our own Milky Way galaxy are due for a collision in the distant future. This is somewhat the exception however as most tend to be travelling away from us and when we look out into the universe most of the galaxies seem to be moving away from each other period. Yet not always and this is what has astronomers and scientists so puzzled as collections of galaxies which are known as galaxy clusters group together and stay together sort of like a family. The mutual attraction of gravity between them keeps them held in a close knit bunch cosmically speaking of course. What has puzzled astronomers for years though is that when the mass of the galaxies is calculated individually and then as the sum total of all the galaxies within a single galaxy cluster it appears as though they do not possess enough mass to stay together. In other words it appears as though they do not have enough matter necessary to generate the levels of gravitational attraction that would be needed to keep them from flying away from one another.

First noticed in the 1930s by Zwicky this led him as we have seen to coin the term dark matter as a way of explaining where the missing matter and therefore gravity was coming from. He reasoned that if we cannot see the matter it must still be there so that normal laws of physics would not be broken and the normal force of gravity could work its weak but far reaching power over the entire cluster and all the resident galaxies within it. The problem was further compounded when it was noticed that not only galaxy clusters did seemingly not possess enough matter to hold themselves together but also galaxies themselves too. It was noticed as we have mentioned in the flat rotation problem that the stars of galaxies move faster than one might expect based on their orbital distances from the local galactic center. More speed means they need more total gravity to keep them in such high rates of motion and without the undesirable effect of flinging their individual stars out into the cosmos. Again not enough matter found in galaxies had people wondering where all the missing mass was. The common thought was that Zwicky was right and there simply must be, lurking somewhere although no one knew where or what to look for, more matter somewhere in the universe. Recent decades have focused heavily on this search for the missing dark matter as it came to be commonly known amongst cosmological circles. Upon even more exhaustive investigation newer more accurate measurements from multiple studies revealed that even more matter was missing than originally thought. This was no fault of the early scientists working on dark matter simply that more modern instruments could take far more accurate measurements from much farther away than before and like many studies we have the Hubble Space Telescope providing a very distant view of the universe.

In fact it was many of the Hubble deep field images that not only showed that the universe was basically never ending and in fact might be never ending but also that it was populated entirely with galaxies. The next step besides the simply eye opening realization that the universe was far larger than we originally thought was to measure these objects with respect to currently accepted theories to see how they stacked up against the observations. The theory that the universe was expanding continually and that it did not seem to possess enough matter to hold itself together held firm. With this gradual understanding of the structure of galaxies within the universe and their motions when compared to one another and to us a new problem was found which was even more matter was missing. This time it was not a lack of gravity that was found but a lack of not gravity that was found. Enter the idea of dark energy the reverse of dark matter which produces a reverse effect as well.

Matter is thought to be responsible for gravity which causes things to move together and stick together and the opposite is dark energy which is thought to push things apart and make them fly apart. Careful analysis of all the observations made showed that something was pushing the galaxies and the galaxy clusters away from one another and it was doing so with surprising strength which meant even more mass was missing from the universe. Here I need to clarify a concept used by scientists when discussing the problem of missing mass in the universe as well as dark matter and dark energy as possible explanations for them. When a scientist says the words missing mass what they are really saying is something that would produce the same effect as normal mass or matter as modern science currently understands it today. This distinction is of the utmost importance in understanding the universe as nowhere does the universe care one way or the other whether or not humans understand something other than matter. What do I mean precisely by this? Simple, the missing matter refers to unexplained observations concerning the motion of large scale massive objects in the universe and not necessarily in conjunction with missing mass in the form of dark matter or dark energy. In other words scientists say that if the forces that appear to be missing were derived from matter and not something else then about 90% to 95 % of the matter in the universe is missing.

The rough mathematics breaks down to what we think of as normal visible matter making up about only five percent of the universe. This includes everything we can see or touch with the human senses such as everyday objects like apple trees that smart people sit under and the things they look at whilst under these trees like moons and stars. Take any of those things and group them in large enough quantities and you have solar systems, galaxies and galaxy clusters which are some of the biggest single types of objects in the known universe. This type of visible or light matter is made of real solid particles that constitute things that give rise one way or another to gravity. The invisible or dark matter is thought to be made of exactly the same stuff in so much as it is made of real physical particles only these particles are like nothing we have ever seen before should they in fact exist. The dark matter will make up five to six times more physical mass of the universe than normal or light matter meaning dark matter should weigh in at about 30% of all the particles in the universe. This leaves us with about 65% to 70 % depending on who you talk to, of the mass in the universe unaccounted for and doing the exact opposite of giving rise to gravity. The dark energy is not making gravity it is repelling things as fast and hard as it can and so much so that seemingly 70% of the universe is governed by it alone. The dark energy just like the dark matter is a form of energy that modern science has never seen before and does not at all understand.

This is where I will restate with several layers of underlining that nothing says dark matter or dark energy even exists at all. Once again the effects of what we have measured seem to be missing mass and so not giving us enough gravity but only based on modern sciences current understanding of the universe, particles within it and the mathematics used to derive all related values from it. As for the dark energy the exact same can be said and with much more vehemence as dark matter itself is poorly if at all understood by modern science and so the opposite which gives rise to a pushing force, the dark energy, was simply plucked out of thin air to suggest a possible mechanism for the apparent spreading out of matter in the cosmos. Here again you can also see the failure of calling everything energy as dark energy is in no way supposed to be some sort of dark photon.

Zero and precisely zero theoretical evidence exists for dark energy other than calculating how much mass it could represent in order to generate the pushing forces observed which are of coursed based off of the same calculations that gave rise to the mass of galaxy clusters and what not plus the maybe missing dark matter they theoretically possess. As you can plainly see what modern science is really saying is that it does not know what is going on in terms of total matter in the universe and that suggesting more matter in the form of dark matter might help solve the problem. Afterall all the visible matter meaning the only matter we know to be real is only five percent of the universe? That sounds an awful lot more like there is something missing in the theory rather than 95% more stuff out there that golly we have just never even been able to detect on any level whatsoever or even see interacting with in any way any of the known particles, force carriers or forces.

The dark energy is suggested in much the same way a mysterious force of unknown energy that conveniently pushes the universe apart. I would like to think these theories of dark matter and dark energy were dreamt up out of the most pure scientific intentions and I believe that since the days of Zwicky in the 1930s they have been. The problem stems from the fact that it seems as though dark matter and dark energy are trying to be fit into existing concepts like the Standard Model of particle physics when if they represent new and exotic forms of matter and energy they might not fit at all. This is what most modern scientists working with experiments designed to find dark matter or energy are aiming to do and if successful it might complete the family of particles within the Standard Model once and for all. The efforts being made to find dark matter and dark energy are nothing short of heroic as every possible experimental avenue is being followed in an effort to seek out any new type of particle. Yet what if the matter does not come from a particle at all? Here we see the great debate of the dark matter come into full swing as two different camps emerged with opposite approaches to explaining dark matter and it all boils down to whether you are feeling macho or wimpy.

MACHOS and WIMPS

The great dark matter search began with MACHOS on one side and WIMPS on the other and while the WIMPS seem to be winning these days the MACHOS might yet make a comeback and knockout the wimps. So what are machos and wimps you ask? They are the two various umbrella terms for the competing theories of what dark matter really is should it exist at all. In truth there are dozens of various theories for what dark matter might be but when you reduce them to their simplest ideas you find that they will belong to one of these two groups and as such are commonly referred to as one or the other. The majority of all dark matter theories are derived from baryonic matter which is made from baryons and include commonplace objects such as protons or neutrons. Here we find that size does not matter when it comes to which dark matter theory is being examined as both will use baryonic matter and this could be a hefty macho or a miniscule wimp. So first let us look at the machos and what the name means.

The term macho is an acronym for a much longer name which is massive astrophysical compact halo object. If that sounds scary it should not as all it really means is large quantities of regular matter that we have been unable to detect. The Earth for example is a macho and so is any piece of solid matter floating or orbiting around our Sun. The reason for this is simple an alien race cannot directly detect our planet through the use of any telescope. Think for a minute how we view the cosmos here on Earth and realize that all we use is the electromagnetic spectrum. Do we examine the universe using the three other forces? No we do not even, the signature of gravity we think we see is just more electromagnetic radiation that is altered by a gravitational field and so we are still learning everything about the cosmos from only one source of information. We do not measure gravity particles or gravity force just the light that is altered by it.

So is any alien race looking at us is bound by the same limits of electromagnetic radiation as a tool for measurement. Radio waves, infrared, microwave and so on up to gamma rays are still only photons. Since Earth does not emit or reflect any appreciable amounts of radiation in any wavelength beyond our own solar system we and all the mass of the Earth are invisible to the astronomers situated on other systems throughout the galaxy. This is the general thinking behind most macho theory as a great number of objects will possess significant mass and yet not emit significant amounts of electromagnetic radiation for us to detect. The most obvious candidates are black holes which not only refuse to emit electromagnetic radiation but there are easily capable of absorbing all they come across rendering them completely invisible to normal telescopes. A black hole of course was formed from a star with at least ten times the solar mass of our own Sun and so possesses a huge amount of matter and gravity.

Black holes of various kinds are good candidates for machos in the universe but are not the only kind which lends much support to the macho theory. The key to many of these objects is that they must be some form of large scale structure and when I say large scale I mean anything with an appreciable size larger than a single particle as that would constitute a form of wimp. Black holes are of course a perfect candidate for a macho that contains high mass, contributes significantly to gravitational effects and yet does not emit visible radiation but there are many others. Things like brown dwarf stars and Jupiter sized planets are good examples of these objects because they can do two things necessary for supporting the machos theory of dark matter. The first is that they contain of course significant mass and simply finding large amounts of raw mass is key to solving where much of the missing matter seems to have gone. Just like black holes objects with lots of mass means lots of gravity to help hold things like galaxies and galaxy clusters together.

The second advantage that they possess is large reserves of deuterium which is thought to have been produced in large quantities during Big Bang nucleosynthesis if the missing dark matter comes from baryonic matter. When astronomers look into the sky they see less deuterium than they should according to theory and objects like brown dwarf stars have not ignited their fusion cores and therefore do not burn brightly enough to see but they do burn deuterium. The Jupiter sized objects of course burn nothing but contain large amounts of hydrogen, helium and deuterium making them ideal reservoirs for these elements. We will return to discussions of machos later but for now let us introduce their counterparts the wimps which also have their share of strengths for a theory of dark matter.

Like machos wimps also have a much longer name stemming from their acronym which is weakly interacting massive particle. The main appeal of the wimp argument is that unlike machos you can solve the problem with many, many, many small objects of the sub-atomic variety enabling them to total the mass of the missing gravity by sheer number. Think for a minute of the humble neutrino which is an almost totally massless particle meaning it weighs essentially nothing and yet they exist in fantastic quantities throughout the universe. A neutrinos mass is cause for some discussion as many values have been reported from varying studies but an extremely rough average for the three neutrino flavor states would be around 0.25 eV and yes I realize many people reading this are disagreeing about the exact number and I hope that even more people reading this are relaxed enough to know I am simply giving a rough figure. The point of the neutrinos low mass is that even though it is far lighter than most other particles it does have a mass and this means it does contribute to gravity. In addition to this the numbers of neutrinos present in the universe is so great that adding up all their little masses means they by themselves can contribute directly to the total mass of the universe itself.

The idea behind wimps is pretty much the same as it involves copious amounts of tiny particles that possess mass and by adding these up we get the missing matter of the universe. If the wimps have an advantage over the humble neutrino it is size and size alone. The rough values for the various wimp masses expected by theory can range by an order of three magnitudes from 1 GeV to 1 TeV. In this case the G means giga or billion and the T means Terra or trillion so quite a difference in the overall masses of predicted wimps. To put this in perspective with the humble neutrino at 0.25 eV a single 1 TeV neutrino would look like this: 1 000 000 000 000 eV making it clearly heavier than a regular flavor neutrino. The hope of the wimps backers is that if you have a particle with such large mass that the huge amounts of missing matter can be made up with the tiny particles in vast numbers. Since the mass of a theoretical dark matter wimp is so high when compared to a single neutrino it is easy to see how the dark matter problem could be solved using wimps. Neutrinos are already known to be present in the universe in obscene numbers and so any particle present in the same amounts but with a vastly superior mass could solve the missing matter problem.

There is a catch to the wimp argument that is unsolvable despite many thorough and sometimes desperate attempts and this is the interacting part of the wimp acronym. The first part of the acronym stands for weakly interacting and it means exactly what it says. Particles in the wimp camp would be very, very unlikely to interact with anything as they pass through most things in the universe without any hindrance whatsoever. This includes normal matter and energy in small or large amounts it does not seem to matter much as the wimps do not interact with them in any significant way. This is not to say that no interactions would be occurring only that they would be so infrequent as to be almost non existent. This is akin to neutrinos which hardly interact with matter at all and freely pass straight through the Earth everyday. It is thought that dark matter wimps would behave much the same way with countless wimps passing straight through the Earth all the time and yet hardly causing any interactions.

The plus side to dark matter wimps if they exist is there much larger mass than a neutrino which should make them literally billions of times easier to detect. The reasoning goes something like if you have a small pebble and you throw it into a pond filled with perfectly still water you will see small ripples emanating from the place where the pebble hit the surface of the water. This is only a small pebble and so the ripples are small without much amplitude and will not carry much energy with them and so not travel very far from the point of impact. Now if we drop a huge boulder into the water it will create a huge splash that has incredible energy and power meaning the waves have very high amplitude and travel very far from the point of impact. If we are hovering above the pond looking down from a height of say 100 meters the pebble impact is tiny and difficult to see but the boulder impact is massive and very easy to see because it carried so much more energy with it.

A dark matter wimp particle would carry a boulder amount of mass and therefore energy with it meaning that any detector looking for them should produce very easy to see flashes that are far brighter, with more energy and completely different from a neutrino signature. That is of course the theory only. In reality things are quite different and so far no dark matter wimps have been detected in any experiment. The approach to finding wimps has been carefully thought out though and no stone has been left unturned. Two approaches have been employed and they both go about looking for the little particles in drastically different ways. The first of course is passive and gentle and involves the use of large ultra-sensitive detectors buried deep with the Earth's surface. These detectors operate in much the same way as a normal neutrino detector and use various substances to hopefully catch one of the extremely rare interactions with dark matter. The interaction would then produce a flash of normal energy from the normal matter present and the properties of the flash can easily be recorded and analyzed using conventional equipment. The facilities themselves are always placed deep underground, sometimes up to a few kilometers underground in order to shield the test samples from any radiation sources that might contaminate the experimental results.

The second approach to finding the dark matter particles is to use a rather violent means of detection which is to smash things together in large and powerful super colliders. This of course has been performed all over the world with the same expected outcome of creating a shower of fundamental particles which all have their own unique signatures. Since they are placed within conventional detectors at the time of the collisions any information recovered can be easily interpreted. The hope is that one of the signatures recorded will be new and unknown from all the known particles and their various properties. Anything new would be a potential dark matter candidate that would allow the experimentalists to narrow their searches and acquire more information on these new particles. In either experiment any new data would be used to check the current dark matter theories to see if anything fits the expected values.

This leads to another important aspect of the search for dark matter which is called super symmetry. The super symmetry theory has been searched for using things like particle colliders and so far has not yielded any positive results however the theory still presents a possible explanation for the missing matter in the universe. According to the super symmetry theory every known type of fermion and boson particle would have a corresponding super partner with varying quantum values. Sometimes these values are the same such as mass but differ by the spin value which is a half integer different. A fermionic particle will have a bosonic super symmetrical particle and will be denoted by a particle with an s prefix in front of it. For example a selectron or a squark would be two examples of such theoretical particles and yes it is fun to say squark. A bosonic particle will have a fermionic super partner with a suffix containing the ino description and might be things like a photino or a higgsino. The obvious advantage of such a theory is that it instantly creates a whole new host of particles which can lend matter to the missing mass problem.

A second advantage is that they would neatly fit into the Standard Model as they would in essence be normal well understood classes of particles and so other than not being detected would not create problems in meshing with the known particles. A universe filled with super symmetric particles would have significantly more mass available to create gravity with and solve the missing matter problem. The trouble with super symmetry theory is that the predicted masses of particles have not been detected in any collider experiments including those run at the most currently powerful collider the LHC. Hope still remains for the super symmetry theory though as new experiments might yield the long desired particles even if their respective values differ slightly from theoretical predictions.

So these are the basic ideas behind what dark matter might be should it exist at all and should it exist in the quantities theorized, but how do they apply to the three main views of the universe? The problem is missing matter and therefore missing gravity from the universe and the quantities are large. The possible candidate theories fall into two main categories the first being the machos and the second being the wimps. First let us examine the machos and how each of the three views would respond to it as a solution to the problem of missing mass. First up is classical mechanics which would fair the best if the missing matter was made up of machos or at least mainly made up of machos. This point needs to be stressed here as it is entirely possible that a combination of machos and wimps will in fact lead to the missing matter. Are there large scale objects with significant mass that we cannot observe here on Earth because they are too faint or do not emit any electromagnetic radiation at all? Yes there are and there is also a good chance that new particles or at least greater amounts of known particles like neutrinos or heavier neutrinos exist elsewhere that have escaped detection as well. The net result is that a combination of both machos and wimps will likely be needed to solve or at least partially solve the missing matter problem. For now we will assume that most of the missing matter is made from machos and the classical mechanics view of the universe has no problem dealing with this sort of object. The reason being is that all large non-quantum scale objects obey the laws of classical mechanics pretty much perfectly in terms of there interactions with each and how they move. So adding more of what classical mechanics can understand would be a perfect solution to the missing matter problem.

Next up we will look at general relativity which as we have all seen is a derivative of classical mechanics and uses many of its concepts and equations at its core. General relativity therefore would also have no problem dealing with more machos in the universe as it follows the same principles of large scale non-quantum objects and their interactions. General relativity also scores high here because it makes better predictions than its predecessor classical mechanics when it comes to things like gravity. Adding more matter means more gravity and this should fit well with the predictions made from general relativity so that the universe balances out all the missing matter with observation.

Lastly we come to quantum mechanics which will score the lowest here as it has no framework for incorporating the large scale effects of gravity with the small scale interactions of fundamental particles on quantum scales and machos are not particles just more large scale masses. The real shining point of quantum mechanics here however is that the matter making up the machos is completely normal baryonic matter. The reason this is so important to quantum mechanics and science in general is because all the matter making up the machos already falls perfectly into the Standard Model of particle physics. With the machos being made of regular particles, atoms and forming into regular cosmic objects no new particles with strange properties or dimensions need be incorporated into the Standard Model. This is of course one of the huge advantages that the machos argument has over the wimps dark matter theory. The downside to many of the wimp candidates is that they exist only on paper and in mathematical equations and to make matters worse many of these are derived particles based off of already derived mathematics or assumptions about conditions that existed earlier in the universe.

In many cases predictions and conditions necessary to make various wimp theories work are based off of a number of assumptions, sound assumptions but still unknowns nonetheless. Especially those involving initial conditions of the Big Bang itself and the state of field strengths and energy levels available for particle formation shortly after the Big Bang. Numerous hypothetical thermal limits are suggested for when wimp particles would become significant players in the overall mass and gravity components of the universe. The Big Bang itself is under debate these days and the nature of the conditions of the universe shortly after it should it have existed as current theory suggests is poorly understood. For this reason the appeal of machos is undeniable as it simply involves more stuff so to speak in the universe and more stuff of a well understood variety. The fact that we cannot detect it with our level of technology is actually a good thing because it means that in decades to come our instrumentation will more than likely be sensitive enough to detect significant amounts of machos that are today hidden from us. The worst case scenario is that the amount of early elemental nucleosynthesis currently accepted for the Big Bang models is not fully understood and that future observations and fine tunings will allow for the necessary amounts of things like helium 4 and deuterium to be present for the machos theory to be more viable. All in all many astronomers would probably be happy to have machos as the answer simply because it gives them far more material to study and as for the rest of the scientific community it presents the simplest explanation with the least headaches in terms of figuring out where all the missing matter went.

Now we will examine the wimps theory and how it compares with the three main views of how the universe functions. Once again the basic understanding of wimps as they are broadly associated under one category is that of small particles which barely interact with normal matter and yet have extremely high masses such that their combined numbers will account for the missing matter. First up is classical mechanics and the idea of multitudes of little particles moving quickly and pretty much evenly throughout the universe whilst not interacting with normal matter. To this end classical mechanics is divided in how favorable the wimps would be in terms of being classically explained. On the one hand classical mechanics is not quantum mechanics and so has difficulty dealing with the small finite interactions of fundamental particles like the wimps would be. Classical mechanics however and some of its laws still work perfectly at the quantum level so the inclusion of a new particle should still obey some classical laws although overall quantum laws will describe them better. The large scale effects that wimps would create in the universe are more favored by classical mechanics as they translate into normal large scale motions of bigger objects like planets, stars and galaxies. So classical mechanics is average in understanding a universe populated by strange wimp particles should they exist.

Next up is quantum mechanics which obviously scores the highest when it comes to understanding a universe full of wimps as at their core wimps are fundamental quantum objects. In fact the wimps should obey all the known laws of quantum mechanics straight out of the gate as they would be nothing more than a new member to the particle zoo. The wimps would even be able to be measured successfully in experiments here on Earth such as those carried out at the Large Hadron Collider assuming the LHC has enough power to create them. The difficulty in studying wimps is that they are thought to only interact through the weak force and even then only sometimes. Many factors effect whether a theoretical wimp particle exchanges any sort of information with normal matter. To start with a wimp is very small and can only interact with the nucleus of an atom which is itself only a tiny percentage of the volume of an atom proper. What this means is that all the normal matter that we see in the universe and that we are currently measuring and wondering why it does not appear to be making enough gravity is mainly empty space and therefore not likely to encounter a single wimp. Remember that even though the number of theorized wimps said to be passing through our bodies for example every second measures easily in the millions or higher that is because we are massive compared to all fundamental particles. Take a single atom from our bodies and the numbers tell a different story which is that possibly there is not even remotely close to 1 wimp per atom inside a human. Meaning the same is true for all objects and therefore the chances of an interaction are extremely small between an atom and a wimp let a lone the tiny nucleus of an atom and a wimp. Should an interaction occur however quantum mechanics can map out exactly what took place in great detail at least in theory and would give particle physicists years of new material to study.

The last up is general relativity which has a love hate relationship with the wimp theory. It absolutely hates wimps because it hates all fundamental particles and quantum objects as general relativity fails utterly at such small scales. On the other hand general relativity loves the overall potential effect of a wimp as they are thought to contribute to the mass of large scale objects such as planets, stars and galaxies. Objects of this size scale are where general relativity makes its best predictions and so the balancing of any theoretical missing matter with the observed matter would allow relativity to behave normally. The difficulty encountered with any of the three views explaining how wimps interact with the universe still stems from the fact that wimps have never been found. It was mentioned earlier that the wimps are poorly understood outside of the derived mathematics from which they were born and that the conditions necessary to narrow down their possible values and thermal limits is also poorly understood. This includes wimp theory in general, wimp interactions with normal matter and the initial conditions of the universe shortly after the Big Bang. All of these are poorly understood and result in a wide variety of wimps that have all categorically failed detection attempts whether those attempts were passive listening or active experimentation. While it might look hopeless for the wimps due to these factors it is actually a good thing in some respects. The fact that all these conditions are not well understood means that there is still room for the wimps to be found. Once our understanding of conditions immediately after the Big Bang are better understood or once we understand better the possible weak interactions of particles newer more refined wimps could be theorized. These new particles might exist and would allow for the possibility of detection once again meaning that wimps could indeed make up some of the missing matter in the universe.

Rate of Expansion of the Universe

The next unsolved problem continues on from this apparent lack of matter in the universe and can be thought of as a lack of energy or a sort of energy crisis. The universe as we have learned is accelerating its rate of expansion and this is not something that makes sense according to the three main views of how the universe behaves. First of all this problem has multiple parts and is not simply a reasoning behind the currently observed increased rate of expansion but has to do with all forms of expansion throughout the history of the cosmos. Obviously when one discusses the idea of an accelerating expansion to the universe dark energy comes to mind as the driving force behind this occurrence. Dark energy as we have also learned is one hundred percent not understood by modern science and evolved solely out of a need to explain this increased rate of expansion. Matter gives rise to gravity which would slow the rate of expansion down as it is believed to have done so in the past. The opposite side of the coin means that energy must therefore be responsible for a runaway rate of expansion and so based off of the idea of dark matter scientists coined the term dark energy. So dark energy is responsible it is thought for the driving apart of galaxies from one another and so on and so forth throughout the universe and over time is getting stronger so therefore must be causing the acceleration of the universes expansion. All of that is theory only and no proof so what else can we look at in order to understand how the universe expands?

To begin with let us look at the Big Bang and the inflationary period that is thought to have rapidly expanded the universe many trillions of times and much faster than the speed of light. This is a form of expansion that we will look at in a bit more detail later as it too addresses one of the unsolved problems of the universe but for now it seems to have been a driving force behind cosmic expansion. The thing is the driving forces behind the theoretical inflationary period are not at all the same as those that are believed to be causing today's increased expansion and this is of course the dark energy. So this creates a simple question which is: why would a universe need two different mechanisms for expansion and why could it not simply use one for all expansion over its entire history? Remember all that heating and cooling and turning on and off given critical energy levels of the inflation period in the early universe which states that inflation cannot exist after the period ended? For this modern science does not know but what modern science does know is that the mechanics behind the inflationary mechanism and those of dark energy as they are described today are completely incompatible.

What modern science does know very well is that the universe is accelerating in its current state of expansion and so an explanation must be found. The last puzzling piece of the missing energy problem is why it seems to be increasing as this would mean that somehow the universe is creating dark energy out of thin air so to speak. The amount of other stuff in the universe has not been decreasing over time it has remained static since the Big Bang so how can the universe just make more energy? Again something is rotten in the state of the cosmic Denmark. Since dark energy is so poorly understood let us take a quick look at the three views and how they might be able to accommodate it. First up is classical mechanics which has a difficult time incorporating quantum effects such as those used to describe energy in discrete packets. Classical mechanics however does quite happily allow for the idea of forces acting upon objects and altering their states of motion. After all an object remains at rest until acted upon by an outside force and so the idea of an outside dark energy force acting upon objects like galaxies makes on a very macro scale perfect sense to classical mechanics.

Quantum mechanics of course does not understand the large scale of the universe but can most easily out of the three incorporate a new form of energy to interact with known quantum objects. If dark energy exists then quantum mechanics will welcome it with open arms and a cozy home in the Standard Model. Lastly is general relativity which utterly fails at incorporating dark energy as really general relativity only deals with gravity and matter and not the reverse. The concept of expansion can be incorporated into relativity equations but on the very base level of the theory of relativity the universe should be slowing down as the initial velocity of the Big Bang is lost over time and gravity creates a curvature which slows the progress of expansion. The complications of dark energy to relativistic predictions of the fate of the universe mean that general relativity has innate inconsistencies within itself. According to relativity finite mass creates finite gravity no matter how you treat the matter making the gravity and so a universe filled with finite mass and space should not expand forever as the finite mass will halt the expansion of the finite volume of space.

Remember according to the Big Bang theory all space was created at once and so the universe is not making more space as it ages only the finite space is getting more spread out. Since the universe is changing its rate of acceleration at will it seems as though fixed finite values do not exist and therefore another mechanism beyond general relativity must be at work. An expanding universe further demonstrates the failure of relativity at universal scales as a curved spacetime with something like a galaxy should be destroyed over time as the rate of expansion increases in the universe. Since the expansion permeates all of space including that which exists inside a galaxy it should be tearing the spacetime curvature of the galaxy apart instead we see that galaxies are stable for billions and billions of years. This demonstrates that something other than pure relativistic effects are at work in the universe as the dark matter holding galaxies together would lose out to the dark energy pushing the universe apart. To this end relativity is not at all useless in examining large scale objects like galaxies but simply that for the inclusion of dark energy and on size scales of the whole universe it cannot provide an explanation for what is really going on. So we see here that the three views have different levels of acceptance to the concept of dark energy but that at their cores none can offer even a remote explanation as to the why of the problem.

Next we will examine another expansion related problem of the universe and this one begins at the very beginning of the universe according to the theory of the Big Bang. It is of course the problem of inflationary theory that we mentioned earlier and the phase of rapid growth in the early cosmos. The problem itself is not with inflationary theory as the theory is just that a theory and is still unproven. The problem lies instead with the observation of a homogenous universe which apparently has no large scale anisotropies in it and is known also as the horizon problem. Anisotropy is simply any directionally dependant phenomena and in the universe could be thought of as any large scale structures that differ from region to region. For example if we look in one direction from Earth we might see millions of galaxies clumped together while in the opposite direction the universe is sparse and almost void of matter and this would demonstrate a directional dependence of matter distribution within the observable universe. The Big Bang predicts the existence of these large scale anisotropies and much of the Big Bang theory has been born out by good theory and observation. So if the Big Bang is correct why is the universe so smooth and uniform no matter where you look in the sky? Why is the horizon problem not solved?

The answer is simple and if we think about the way the universe looks to us here on Earth it is uniform and equal in all directions and for us that means approximately 14 billion light years all around us. This is because the universe is 14 billion years old and the maximum amount of time light or any other information could travel in the universe if the speed of light is the limit is 14 billion light years. So far so good and we do not have any problems because the light from a galaxy let us say on our left takes the same amount of time to reach us as a galaxy on our right and both situated at the edge of the observable universe. The horizon problem stems from the fact that the universe and therefore all the information in it is uniform everywhere we look and yet a galaxy on the left of us and the right of us are not 14 billion light years apart from each other but are 28 billion light years apart. What this means is that the information for the left and right galaxies cannot have possible reached each other as the universe is too far apart from each other.

In other words the visible horizon of the left and right galaxies does not over lap their spheres of causality and yet they appear to be uniform. The Big Bang should not allow something like this to happen as you might reasonably expect their local properties such as matter density or distribution to be vastly different from one another. Instead they do not possess statistically significant differences from one another so the universe must have had a mechanism for setting up a uniform sharing of information when it was young that has grown to what we see today. The Big Bang does not predict such observations and yet of course the observations as I have repeatedly said are King and simply never wrong. So inflation theory seeks to solve the horizon problem by rapidly spreading out the entire universe early on in its life so that today the small quantum differences that existed are now spread out evenly in all directions. Inflation theory is a neat way of cleaning up some of the mess left behind by the Big Bang but it does create some more unsolved problems along the way.

First up is the mechanism for the inflation theory to work which is the phase of rapid expansion of the early universe. The inflation theory predicts an inflation field with an inflaton force carrier for the field in order to drive the expansion and fit in with known physics. The main snag with this idea is that according to science fields do not simply disappear and the idea that a field of such power could exist as a completely separate fundamental force and overpower the universe itself to expand could simply vanish as if turned off. Not only turned off but then destroyed so that the field is no longer in our

Standard Model of particle physics. The mathematical descriptions of such a field and such a force carrier are of course only purely hypothetical at best as no direct observations were ever made by humans of the expansion of the universe. Therefore even on a theoretical basis the idea of inflation fields and inflatons is difficult to back up and yet the overall theory does seem to solve a number of nagging problems. This leads to the second part of the early expansion problem of the universe which is if inflation by any means is in fact correct then why did it stop and can it still be going on elsewhere in the universe or was an entirely different mechanism responsible for the rapid early expansion of the universe.

The why did it stop is unknown as we have mentioned before the idea of an inflation field would not simply go away and so should not have lost all strength today. A curious prediction of inflation theory and the early rapid expansion of the universe is that at this point it appears as though gravity flowed in reverse and for a time was a form of anti-gravity that pushed the universe apart. If true then general relativity is again out the window as gravity according to Einstein works only one way and in fairness to relativity its predictions of gravitational forces is well researched. So how then can gravity have worked backwards in the early inflationary universe? Could it therefore be another mechanism altogether that was driving the expansion of the universe? The answer is yes it very well could have as a theory that can solve the horizon and uniformity problems of the universe does not have to exclusively be solved by an inflationary field and inflaton force carrier. A new theory of the rapid expansion of the early universe might also explain whether or not the forces of expansion could still be going on somewhere in the universe today. So for the three views classical will accept any new force pushing the universe apart as will quantum mechanics and of course quantum mechanics will score higher hear as it can incorporate a new inflationary field or inflaton particle should they exist. General relativity as we have seen cannot incorporate a new quantum field or particle as not only does relativity not mesh with quantum mechanics but the large scale predictions of relativity fail under reverse gravity inflationary epochs.

Life of the Universe and Death?

This leads us to a discussion of another problem facing the universe which is what is the cosmos ultimate fate given these discrepancies in rates of expansion that exist and that they seem to have changed over time? The answer to this question is basically the fate of the universe. If all the theories of how the universe began are correct then everything started in one mind-bending explosion and has progressed from there to create the universe we now view today. The starting conditions for the universe appear to be very different than what we observe now and this means going from a super small, dense and hot singularity to a very cool and spread out cosmos. Since the expansion of the universe seems to be accelerating it is likely to get much colder and less dense than it is today. But how far can that go and will some other mechanism take over to stop the expansion which after all is plausible since we know that the acceleration has sped up why can it not slow down? To begin with let us examine what would happen if dark energy were to win out, nothing else emerged to stop it and the time span was infinite.

Essentially you get what has been rather un-originally dubbed the Big Rip. To be fair however all the theories currently accepted by modern science as they pertain to the fate of the universe are derivatives of the original Big Bang moniker. We will look at the Big Rip, the Big Crunch, the Big Freeze and the Big Bounce. For now back to the Big Rip and what it essentially states is that the dark energy currently pushing the universe apart will literally push everything apart. What this means is that the large scale objects we see today are only the beginning and when one looks at a distant galaxy and how it is flying away from us ever faster this same logic can be applied to much smaller objects. To begin with let us disassemble a galaxy in the far future as the dark energy pushes galaxies apart and their stars go flying off into space. Next the solar systems of those stars will break up as they too are torn apart. Now we get down to individual stars and planets which according to the Big Rip theory will also break apart until only atoms are left in the universe. At this point even the atoms themselves and the fundamental particles that they are built from will be pushed apart from one another and none of the four known fundamental forces will ever be stronger than the dark energy again.

The net result is literally everything in the universe being ripped apart from the inside by the outward pressure generated by the dark energy and hence the theoretical fate of the universe from this perspective is called the Big Rip. The Big Rip also includes the ripping apart of spacetime as well and so will truly be the complete destruction of the universe with no mechanism available to recycle itself and so this will be the end of literally everything according to the theory. This has been roughly calculated to happen some 22 billion years into the future with most of the drastic effects taking place with the last few months for solar system collapse and the last few minutes for atomic collapse.

Now let us look at a scenario that is a little less extreme than this but still involves copious amounts of dark energy running amok freely within the universe. This theory is called the Big Freeze and deals with the total energy and work capability of everything within the universe and so in a sense is a kind of ultimate entropy death for the cosmos. Nothing bizarrely strange like atoms ripping themselves to pieces will happen but eventually the universe will have spread itself so thin that stars will become isolated from one another some 100 trillion years or so into the future. This is due to the dark energy pushing everything apart from everything else just as in the previous theory only this time it does not posses the run away energy necessary to literally rip fundamental particles and spacetime apart too. This can be seen as a critical dark energy density to the universe which is likened to the early universe and its critical density phases of being too strong or too weak. What this means is that new star formation from older stars going to supernovae will not occur and therefore a finite amount of time and energy exists in the universe.

As time progresses stars will run out of fuel to burn and therefore stop generating heat. Given enough time the stars will all grow cold and the only thing remaining will be black holes. Given even more time these black holes will run out of energy too as they emit Hawking radiation leaving them to simply fizzle out and disappear. At this point the temperature in the universe has reached thermodynamic equilibrium with all of space and is now a uniform value everywhere. Anytime this occurs not only will energy be at a minimum and entropy will be at a maximum but the ability to do useful work is gone from the universe as well. In other words no chance to restart things from the beginning and all you are left with is a cold, utterly dark and dead universe which has essentially frozen in place and hence gives us the theories name of the Big Freeze.

Remember from the quantum potential well that once everything has reached the lowest possible energy state no more work can be done. This means no work available to make new stars from existing particles to reheat the universe and begin again. In addition to this since the universe is so cold and spread out quantum tunneling will not allow for more work to be done as any particle that manages the low probability of theoretically tunneling out of its potential well is no where near another particle to do work with. Many factors can play into the various Big Freeze scenarios such as protons decaying or not at all, the existence of significant amounts of quantum tunneling on remnant matter or whether or not the vacuum energy remains constant or decays to a lower state. The time scale for this theory varies somewhat as different views on the rate of energy loss exist and via what pathways but ultimately it will take untold trillions of years and beyond that the universe will simply remain cold and dark.

So far we have been examining fates of the universe or at least possible fates of our beloved universe that involve it growing ever larger and eventually dying out through some uncontrolled expansion mechanism. There are two other scenarios that might play out in the distant future and these both are quite the opposite to the idea of an expanding and popping universe. The first we will look at is the Big Crunch and from this name we can rightly conclude that the universe will end up very small again. Basically instead of forever exploding outwards as with the previous two theories it will rush inward on itself and eventually smash back together in the middle. This idea was of course one of the first proposed by modern science as a way that the cosmos might ultimately end and was conceived before the observations were made that the universe is still expanding. This does not mean that it is wrong or that it cannot happen and indeed it is no more impossible for this to happen than for the two unproven expansion theories to happen either. What it all comes down to is the Hubble Constant which measures how fast the universe is expanding.

Named of course for Hubble who discovered that the universe was not static but in fact expanding and whose initial insights into an expanding universe gave science the idea that it must have originated in a sort of central point we now call the Big Bang theory. Coupled with this is the overall density of the universe which is made up of the force of gravity and the amount of matter in the universe. It can be thought of as a sort of pressure which will give you different values based on the density of matter, its distribution in micro and macro scales and how much force gravity can exert on it. Summing all of this up and you get what has been called the critical density of the universe and simply put if the density is higher than this value the force of gravity will eventually slow everything down and reverse the expansion of the universe. If the density is lower than the critical density then the universe will expand forever as gravity will not be strong enough to halt the expansion.

In the Big Crunch theory the density of the universe is high enough to prevent it from expanding forever and what this means is that at some far distant point in the future the universe will slowly grind to a halt. After which it will begin contracting ever so slowly under the force of gravity and as it continues to contract it also accelerates. Towards the end of contraction the matter in the universe will be compressed rather haphazardly into black holes and eventually be travelling at or much faster than the speed of light towards the center of the universe. Even if matter is not travelling faster than light inside the universe the universe itself might be able to far exceed the speed of light as it contracts in a sort of reverse inflation theory fashion. The net result of this is a single black hole containing all matter in the universe which is itself crushed down to a single Big Crunch singularity. This smashing of everything back together into a single point would be the ultimate fate of the universe and the end of the line so to speak.

The next question one might ask is what happens after a Big Crunch? There are two possible answers and the first is that nothing at all happens and the universe dies in the process. This idea must be accepted with as much validity as the idea of the Big Rip where the universe dies after being torn apart and there is not a what next to consider, the universe is simply dead and gone. The second answer to the question is that the universe is somehow reborn with another Big Bang. This idea has had many names over the years starting with simply saying a Big Bang follows a Big Crunch, the Oscillatory theory and now the Big Bounce. I warned you about those uncreative names. Yes the Big Bounce is what some theorists believe will happen at the end of our universe and conversely the beginning of another. A cyclical system that states the universe crunching in on itself might contain the necessary quantum deviations necessary to create a new universe. I say deviations because it is assumed given certain unusual conditions of a Big Crunch that space will change enough to allow things like the speed of light in a vacuum to be altered. In other words the laws of physics can be bent to solve the problem of creating a Big Bounce and there are many ways this can happen but in general things happen in intervals smaller than a Planck time.

The theories all for new quantum effects to take over and basically generate a new explosion complete with inflationary period thusly making a new universe. The ideas behind this are actually somewhat solid as general relativity fails utterly at a singularity where energy becomes infinite and volume becomes zero. This is of course impossible and everyone knows this so the idea that relativity cannot explain what happens at these scales is perfectly correct. The quantum world would then become far more important in dictating what is allowed to happen to the matter, energy and space in the now highly compactified universe. It has been suggested that if quantum gravity is allowed to explore all possible probabilities at these extreme conditions then it can in fact become repulsive and create a pushing effect. This pushing effect is of course very rapid and results in what cosmologists understand today as a Big Bang. The net result is a quantum mechanical explanation for the Big Bounce theory that does away with the limitations of general relativity. Theoretically anyway and given a few degrees of freedom with the known laws of physics as they would exist inside a Big Crunch. Nevertheless the Big Bounce is another of the possible fates of the universe equally valid alongside its counterparts and the beautiful thing about a Big Bounce model is that it explains why there is always a universe present which after all is how we observe things today.

So by examining these four fates of the universe is it possible for the three main views of how the universe functions to cope with any of these possible fates? To start with each of these separate fates is generated by significantly different mechanisms and so produces very different outcomes. The way in which each of the three main views will deal with them however is at best fairly broad spectrum as much of the four theories themselves rely heavily on purely hypothetical or undiscovered laws of physics. Nevertheless as bizarre as some of the fates may seem they are based off of decent assumptions and to some extent fit in neatly with the Standard Model as it is understood today. First up is classical mechanics which is most at home with the Big Freeze and the Big Crunch as these both involve known forces acting on objects well understood by modern science. The gradual cooling of the universe fits neatly with normal thermodynamics and the gradual collapse of the universe fits neatly with Newton's laws of motion as gravity is simply a force acting on matter. Classical mechanics has difficulty however with the Big Rip and the Big Bounce as both of those involve quantum influences at their extremes.

General relativity has the most difficulty with all the theories as it is unable to deal with quantum effects such as those present during the dissociation of all fundamental particles from one another as well as spacetime during the Big Rip. The Big Freeze is probably the most Einstein friendly of the bunch as it involves nothing more than the universe cooling off due to very low temperature with no future work possible due to entropy changes. The Big Crunch results in a singularity at the center of whatever is left of the universe which is of course the death knell of all relativity as it cannot remotely understand singularities or the physics behind any similar type of object. The Big Bounce is also utterly unknown to general relativity as it is purely quantum mechanical effects that would be responsible for the formation of a new Big Bang or inflationary event.

Lastly is quantum mechanics which is the big winner when it comes to dealing with the four possible fates of the universe many billions to trillions of years from now. At the core of the Big Freeze and the Big Rip is dark energy or phantom energy which acts upon the smallest quantum objects slowly overtime. The final results are quantum interactions that eventually cool off the entire universe as with the Big Freeze or more quantum interactions that eventually unravel matter itself theoretically as seen in the Big Rip. The most curious quantum effect and undoubtedly the one that quantum mechanics fails to understand completely as do the other two views as well is the destruction of space itself during the very last moments of the Big Rip. No known mechanism can explain what is happening at that time to space itself as nothing in modern science today understands what space really is. However quantum mechanics does understand what would be necessary to make matter itself dissociate and so has the best chance of the three main views in coming to grips with the full destruction of the universe. As for the two universe contracting theories the Big Crunch and the Big Bounce quantum mechanics has some difficulty at the start but no difficulty at the finish. The start involves not so much a problem for quantum mechanics but rather a problem in unknown forces or energies.

The exact mechanism by which the universe would start to contract is unknown to modern science as is the mechanism that would cause it to stop expanding in the very first place. This is where quantum mechanics runs into some problems as none of the currently accepted forces in the universe can stop the overall expansion. The good news for quantum theorists however is that if a new force or energy were presented it would fit perfectly into quantum mechanics and simply provide the necessary power to reverse the expansion of the universe. From this point of view quantum mechanics has no trouble dealing with the two contracting universe fate theories. At the end of both theories is where quantum mechanics fairs best as the Big Crunch ends in a singularity where general relativity breaks down but quantum mechanics does not. Following on from this quantum mechanics can provide a natural explanation for why the last theory the Big Bounce would even take place. For this reason quantum mechanics is probably the best of the three views currently accepted by modern science to provide a reasonable explanation for the ultimate fate of the universe regardless of which fate is examined.

Absence of Light Voids and Information Loss

These unsolved problems of rates of expansion, homogeneity and fate are the largest scale unsolved problems possible as they involve the entire universe as a whole. There are many other problems that exist within the universe itself that can be treated as more isolated systems. That being said some of these problems are in no way small as they can involve huge objects such as super massive black holes or galaxy cluster collisions. To begin with let us examine black holes for a moment and what they bring to the table of unsolved problems. The first thing they bring is of course themselves as black holes are known to exist but they are not understood as to what they really are. The descriptions of black holes as given by modern science deal more with how they affect the objects in their local neighborhood and less with what a black hole really is. The closest description of a black hole comes from general relativity which describes it as a place in space where volume is zero, gravity is infinite and time stops completely. This is a description of a singularity which is the very heart of a black hole and is the best description mathematically available to modern science for what these objects actually are. It is also totally incorrect.

The reason being is that a singularity cannot exist with the properties described by general relativity and in fact no object can. This is not even a point of contention amongst modern scientists they simply admit that a singularity is a way of saying that they do not understand what is really happening to matter and space under those extreme conditions. This realization that general relativity cannot hope to explain what a black hole is becomes one of the strongest motivators behind research into things like a theory of quantum gravity.

To this end we can begin to understand that relativity has no chance of understanding black holes while quantum mechanics has the best as the black hole at its core will still be made of quantum objects that will obey quantum rules even if those of relativity breakdown. Classical mechanics is somewhere in the middle as it is in between quantum mechanics and general relativity and in fact quantum mechanics and general relativity would both not exist without their classical ancestor. Black holes and the problems facing modern science in terms of reaching a full understanding of them is only one of the unsolved problems in this list of smaller systems. Others that are related to black holes can include things like information paradoxes where the rules of quantum mechanics appear to break down.

One of the fundamental laws of quantum mechanics is that information about physical systems cannot be lost from the universe under any circumstances. This is not to say that information cannot be lost from the universe only that it is a violation of the laws of quantum mechanics and it might be a reality for modern science that information can in fact be lost. It was Stephen Hawking himself who first proposed that information could be lost inside of a black hole. The argument runs something like this; you start with a black hole which is separated from the rest of the universe by its event horizon. From there you have empty space just outside the event horizon which even if totally devoid of matter still possess quantum energy. Quantum space theoretically creates pairs of particles which more or less pop into existence in accordance with quantum probabilities. Normally these pair production particles encounter nothing in their short lives other than each other with whom they promptly undergo mutual annihilation events. The information about these particles has so far not been removed from the universe and nothing has been violated according to quantum mechanics.

Hawking argued that just above an event horizon pair production would still occur but with a slight difference as one of the two particles would be projected into the black holes event horizon where it would become trapped. The second particle would therefore have nothing with which it could annihilate itself and thus follow the normal information retaining quantum mechanical chain of events. In this way a single particle from any pair production event would be lost inside a black hole and therefore take the quantum information it carried with it. This would allow for a possible mechanism of information being lost from the universe which would violate the human constructed laws of quantum mechanics but not obviously be impossible as it occurred naturally in the universe. Recent developments in theoretical physics have hypothesized mathematical pathways by which the information is not lost however they are unproven and without understanding what black holes are completely are most likely false. Simply put not everyone agrees with these new ideas concerning the preservation of information no matter what throughout the universe and until black holes are fully understood or at least much better understood no one can say definitively that information is preserved.

The information paradox is not the only unsolved problem ascribed to black holes and others include black hole radiation which follows from the information paradox. A consequence of the pair production event close to a black holes event horizon is the occurrence of something called Hawking radiation. Hawking radiation refers to the energy given off from a black hole in this process which occurs because of the black holes gravitation just above the event horizon. A simplified version of this process involves the standard pair production of quantum particles we have seen with the exception that the gravity of the black hole itself is responsible for the creation of the pair. The fact that the black hole itself uses its own gravitational energy to create a pair of particles is why the black hole emits radiation in the form of the one particle that survives not falling into the event horizon and obviously not annihilating with its partner. The unformed particles which did not yet exist in our universe are brought into existence as real particles through the black holes gravitational field perturbing the vacuum. The effect is strong enough to cause the vacuum to create two new particles one of which is absorbed by the black hole and one which is lost from it. So in this way a very, very small amount of energy is radiated away from the black hole as a form of black body radiation.

The problem here is that the black hole must actually lose some mass in the process in order to obey the laws of conservation of energy. One might think that the black hole should gain mass as it absorbs one of the particles while the other flies off into space but this is not so. The particles were brought into reality by the gravitational energy of the black hole and so we can think of this as two units of energy spent by the black hole in order to make a pair of particles. The pair of particles does not annihilate which would return the energy to the black hole and instead one is lost from the system. So although the black hole does absorb one particle or one unit of energy spent the black hole also loses one particle or one energy spent. The net result is that the black hole only gains back half the units of energy spent needed to create the pair of particles and so has lost energy which can be thought of as mass. In this way the black hole has done two things the first of which is emit radiation and the second is to lose mass in the process. What we see here is a new problem arising which is the information paradox and it could have been discussed before or after the radiation as the two are interlinked together and so we simply picked one to talk about before the other.

The reason why information seems to be lost is that the radiation can be thought of in two ways the first being the traditional method of thermal radiation where information about the body emitting the radiation is contained within it. The second way is the Hawking radiation which does not contain this information but has the advantage of better suiting the data at hand. So do we trust the less well fit traditional method or the better fitting Hawking method? Any time a new theory comes along that better fits the data it is usually adopted as quantum mechanics were over classical mechanics and so like it or not the explanation put forth by Hawking seems to be the one to use. In this way information can be lost from the universe inside a black hole and to make matters worse there is more than one way that a black hole can be theorized to lose mass.

Another method by which the black hole can be seen to lose mass as emitting radiation is by strong quantum effects near the event horizon and resembles those believed to be at work in quantum tunneling events. Quantum tunneling as you might recall is the low probability that a particle can exist outside of an area where it could normally not travel. The tricky bit about quantum tunneling is that the exact mechanism is not known at all to modern science as it exists only within the mathematics and the process by which a solid particle can move through things that it should not be able to or exist seemingly in two places at once is impossible in reality. Although the mechanism by which quantum tunneling appears to take effect is unknown to modern science the fact that it seems to occur in certain conditions is undeniable. The fact that it only seems to be a real effect might again be an error in observation or some unknown mechanism that modern science does not know. The event horizon of a black hole is one such place where pair production of particles is thought to occur just below the surface of the event horizon in an area where normally nothing is thought to be able to escape. What happens here is one of the particles falls into the black hole and is lost to the singularity while the other through quantum tunneling is able to escape the event horizon of the black hole and be radiated outward into space. Again we have the black hole creating two particles but only retaining one which can lead to the black hole both emitting radiation but also losing mass in the process.

One last process by which a black hole is thought to lose mass while emitting radiation is through the energies associated with the quantum particles produced just beyond the event horizon. If the black hole creates two particles from the vacuum and one escapes and one is trapped the type of energy the particles carry with them can affect the black hole in terms of its overall mass. If the particles can annihilate with each other this means they must be opposites with one particle containing normal energy and the other containing negative energy. If the negative energy particle is lost to the black hole it will decrease the black holes overall mass and thus allow the black hole to once again emit radiation while losing mass. No matter the scenario it seems that black holes given the right conditions will emit some form of radiation and lose some mass in the process. The unsolved problem of black holes and radiation here stems from the fact that no one knows which of these scenarios the correct one is or if any of them are correct at all.

At the core of this conundrum is the reality that scientists know that black holes or at least something akin to them exists but not at all how they are made or what they are made of. It is hoped that the identification of black hole radiation directly will shed some light on what they are doing behind the closed doors of their event horizons. The theoretical models tell us of another fly in the ointment when it comes to black holes and black hole radiation. That fly is the fact that despite black holes apparent loss of mass the black hole will only lose mass at an incredibly slow rate and it will take untold ages for one to finally evaporate. Any normal sized black hole made from a star of around ten or more solar masses will actually not evaporate as it loses mass very slowly. It turns out in fact that the smaller the black hole the hotter it should be and the faster it should evaporate. So from this logic a black hole of normal size will evaporate slowly and be relatively cool meaning that it is not hot enough to be emitting that much radiation. The more radiation it emits the more heat it has and therefore the more mass it is losing.

The trouble is a black hole of normal size is so cold only a few tens of nano-Kelvins or so that it will be absorbing far more radiation than it will ever emit from the free energy of the Cosmic Microwave Background radiation. So the black hole is in a sense growing ever bigger and the possibility of seeing any of the black hole radiation emitted drops to near zero even with the best equipment. If you could make a black hole out of a mass of something like a small planet, say perhaps the moon it would have a temperature roughly equal to the Cosmic Microwave Background radiation allowing it to absorb as much energy as it emits and therefore keeps its mass stable. A much smaller black hole and these are mostly hypothetical primordial black holes are so small and microscopic that they will be able to evaporate as they will give off more radiation than they absorb from the Cosmic Microwave Background. So now we have the troubles of not only understanding what a black hole really is but whether or not it is losing mass via emitted radiation. All of these factors must then be weighed on whether or not information can be lost from the universe inside a black hole and does this change given their different sizes, temperatures and behaviors?

Here we see a definite problem arise in the form of information loss inside black holes as it seems to violate the principle of unitarity. Unitarity is a quantum mathematical statement which says that the sum of all possible outcomes of an event is one. The reason for this is that quantum mechanics is based on probabilities and any event that is guaranteed to occur is given a 1:1 probability of succeeding. In other words it will occur one time out of every one time it happens or always 100% of the time. Since the act making an observation reveals the systems true state at any given time whatever you observe is the correct outcome that had to happen so it happened 1:1 times. Behind this 1:1 outcome are all the possible states the quantum system could have possessed prior to the observation being made and so if you sum up all of the near infinite possibilities for any given quantum system it must add up to the 1:1 you did observe. In quantum mechanics the information for a system is inside the wave function describing that system and the wave function is unitary as it evolves over time. Since it is unitary it must contain all information so that it can sum to 1:1 at the end observation but a black hole seems to violate this. If information could be lost into a black hole this means that some of the information inside the wave function used to describe a physical quantum object would be lost. A loss of information in the wave function means a loss of unitarity and so a loss of the 1:1 so desired by quantum mathematics. This violation of beloved quantum beliefs is the source of not only great concern but also one of the unsolved problems of the universe and just one of the unsolved problems relating to black holes.

So how do the three main views of science stack up against this type of problem? Well none too good really as each is unable to deal with the loss of information at their own levels. Starting with classical mechanics we see that a particular set of information describing a system usually deals with forces and changes in the behavior of the object as a result. When it comes to black holes what forces are acting on the object besides gravity are unknown and once the object passes into the event horizon the forces acting on it conceal what they in fact do to the object and so describing the details of a system in classical terms is difficult at best inside a black hole. The net result is classical mechanics might be able to deal with a loss of information best of the three if it knew what was happening to the object in terms of the forces acting on it inside the black hole.

Quantum mechanics on the other hand can deal with information that classical mechanics cannot such as quantum properties like spin and charge. To this end quantum mechanics has the best chance of describing what is happening to the object in terms of information but once the object passes into the black hole quantum mechanics fails. The reason is that one of the base principles of quantum mechanics is that no matter what you do to a system you can back track the events if you have all the information. So no matter how jumbled up something might be quantum mechanics dictates that it can always figure out where the order of events came from and how they unfolded. This is of course assuming that quantum mechanics is in fact correct given any situation and nothing says this is the case when dealing with the new type of physics that exists within a black hole. If it is possible for information to be lost or at least diminished from the universe then quantum mechanics has no hope of answering this problem whatsoever as it is simply not built to do so.

This leaves relativity which has the least chance of explaining what is going on for two reasons the first being that relativity tends not to deal with ultra small fundamental particles at all. It is true that there is much information that relativity can use to describe objects as we have seen with things like the light bending in a high gravitational field. The trouble is general relativity never says why this happens only that it does and it can calculate how much. The second trouble with relativity and the black hole information paradox comes from the fact that relativity itself fails one hundred percent inside a black hole and so can only work outside of it in regular space. This means that information about the system can be garnered from anything that happens above or outside the event horizon but once the particle travels past this barrier and into the black hole itself relativity fails completely. The thing of it is that the information loss and whatever is happening to the particle that might reveal any answers does not occur above the event horizon but below it and so we see that relativity stands no chance of answering this problem either. So all in all with the nature of black holes and not only how they really do work and what they are made of still eluding modern science the question of whether or not information can be lost within a black hole is still unknown for sure.

While we are thinking about the three main views of science and which fits these problems best let us look for a moment at where this information paradox comes from. It comes from the presupposition made by quantum mechanics that all information must be preserved in order to maintain unitarity of the wave function as it evolves over time. So at the heart of this is the idea of the quantum mathematical wave function which describes all possible states that a system can exist in at any given moment. The exact state of the system or in other words its exact reality is not known until it has been measured. The act of measuring therefore collapses the wave function and reveals which of the probabilities was in fact the correct one. Seems simple enough except for one small problem with huge implications which is what really constitutes a measurement anyway?

Observing Quantum Systems and Implications

Traditionally any quantum system must be isolated in order to study it properly and this comes back to information. The nature of a quantum system contains all the information necessary to describe what is happening to a particular object at a particular time. What is happening to the object also deals with absolutely all of the known properties of said object and this is everything that classical mechanics and general relativity do not possess. Things of this type are spin, charge and the like which may be influenced on some scale by classical laws but classical mechanics do not contain these variables. From here we can see that the most base level interactions of the most fundamental particles are being described and in describing them we use quanta which are thought to be discrete jumps of whatever we are talking about. The idea of discrete jumps or packets of information is key to making any sort of calculation in quantum mechanics as it denotes the smallest and indivisible changes to the system. All of what modern science knows about the universe and every piece of matter and energy within it obey these tiny little jumps. To this end any sort of measurement must in a sense investigate what state the quantum system is in at a particular time and in order to do this some sort of interaction must be made with the system. This means any sort of interaction as an outside observer and for this we will say a scientist conducting an experiment must collect data or information in order to see what is happening in the system. So in order for the scientist to learn about what is going on with the system some information must come from the system in order for the scientist to collect data and make observations.

In other words something must quantumly interact with the scientists instruments from the system being studied in order to make an observation. This on the surface does not seem to be such a big problem as scientists routinely interact with all sorts of experiments on all sorts of scales to collect data. Think for a moment of a scientist simply sticking a thermometer into a solution to read the temperature and we have an example of an instrument interacting with the system in order to learn about it. The thing is this is fine for a large scale system where many particles are involved and the energies are so large that the small fluctuations caused by the thermometers own temperature is negligible when compared to the overall temperature of the system. Yet no matter what the temperature of the thermometer will not be perfectly calibrated to the solution and how could it be the scientists needed the thermometer in order to check the systems temperature anyway. So in this way the scientists are changing the nature of the system and the information within it only in this case the system is a macro system where the change is tiny and can be ignored. The fact of the matter is that a change does happen though and this is seen in the laws of thermodynamics where we have seen that the two objects the system and the thermometer will now reach equilibrium with each other and as such alter the system.

On a quantum scale which is a very much so micro system the interaction of anything with the system can have huge consequences. A thermometer is made of countless quantum objects and so the behavior of them can be calculated as an average that perfectly obeys classical mechanics. On a quantum scale the average cannot be used and instead only absolute values may be taken as each value in and of itself is in fact quantized. Here the information of the system is greatly affected by any sort of interaction with an instrument as the interaction must involve direct quantum contact if you will of the two objects being the system and the instrument.

How this causes a problem for observation is simple as the nature of the system is unknown and therefore something we want to learn about. The nature of the instrument is known as it must be in order for it to be useful to us. So if we combine the known properties of the instrument with the unknown system we actually bias the outcome as we disrupt the wave equation by destroying some of the possibilities within it. This means the quantum system was not allowed to evolve over time freely as it wanted to but was stopped at some point by the scientist who collapsed the wave function. In this way it is difficult to know what is truly happening in the system and therefore truly know the behavior of quantum objects and this is a puzzle that has existed for over a hundred years in modern science. So we are left with a problem of what makes an observation an observation anyway?

An observation might be from an instrument that is used to study a system and this instrument is of course inanimate in nature and so does not create a reality of the universe through perception as has been argued in the past. The observation could be made of a single external quantum particle or object that was used to probe the system and is probably the least intrusive interaction that could be made. The observation could also be made by an intelligent creature such as Schrodinger or his cat both of which are living breathing thinking animals that can observe the world around them and therefore take note of quantum systems. The system being studied however does not give up its secrets easily and no matter which of these observations are used appears to give different results depending on whether or not we look at the system so to speak. According to quantum mechanics objects can seemingly do strange things that defy common sense such as be in two places at once or two states at once. The wave function does not provide an answer as to which is which until it collapses through observation. This act of observation seems to affect the results of whatever experiment is being conducted and has caused many paradoxes to become apparent.

The biggest paradox of course has been described as Schrodinger's Cat which we have discussed before and sits at the core of this unsolved problem. The Copenhagen interpretation of Bohr and Heisenberg does not sit well with many and is known to create questions about quantum mechanics such as the Schrodinger's Cat paradox which was an attempt to illustrate this observation problem. To this day the very nature of wave function collapse and wave-particle duality remain as unsolved problems for modern science. The trouble with this problem is that it covers a great many things when it comes to the behavior of matter at the fundamental level. First up of course is the true nature of matter which is a particle, wave or both and each of these concepts has received much scrutiny over the decades. At some points in both experiments and mathematics matter is better suited to behaving like a particle or at least it is easier to describe it that way given the conditions of a certain system. Of course the same can be said to be true for matter as a wave in certain conditions and as a wave-particle.

What is obviously missing from modern science and specifically quantum mechanics is a better understanding of what is really happening down at those tiny microscopic fundamental levels. Matter is doing things that seem strange and yet these odd behaviors do not manifest themselves for everyday real world objects of the macro variety. Many have argued that quantum effects should appear at macro scales even if the probabilities are extraordinarily low as this would allow the observations made of micro scale objects to remain valid. That being said people for example or cats do not exhibit things like superposition and so some have postulated that this is the true nature of matter and it is the quantum mechanics that has got it wrong.

The apparent collapse of the wave function demonstrates which state the superposition was in and gives us the information of the system but only after we observe it. This is of course impossible as everyone knows as the object being studied really only was in one position it only appeared to be in many due to the way quantum equations are calculated. So here we have the problem full in front of us as matter really is only behaving normally but modern science lacks the tools or equations to see it. This lack of understanding goes beyond the already existing problems of observation, wave-particle duality and superposition to include things like quantum entanglement as well. It is known that two objects can become entangled so as to share information with each other even when separated by a significant distance. The distance is such that no information exchange should normally be allowed to take place between them and yet information exchange is occurring. When a measurement of one particle is made and its true state revealed the state of the second particle is revealed simultaneously. Even though the true state of both was not known before and the two particles are dependent of one another for the behaviors of their observed states. The information is not travelling at the speed of light the information is travelling infinitely fast or very close to it in order to be revealed at the literal exact same time. This has been proven time and time again which brings up the problems of how the particles know what the other is up to and how they can share information so quickly.

So what is left after looking at the various aspects of quantum mechanics is a greater understanding of what is not understood by modern science with respect to the most base level interactions in the universe. This confusion of the wave-particle duality, wave function collapse, superposition and quantum entanglement to name just a few gives us another unsolved problem of the universe. A brief glance at how the three main views might deal with this is all that is needed as there is so much quantum ground to cover. Looking at classical mechanics we see it offers no explanation directly as to what is happening only that the objects in question are acting as though they are being affected by a force which would be able to be explained through classical and quantum means. Unfortunately classical mechanics does not have the ability to tell us what that new force is only that if it is real then it can behave normally with everything else in the universe as nothing in classical mechanics is limited by the speed of light

Quantum mechanics fairs oddly the best and the worst at explaining what is going on with these unknown phenomena as they belong firmly in the realm of the quantum world. Obviously whatever is happening to the objects is occurring on a quantum scale as the interactions are seemingly of the smallest order and take place in set or quantized amounts. Any new insights into the nature of these interactions will no doubt be described at some point in quantum terms and therefore can be explained using quantum and some classical mechanics. However quantum mechanics also fails the worst in explaining these strange effects as it is quantum mechanics itself that seems to have generated many of them. In fairness some of the effects appear to be simply a part of the cosmos that quantum mechanics is just observing passively and that is fine but some appear to have been created by quantum mechanics itself.

For example we have the problem of the wave function collapse which describes real solid objects found in the universe but in a human mathematical sense so the problem may not be with the object but the way humans are trying to map it through equations which would make this problem one created by quantum mechanics. Lastly we have general relativity which cannot hope to explain any small scale problems at all, however this does not really matter as something fundamental is missing from modern sciences understanding of how these phenomena truly manifest themselves in the universe. The problems that remain unsolved in quantum mechanics so behoove modern science and anything that the three main views can comprehend that it appears as though something unknown and new must be needed to end these questions.

Baryon Asymmetry

Now from the very, very small to the very small and very large all at the same time we find the next truly unsolved problem in modern science today which is of course Baryon asymmetry. This problem is one that many scientists both love and hate as it seems to be something relatively tangible to humans. They of course hate it because it is not understood and modern science has no models or equations with which to explain it but they also love it because experiments concerning it can actually be carried out in small scale here on Earth. This is finally something exciting for the experimentalists to sink their teeth into as unlike large scale phenomena such as the inflation of the universe baryons can be studied everyday here in the laboratory. It is baryons which make up all of us and everything around us after all and consist of the familiar proton and neutron and the like. Comprised of three quarks in various numbers and with various things like flavor or color we have seen them before and know that they make up normal visible matter throughout the universe. We have studied them for decades and centuries directly and been aware of them at some level for countless millennia and so why do they create a problem for modern day scientists? The reason is simple we find only one kind of them when we should find two or at least be able to see a lot more of the second kind.

This second kind is of course the anti-baryon which would make up the bulk of what we think of as anti-matter. Anti-matter can be made of more things than just baryons of course such as a positron which is the antiparticle to an electron which is a lepton itself. The reason it is called a baryon asymmetry by some is because the weight and therefore mass contribution to the universe from things like protons and neutrons is thousands of times greater than that of say the electron. This is also where we get the idea of the problem being both big and small at the same time as it is small due to the fact that we are dealing with tiny subatomic particles but big because they fill the entire universe. In addition to this no corner of the universe is truly devoid of matter as everywhere we look we see more of it with interstellar and intergalactic mediums between them filled with gas, dust and other small objects all made of baryons and such. It is also a big problem because from the very beginning of the universe the fate of the cosmos as we see it today was determined.

This refers to the idea that during the Big Bang and the short time span afterwards when matter and energy popped into existence the baryogenesis that occurred should have according to currently accepted theories created equal parts matter and anti-matter during that time. If this were so then the early hot and very, very dense universe would have provided ample opportunities for baryons and anti-baryons to meet up and annihilate each other destroying them both. This would lead to a veritable sea of nothing but high energy photons which over the billions of years would red-shift some of their energy to lower energy and longer wavelength photons. This would create a universe of pure energy and no matter which is obviously not the case today for two reasons the first being when we look at the cosmos we do not see only energy and the second being of course is that we are there to not see it. We are made of baryons and so they must have survived this theoretical period of annihilation or we would not be here to ponder why it did not happen in the first place.

To date there is no currently accepted model to explain why the universe ended up with far more regular baryons than it did anti-baryons but a few oddities of such a problem have been identified. The first is where is all the anti-matter? This may sound straight forward and it is for good reason which is namely the fact that the equations predict equal amounts of matter and anti-matter. Since the equations which may or may not be right say there should be anti-matter we should at least assume they are right in order to prove them wrong and to this end look for anti-matter. Some scientists are doing just that and have postulated that entire anti-matter objects should physically exist in the universe and not just little grains of dust but whole stars or galaxies made of pure anti-matter.

The universe is big and we are still in the infancy of exploring it as we have yet to leave our planet and journey out among the stars to examine them up close so the idea that whole anti-matter galaxies could be out there is actually not so far fetched. The thing of it is that anti-matter behaves identically to regular matter and so a galaxy comprised solely of anti-matter baryons and what not would produce the same gravitational effects or red-shifting of its photons that we would see from a normal galaxy. In other words we could in theory be staring at anti-matter galaxies all around us and not even know it and in fact we might be the only matter galaxy in the entire universe of anti-matter galaxies. Perhaps other species' astronomers are looking through the universe right now trying to prove we exist. Even if they are not and even if the most probable answer is the correct one which of course that the entire universe is populated with a majority of regular matter it does leave room for large areas of anti-matter.

During the theoretical inflationary period quantum fluctuations grew in size faster than the speed of light and were spread across and throughout the universe beyond our observable horizon. This means that if a quantum fluctuation occurred where anti-matter was dominant over matter it could have after the inflationary period ended created a whole part of the universe made from pure anti-matter. What scientists are looking for here on Earth to prove something even remotely similar to this happened is by looking for areas of high gamma ray production. The annihilation of matter and anti-matter inevitably produces high energy short wavelength gamma rays. The goal is to see if we can find areas of high gamma ray production where no known sources of gamma ray production exist.

Think of active galactic nuclei or super massive black holes which are thought to produce anti-matter relativistic jets from their poles. The anti-matter flies away from the black holes and annihilates with regular matter in the interstellar or intergalactic mediums producing what we detect as gamma rays. This is the sort of cosmic interaction astronomers are looking for somewhere in the universe. You might expect to see unusually high levels of gamma rays in otherwise average areas of space. They might be found at the edge of the universe where one area of cosmic inflation from our own matter rich area of the universe mixes with another part that was filled with anti-matter and separated by cosmic inflation billions of years ago. This sort of occurrence might be possible and the reason it has yet to be detected is that the light simply has not had enough time to travel all the way to Earth yet. If for example the universe is older than 13.8 billion years or the light is coming from an area of the universe removed from us farther than 13.8 billion light years the photons produced in annihilation events simply are not old enough to have reached us.

Other possible theories for the matter over anti-matter problem involve matter and anti-matter being created equally in the Big Bang quite normally. What happens afterwards is also fairly normal as it involves the anti-matter and matter decaying at different rates such that the matter wins out. In other words the matter takes longer to decay than the anti-matter so that the dominance of regular matter is assured simply by it not having to annihilate with as much anti-matter. Since the anti-matter decays faster it simply goes away before significant or catastrophic levels of annihilation can occur. This would leave a very small fraction of the matter in the universe as regular matter and therefore no matter how small this might be it would provide for the stars, galaxies and visible universe we see today. It seems to make some level of immediate sense as not everything in the universe progresses at the same rate of change. Take for instance normal heavy elements around the uranium range and especially higher and look at their respective half-lives. While the graph for plotting elemental weight and half life decay time is not a smooth one it does not have to be it only has to be different. What this shows is that any element or any isotope of an element will possess a different half life from the next including isotopes of itself. This causes the test samples to decay at different rates and thus produce very different half lives on the order of fractions of seconds to millions of years or more.

The decay process that varies in normal elements might exist also in things like matter and anti-matter even at the fundamental level so that while most of the CPT symmetry of the particles and their antiparticles is preserved slight delays in decay can be real. This would produce large scale effects when taken into account for all matter created at the time of the Big Bang and the entire size of the universe. CPT symmetry refers to what is known as charge, parity and time symmetry of a particle. This is a theory which states that any particle should be identical to its mirror image if transposed through any of the CPT variables. The concept is simple if we make a mirror image of our universe everything should evolve the same as it has for us here and in order for that to happen CPT symmetry cannot be broken. The particles of that mirror universe would be identical if they were charge reversed or in other words made up of anti-matter. If they were physically swapped in terms of spatial coordinates say left for right their parity would be preserved and reversal of momentum or time would also not affect them. Commonly known as CP symmetry as time is almost always involved with them it was noted in 1964 that this truism of theory did not always hold.

The two American scientists James Cronin and Val Logsdon Fitch demonstrated that small subatomic particles called kaons and specifically neutral kaons violated CP symmetry. What they demonstrated was that the decay of these particles and the subsequent reaction involved were not bound by the time parameter. Meaning that if the experiment or reaction could be run in reverse you would not end up with the same point from which you started the original experiment and thus it was time invariant and therefore broke CP symmetry. The pair were awarded the Nobel Prize in 1980 and since that time newer experiments have backed up there groundbreaking findings. Before this time it was a long held belief in physics that the universe and everything in it was ruled by symmetrical values this proof that the symmetry of the universe did not hold meant that a reworking and rethinking of the basic way in which the universe works was in order. The problem remains to this day why the CP violation occurs commonly in the weak force but not the strong nuclear force. Basically after the Big Bang if the universe conserved CP symmetry the matter and anti-matter created should have annihilated perfectly with each other but since it did not some other events, laws of physics or something more interesting must have happened in order for the basic laws of the universe to not apply equally to matter and anti-matter. Why any of this happened in literally the very first place after the formation of the universe is one of the unsolved problems still facing modern science.

Now let us see quickly how the three main views will stack up against it. First is classical mechanics which although it deals with the very small and the very large at the same time does not incorporate things like decay traditionally which is seen in the weak CP violation. If a force is responsible for causing some CP violation and not others then a possible classical explanation would be possible using the strength of the force acting on the objects to allow for breaking symmetry in some cases but not in all. Classical mechanics however does not deal with the idea of mirror universes and so most likely will not be able to provide a full explanation. General relativity deals with the larger things but not at all with the smaller things and since relativistic effects rarely play out on quantum scales it is impossible for anything offered by general relativity to explain this anomaly. Quantum mechanics is the winner here although it cannot truly win. What this means is that despite quantum mechanics offering a decent explanation for mirror properties and can be used to experimentally show that the weak force is susceptible to CP violation it has no answer for why the strong force does not cause CP violation. This is an inherent problem with quantum mechanics and the Standard Model which along with quantum chromodynamics as the outcomes of behaviors of matter within them both allow for some but not all CP violation to occur. Nevertheless quantum mechanics and the like offer the most in depth explanation of the objects in question available to modern science and so it stands to reason that if more information can be given or if certain fine tuning problems concerning the theories can be overcome then it stands the best chance of incorporating CP violations.

All of this of course stems from the fact that the direct cause of the amount of normal baryons outweighing the amount of anti-baryons is still not accounted for. The Big Bang has produced a universe which is real and we can touch, see and hear and so on rather than one where all matter has annihilated and only energy pervades the cosmos. The fact that matter according to modern theory should have been created in equal parts yet does not appear to have sustained itself in equal parts puts the idea of baryogenesis during and shortly after the Big Bang into question. What if the currently accepted theories of matter creation are in fact all wrong? After all nothing says they have to be right in the first place and this would allow the universe to create whatever kinds and amounts of matter or anti-matter it desired. In fact the universe did create exactly what it wanted and in the quantities that it wanted meaning once again we have to get the models right and stop trying to get the universe to fit the equations. To this end a look at what went on early in the first Big Bang moments and the creation of matter from the three views shows that general relativity fails again, classical mechanics is a solid second and the winner is quantum mechanics.

General relativity breaks down under any sort of condition resembling those of the early universe and is as such useless for early examinations of the matter over anti-matter problem. Classical mechanics describes how objects interact with each other or to externally applied forces and since the exact forces during the Big Bang are unknown to modern science it might be possible that classical mechanics can offer a solution. This solution will be seen at the fundamental level but if an unknown force existed during the early moments of the universe such as the inflationary force then perhaps it could create simple classical forces acting on early matter such that it simply favors matter. While a true unknown at this point as the inflationary force has yet to be proven if ever it is still worthy of contemplation.

The quantum world does not understand what happened at this early stage of the universe either as the extreme conditions present cause normal quantum interactions to behave very differently or not at all. Remember the quark-gluon plasma and how it is dominated solely by the strong nuclear force on scales far exceeding today's atomic nuclei. Yet if a better understanding of quantum forces under extreme conditions or the quantum inclusion of undiscovered forces like the temporary inflationary force it might like the classical mechanics approach offer a simple cause and effect for why matter dominates the universe today. Does the rate at which CP violation takes place change under extreme primordial conditions such that it favors matter over anti-matter exponentially more? This is still obviously one of the key unsolved problems facing modern science.

Number of Dimensions

The next unsolved problem facing modern science is something that has at its core a very empirical and somewhat philosophical as well as scientific beginning and it is how many dimensions are there in the universe? Like all of the biggest problems they sometimes begin from the smallest questions and this one is no different than the dreaded question faced by many parents. This question of course is posed by a child usually of a curious mind that upon seeing something unfold before them simply asks why? The parent for instance might be looking at the newly fallen apple in the families backyard which has just honored Newton by falling to the ground as everyone knows it will. The parent is now also being faced with their child who looks to them for answers to everything and tells the child the apple fell because it was ripe and ready to fall. The child says but why did it fall down and not up or sideways? The parent dutifully explains that the apple fell because of a magical invisible force called gravity which the Earth has. The child undaunted and armed with a never ending supply of why powered ammunition asks again why? The parent says because the Earth is a big mass and little masses like the apple are attracted to it and so the apple fell down. Once more and this time for the finishing blow the child asks why? The parent unless they feel like lying to their own offspring will now say because it just does and that will be that.

In truth however we are all faced with the reality that the parent also does not know why the apple fell as no one in modern science knows what gravity is, how it works and where it really comes from and so it is impossible for the parent to know why either. Perhaps the parent should go up to the scientists and in the same monotone but innocent voice ask why? Gravity as we can see from this universal example is a familiar quandary of the scientific world as it is much studied and measured but little understood. Another problem that is thought to be well understood is the idea of dimensions in the universe and just as gravity creates a huge discussion from a simple little why? So does the number of dimensions spring forth as a preposterously large problem from a very tiny and simple question.

Here we are faced with an old but new problem as in the olden days of centuries to thousands of years ago it was reasoned that three dimensions existed. This is of course an empirical measurement that can be made using your own senses and from everyday to day perspective makes perfect sense. Our senses are keen and honed over billions of years of evolution to scrutinize the world and universe around us to our advantage. Yet our senses would never be aware on their own of things like atoms, protons, neutrons and the like yet they not only reveal themselves to us everyday but we ourselves and our senses are actually powered by them. To this end we must be critical of everything we encounter in the world as much is not as it first seems and dimensions are fast becoming one of those things.

More accurately and like it or not the massive problem of the number of dimensions in the universe is here already and no clear answer is at hand to modern science. Traditionally the answer is three with length, width and height being the common names used to describe them and perhaps if you are more mathematically bent you will say the X, Y and Z axis. The ancients of numerous cultures world wide independently came to the conclusion that the world was built on three dimensions. Much architecture, mathematics and even science was built off of this simple assumption. Some time later a fourth dimension was added and no one can truly say when as vagaries abound when listing all the times time was mentioned as a thing so to speak. In general however time can be seen as the fourth dimension as the dimension through which objects with three dimensions move or persist in. If you are new to physics and especially theoretical physics you may have heard the term 3+1 space and wondered what this could refer to.

What this means is simply 3 spatial dimensions and 1 temporal dimension used to describe a system that exists anywhere at any scale in the universe. This addition of time is the newest apparently non-physical dimension that humans have commonly added to theories concerning the universe. It is said to be non-physical as it does not refer to size or shape but rather as has been stated for centuries a dimension through which things with size and shape move or persist. This is a very important distinction to make as it raises the questions of why does the universe have three times as many spatial dimensions and if it can have more than one spatial dimension can it have more than one temporal dimension?

The answer to the first question is that maybe the universe does not have three times as many spatial dimensions as temporal ones as it might be that it has in fact zero temporal dimensions. The idea of zero temporal dimensions means that time is not real and only a figment of the intelligent mind and one that is used as a measurement tool to track the progress of events. This argument we have previously examined and in some sense makes a lot of sense as everything else in the universe can be reversed except for time. Why would the universe create literally everything that works both ways forwards and backwards like particles, forces and every combination of reaction imaginable but leave time as a special separate entity that is exempt from these rules? The answer is it probably did not and the most likely explanation is that time is not real and only something humans use to explain things like causality just the way humans use entropy to describe things but there is no entropy dimension either.

Causality is also known more commonly as cause and effect and refers to the link between one process and another such that the second is brought about by means of the first. Examples of causality exist everywhere in the universe at all times and are not confined to specific fields such as physics alone. Take for instance the idea of creating energy and you have a reaction process whereby the mixing or interacting of two chemicals creates a product. The cause and effect here is the product being a result of the reactants and there is a before and after which is described as taking place over time but does not involve time in the reactants or products. This is a chemistry example of causality that is mediated by physics causality of interaction of quantum systems through some classical means. If that seems complicated here is a more straight forward example. You see someone eating a delicious juicy cheeseburger and drinking an ice cold beer and this causes you to get hungry and jealous and also you have now decided what to have for dinner.

We talked earlier about all intelligent creatures understanding time in order to make sense of the universe around them and we used the idea of lions on the hunt calculating distances, speeds and probabilities of success and therefore proving they understand time as well as humans do so time is a universal measuring tool. Now if we look at causality and lions we find that they inevitably get hungry with the passage of time and even more causality as they choose to hunt and eat. So we have biology working with causality too as lions have a cause and effect of getting hungry and hunting which results in physics and chemistry as they burn food for energy like our reactants and products. In other words causality is everywhere in the universe showing us that things happen in an order that has a before and an after which tends to indicate that time is simply not real. Sure the cause and effect can be measured to have progressed from one instance to another but at no point is time involved in it.

The idea of extra dimensions of time has been considered but would always raise the question of causality being satisfied and since the entire universe seems to use it and has since its creation some 13.8 billion years ago it seems highly unlikely that extra time dimensions exist. In addition to this even the concept of extra time dimensions exists only theoretically and with a massive number of unproven assumptions about how the universe works. So the idea that there are more than the 3+1 space dimensions in terms of a +2 for the temporal one seems to be out at the moment and likely never to occur. This is of course further disproven at least theoretically by the fact that many feel time does not at all exist so you are left with a 3+0 space dimension system. Perhaps other dimensions exist but spatially instead of temporally and theories of this type abound.

The simplest extra dimension to add to the universe is one more dimension to space itself and would result in there being a 4+1 number to the cosmos. The temporal dimension remains unchanged of course and the three dimensions in which objects are made and move also remain the same. The basic idea of the fourth dimension is that space itself is warped into a kind of folded four dimensional structure. This model is called the RS model or Randall-Sundrum model after its contributors and is akin to the ADD model or Arkani-Hamed, Dimopoulos and Dvali model named for its contributors. Both models deal with using extra dimension beyond the 3+1 in order to do one thing which is solve the hierarchy problem. The hierarchy problem refers to modern sciences failure to explain why there is such a large discrepancy in the strengths of the fundamental forces. Specifically why is gravity so many orders of magnitude weaker than the other three?

To this end we will look briefly at both of these two theories to see how they deal with the problem. First is the RS model which uses branes as a way of explaining energy discrepancies between various particles and fields. Simply put there are two branes in the universe according to this theory one is the gravitybrane and the other the weakbrane. Normal everyday particles and objects like those belonging to the Standard Model exist in the weakbrane in a 3+1 space and obey normal laws and exhibit normal energies. As you probably guessed the gravitybrane contains the extra dimension of 4+1 space and is the home of the theoretical graviton. The gravity brane has positive energy and the weakbrane negative energy which causes the warping of space to create a 4+1 system according to the theory. The difference in energy scales between the branes and the effort it takes the 4+1 dimensional graviton to travel to and interact with our 3+1 universe creates the imbalance seen in the strengths of the fundamental forces according to the RS theory.

Next up is the ADD model which also uses branes to solve the hierarchy problem and attempt to understand gravities interaction with Standard Model particles. This theory involves normal matter as we encounter everyday and the forces that bind them to also exist in a 3+1 dimensional space but that gravity can exist in more dimensions. In fact gravity according to this theory does not have to be bound to only one extra dimension and can exist outside of our 3+1 space in multiple extra dimensions. The reason this was adopted was extra dimensions allow for some theories of gravity being freed up from the confines of the Planck scale. Essentially the scale that can now be reached is lower than that of the Planck scale and these extra dimensions provide for differences in the strength of gravity when compared to the other three forces. Leaving the mathematics of both of these theories aside we see that the addition of extra dimensions at least theoretically can be used to solve the hierarchy problem and solve one problem in physics. After these of course we encounter still more theories that use extra dimensions to explain not just the hierarchy problem but also the gravity problem in general. These we have seen already but need mentioning again as they go about adding extra dimensions in different ways then previous theories. We are of course talking about string theories that use multiple dimensions to explain the incorporation of gravity into the Standard Model of particle physics.

The previous theories used extra dimensions of space itself called branes which are very large theoretical dimensions onto which things like matter can be bound. String theories sometimes use branes as attachment points for their strings specifically in differentiating between the open and closed one dimensional variety. However the real use of multiple dimensions in string theories is not on the very large scale of the universe but the very small scale of the fundamental particle. String theories add extra dimensions to the universe but do so with respect to matter which they believe to be made up of strings. The strings can have multiple extra dimensions which are not normally seen by us at any scale in the everyday universe. In fact many of these extra dimensions would be impossible to see even experimentally at all as they undergo a mathematical process called compactification. Compactification involves the folding up of these extra dimensions into essentially a single point object that can then be used to interact with the 3+1 space we are familiar with. String theories are trying to unify the force of gravity with the other three forces and solve the hierarchy problem at the same time from a kind of miniature approach. If correct it too provides evidence for the existence of multiple dimensions in the universe beyond the familiar 3+1 that modern science currently believes to exist. So now that we have seen the various theories vying for top spot in the problem of how many dimensions are in the universe let us see if they are compatible with the three main views.

First up is classical mechanics that would be perfectly happy with extra dimensions either on the large or small scale as the objects within them still obey classic laws of motion. If you have classical objects moving through higher dimensions or extra dimensions compactified into small spaces moving through normal space their respective motions can still be calculated classically. In addition to this the driving mechanisms behind these extra dimensions represent not curvatures of spacetime but real forces which can be applied in a classical way albeit through quantum interpretive mathematics. On that note general relativity fails at most of the theories except maybe for the RS model that uses a combination of negative and positive energy to warp the extra dimension of space. It might be possible to incorporate relativity into the warped extra dimension of space if the exotic forms of brane energies can be integrated into a general relativity framework should the theory be found correct. Lastly we find quantum mechanics surprisingly struggling with these theories for the most part except of course for string theories. String theory is one where quantum mechanics is right at home and can be used to explain the compactified dimensions of vibrating open or closed strings easily. Where quantum mechanics falters is at the large scale and specifically the brane systems used in many of these theories and the reason is quantum mechanics cannot deal with large scale objects. The interactions of the quantum objects with the branes can be equated somewhat but that is not where the problem lies.

The main crux of the difficulty in employing pure quantum mechanics to extra dimensions on large scales is that the concept of a brane which exists in normal 3+1 space or in multiple higher say 4+1 or 5+1 space configurations is that quantum mechanics and the Standard Model which go hand in hand do not deal with exotic physics. A brane is and of by itself a new type of thing so to say in the universe and nothing existing physically or experimentally within quantum mechanics or the Standard Model has ever been identified or measured. Only the most extrapolated, convoluted and derived mathematical pure theories are able to dream up these new objects and as such the nature of their behavior over the scale of the universe is unknown to modern science. A quantum interaction might work between a single string and a brane but how these become large scale billions of light years across effects where the universe seems to ebb and flow is unknown. The best possible outcome would be if these theories are correct and multiple static branes do exist in the universe by which all observed effects are created then a classical mechanics explanation of the large scale movements through quantum interactions could be used to account for things like motion of normal objects in 3+1 space and a plausible explanation of the hierarchy problem. This is of course assuming that any of these theories are even remotely correct as the inclusion of things like multiple dimensions, branes and compactified objects are let us face it, fairly outlandish and were only ever theorized in the first place as an attempt to solve the huge gravity problem. While based in theory and mathematics and of course created with the best of intentions no doubt there is one serious trouble with all of these theories which is proof.

The lack of proof takes two forms and in both cases destroys these theories. The first form is the fact that despite certain energy boundaries being hypothesized for various particles, branes or what not none have ever been discovered. Even the mighty Large Hadron Collider has not found them and in some cases it is not a lack of power for once. Normally insufficient power is the stumbling block by which theories cannot be tested in the laboratory but in the case of these modern multiple dimension theories for once the energy ranges lie within what the LHC can create. There is also the fact that many of these theories at their core and by the admission of their champions are untestable in the first place. This is falling directly into the trap of theories that cannot even be proven wrong and so are at their core worthless.

A perfect example of this is the over-exuberance of many to use philosophical mathematics and derivatives of derivatives to create objects that can only be realized in theory and never actually physically. In other words do not trust the mathematics too much. The example of the strongest of the theories being string theory illustrates this perfectly as it has been argued that strings may never be proven to exist as it is impossible to use 3+1 laboratory equipment to examine multiple dimensional compactified objects. How could one build any experiment in any laboratory that would be able to probe the depths of these exotic objects? Without direct proof this is akin to string theorists awarding themselves the Nobel Prize for being right based off of their own votes. Concepts of things like branes are subject to similar impossibilities that result in many stating they cannot ever be proven. This does not mean for a single instance that the lack of ability to prove them wrong by default will therefore prove them correct.

This is a topic of much debate that has yet to be resolved in the scientific community but one that must be in order to make any form of meaningful progress beyond where modern science is currently mired. This brings us to the second form of lack of proof where these theories are concerned and that is empirical proof. When I say empirical proof I mean any form of proof that can be derived from the senses or instrumentation in a laboratory. The fact that some say purely theoretical objects like dark matter, dark energy, strings and multiple dimensions to name a few of them cannot be ever detected directly or indirectly either is proof that they do not in fact exist. The simplest explanation for why these things cannot be detected and have not been detected for thousands of years of science and billions of years of evolution by any species or in any laboratory is because they do not in fact exist. They have not been found and will not be found because there is nothing to find. Simply creating an exotic theory which promises much does not make it a reality more powerful than the humdrum alternative which is it is false. Perhaps the reason the universe is exactly as it appears is because the universe really is exactly as it appears. We have space which matter and energy move through and that is all. The key is to simply find out how these things interact with one another rather than create more things to interact with the existing things that modern science does not fully understand yet anyway. So in summation while these theories do offer a possible explanation for how the universe interacts the fact that many are simply untestable in normal conditions is a serious setback for them. They cannot even be proven wrong is a powerful statement and one upon which most modern science was founded and more importantly proved.

Is their More to the Universe?

One more area of interest to the study of the cosmos as a whole is what really constitutes the universe in the first place? This topic is one that is much more theoretical and philosophical at the same time and may never be solved but since we have been looking at theories which are unsolvable themselves let us simply dive straight into it. So the basic problem is modern science does not know what the universe is in some of its most broad senses. For example what is the size of the universe, what is its shape and can it be possible for something to exist beyond the universe or does that sort of question not even make sense? The first aspect of these problems lies in the simple size of the universe and to this end we only know definitively what we can see. This is what the Hubble Space Telescope can see which is roughly 14 billion light years away. If by the time you read this book the James

Webb Space Telescope has been launched and is returning useable data then that distance will have been pushed back a few more billion light years. This makes the universe somewhere around let us say 15 billion light years old on average that we situated here on Earth can directly observe through all known means. The total size of the universe is therefore 30 billion light years across and since each light year is approximately 10 trillion kilometers we can come to one universal conclusion which is the cosmos is really big. How is that for scientific accuracy? It is really big.

Well if this may sound a trifle curt it is meant to be so as modern science lacks the correct model to effectively predict what lies beyond and we know something must be based on a huge coincidence of the supposition that it is exactly 30 billion light years across. This is of course the rather arrogant or dimwitted assumption that the Earth is at the precise center of the entire universe. Mathematically speaking the odds of this fact are to pardon the pun, astronomical. So we can now say that the universe must be larger than 30 billion light years across and this brings up the problem of how big is it really?

Since we are not at the center the universe therefore extends outwards in some number of unknown directions far from Earth. I say unknown directions because if we think of the placement of our solar system within the Milky Way galaxy we see that we are near the outer edge of one of the spiral arms. This means that some directions from Earth within the galaxy we find more or less space. To think of this another way if we travel outwards from the Earth away from the center of the galaxy we would run out of galaxy long before we would if we had tried to travel to the center and out the other side. The universe might be very much the same kind of construction as the galaxy with some directions going on longer than others. Certain cosmic microwave background anisotropies tend to align themselves with the ecliptic place of Earth and the solar system. Just as one would expect stars to do around a galaxy so could the universe itself have a kind of spiral galaxy shape? If so how does this affect the total size of the universe?

Here we can see size and shape of the universe possibly merging and whether they share a type of galactic resemblance is not that important only that the two are inexorably linked. Since they are linked we are faced with the second major problem which is what is the shape of the universe? Again from where we sit here on Earth and what we can observe directly the universe is a perfect sphere 30 billion light years across. This is only how it appears as our detection tools have a limit to their range and that limit is uniform no matter the direction we look and therefore constructing a sphere from these limitations will always be uniform and perfect. This is mathematically pleasing but practically false as the universe is most likely not a perfect sphere. Show me one truly perfect sphere occurring in nature right down to the quantum particle that makes up its outermost radius.

Of course this is a rhetorical challenge as you cannot and the odds that something as large and complex as the entire cosmos is also perfectly spherical is again at best a vastly remote possibility. No matter what though the universe is real and a tangible physical object so it does have a shape. Starting from the Big Bang as a potential theory for how the universe began its life what shape would that beginning yield? In the initial conditions of the Big Bang hypotheses it is believed that the universe had no shape as there was no universe before the Big Bang. The very first instant after the explosion the singularity that is thought to have given rise to the cosmos would have expanded outwards in all directions fairly uniformly. The actual singularity is an object that if real has unknown properties and one of these properties would be dimensions. So the starting shape of the universe is not known for certain as the degree to which the singularity propagated in each direction is not known and this propagation is what represented the spread of space and matter.

The next phase is thought to have been the inflationary period which would roughly blow up the universe many trillions upon trillions of times in as few fractions of a second. This creates a smoothing effect to the early universe as the rapid expansion means a rapid increase in volume of the universe. As any volume increases so does the surface area encapsulating that volume. Increasing surface area tends to smooth out irregularities over time. This is not true of the size of the actually irregularity itself but of the comparative size of the irregularity to the total area available. So think of it like this if you have a sphere of ten centimeters in diameter and there is a one centimeter high bump on it the bump is one tenth the size of the sphere. If you blow that sphere up to the size of one meter or a hundred centimeters the size of the bump is now only one hundredth the size of the sphere. The comparative size of the bump to the sphere has decreased greatly while the actual size of the bump has not as space has simply pushed the bump farther away from the center of the explosion.

An identical effect would have been present throughout the surface of the universe at the time of inflation such that the internal volume of the cosmos grew rapidly while surface imperfections essentially maintained there size as they were at the edge of the universe where no inflation occurred outside of space. Taken over many doublings of the size of the universe and the comparative size of any imperfections falls to near zero. However no matter how large the universe gets the comparative size never truly falls to zero. To this end the surface of the universe must be incredibly smooth when viewed from a great distance and appear to have a near zero amount of imperfections on it. This translates into a nearly but not perfect spherical shape assuming the explosion and inflation of the early cosmos was uniform in all directions. So there is no escaping the fact that the universe is likely not a perfect sphere as was often thought by many ancients.

Some apparent evidence can be found for this in the idea of dark flow. Dark flow is the unoriginal dark name given to anything not yet understood by modern science about the universe and refers to the apparent flow of large amounts of matter against normal universal motions. Essentially scientists have observed massive multi light year tracks of space and the matter within it moving against the normal motion of all other objects in their surroundings. These flows occur as if a force where acting upon them from outside the visible universe. How this can be interpreted as a non-spherical universe is simple the universe on its largest scales must have huge irregularities that can account for forces strong enough to move whole swaths of the cosmos at will. Irregularities of this nature would include large amounts of matter or space such that the combined effect on visible matter allows for odd motions in what we can view of the universe. Since it is believed that space and matter are linked the space holding the matter responsible for pulling on the visible matter would be distorted to a massive degree preventing that part of the universe from having a smooth spherical surface. This is of course assuming the dark flow is occurring due to something at the edge of the universe.

However the universe no matter how big would have an edge from the Big Bang concept and any dark flow events near the edge would distort it to a huge degree resulting in a lump or maybe dimple in the surface of the universe. The size of the lump or dimple needed to cause billions of light years worth of cosmos to flow must be greater in size than the flow itself in order to move that much matter and space at will against the forces of the surrounding universe. So any lumps or dimples would be themselves many billions of light years or more across. This means that the universe no matter how large it is must have a roughly spherical but slightly imperfect shape at its greatest sizes. This is based on a neat spherical explosion to the Big Bang however if the explosion looked anything like a conventional explosion as we see on Earth then the irregularities would create huge plumes of universe stretching in random directions some much larger and faster growing than the others. Even with inflation we would only see the plumes growing rapidly and the end result would be a massively non-spherical and random universe. Plumes might exist however as the small quantum irregularities believed to have been inside the singularity and made manifest beyond the observable horizon would create different areas of the universe information isolated from our own. Other ideas can abound for the shape of the universe such as cubes, flat planes or complex polygonal structures but these are rooted in mathematics only and are not backed up by theories such as the Big Bang, inflation or general empirical observations of the visible universe.

The third constituent of the universe after size and shape is the possible end of the universe. No I do not mean the temporal end of the universe if that even exists but instead the physical end to the universe. This can be thought of as the edge of the world as many ancient cultures or perhaps all believed at some point in time whether this was based off of empirical observation or cultural beliefs of the Earth, the celestial abode where the gods lived or some other form of legend. The comparison to the edge of the world is very apt however as the ancients thought if you walked long enough on this flat plane of the world you would come to an edge over which nothing existed. At least nothing of the world as they knew it would have existed and so we are left with the same question but on a much larger scale here. Does the universe have an edge and what lies beyond it as one would be leaving the universe as we know it today?

To start off one must define what universe really means and then we can examine possibilities from there on out. The universe means everything and nothing can possibly exist outside of it. I have long disliked the idea or terminology of the multiverse and known it to be quite wrong as even multiple universes would be contained in the whole largest universe. The idea of multiple universes should really be called multiple visible universes just as we see a universe around us that extends some 14 billion years from Earth and light beyond this would not have had time to travel the immense distances from the beginning of the Big Bang. These multiple universes as they are thought to exist whether in multiple dimensions or realities would still just be a small part of the whole universe that encapsulates them all.

So first off the universe really is everything and this includes theoretical multiple universes, dimensions and areas outside of our visible universe. No matter how big, convoluted or remote any structure in the universe might be it is still inside the single whole universe and so eventually is incorporated into the one cosmos. That being said let us travel any necessary distance to get to the very edge of it, the very boundary of the one whole universe and take a look at what we find. Well what we find from a logical stand point is that there is a nothingness that exists all around the universe. This is not to say that even nothing is something or any other type of circuitous wordplay applies whatsoever. Here nothing really is nothing and we know this logically because anything that is anything is part of the universe. Since the universe is everything as we have seen there is nothing outside of it because there is no universe there for it to be in. No you cannot even hypothetically place an object outside of the universe and tritely ask but what if? The universe instantly encapsulates it before you even move it outside and even the idea of postulating a hypothesis in a region outside the universe increases the universes domain to extend to that point and beyond before you even had the thought. Why?

Simple, the universe is everything and nothing can be outside it no matter how hard you try no matter how impossibly hypothetical your thought device it fails as the universe is where everything lives and all you have managed to do is help the universe in creating more of itself by creating a hypothetical area beyond its reach. No matter how many hypothetical and quite impossible scenarios you attempt to concoct the universe is and always will be one step ahead of you so just get used to it. To this end logic and rational thought dictate that there is an edge to the universe and there is nothing past it so you need not worry about falling off. You would be all dressed up with no where to go as the saying goes. Do we have scientific proof of this fact or at least a workable theory for it?

Yes we do and whether or not it is proven correct through fine tuning or not makes little difference the Big Bang theory has had much success and is based off of the idea of nothing outside the universe. In fact the Big Bang theory needs nothing to be outside the universe because in order for the Big Bang theory to work the universe must be created by it. This means that there was no universe for an explosion to take place in and spread out through as the explosion literally was the universe itself being created. To talk about outside the universe in terms of the Big Bang is to talk about outside the singularity which started it all off and before the singularity there was nothing. After the explosion began the universe was created and not just the matter and energy that modern science is fairly cozy with but the space through which it all moves. The fastest expanding part of the explosion was the background of space which is capable of moving much faster than light itself. This space expanded ahead of and still does presumably of all matter and energy within it meaning it would be impossible for any matter or energy to pass beyond the boundaries of space and therefore the universe itself. To make matters even more complicated each of the three main views of modern science have there own ideas of whether or not you can ever go past the edge of the universe. These ideas are for a moment ignoring the simple rational and logical fact that the universe as we mentioned earlier constitutes everything and therefore anything moving outside it simply expands the universes realm. To this end we will quickly look at the three main views and how they handle the first two problems of space.

First off is the size of the universe problem and what we can see here is that classical mechanics and quantum mechanics have nothing much to say about the overall size of the cosmos really. This is because neither of them is involved intimately with the actual laws governing the size and in fact they are both happy with a universe of any size so long as the objects within it can be described using their laws. As for general relativity much is the same and a universe of any size is acceptable too except for the idea of the speed of light being the fastest possible in the universe. I must point out that no where ever in the rules of the universe does it say that electromagnetic radiation, photons or simply light as you might like to call it need be the fastest thing ever.

This is something that modern science and its followers have decided to accept without checking with the universe before hand. Many of Einstein's rules are built off of the assumption that the speed of light in a vacuum which is approximately 300 000 km/s is the speed limit to the universe. The fact that he incorporated this belief into his calculations and that his calculations appear to give good measurement of various phenomena while compelling does not make them necessarily true. Remember that special and general relativity are not attempts at making a grand unified theory and cannot nor will they ever be able to solve all the unsolved problems in physics and these problems do exist so to solve them you need to break the rules of both special and general relativity. Afterall in the last one hundred years no one in modern science has really broken the rules of Einstein and look where that has gotten us, nowhere fast and maybe even faster than light. For some of you reading this you are about to burn my book for others a small smile is spreading slowly across your face nanometer by nanometer but it is there. I will dive fully into the shortcomings of general relativity and all of the three views later as it should be fairly obvious to everyone not just the scientifically minded that they are of course wrong in some or many respects. Think about it for a moment if they were not no one would be discussing the unsolved problems in the universe would they?

For now however let us look at a relativity based view of the size of the universe. According to relativity light has not had time to travel from areas of the universe farther away than 14 billion light years as the universe has not had time to exist long enough for the information to travel that distance. If things can travel faster than the speed of light however we see that the absolute size of the universe according to relativity is wrong. This is not because light has suddenly changed the speed at which it travels or the laws that we use to measure the speed of light and distance are suddenly wrong only that a universe can now be much larger or smaller, older or younger than need be dictated by relativity alone. Unseen forces responsible for some of the unsolved problems of science might be perturbing the matter and energy we do see to a point where space can behave quite differently than what we might expect. As I stated earlier the relativity view can work with a universe of any size I only cautioned against assuming that light was the fastest and therefore best marker of distance and time available in measuring the cosmos. To this end we do know that things can travel faster than light and have seen this numerous times already whether it be the inflationary epoch of the universe or more modern realizations like quantum entanglement. What this means is that the transfer of information is in fact quite different than general relativity predicts and so using purely relativity as a measurement tool for both size and age of the universe is inaccurate at large scales.

Moving on to the shape of the universe and the three views interpretation of it we find that there really is nothing to stop any of the views from working decently here. The reason is simple as we find that the shape of space is on a small and large scale compatible with the accepted ideas of modern science for the most part. Take for instance classical mechanics and a very young universe and you find that the shape of this early hot and dense cosmos still obeys classical laws for the most part in terms of interaction between particles and the distribution of thermodynamic energy. Even a small scale universe of irregular or perfectly spherical shape would still allow for the classical laws to be obeyed. Similarly a large scale universe works equally well for classical mechanics albeit that the interactions are far less frequent as things are so spread out. The governing classical mechanics of either a small or large scale cosmos can operate normally despite the cosmos possibly having a variety of shapes as the forces dictating the motions of matter and energy within it are not affected by the shape of the system in which they are contained. Think for a moment about isolated systems in thermodynamics and realize that the increase in entropy for example will be no different if the system were a sphere or a cube at the end of the day.

This leads us directly into quantum mechanics which is after all merely a more detailed and information rich application of classical mechanics with a few additions to the specific details of fundamental particles and forces. Whereas classical mechanics tends to deal with averages of macroscopic constituents to a system quantum mechanics deals with individual microscopic particles of the same systems but at the end of the day both views are on the same side. From a quantum mechanics viewpoint we therefore see that universe shape is of little consequence yet again as small or large scale universe size does not affect the interactions of quantum objects within the system. Small scale sees an increase of course in the number and frequency of interactions and large scale sees these interactions occurring far less frequently as a result. Shape is in all of the three views smoothed out at very large scales and as we have seen the curvature of the edge of the universe at large scales is near zero making it relatively flat and uninteresting. This dictates that no unusual quantum effects should be present at very large scales despite the microscopic nature of quantum systems. On small scales any significant shape changes in the cosmos would most likely produce unusual effects for individual particles as the density of the early universe would exert a far greater effect on any single object.

Yet even at these small scales the quantum interactions are bound by the fundamental forces present throughout space which cannot exceed the dimensions of space or therefore the shape of space and so still operate normally. A shape difference such that one part of the universe might for example protrude away farther from the rest would only translate into a difference in density for localized regions of the cosmos and so only change the rate of quantum reactions in that area but not the overall nature of said reactions. Remember that a large scale gravitational change such as those thought to be at work in dark flow do not occur in a small quantum system where space is so small gravity is essentially ignored making all interactions perfectly normal quantum actions. So as a result shape does not bother quantum mechanics when referring to the overall cosmos at small and young or large and older scales.

General relativity Deals less with matter and energy, particles and forces and more with the large scale deformations if you will of space so the effects of irregularities in the shape of the universe will be felt more strongly at least theoretically by relativity. A small universe will have denser space with more significant deformations if they ever existed than a larger more uniformly flattened out space and this is where relativity is most affected. A small scale universe would allow for greater bending of theoretical spacetime as the densities of matter are far higher and according to relativity it is matter that bends spacetime and produces gravity. So some areas of an early irregularly shaped universe might have much higher gravities than others which would affect the overall evolution of the universe to its present day large scale state. Areas with higher gravity might produce areas of more black holes than others earlier on in the universe and therefore create large scale anisotropies later in the life of the cosmos.

I say might because it is possible that these mechanics could have worked but are also impossible due to inflationary theory. The inflationary theory expanded the size of the universe many times in far fewer seconds to create a uniform distribution of matter and energy throughout the universe as well as giving rise to an epoch of so called negative gravity. This means that long before any relativistic effects could manifest themselves into the universe the cosmos had already spread out uniformly and denied these effects a foothold in the early universe. All in all it seems that no matter which view we adopt as our favorite from modern sciences tool kit we find that none can dictate the shape of the universe at either small or large scales. In addition to this none can therefore provide us with any useful information of the universe at small or large scales meaning we cannot hope to learn anything about the shape of the cosmos based on these three views. They simply lack the largest of scales interpretations of space, matter and energy necessary to look at the biggest of the big pictures.

From here we enter some of the most interesting discussions of the three views and the size, shape and edge of the universe problems as the last problem the edge of the universe finally allows us to sink our teeth into much more factual evidence concerning how each of these views deals with the very boundary of the universe. The edge of the universe as we have discovered by definition can never be crossed as this would simply expand the borders of the universe itself. However what we can look at is how these borders might be expanded in other words we can reach the edge of the universe and not cross it but push it farther out. This is of course some of the most unknown territory in all of modern science but surprisingly a solid background for mechanisms that might be able to push the border back a bit are theoretically possible. That being said it is only slightly probable and still fully impossible if that makes any sense at all to you.

Nevertheless trifles such as the known laws of physics have never stopped science before from attempting to push back the curtain on what is possible and what is not and we are not going to stop it now. So without holding back at all let us look at each of the three main views of modern science and the possibility each has to theoretically push back the border of the universe and technically go beyond the edge.

The first up is quantum and while it was not the first to appear on the theoretical scene we will examine it first for a reason and this reason is that it can never cross the boundary of the universe. Sadly it is true for quantum mechanics but it is fully impossible for any quantum mechanism to exist at all anywhere that could ever conceivably cross over the edge of the universe. The reason this is so is because of the fact that at the core of quantum mechanics is the idea that particles are simply excitations of the fields of the fundamental forces in the universe. The ability to manifest a particle is due to perturbations in the fields that for reasons unknown to modern science simply choose to occur. A particle only appears to move freely in a quantum universe when in fact it is merely travelling along preexisting paths of the fields in space. Think of a small pile of dirt swept under a carpet as the classic image goes and the little bump it creates as a result. The bump only exists because the carpet is there to bulge up and create it in the first place. Move the dirt pile under the carpet a ways and all you have done is make the existing carpet bulge in a new location. The carpet is what creates the bump and the pile of dirt is the perturbation in the smooth fabric of the carpet. A particle is the perturbation in the smooth fields of the quantum space and so is linked permanently to the fields. What does this mean for the edge of the universe?

Simple, since the universe consists roughly speaking of space and matter the edge of the universe would be the edge of space too. According to quantum theory the fields making up particles are a part of space and so end wherever space ends. The particles are only perturbations of the fields and so can not travel anywhere the fields cannot as well. Since the fields are held by space no particle can go beyond the field which ends rather abruptly at the edge of space itself. No matter how you slice it is impossible to create a scenario based on quantum mechanics that allows for a particle to cross over the edge of the universe. In addition the edge of the universe is also the edge of space and this contains the quantum fields so one cannot move the fields beyond the edge of space as they are embedded in it and this includes the theoretical trick of quantum tunneling too.

We can see here that despite the lack of quantum mechanics being able to even theoretically cross over the edge of the universe we do have much more solid ground to stand on when discussing an unsolved problem of the universe. The key to any of these views is their model strength which is essentially how they see the universe being put together and how well that can be used to explain the goings on of everything in that universe. Things like the proof and the mathematics necessary to describe a good model always come second as the concept is all important. The fact that this is true allows us to forgo the usual necessity for proof and mathematics and look at the fundamental teachings of each view and apply this raw knowledge to the unsolved problems of the universe. Whether they succeed or inevitably fail is of little consequence as it permits far greater insight into the workings of the universe than simply plodding on and on with little niggling details as the theories are currently stagnated in. To that end the failure of quantum mechanics to cross over the boundary or edge of the universe should actually be lauded as a triumph for the theory as it has such firm rules a definite answer can be gleaned from a simple analysis of the problem through use of its model.

Now we will look at the last theory to emerge which is of course general relativity and like quantum mechanics has very definite rules as to how it behaves as an idea. The first rule one must understand when thinking about general relativity is that at its very core space and time are one and the same thing intricately woven into what Einstein called the fabric of spacetime. Spacetime are not separate entities as they are in classical mechanics and quantum mechanics but are one and the same and how they behave dictate according to relativity how the universe also behaves most of the time. As we all know relativity does not work at all on the small particle scale but it does seem to work on the larger scales of things like planets, stars and the like so let us look at the largest scale possible the universe and see if relativity can go beyond the edge.

First of all we know that spacetime cannot go over the edge of the universe as it is the boundary of the universe itself according to general relativity and instead we must look at anything moving in spacetime. This means an ordinary object with mass such as Newton has described in classical mechanics. According to classical mechanics objects simply move where they want to and developing this idea further general relativity states that objects move with respect to each other and spacetime too. Objects can be said to fall along the curvature of spacetime and so at the edge of the universe as long as an object has sufficient escape velocity to not fall backwards into the curvature of spacetime there is no logical reason why according to relativity that it cannot leave space time. This is also due to the fact that general relativity does not bind matter and particles to spacetime in a quantum way or does not view them as excitations of a quantum field.

In general relativity as in classical mechanics before it matter is independent of space and can move freely. This is just the same kind of thinking that also permeates classical mechanics and general relativity when talking about time as time is integral to the existence of matter but matter can move at different rates of time in relativity whereas this is not true for classical mechanics. Nevertheless time is a part of matter according to general relativity and you cannot describe any system using general relativity without using time and in fact the very nature of matter needs time as a fourth dimension in relativity along with the three spatial ones. This is afterall how general relativity was born by tracing the world lines of an object which can only be done from position to position when described time from time.

This however creates a huge flaw in relativity when talking about leaving the universe by crossing over the edge of the cosmos. While the curvature of spacetime at the very edge of the universe would be so incredibly shallow due to the total size of the universe spreading all matter out so far and the sheer size of the edge being a near zero curvature geometry anyway we find a flaw in how general relativity describes the universe that prevents matter from leaving the cosmos. The flaw is that despite the curvature being so small as to require little energy of momentum in order to travel over the edge the nature of spacetime itself prevents any matter from leaving the universe by weaving time into the fabric of space. According to relativity remember matter can move independently of space and this is how objects move freely in spacetime but objects cannot also exist without time according to relativity and so since space and time are linked and time itself is a part of space matter cannot leave spacetime as it would lose its fourth temporal dimension. Matter might be able to move freely in space but it requires time to exist and so leaving spacetime is forbidden as matter cannot exist without the time component of spacetime. Since spacetime stops at the edge of the universe and in fact is the edge of the universe nothing can escape the universe according to relativity no matter how hard one tries. There would be no temporal frames of reference to plot the free fall motion of matter without all of spacetime and so the information necessary to describe an object evaporates for relativity at this point.

This is also true at the most microscopic and fundamental levels where even though relativity does not have a mechanism to describe these tiny details the same relativistic laws must apply to everything in order to make general relativity a workable answer to describe the universe and so all of the very smallest constituents of matter cannot exist without time as well according to the theory of general relativity. The fundamental particles in matter would most likely dissolve at the edge of space and never cross it as the time frames needed for them to exist as complex systems like atoms would be lost. Not only can matter never travel outside the universe but it would also cease to be matter completely without time and so further negates the possibility of matter travelling beyond the edge of the universe according to general relativity. Again we see the inability for one of the three views to escape the universe and again we see the carved in stone rules of the theory laid out so as to explain why it cannot leave the universe which is overall a success of the theory. By stating what it can and cannot describe at the level of theoretical ideas versus mathematics the idea behind general relativity can be explored even if the edge of the universe cannot.

Lastly we come to classical mechanics and our last hopes for the three main views of crossing over the edge of the universe. Despite the fact that classical mechanics was the first of the three views to be developed chronologically we are dealing with it last as it has believe it or not the best chance of crossing over the edge of the universe. An object in classical mechanics is independent of the rest of the universe and this includes time and space. In fact according to Newton's ideas of motion and matter all objects have their own relative space and time when compared to the absolute time and space of the universe. The absolute time and space are sort of the mathematical starting points from which humans make relative observations for other objects such as two objects in motion relative to one another.

This has been explored previously with respect to both space and time but must be restated here as this notion provides for the interesting effect of normal objects being able to move independent of space and time. Meaning that when one takes a normal object to the edge of the universe and applies a force to it sufficient to propel it beyond the edge of the universe the object will encounter no resistance upon reaching the edge of the universe. The object is therefore free to leave the universe and even leave space itself as a classical object can and does exist independently of external variables as you might recall the variable space inside the object is independent of the absolute space and so the matter of the classical object takes is own internal space with it over the edge.

This is of important note because the idea of particles themselves in quantum mechanics forbids them from detaching from the fields of space that created them and therefore leaving the universe. This is important because in general relativity objects cannot be unbound from spacetime as they need both of these in order to exist in three plus one space. In both quantum mechanics and general relativity theories we find the laws of physics breaking down at the edge of the universe. This is not so for classical mechanics and so provides the only explainable framework for anything leaving the universe at all. Think about this for a moment as a quantum or relative object tries to leave the universe and realize that one of two things and possibly both happen to it upon reaching the edge. The first is that the idea behind the quantum mechanical theory and general relativity theory is flawed and therefore no amount of fine tuning of the mathematics or theories will correct these problems. The result is nothing can leave the universe according to those two ideas.

The second conclusion is that they are still incorrect as it is impossible even under their own sets of rules for matter to spontaneously vanish upon reaching the edge. The breakdown of quantum fields or the end of spacetime at the edge of the universe provides no method of recycling the energy carried by an object at the edge of the universe because there is an actual edge to the universe and not a closed loop which would allow for the energy and fields or what not to be recycled back into the cosmos. The information about the fields, spacetime and matter would be lost from the universe which is believed to be forbidden by modern science. The only theory that does not forbid this is classical mechanics which allows for any object in motion to keep travelling unless acted upon by another force. All one needs to do is set an object in motion headed for the edge of the universe and then not supply an external force to keep it inside the universe and just let it float away. Does this mean that the classical view of objects leaving the universe is therefore one hundred percent correct?

Well maybe yes and maybe no. If classical mechanics is proven to be right then yes it can do whatever it so chooses at the edge of the universe and nuts to the rest of us. If no then we get to keep all objects within the cosmos and try something different in our quest to charge over the edge of the universe. The reason why I make this distinction between the failures of quantum mechanics and general relativity to cross over the edge and the success of classical mechanics to do just that is because classical mechanics is one of the longest standing and best fitting views we have of the universe to date. Before everyone out there in the science community laughs themselves silly let us think about the much bigger picture for a moment in space, time and most importantly ideas. First of all quantum mechanics fails at the larger scales utterly and general relativity fails at the smaller scales utterly whereas classical mechanics tends to work decently at both.

Yes I said decently and not perfectly because it does not and to end this right now I never said nor will I ever say classical mechanics is the true answer to the problem of how the universe works as it most certainly is not. What I will say though is that when I see a theory that offers a glimmer of hope for both small and large scale systems I think it should at least be investigated further. Therefore we must remind ourselves of one more crucial factor and that is both quantum mechanics and general relativity are just as dead wrong about solving the problem of how the universe works as is classical mechanics. This should be obvious to even the most scientifically lackluster out there as if either of these theories was correct then they would simultaneously not fail utterly at one extreme or the other and they would also have answered the entire list of unsolved problems about the universe already.

Since we all know that to be impossible for quantum mechanics and general relativity we can at least entertain the idea that classical mechanics is still correct and afterall is the root of all quantum mechanical and relativity theories. More observable and measurable systems in the universe follow classical mechanical rules and have done so from the inception of the classical mechanical idea up until today for modern science. Does classical mechanics offer up perfect predictions about certain outcomes when compared to other theories? No but that is simply because it was the first rough draft of the theory and centuries later insights into the world of subatomic particles and large scale objects in the universe proved it needed to be refined but not replaced. To this end the classical mechanics view of the universe is a valid one and can therefore offer a valid mechanism by which an object and therefore information might leave the universe. This idea is actually favored by some new theories which suggest that other universes exist close to ours but information travelling through them takes a one way ride to the other side and so depletes our universe slightly. No need to name these theories here as you are probably familiar with some and they involve exotic, unproven and usually untestable theories of black holes, multiverse, extra dimensions and so on and so forth.

Lastly on the subject do remember that while I am having some investigative fun with these three main views I am also simply posing the what if scenario as to whether anything can in fact leave our universe and therefore be said to have crossed over the edge. Since the universe encompasses everything anyway we know that leaving the universe means simply creating more normal universe and so defeats the whole purpose of the question of whether or not you can leave the universe. What I do want you to take away from all of this is the deeper look into the ideas behind the three main views of modern science as it is the basic concepts of how these models work that is of the utmost importance and forever supersedes the mathematical rigidities they possess which were created afterwards. Even today people tweak and re-tweak the parameters and variables in an attempt to fit the mathematics of each of the three views to the observations of the universe. What this means is that the observations are obviously correct and the three main views are flawed because they require the addition of, creation of or tweaking of some of their parts or variables. In other words the mathematics behind them are not as important as the key concepts as the mathematics are but fleeting things which can be fine tuned when needed but the core ideas of the models are not to be tampered with as they set out the rules by which the mathematics are derived.

Quantum Entanglement and Non-Locality

This idea of the models core concepts being of the utmost importance in determining its success has severe implications for all the three main views and can be illustrated by examining an unsolved problem that would shake up the rules of modern science and physics. This is of course the problem of locality and specifically are non-local events a reality in the universe whether in the laboratory or the cosmos in general. First up let us examine what a local event is as it is described and understood by modern science and then we can give examples and finally talk about whether or not non-local phenomena exist. A local event is one that influences an object through interactions with its surroundings. This can be thought of as a single quantum object and the entire quantum system used to describe it and would constitute a micro scale local event.

A macro scale local event can be anything much more ordinary such as a cup of tea sitting on a table where the tea experiences effects from its surroundings. The most notable of course would be that the surroundings are much cooler than the tea and so heat is lost from the cup of tea to the room which would play the role of the system. The information contained in the cup of tea with respects to its thermal energy can be transmitted to the room and is therefore said to maintain the principle of locality. A quantum system is also said to maintain the principle of locality by transmitting information at or below the speed of light with its surroundings or by receiving information at or below the speed of light from the surroundings of the system. The general idea is that for something to influence something else at a distance meaning that they are not in direct contact with one another then something must be controlling that action. Such controlling actions could be mediated by a field, particle or force carrier as any of these could travel from one object or place to the next and carry information with them which when added to the destination object causes some action to occur.

Think of the exchange of information from two atoms and the electrons they share or the forces binding them generated by the underlying fields of quantum mechanics. The reason I mentioned the actions or information exchange is said to occur at or below the speed of light is that modern science bases the idea of locality off of special relativity which forbids anything in the universe from going faster than the speed of light. If we assume that relativity is correct then that means the total travel time of actions or information from two separate objects or points cannot exceed the speed of light and also that the time it takes for these actions to occur is a non-zero timeframe. This is because even light travelling from one point or object to the next will still require some length of duration to pass in order for it to carry the information along and thus the speed of light creates a travel time greater than zero no matter how close to zero it might seem. Quantum mechanics however seems to violate this law of locality pretty much every time you make a quantum measurement as information is resolved from the system in perfect zero time meaning instantly and with no delay. So in other words information has travelled much faster than the speed of light which is thought to be impossible by modern science.

A possible paradox was evolved in 1935 known as the EPR paradox which would seem to prove that locality must not always be obeyed in the universe and under certain special conditions information can travel faster than light. The EPR paradox was named after Albert Einstein, Boris Podolsky and Nathan Rosen who recognized that something in the universe was able to break the law of special relativity that light was the fastest speed possible in the universe. To do this they looked at quantum entanglement and the resulting information that is collected upon making an observation of one of the entangled objects. In such an experiment as one where two objects are made to be quantum entangled they are also separated by some distance and even though any small distance will do it has been found that the distances can actually be quite great as well which only further confounds the problem. The measurement of one of the objects causes the properties of the other to be instantly known as the wave functions of both particles collapse simultaneously. This means the accumulation of information concerning the state of one particle instantly allows for the information of the second particle to be known as well and since the particles where entangled some action must have passed between them in zero time which is of course infinitely faster than the speed of light. In the laboratory one physical particle is measured and its state revealed, the entangled particle has its own physical properties revealed at exactly the same time.

This EPR paradox should be taken seriously for a couple of reasons the first of which is that it was partly developed by Einstein who himself said from his own work nothing can travel faster than light and yet there he was actively saying I believe that something spooky is happening. This is of course where we get Einstein's famous joke about quantum entanglement being nothing more than spooky actions at a distance. So we have the approval of the developer of relativity saying that something unexplainable that violates his own laws is happening. The second reason we should take this seriously is that it is real as in quantum entanglement is not some fanciful effect dreamt up in the equations of a mad scientist and instead is one of the most studied, scrutinized and debated aspects of physics to date. Quantum entanglement is real and verifiable and repeatable which denotes all the hallmarks of a fact in modern science and so the fact that information is shared immediately is of major significance. Why this is so important is that something unknown to modern science is occurring and this can be inferred from the idea of information as understood by modern science.

At present modern science understands information essentially as anything that can alter a known physical object which is usually a quantum object by a set amount. So what is a known object and what is a set amount? Well everything as modern sciences knows it and a simple example would be an atom which can be viewed as a quantum object which is well understood. A second atom might emit a photon which is of course another quantum object towards the first atom. The first atom will absorb the photon and the electron orbiting it has gained a quantized amount of energy and transitions to a higher orbital energy state. In this way we can say that information was shared between the two quantum objects because the information carrying the loss of energy of the second atom is placed into the first atom and so by knowing how much energy the first atom has absorbed we know how much the second has lost.

In other words the information necessary to describe what just occurred between the two quantum objects has been preserved and understood nicely by modern science. This ties directly into the idea that we have touched on before and will again of total information in the universe and whether or not it can be lost either completely or partially forever. The very idea that information is never lost is also a firmly held belief of modern science that underlines a number of key theories currently accepted by modern science. So the fact that something like quantum entanglement can seem to violate that most beloved of principles is a concern for modern theories and one that must be addressed if a complete picture of the universe is to be had. Hence we have the creation of the EPR paradox as a way of realizing the fact that something is allowing information to travel faster than light to occur.

From this EPR paradox we arrive at the work of an Irish physicist by the name of John Stewart Bell who came up with a way of determining whether or not a particular system violated the known laws of the universe. Named after him and now called Bell's theorem it is a set of rules that must be obeyed or broken during an experiment and the consequences of which result in a system and here we specifically mean a quantum system either maintaining locality or breaking the laws of physics. The obeying aspect is of course quite boring by all accounts as it results in the usual and fully expected outcomes but that does not mean it is unimportant as the core beliefs of science technically hang in the balance. Should the laws be obeyed and it must be the case in every experiment always and under every possible parameter then modern science can for now at least hang on to the ideas of the Standard model and relativity. If the rules are broken and therefore not obeyed then we find that something is flawed in relativity as well as quantum mechanics too.

This flaw in quantum mechanics is not really a flaw just a lack of understanding based on the current models of quantum systems a newer and more complete model of quantum systems might account for the abhorrent behaviors found during experiments involving quantum entanglement. It would mean however that current quantum models used by modern science are incomplete and do not fully explain the systems they are designed to describe which opens the door for new ideas to fill in the gaps. So in order to examine what is being tested in Bell's theorem one must be acquainted with two words and how they apply to physics the first is locality which we have seen and the second is realism.

When thinking of realism many meanings can potentially spring to mind the most obvious being that if you can interact with something on the empirical level then it is real. This is of course how our ancestors understood the world and as such still holds true today with the caveat that we now have the luxury of examining things in much more minutia than ever before. Take for example the humble atom once again and realize that while we are empirically aware of the electromagnetic forces at work in keeping electrons happily orbiting their atomic nuclei through the act of touch and encountering resistance we are not aware in any empirical way of the nuclear forces that bind atomic nuclei. They are working unseen and unfelt by the senses available to biological organisms. They are however still very real and therefore a layer of the universe must and does exist that is real but beyond empirical realism. A second definition of realism is a more philosophical one which we have touched on before that states we create a reality of the universe by examining it and this we have seen in discussions of time and space.

What science means by realism is yet a third definition and can be summarized simply as the results of a measurement exist without the need for an observer to make them real. In other words when someone looks at a quantum system and sees multiple possibilities that are unknown probabilities until the act of measurement it is not the observer that materializes the measurement into being. Once again what this breaks down to is it is not so simply because you looked at, it was going to be so anyway. Without going into further detail here about locality, realism and local realism as it is known to physicists let us take what we have learned and examine Bell's theorem. Essentially Bell addressed the EPR paradox in such a way that he demonstrated no matter how one changed the system or made a measurement and with local hidden variables or not the results would disagree with quantum mechanical theory. So in other words he came up with a proof that spooky actions at a distance must be real and locality and realism were in certain cases null.

His basic experimental idea revolved around entangled particles linked by spin for example which can have different outcomes for predictions made using quantum mechanics. This is of course the basic idea behind probabilities of a particular function in quantum mechanics and all the possible states it can occupy. The idea that a normal non-spooky action at a distance phenomena was occurring in the system would be explained by a hidden variable. If hidden variables were responsible for the predicted results Bell argued they would be bound by certain limitations he called inequalities and hence we have the term Bell's inequalities. If the hidden variables cannot satisfy the inequalities as is the case with quantum phenomena such as entanglement this means that no hidden variables are present and something non-local is happening. Put another way locality is lost during quantum operations of certain kinds, spooky actions at a distance are real and something in the universe is capable of transmitting information or actions faster than the speed of light thus making relativity partially or wholly invalid. So if one incorporates local realism which is said to be a combination of locality and realism obviously a physical theory cannot be derived that accounts for all the possible predictions of quantum mechanics. Even the concept of local hidden variables cannot account for the inconsistencies encountered during quantum entanglement experiments.

A hidden variable is one that is simply believed to be at work but not fully understood or incorporated into the framework of quantum mechanical systems and it was hoped by many for at least as many decades that hidden variables would emerge to resolve the spooky actions at a distance. None have been found and even Bell himself acknowledged that quantum mechanics must contain no hidden variables and therefore is a non-local event in certain cases. A hidden variable can be further thought of as anything that fits into the Standard Model or relativity as it must be a field, energy or matter that is non-superluminal meaning does not exceed the speed of light. Thus the EPR paradox remains along with Bell's theorem as proof that quantum systems do not always obey the known laws of physics.

Here I must reveal the stark fact that paradox is the wrong word or way of describing what is in fact going on here. Remember the universe does not and should not care what humans or any other intelligent species thinks about how it should work. The fact of the matter is that the universe works the correct way and ideas trying to understand the universe are working to learn that way in modern science. There is no paradox occurring because a paradox essentially means an impossibility and since the quantum entanglement for example is happening and real it means the universe is working just fine and not doing what is impossible as that would be, well impossible. So this leaves us with the realization that nothing wrong or bad is happening with quantum entanglement events and instead another unknown to modern science mechanism is at work which allows for the transfer of information faster than light.

This in turn leaves us to further realize that information indeed is not confined to the current understanding that modern science has given it. Which is to say that information need not be made of something that is bound by the slow speeds of special relativity and therefore need not be a field, energy or matter. Since one quantum object is made immediately aware of the behavior of the other it is obvious that the information being shared is not made of relativistic objects or fields. It should furthermore be fairly obvious to even the most lethargic of thinkers that this is indeed true as relativity, quantum mechanics and classical mechanics in any combination have not and will not solve how the universe works as they are not complete pictures of the universe and its parts. Something great is missing from them and therefore forever prevents them from solving the problem of how the universe works. The missing link to this problem is at least partly hidden in the world of quantum entanglement and events which exhibit non-locality. So the fact that information of some kind currently unknown to modern science is transferring an unknown to modern science type of information is not at all a paradox but a reality as solid as any other.

Now that we have looked at a problem which truly is unsolved by modern physics we must look at what it means for the three main views and their respective chances of solving it should a solution present itself. The first thing to remember is that this is not the sort of problem that has ever really cropped up in modern physics before as it involves an idea that is completely outside the realm of anything previously encountered. Is this sort of thing unsolvable? Well of course not as science has faced numerous challenges in the past and with time and hard work have fought to overcome them. The question is whether or not anything in the current arsenal of modern science has what it takes to bring down this new opponent as it were.

The first main theory employed by modern science is classical mechanics which despite being significantly tweaked to produce both quantum mechanics and relativity is still unchallenged for most everyday applications. Afterall the laws of classical thermodynamics for example are one hundred percent accurate for a larger non-quantum system and so represent an example of their continued usefulness. As for effects that are abandoning the familiar local-realism classical mechanics falls short as it is almost impossible to describe these effects in macroscopic more classical systemics. The idea that an effect is caused by something other than a traditional force cannot be handled by classical mechanics as it relies on every action being mediated by a force. This can be a force on an object, a force on a force or an object on an object but something known must be occurring for which classical mechanics has no answer. It is possible that an unknown force acting on quantum objects would in the general sense satisfy classical mechanics by remaining a traditional force creating an action on an object's behavior but it would be happening in a new and non-traditional classical way. If this is somewhat foggy it is not your fault remember that classical mechanics as well as the other two main views are truly dealing with territory unknown to modern science.

This brings us to the second of the main views which normally fairs the best with the unsolved problems and that is quantum mechanics. Well here the usual favorite also falls short as nothing in quantum mechanics explains what is happening here. It is in fact because of quantum mechanics that the problem has become known and unsolved in the first place. This is somewhat akin to quantum mechanics admitting it needs a helping hand from the other two main views just as relativity admits it needs help from the other two in solving the problem it created itself of the singularity. So quantum mechanics can best describe what is going on and why it is unsolved but it cannot offer a solution using its own mathematics as they contain one major problem themselves which is probability. Quantum mechanics is in some ways nothing more than a good set of probabilities used to explain observed phenomena and in so doing fails at the idea of an entanglement. The reason is simple if you look at the classic entangled pair experiment as carried out by hundreds of scientists over the years and not just Bell.

A pair of entangled particles could have any normal probability distribution of let us say spin based on normal quantum mechanics that should be unknown and therefore not correlated with the other particle. When the wave function is collapsed that predicts these probabilities we find the particles always knowing what the other was up to and this does not make sense from a normal scientific standpoint as modern science understands the universe. Think for a minute of flipping a coin for a very simply analogy to this problem and you quickly realize you have a one half chance of getting heads or tails. If you flip two coins you have a one half chance per coin and they are not linked at all and the probabilities are said to be independent of one another.

In the world of quantum entanglement however the coins become linked by some as yet unknown to modern science mechanism that makes one coin always land heads or tails based on the other and can never again exhibit the random heads or tails probabilities it did before they were linked. So quantum entanglement forces a very real effect upon what should be non-entangled events. Never again can your coin tosses be random and normally that makes no sense. Since the entanglement phenomena is real and is not created by any hidden variables which would be explainable using quantum mechanics and quantum equations it defies any quantum explanation. Quantum mechanics therefore cannot solve this problem and even if a reason for the effect is found and given to modern science it will require a rewrite of some of the core equations of quantum mechanics itself.

This leaves relativity as the hopeful savior of the problem of these non-locality effects and unfortunately for modern science it fails too. This is due to more than one reason the first of which is that it is relativity which fails as well at the quantum entanglement problem in the first place not just quantum mechanics. The theory of relativity states that nothing in the universe can exceed the speed of light in a vacuum and so the fact that something is travelling faster than light during quantum entanglement events already excludes relativity from ever solving the problem. The mathematical framework of relativity is limited by its assumption that nothing can travel faster than light and that nothing can therefore transmit information faster than light in a vacuum. This fails utterly at a quantum level with quantum entanglement events and is compounded by the fact that the mathematics of relativity fail more than once in the universe.

Two more examples are during the faster than light inflationary period of the universe where it is known that space itself can travel much, much faster than light and therefore exposes the reality that maybe other things in the universe are not bound by the speed of light limit. The second place it fails is inside a black hole and specifically at the point of a singularity which is of course a physical impossibility which casts even more doubt onto the usefulness of relativity and all of its equations as the amount of trust that should be placed in them wanes. Any new mechanism for the quantum entanglement effect would include forces, objects or mathematics that fall outside the realm of the speed of light being the limit to the universe and so by their very nature cannot be incorporated into relativity at all.

The stark fact of the matter is that all three main views cannot deal with the realities of non-local events such as quantum entanglement. This is a bit of a shocker for modern science as up until now at least one of the three main views has showed promise for at least chipping away at the unsolved problems of the cosmos. If this is where modern science is headed then the reality occurs that more than one problem might be completely unsolvable and unknown to modern science.

Now we enter some of the most fascinating and uncomfortable territory fully unexplored and uncharted in its apparent weirdness in all of modern science which is the idea that some of the core beliefs held most dear by modern science are in fact wrong. This is of course fascinating because it is a chance for new and therefore correct models to take hold and flourish in the wake of the mess left behind by the last century of physics. Mess is of course a very general word as some of what has come out of the last century is some of the best work modern science has ever done but mess is still applicable because starting from the ground up no one in modern science can still answer that question of how the universe works. In order to do that the core beliefs and basis for understanding must be correct first and this is where I have been going in examining the ideas of the three main views of modern science being classical mechanics, quantum mechanics and general relativity. We will return to the fascinating aspect later but for now let us examine the uncomfortable aspect which is of course the fact that the three main views cannot handle the universe head on and are to different degrees trod upon by the cosmos.

Take for instance the idea of relativity and when I say relativity I mean all of it rolled into one and this is both special relativity and general relativity. Special relativity states that nothing can move faster than the speed of light which works well for some aspects of the universe but not all. Matter seems to obey this but space itself during the theoretical inflationary period does not. The universe is made of both space and matter so right from the get go we know that special relativity is wrong or at least partially wrong as it cannot fully describe the universe as a whole. Of interest here is also the fact that for apparently no one hundred percent proven reason the modern scientific community of both yesteryear and today take at face value something that Einstein simply made up which is that light in a vacuum must be the fastest possible speed in the universe. Einstein does not nor does his work understand or explain how the universe works and so his assertion that the speed of light in a vacuum is the fastest speed possible is untrue.

Think of this another way and realize that nowhere in the universe is a signpost stating one shall not exceed the speed limit of light. General relativity which builds on the core beliefs of special relativity attempts to explain the effect of gravity using the speed of light as a limit as well. Modern science often applauds Einstein for his equations which better fit observation than those of Newton but I need to remind everyone here that Einstein's equations are only calculators and do not explain why the universe behaves as it does. Newton's own work led to insights such as light bending around massive gravitational objects such as stars centuries before Einstein and without a thing such as curved spacetime. If given centuries better observations and scientific equipment the genius of Newton would certainly have improved his calculations to match those observed. To this end taking a different entire view of gravity from Einstein simply because his numbers match up better does not usurp the core ideals of how the universe works in terms of gravity. All other forces in the universe are forces so why would gravity be different?

If gravity is a force and not a result of bending of spacetime then relativity is out the window again. For now I will forgo such discussion and focus on simply relativity and the universe as a whole with respect to curvatures and gravity. The basic premise of Einstein's equations is to match different tensors with different masses of observables to produce a curvature of a set amount. That is it and nothing more but a set of rules to say if you have this much mass in this much volume you get this much curvature of what relativity calls spacetime. The resultant effect according to relativity is gravity but nowhere in any of this does it ever explain what and why it is happening. No stop thinking it does as I can hear your thoughts already saying curvature is the why and the what. Use Einstein's work and show the modern scientific community what is happening at the most fundamental level to make these things happen and of more importance why. What is matter doing to space to warp it and what is it doing from a quantum mechanical perspective? What is space anyway? Why should an object fall into anything if not worked upon by a force or force carrier? Spacetime can be curved all it wants if an object is minding its own business what influences it to start moving towards the area of highest gravity? The list of questions that Einstein and his body of work cannot answer is longer still but these are the very real questions that need answering. The theory of relativity itself proves itself false through one of its own predications.

The ever interesting black hole makes yet another appearance and was first predicted by the equations of relativity. I do not doubt that relativity has its place and has its uses but I am simply saying it is not correct in explaining the universe and this goes for even gravity which it is aimed squarely at and not things like fundamental particles and the like. Relativity predicted correctly the existence of black holes or completely gravitationally collapsed objects and in so doing proved that as a theory it was itself incomplete, wrong or both. At the core of a black hole is a singularity according to relativity which essentially when one solves the equations gives you a value of infinity. While infinity is a number that gets tossed around quite a bit most people do not stop to think of what it really means. In most peoples mind it means simply something without end or anything that no matter how big you can always keep adding a one to forever. In the real universe and physics of course infinity means a mistake as nothing in the real universe ever has the value of infinity.

So at the point of a singularity we find three such infinities as predicted by relativity and they are in no particular order gravity, volume and time. At the point of a singularity we see infinite gravity which is of course absurd as an infinite gravity would produce an infinite curvature of spacetime if it were real and destroy the universe. We also see an infinitely small volume which is of course also absurd as nothing can have zero dimensions as you need dimensions to describe anything therefore something with no dimensions is nothing and therefore false. Lastly time is infinitely small or in other words completely stopped and this means truly stopped not simply slowed down to imperceptibly short lengths. Time is not necessarily real as we have discussed earlier for various reasons and furthermore it is impossible to have zero time because Einstein's whole idea of special and general relativity relies solely on the passage of time from reference frame to reference frame. Without time you have no relativity which was the theory that gave you this impossible singularity in the first place so once again relativity proves itself wrong.

This is sadly for modern science only the beginning of the failures of relativity to which so much faith has been placed and to which so many answers will never come. The practical application of this from a theoretical physics stand point is that general relativity is also wrong and therefore despite its calculations being successful its core beliefs are incorrect. Thus using general relativity or any other relativity to explain the universe even just the small gravitational part of it is false. So this is where we see the uncomfortableness of modern science growing from and in fairness to relativity quantum mechanics does not fair much better.

In quantum mechanics we find a decent description of what the universe is up to on the very, very small scale which involves three of the four known fundamental forces. These forces are the electromagnetic force which is what holds electrons to their partner nuclei and of course help to transfer energy from quantum object to quantum object using photons. The nucleus contains two more of the fundamental forces the first of which is the strong nuclear force or just the strong force. The strong force is what binds things like hadrons together and two common hadrons are of course the proton and neutron which make up the nucleus of all atoms. The third of the fundamental forces described by quantum mechanics in detail is the weak nuclear force or the weak force. This force is the one responsible for the decay phenomena seen in various atomic nuclei and many subatomic particles. This is the basic domain of quantum mechanics as it seeks to more or less map out and measure the values of each particle, the interaction force between them and how they combine or decay to give rise to new end products. What quantum mechanics does not do in any sense is incorporate gravity into its explanatory framework of known particles and the three remaining known fundamental forces.

The idea of warped space is essentially non-existent at the quantum scale as it is currently believed by modern science that the amount of change in space of the quantum system is zero or so close to it that it does not matter significantly. So the net result of this is a beautiful mapping of everything by quantum mechanics but a complete failure to incorporate any apparent relativistic effects. So quantum mechanics does not explain any of the large scale observations made of the universe from gravity of things like stars and galaxies to unseen forces like the acceleration of the expansion of the known universe. Yet we know that quantum mechanics must be at least partially correct because it is made of repeatable and testable objects that obey certain known probabilities. So what happens if you simply incorporate the known laws of quantum mechanics into the theory of general relativity for large scale objects in the universe such as black holes? Well instead of one infinity value which as we have discussed is impossible in physical reality you get an infinite series of infinities each more impossible than the last. Remember that quantum mechanics is based off of the linear equations of probabilities existing for every conceivable possibility of a known system and when applied to the already infinite values of an object like a black hole produces the biggest impossibilities of all.

In other words you cannot get quantum mechanics and any form of relativity to mesh and explain the universe as the theories are described today by modern science. While one works well for the large scale and one for the small scale simply mashing them together fails utterly and this means that neither can hope to explain the universe as they both fail from different directions. If you look at relativity as the big scale top down approach to understanding the universe we get a failure to explain the cosmos. If we look at quantum mechanics as a small scale bottom up approach we get another failure in explaining the cosmos. To make matters worse no where in the middle of this top down and bottom up world do we find a happy middle ground. This means that both theories are incomplete or unfortunately just plain wrong. Now does the uncomfortableness of the current situation faced by modern science make more sense?

If it does not I will explain one more problem cropping up from both of the competing theories that details why they both fall short. This last problem both quantum mechanics and relativity share is that at their very cores they explain nothing of why the things observed in the universe do what they do and why. Let me put this another way and ask you to explain rationally and with known facts only why any particle does anything the way we see it using either theories. Take for example what we examined earlier using relativity and its thin explanation or rather its general glossing over of the why things fall in the universe. How does anything using a curved spacetime due to the presence of matter explain why nuclear decay occurs for example? It is okay to not answer as Einstein who invented the theory did not know either. Now ask a quantum mechanics perspective question of the same nature and inquire why nuclear decay happens as well? Quantum mechanics cannot answer this but it can give you a very detailed and precise account of the step by step chain of events. Only it has no idea why these events started in the first place and this is the crux of the problem once again as neither theory explains why things are happening and offer no mechanisms for why they might be too.

Lastly we are left with classical mechanics which is the third and final main view accepted by modern science. Some might be scoffing at the idea of classical mechanics being an accepted view of modern science as they believe it has long since been supplanted by both quantum mechanics and general relativity. This would be true right up until one points out that classical mechanics has had a far longer and more successful run than quantum mechanics and general relativity combined. Many of the rules and laws put in place by classical mechanics have stood the test of time and are still in use today and in fact efforts have been made to update them to fit with modern observations and unsolved problems. Classical mechanics has also enjoyed one other major benefit that both quantum mechanics nor general relativity can enjoy and that is a lack of newly created problems. Both quantum mechanics and general relativity have brought forth fresh unsolved problems to the forefront of theoretical science as we have seen with things like entanglement and singularities respectively. Answers to both of these phenomena do exist but are impossible to reach using the parent theories that discovered them. Classical mechanics on the other hand gave birth to no such unsolved problems and instead was aimed squarely at solving unknown problems by way of explanation through a sound model. True classical mechanics has not been able to fully explain all the unknowns but it at least created no new ones.

This leads us to the third plus in favor of classical mechanics which is the laws of classical mechanics offer the best explanation of why things happen as they do. Like it or not classical laws work best for real objects that we can experiment on and measure with ease and that you and I and the Earth are made of. Take for instance this book you are reading and think about its dimensions and mass for a moment. The book is far too large to ever be described by a purely quantum mechanical system and far too small to ever demonstrate and relativistic effects. This leaves the classical laws to explain why it behaves as it does and these laws fit better than any at this scale which is the scale which we as humans live at. Once modern science strays into the territory of the very small or the very large the two views remaining which are quantum mechanics and general relativity shine briefly before failing altogether.

There might be something more to the simple classical laws than modern science is giving them credit for as even the standard model works best with classical mechanics. The standard model is afterall a small scale look at classical macro systems whereby everything in the universe is acted upon by a force and that force is real and not the result of something like a bent spacetime. While all of what I have said is fact it must still be reflected upon however that classical mechanics does not answer the unsolved problems in the universe only that it offers the best balance between the worlds of the very small and the worlds of the very large. To this end it cannot be fully discarded as a better application of both quantum mechanics and general relativity to classical mechanics might fair better than any of the three alone in solving unknown problems in the cosmos.

Where Human Knowledge Stands According to Modern Science

So there you have it once and for all the final product of human sciences culmination of rational thought, experiments and theories for the last few thousand years or so. Were you expecting something more grandiose perhaps? In truth the current state of our species journey from pure questions to at least some answers as summed up by modern science is sadly a bit of a fizzle. Great progress has been made as we have adopted, tested and discarded old theories for new but the stark reality is that the three main views which represent the culmination of human scientific thought are incomplete or wrong. Furthermore they are mostly incompatible and in disagreement with one another meaning they have all at least partly followed the incorrect path to the truth. Since modern science has built its sole purpose to using these three views to understand the universe humans are now faced with the loss of a new direction to go in.

The term ad hoc gets thrown around a lot these days in science and for good reason as ad hoc is pretty much what has been done to all the three main views for at least the last century. The trouble is that the base theories themselves do not remotely come close to explaining the universe for starters and so great faith was placed in the newer ad hoc versions. These newer versions still do not explain the universe even with any caveats we see fit to grant them and this is of supreme importance in reflecting upon the three main views. The three main views in their original forms deal with very few and sometimes no unknown variables or circumstances in their attempts to explain the universe. This is not true however of the more ad hoc models as they all include multiple new mathematical inventions in order to work and even still with whatever is dreamt up for them is insufficient to explain the universe still.

More particles and more dimensions? Sure no problem invent whatever you want to explain the universe using whatever of the three main views you currently backup. The problem there is that these new variables just create more unsolved problems than existed before and now you have people attempting to get the universe to fit the mathematics rather than get the theory to fit the universe. If you need proof of this look at one of the most adored in modern science theories of the twentieth century, relativity and you see what I am getting at. With no proof experimentally whatsoever Einstein simply plucked curved spacetime from thin air and used it to explain gravitation in the universe. But you argue relativity does a good job of predicting gravitational effects in the universe and so it must be right on all counts. Except for the small problem that relativity predicts the existence of an impossible structure the singularity through its own curvatures and so has destroyed itself. One might quibble over such notions but that someone would only demonstrate their lack of creativity or worse their inability to accept new ideas as their minds have long since stagnated in the past.

Again you need proof of this? Hardly a problem as we all remember the fuss made when it was suggested that the Earth not be the centre of the solar system and indeed the whole universe. Or how about the fact that planets do not revolve around the Sun in perfect circles but rather imperfect ellipses? When the alternate and also correct new models were presented to explain the universe they were not met with love and adoration from a sea of open scientific or philosophical arms but rather with revulsion and barbed attacks attempting futilely to prove them wrong. Yet history has lumbered ever forward as does the arrow of time to which its yoke is affixed and bore out the newer models proving that the Earth rotates in an ellipse around the Sun and that both of them are quite nearly spherical and not at all flat.

Now back in the present time we are faced with models that cannot explain the observations of the universe once again just as old models or perfect circles could not account for the elliptical paths which the planets take. Today the old models being the three main views cannot account for the observations of the universe on its most grand and miniscule scales and just as attempts were made to ad hoc perfect circles we now have new attempts being made to ad hoc the three main views. This I do not see as a problem rather as perfect timing as modern science is keenly aware of this fact or at least a few of the keenest minds in modern science are aware of it. Seeing the proof that the three main views are incapable in any fashion of explaining the universe is a positive step forward that will hopefully ready modern science for a new model that fills all the holes in the current understanding of the universe and solves the rather lengthy list of unsolved problems while it's at it. Not to mention it will incorporate the good points of the three main views whilst doing away with the niggling troublesome aspects of them as well. A purely traditional model of the universe using the three main views will never explain the small scale and the large in one nor will it allow for the tried and true laws of things like classical mechanics, relativistic effects and proven quantum experimentation to play nice together. So the time is ripe for a new model that will finally solve the problem of how the universe works literally for good and be able to account for all observations no matter where or when we choose to look.

What is it up against? Simple, the big problems that remain unsolved in the universe and some of the biggest questions that have been mainly ignored in an attempt to make quick progress. Unsolved problems are one thing and they can be made of observations or experimental results but the other set of challenges which are the neglected answers are another matter of interest as well. One quick illustration of such neglected answers can be found in asking and providing a reasonable answer for what is space really? This may sound to some unimportant and to others it is of the most profound significance imaginable. Think about this for a moment and in truth ask yourself if anyone has ever provided an accurate description of what space really is? No I do not mean an aether, a spacetime or an immutable sphere what I mean is what is space really? None of these previous descriptions have actually said anything really as none have even remotely attempted to address what the physical properties of space is itself, what it is made of and how it interacts with matter or matter interacts with it. This topic will be discussed and answered later but for now it easily illustrates an area of debate in modern science that has been neglected in order for quick gains.

General relativity for example says that spacetime is a real structure but never describes what it is or how it works only that since it is curved we can ignore those trifling details and charge full speed ahead using spacetime for whatever we like. Yet we know spacetime does not work for everything in the universe such as quantum systems or singularities so it is not correct and yet it has been ignored in order to use relativity as a tool for getting some quick results scientifically. When Newton's ideas of space were found to be lacking in scope to describe the universe they were discarded and replaced by new ones and so relativity is in desperate need of replacing. I know as I say this a great number of scientific minds are in agreement with me and would welcome a new model to explain it all so to speak. Quantum mechanics is in equal need of an upgrade if you like as it too neglects what something as simple and important as what space is as well. Quantum mechanics provided a wealth of new information concerning matter, energy and how they interact with one another but never addressed from the very beginning how they interact with space. To this regrettable end we again come across an incomplete theory of how the universe works that was adopted and exploited as it provided a great number of ready to be plucked answers. No one stopped and said for a minute that before we go whole hog into utilizing both quantum mechanics and relativity we should really address unanswered fundamental questions such as what is space really? Remember that the idea that strings are real, that quantum field theory or the theory of general relativity are correct is false as they are only theories. Let go of these assumptions and entertain a new way of thinking, a new way of looking at the universe that is built from the ground up and aims to answer what something like space truly is. Decades of progress and ad hoc theories have passed and now the rift between the small and the large has blossomed into a wholly beautiful and unsolvable flower.

Chapter 13
The Flower, The Net and The Spheres

The Future or What we need to overcome and do to get there

Beautiful in that it screams out a need for something new and delineates the scientific eras of past ad hoc with much needed future new models and proofs. In order for any single new model to help modern science it must address all these concerns and more and do so at once. Gone into the past are the days when simply trying to explain gravity were enough to warrant a paradigm shift in scientific thinking. That truism is most profound today as new theories once again tend to focus on only one problem at a time and much like trying to hold an armful of perfect frictionless spheres aloft by focusing on one the rest slip from your grasp. In order to contain them all you must face this challenge head on and snare all the spheres in one all encompassing answer of a net where the spheres and their secrets of the universe cannot escape resolution. These spheres range in size and scale of description from small missing pieces of our current picture of the universe to massive and sometimes seemingly unanswerable question that go well beyond the scope of anything we have ever been able to measure such as what happened before the big bang, what will happen at the end of the universe and so on and so forth. The spheres however when weighted against one another are found to be completely equal in value as in order to answer the problem of how the universe works each one must be addressed simultaneously within the net of answers. This will allow for one more incredible possibility to occur with a one to one probability which is in doing so the questions that seem beyond scope such as what happened before the big bang will materialize with not close to approaching but one hundred percent certainty. This is due to the simple fact that all the spheres are now held within the same net of answers and in order for the net to work for one sphere it must work for all so in essence the rules that are used by the model to govern the actions of the smaller and more measurable spheres will enable us to know for certain the actions of the seemingly large and impossible ones too.

Space

Want to see what is in the net? I will list only the necessary spheres although any model that can reasonably contain them will definitely hold the answers to other spheres as well. The first sphere is the unsolved problem of what space really is and while that is a simple answer the underlying mechanism is a must for understanding how the universe works. The reason is simple as space is the backdrop if you will or the sort of invisible glow that everything else exists in. Space must be defined and understood in a new model system as it is important in how everything we see and can measure or observe works. If space were so unimportant then everything from the big bang up until anything we observe in the cosmos today would progress according to laws of the Standard Model and that model alone, however we know better. So space is the first thing that must be revealed and the sphere that follows is one that directly ties into space as do all things but in this case is a somewhat larger scale application of it.

Space and Matter Interactions

The next sphere shows how matter interacts with space in the first place. This is of course a much larger application of just understanding space because all matter and energy as well interact with space and does so across the entire universe. Thus the need to explain how matter and space share the universe together is integral in constructing a cosmos filled with all the structures we see from the Earth to the Milky Way and beyond.

Gravity

Now we come to sphere three which for many is a very much exciting and confusing one as it involves the final explanation of what gravity really is and in some cases more importantly how it works. Yes the answer to gravity is finally given as the interaction of matter and empty space is understood and this is obvious to everyone no matter which of the three main views they might currently back up as we all know that matter can be thought of as making gravity. This is technically true and I will explain exactly how gravity must work in order to produce the large scale observations we see everyday throughout the cosmos but be created from quantum level particles. I told you sphere three was going to be interesting and it leads right into the fourth sphere which reveals some unexpected effects of gravity.

Black Holes

This is illustrated by using our long time cosmic friends the black holes which as many have suspected for decades and certainly in recent years hold the key to many secrets of the universe. Well a secret untold is a secret still, a secret told is one no more and the black hole is telling us some of the secrets of the universe rendering them no longer mysteries but truths. Thus an understanding of the once strange workings of black holes is needed to answer how the universe works as they make up a decent portion of it.

Expansion Phase

Sphere five now takes us to a particular application of gravity now that we understand space and matter interactions perfectly. This application is the early expansion of the universe after the big bang which seeks to smooth out the universe to produce the roughly homogenous one we see today. That fifth sphere moves quickly into the sixth sphere which looks at how the interplay of space, matter and the resultant gravity can be used to explain the deceleration and acceleration phases of the universe following the big bang and leading up to today.

Rate of Acceleration of the Expansion of the Universe

This is a key sphere as it provides the mechanism for the history of the universe which we have been able to surmise and couple it with space and matter. In other words the explanation given by the sixth sphere is observational proof that space and matter must interact in a set way. Since this deals with a large scale application of the space and matter interaction mechanism it leads to the number seven sphere which incorporates the infuriatingly missing matter and energy we believe we are indirectly observing in the universe.

Missing Matter and Energy

So this will address missing components of the universe in a way that does not introduce new particles or energies as that would only complicate the universe further than is needed. Like tracing a path from A to B one can measure distance or displacement and realize you do not need extra side trips to go from A to B. The shortest route is the one favored by the universe in terms of the smallest number of variables and so an explanation of missing matter using less stuff is better than one using more stuff as will be illustrated.

Fields

Sphere eight contains a more in depth look at what forces and fields are as they permeate everything in the universe yet are not well described by the three main views. Any new model must provide an explanation of what these fields such as gravity and magnetism for example are and how they work. Gravity pulls stuff from areas of little mass to areas of high mass; magnetism attracts or repels based on arrangements of magnetic poles. We have all heard that for quite some time but ask anyone in modern science today to explain what is going on in terms of the actually workings of fields and you get nothing but vagaries or a total glossing over of the question just posed. This is therefore one of the important and unanswered questions that need to be addressed by any new model if it is to encapsulate everything in the universe into a single solution. The exact workings or at least much more specific workings of things like gravity and magnetism have been largely side stepped in order to focus on making measurement or harnessing them for useful applications. This is all well and good and has undoubtedly led to many important areas of science making rapid progress but have left other areas such as a fundamental understanding of the universe at its most base levels a back burner priority. The unsolved problems of the universe are not going to be solved unless sphere eight can be resolved and the more exact workings of invisible fields resolved. From sphere eight we can try to apply what we have learned about fields into more Loki like aspects of the unsolved problems list and enter sphere nine.

Quantum Entanglement

Sphere nine must hold the key to very real and very unanswered problems as faced by modern science such as quantum entanglement. Not only is quantum entanglement a real phenomena and not only is it one being exploited but not understood in laboratories today but it holds the key to the idea of non-local events and actions. Solving quantum entanglement by itself is necessary for any theory to truly answer the biggest question possible of how the universe works but will also answer locality and this is one of the biggest concerns in modern science. The entire basis of modern science and the three main views relies on non-local actions not being real and yet they are so they must be addressed before any further progress into solving the universe can be made. Remember that just because the three main views do not have room for such things does not mean they are imaginary. In fact we do know they are real and so must stop trying to make the universe fit the three main views and adopt a model that explains these events even if it is only a first step in understanding them.

From here we come to a series of spheres that will solve numerous unsolved problems more or less one at a time but will do so by utilizing the rules laid down by the previous ones. In truth a general grey area surrounds all the spheres as after all they can all be described as a part of the whole which is of course a single model that explains the universe. I will differentiate the spheres into technically two groups the ones that tend to introduce the big concepts and the ones that tend to explain the big concepts by way of observation. That being said they are fairly universally interchangeable but must be separated so that they present the solutions in something of a reasonable order. This therefore brings us to sphere ten which contains the solution to the matter over anti-matter problem faced by modern science and is really not a problem.

Matter over Anti-matter

The universe has created itself and evolved as we see it because that is the way things work and so no problem exists only a hole in the knowledge modern science possess and sphere ten has the answer. The solution to the matter over anti-matter problem for any new model must be explained using the rules laid out that govern the cosmos on its most grand scales. So the laws of space and matter and how they interact provide the answer to why the universe chose matter over anti-matter. To this end applying these rules will solve the problem and be even more proof that this new model does in fact explain everything as all the spheres are proofs themselves of the entire model being correct. Since this is a large problem to solve, the entire universe, far more examples of proof will be useful and demanded in establishing any new model as fully correct. So any new model must incorporate a whole plethora of current observations and unsolved problems faced by modern science in order to fully cover the cosmos.

Dimensions

Further problems found and answered in sphere eleven will also provide more proof of any new model being correct as it deals with the issue of dimensions in the cosmos. The specific aspect of sphere eleven that is most interesting is the number of total dimensions to the universe. The word total must be used here as modern science currently understands dimensions to be made of various different types. There are the normal three spatial dimensions that everyone is familiar with, the dimension of time which is unique according to modern science and made of a different material than spatial dimensions. Two more types of dimensions have been proposed as well by modern scientific theories and these are multiple hidden dimensions such as those found in various string theories and usually take on the typical compactified shape as well as brane dimensions which appear to be some sort of ethereal backdrop separate from space, matter and time and yet interacting with them again depending on the theory.

Needless to say the total number of dimensions must be summed if one is to understand the universe and in addition to that the number of and types of theoretical dimensions in the universe also must be rectified. The correct theory will be the one that uses the least number of dimensions possible as the universe never tends towards more complicated explanations of itself. So any new model that is correct will have to nail down the dimensions problem once and for all and be able to explain universe in as few dimensions as possible. Once we have the number of dimensions neatly tucked away we can examine the universe on a far off future level at what some would say is the end of all dimensions.

Fate of the Cosmos

In other words sphere twelve will contain the answer to how the universe will finally end its life or start a new one. The fate of the universe has been proposed multiple times in multiple ways and yet scientists know that it really will ultimately only have one end and so any model that explains the universe will find that one true end. A very exciting sphere to be sure as much debate over the fate of the universe has raged since the dawn of human time and been fought back and forth theologically, philosophically and scientifically for thousands of years. So this new model will have the answer that cannot be refuted despite what the end of the world myths predict and that is rational music to scientific ears.

Information Loss 1

Another particularly interesting sphere for the truly science minded is contained in sphere thirteen and involves the idea of information loss. This is a fantastic issue of debate and discussion in modern science as there are those who regard the sanctity of preservation of information as a definite property of the universe and yet nothing says that this must be so. Like non-local events and spooky actions at a distance not everything fits into modern science the way modern science wants. It is only currently assumed by modern science that the speed of light is the fastest in the universe, non-local events are impossible and that information cannot ever be lost. Long before our time people thought it was impossible for the Sun to be the centre of the solar system and that anything else was plain wrong as it did not fit into our view of what was then modern scientific thinking. Sphere thirteen will answer this similar problem of information loss and reveal whether or not we can hold onto what modern science cherishes as resolute truth or open our minds and breathe the fresh air of a new and final truth. Regardless of what the answer is any new model must deal head on with this question of information loss if it is to explain truly everything in the cosmos.

About Time

Sphere fourteen is just as much a paradigm shift as its predecessor sphere in that it will also reveal a great truth and a very fundamental one about the universe which is of course time. We have discussed before the idea of time and whether or not it is real and the most I was willing to say in each of those instances was that the question remains unanswered. This is the truth about time for modern science as it remains an open question and therefore an unsolved problem. Time may be a part of space as is believed to be true by one of the three main views relativity. Time may not necessarily be true as is believed by certain quantum mechanics and many philosophical arguments. Time may yet also be some sort of combination as seen by Classical mechanics with time possessing two different modalities. All three main views and modern science are stumped as to what time really is so any new model of the universe must finally pick a side in the time is real or fake debate. This choosing of a side will not only reveal many of the inner workings of the cosmos in accurate detail but also represent another shift in thinking towards new futures as time will once and for all be held in a fixed place in science or it will be discarded. Like local events, the speed of light and so on time represents one of the key concepts that modern science has had to take for granted in order to move forward and is another good example of a basic fundamental problem that was never resolved properly before moving on. Thus sphere fourteen will answer this problem and once and for all and present time in such a way that it can be said to hold to one side or the other while always fitting with the observations of the universe.

Entropy

We can now move forward like the arrow of time to the next problem in our list which also seeks to answer one of the most basic and yet as poorly understood concepts by modern science which is entropy found in sphere fifteen. A very basic evaluation of time is that it always flows from the past to the present and into the future and inexorably linked to this is the concept of entropy. It is believed by modern science that the entropy of the universe very much enjoys a similar arrow of entropy that always moves forwards. This arrow increases in entropy and not time and as we all know is a basic measure of the distribution and disorder of things and events within the universe. The massively head scratching question here is can entropy be reversed or to put it another way can we get the universe back to some more ordered and lower entropy state as when it began? The laws of physics do not dictate anywhere that the flow of entropy could not move into reverse only that the probability of such a thing happening is so mind bogglingly remote it approaches zero. Yet the universe began at a very low state of entropy and so a very simple and innocent question is can it return to such a state perhaps through some exotic balancing mechanism involving reverse time or alternate universes? Both of these concepts have been explored and are virtually untestable so any new model that wants to be accepted must accommodate a testable idea of recycling entropy even if that idea is only mathematically testable. Sphere fifteen will therefore contain an answer to the entropy question that uses the least number of variables possible which mesh seamlessly with the same rules that explain the rest of the universe and can be at least mathematically proven. The answers found to both time and entropy might see their arrows left alone, reversed, formed into circles or discarded altogether but no matter the outcome a new way of thinking about the cosmos will be proven to and thrust upon science the likes of which cannot be ignored.

Matter

Sphere sixteen will be a small departure from the concept of wake up calls in thinking but it will still answer a fundamental question that has been asked for hundreds to thousands of years or more and never answered. It is the problem of the true nature of matter and in terms of an unsolved problem that science has glossed over in an effort to make quick gains somewhere else it has no equal. The reason is quite simple the true nature of matter is probably one of the earliest questions asked right after the biggest of all questions which is of course how does the universe work? Humans have been looking at all the stuff around us and that includes ourselves in an attempt to understand what it is that we and everything else is made of. We touched on this far back at the beginning of the book with the ancients and their various rational explanations for matter as being something akin to the four elements. That is how far back the question goes and as you can see predates pretty much all the other questions by a good several thousand years or so. Yet this is one of the key unanswered questions that science has been forced to bypass in order to make more progress. Science and this includes all modern science has not bypassed these questions as some sort of information get rich quick scheme but rather out of necessity as the question to this day remains unanswered by modern science itself.

Yet here we are at the end of the road for the three main theories which are the culmination of thought about this very problem over the last few thousand years and most likely beyond. We have access to tools never dreamt of by our ancient ancestors and have probed matter to its very limits only moments after the big bang. This is the end of the road for the three main views as none can explain what matter truly is and certainly not in a way that agrees with the other views. So this question must be properly addressed with a new idea on what matter is if any model is to attempt to explain the universe. Indeed this problem must be answered before most of science can progress any further as we have unraveled the atom down to sub-atomic fundamental particles and even some that may not ever be real and exist only in theory. So a new model for the nature of matter must be proposed and it first and foremost must be compatible with space and in fact its interactions with space must be the number one concern for any new model as it is known that the universe is neatly made up of both matter and space. Sphere sixteen will surely hold new ideas that may not be complete departures in thinking but will certainly prove to be of terrible and sleepless fascination to many science minded individuals.

Light

Sphere seventeen will also prove to be very interesting and deal with another key aspect of the universe that has been heavily studied like matter but not fully understood also like matter and it is of course light. Yes the humble and ever important photon or light wave is well measured and scrutinized but at its core not fully understood. Think about it for a moment and you realize that everything in the universe which is matter is amazingly complex and bound by four forces at least. From it we can construct planets and people, stars and galaxies all with massively different properties and characteristics. Yet light is only ever light. Sure it can be more energetic and sometimes more nonchalant in its approach to how much energy it carries but that is about all. You can never combine light to form light or energy atoms and this means no light molecules no light galaxies no light anything and just light. So why just light? What is it about light that makes it always travel at the speed of light and accelerate instantaneously to this speed regardless of anything else like motion?

All other substances in the universe are bound by countless rules and laws even forbidding them to travel at light speed. Relativity we know by now to be incorrect in a great number of aspects and so we should not look to it for answers concerning light and yet light does seem to obey things like the presence of large gravitational fields. It is as if light is something separate from everything else in the universe as energy is not matter or we would find it behaving like matter. This point is fact and not opinion or light would be dancing all around the Higgs field and yet it does not. This is a clear proof that whatever matter is it is something different as it slows down enough to interact with the Higgs boson and do some neat tricks whereas light is too hurried to be bothered it would seem. Therefore the nature of light or at the very least a better understanding of why space and matter affect it they way they do must be given in sphere seventeen before any new model can once again incorporate everything in the universe properly.

Electron Drop Down

Another curious aspect of physics that has so far eluded modern science is some of the additional strange properties of quantum systems and these can be found in sphere eighteen. A very simple one and one that has been troublesome for a while is the simple act of an electron dropping down an orbital. Yes that really is it to be exact an atom absorbs a photon and gains more energy the result of which is the electron or electrons orbiting it jump to a higher orbital and then jump back down again. As the electron jumps back down a photon is emitted from the nucleus and everything is pretty much the way it was just a quantum instance before. Modern science understands some of what is going on here but not all and in the case of the atom on a quantum scale no solid answer has yet been given or accepted. Certainly not one that would be able to fit with the rest of the unsolved problems of the universe. So the atom gains energy and increases the orbital of the electron in orbit around it and modern science has measured the quantum amount to a tee. What modern science does not understand is why the electron chooses to suddenly drop back down when it appears if nothing else has interacted with the atom. In other words a visible action is recorded for why the orbital energy increases but when the orbital energy decreases no visible action is seen.

This simple act which occurs almost an infinite number of times every second throughout the universe seems to be doing the impossible which is violating the known laws of physics. An object is not remaining at rest until another object or force interacts with it but rather it is doing what it wants on its own with no external influence. It is as if another spooky action at a distance was occurring and to make matters worse it is very well documented using one of the most accepted of the three views which is quantum mechanics. So we have another unsolved problem of modern science that is simple in nature and yet requires some new model to explain as the three main views have once again come up short. Sphere eighteen must be accounted for if any new model is to explain everything in the universe and do so in a way that can still be measured in quantum values as part of the action here has already been measured using quantum amounts. This is a very strange puzzle for modern science as part of the homework appears to be done already but something fundamental and profound is clearly missing only modern science cannot see it.

Information 2

Another problem modern science cannot see and is as yet aware of its existence is information in the universe. Sphere nineteen will revisit sphere thirteen as well as information and not the typical kind you encounter everyday but physical information of fundamental objects. This is the kind of information that scientists use to essentially catalogue everything they can about the cosmos. It might be the particular properties of a fundamental particle such as charge or spin and it could also be the way two such particles interact with one another. These interactions will undoubtedly create a change in the basic properties of the two particles as they might do something as simple as bounce off each other and change direction. The change in direction means the information needed to describe the behavior of those particles in the universe has now changed. Add this to the fact that the universe is made of nearly countless particles and countless interactions and the amount of information involved is staggering.

The problem that remains unsolved here is whether or not in that huge jumble of interactions and information flow something can occur that will cause the universe to lose information. Information loss in the universe is currently thought by modern science to be quite impossible and any ideas that crop up suggesting information can be lost are usually attacked with extreme prejudice. This does not mean that just because the current winds of favor blow towards keeping information preserved in the universe information cannot be lost. In fact there is a strong possibility that information can be lost at some point in the universe and a new model explaining how this can be done would most assuredly shake the foundations of some scientific pillars.

Currently it is thought that if given enough computing power and time the previous conditions for everything in the universe can be calculated meaning that you can always back track to recover your information and recreate the initial conditions of the system in question. If it can be shown that information can in certain cases be lost from the universe it must be answered in sphere nineteen and must be done so using the same framework that explains the rest of the universe. If all of this can be done at least a little bit so to speak any new model would have made significant progress in proving itself correct. So the combined answers from sphere thirteen and nineteen will finally show the complete truth to information loss in the universe, entropy and do so in a way that meshes not just these two spheres but the explanation of these to things like black holes, quantum mechanics and space physics.

Vacuum Catastrophe

Sphere twenty is a big one for sure as it deals with a very big mistake and that is of course the vacuum catastrophe. Science is no stranger to making mistakes and in fact many of the best breakthroughs come from things not working out as predicted and giving unexpected results. Rather than be a hindrance to the forward progress of human understanding of the cosmos it turns out that it is quite a serendipitous occurrence. This is due to the fact that every time something does not work as expected it usually reveals new clues about the correct path to follow and if not it at least tells you that the one you are following is a dead end and should be abandoned. This is very much what is happening with the discovery of the vacuum catastrophe in modern science today. The vacuum catastrophe arises because the predicted value of energy in the universe is around 125 orders of magnitude above the measured value for the vacuum energy. This massive gap in values is of course a catastrophe and lets us know with perfect certainty that no fine tuning of the models or data is needed but rather that with such a large gap something is dead wrong and must be dropped like a hot coal.

The very high value for the vacuum energy comes from quantum field theory which is theory only and so is of very low weight. Quantum field theory has made some useful predictions however and so it cannot for now at least be fully ignored as it might still be right. The measured value comes from real experiments here on Earth in laboratories which can be closely monitored and repeated so they carry great weight. The experiment is the now famous Casimir effect whereby two metal plates are positioned parallel to one another and with a gap between them of only a few nanometers or so. The idea is that the vacuum energy in space fluctuates according to quantum mechanical laws and resembles familiar waves where some can be longer and some shorter in length. The closeness of the plates can be generally thought of as excluding the longer wavelengths of quantum field energy stored in the vacuum and as such only the shorter waves are in between the plates. Since you still have both kinds of waves outside the plates you have more energy and therefore more push to be applied to the plates and this should result in the two plates being moved together. In theory at least this is how it works and while the Casimir effect has been conducted several times in laboratories over the decades since its inception in 1947 by Dutch physicist Hendrik Casimir the exact reason why the plates move together is still debated. The problem is that it is unclear as to which is correct. Is the prediction of quantum field theory so high as to be absurdly wrong or is something else causing the experimental results that is not vacuum energy?

The Casimir effect as measured if it is indeed the correct mechanism at work would prove that the vacuum energy of the universe is the much, much lower of the two values. If the idea behind quantum field theory is correct that every quantized volume of space has stored energy for all the normal forces in it then the much larger value must be accepted. This is because the energy is stored in the quantum field theory in the vacuum of space even if it is not measurable at that time because there is no change in its distribution. In other words just because nothing is present in that space it still possess all the necessary energy to be measured properly if something where there. This results in the hugely large number of vacuum energy and the catastrophe is born. While a lot of speculation still surrounds this failure of modern science and it is hazy as to which view is more correct any new model of the universe could at least offer a helpful hint as to which direction further investigation should go whether that be the big value or the small value. To this end sphere twenty will demonstrate an explanation for which side is correct in the catastrophe wars and move modern science towards the truth with the result that any proof made in that direction will lend validity to the model that created it. By now we have reached a point where most of the major problems have been dealt with and are now faced with ones that are somewhat more specific in nature.

Supernova

Sphere twenty-one holds the key to most of the remaining unsolved problems in that there is no particular order to put them in. The exact mechanism that drives supernovae might be found here as modern science still does not fully understand the full sequence of events that leads to a violent supernovae or hypernovae. The three main views are at a loss for providing a reasonable walkthrough of the start to finish of the whole explosion process as can be explained using small quantum sized fundamental objects that result in a very large macro scale event. In other words we have another illustration of a need for a correct model to explain the linking of the small and the large in the universe and the example of a supernova is simply one more way to demonstrate it. Also by now we have looked at matter more closely as a proper understanding of what matter really is is necessary to fully understand the universe. The exact nature of matter will also help explain things like supernova and this is only one small example of the positive feedback loop that will prove any new model as one that truly explains the cosmos.

So as far as matter goes we can now look at smaller details such as the generations of matter which appears to come in threes. This is of course especially apt for things like electrons and quarks which have three generations of particles. Each generation has a heavier version of itself and retains the same properties such as charge in for example electrons. An electron is negatively charged and has a certain mass the next heaviest generation of the electron is the muon which is still negatively charged but is heavier. After the muon comes the tau which is also negatively charged and heavier still than the muon and the tau decays the fastest after which the muon decays and lastly the electron which is exceedingly long lived and stable. So it is easy to see for example using electrons three generations of matter that go up in mass while going down in lifespan. Why is this so? A simple question but one that modern science cannot answer using the three main views as these exotic versions of the electron the muon and tau respectively seem to serve no real purpose in the universe. They are created and then quickly disappear so why does the universe need them and if it really does not need them then what makes them the way they are?

This is the sort of problem that is faced once again in sphere twenty-one as a more specific problem to a larger one that has been solved before. The solving of these smaller but no less important specific ones will further help to prove the validity of any new model of the universe. By now this kind of thinking should be obvious to anyone as a new model of the universe that is successful in solving the big sweeping problems will no doubt help solve the smaller ones too. Sphere twenty-one therefore is a veritable buffet of scientific tidbits for the list of unsolved problems to snack on.

The Answer

After sphere twenty-one the last sphere is twenty-two and while many more could be added as they would likewise contain many more specific problems like sphere twenty-one we can now move on to the last sphere in the net which is essentially a summation of the achievements of the previous spheres. Sphere twenty-two will therefore draw a common thread between all the solved problems of the previous spheres in such a way that it demonstrates the proof of the new model. A core set of ideas and rules will be made present so that the solution the new model provides to how the universe works can be summed up into a simple beautiful answer. In many ways the spheres preceding this last sphere are solutions to unsolved problems that are also proofs of what this last sphere is saying about how the model explains the cosmos. The reason the models core system is fully explained last is due to the fact that it is a new model and as such will face may preconceived oppositions to its truths. Like any new model of the universe the status quo must be overcome and the gentlest way to do this is to show how it can solve problems that other theories such as the three main views cannot either partially or fully. Once all proof has been shown even if it is in what many might consider a rough form sphere twenty-two will assuredly show that this new model is actually a diamond in the rough instead. A beautiful answer to the oldest question to ever occupy the curious minds of our species and all intelligent minds in the galaxy and beyond which is simply put how does the universe work?

Chapter 14
New Universe Order

What I have done

The universe is awesome I love it and it is where I keep all my stuff. Universes are so cool when you think about it for a moment that everyone is going to want their own pretty soon. They can make life from lifelessness and they seem to appear out of nowhere which is kind of like making their very own little universe lives out of their own little universe nothingness too. They are naturally litter box trained and make for the perfect pet which is why I think they have been so wildly successful a hit with all intelligent life in the cosmos. Yet the cute little darlings are still tricky to understand and if we want to be proper owners of our universe we need to understand it fully. Now we must find a way to answer the oldest and most difficult of questions which is how does the universe work? The problem of how the universe works is made easier by looking at what a universe is. A universe is as we have seen already is everything there is with nothing left out and this means any alternate dimensions or alternate multiverses should they exist are still within the domain of the one true all encompassing universe. The trouble that humanity has faced since it first asked this question is that it does not know how to put a universe together only that we live in one.

Take for example the two things you need to build a universe and you end up with space and stuff. At the beginning of this book I pointed out that humanities quest to understand the universe comes from its desire to make sense of the stuff around itself. This stuff is the matter and energy that we experience everyday and are made of ourselves. This is one part of the universe and has received the vast majority of the attention of study over the millennia. The other part of the universe is the space that the stuff lives in and it has easily received next to no attention whatsoever when compared to the stuff. Have people tried to come up with ideas about space before and the answer is of course yes. The trouble is most are exceedingly superficial and usually fairly easy to find flaws with and so people give up on working on the space part of the universe and settle back into more comfortable environs of looking at the stuff again. This cozy armchair view of the universe has resulted in a lopsided understanding of the universe as a whole and led to the failing of any one model to explain the universe. The key without a doubt to solving how the universe works is to understand space far better than ever before and to insert that knowledge into the building of a universe. Yes I am aware of how many times I am saying universe in each of these sentences and I feel that it is warranted because we are at the end of the scientific journey as the understanding of everything really is a terminus point and that point is the entire universe all at once.

It is time to finally face the entire universe head on in order to create a new model that can adequately explain it as we have assembled all the pieces we need to do so. The stuff side of the universe is half of the pieces and as recent non-developments of the last century or so have shown no amount of ad hoc to the existing theories is going to magically fill in the holes of our knowledge. There are holes in the knowledge of stuff and we can fill them but we need space to do that. This is not just because modern sciences understanding of space is so limited or blinded by trying to fit space to fit with current theories but because in order to understand matter and the universe we need all the parts and space is missing.

Think about this for a moment and you will remember that matter appears to affect the behavior of space and likewise space affects the behavior of matter and in far more profound ways than many might expect. We need more than ever the concept of space to be presented in its truthful form so that the interaction of space with the stuff of the universe can solve our matter problems as well. The missing details of matter are there because matter interacts with space and without fully understanding space one can never hope to fill those details. This is the single biggest challenge in solving the universe in one single model over the ages and it is the interaction of space and matter.

They do not exist separate from one another and they do not go about their business very much indifferent to the other. Instead they constantly interact with one another every conceivable instant forever and they will never stop. The correct fusing of the two elements of the universe being space and matter will explain three things perfectly. The first is the nature of space and matter as the behaviors of both are affected by both. The second is the interaction of the two to create a universe as we see it now and as it has existed in the past and how it will exist in the future. The third thing it will explain is all of the unsolved problems currently faced by modern science and by solving these problems using a single mechanism it will prove itself correct. The universe works only one way and any model that can explain the phenomena seen within it is by definition correct as the phenomena themselves are physical proof existing within that same universe.

To this end there is no real way to present this in a start to finish kind of order as the universe is everything and starting at any one point is as good as starting at any other. What for example is the start point on the surface of a sphere? Starting from anywhere will easily lead you to any other equal anywhere and so on and so forth. Simply jumping in somewhere and beginning to explain the space and matter interactions will eventually lead to a complete understanding of the universe. These interactions take place throughout the age of the universe with some being more relevant earlier on in the age of the universe and some later on. The truth is no one interaction is more relevant than any other so we could just as well start at the end of the universe or the middle and still easily explain the beginning. Nevertheless explaining some of the interactions will explain the rest so starting roughly at the beginning of the universe and moving forward will be helpful. At any time necessary the various middle bits can also be included to complete the picture.

What modern science is lacking and what it has been looking for actively for the last century and more is a complete model that does finally solve the problem of how the universe works. One single model that can incorporate the whole of the universe and finally merge the realm of the ultra small with that of the ultra large and this means finding a way to reconcile the quantum mechanical world so that it can account for universe sized observations. To date modern science has not been able to do this in a way that allows a single underlying mechanism to solve every problem in the list of unsolved problems and without creating copious amounts of unknown variables. Various theories have been proposed to deal with a single problem at a time but none have been able to solve them. In addition to this each of these individual theories is not designed to and cannot solve any of the other problems on the list. So what you are left with is multiple theories for multiple problems and in some cases the problems persist with no viable theories to explain them at all. This is the sad truth of the state of affairs concerning modern science and the unsolved problems in the universe. It is also the sad holding pattern to which all of human understanding of the universe has been forced to uncomfortably settle in.

Here is where I begin explaining my own work and as I said at the beginning of my book I will be using a gentle conversational tone to everything I explain. This means I will be slowly building upon the explanation of my work section by section and for the absolute last, smallest and minute inner workings of the universe you may have to be patient and wait till the very end. I promise though, that as my work progresses I will be adding more and deeper details to how everything in the universe works until all unknowns are simply gone. When I say unknowns I mean that when I am finished there will be nothing left to chance and nothing left unexplained. Modern science is too full of things that are simply stated to occur but no reason as to why. Take for example the act of something getting hot and you have modern science saying that as things get hot they emit photons. Seems harmless enough but if you ask modern science the exact why of this fact it will come up short. Explain modern science, exactly where a photon comes from inside a hadron or from the quarks and gluons inside of it? Why a photon? See where I am going? Modern science is great at saying yes we measure a photon being emitted and we know to the quantum level what the strength of that photon is but that is about it. So by the time I have finished explaining here at least my work you will see why something as simple as this occurs. In fact you will see using my work why everything happens in the universe. This is the point of my casual approach to explaining how the universe works as this will allow me to gently introduce each new concept as needed to build towards the final answer. So once again thank you for your patience up until this point and thank you for your future patience for my work to come. I assure you though that for those who can be patient and understand what I have written you will be rewarded with the true and only way the universe works.

So I will now propose my own model to explain the universe and once and for all solve the problem of how the universe works. I have been hinting at this for some time throughout this book and the perceptive of you will no doubt have clued into this fact but now it is time to fully reveal my work. I will begin my journey here and ask that you come along with me as an observer to my work whereby I will explain each and every working detail of the universe thoroughly. I will explain every unsolved problem fully not just some and also I will walk you through the universe from not just its creation, life and death but also explain how it and everything existed both before it was born and after it dies. I will tie all of this together into one single model system that I have discovered that I fully believe, in fact that I know to be the true way in which the universe works. I make this statement due to the fact that I am able to use my one model to explain each problem on the list of unsolved problems. In other words instead of using multiple theories for multiple problems I can show how all can be solved using a single model. This same model will tie the small to the big and everything in between as well.

It is a long journey where I will take time to carefully detail how we know the universe to work in a certain situation given the king of all facts which are observations and then I will show how my work can explain each of these observations in turn. What I will do is finally reveal what space really is and in so doing answer that ancient question of what makes the universe besides the stuff that we see within it. A successful fusing of space and matter has been elusive throughout history but now I can show how space and matter can be made one. I will then show from this how the interaction of matter with space creates the universe as we see it today and as we have observed it billions of years ago and how all of this fits perfectly with our understanding of the nature of the stuff that makes up matter including us. This model explaining the fusing of space and matter and demonstrating the interactions that they share will also result in an explanation of testable phenomena. It will explain the apparent anomalies of things like quantum mechanics as well as measureable observations being made here on Earth and in space. In other words it will be falsifiable as I will lay out the basic rules of how space and matter interact such that these rules can be scrutinized and tested.

◄New Universe Order►

What I will show with my work might represent systems that will require years of research before the apparatus necessary to test them fully can be developed but they will be testable in full eventually.

I invite everyone reading this and examining my work to look at it with an open mind but also a scientific one that will not accept it unless it is proven one hundred percent correct. If you follow me on this journey I would like to show you all the ways and all the problems it can solve in the universe and I hope that by the time I am through explaining them all you can also see that this one answer is in fact correct. I say this because even before the journey begins as a single model that can rationally and logically explain all the unsolved problems in the universe using the same mechanism each time cannot be incorrect. There simply must be something to it. No matter how it is first received I believe the brightest and most forward thinking minds will see in it a truth that I hope moves us from a stagnant understanding of the universe to one where all the secrets of the universe are laid bare before us.

We have seen the steady advance in our understanding of the world around us from our ancient ancestors until today and the failures of the three main views. From here the smooth flow of our knowledge will not cease with the three main views but instead it is time to move beyond them. It is time for humanity to once again move forward in great strides and change its thinking as we have done countless times before. No longer do we believe the Earth to be flat, isolated and at the centre of the universe and now it is time for us to no longer believe that the answers to the problems of the universe are beyond our understanding. The theories of quantum mechanics and general relativity came about as a need to understand the universe and using their respective tools cannot solve said universe. So now I would like to keep the observations and experiments of the last hundred years that do seem to work but start again and build a universe from a new perspective. Let me introduce new concepts that allow for the universe to be created as we see today but are not bound by quantum mechanics or general relativity. Come with me on my journey and I will show you the solution to the greatest problem of all: how the universe works.

Chapter 15
Starting Off Small

Space

The first thing I need to discuss with respect to my work is space. Space has many meanings and there is considerable overlap and confusion when distinguishing between the various usages. This is due to two factors the first being multiple acceptances of the word space come into and out of favor over time. The second is that as I have said before the true idea and nature of space has eluded modern science and a degree of severe ambiguity surrounds how space is described. To these ends I will describe what space means in general to most people and what space means to me as well as define the true version of space. This true version of space is of course part of the major framework of the universe with the other being matter or the stuff we see all around us. When most people think of space they think of the cosmos and basically the region that lies between things like stars, comets, planets and the like. This region is seen as cold, harsh and a vacuum that exists devoid of matter and energy unless something moves through it. Think for a moment of absolutely nothing and a single photon or hydrogen atom drifting lazily by and you have an idea of what most people think of as a cold space vacuum.

As for the three main views each has a slightly different idea of space and we will start with classical mechanics. Classical mechanics views space as a backdrop to which everything else moves through and these objects obey classical mechanical interactions. The classical space is also known sometimes as absolute space and is seen at its limits to never move or change shape. This is perhaps the familiar and warm view of space most people first have even as children as the universe makes the most sense to them if organized this way. The second view is relativity which sees space exactly the same as classical mechanics whereby space is a backdrop that things move through but here space is not absolute. In general relativity space is actually flexible and can be bent or curved depending on the amount of matter present in it. This still allows for some classical laws to hold but introduces the idea of a non-rigid spacetime. Time is of course the second way that relativity differs from classical mechanics in that space and time are one and the same material which is the opposite of classical mechanics. Those who learn about the cosmos are often taught and accept this theory of space second.

The last of the three main views which is quantum mechanics most people are never taught and the ideas are quite different from both classical mechanics and general relativity. In quantum mechanics the idea of space is that it is a vacuum filled with energy that can be excited in order to create particles like force carriers but it cannot be flexible. This background vacuum energy or field energy as it is known in quantum mechanics is thought to permeate all space with the areas devoid of matter being truly empty. Some versions of quantum theory allow for particles to spontaneously pop into and out of existence in accordance with laws of quantum uncertainty concerning the vacuum fields. This is in direct contrast to classical mechanics and general relativity which see space as empty with nothing in it and this explains why general relativity in particular fails to incorporate quantum systems completely. On the other hand the quantum mechanics view of space is unable to explain the large scale bending of space as it deals solely with the ultra microscopic scales of the cosmos. We have now seen the three main views and how they think of space each with some advantages and disadvantages and what the word space means to each of them in particular.

My work deals with a different kind of space altogether that can incorporate the small scale effects and still be scaled up to explain the large scale ones too. In order to do this I describe space as Empty Space. The reason I do this and what I mean by empty space will be made more clear as I progress through the various unsolved problems of the universe but for now I will give a brief description of what I mean by empty space. Empty space is the true type of space in the universe and it is one that is totally devoid of all matter, quantum effects and time. This type of space can and does exist under certain conditions which I will explain later and occurs somewhat frequently in our universe. Matter can exist independently of space under extreme conditions to the extent that a field or vacuum energy explanation of space does not work. Likewise spacetime is also equally unnecessary as the apparent effects of time can be removed from certain systems given the right and again rare conditions. These conditions however are not so rare that they exist say only at the beginning and end of the universe instead I insist that they can exist in observable ways even today and we have observed them. For now however I will focus on the idea of Empty Space and show you what the general build of empty space is without matter and from there I will of course build up to a fully interacting universe of empty space and matter so that you get the familiar cosmos we see today.

ESS

The largest part of the universe which is space is made of the smallest parts of the universe. Empty space is not one large contiguous fabric of space and time or one set of fields as currently believed by many but rather it is made of the smallest bits imaginable. Empty space is made of a near infinite number of small structures which I will be referring to as Empty Space Structures. These empty space structures or ESS will be useful when explaining interactions of space and matter as well as interactions between individual structures and their neighbors. The basic geometry of an ESS is a sphere and can be pictured as a basically hollow sphere with the outer shell possessing a faint glow. This is simply a picture I want all of you to familiarize yourself with now so that I can move forward from this point and explain all the various unsolved problems using an easy to recognize object. I should point out here that the first part of my explanation of the universe will start out somewhat basic and become progressively more complex so as not to overwhelm the reader with parts of the model.

To this end just think of a small hollow glowing sphere for now and you can think of whatever color helps you to visualize the little ESS better. So now that we have our little spherical ESS I will explain why they are round in the first place. The reason they are round is the universe is lazy or efficient however you choose to see it as the universe always takes the easiest route. A sphere is the simplest structure with which to construct something especially down at the very fundamental level of everything. The reason this is so is that anything at the very smallest scales would require an additional set of energy and information put into it in order to create a design for a geometrical shape that deviates from a sphere. In other words it takes work to make something non-spherical such as the information and energy necessary to shape the ESS into something like a cube, triangle or the like. A sphere is a natural geometrical shape that involves the least amount of energy for the universe to maintain. As we know the universe always tends to the lowest energy expenditure possible and the least number of variables so introducing a shape other than a sphere would tend to involve more energy and require additional external variables to the ESS to shape them into something other than a sphere. So a sphere is the shape of all ESS and they are hollow with the spaces between themselves and each other being truly devoid of everything.

This space that exists between them is made of nothingness which is the undisputed lack of everything meaning it has no structure, no energy, no matter, no fields no anything. It exists within the universe but it has nothing in it and has zero physical characteristics and it also does not require any. There are no scientific or philosophical debates here about what this nothing is made of; it is simply nothing. If you need a more complete understanding of it remember that these little spaces contain as much nothingness as going past the edge of the universe. What I am trying to get across is limited by the descriptive powers of language which tend to always be bound to some preconceived idea of what they are describing. In this case I am appealing to the excusal of preformed linguistic ideals in order to describe the utterly nothing that exists in the universe when one excludes both matter and empty space. These nothings are still useful however because they allow for a variety of interactions to take place all the time in the universe between ESS and matter as well as between ESS themselves. Later answers to unsolved problems such as the Big Bang will illustrate the usefulness of these nothings and in fact their necessity but it still does not mean that they include anything at all they remain true nothingness. This is enough description of the nothing for now and we will return to it later allowing us to once again focus on the ESS that covers the entire universe.

The ESS do in fact cover the entire universe unlike matter or energy which does not. The ESS extend to the boundaries of the universe or in other words the edge or end of the universe which extends far beyond the visual detection range of any of our instruments. They exist obviously where matter and energy are found or rather matter and energy exist where ESS are found which is everywhere. The ESS will interact with each other and with matter as well which can produce a variety of effects. These effects will be discussed when the appropriate unsolved problem calls upon them but for now I can at least list the interactions. The first and simplest interaction is that of one ESS and its neighbors and takes place at the absolute most fundamental levels of the universes construction. These interactions are the type described earlier which include Big Bang events but are not solely confined to the creation of the universe. A second type of interaction is between ESS and any matter that happens to pass close to them. This type of interaction is a standard matter and ESS type that we experience everyday and results in things like quantum interactions of the smallest subatomic particles. These interactions also go on to make up any larger macro scale object such as a person, a planet or even a whole star. The third type of interaction is that of multiple areas of matter occupied ESS and can be though of as the interaction between any two objects. A simple example of this would be say, you and the book you are holding.

The interactions can become extremely complex but all are rooted in the single same matter to ESS interaction that happens at and below the quantum level. The increasing number of these interactions results in larger constructions such as the Earth and what not. The behavior of both ESS and matter will be dealt with in greater detail in the next section for now we can sum up the basic look of the ESS. They are small hollow spheres that make up space as it is commonly known and they can interact with each other as well as matter. They fill the entire universe although they do not constitute the whole of the universe as matter is the other necessary piece. This general picture is all that is needed for now concerning ESS and will serve as a starting point for anyone trying to understand how the universe works. I will be using ESS constantly throughout my work and by giving you this early description of them I can now build upon the workings of the universe such that each newly solved problem solidifies the shape of the ESS. The more you see them in action the more this basic shape will help you to understand the interactions of matter and ESS so that we build a universe as we see it today. This picture of little hollow glowing spheres is only the beginning of the true nature and shape of space and much more about the general shape will be revealed by examining the ESS and matter interactions. The detail needed to describe these interactions is more in depth and lengthy so will be contained in its own section.

ESS and Matter Interactions

Now that you have an understanding of the basic shape of space we can begin to build on how matter interacts with space and by doing so learn more about space as it is constructed of ESS. Remember we are still building upon our model of what space really is and I must underline this point again as the true nature of space is the missing piece of the solution that has eluded science in solving how the universe works. There is far more to the nature of space than just little hollow spheres that glow slightly. Before we get to that we must first talk about how matter passes close to ESS in the first place. To this end I must once again deal with the limitations of language as no word exists to describe how matter exists around the ESS. For example matter does not come close to the ESS and the closer it gets to the ESS the stronger the interactions or any such idea as that. Nor does matter float over the ESS as if it is some ball awash on an ocean of empty space. Instead the matter exists around the ESS as ESS are smaller than matter. This can be thought of as any fundamental particle such as an electron or quark or any composite particle such as a proton or neutron. The ESS instead exist as a series of spheres one on top of each other in rows and columns stretching as far as the eye can see in all directions. You should by now have a picture of the ESS as stacks of spheres such that a line can pass through the very centers of each along the X, Y and Z axis. Another way to build this mental picture of the ESS is to imagine a floor covered with these little hollow glowing orbs where the orbs are aligned parallel and perpendicular to each other in a grid type fashion. This allows you to draw a line through all of their centers along any given row or column and means that they are not placed next to each other in a staggered fashion which would prevent the drawing of such lines. After you have mentally constructed one of these ESS mats simply create another in your mind on top of that identical to the first. Repeat this process until you have a solid room full of little hollow glowing ESS spheres and you can now draw a line vertically through any of their centers as well.

This is the basic shape of all ESS and how they align themselves in the universe under normal conditions. Conditions can and do exist that this neat little alignment is gently nudged in certain directions and we will examine those later but for now we have the shape of space in its matter removed form. This is to say ESS as they align themselves when matter is not present to any remote degree. I am however explaining about matter and ESS here so we must once again describe how matter moves through, around or over ESS. Again limitations in language not withstanding matter can be loosely thought of for now as a larger sphere that can envelop a great number of ESS at once. Picture your room full of ESS and see all the cute little spheres glowing gently before you and now place a larger mostly opaque orb over a bunch of them in the middle of the room. What you can now see is that a single particle which is the mostly opaque orb holds many ESS within it. The orb is mostly opaque however and what this reveals to you is that the ESS are still dimly visible inside the orb. In other words the matter does not move close to, hover over or anything else to the ESS but instead it sits all around it at once. The ESS are in effect like a nebel or mist that normal matter can move through at the same time as it moves around and envelopes it. Now if we move that mental orb with our minds through the room from say right to left in our field of view we see that the ESS inside of and outside of it are unmoved. So the nebel is undisturbed by the passage of matter through the room. In other words this allows for matter to still move through space as we normally perceive it and allow space to remain as the steady backdrop of the universe.

This illustrates one of the main problems with the three main views as held by modern science which is a lack of understanding of the true nature of space. According to the three main views space should be both rigid and flexible at once. Classical mechanics and quantum mechanics favor a rigid space structure where effects are so small since they are quantized this means that space cannot by made to move as this is unnecessary at such small scales. Objects in the real world move as classically described and this includes you, me and everything around us. So in other words both classically and in terms of quantum mechanics matter need only move through a backdrop of space and that is that. General relativity on the other hand allows for space to be bent and so implies a degree of non rigidity but general relativity fails one hundred percent at describing anything of a fundamental nature. In addition to this the apparent motion of freefall objects while calculated relative to one another cannot be fully correct because they must also be able to be calculated relative to another object which is any random point in space. Any random point in space is not bent and therefore relativity is once again incorrect because it cannot account for a single rigid point of immoveable space. So both ideas touted by the three main views of a rigid space and a flexible space seem equally right and equally wrong. In other words what is needed is the long sought after reconciliation of the two different vantage points on what space really is. The ESS and matter interaction directly solves this problem by allowing space to be both rigid and flexible at the same time.

If we look back at the room full of little ESS spheres and the mostly opaque orb moving in it we see that the orb can move freely through the room and therefore leave the ESS right where it found them. So the orb will be encountering ESS that are not yet inside of it as well as those that are and the ESS that it has left behind as it moves from right to left. If the orb moves out of the room we have the ESS room filled with its neat and structured series of spheres just like before. In this way the ESS can provide for a rigid type of space that fits with classical mechanics and quantum mechanics as well. We will be looking at rigid quantum type interactions of ESS and matter later but for now I will describe the second type of interactions for there are two.

The second type of interactions comes in two forms but both allow for space to be non-rigid. The most basic type of this interaction is a new concept that I must now introduce called spatial compression. Spatial compression is one of the most important aspects of ESS as it allows them to effectively change their shape. Changing their shape allows them to also behave like a non-rigid space that can therefore solve the limitations of classical mechanics and quantum mechanics in terms of using only a rigid space system. What spatial compression means is exactly how it sounds and I am keeping it simple so as to demonstrate how my work is keeping the entire universe as simple as possible. The simplicity of spatial compression means that given certain circumstances the ESS can as I said change their shape. The shape change is only a volume change and can effectively be thought of and calculated as the volume of a sphere whose radius has decreased. So in other words the room full of little hollow glowing spheres stays full of perfect spheres it is just that some may be smaller or larger than others.

If we think back to the room with the mostly opaque orb floating in it we see that the ESS most closely associated with the orb are spatially compressed. In other words they are slightly smaller than their farther away more distant neighbors. As the orb moves from right to left in our room we see the ESS compress and become smaller as the orb approaches and passes through them. The enveloped ESS can be seen to be smaller than those not yet in the orb and those left behind by the orb. This also brings up another aspect of the ESS which helps to explain two things at once. The first is more of the true nature of empty space and the second is the interaction of matter and ESS. The space that is left behind by the orb returns to its normal size and so it is no longer spatially compressed and instead it has rebounded to its normal volume. This demonstrates the ESS ability to not only spatially compress but to spatially rebound as well. The interaction of ESS and matter is also further developed here as the obvious conclusion is drawn that if matter causes the ESS to compress then the absence or removal of matter allows for the ESS to rebound to their normal volumes. There are other interactions that the ESS can experience with matter as well as a more detailed explanation of the ESS spatial compression and spatial rebound but they will be dealt with individually in later unsolved problems. Each of the unsolved problems is detailed enough to fully explain an interaction type on its own and for now simply grasping the general ESS and matter interactions is enough.

What we can learn from this however is that the ESS can act both as a rigid space when needed and as a non-rigid space at the same time. If we also look at the spheres in the room we notice that in the room without the orb the ESS are all of equal sizes. Now when we place the orb in the room we see that some of the ESS are normal size and some are smaller but the way the size progression is viewed by each individual ESS is interesting. Instead of the traditional view of a smooth spacetime that can be bent smoothly and with infinite degrees of curvature ESS follow a more structured shape change. Each ESS will decrease or increase in volume as the orb approaches and envelops or moves on and leaves behind respectively the ESS it encounters. So as we examine the volume of the most distant ESS they are normal, the ESS closer to the orb are smaller and the ones inside are smallest still. However volumetric analyses of the individual ESS reveals that the spheres all have equal volumes in relation to their distance to the orb.

So an ESS just outside the orb on the left where the orb is about to envelop it will have a slightly smaller volume than its neighbor. This ESS is in a sense perfectly touching its sphere to the orb and is therefore as close as it can possibly get before being inside the orb proper. The ESS on the right side of the orb that has been just left behind is also perfectly touching its sphere to the orb too. These two ESS will have identical volumes given their proximity to the orb as will all the ESS that are just on the outside of the orb. The ESS inside the immediate wall of the orb will have a slightly smaller volume still but all ESS inside the orb that are touching its wall will have an identical volume. The volume of any ESS one sphere farther away or closer to the center of the orb will all share identical volumes to their respective neighbors whether outside or in.

The mostly opaque matter orb will not slowly move over a single ESS but move by a set amount until the next ESS spheres are inside it making the movement quantized and not perfectly smooth. The volume changes are equivalent and important because they always occur in set amounts in other words they are quantized to a certain extent. I say certain extent because quantum mechanics deals with particles and their forces and not ESS but the idea remains the same. ESS share a different sort of information exchange that I will discuss in a future section but the important aspect of this system is that although it is governed by different properties than normal quantum forces it does share a direct exchange with all known quantum particles. These aspects will be dealt with in numerous unsolved problems and will help explain how space can once again be seen as both rigid and non-rigid as is needed.

A quantized jump in volume is made by each individual ESS as needed and separate from all the rest allowing the space to be both rigid and non-rigid at the same time. The volume changes can be calculated in set amounts which does away with the problem of an infinitely smooth curving of space which does not mesh with the quantum world. The interactions of ESS and matter in what can be thought of as a quantized way will be dealt with in extreme detail when warranted by individual problems and again I apologize for the necessity of a basic description as it is still necessary to provide a general picture of the ESS nature of space and its interactions with matter. We are providing a solid stable base from which to build upon in later sections.

Once again it is important here to revisit the failures of quantum mechanics to understand space in a way that can make it non-rigid at large scales as well as general relativities failure to quantize space at small scales. Space on a large scale clearly does not behave in a set quantized manner as we can see many of these effects everyday in phenomena like the bending of light around large gravitational objects. Thus quantum mechanics cannot hope to describe what space is or calculate using its equations how it will behave at cosmic scales. On the other hand general relativity is even vaguer as it never describes anywhere what space really is. The equations of general relativity are based off of Newtonian thought which does tell us that certain things like the bending of light can occur given high enough gravitational forces. It is important here to remind the reader that Newton said the word force or in other words gravity is not a result of a simple bending of spacetime as Einstein believed. The fact that Einstein usurped Newton's equations such that he was able to create more accurate predictions to things like light bending around the Sun does not give him or his work on relativity carte blanche to simply create spacetime.

What I mean by this is simply saying that space and time are woven into the same fabric as he called it and that it bends which is the sole reason gravity pulls two objects together is incorrect. Quantum mechanics and its supporters should be front and center in attacking the idea of gravity as a sole result of bending space and time as it represents a fundamental force in the universe. All of the other fundamental forces are governed by quantum mechanics and force carriers which carry real forces and not curvatures. For example there is no strong nuclear force curvature of space, nor is there an electromagnetic curvature of space they are forces pure and simple. The idea that gravity is somehow and I dare say magically different than all the other forces is incorrect as this would involve the universe creating an unnecessary extra property of gravity with its own rules just to satisfy the theory of general relativity. Once again we are faced with the idea of science attempting to make the universe conform to a theory rather than find the true way in which the universe works and develop a model to explain it.

If this point is not sufficiently understood I shall dismantle spacetime in one additional way here and that is by asking the simple question what is a spacetime anyways? Think about this for a moment and you realize that Einstein never once said what spacetime really is only that it curves. Curving spacetime was just Einstein's way of saying he did not understand what space really was as it offers no explanation of the properties of space. Everything in the universe no matter how illusive it might seem at first is actually made of something tangible whether you think of light or energy for example they are all real physical objects in the sense that the can be measured. Yes even light is physically real as it has an effect on matter that is measurable but according to Einstein space is simply non-existent. It is a non-existent forceless fabric that only needs to curve in order to create gravity and for some reason it needs no quantum mechanism with which to interact with every known particle or force carrier which are all quantum in nature. Space is part of our universe and so must be made of something and cannot just be curved to fit an equation and that is that.

Space must be made of something and it must be made of something tangible even in the most indirect sense as it interacts directly with very real and physical objects like matter. Einstein never explained as he did not understand what spacetime was and so his spacetime does not explain what space is at a quantum level. Since all matter and energy are governed and we know this to be one hundred percent correct by quantum forces their interaction with space must be at some level quantized as well. The idea of spacetime and it being curved in order to have objects spontaneously fall into these curvatures explains precisely nothing about the quantum interactions of space and matter. The idea first put forth by Newton that gravity is a force no different than any other is most likely correct and backed by quantum mechanics which search for a real quantized force carrier for gravity. In fact I know it is at least partially correct as I have discovered with my work and it is used directly in combining both quantum mechanics and the large scale observations that are calculated by but not explained by general relativity.

The interaction of matter and ESS is the direct link to how you can produce quantized changes in space as seen in variations of ESS volumes and still account for large scale changes in space while satisfying the need for gravity to be a force. An introduction to ESS gravity will be introduced in the next section and followed as needed by further explanations of unsolved problems that help to prove this model correct.

Gravity and Basic Interactions

The previous section outlined the extremely important ability of space through ESS to be both rigid and non-rigid as needed by the interaction of matter with it. It also showed that ESS are smaller than matter and that matter can move freely in it or around it however you choose to look at it. The changes in ESS volume due to its interactions with matter are made in set jumps up or down as the ESS rebound or spatially compress as needed. The fact that ESS can both spatially compress and naturally spatially rebound to alter their volumes is also of importance here. The mechanism by which ESS rebound naturally will be dealt with later for now we will focus on the spatial compression of ESS. Spatial compression of ESS occurs in the presence of matter and we associate matter no matter which of the three views you choose with gravity.

Classical mechanics saw gravity as the force emanating from any object with mass and therefore two known masses will through the gravitational constant which represents the amount of attraction for the force of gravity be drawn together. The force here is generated from the matter and like a string tethering two spheres holds the objects together. General relativity states that matter does not generate the force but that the presence of matter in space will curve spacetime and the curvature of spacetime is what creates the gravity which causes the two objects to fall towards one another and therefore not be attracted to each other but rather fall towards spacetime. In other words gravity is not emanating from matter but from space according to this theory. Quantum mechanics does not deal directly with gravity but it does deal directly with matter that is obviously linked to gravity and measures matter in set quantized amounts. There appear to be some truth to all three views but the correct combination of space and matter is unknown to modern science.

I will explain using ESS how the universe combines both space and matter such that it can create the force of gravity as we observe it today. I say observe as the ESS explanation of gravity will satisfy all three main views at least where they appear to be correct concerning attraction between two objects. The idea of gravity emanating from matter will be satisfied, the idea of gravity emanating from space will be satisfied and the need to measure this in a quantized way will be satisfied. The description I will now give is of course a basic description as we are still building and will be for quite some time the structure, nature and properties of space as the need to fully explain what space is remains of the utmost importance in establishing how space and matter interact and therefore solving how the universe works.

So let us return once again to our familiar room full of little ESS spheres and once again notice they are identical in size to each other. This is space in the absence of matter and the introduction of our mostly opaque orb will serve as a typical source of gravity. When I say source I mean the basic idea humans have understood for centuries that matter is associated with gravity regardless of the theory discussing it. Now that the orb is floating in the middle of the room of ESS we can see that some are now smaller than they used to be and this is of course a sign that matter is interacting directly with the space around it. The ESS closest to the orb and obviously those inside of it are smaller than the rest and the very center ESS in the orb is the smallest ESS in the entire room. So far so good we have space and we have matter which we know the universe to be made of in the same system. The mass of the object which is in this case the orb will of course determine the strength of the gravity generated in the system.

This is once again something people have known for centuries as something like the Sun produces much more gravity than the Earth which in turn produces more gravity than an apple that falls to its surface. All however are objects made of matter and all will produce gravity no matter how small; in a very minute way the apple is causing the Sun to be attracted to it too. Even a single sub-atomic fundamental particle will create gravity and the orb can be thought of as a single sub-atomic particle. Multiple particles exist and have various masses but for our example the orb will have any arbitrary mass as we are demonstrating the mechanism by which gravity is created and not trying to measure the value for a specific particle although this can be done. Any object such as the orb is a particle which is made of matter or energy or a combination of both and the total energy or mass of the object can be calculated for now using conventional means. What this means is any object that modern science can measure in a lab is made of the stuff of the universe which is one of the two parts needed to make the universe. This stuff is governed by quantum mechanics which we know and quantum mechanics is solely involved in the stuff side of the universe and not the space side at all. The way the matter interacts with space is far more than simply just saying mass makes gravity the more you have the more you get.

Macht

Part of the actual particle that we think of as normal matter is being given to the ESS. Yes that is right I said part of the particle is actually being given to a degree to space. The way matter and space interact that has been missing for all these years is simple the two can exchange information with one another but not in the typical way you might think. The energy being given to the ESS by matter is not used by the ESS as energy. What I mean is once the ESS gets a hold of the energy it does not use it in a way that can be described using quantum mechanics anymore as it is no longer normal energy. The ESS have something that is known as macht which is the way ESS use the energy and can share it as well. What the ESS share is something called macht which means power in German.

Once again you can see I am using a German word to describe something happening in the universe because I realized that all the words used in English to describe energy, work, force or what not have been used. In addition science already has a very well defined explanation for each of these words that is rooted firmly in what can be thought of as matter or energy science. In other words not space science as there is no vocabulary left to use to describe actions and forces inside space and space alone I decided to use a word that I feel describes what is happening in the ESS. They are possessing power and so to avoid confusion with the English usage in science of the word power I used macht to differentiate between matter and energy power and ESS power. In addition to this much of the early works published and indeed most of the early work done in physics in terms of quantum mechanics, relativity and particle research was all done in German around the 1900s or so and so after reading and learning many new words in German used to describe all manner of scientific discovery I thought it fitting to honor those past scientists with their words once again. Once again I apologize to my German readers as to you the word macht still means power but I do thank you for providing me with another chance to create a term based off of historical scientific influence from the last few centuries. So now that I have explained the term macht you will be hearing it continuously as it applies to everything that the ESS do as well as everything that the ESS and matter need to interact with each other and ultimately ESS macht is needed to construct a universe as we see it and solve all of the unsolved problems currently facing modern science.

Looking directly at the little ESS we see in the room we notice that all the ESS have a faint glow which comes from the amount of macht each possesses. The ESS closest to the orb have a brighter glow meaning they have more macht than their neighbors more distant from the orb. The ESS inside the orb are brighter still with even more macht and the ESS at the very center is the brightest with the most macht of all the ESS in the room. So what we have learned is that the ESS experiences a decrease in volume relative to their position in space to the center of the orb while they experience an increase in macht. This is the first relationship of significance between matter and space interactions and the two of them together both decreases ESS in volume along with increases in macht and are what create gravity.

The macht I have been describing for the ESS I have not yet explained what it exactly is only that it comes from the ESS interactions with any object of mass. The true explanation of ESS macht is not simple and not relegated to only one variable but since we are discussing gravity here I will at least deal with the basics of that. For the example of the room of ESS spheres and the orb the macht that we see as a faint glow in the surface of the spheres can be thought of as gravity. Although gravity is not solely the result of just ESS macht and by that I mean that it is. If this seems confusing it is not but it does involve both properties of the ESS as it interacts with matter and this is the missing key to gravity and why it behaves as it does. Remember that the ESS touch each other too and that the glow appears brightest inside the very center of the orb and then fades as you move outward from its surface. What this means is the ESS outside of the orb are not actually touching the matter that makes up the orb and so they are not getting any macht glow from it.

The only place they are getting macht from is the ESS that are inside the orb and so we have another fundamental property of space revealed which is that space shares macht. In other words ESS can be spatially compressed by its interaction with or proximity to matter but ESS can also share macht amongst themselves from that matter. So all the ESS in the room feel a gradation of macht which is strongest in the center of the orb and diminishes as you travel outward from the center of the orb and this is accompanied by a spatial compression of the ESS which is greatest inside the very center of the orb and diminishes as you travel out from it. The spatial compression of the ESS is a direct result of the macht of the ESS and in other words the ESS macht spatially compresses the ESS itself. The interaction with matter provides energy to create macht which is where the ESS gets the necessary ability to spatially compress themselves. If this seems a bit overwhelming let us quickly sum things up so far starting with the universe. The universe is made of space and stuff and the stuff is made of matter and energy. Part of the matter and energy of all stuff in the universe interacts with ESS and increases the macht of the ESS. The more ESS macht that is present the more the macht cause the ESS to spatially compress which results in a decrease in volume of the ESS. The combination of decreased ESS volume and increased ESS macht creates what is known in modern science as the force of gravity. I will explain shortly how gravity works in ways that we see around us but for now let us quickly look at the energy and macht used by the matter and ESS in these interactions.

The matter gives some of its energy to the ESS and the ESS in turn powers up its macht as a result. The energy as I have been calling it that comes from matter need not be thought of as the standard definition of energy which often people will think of as electromagnetic energy. This is the most common type of energy most people think of as it is popularized in relativities mass to energy equivalence theorem. A particle of matter in the form of an electron will annihilate with a particle of matter in the form of a positron producing two high energy photons as a result and the energies of these photons are equivalent to the rest mass energy of the two particles. This kind of energy the electromagnetic type is not however the only type of energy in the universe and in fact many other kinds existed to human knowledge before we understood light.

Take for example kinetic energy, potential energy, chemical energy or quantum energy just to name a few. So when I say energy is being given from matter to ESS I only mean for now that we can think of some form of useable or workable force being given to the ESS. This basic giving of energy is sufficient for now in that it allows me to explain what is happening between matter and ESS or in other words how space and matter interact. Remember the interactions taking place between space and matter are the key here and the details can wait for later as the interplay between the two is the solution to the problem of how the universe works and for now I am still building a picture of all these occurrences. To this end simply accept for now that matter is in a sense donating some of its abilities if you will to be used by ESS for various purposes.

What is donated from the matter however is donated in a set amount or as one might expect a quantized amount. This is true for two reasons the first of which is that matter as we know is governed by quantum mechanical rules and so it must continue to obey these rules when interacting with space in the form of individual ESS. The second reason is that the ESS are unlike the smooth continuous type of space described by general relativity and are instead made out of small spheres which as we have seen are identical to each other under normal amounts of spatial compression or rebound. In other words space is made of set amounts in the form of individual ESS and so space will not absorb anything whatsoever from matter in a smooth contiguous fashion but rather only in small set amounts that can be handled by each ESS. So what we see here is that the ESS of space are themselves measurable in what can be thought of as a quantized way. Therefore both the energy being given by matter to the ESS and the macht used by the ESS plus all changes to and from the two constituents of the system are determined by quantized amounts. This explains a way that the matter and space interactions can be made measureable in a useful quantum mechanical fashion and leads directly to one aspect of this energy to macht sharing between the two.

The energy given by the matter is of course one set amount given the local spatial compression and other factors which I will explain later as changes in things like spatial compression can produce profound effects on matter. In addition the way and rate that matter interacts with ESS changes based on such factors as local spatial compression which again I will clear up in a later section. For now the mostly opaque orb in our room donates one set amount of energy to the ESS that it sits around and so we register one drop of a quantized amount from the matter. The orb however sits atop multiple ESS and it cannot obviously give one piece of itself to every ESS as the orb is far larger than multiple ESS. The ESS are smaller and so can only take a portion of the energy given. What happens in matter and space interactions is that the ESS each take up a small amount of the energy donated by the orb which is of course a small quantized amount per individual ESS but will be summed to the total energy given by the matter.

So in other words if the orb is able to hold 100 ESS for arguments sake as this is not to scale, then the ESS will each have their macht increased by $1/100^{th}$ of the energy donated by the matter. This is a form of conservation of exchanged energy and macht between the two different parts of the universe and means you do not lose anything permanently from the matter such that continued exposure to the ESS drains the matter down to nothingness. This is because the ESS under normal spatial compression can only absorb so much energy from the matter and this prevents them from absorbing more. The only way to absorb more energy is to change the amount of local spatial compression which again we will examine later as it solves a great number of the unsolved problems in physics. As for the fact that the ESS are smaller than matter it is possible here that the scale of ESS are below Planck scales, meaning the amount of energy donated by matter might be a normal quantum Planck quantity but that ESS can use smaller values of macht than anything in the Planck scale.

Schales

Back to a simple sharing of energy and macht we find that the ESS split the energy roughly equally amongst themselves inside the orb but they also share their macht with ESS not inside the orb. This is something I have already explained and you will remember the faint macht glow coming from the ESS spheres in our little room. So the ESS in our room were at first simply sitting there in the room with no matter present. The orb appears and the ESS inside the orb will now absorb and share the quantized amount of energy given to them by the matter. This occurs only for a single moment as the next moment involves no further energy being given from the matter to the ESS but instead involves the ESS sharing macht with their neighboring ESS. The first ESS to receive this gifted macht are the ESS spheres sitting directly outside of the orb. The amount of spheres surrounding the orb is thought of as a larger matter free ESS schale or shell made of multiple ESS that takes energy from the

ESS that are inside the orb proper. I chose to simply use once again a German word here for the shell that surrounds the ESS which are inside any matter as the word shell obviously has already been used numerous times scientifically and so to separate those usages from purely ones relating to ESS and space physics I will now use schale.

The ESS inside the orb will share some of their macht with the schale of ESS spheres close to them and this sharing of macht continues from moment to moment as the macht is further shared with ever more distant and ever larger ESS schales. The amount of energy taken from the matter is still unchanged but the amount of macht now possessed by the ESS inside of it has decreased as they have shared some of their macht with the ESS occupying the schales surrounding the orb. No matter how far you care to look at the macht which is shared from the ESS inside the orb the total amount of macht possessed by all ESS inside the orb and those in the surrounding schales will always sum to the amount of energy given by the matter in the first moment. The increase in volume of the ESS schales around the orb will also correspond with a decrease in macht power as you travel from the orbs center outward and this is due to the fact that the ESS inside the orb are still in direct contact from a space to matter perspective. Therefore these internal ESS are still receiving the energy directly and therefore holding the Lion's share of the macht.

You can easily calculate the drop in macht with increasing numbers of schales outside the orb using basic mathematics and this can then be incorporated with the quantum amount of energy given from the matter. What this means is that you can not only calculate the drop off in ESS macht as distance from the center of the orb and the surface of the orb increases but also that if you know the amount of macht present in a given area and the amount of spatial compression you can calculate the amount of energy given to the local ESS and therefore the amount and type of matter that donated it. You do not have to ever see the matter in order to know that it is there if you know both the macht values and spatial compression of the local ESS.

In other words you never have to see the matter if you can see the gravity. Experiments that can measure ESS can always determine what type and how much matter is present in that local area and the reverse is also true as the matter in a certain region of the universe will tell you about the ESS and macht present as well. More on this later as more detailed discussions of space and matter interactions will explain types of macht, spatial compression and the effects on both ESS and matter as we reach those unsolved problems and solve them.

Another facet of ESS and matter interactions comes from the quantum nature of matter itself. One of the inherent properties of matter is that according to quantum mechanics it never quite sits still and will be seen in many places at once or in many probabilities at once within a given volume of space. The exact mechanism by which this occurs will be explained later but for now we will once again examine the general volume of the system and the ESS macht glow we see in it as it relates to matter. It is known that ESS gains some of its macht from interactions with matter and that matter gives this in the form of donated energy. It is also known that the ESS will share this energy amongst each other by sharing macht and that the macht is strongest in the dead center of whatever matter we are dealing with. However matter tends to jump around in space as believed by quantum mechanics making its position in space seemingly random. This random motion is centered around a probability field which attempts to predict the possible location of the matter at any given time. To this end we see that matter will be vibrating almost randomly in three dimensions in space. The orb in our room therefore is vibrating randomly in three dimensions of the

ESS spheres it is enveloping and what this means is the very center of the orb does not always exist at the very center of the orb.

Instead the orb will make small quantum motions that move the center here and there throughout the ESS. This motion of the orb also means the surface of the orb is moving slightly in the ESS spheres and will sometimes be enveloping various ESS and sometimes not. The effect on the ESS is that some are inside the orb and some are existing within a schale and a moment or so later as the orb undergoes more quantum vibrations the ESS that were part of a schale might now randomly find themselves inside the orb. ESS inside the orb are in direct contact with the energy being donated by the matter and so possess more immediate macht than ESS in schales. ESS in schales must get a boost in macht from their fellow ESS inside the orb and so it is obvious that the quantum motion of the matter results in complex patterns of sharing macht amongst ESS and donating energy from the matter to specific ESS. These patterns can be seen in the gentle increases and decreases in the macht glow of the ESS inside the room both inside the mostly opaque orb and in the ESS spheres surrounding it in schales. The ESS will always share the macht amongst each other and so one pattern emerges which is the brightest ESS glow and therefore the ESS with the most macht will always be seen at the center of the orb although the orb appears to move around the center of the ESS glow.

The sharing of macht amongst ESS is so rapid that the various quantum motions of the matter while rapid and real will appear to be almost non-existent when viewed in terms of ESS macht glow alone. What this means is that the overall macht will move with the ESS more rapidly than the quantum motion will move the matter. Think for a moment about the orb being perfectly stationary and not vibrating quantumly at all and also see the ESS inside the orb. Now look at the macht glow of the ESS and notice that the very center ESS is the single brightest ESS in the room. This represents the perfect center of the ESS spheres and of the orb as shown in a single moment devoid of changes in time. Now allow the orb to vibrate quantumly again and you see it hurriedly moving in all three dimensions around its center the surface of the orb becomes slightly fuzzy or hazy as the rapid quantum motions blurs the crisp surface of the orb.

If you look inside the orb and see the ESS inside you will notice that the center of the ESS glow is not moving with respect to the surface of the orb but that it has grown slightly fuzzy or hazy as well. The exact center ESS we saw when the orb was motionless is still the center and brightest ESS but the ESS surrounding it are all now glowing much more brightly than before. You should have the illusion that the center is not moving of the orb while the surfaces are vibrating in all three dimensions. The ESS glow at the center of the orb is the pictorial representation of the orbs center in this view but is not the true center of the orb. This illustrates that the orb can and is subject to quantum motions but that the quantum motions are much slower than the transfer of macht between ESS. The net result is the ESS remain motionless in the room and share a common center of macht while the orb or matter is free to move around the ESS throughout the room. This demonstrates how quantum motions of matter are both accounted for by ESS and at the same time dealt with by ESS through the sharing of macht. Viewing the room at the time scale of matter produces this picture as the quantum motions are being shown in there real time speed. Thus we see a steady state glow from the ESS while the orb which represents the matter moves rapidly.

If we view the changes from the ESS sphere time scale we see that the matter is essentially stopped in its tracks and frozen while the ESS is able to share macht. The sharing of macht will be a smooth flow of macht from the area of space where the matter used to be to where it is now. In other words if the matter made a quantum motion from right to left in our room we see the ESS at the new left centered position as being slightly dimmer than those at the right where the center used to be. As a moment passes the shift in macht increases the glow towards the left and the matter itself is still motionless. The reason this happens is the ESS macht sharing though much faster than quantum motions still is bound to the energy donated from the matter and thus the center of the matter itself. Since the matter is physically moving in quantized amounts the ESS will no matter how fast they share macht have to wait for that motion to happen. Once it has happened however the macht flows so fast that the glow is balanced amongst ESS long before the next quantum motion can occur.

This is what produces a steady state glow in ESS macht sharing when the film is sped up to the speed of the quantum motions. It is a bit like your finger turning on a flashlight pointed at a wall as the light transmitted from the flashlight reaches the wall so quickly compared to your finger moving the switch on the light source. From the point of view of the light your finger has stopped in time but no matter how fast light is it still must wait for the actual action input by you to the flashlight. Once the action is input however the light compensates for your sluggishness by reaching the wall almost instantly. The analogy here is for demonstrative purposes only and does not necessarily represent a scale example in action. To that end however ESS macht sharing speeds will be dealt with later in some of the unsolved problems proving them faster than quantum action or to be more direct explaining the apparent quantum behavior.

Now it is time for us to look at the ESS macht glow a little more in depth as it travels through the room of little hollow ESS spheres and the mostly opaque orb inside the room. The first thing we need to realize is the fact that the ESS macht is glowing not just inside the orb proper which is to be expected as matter and space interact to form gravity but also outside the orb. The problem of gravity which is what the ESS macht is solving we all know extends beyond any piece of matter in the universe. In fact it is the farthest reaching of all the fundamental forces in the universe by far.

Take for example the extremely powerful force inside the nuclei of atoms known as the strong nuclear force and see that its range is roughly a few femtometers or the diameter of the nucleus itself. It may be many orders of magnitude more powerful than gravity but it has essentially no range whatsoever where as by comparison gravity is far, far weaker than the strong nuclear force but extends throughout the entire universe. Literally every part of the universe in existence is under the influence of gravity almost all of the time. I said almost all of the time because there are some rare situations and conditions that allow for gravity to momentarily behave unusually to how it normally does in our everyday lives. The explanation for the disparity in strengths of fundamental forces and for any seemingly peculiar behaviors of gravity will be dealt with later for now I will continue to explain how ESS and the macht it possesses creates the force we know as gravity.

So it is understood that gravity extends out a long ways from any matter that it originates from and the ESS macht is why this happens. The glow of macht coming from the ESS within the orb is shared amongst close-by ESS contained in the schales surrounding the orb. If we examine the schales more closely however we see that the glow is far beyond the few closest schales and in fact extends far out into space. This is the space that is thought of as having nothing in it as in no matter or energy. The space here is in a sense empty of normal things in the universe which is of course the stuff which makes up one half of it. However the other half of the universe which is space is there and is not empty it in fact contains near infinite numbers of ESS which all contain macht. The space here might appear to be empty to the conventional views of modern science but it is in fact quite alive and well.

The ESS at these remote distances from the orb are all affected by the increase in macht and the subsequent sharing of macht by the ESS inside the orb. Remember that the ESS are now splitting their donated energy from the matter with a huge number of ESS in all the surrounding schales and this splitting of macht is what partially explains the weakness of the gravitational force in the set jumps of ESS space. So if we go back to the idea of 1/100th of the energy being given to ESS inside the orb and now we extend the macht sharing through schales far away from the orb it might be that each ESS now only has 1/1 000 000 000th of the donated energy and so the much more powerful forces inside the matter from say the other three forces if you will is diminished into a weak gravity. This is a very rough explanation that I will deal with later and so there is no need to run with this idea yet but you will be getting a fuller look at these facts later.

This extension of the effects of the macht sharing is what creates the attractive force of gravity that we understand today and did all manner of wonderful things like drop apples on to Newton's head. Remember we are looking at gravity macht here which is a part of the ESS that makes up empty space and we know that matter is responsible for creating gravity too. The more matter you have the more gravity you get and this is done by increasing the macht gravity of any ESS the matter encounters. The ESS macht travels outward and is shared by other ESS with its strength diminishing with each ESS it encounters and therefore creating a nice gentle decrease in field strength as you move outward from any source of matter.

Think of the Earth, Sun or any other large object and you see gravity is strongest at its surface and gets weaker as you travel outward 90 degrees from that point into space. Thus we see that ESS macht similarly drops off as you move away from the source of matter and this means the amount of spatial compression changes too. The ESS inside the orb are under highest spatial compression while the immediate schales experience less but still very significant levels of spatial compression. The amount of spatial compression which is also reflected in a decrease in ESS volume gradually lessens as you move to ever more distance schales from the orb. The effects however are very far reaching as is known for gravity so that even at great distances the ESS are not unaffected by the orbs presence. In other words the ESS are not as they were before the orb entered the room and will all experience some of the effects of the energy exchange from the matter to the ESS happening far away. The ESS will even here be slightly compressed and have slightly elevated macht gravity inside them.

Now it is time to look at a natural property of matter which is this: matter is always attracted to a gravitational field. This may come as no surprise to most and it should not as this is an undeniable fact of the universe. How matter is attracted to the force of gravity however can be explained step by step by looking at the little hollow glowing spheres known as ESS. We are now going to introduce a second and much smaller orb into the room and for your mental reference picture we will make it appear on the far left side of the room some distance from the wall. This orb if it were the only object in the room would remain motionless in place except for its small quantum vibrations around a central point like the original larger orb. This smaller orb is likewise mostly opaque and likewise affects the ESS in all the same ways as before.

The key of matter being attracted naturally to gravity can be seen perfectly if we release the second orb and allow it to interact freely with the system of the room. This includes the ESS in the room and any effects being created by the first orb. The second orb is made of particles of matter which are naturally attracted to the gravity macht of the ESS around them and this is the same thing we have seen with the first orb. It is known that stronger gravitational fields attract matter in a more significant way and this is why the Sun can accelerate objects faster towards its surface than the Earth can.

So this means the second orb if it were left alone in the room without the first orb would experience a normal level of attraction to the ESS macht of gravity present. If the amount of ESS macht is increased however the second orb will feel a greater attraction to the ESS. The first orb though distant from the second orb has a larger mass that has increased the ESS macht of all the space in the room and caused them all to become slightly spatially compressed. The second orb is therefore encountering more ESS as they are now smaller in volume and remember the orbs which represent matter sit over and around or envelop the ESS inside them. Therefore the ESS present inside the second orb is greater than if the second orb were alone in the room. In addition to this the ESS macht has been increased by the first orb and so the macht gravity force of the ESS is now also stronger.

The second orb is of course made of matter which is naturally attracted to space and gravity which we now know to be ESS macht. The result is the perfectly motionless second orb will continue to be attracted to ever higher ESS macht and ESS concentrations. So we see from moment to moment and these time scales are of the shortest order possible which are ESS governed and not quantum timescales which are longer the second orb gradual shift ESS by ESS towards the first orb. The timescales take place in two steps as the matter can only exist in a particular quantum position at a time in the universe and so quantum shifts in position are one time scale. The sharing of ESS macht as we have seen is much faster than this and so the ESS macht sharing effects of the second orb over its local spatially compressed ESS are on another. Combining the two gives an overall timescale which shows the motion moment to moment of the second orb as it travels inside the room.

What we see close up inside the second orb is that the ESS here seek to share more power with each other and so share the macht between them and their neighboring schales of ESS that extend all the way to the neighboring schales of the first orb. The ESS are also getting smaller in size as the second orb moves closer to the first orb as the first orb is much larger, has more mass and can therefore give more energy to the ESS and the ESS can therefore generate greater levels of macht. This reflects the observation of a larger object attracting a smaller object to it but also that the larger object is still slightly moving towards the second smaller one. This sharing of macht extends in a straight line path from the very center of each orb to the other. Each orb seeks to share the ESS and is attracted to ever more spatially compressed ESS and higher ESS macht. So both orbs move towards the common center of gravity they now share although the second orb does overall most of the moving through space in terms of raw distance.

In terms of moving through space by sharing energy both orbs are sharing equally because even though the larger orb is not moving much right to left it is moving more matter and macht with it. If we look once again at the smaller orb and track its progress towards the first we see that it will continue to encounter ever more spatially compressed ESS and ever higher ESS macht strengths. It will in other words accelerate until it has reached the maximum possible velocity allowed under the amount of ESS spatial compression and ESS macht glow produced by the first orb. This is how objects of different size have different gravitational strengths as an object falling towards the Sun for example will always fall faster than an object towards the Earth or can be thought of in reverse as escape velocity.

The attraction of the second orb to the first is also compounded by the fact that since matter is attracted to the macht of ESS if more ESS are present the matter will be attracted even more strongly. So there are two and not one effects working to produce gravity which are the number of ESS the object with mass encounters and the strength of the ESS macht possessed by those same ESS in what is known as local spatial compression. The local spatial compression of an area of space differs with the mass of any matter within it. The Sun is much better at spatially compressing the ESS in its local area and increasing the ESS macht they each possess than the Earth is. The Earth is likewise much better at spatially compressing its local ESS and increasing their macht glow than we are and so this is why we do not simply fly or float off the surface of the Earth and nor does the Earth depart from the Sun.

This attraction between matter and the ESS macht means the second orb will always move towards the larger first orb as the matter in the second orb encounters more ESS as they are smaller in volume and feels more pull from the macht of those ESS. No matter how many objects you have in any volume of space they will all donate some of their energy to the ESS of space based on their own total mass and the ESS within the volume of space will seek to share and balance out the macht they possess. This means the objects will naturally follow a path in space dictated by the ESS and the macht they carry but generated in a sense from the matter they themselves are made of. It after all takes two things to make gravity which are matter and space not just one or the other.

Earlier I discussed the three main views and the different ways they believed gravity to exist in the universe. The earliest view which is classical mechanics treats gravity as a pure force that acts upon objects following the classical rules of motion. Quantum mechanics treats gravity as a force as well that must be able to fit into the Standard Model of particle physics and as such will have a force carrier particle such as the graviton. The quantum mechanical view treats space as small systems that do not feel the effects of large scale objects like whole stars or galaxies whereas general relativity does. General relativity deals only with large scale effects of gravity and attributes them not to particles or to forces but only to spacetime which it says is curved so that objects fall into gravity wells. It is well known that none of these three views is correct as should be clear by now but that certain parts of them do seem to fit observation well. To this end the mechanisms by which they are incorrect but the assertion made by each that space must at least partly behave in a manner that they each champion is correct.

I said earlier that I would be able to solve the mystery of gravity using ESS and macht in such a way that it allowed for inclusion of what is observationally correct about both the small scale world and the large scale universe. This is accomplished by treating space exactly as I have stated with ESS that can be spatially compressed or allowed to rebound, macht sharing between ESS and energy or ability if you will being donated from matter to space. If you examine the actions of gravity of the two orbs I have used for demonstration purposes before and the ESS they are enveloped by and surrounded by in the room you will see that all the small and large scale observations are met without violating each other.

Take for instance the need for gravity to arise from matter as is believed by both classical mechanics and quantum mechanics as both deal with objects as either classical or quantum particles. From the point of view of these two theories matter is necessary for the generation of the forces whether that is a simple classical force or a force carrier driven one, matter is always making the gravity according to them. This is as we know partially true as these two theories are not completely incorrect with respect to gravity and the way that matter is still used to generate gravity is by donation of energy. The orb or orbs in the room depending on when we talked about gravity will donate some of their energy to ESS and this results in the turning on of gravity if you will. The energy donated is necessary for the ESS macht to become higher in the local area of space and so allows the ESS existing around the orb in schales to share macht. The macht is also one of the working mechanisms for gravity and so is partly being generated from matter satisfying the force of gravity originating from matter observations.

Newton can still be correct in saying that the two bodies will create an attractive force that pulls them together and quantum mechanics can still say that the force is governed from particles that fit into the Standard Model of particle physics. So we have gravity partly originating from matter in the universe in order to create gravitational effects on objects of all size as we see them today. The quantum scale space will still be seen as unaffected by gravity because the transfer of macht from ESS to ESS exists within any quantum system at a scale far smaller than the particles being studied. The gravitational effects are there simply too small to be detected by quantum mechanics and will not become significant until other conditions are met that I will explain when examining objects like black holes. I will say this for now that at black hole scales ESS effects are present in quantum systems and manifest themselves in measureable ways even for single fundamental particle systems.

Looking at the large scale idea that gravity comes from space and not matter is also satisfied in this way of using ESS and macht as space is obviously made of ESS. As I have stated before the biggest stumbling block to progress and filling in of the holes of knowledge of modern science concerning the workings of the universe comes from not understanding the true structure and nature of space. Knowing the true structure will solve all these problems and the true structure and nature of space are ESS carrying macht within them. The ESS carrying macht is the reason it appears as space is responsible for the generation of gravity and explains why space appears to be curved in large gravitational fields. Space does not bend or curve it compresses and that is how it changes shape on large scales and this mechanism is brought about by macht sharing from ESS to ESS within space itself. The ESS as you now know undergo changes in volume in the presence of direct matter contact or macht sharing between schales. The natural behavior of matter is to be attracted to space which is obvious from observations made throughout the universe whether here on Earth in a lab or deep in space surrounding a galaxy cluster. Since matter is attracted to space it can be attracted to greater or lesser degrees depending on the local area of spatial compression of the ESS. Under normal conditions the spatial compression is minimal and so the number of ESS per volume of matter tends to not be of major importance in determining gravitational effects although it is still at work.

What is significant at this scale which can be thought of as the scale of say an astronaut and his spaceship or the Earth is the macht sharing between ESS. Since the spatial compression is minimal here the macht is used more for sharing between ESS and thus increasing the gravity macht of distant ESS which are not inside matter. Again a detailed description of when and why spatial compression and macht sharing vary with respect to gravitational effects will be dealt with later in black holes. Here however this can be seen why the Earth or the astronaut does not cause significant changes in ESS spatial compression of space but the Sun for example does.

For now however we see that both spatially compressing ESS and the sharing of ESS macht between themselves is the mechanism by which space creates gravity. Any object in space feels the effect of gravity and will therefore be pulled towards the strongest source of gravity. Gravity arising from space is made possible by the ESS being a part of space and so their compression to allow more ESS per volume of matter helps to explain gravity. The ability of ESS to share macht at great distances from the object that gave them energy also explains gravity. The second orb for example feels the increased macht glow in the ESS it encounters in the room and is attracted towards the first orb by feeling gravity emanating from space. At large non quantum scales space therefore appears to be the source of gravity even though it is a combination of small and large scale effects that makes gravity work as we see it today. The large scale effects are carried by ESS macht glow and so can travel outwards a great distance from any source of matter that created them which is far outside the range of any quantum explanation.

The source of the increased macht glow comes from matter however which happens on a quantum scale and in terms of ESS on a smaller than quantum scale so quantum systems are hugely important in explaining gravity as well. In other words both the small and large scale effects of quantum mechanics and things like bending light on cosmic scales can be incorporated into one model. This still allows everyday objects to behave classically as makes empirical sense using classical mechanics and returns gravity to a force rather than the curving of some imaginary non-quantum spacetime. A quick note here on anyone saying that general relativity must still be correct and that spacetime must exist due to apparent changes in time dilation in the cosmos as I will be explaining what time is later and why it only seems to be doing what relativity says.

The model of gravity using ESS and the macht it carries solves many of the problems faced when looking at how gravity behaves in the universe. Gravity works weakly on very large scales that are infinite across the entire universe. Gravity also influences all matter and energy in the universe and since all matter and energy are governed by quantum mechanics there has long been a strong suspicion that gravity is somewhat quantum in nature as well. Now it is known what and how gravity works in such a way that allows for the classical mechanics idea of gravity being a force emanating from an object or between two objects to be real. The general relativity theory of gravity being a property of space alone is incorrect but it has been shown that gravity is in part carried by space itself. Large scale gravitational effects ruling the cosmos is not a new idea and goes back to Newton but until now the workings of these effects have been unknown or misunderstood. Yet the reality is that large scale gravitational effects are of course real and observable all around us since ancient times.

It is somewhat surprising that our knowledge as humans of everything around us has made relatively steady progress since pre-history but gravity has not. Even the most advanced theories and interpretations of general relativity and quantum mechanics have failed in their ad hoc attempts to solve the problem. Once again I gently remind the reader that it was the lack of understanding of space at its most fundamental levels that led to this inevitable failure in understanding gravity. The reason of course is that gravity is the result of the interaction of matter and space and not one or the other alone so an incorporation of the two was necessary in order to finally solve the problem of what gravity is and how it really works.

The most useful aspect of the ESS macht explanation of gravity as I will be discussing its relevance to gravity here only and save other uses for later is that gravity can finally be applied to both the small and large scales. The quantum world which sees matter and energy as particles or waves that are naught but quantum objects following set quantum rules can be kept in the ESS macht model of space. Likewise all large scale observations of the universe seen through either classical mechanical or general relativistic eyes can also be kept and now built from the ground up. In other words large scale space behavior can now be explained using small scale fundamental physics.

This is the marriage of both quantum physics and what I have named space physics. Space physics is of course the physical realm of space or empty space as some people call it and is dictated by the behavior of ESS and the macht they carry along with all interactions arising from those two things. These interactions can be from ESS to ESS contact or ESS to matter contact or combinations of effects between the two or multiple groups of matter and space interactions. If one so chooses to measure space in quantum mechanical terms then failure is inevitable because in the world of small scale quantum systems space is seen as static. If one tries to view the universe through classical mechanics or general relativity then failure is once again inevitable as these two theories cannot hope to ever explain the small scale and instead they merely catalogue the large scale. Space physics using ESS macht can merge the two by demonstrating how quantum particles and forces interact with ESS to alter macht levels and therefore alter space. The alteration of space can then be measured as a normal system although it is not a purely quantum mechanical one. Quantum mechanics explain the matter and energy or the stuff in the universe fairly well and space physics explains space as well as all space and matter interactions which is the second half of the ingredients you need to make a universe.

I have stated before that the amount of energy given to ESS by matter is of a set amount based on the local amount of spatial compression of the ESS and the macht they carry to name two variables. This donated energy is of course a quantum amount as the object donating it is quantum in nature and must obey quantum rules. So just as an electron will move up or down orbitals as the parent nucleus it orbits gains or loses energy in set quantized amounts so does the energy donated to ESS come in set amounts. The ESS use space physics which is different than quantum physics in many respects but it is also similar in many respects such that changes in macht between ESS also occur in what can be thought of as set quantized amounts. The scale of ESS and the macht they carry and share occurs at the level of space physics which is far more sub-divided than quantum physics and results in much smaller scales. These smaller scales are what results in a seemingly smooth and liquid flow to space as space and the way it behaves has always been observed to be much more refined than quantum systems.

Nevertheless the macht shared between ESS still follow rules to how they are shared and how much of them go to each ESS. The ESS inside matter such as the ESS spheres inside our mostly opaque orb as well as the ESS located in the schales outside the orb and extending into space will be following these rules. The macht shared will be transferred in again what can be thought of as set discrete amounts from ESS to ESS. In other words the ESS and the macht they carry are measureable at the scale of space physics.

With both quantum systems and space systems being made measureable and with the two systems interacting it is possible to measure the total interactions between quantum physics and space physics. The interactions between the two types of physics results in gravity so by measuring the two together you are actually measuring gravity for the first time at the small scale. Combining the rules for quantum mechanics and the rules for space physics will give you a single set of rules that dictates how matter and space interact. It is possible to then create a single set of equations that can freely exchange information between the two such that ESS macht given local spatial compression can be accounted for. The scaling up of this is able to move the equations from the small scale to the large scale.

Variables such as ESS number, macht strength and local spatial compression are all incorporated into space physics and need to be fused with the energy exchange from quantum systems. Once done however it is possible to zoom out if you will from the ultra small scale quantum world that modern science is used to and travel to any larger scale observed in the universe. So in this way a mathematically workable equation exists from the small ESS scale to the larger quantum mechanics scale and finally to the universe scale itself. You can use the ESS model of space to explain gravity on any scale which bridges the gap between small systems and massive objects such as stars, galaxies and the entire universe itself.

Facts

I have discussed here the reality of such equations existing between quantum mechanics and space physics and the fact that space physics governs how space behaves. Before I cover any more of these interactions or equations I would like to sum up some of the key points of ESS macht gravity that you have learned so far. The first fact is that matter donates or gives part of its energy or abilities if you will to space. This is massively obvious to all as we know that matter does alter the geometry or shape of space and this can be observed regularly in the cosmos. The second fact is that the energy donated to the ESS affects them by way of decreasing their total volume per ESS. The amount of ESS volume change is dependant on multiple variables or more interestingly can produce multiple effects that I will be dealing with in future sections. For now however a simple matter and space interaction produces a decrease in volume of the ESS inside the matter.

The third fact is that as the ESS interacts with matter they transform the donated energy into ESS macht which is the life force of all ESS. The ESS macht level will increase for any ESS inside matter and the ESS macht level will increase in any nearby schales as well. This is due to sharing of macht from the ESS in contact with matter and the surrounding ESS outside of the matter. The fourth fact is that the ESS macht glow extends outwards a great distance from the point of matter and space interaction which saw the increase in macht in the initial moments of contact between matter and space in the system. This glow will also affect the size of the ESS farther away from the area of ESS and matter interaction.

The fifth fact is that matter is naturally attracted to space and particularly to the ESS macht gravity of space. This means any other object close to the first object which can be thought of as producing the gravity will react to the changes in ESS volume and macht levels in the space the second object occupies. The sixth fact is that it is combination of the number of ESS and the strength of the ESS macht glow that attracts the second object to the first. This attraction explains why any smaller object will travel towards a larger one as the object is really following gravity. Fact seven is that gravity is now revealed as both a force and a product of the shape of space as it takes both the macht carried by the ESS and volume changes in ESS to create a gravitational attraction between two masses.

The eighth fact is that all of these interactions and the energies or macht involved whether it is coming from matter or ESS respectively can be measured in real and meaningful set amounts producing a way to incorporate both the physics of matter and space as one. This results in an explanation of gravity such that small scale interactions of known fundamental particles can be made with space and it's ESS in order to explain macht changes that can be scaled upwards to any large scale object including the universe in a seamless manner.

This is a general summation of the facts that govern the interactions of matter with space and as some of the brighter readers have no doubt noticed I have only dealt with these interactions in a fairly simple way. For example I have only mentioned what happens when space compresses in great detail and listed some of the rules governing the effects of matter and space when spatial compression increases. I have not dealt with here all the forms of spatial compression possible in the universe when talking about interactions between matter and space or just ESS alone. In addition I have only briefly touched on the idea of ESS macht glow and ESS number per matter volume in explaining gravity but have not explored what happens when these conditions are taken to greater levels. In short there are a number of variables that I have left alone for now as I was aiming to only explain the basic interaction of matter and space and also to continue my explanation of ESS and how space is really made. In order to more fully develop my work and to prove that space is in fact made of ESS and ESS macht glow I will need to apply some of the unsolved problems of the universe to the variables I have not fully discussed here. In some sense the next many sections of my book will be focused not so much on developing the properties of space although I will continue to do so but I will be shifting my focus now on pure proofs of my work. If I can reasonably explain the unsolved problems of the universe using my work I should be able to prove it is correct. To this end I would like you now to keep an open mind and push out any limiting thoughts or theories that might exist in your minds eye. I would also like you to accept what I tell you as truth only as far as you need to understand how I am solving any of the problems I present. I hope in this open minded way you will see that the universe is constructed of ESS carrying macht and that any of modern sciences unsolved problems can in fact be solved using them.

Chapter 16
Black Holes

Black Holes and ESS macht

I hope by now that some of you are well and truly interested in ESS and the macht glow they carry as the way of explaining space such that gravity can be incorporated into known quantum mechanical systems and still produce the large scale observations we see in the universe. I also realize that many of you will no doubt want me to explore further many of the space and matter interactions in an effort to further prove my work to you. I fully understand this and I am well aware of the number of times that I have said I will deal with a particular topic in a later section and to that end I will now apply my ESS macht space to a direct unsolved problem in modern science which is the black hole. The concept of a black hole is a very lengthy one and as such I will only cover a general application of ESS macht and ESS volume changes with respect to matter to show how a seemingly static view of space from the quantum mechanical world can be made to feel large scale relativistic effects still. The absolute exact mechanism by which all black holes are made, what they are made out of and how they behave I will leave until later sections as the problem is not one that can be encapsulated in a single sweeping stroke.

This first explanation of a black hole will deal very little to nothing with the matter that a black hole is made of and instead focus purely on the space a black hole is made of. The first thing you must understand about black holes is that yes they are in fact made of space as well as matter. Black holes are directly constructed from a combination of matter and space unlike the general relativity theory that they are merely a product of bent spacetime. A black hole according to general relativity is merely an area of space that has become so curved that its gravity well is infinite and this is achieved by squeezing the matter in it into an infinitely small space and as such time stops. The general relativity view of black holes is one hundred percent wrong.

It cannot be said more bluntly than that and it must be said. The whole idea of a black hole arising from general relativity is a rather childish run away one whereby matter was crushed to impossible degrees just for fun so that space could be curved to impossible degrees just for fun. It was nothing more than a crude thought experiment involving simple what if scenarios that allowed people to play with the mathematics. It is impossible for a black hole to exist in any form as dictated by general relativity for numerous reasons the first of which is that the theory of general relativity itself is of course wrong. If this were not so then no one in modern science would be bothered trying to solve the unsolved problems of the universe because there would be no unsolved problems as general relativity would have already answered them. This obviously has not happened and partly because general relativity has created some of those unsolved problems itself telling us that it is incorrect.

The second reason that general relativity is incorrect in how it views black holes comes from the fact that once again relativity offers no clear explanation of what is actually happening in the universe to make a black hole in the first place. Once again I must remind the reader that never once did Einstein propose a solid explanation of what space was or why it behaved as it did. Einstein simply said my formulas account for certain masses in certain volumes that match gravitational effects and in fairness to Einstein many people who came after him simply ran with his theory and claimed that general relativity accounts for numerous strange things in the universe. The bottom line here is that general relativity never explains what space is or why it does anything and so is rendered useless. Another reason that general relativity is incorrect in its theory of a black hole comes from the infinities contained in its own equations. These are of course the infinite gravity produced by a black hole, the zero volume of the matter in the black hole and the stoppage of time in the black hole. All of these are encapsulated in one of physics biggest mistakes or rather in one of Einstein's biggest mistakes which is the singularity. I hate singularities.

A singularity is a point of dimensionless space that contains all three of the impossible infinites described and predicted by general relativity. A singularity is also modern sciences way of saying it does not understand what is happening in the universe at that point and under those conditions. Regrettably or perhaps innocently depending on how you view this phenomenon popular culture has led many to believe that singularities actually exist. To this end it is often wrongly thought that singularities are real when in truth they exist only in the mathematical sense and even then are seen as giant mistakes. The reason for this is simple you cannot possess any infinite value in the universe no matter what or how you try. You can never have infinite density which yields zero dimensions to all the matter crushed into a black holes singularity. Nor can you have infinite gravity as a result which inversely slows time to a dead stop.

To prove this concept rather quickly and simply let us assume for a moment that the impossible singularity is real and that black holes behave exactly as general relativity predicts. What you have is a single point in space with infinite gravity and for arguments sake let us say that we are talking about the very first black hole to exist in our universe which would be about 200-300 hundred millions years after the big bang. Yes these are rough estimates just to prove a quick point but nonetheless they will prove my point here. So this single black hole has infinite gravity because we are suspending any sort of rational thinking about cosmological objects for a moment. This means it really does possess the impossible which is an infinity and so can do infinite things like have an infinitely strong gravitational pull which in turn has an infinite reach with infinite strength. You can probably see where I am headed with this as any single object with an infinite gravity would have long consumed our universe. The curvature produced by the black hole would curve all of spacetime meaning it would curve the entire universe onto itself and very quickly kill the infant universe which remember was only expanding for a few hundred million years and not the billions we see today. So the universe is collapsed on itself and we are all dead and so how are we discussing this point? Exactly we are discussing this point and so through a quick rational empirical evaluation of the theory of infinities and general relativity we can see that infinities are impossible and therefore general relativity is wrong.

Space does change its shape though and black holes do exist and they must be governed by quantum mechanics to some extent because that is what rules matter which made the black hole in the first place. In order to therefore understand what a black hole is which is of course not a hole at all but a rather dense sphere of crushed matter spinning in space we must look first at the space it is made of. This examination of the space inside a black hole will do two things the first of which is help to further develop the model of ESS macht and how it behaves in the universe. The second is of course to begin explaining at least from a space physics standpoint what black holes really and truly are. To begin with let us remember that space as viewed in quantum mechanics is static with no room to incorporate changes in the shape of space which up until now has been called relativistic effects. The failure of quantum mechanics and relativity to play nice comes from the failure of general relativity to bring anything useful to the table as it were. Quantum mechanics brings many solid numbers and facts and such but general relativity only says space curves now deal with it. That might sound a bit short and it was meant to as the shackles of general relativity must now be cast off in order to make progress towards solving the problem of how the universe works.

I will start with solving how gravity works by way of using black holes to demonstrate how space is not static in quantum systems. The static view of space can only be realized in what is known as normal spatial compression and this refers to once again a scientifically Earth-centric view of the universe and physics in general. Nevertheless it must be used as this is where we as humans have conducted all of our experiments. So we can now look at quantum space and see that normal spatial compression reigns and explains why space appears static as the local spatial compression of Earth does not change so we can never record changes in actions between fundamental particles in these ways. If we could have three laboratories set up in three different areas of spatial compression we would in fact notice these changes. Gravity still holds matter together and causes it to react by action of a force under normal spatial compression so we do know it is here and working. A curvature of spacetime does not explain gravity because it is only one factor used in normal spatial compression around an object like the Earth and a dreadnought of an object like a super massive black hole. Modern science has been looking for only one gravitational factor to explain all gravity related effects in the universe and this will never work especially one based off of general relativity. Quantum mechanics on the other hand has been begging for a better explanation of gravity and ESS macht is the answer it has been searching for all its life.

ESS macht gravity uses two factors to explain how the normal force of gravity as we perceive can be used to construct the Earth and the super massive black hole at the center of the Milky Way galaxy our home. To begin with let me remind you that gravity is a summation of ESS volume or spatial compression of local space and the macht that ESS share with each other from ESS and matter contact to ESS schales around the ESS matter interaction area. Why does gravity work on Earth when it appears to not affect changes in quantum systems? The answer is the ESS macht which is far more effective at transferring gravitational force through space than can be achieved by spatial compression in Earth like objects. In other words general relativity is next to useless at quantum scales which have been well known for a century or more. This is why it appears that space does not behave in any malleable way at quantum scales. The effects of gravity on matter right down to the fundamental level are due to ESS macht glow which we know matter is attracted to. If you increase the macht strength matter is attracted to space even more and this is fairly obvious when dropping objects from different heights. If you drop an object from 1000 kilometers above the surface of the Earth it will fall but slowly and accelerate to its terminal velocity before impacting the surface of the planet. If you drop an object from one kilometer above the surface of the Earth it will reach its terminal velocity much faster than the space dropped object. How does this prove ESS macht correct? Simple a quantum laboratory placed at 1000 kilometers will not notice significant changes in the amount of spatial compression of the system when compared to a laboratory one kilometer above the Earth or one more on the surface of the planet. All quantum measurements will behave as though space were not curving according to relativity and this is perfectly correct. So if spacetime curvature is not responsible for gravitational effects why does one object suddenly fall much faster when released?

Space has not changed shape at different altitudes to any appreciable degree around an object like the Earth but the transmission of macht energy has. An ESS located at the 1000 kilometer schale level will have a much weaker glow than one at one kilometer and weaker still than ESS located at the surface of the planet. Matter is attracted to gravity, all theories know this and this is not under debate but if gravity is not the result of curvatures it must be the result of forces. Quantum measurable forces of gravity such as those found in the ESS macht glow of space. In this way the gravitational constant can remain the same and satisfy physics when needed but changes in quantum space are not needed and instead increases in quantum force are included in calculations of particle actions in quantum systems. This removes the theory of curved spacetime from quantum systems at normal spatial compressions but allows for observational gravitational effects to hold true. The fact that gravity still appears to behave as predicted is because space is made of both ESS which undergo changes in volume as well as ESS macht which undergo changes in glow strength. Quantum mechanics and classical mechanics which both use forces can now be satisfied mathematically. General relativity which is based off of classical mechanics still gets to assume changes in space geometry but the attributing of gravity to solely changes in space geometry is now removed. This demonstration is only a quick look at gravity as it works under normal spatial compression through ESS macht glow and the values used are for illustrative purposes only however the idea holds true.

If you are still somewhat uncomfortable with the idea of abandoning general relativity as a way of explaining gravity especially because its equations produce good results let me put it to you this way: general relativity can still be used to calculate motions due to gravitational fields only the reason these motions occur is not explained correctly by general relativity. I do acknowledge that the predictions of gravitational effects made by relativity are good ones only the reason is not due to spacetime bending this is simply a mistake that Einstein made as he was unable to understand how the universe worked. Two factors are needed to explain gravity and both will be more useful or less useful based on the amount of spatial compression of local space and other factors as you will learn. Under normal spatial compression gravity behaves as modern science predicts using known equations but it occurs for two reasons explained by space physics.

Now to explain why spacetime curvature is not the sole reason gravity exists and to explain the nature of space in a black hole, yes you have all been very patient and I thank you for this, we can now look at space inside a black hole. Both factors of gravity, the ESS shape and ESS macht glow are important here and can easily be incorporated into quantum mechanics. To begin with it must be realized that the conditions surrounding black holes are of course extreme but they are not infinite. This includes both the conditions outside the black hole, at or below the event horizon and at the very heart of the black hole where the singularity does not exist. The use of two factors in explaining gravity also allows for gravitational predictions made by general relativity in terms of motions on objects due to gravity to still be made but eliminates the failure of general relativity when it comes to predicting a singularity. This is of course due to the fact the general relativity uses changes in space geometry as the sole means of explaining gravity and so creates an impossible object the singularity but if you remove this spacetime dependency of gravity you can eliminate the singularity from the equations. In order to do this we will return to our room of ESS spheres and the matter orb and of course the matter orb is not to scale but represents a black hole.

When we first looked inside our original regular orb we noticed that the ESS spheres had decreased in volume but only slightly and even this effect was exaggerated to illustrate the nature of ESS and matter interactions. In reality that first orb was a small object like a proton or a fundamental particle and so would never cause very significant spatial compression of the local area but it would cause a significant increase in ESS macht. If we look at a black hole orb in our room we see right away that the ESS are now incredibly compressed to the point where the little hollow spheres that we first saw are now more like little hollow dots. They are still however hollow and this fact will become unbelievably important when answering other unsolved problems but for now we need only know they are still hollow.

The volume inside each ESS orb however is so reduced that they are a bare fraction of their original size and this is of course due to the extremely large interaction they are experiencing from the matter of the black hole. Now remember also that ESS will decrease in volume due to macht sharing as we saw the ESS in the surrounding schales experience a decrease in volume as well even though they were not directly receiving any energy from the matter. So in truth it is macht strength that cause ESS volume to decrease and in the case of a black hole so much energy is being received from the black hole matter that the macht glow is incredibly strong. If you have to imagine the little ESS hollow spheres as glowing almost white blindingly bright then do so as long as you understand that this corresponds to an increase in macht strength coupled with a decrease in ESS volume.

Now remember that gravity is also due to ESS volume or spatial compression of the local area and you see that the ESS are highly spatially compressed. What this means is that the total volume of the orb can have many more ESS in it than an object of similar size under normal spatial compression. Imagine the Earth as a standard volume and picture for example and definitely not to scale 100 ESS spheres inside it and now picture a black hole the same diameter as the Earth with 100 million ESS spheres inside it. This should give you an idea of ESS number per volume of matter in an area of low spatial compression and high spatial compression respectively. Since it is now understood that high spatial compression allows for many more ESS to be inside a given volume of matter we can also understand that matter is attracted to ESS macht. Since matter is attracted to ESS macht which is what space is made of we see that the matter of a black hole is filled with many more ESS and so can be attracted to a greater number of ESS at once. Therefore the matter is now even more highly attracted to space which in turn generates the high macht levels observed.

The room of little hollow ESS spheres however now looks very different than before as the spheres inside the black hole matter volume appear to be little hollow dots. These dots came from the ESS in the room and even though the volume of matter in the black hole orb has not changed the number of ESS has. The only place they came from was outside the orb and so we see that more space has actually moved into the same volume of matter which is the black hole. This is the very first point I mentioned about black holes which is that they are also made up of space. You cannot create a black hole without space and matter so a black hole is not just very dense matter it is also increased amounts of space. We know this to be increased numbers of ESS and without these ESS being a part of the black hole under extremely high levels of spatial compression the black hole would immediately explode outward as one of its two key components would be lost.

This also differentiates from the misconception that black holes have space flowing into them like a river as this is quite wrong. The space does not forever flow into a black hole with the black hole gobbling up all space but rather the black hole has a very regulated amount of space permitted inside of it and that is all that is allowed. There will be a longer discussion of black hole workings later as I mentioned before this is only a basic introduction to black hole physics. These later topics will include black hole matter, black hole matter and space interactions and black hole life history data and behaviors. For now however just focus on seeing that there are many, many more ESS per volume of the black hole than there were before or for an object the same size as the black hole but under normal local spatial compression.

What we can learn from black hole construction here using increased numbers of ESS is a quantum explanation of gravity as a product of two gravitational factors. Since quantum systems deal with matter as particles we can see that the normally static space of a quantum system is quite different. In our earthbound laboratory we notice no significant change in spatial compression when we moved up or down from the Earth's surface. The reason for this we have seen but in a black hole we also know that many more ESS are spatially compressed into the same volume of space. So now picture a quantum system where in the background of whatever particle you might be picturing you see hundreds of little spheres. The scales present are of course greatly exaggerated but I hope you are seeing the picture fully here.

Normal quantum systems cannot see the spheres to any real degree because not only are they not noticing changes in spatial compression for comparative purposes but also because the ESS are large enough here to not really affect anything. In the quantum system of the black hole the exact same setup of particles is present but since space is under such high levels of compression the number of ESS pictured in the quantum system suddenly jumps up and becomes apparent. ESS macht glow is still present and working hard to influence the matter and space interactions to produce gravity as we observe it but the number of ESS are now also finally a factor. In this way the space of any quantum system can finally incorporate what are thought of as changes to space geometry at a small scale. The full effect of this is that changes of gravitational force at the quantum scale are made of both the macht glow and the number of ESS due to spatial compression.

Any picture of a quantum system in your mind should now be one where you can see dimly glowing hollow spheres in the background of the system. Under normal spatial compression they do not affect the actions of particles to any great degree because gravity is so weak here and this we now know is because the ESS are too large and the ESS macht glow too weak to overcome actions affecting the particles from the other three fundamental forces. In a black hole however you should now picture hundreds of blindingly bright white hollow dots which exhibit such a high degree of influence on the particles as to now have the ability to alter quantum interactions even over the other three forces. There will be a much more detailed explanation of ESS and ESS macht glow on quantum systems and all fundamental forces later that will cover every known force in the universe and their behavior with matter. For now however you can begin to see how space becomes relevant for the first time in quantum scale systems and how the incorporation of large scale observations of the universe can be explained using small scale interactions.

This inclusion of two factors of gravity should be easily accepted as multiple factors are already understood to explain other forces such as the electromagnetic force. Using two factors to explain gravity also directly does away with the impossible singularity as gravity is no longer explained using space geometry only. In a black hole we know that ESS density per matter volume increases and that macht glow shared by the ESS also increase. Combining these two factors allows for the full effect of black hole gravitation to be realized in a way that agrees with calculations but keeps space from being infinite at any one point and creating a singularity. Since the black hole derives its gravity from two factors space no longer needs to be crushed into a point of zero dimensions and in fact the ESS forbid this naturally which I have touched on when I explained that they are still hollow dots.

Since the burden of generating gravity has been removed from space geometry alone we can now give part of the workload to the ESS macht. The black hole will therefore have a finite volume and finite density even at its very center which prevents the gravitational force from becoming infinite as predicted by general relativity. It also prevents the matter from being squeezed into a point of zero dimensions and stops time from well, stopping. The way gravity can still be so high in a black hole and match predicted values for a given mass and volume is by sharing the force between ESS volume and density and ESS macht glow. I will remind the reader that ESS share macht glow amongst them such that macht generated from ESS inside the black hole but not at the center will still be transmitted to the center as well as outwards. The overall sharing properties of ESS macht amongst themselves will also be covered a little later but for now understand that ESS macht glow can and is shared to some extent at a central point in the black hole or any object.

What modern science views as a single point in space that contains the highest level of gravity for a black hole or any object is maintained but in the case of a black hole the necessary and impossible condition of general relativity of zero dimensions and infinite gravity are now removed. This is directly due to the fact that in general relativity all the force of gravity is contained in only a physical curvature of spacetime and therefore a point exists according to these equations where space can be bent until it is infinitely small. Using ESS and macht allows for a second factor to be used in explaining gravity which stops a point of zero dimensions and infinite curvature from existing. The curvature of spacetime will never reach infinity here for two reasons the first being spacetime is not actually real and I will explain this in full later on. The second reason is that the macht glow which accounts for some of the gravity acts as a stopper to the continual compression of space at this central point in the black hole.

Since gravity is made of two factors you can never mathematically bend space enough to cause it to become infinite because no matter how much mass or curving of space you use some of the force of gravity comes from macht. So increasing mass or decreasing volume in an attempt to curve space more will only ever help to ensure that more macht is present as the overall gravitation of the object increases. Some macht will always exist and so always prevent space from curving fully as a result no singularity can ever exist and so the impossibility of general relativity is destroyed using ESS and macht glow of space physics. I will now make a bold statement that I hope most recognize as acceptable as it has already been done before in the history of physics and it is this: In order to demonstrate that his work was a better fit for the universe Einstein had to prove Newton's own work was less able to fit observations and show the flaws in his theory and now I am demonstrating that my work better fits the universe than Einstein's by eliminating the singularity which is one flaw of general relativity and explaining the observables of the universe such as black holes and gravity.

Let us now return to further aspects of black hole physics as they relate to space and gravity. We will once again examine the room filled with ESS spheres and as before note that many of the ESS once outside the black hole have been spatially compressed and moved inside the black hole proper. The space in the room where the ESS used to be outside the black hole does not remain empty as adjacent ESS move towards the black hole to fill the gap. Space however does not flow like a continuous fluid into the black hole nor does it behave in any way like a river, stream or waterfall as you approach the black hole. Instead space calmly and smoothly fills in the gaps that would have been created by the ESS that moved into the black hole. What you now see is a room where the ESS are largest at the walls and decrease in size as you move towards the black hole and as you get very close to the black hole they decrease in size drastically. The drastic decrease in size as you remember is also facilitated by the mathematical decrease in the number of ESS available to each smaller schale and therefore a concentration of macht in these smaller schales. This continues until you reach the interior of the black hole and the spheres are now hollow little dots of ESS. In this way the confusion of space flowing into a black hole and being forever captured by it is done away with quite nicely.

Instead space is still very much held in place and finite and in no way must flow forever. The center ESS in the black hole acts as a sort of stationary piece of space and the other ESS bunch up around it. This is of course assuming the black hole remains stationary if it were to move the central ESS would of course cease to be the center of space in the black hole as a new ESS becomes the center of space instead. The ESS outside the black hole are not forced to increase in size in some desperate attempt to accommodate the loss of other ESS into the black hole as more ESS from outside simply move inward to fill the gaps. Remember the ESS macht extends beyond the black hole for many schales and so will affect those ESS such that they experience a decrease in volume as well. Thus all ESS maintain a gentle spatial compression when farthest away from the black hole until they are greatly spatially compressed inside the black hole proper.

One more note on black holes and space as it applies to space and matter interactions as well as gravity in general before we move on. Another aspect I touched briefly on before and I will alight on once more is the idea of increased ESS in side matter volume. This idea you have been made aware of by now and as the black hole demonstrates the number of ESS per volume of matter is variable depending on the local spatial compression in the area. What I mentioned earlier was the fact that increased ESS inside of matter also allows for increased energy to be donated from that matter back into the ESS. This is one of the most basic matter and space interactions possible as you learned some time ago. Matter is attracted to space and will donate some of its energy or abilities to the ESS it envelops. The ESS here will then use that energy and transform it into ESS macht which it will share with neighboring schale macht. The ESS macht glow of all ESS in the area becomes stronger and this causes the ESS to both decrease in volume and to extend the reach of the macht glow through sharing. In a black hole this effect is greatly exaggerated as is not the case for normal spatial compression on Earth which if you remember is our standard value. I will explain later the exact details of why these effects happen given various levels of spatial compression later and in complete detail but for now I will simply tell you what does in fact happen.

In a black hole the gravity actually increases in strength as a result of this effect. The number of ESS per volume of matter has increased greatly and this in turn means that the matter is even more attracted to space. Since matter is more attracted to space it gives even more energy to the space and specifically to the ESS in that space. This allows the ESS to greatly increase their macht and share the macht with even more schales and even stronger with each schale than before. In other words given the right conditions gravity is variable and not a fixed constant. To put this another way if I have one tonne of matter and I leave it in its non black hole state it will produce for example one unit of gravity. According to modern science gravity inside a black hole will not exceed this limit and remain at the strength of one unit. However because we now know that gravity is made of two factors and we know how gravity is created in the interaction of matter with space we can see that inside an object like a black hole one tonne of matter will produce greater than one unit of gravity. I am not suggesting anything outlandish such as one tonne of matter will produce one trillion units of gravity when in its black hole form only that gravity created by exceptional objects like black holes is not fixed.

The gravity of a black hole is greater than what would be produced if the mass of the black hole were to be unpacked into the form of a star for example. This idea will become perfectly clear in later sections as I explain more about black holes and deal with other unsolved problems to show you their true answers. I only want to place this concept into your minds as it will become relevant in the future. For now just realize that gravity is not always a fixed constant in the universe and can behave quite differently than predicted by the three main views but in reality behaves perfectly normally for the universe once understood using space physics. A simple example of this can be found in variable star systems such as the one that produced Cygnus X-1.

The star system Cygnus X-1 is of course a well known binary star system that contains one normal star and one black hole. The key interesting feature of Cygnus X-1 comes from the X in its name which scientists gave it because it was a strong source of x-ray emissions in space at a time when x-ray astronomy was still being developed. The astronomical significance of x-ray astronomy aside we will focus on the source of the x-rays which is the black hole part of the binary system. Well in fairness it is also from the normal companion star in the system as well as without it the black hole could not make x-rays at all. So what is going on here and why is this star system significant in demonstrating the variable power of gravity as it relates to black holes? First off the what is going on here part is that gas and other upper atmosphere materials from around the photosphere of the companion star are being slowly drawn off by the black hole and consumed by it. As the hot infalling gas and charged particles meet the black hole they are accelerated to very high speeds and in some cases near the speed of light before crossing over the event horizon to be added to the mass of the black hole. Before plunging into the event horizon some of the gas is moving so quickly it emits x-rays which are able to escape the black hole because they are not yet over the event horizon. These x-rays of course travel through space to Earth and are what astronomers of the time were detecting. The overall star system Cygnus X-1 became famous as it was the first solid candidate system to contain and therefore possibly prove the existence of black holes. That is the basic mechanics of how this system is producing x-rays and now we will examine the why it is significant in explaining and proving variable gravity.

Think of the star system before one of the two stars became a black hole and you have a normal binary star system. One of the stars is fairly average and the other is a super giant star with a mass greatly in excess of the Sun probably at least ten solar masses as this is thought to be a fairly safe estimate of the kind of stellar mass that is needed to reliably produce a black hole upon going supernovae. This larger super giant star during the fusion phase of its life was not able to pull material away from its smaller companion and consume it. This is due to the fact that it did not possess sufficient gravity to accomplish this task. We know this to be true because we have observed countless double or binary star systems in the universe and have never once observed a single star in any of those systems drawing material off of their companion. This is the first clue about variable gravity as it shows that normal gravitational effects even those created by a super giant star are not sufficient to in a sense eat or absorb another star. In other words the spatial compression generated by that star and the space around it are not high enough to alter gravity in any non-standard ways.

This topic will be dealt with in complete detail later as the composition of objects like stars, neutron stars and black holes all differ and all affect spatial compression in different ways. For now simply know that the super giant star was not massive enough to get the job done.

The next thing that happened in the Cygnus system was that the super giant star depleted its supply of fusible materials and rapidly underwent a supernova event that resulted in the formation of a black hole. Like a typical super giant star it produced a violent explosion whereby most of the material, up to around 90 percent of the star was blown off and the incredible explosion created a black hole. Fast forward many millions or billions of years since the event and you have a nice inconspicuous star system emitting x-rays for some unknown reason. The reason for x-ray emission has since been determined and the Cygnus system earned its X-1 title to boot. So what we now have is a normal star still burning brightly in the system and a black hole eating it small portions of gas at a time. The gas forms a glowing hot accretion disc which produces the x-rays as we have seen and yet according to a traditional standard non-variable gravity view of the universe this should not be possible. If you think about the overall mass of the black hole and how general relativity theorizes gravity to work it is immediately obvious why this is so. The black hole according to general relativity produces the same gravity as the star that produced it before only because it is a more compact object the curvature of spacetime is increased so the strength of the escape velocity of the black hole is much higher but the reach of the field is less.

There are two flaws with this reasoning the first being that a similar mass will produce similar gravity before and after the supernova event. The star before the supernova event contained 100 percent of its mass and so according to general relativity should produce a normal 100 percent of its gravity. After the supernova event most of the stars mass was lost due to ejected material or neutrino emissions and so it is possible that only ten percent of the original mass of the star remains. This means the black hole should produce only ten percent of the gravity it did as when it was a normal star. Furthermore general relativity states that the overall reach of the gravitational field of the black hole is now greatly diminished because there is a relationship to be maintained between strength and distance of gravity according to Einstein's theory. If the spacetime creating the gravity is less curved according to relativity then the strength of the gravity will be weaker but reach farther out from the object. If you crush the object so that it is a tiny object like a black hole the gravity is now stronger but the reach is vastly diminished. In this way it is easy to see that any star such as the one that produced the black hole in Cygnus X-1 cannot possibly possess the gravitational laws stated by the theory of general relativity.

A super giant star has a diameter of hundreds of millions of kilometers and the binary system would require that the two stars are at tens of millions if not hundreds of millions of kilometers apart. This means the tiny tens of kilometers wide black hole left after the super giant star went super nova would be so far distant from its companion that now the general relativity explanation of gravity could never account for the black holes influence over the companion star. The reach of the black holes gravity would fall miserably short of reaching the companion star and would certainly have no where near the strength to actively draw material off the surface of it for future consumption and production of x-ray emissions. The star after the explosion has lost mass and so if before it did not have strength enough to pull materials off the companion star it certainly does not now. In other words a star system such as Cygnus X-1 proves the theory of general relativity incorrect in explaining how gravity works and how black holes are constructed in terms of local space.

To explain how gravity does work in such a star system one must use the ESS macht description of space as I have shown before. The limitations of gravitational strength and gravitational reach as stated by the theory of general relativity are negated using space physics. In space physics both the overall strength of the gravity and the reach of the gravity are greatly increased inside an object such as a black hole. This is due to both increased spatial compression of ESS inside the local area of space and inside the volume of black hole matter. It is also due to increased macht glow inside the space of the black hole matter and the schales surrounding it. The mechanism of matter donating more energy to ESS which we have examined before as an explanation of how gravity works in general is in use here. The black hole causes the matter to donate more energy to the ESS inside of it and therefore increases the ESS macht strength which results in variable gravity. The gravity of a kilogram of black hole using space physics is more than the gravity of a kilogram of normal stellar matter.

This includes both the strength and reach of gravity as generated by a specific object. In this way a black hole possessing only as little as ten percent of its original stellar mass and now hundreds of millions of kilometers farther from its companion star can in fact do what it could not during its fusion life cycle which is actively and not passively draw material off of the companion star for consumption. This is only one proof of variable gravity and the workings of gravity as explained by ESS macht and space physics. If you find the idea of variable gravity to be a grievous violation of the laws of known physics, don't; variable gravity is the only explanation for two of the biggest unsolved problems in modern physics. I will be showing you further observational proofs of this fact from the universe in later sections and these will also include a more detailed explanation of spatial rebound which I have mentioned briefly before.

Facts

To end this introduction to black hole physics I will sum up some of the key points that I have shown you. The first of the key facts about black holes is that they are made up of space just as much as they are made of matter. Now that it is known that space has a real and definite structure with properties and laws that govern them it is also known that objects such as black holes cannot be without space as well as matter. Space is actually a part of all objects in the universe as we know even down to a single fundamental particle nevertheless the degree to which space is present inside a black hole and to which it contributes to the overall workings of a black hole are simply more prevalent than in other types of objects.

The second fact of black holes is that there is no such thing as a singularity in the universe and this includes anywhere in the cosmos not just the heart of black holes. Physical impossibilities of the infinities generated from the mathematics of general relativity aside it is the combination of ESS and macht that prevent singularities from ever existing. The calculable amounts of gravity from a known mass will still be possible using both ESS and macht to explain gravity only the inclusion of macht into gravity forbids an infinite bending of space which would result in a singularity.

The third fact of black hole physics is that gravity is constructed of two factors which both become significant inside a black hole. I have already explained that gravity is made of two factors which are ESS and macht glow but that in areas of the universe like around the Earth it is macht that contributes to more of the gravitational force than total ESS per volume of matter. Inside a black hole however the total number of ESS per volume of matter is greatly increased and will contribute far more substantially to the overall gravitational force of the black hole than just macht glow alone.

This leads directly to the fourth fact of black holes which is they exhibit levels of elevated gravity and are a source of proof that gravity is a variable force in the universe. Examples of this can be found in star systems such as Cygnus X-1 and the like where mass depleted super giants through supernova mechanisms produce black holes that are hundreds of millions of kilometers farther from their companion star and yet are able to extend their gravitational field far beyond the reach of the theory of general relativity and draw matter into their accretion discs eventually producing detectable x-rays. Likewise I will show later how the other forces can in certain cases be variable as well and you might be comfortable with the idea of the strong force and quark-gluon plasma.

The fifth fact of black hole physics is the increased ESS inside the black hole matter demonstrates the applicability of volume changes of space to quantum systems in a meaningful and non-trivial way. Black holes are good proofs of the fact that space becomes a significant contributor to quantum level systems when spatially compressed to these scales. Significant changes to quantum actions will result for the first time in measurable and calculable ways inside black hole space as compared to Earth space which is under normal spatial compression. This marker of normal spatial compression can be added to the list of scientific standards such as temperature and pressure and the like if relevant for a given experiment or calculation.

The sixth fact of black hole physics is simply one that helps explain variable gravity which is that matter can be made to give more or less of its donated energy to ESS depending on a given set of conditions. This fact I have left for the end even though it was not the last I mentioned in this section because it will tie in directly to the next section which deals with another unsolved problem facing modern science.

I would like to add a quick and fun little note here at the end of this section as I have just introduced some of the basic laws governing black holes I thought it would be nice to include a harmless guess about a wandering black hole. This wandering black hole is one that I had the notion of the other day as circling our own Sun at some great distance in the region of space beyond the Kuiper belt. A purely fun thought about why the number of objects drops off suddenly and for no reason in accordance with known laws of matter distribution and gravitational effects of the Sun in the Kuiper belt would be because a little black hole from a long ago supernova is or was slowly orbiting in this region of space.

It would be easily capable of clearing its orbital path leaving no objects for us to detect. In addition the usual astronomical tool of looking night after night for little dots that seem to be moving against the back drop of stars would falter as the black hole would be all but invisible. Unless a sufficiently bright and distant object were to pass directly behind the black hole as we look out at it we would not be able to see any lensing effects due to gravity either. The drop off edge of the Kuiper belt might be the rough distance of the Lagrange point between the Sun and this wandering black hole that is now our neighbor. Neither body the Sun or the black hole can attract these objects to any significant degree so they are relatively stationary while the orbital path of the black hole is free of detectable materials. Again I just thought of this interesting possibility the other day and thought it might be a fun little tidbit to include even if only in the hopes of getting some bored astronomer on the hunt for gravitational lensing in our solar system's backyard or perhaps some other mind's mental juices flowing.

Chapter 17
Early Universe Behavior

The Big Bang

The Big Bang eh? Where should I start? Well first off I will not be explaining the what banged, why it banged or what was there before it banged until a later section I will tease you for now and let you know that a description of spaced incorporating ESS and macht sharing amongst ESS reveals the answers to these three questions. I also will not be dealing with the intricacies of matter space interactions that exist as a result of these three questions here but I will be giving an explanation of the Big Bang from the moment it exploded and for a short time after it rapidly expanded to form our baby universe. You will be introduced here to spatial rebound and further examples of variable gravity. The main focus will once again be on basic space and matter interactions such that they are necessary for still building upon the behavior of empty space. In short order let us blow up a universe shall we?

So there you have it, a Big Bang event has just taken place and we have a baby universe being born creating space and matter with it. First of all it did not create space and matter with it this would be the assumption of the bang arising from a singularity which by now you should all know is a no-no. These mechanisms will be explained in full later as I mentioned but for now I will use the traditional very small point of explosion to create a universe theory. So the universe is exploding outward in all directions from a very small area compared to the size it is today and that might be anything because even something as large as our solar system is basically a nothingth of the volume of the total universe. So even something that large would still be at the present scale of the universe today basically a tiny point in space from which one might say everything else sprang up. If space and matter were not created in the Big Bang then they must have been present in some form at least shortly after the bang. So you can now picture a small universe rapidly expanding outward in all directions from a single point and that space and matter having come from somewhere are spreading out with it.

The first thing you must realize is that space has not changed its properties since it was created anymore than an electron or any other fundamental particle has. Scientists love to say that the original electrons created in the Big Bang are still with us today totally unchanged and this little tidbit is meant to blow your mind. Whether your mind is blown or not is of little consequence as the fact that electrons have not changed is very much correct. That being said space itself has not changed either and just as an electron retains its base properties over time so does space and specifically empty space. What this means is that similar to the way all the galaxies in the universe were back tracked after Hubble's remarkable discovery to a single point of the Big Bang so is all the space in the universe back tracked to this same starting point.

As we know space is susceptible to compression and individual ESS can be made to decrease in volume. Multiple ESS decreasing in volume in a local area will make the space there seem to shrink and this is what one finds around any large object like the Earth or the Sun for example. So if we back track space to the beginning of the universe we find that all the ESS have become quite small and do not resemble little hollow spheres the likes of which we first saw in our little room but instead look like little hollow dots which are more familiar to us from our black hole orb. The ESS can be rewound from today until we reach the point of the Big Bang and we see that all the ESS are massively spatially compressed into a tiny sphere along with all the matter in the universe. Again I must remind the reader that I will explain fully what really happened at the beginning of the universe and many of these traditional tiny points of near dimensionless space are simply and absurdly wrong in the Big Bang theories of modern science.

Nevertheless we have now travelled backward in events, this is more accurate than saying backwards in time as I will explain in a later section, to the very beginning and in fact the very moment of the Big Bang. Now we can finally understand that space here is made of the most highly spatially compressed ESS possible in the universe and all the matter is packed up in that tiny space too. So when we now run events forward again and we see our baby universe exploding into life we can work with the fact that space is made of all the highly spatially compressed ESS that will ever exist in our universe. More importantly is that just like the electron which has remained unchanged so have the ESS and this means that all the laws governing the behaviors of ESS are still present too.

The first behavior of ESS that I ever taught you is that they compress due to interactions with matter where a part of the matter is donated to the ESS and used in its macht glow. Time for rule number one in exploding a universe which is this: there is no chemical or nuclear explosive used. You might find this strange but I assure you there is a dead serious reason for saying this. Most people think of an explosion as an outward movement of materials and energy from a region of high density and low entropy to a region of low density and high entropy. They are not wrong if you are looking at a conventional explosion such as any normal chemical or even a nuclear detonation. Leaving the entropy out for a minute because in truth most people do not think of that when they think of an explosion we can focus on the main fireball that most people are picturing. This fireball is propelled outward by the chemical or nuclear explosive force but in a Big Bang no one has lit a little wick to cause it to explode. Nor has the universe been powered apart in all directions by a means of chemical or nuclear energy.

Think about this for a minute there is no known driving force in all of modern science for the physical force moving the universe apart from a region of highly compressed material to a region of nice uniform spread out matter as we see today. What powered the explosion? What was the active pushing mechanism that literally expanded the universe? Even modern science knows that it could not be anything quantum either of matter or energy as it is known that space itself is what dragged the matter along with it as it moved outward from the center of the Big Bang. Since this is the case no explanation arising from matter or energy based perspectives can drive the universe apart in the explosion. But the universe did explode and it did expand until it has taken the shape we see today and so what did this? What was the fundamental force that supersedes all matter and energy such that it could carry them along with it in the expansion of the universe?

The answer is of course the only thing left in the universe which is space itself and as I said truly empty space. I did not choose the name Empty Space Structures or ESS for short for no reason as I meant space that is truly devoid of all matter and energy as is known by the Standard Model of particle physics and modern science. The only thing in ESS therefore is ESS and of course the macht it carries. Remember macht is only amplified by the presence of and subsequent interactions of ESS with matter but macht exists as part of the fundamental properties of ESS regardless of matter being there or not. The key to powering the universe apart in all directions at the most fantastic of speeds can only come from one power source and that is the ESS macht that makes up space. The simple interactions of matter with ESS explain how this is so as we can once again make use of our ever helpful room full of little hollow ESS spheres and mostly opaque orbs that we place in it.

Think back to when you first learned how matter causes ESS to decrease in volume from their original size. The matter interacting with the ESS causes it to shrink its volume slightly due to the ESS macht being increased. As the orb moves from right to left in the room we see the ESS that the orb leaves behind rebounding again to their original volume once they are far enough away from the orb. You will also remember that the ESS in the room will never be able to rebound back to their original size as long as any matter exists within the room as the ESS will share macht amongst themselves and therefore they will all have a slightly higher than baseline macht glow. This prevents them from ever rebounding to their original size meaning they will always be slightly smaller than if they were completely left alone with no matter interacting with them at all. So in the very beginning of the universe we find that space is flying away from the center of the Big Bang in every direction and in other words it is rebounding. We also know that spatial rebound is caused by the decrease in macht amongst ESS which is in turn caused by a donation of energy from matter during space and matter interactions.

Therefore during the Big Bang the ESS are obviously experiencing a decrease in the amount of donated energy coming from matter and this allows them to rebound. The state of matter is also somewhat known to modern science at stages of the universes birth just after this phase and matter does not at all resemble what we see today. There are no large structures such as planets, stars or galaxies and in fact there are no small structures either. This means no molecules, crystals or even atoms and in fact there are no hadrons either so no protons or neutrons. At this stage matter is only in its most fundamental forms and exists as a tiny assortment of particles. The early stages of the universe are likewise devoid of certain fundamental forces as well or more accurately these forces have not yet broken apart from each other to become the separate forces we see today. So in essence these forces do not exist as they have not been turned on yet and this is because matter itself cannot even congregate with itself yet. This is what many people know as the quark-gluon plasma of the early universe which is dominated exclusively by the strong nuclear force. Looking back before this and even closer to the Big Bang it is theorized that only space and pure energy existed which in turn would lead one to think that even more forces where turned off. The truth is all the forces in the universe were turned off.

At the very moment and shortly after the Big Bang exploded the four fundamental forces of the universe did not exist not because they had yet to be created but because they had yet to be turned on. Think for a moment of a free quark which finds a couple more friends and becomes a proton. While capable of carrying a positive charge once assembled into a proton the individual quarks do not. The ability to use the charge is carried by the electromagnetic force as anyone knows from interactions of positive protons with negative electrons but in the free quark form it is impossible to use the electromagnetic force. So in this form of matter the electromagnetic force does not yet exist in the universe because it has not yet been turned on. Modern science believes that matter gains its mass from interacting with the Higg's field and that if it does not interact with it its mass has not yet been turned on. These are two examples of fields or properties of matter not existing given certain conditions and the same was very true of the early universe just after the Big Bang. Gravity was not turned on. The ESS structures were in the very first hot moments of the Big Bang free as in completely free of matter just as the little hollow spheres were in our room before we placed the mostly opaque orb in it. Gravity as we know is the result of two factors one of which is

ESS spatial compression and the other is the ESS macht shared inside space. If matter is removed completely from interacting with the ESS as was the case in the very first moments of the Big Bang you remove all donated energy from matter to the ESS macht glow and so as we have learned the ESS will rebound to their original volume.

This reveals another obvious property of space which is that it naturally tends to increase in volume when matter is removed from it. This can be observed for a fact everyday when we look at large gravitational objects in space. Look at the Sun and see that it deforms space around it such that light can be bent as it passes from behind the Sun and reaches Earth where we detect this deflection. The Sun is deforming space through spatial compression but the Sun does not stay permanently in one place, it orbits the Milky Way's galactic center at a rate of 250 million years. So what happens to the spatially compressed ESS that the Sun no longer occupies? Well they do not stay spatially compressed of course and they rebound to their previous size before the Sun orbited over them and we can prove this by looking in the region of space the Sun used to occupy and seeing that there are no light bending effects present. So we know for a fact that space rebounds if matter leaves its local area and in the case of the early Big Bang universe matter has left it completely.

This is the stage of the universes life that is often described as being space filled with pure energy. In this stage the universe is expanding at such a rapid pace that it will smooth itself out in a second or so. The ESS are free to rebound as quickly as they can because matter is not giving them any energy to activate gravity through the macht the ESS carry and share. The rebounding of each ESS occurs at a speed many, many, many times faster than the speed of light and the universe goes from a tiny dot to a massive ball in a fraction of a fraction of a fraction of a second. This is due to the fact that ESS share macht and operate all of their actions at speeds far faster than those of the quantum world which are light speed or slower. Space will always possess an exact center in any Big Bang explosion as there will always be one ESS dead center in the universe because the smallest structures in the universe are ESS. Since space is the backdrop of the universe it creates its own dead center always and this provides the pushing off point for all expanding ESS around it to rebound as well. The fundamental geometry of the universe can be generated from this central ESS as a single sphere where the geometry grows from rings of spheres packed around the first. The interlocking of spheres forms a recognizable pattern that is then repeated through a near infinite number of schales from the very center of the universe to its furthest most edges. Since space tends to optimize its arrangement of ESS it will always form a near perfect sphere as this is the geometrical arrangement of schale after schale of ESS from the center.

Think of a 10 centimeter ball surrounded by a layer of balls of equal size and repeat this until the ball is as big as the Earth. Viewed all at once the Earth appears to be a near perfect sphere but up close the surface is not quite smooth as it is made of 10 centimeter balls which will jut upwards a bit from the overall surface. Any object the size of the Earth will naturally pull itself into a ball following the rule that each ball must touch its neighbor. The universe is built the exact same way as all ESS will seek to touch each other equally so that they can share the macht they carry equally. The macht holds the ESS next to each other and so will always pull the ESS in the universe into a sphere shape which if viewed up close is slightly imperfect at the surface due to the outer most ESS jutting upward slightly. This is a property of how the universe generates itself and how ESS interact with each other. The shape of the entire universe is in fact a sphere with a fine layer of imperfection to that sphere on its outer surface. ESS and the properties they posses explain the shape of the universe and this can be determined from the initial moments of the Big Bang.

The other property of rebound in explaining the Big Bang and how it worked is that it provides the explosive outward force necessary to explain why space and the universe expanded in the first place. With no conventional chemical, nuclear or quantum explosive mechanism available the answer must lie with and in space itself. The natural property of ESS to rebound is the driving force that causes the universe to explode outwards. The rebound of empty space as explained by the absence of matter also provides the rapid expansion mechanism necessary to produce a smooth and uniform distribution of matter as seen today anywhere you look in the universe. The traditional inflationary mechanism involves a few odd things which include the reversal of gravity to push instead of pull, the existence of an inflationary field and the necessity of a force carrier called the inflaton for that field.

In addition it also calls for the field to only exist and be active for a short time in the beginning of the universe which means the universe went to all the extraneous work of creating another fundamental force and force carrier to be used for a trillionth of a second and then never again. All fields and force carriers fall under the restrictions of the Standard Model and quantum mechanics and as such can never travel faster than the speed of light and therefore never expand space faster than the speed of light either. This is of course highly unlikely to the point of being completely unlikely but inflation theory does provide for a reasonable explanation for why the universe is so uniform no matter where we look. To this I say the idea behind a rapid expansion phase of the universe is a just one only the machinations behind it are not what inflationary theory say they are.

Instead the natural properties of ESS offer a simpler way of getting the universe to spread itself out so evenly. We have seen that ESS and ESS macht are responsible for gravity and used them to eliminate the singularity postulated by the theory of general relativity and now we see them able to explain the rapid expansion stage of the universe. Remember again that the rate at which actions take place on the ESS scale are far faster than quantum scales and far faster than the speed of light. This means that a brief rapid expansion stage of the universe where space rebounds many times faster than light could ever travel is real and provides for a uniform distribution of matter in our universe today. The speed of light is the speed limit for all Standard Model and quantum physics and this includes matter and energy but it is not the speed limit for the universe. The universe is made of two parts space and the stuff in space as we know. The space part of the universe is made of ESS and ESS macht glow and these are not bound by relativity, do not obey the speed limit of light and move outside of and much faster than matter and energy physics as understood through quantum mechanics. The rapid expansion stage of the universe proves this as space is allowed to expand many times faster than light and therefore simply carry the matter and energy in the universe with it as it expands.

The mechanism for this is a natural one that we all know is real which is space rebounds in the absence of matter. As I said earlier space will always carry a macht glow if matter is interacting with it anywhere but after the Big Bang when the two are separate as in the space and pure energy stage of the universe the macht glow is at its minimum. In the case of the macht glow of gravity it is in fact off and so the ESS are allowed to behave as though they were the sole things in our little room with no matter orbs present. They will naturally rebound to their pre-matter volumes and all of them doing this at once much faster than the speed of light and in every part of the baby universe at once means they drive all of space and all of the universe apart in every direction quickly and uniformly.

Remember the law of conservation of macht and energy? The only reason space cannot behave in a faster than light way normally is because the slow quantum speed of matter and energy are bound to it through macht. In the early universe with the matter gone this limit is removed and the true macht speed of ESS can be realized. No new fields or fundamental forces are needed, no new carrier particles are needed and gravity is not in fact working bizarrely in reverse. Instead space is doing nothing new and nothing unusual it is only behaving exactly as it should and this explains why gravity appears to work in reverse and why the universe does undergo a brief rapid stage of expansion that will smooth itself out quickly.

What turns this stage of the universe off? Easy, gravity. Gravity from the matter that is now existing in the most primitive forms in the universe. This newly available matter will as we have seen before and observe everyday interact with space and this causes energy to be donated to the ESS. The ESS in turn uses this to fuel the macht glow of gravity which everyone knows causes the ESS to decrease in volume and spatially compress. It also returns the familiar law of conservation of macht and energy meaning motions in the universe are now once again slowed to sub-light speed. The same simple force that makes gravity work on the scale of an apple and the Earth, the Earth and the Sun or in the heart of a black hole can be used to explain how the universe powered its Big Bang explosion, how it expanded rapidly with no new or bizarre properties and then once again how it naturally ended this period of expansion without the need for new force carriers or new fundamental fields to be created for the sole purpose of being used for a trillionth of a second and then never again. All of this is brought about by the simple fact that gravity is variable depending on the amount of interaction matter has with ESS and the ESS macht glow and how much spatial compression is present in the local area which at the time of the Big Bang the local area was the entire universe. You should be recalling the total ESS per volume of matter right about now and macht glow strength. A variable gravity system using ESS and ESS macht can explain everything in the universe from a simple apple falling to the Earth, the interiors of black holes and the explosive force of the Big Bang and the rapid expansion stage of the universe right down to where the expansion stops.

The key facts for this section therefore can be started off with the first which is ESS undergo rebound when the interaction of space and matter is decreased. The second is ESS and ESS macht travel many times faster than the speed of light. Three, the speed of light is not the speed limit for the universe only for matter and energy. Fourth would be that in the total absence of interaction with matter an expansion stage is then produced. The fifth fact is that this expansion phase involves no new mechanisms and can be explained using the exact same ones for all gravitational effects. The sixth is of course that this is further proof that gravity is variable. The seventh fact is that the nature of the explosion of the universe and the properties and behaviors of ESS means the universe will always have a near spherical shape due to ESS seeking to attach themselves to each other in schales. The eighth fact is that the natural interactions of space with matter will turn gravity back on as ESS macht of gravity is once again shared with all ESS and thus end the rapid expansion stage of the universe. The next section will once again explain the use of variable gravity to answer some of the unsolved problems facing modern science and will continue the evolution of the universe as a general way of keeping things in order with respect to explaining space and matter interactions.

Deceleration and Acceleration

The next unsolved problem in physics relates to the expansion of the universe after the Big Bang has taken place and matter has settled into the familiar forms we know today. The unsolved problem is why the expansion of the universe after the initial Big Bang explosion decelerated for a period of time and has slowly started to accelerate again. This is not to say that the universe ever stopped expanding after the explosion that created it only that the rate of expansion slowed and then increased again but at all times the universe was still moving outward from the central point of the blast. This change in rates of expansion of the universe is commonly known and well studied but what has eluded modern science is why it has happened in the first place. After all conventional explosions always start out moving as fast as they possibly can and then begin to slow down afterwards until they eventually stop. The universe is slightly different as it is unclear as to whether or not it will ever stop but that is a different unsolved problem for later. No matter what the universe is not behaving like a conventional explosion whether chemical or nuclear in nature.

We do know for a fact from Newton's work that everything in the universe usually tends to remain just as it is forever unless acted upon by an outside force. In other words in order to cause an explosion to decelerate and then accelerate again would normally take the outside influence of an additional force to change the rate of the expanding fireball for example. This tends to make people think that something is acting upon space such that it has slowed and sped up its rate of expansion over the billions of years since the Big Bang. The typical two candidate answers to this problem as is currently favored by modern science are dark matter and dark energy. Dark matter is simply a name given to any matter that cannot be seen as we know from the 1930s and Zwicky's work. Since then both WIMPS and MACHOS of different types have been reasonably proposed as the true dark matter of the universe. That is if dark matter even exists.

You see no one has ever found dark matter and certainly not in sufficient amounts to explain observed phenomena. However we will leave out dark matter discussions in full for now as I only need remind the reader of its presence here for its contribution to gravity. That is all dark matter is really used for anyway it is simply to provide more mass and therefore more gravity where people think it should go. As for the deceleration of the universe most dark matter proponents are willing to simply say that the universe did behave like a normal explosion right up until the point it started to accelerate again.

The acceleration is then attributed to dark matters opposite twin which is of course dark energy. Real creative naming I know but that is what it has been called as again it simply refers to energy that cannot be seen normally. It strikes me as odd anyway because anytime someone says there is something that cannot be seen it is because there is nothing there to see. Nevertheless dark energy is completely unknown to modern science in every form whatsoever. If you meet someone who tries to describe the properties of dark energy to you simply smile and nod then excuse yourself as anyone worth their reputation in science will tell you that dark energy is a purely hypothetical theory at best. What most people who do support dark energy do say is that it provides a pushing force to space. This much is in my opinion fine as something does seem to be making space push apart and I hold nothing against dark energy proponents for it. Again I will be returning to the dark matter and dark energy discussions later, for now we will look at dark energy as a pushing force in the universe driving space to accelerate its rate of expansion once more.

So we have the universe going from fast to slow to fast again in terms of its rate of expansion and we have dark matter being credited for the slow down as it contributes to gravity which pulls the universe together and then we have dark energy being credited for pushing the universe apart again. This would seem to make sense right up until you think about this more closely. If the universe created all matter in the Big Bang and this matter has remained unchanged since the beginning of the universe then why did the universe not collapse back on itself due to gravity? In other words look at fundamental particles like the electron or quark and of which dark matter is also theorized to be fundamental and you see that no matter what you do to these electrons or quarks you cannot destroy them.

Create a neutron star you say and bind up the electrons in the proton electron capture mechanism? Well you have technically gotten rid of an electron but in reality you have not as the mass of the electron and its charge have been conserved in side the neutron. This is why the neutron is heavier than an electron and proton and has no charge so you have not actually destroyed anything and the gravity of the matter remains. In other words the universe has not lost any matter and therefore not lost gravity so should still slow down. Dark matter is the same as a theoretical fundamental particle, you will always be able to conserve its mass somewhere and therefore conserve its gravity such that the universe does not lose any gravity and therefore should still slow down.

What about annihilating an electron with a positron you ask surely that gets rid of its mass? Well you are almost correct as the electron is now a high energy gamma ray photon and so is the positron but the equivalence principle of relativity tells you that matter and energy are convertible mathematically and so the matter remains. If you are having trouble understanding this think of a hydrogen atom that is floating free in space and has a mass of one unit for arguments sake. If you introduce a photon to the system such that the nucleus of the atom in this case the proton captures it you will have now a mass of one plus one photon. The equivalence principle states that since the atom has absorbed the photon and since matter and energy are the same according to Einstein the actual mass of the hydrogen atom has increased by one photons worth despite the photon being a massless particle. So here you can see that even in apparent death the electron still retains its mass and therefore contributes its equivalent amount of energy to gravity in the universe and most likely by being swallowed up by some normal or dark matter should it exist. So this means that the gravity of the universe has not decreased and it should still slow down.

Now we see the universe increasing its rate of expansion in the acceleration phase and this has been theorized as due to dark energy. The trouble is where did the dark energy come from? All energy and matter as we know it was created in the Big Bang and so therefore was the dark energy as in all of it. If this again were the case how could the universe start to accelerate its rate of expansion again as this means the dark energy content of the universe is increasing. Creating something from nothing is of course impossible and so ruled out as a possible explanation. This means that if dark energy is correct as the answer for why the universe is accelerating in its expansion then huge quantities of the stuff must be being created even as we speak. How much must be created? Well according to the breakdown of modern science more than all the visible, neutrino and dark matter in the universe combined. In fact about 70 percent of the universe is theorized to be made of dark energy which is logically impossible. The reason this is so is that if it were matter and therefore gravity would have little to no consequence whatsoever on the fate of the universe in terms of expansion or even total life cycle. People have often said that if the universe had a bit more matter or was a bit denser then we would be long dead from a Big Crunch well before any stars or life could have evolved. Well the same is true of too little matter and too little density or too much dark energy push as the universe would have long ago torn itself asunder in a Big Rip. Again not enough time for stars or life to evolve and therefore for any of us to sit around at leisure sipping tea and contemplating these what if scenarios.

So from a logical critical universe density perspective the chances of dark energy being this prevalent if at all is highly improbable and the mere notion of dark energy increasing is not only impossible but would lead to such a rapid runaway effect that the universe would be destroyed before I finished this sentence. Well since I have placed my period thusly at the tail of my previous sentence I believe I have warded off the evil unseen spirits. Frivolity of semantics aside dark energy is thought to reside inside the vacuum energy of space and since all matter inhabits space a very real destructive mechanism for pulling atoms and the like to pieces exists. A subsequent increase in the pressure exerted internally on any quantum system would have already upset the carefully balanced universe such that again life and matter would be long since destroyed. This has to do with the very careful balancing act that modern science believes to be at work at precisely this moment in the universe as I have stated too little or too much of something kills us instantly because destruction is wrought at the quantum level and therefore the quantum timescale. You cannot perceive this death because your thought processes work at a speed below quantum scales so you would not be aware of anything anyway.

This perfect balance just as we happen to perceive the universe today is folly for another reason and that is the universe is old and has behaved just as we observe it for billions of billions of years. If the critical balance was only achieved recently then why was the universe not destroyed billions of years ago? The conditions for stellar life as well as the following biological life would not have been possible in this distant past and so no stellar nucleosynthesis to eventually make us possible. For the record I am only using these human-centric arguments of the universe because I disagree with their use wholeheartedly in science and philosophy. Humans have long contained the arrogance necessary to make such and I will say it bluntly, stupid claims that things are only as they are because we perceive them that way. I will ask the universe to make trees fall in forests the universe over and you had better get over the fact that they all make a sound. The universe as I have said before does not care whether or not we understand it and does not change itself to fit theories. Just because something exists in our own fictional landscape of mathematics and ad hoc theories run wild is no excuse for trying to make the universe conform to us rather than the other way around.

I apologize to any that are theoretically offended by this but I do so for your benefit the sooner this fact is realized the sooner the truth can be made what it should be which is a simple fact like all the rest. To those who are reading this and not offended I believe that you are the bright ones ready to accept a new order in understanding such that we can finally reveal how the universe works. This is the dream claimed by all who study the cosmos so how can we solve the problem of the decelerating and accelerating universe if we do not use more matter and more energy? In other words can the problem be solved without violating any of the known laws of science such that we do not need to decrease matter or create new matter after the Big Bang? In other words how can we explain how the universe behaved differently in the past billions of years, how it behaves today and how it will behave in the billions to trillions of years to come in such a way that it does not upset any critical values and destroy itself?

The answer is the same I have been giving you since the start of my work and it is changes in ESS macht due to spatial compression and interactions with matter. The universe was created with a set amount of space and stuff and yet it has persisted over its entire life without tipping any critical values. The changes in the universe and its appearance are as big as they get involving whole galaxy clusters and filaments as more energy and matter than humans have ever been able to conceive of in our entire history. So with these massive changes one might expect the universe to behave differently and if not, which it has not really then what keeps it in check? The variable gravity I have mentioned before is again going to be proved as the mechanism by which this decrease and increase in the acceleration of the expansion of the universe is made possible without upsetting balances, creating more dark anythings or relying on the perfect timing of how we perceive things. It will also explain the way this occurs such that no additional variables such as new matter or energy is introduced and instead will be made possible by once again looking at what empty space really is and how it works with matter. I like making the universe simpler and removing variables as it tells me that the universe is behaving not as I think it should but as it wants too which is always by choosing the easiest path. Matter always exists at its rest and not excited state if you allow it whether that is classical mechanics, general relativity or quantum mechanics as the method of explaining what is going on.

So the least number of variables for this unsolved problem is using only space itself to explain why it has decided to slow down and speed up again. The first thing we must remember is that the universe began as a massive almost infinitely faster than light explosion that pushed everything outwards from a central point. This point is of course the central ESS of the universe in terms of the Big Bang and set everything moving outwards creating the universe as it went. Like any force as we have known since the birth of classical mechanics the universe does not want to stop expanding unless something makes it so and in this case it is the universe itself that is making it change speeds. The expansion of the universe is going to progress in multiple stages and I will only discuss three of them here as all others will be handled in another section and another unsolved problem. The first stage is the initial explosion which we know to have started everything moving outward and will continue normally unless acted upon.

The second stage is the deceleration of the expansion phase of the universe and this was due to increases in gravity in the universe. Extra matter was not needed to make more gravity as we know gravity to be variable you only need to turn up the dial a little. We do know that after the rapid expansion stage of the universe when space was rebounding faster than light that it was the introduction of matter back with the ESS that caused this stage to stop. Matter all through the universe was able to once again interact with the ESS and power the macht glow of gravity to halt the rapid expansion rebound. But matter was not in its present form as we see it today and existed as a collection of particles only it would be much later that complex assemblies like gas clouds and stars would be born. When these assemblies were made possible the density of matter in the universe increased which as we have seen increases spatial compression. The increase in spatial compression is denoted by a decrease in the volume of ESS and an increase in the macht glow they share with each other. As we have seen the more ESS are available per volume of matter the greater the interaction of space and matter will be. Any increase in interactions of space and matter will lead directly to more energy being donated to the ESS by the matter. The ESS in turn will increase their macht and share this throughout the local area of space in ever expanding schales. The combination of ESS and macht are what cause the force of gravity to be made and so increasing the two factors increases gravity. In other words as we have seen before gravity is not constant for a set amount of matter and is variable.

The second stage of the universe is seen to exist during and after the formation of large numbers of stars and other assemblies in the universe. The earliest stars in the universe were all formed from clouds of gas much denser than is available today and so as a result most early stars in the universe were massive super giants with masses many times that of the Sun. The inevitable supernovas and black holes to follow were produced in copious amounts evenly throughout the universe. The distribution was very uniform as the distribution of gas from which their stars were formed was uniform although not perfectly uniform of course. This means that a uniform distribution of black holes from supergiant stars was made available evenly throughout the cosmos and with them increased levels of variable gravity. The highly spatially compressed nature of space in black holes as has already been learned is one mechanism of turning up the force of gravity. This caused the overall expansion of the universe to slow down but not stop obviously as the initial outward momentum was still present.

The universe without variable gravity should never have slowed down in the first place as a constant gravity would simply ride along with the outward momentum from the initial explosion and take regular stars, planets and so on along with it as matter continued to be dragged with expanding space but not slowing space down. Anything beyond the light cone according to modern science cannot affect anything else and so pockets of universe would move outward regardless of whether or not gravity was trying to stop it. This is of course only assuming a non-variable force of gravity and would be limited to light or sub-light speed physics. This is also due to the fact that according to general relativity gravity the effects of a gravitational field are limited for a set mass and therefore cannot ever bend space enough to halt the expansion of the universe anyway as space will not curve enough between isolated objects. Creating more matter or energy from after the Big Bang is impossible and so you cannot use those theories to explain the observed phenomena. In short you cannot build a universe without variable gravity.

So now we have seen how the usage of variable gravity and ESS have allowed the universe to explode, rebound faster than light and then decelerate its rate of expansion but how does it permit the universe to accelerate again? It might seem that the variable gravity would cause the universe to collapse again and for this I will leave all discussion of the ultimate fate of the universe until later I will focus now only on a mechanism for speeding up the expansion again. What you must remember now is that gravity is as I said variable we can turn it on, turn it up and of course do the opposite we can turn it down and even off. Turning it off is simple as the mechanism for this has already been demonstrated albeit under specific conditions by using unrestrained spatial rebound during the Big Bang. However turning it down is much more easily done than said. This is no joke just wait a second and gravity has been turned down as I have shown before by something like the Sun leaving some local space behind as it orbits the galaxy. This space now removed from direct interactions with matter has its macht glow of gravity decreased as it becomes an external schale to the ESS inside the Sun. The decrease in macht glow has made the volume of the ESS in that schale increase and therefore the force of gravity in that local area of space has decreased. I told you it was easier done than said and now I will explain how this works on a universe scale and why it appears to us here on Earth as if the entire universe is accelerating apart in all directions.

The first thing you must come to grips with just like trees being noisy drunkards in forests when no one is watching is that not everything is actually moving away from us as many would have you think. Take for example the Andromeda galaxy which will collide with us in the future and you have an example of a massive object that is 200 000 light years across and is coming at us and not going away from us. Throw into this the millions of light years of space between us and you have a non-trivial volume of the universe that despite the apparent accelerating expansion of the universe or the ever increasing theoretical threat of dark energy is not actually getting farther away from us. Scale this up to things like galaxy cluster collisions and you see that even larger structures are shrugging off the so called rule that everything must eventually fly apart. Yet the universe on average does seem to be accelerating away from us. Is this a product of the ever increasing variable gravity of the Milky Way spatially compressing space ever more around us and shifting the light of incoming photons from every direction we look? Partly yes as the local area of spatial compression around the Milky Way is growing with every passing macht second of the cosmos. This effect is of course a demonstration of variable gravity but there is another mechanism at work here also using variable gravity that occurs far from our home galaxy. In truth the effect is happening right here too and throughout the universe but we perceive it only at long range.

What this effect is seen in the universe as it ages beyond the first and second stages and actually began as soon as the second stage did too. As matter began to coalesce in the second stage space compressed in certain areas due to more matter being attracted to it. The areas of space that lay in between these were being depleted of matter and so had their ESS macht glow decreased and this means the ESS that were at the same level of spatial compression as everywhere in the universe after the Big Bang were now rebounding. Rebounding means getting larger in volume and therefore spreading out more as macht glow decreases. Spreading out means increased distance between two regions and this effect simply allowed to continue for billions of years has resulted in the universe increasing its acceleration again.

The current maps of the known universe which cover tens of billions of light years show clear filaments in the structure of the universe and these filaments are as old as the Big Bang itself. Even before the light of recombination was lit the shape of the universe was made not in matter but in ESS and the macht glow they carry. This shape is what we are looking at today and shows all regions of space that have experienced significant levels of increased variable gravity and those that have therefore rebounded with less matter in them. As the space there rebounds the ESS increase in volume which pushes against neighboring ESS and drives them away. The ESS of the universe are all connected and since the ESS are only pushing but not sharing more macht they do not experience a similar decrease in volume as they move against distant macht. Remember it takes both matter interactions and macht sharing between schales to alter ESS volume if one is missing ESS can push on each other freely. In addition gravities reach is unmatched but not infinite and so any areas sufficiently distant from significant sources of matter will not have any elevated gravity and macht glow there. The maximum distance is the one that would break the law of conservation of macht and energy and so is a real boundary to gravities reach from any object with mass.

Where the Earth is and therefore where we are observing this acceleration phase from is inside a filament and when we look outwards we see the combination of two effects at once which makes us believe the universe is going to expand forever and tear itself apart. Areas of space behind us closer to the true center of the universe we are in fact accelerating from and so they appear to be accelerating away from us while areas ahead of us are accelerating away and again we see the same effect both of which are produced by our perspective to spatial rebound and subsequent ESS push in the universe as compared to the true universe center. As the universe continues to age more matter is taken up by the filaments and the objects within them causing more variable gravity to increase. This means the space between filaments continues to rebound more and more and so expands and push farther and faster making it seem as if the acceleration is ever increasing as well.

The first effect is the remnant momentum from the Big Bang which is continuing to carry space, matter and the universe itself outwards from the initial explosion. Remember not to fall prey to human arrogance and assume that we are observing the universe at some special time when the initial momentum should be long gone as it is still with us. This idea should be instantly acceptable to all minds as it would be equally arrogant of a species to make assumptions of the universe perhaps tens of billions of years from now when the initial momentum is gone. Again I will discuss and explain the fate of the universe later.

The second effect that makes the universe seem to accelerate away from us in all directions is the rebounding of space in areas of the universe between filaments. The scale of these spaces are on the orders of billions of light years in some cases and will as they say play tricks on your eyes. In this case play tricks with the photons we are detecting here on Earth. Light will be handled as a topic later and specifically light and ESS interactions but for now we are looking only at the space between filaments and their high levels of spatial rebound. Any other observer in any other system in the universe would notice the same acceleration phenomena as their filament is separated from all the others too. Remember it is a combination of the two factors which are the remnant momentum of the initial explosion and the variable gravity of the space in either the filaments or inter-filament space that is causing this effect. Variable gravity, spatial rebound, remnant momentum and combinations of these effects will create an acceleration of the universe that will last for a while but not forever. The overall filaments are themselves still being carried along with the remnant momentum and are moving away from us too and even accelerating in areas of space even larger than this scale. As galaxies age and accumulate more materials they will also show these effects between galaxies within a filament and so all factors remain the same for apparent acceleration away from us.

From our humble Earth perspective the universe is accelerating its rate of expansion and objects are moving away from us in all directions. The idea that objects are moving away from specifically us in all directions and that this can only be explained by an increase in the tearing apart force of the universe known as vacuum energy or dark energy is again arrogant as that would place us neatly at the dead center of the universe. Did it never occur to some that the Earth, the Sun, the Milky way itself are all moving inside the universe and so the notion that we are at exactly the center from where everything only appears to be moving away from us perfectly is not also arrogant? Not to mention that this is impossible as if we were the very center of the universe and yet still able to move freely in space for the last 14 billion years or so how did we move and yet stay always at the center for those years? The simple answer is that it only appears as though we are at the very center of the universe. Is the universe accelerating its rate of expansion? Yes. Is this a combination of factors? Yes. The observation is real and the variable gravity as well as the remnant momentum explains how this is possible without creating new matter or energy and without upsetting any critical thresholds that would otherwise signal the very rapid demise of the universe.

Facts

The key facts here are first of all that remnant momentum still exists in the universe and will for quite some time. The second is that the increase in ESS macht due to spatial compression allowed for greater space and matter interactions which increased the strength of gravity. The third fact is that the resulting spatial rebound in areas of the universe depleted in matter expands the space there which pushes on the remaining matter filled space. The fourth fact is the push is what we can see between filaments and is also combined with the remnant momentum providing for both factors that account for the acceleration increase. The fifth fact is that all this can be accomplished with a minimum of variables and needs nothing other than space and matter in order to operate. All the same interactions mechanisms between space and matter are here with nothing new or changed simply viewed at the scale of the entire universe and shown in three stages.

The sixth fact here is the very shape of the universe before first light is made possible by looking at filament distribution and filament composition as all future structure to the universe is carried through ESS. Specifically the initial Big Bang explosion can be reverse mapped from all current structures of the universe due to the fact that it is the ESS and the ESS macht that first carried all interactions from individual fundamental particles through to the present day although the necessary computations are quite staggering. The next section will deal with a continuation of these topics by looking again at the limitations of theories such as dark matter and explaining unsolved problems without the use of additional variables such as early universe structure.

Chapter 18
Mass, Galaxy Rotation, Galaxy Formation

Mass

The universe appears to be missing matter from somewhere as the necessary amounts of mass needed to hold the universe together are not fully accounted for. This is of course the topic of much debate in modern science over the last few decades as it has become apparent that large scale objects like galaxies and galaxy clusters seem to not contain enough normal or baryonic matter to hold themselves together. In other words the visible mass of a spiral galaxy for example seems insufficient to provide enough gravity pulling inwards to counteract the motion of stars in orbit around its galactic center such that their orbital velocities should cause the galaxies to simply fly apart. Think for a moment of Newton and his idea that gravity is a force like a tethered string between two objects and imagine one object twirling about around your hand held by a string. If the force of gravity is strong enough the object will remain tethered to your hand by the string. If the object you are twirling is an apple then the string should hold easily but if you are trying to twirl a pumpkin there is no way for the string to hold. The string will break and the pumpkin will fly off away from your hand therefore breaking free of your Newtonian gravitational force. This is similar to what is happening inside galaxies according to the three main views as stars should be able to break the strings holding them to their galaxies centers.

Obviously this is not the case as galaxies have existed for billions of billions of years and have not and are not flying apart today. So why are they not flying apart or more accurately where is all the missing matter in the universe that we cannot see visibly when we look at these galaxies and galaxy clusters? To clarify just because it appears that the matter is missing or the gravity is insufficient does not actually make it true. What I mean by this is that the three main views of modern science cannot understand where the matter appears to have gone but this does not mean that any matter really is gone. This is another example of trying to get the universe to fit theory rather than looking at all factual observations and making a model to fit those and that means all of them at once. In reality the universe is once again behaving just as it should and it is the failings of the three main views that create this unsolved problem of the missing matter. I have already proved this so by showing that the universe has survived quite happily for billions of years since the Big Bang without any strange consequences that would be caused if matter was missing.

The solution that modern science has been supporting for decades concerning this problem is the infamous dark matter that if you are even slightly interested in science have heard far too much about already. Well I do not mean to disappoint you but you are about to hear more as I must discuss the two main types of dark matter and the reasons they cannot explain the missing matter and therefore gravity. The first thing you must realize is that dark matter is only a quick name given back in the 1930s to describe matter that cannot be seen in the normal electromagnetic spectrum by astronomers. Photons make up the electromagnetic spectrum and also light and so we say that matter that gives off photons is light matter if you will. Matter that cannot be detected in any of the electromagnetic wavelengths is said therefore to be dark matter and remember this is only if it exists. Since everything in the universe does emit photons if you poke and prod at it correctly the very idea that some exotic form of matter exists such that it cannot emit any photons directly or indirectly is immediately suspect. Even black holes are theorized to emit radiation in the form of

Hawking radiation and so are not truly black after all. So matter that is dark and cannot emit any light whatsoever yet outnumbers all visible baryonic matter up to 6:1 is now scoffingly suspect.

Here is a quick one, if dark matter makes up six times more mass than visible matter in the universe then how come we do not see gravitational lensing effects literally everywhere we look? Looking out into the galaxy around us the empty space is supposedly filled with scads of this fictional dark matter stuff and so looking at any object should always appear heavily distorted. Yet when we do see lensing effects they are always neatly circular as if a single large object is hiding there such as a black hole or we see only some distortion and at great distances meaning there is ample room for other black hole type objects to obscure what we see. A long passed region of star formation having all become black holes could easily create isolated patches of non-circular lensing effects. People love looking at Hubble Space Telescope images that show areas of star formation in gas and dust and yet never thought that this happened billions of years ago and now all that is left are pockets of black holes in between us and some distant light source. This is especially confounding when it is realized that most modern astronomers accept the fact that black holes must be fairly common in our own galaxy alone. So it is very reasonable that normal events are causing things like gravitational lensing of distant luminous objects as opposed to some theoretical form of missing dark matter.

Yet the theories of missing matter persist in the two dark matter camps and in fact decades old searches by tens of thousands strong or more scientists backed by tens of billions of dollars worth of equipment search for it daily. They have found nothing. Yes that is the truth they have found no dark matter whatsoever and this is a huge red flag as never before in the history of humanity have we failed to find something if it was real. Excluding the years we thought the universe to be our solar system alone and other such nonsense the modern era of science over the last couple hundred years has been wholly successful. We thought atoms were real and in short order we found them. Next we though things like neutrons were real and found them quite quickly afterwards. Harness the power of nuclear energy? Sure we can do that and did. These are but a few examples but they illustrate that at each correct step in science the truth was found in years or maybe, maybe decades versus centuries or millennia.

Dark matter has been looked for since the 1930s and now here we are as I write this some approximate 80 years later and no dark matter found. This is a good indicator that the missing matter and missing gravity is not a problem of the universe but a problem of the three main views and the limitations they place and dictate on how the universe should work. Remember you can use any of three different views and everything they theorize about the universe to solve the problem of missing matter and still cannot find it. We have already seen the limitations of the three main views in the past so why should they with their flaws be correct now? The answer is of course that they are not correct or no such unsolved problems would exist and so looking beyond them is the only hope of solving how the universe works. I reiterate this fact again as it is the restraints on thinking generated by these three views that has led to decades of fruitless experiments and ad hoc theories concerning our universe and its secrets.

To be fair to the dark matter supporters though not all of the dark matter theories should be completely discarded as some do make some sense. Of the two types of dark matter I will look at the MACHOS first which as you all remember are made of larger objects like rogue planets, asteroids, black holes and even simple gas and dust. In other words they are simply more matter of which we are familiar at least only for some reason or other they are not emitting any light for us to detect. Now that being said remember there is a limit to our detection capabilities and just because something is too dim for us to view directly does not mean that it is not there. If I were to look for a small asteroid say one kilometer in diameter randomly placed in the Milky Way it would no matter how hard I tried fail to emit enough light for me to detect it as it is simply too dim and distant. Even if I were to fill the space of our galaxy with trillions of such objects they still would not emit enough light to be seen and yet trillions of kilometer sized asteroids has a definite mass that if held in one object would be much easier to detect. So the idea of dark matter being made of things like this is not so far fetched as even today we have to look very hard in our own backyard just to see Kuiper belt objects let alone Kuiper belt sized objects light years away or even in another galaxy. If you throw into this gas and dust you can account for even more matter as detecting these is near impossible at any great distance.

Now that being said many opponents to the MACHOS camp have suggested that if a uniform distribution of such objects did exist and were present in sufficient densities to account for the missing matter then they would still radiate a faint glow from all the absorbed radiation they receive from the active stars in their parent galaxy. This is the argument made by the WIMPS who are the sworn enemies of the MACHOS and will rumble with them at the drop of a hat. The two camps are not really rival gangs destined to clash only that the cons of one side are the flaws of the other and each side often exploits these in an attempt to lend additional validity to their own theory. Nevertheless the argument is a valid one as copious amounts of evenly distributed dust sufficient to provide all the missing gravity would radiate electromagnetic energy on some wavelength and be detectable here on Earth. However larger MACHOS objects would not do this such as black holes and rogue planets. A single planet sized object hidden in the blackness of interstellar space would receive very little energy from the surrounding stars in the first place and therefore emit even less of that back into space for us to see. In a kind of backwards thinking to most the larger an object is the harder it is to see in space sometimes.

Take black holes as another example and you can see the problem illustrated once again as these emit next to no radiation and are called black for a reason as they abhor giving up energy of any sort. A single rogue black hole can account for cubic light years worth of gas and dust by mass and still go undetected even if we were looking right at it with today's best detectors. It would not take a great number of these wandering black holes to account for a large portion of the missing gravity even by the standards of the three main views as they are simply so massive. The idea that numerous black holes exist in our galaxy as well as every other galaxy in the universe should actually be a simple one to accept. Stars that are larger than our own Sun by a few or more solar masses tend to go supernova upon using up their supply of fusible stellar fuel. The stars that end up as supernova are all massive stars which are known to have a short lifespan meaning they would burn through their supplies of fuel in only a few hundred million years and then become black holes or also neutron stars.

Our galaxy is on the order of 12 billion years old and so 11.5 billion years ago all the supergiant stars destined to become black holes would have already. By the time we or any other intelligent species comes along to question the supposed missing matter the galaxies of the universe would be populated with countless black holes already and all would be undetectable. Couple this fact with the reality that all early stars in the universe tended to burn hot and fast with the dense supply of early hydrogen that many more black holes exist than previously thought. So objects like this are a viable mechanism for explaining where the missing matter has gone. Although the WIMPS camp is not incorrect however in stating that there would not be enough of these hidden black holes to account for the missing matter either. It is true black holes can help the search for an answer but they themselves are not the full answer. Nevertheless the MACHOS side is correct in stating that limitations on human astronomical detection capabilities has prevented us from seeing all of the gas, dust, rocks, planets and black holes that do contribute to a percentage of the missing matter. In this respect I partially support the idea that simply more unseen conventional forms of matter can be useful in explaining where some but not all of the missing gravity appears to have gone.

The opposite side of the argument is the WIMPS camp which we remember is made of swarms of tiny particles as yet undiscovered and flying everywhere at once in the universe. There are numerous WIMP particles theorized and they are all based off of two factors the first being one or more of the three main views and applying their rules to the universe and the second are any number of assumptions made about the way the universe works in general. These two factors are needed by WIMP theorists in order to whittle down the myriad possibilities of candidate particles to a more manageable number. They are also used for a second purpose which is to immediately exclude any far fetched or ridiculous particle candidates that are obviously incorrect. This careful planning by the WIMPS camp means that as far as modern science is concerned they have left nothing to chance and if a particle is not theorized by them it does not exist or exists at such an outlandishly low probability that it is ludicrous to pursue. What it also means is that the general behaviors of the various candidate particles can be summarized easily such that all WIMPS will obey a set number of WIMP properties.

The first is that they are all tiny particles similar in number to neutrinos but can be theorized with different amounts of size or mass or both which might make them like a neutrino and might not. The bottom line is that they are very small and not at all large like a hadron or anything of similar size. The second rule is that they interact rarely with normal matter similar to the way that neutrinos interact infrequently with normal matter. The third and last property of note about WIMPS is that they are everywhere in the universe and by this the WIMP theorists literally mean everywhere. The particles are theorized to be so small, so weakly interacting and so plentiful in order to make up six times the amount of mass in the universe as normal matter that they literally are everywhere in unfathomable numbers. Taking a typical human for example in size and mass and trillions of these dark matter particles would pass through your body every second. All the while you would be perfectly unaware of any of them.

The idea behind the WIMPS dark matter particles is that they do not interact much with normal matter, although they must from time to time like a neutrino but instead contribute more to the gravity of a galaxy or galaxy cluster. I said galaxy or galaxy cluster because dark matter is not needed to hold something like our solar system together as such a cosmically small object as a single solar system obeys perfectly normal relations between known mass, forces of gravity and orbital velocities. In other words the planets closest to their stars such as Mercury orbit the fastest while those much more distant like Neptune or Pluto orbit far, far slower.

It is useful here to point out that the Flat Galaxy Rotation problem is one that dark matter in either of the two camps is trying to answer as the stars in galaxies seem to disobey the normal orbital velocities and distances that are held firm inside our own solar system. In galactic terms it seems as though both the inner Mercury, Venus and outer Neptune, Pluto like stars orbit at the same speeds which contradicts the expected laws of the three main views. Nevertheless the universe once again does not care what modern science or the three main views thinks and happily and contentedly orbits stars around galaxies at nearly the same velocity all the time. So what we have seen here is a contribution to the background gravity of a large object like a galaxy or galaxy cluster by way of numerous WIMP dark matter particles although the particles of course do not do anything as absurd as orbit the galaxies themselves. This is a known fact as no other particle orbits a galaxy and all forms of particles simply move freely through space and this again has been proven countless times by observing neutrinos, photons and the like emitted from distant galaxies all around us in the cosmos and no matter how many billions of light years they travel they move in straight paths right to us.

So now we must look at whether or not dark matter particles can be used to explain such observed phenomena as solving the rotation problem, holding galaxies and galaxy clusters together, dealing with large scale interactions like galaxy cluster collisions and simply accounting for the missing matter in the first place. The first thing we will look at is the success of detecting these particles here on Earth as predicted by dark matter particle theory. The success to date for detecting any particles of the correct dark matter masses has been zero. Yes it is a sad truth for supporters of the WIMP type of dark matter as all searches have been fruitless. The experiments have been of course carefully thought out and carefully planned and as such the actual detectors have been of some of the best if not the best in the world. The problem is that no matter how many detectors are used, no matter how sensitive they are and no matter how long they are allowed to run not a single dark matter particle has ever been detected. This is a huge failure for the WIMP camp as we have seen before that anytime humans have attempted to find a new particle from theory the particle if real is found within years of prediction or maybe a decade or two and found conclusively as well. This is not so the case with dark matter particles as searches of them have been going on since the 1930s when dark matter was first proposed as something that might exist.

Since then and notably around the last forty or so years a great interest and investment of time has been placed on finding dark matter particles specifically and to no avail. The flat out failure to find dark matter particles is bad enough but there is a second problem with the WIMP theory and the history of not detecting dark matter particles and that is they should if they exist be everywhere. This is of course one of the main three criteria for defining any of the WIMP candidate particles as they must exist in simply brain shattering numbers if they are to exist at all and have any hope of accounting for an atrociously large five to six times the matter of all known visible matter in the universe. Think about this for a moment and realize that a single black hole can be 10, 100 a million times more massive than our Sun and think about how many particles are in our Sun. Think about how dense it is and about how dense a black hole must be and then add in every grain of dust, every wisp of gas, every planet, moon and floating rock in space and realize that WIMP particles are supposed to make up at least five times that in mass. If that is the case then WIMPs should be everywhere as in hundreds passing through every atom in your body every second. Even something as tiny as an atom is mainly empty space and can easily hold this many WIMPs and more, so if WIMPS are so plentiful they must have been detected by now. Again no they have not and to make matters worse other equally tiny and rarely interacting particles like neutrinos are easily detected.

When the first neutrino detectors went online they were fairly crude by today's standards in the sense that they lacked today's sensitivity in detecting neutrino events and yet they have been very successful. This goes back to the reality that any real theoretical particle is easily found and not long after neutrinos were hunted for they were cornered and caught so to speak. The first neutrino detectors were turned on and immediately lit up signaling multiple neutrino detections per year and so many so that there really is no need to run the detectors around the clock. Neutrinos are reluctant to interact with normal matter as are the theoretical WIMPs and yet neutrinos make up a tiny percentage of matter in the universe. That being said neutrinos were found to be real and even though they are theoretically insignificant to the number of WIMPs that are theorized to exist they can be detected with mundane ease. On the other hand dark matter detectors have laid silent for decades without so much as a single solid event ever recorded and this is through the whole range of various theoretical WIMP candidates and that is coupled with the fact that WIMPs out number neutrinos by several orders of magnitude according to theory. If WIMPs were real then detectors would have been glowing as brightly as the Sun with all the recorded dark matter interactions events going on inside them. They remain silent and dark, much darker then the matter they were designed to locate. This is one of the most compelling arguments to be made from an experimental stand point on the non-existence of dark matter as WIMPS as by now they should have been detected not ambiguously or with results open to interpretation and opinion but with undeniable solid facts and these facts have simply never materialized.

The total lack of evidence to support the WIMP theory of dark matter is only the first of a series of proofs that dark matter does not exist as WIMPs. A second can be found in the way that WIMPs are thought to work by contributing mass to the universe and so account for the apparent missing gravity. A MACHO contributes to the missing gravity problem simply by being more normal everyday matter that as we know interacts with space to help create gravity. Not finding enough MACHOs to solve the missing matter problem might be due to simple limitations on detection capabilities but in the end they still work the way we might expect as far as gravity is concerned. WIMPs on the other hand even if we assume they exist in the sufficient quantities to account for the missing gravity will not actually solve anything based on how they are theorized to behave. To understand this simple fact look for a moment at a regular supernova and realize that it creates both light and neutrinos as it explodes and that these explosions can come to us from distant galaxies. Why this is important is because despite the incredible distances between us and the supernova events the neutrinos always travel in a straight line from the exploding star straight to us and this is proven by the preceding neutrino detection of supernovas and the following detection of visible photons or light. The photons we know to have travelled in a straight line and we can see this by observing no deflection events due to high gravitational fields that would otherwise distort the incoming light waves. What this illustrates is the light, numerous and weakly interacting neutrinos obedience to the laws of physics such that they are not affected to any great degree by the gravity they encounter inside the parent galaxy where the supernova first occurred or by travelling through millions or more of light years to reach us or by the gravity of the Milky Way either.

Now let us imagine a dark matter WIMP particle making the same journey and realize rather quickly that it cannot do so. This may seem strange to some as one would ask why it cannot make the trip to us? The answer is simple and that is according to WIMP theory dark matter halos exist around galaxies that are a sign that it is dark matter responsible for holding the galaxy together and not flying apart like one might expect from the flat rotation problem analogy. So dark matter is forced to clump as they say around galaxies and this is the massive five to six times more matter than visible matter that keeps the galaxies from flying apart. Well this is obviously wrong as if dark matter were to be forced to clump so closely to any parent galaxy and still be responsible for the huge missing mass discrepancy then it would be impossible for dark matter to ever escape the galaxy. No particle in existence is ever forced to stay bound to a galaxy and the closest WIMP like particle which is the neutrino has the opposite effect of barely interacting with anything and doing whatever it wants. If dark matter were in any way similar to a neutrino only less likely to interact with matter or anything else then it would be completely free of a galaxies hold and wander freely and evenly throughout the universe. No galaxy would be able to hold a dark matter particle and as such no halos or clumps of dark matter can be kept in permanent possession of a galaxy. Dark matter particles by their own theoretical descriptions and base properties forbid them from ever doing something like congregating in a single place and after light would be the most uniform and spread out form of matter in the universe similar to the cosmic microwave background radiations spread throughout the universe.

The only way anything in the universe stays bound to something else is through charge or gravity and since dark matter carries no charge and certainly does not interact with matter through the electromagnetic force it cannot be a charge reason. Also dark matter is thought to interact only with gravity thereby boosting its power and so again would not be interested in interacting in any charge reactions with matter. If gravity is responsible for holding dark matter in clumps around galaxies then normal physical laws must be obeyed and these laws apply to every other known particle or force carrier in existence making dark matter unlikely to break this mould. The laws of physics at work here are simple as a dark matter particle would need to be ultra heavy on the scale similar to gas or dust in order for it to be massive enough to be held by gravity in a clump fashion around a galaxy or it would need to be moving incredibly slowly in order for the minute force of gravity to alter its trajectory over time. It might also be a combination of these two factors but nevertheless both are forbidden by dark matter theory. Dark matter is an ultra light particle that travels at neutrino speeds which is close to that of light and so cannot be heavy enough or move slow enough to clump around a galaxy or galaxy cluster. Remember particles do not orbit planets, stars or galaxies but move freely as one would expect through space. Not to mention if dark matter clumps around galaxies in halos then it is not clumping in between galaxies in galaxy clusters therefore destroying the WIMP theory again.

If this is still not clear enough then consider how having freely roaming WIMPs evenly spread throughout the universe would never solve a problem as it could never locally add to the gravity of anything like a galaxy. In other words galaxies would still fly apart because even if the total amount of matter in the universe were increased it would be done so uniformly meaning that each individual galaxy would still fly apart as through visible matter they appear to not be massive enough to generate the gravitational fields necessary to hold themselves together. Think for a moment about dark matter particles interacting with matter as a way of explaining the local galactic clumping theory and you see that the number of interactions is far too low to make this an effective means of explaining the clumping theory. If dark matter interacted with matter frequently enough and strongly enough to actually cause the local galactic gravity to rise enough to cause the same dark matter particles to clump and not escape the galaxy then detection events in dark matter experiments would literally be red lined. This is not the case and interacting with matter is forbidden in dark matter theory as dark matter is thought to solely interact with gravity such that we only see the effects as changes in gravitational events like bending light or full gravitational lensing. Now if dark matter were to only interact with space and create gravity as a result then you would find uniform dark matter and gravity interactions spread out evenly throughout the universe making dark matter unnecessary at all for the simple fact that gravity would be increased uniformly and therefore a universe with and without WIMPs would be the same. This similarity means that the flat rotation problem, galaxy cluster and galaxy missing mass problems still persist. A galaxy with dark matter flies apart just as easily as a galaxy without dark matter. Dark matter particles would never clump and so never contribute to the local gravity of anything in the universe meaning they are not a solution to any unsolved problems as they simply and rationally cannot be.

I should not need to illustrate this spatial distribution of particles due to weak interactions, relative velocities and abundances again so I will now move on to other flaws in the dark matter particle theory as a solution to apparent missing mass in the universe. Remember it is the central beliefs of dark matter theories that dark matter rarely interacts with anything and yet clumps in halos around galaxies and I will be building on these points here for proving WIMP dark matter theories as false. Dark matter encounters a fatal flaw in its own theory as it tries to explain the motion of objects within the galaxies that we see around us and our own galaxy that we inhabit. This problem is in explaining the rotational speeds of planets around stars in systems and these systems around the galactic center. For a moment let us review how dark matter is theorized to work in contributing extra gravity and then we will show the flaw in this thinking through observations.

Dark matter particles of any WIMP kind are thought to interact only with gravity such that they will then in turn increase the strength of gravity in a local area. In other words if you have a piece of matter in the universe it should according to the theory of general relativity curve spacetime into a gravity well. The gravity is created by the spacetime curvature and adding more matter will curve spacetime even farther according to theory. Dark matter then steps in and interacts with space such that the more space it interacts with the more it will boost the strength of gravity in the local area. So the more you curve spacetime the more dark matter will theoretically contribute to gravity and account for the missing matter problem. At the area of greatest spacetime curvature you have the most gravity boosting due to the most dark matter interactions in that area of space and as you move outward from the region of maximum curvature the dark matter interactions fall off in a nice smooth pattern matching spacetime curvatures predicted by the theory of general relativity. What this means is that you have the most extra gravity at the center of highest mass and the least amount of gravity or to put it another way a weaker gravitational field farther away. This is the normal picture of gravity that we all understand in terms of the farther you are from an object the slower a second object will orbit the first. Think of our own solar system and how fast Mercury orbits when compared to the outer planets like Neptune or Pluto. We know for a fact that the normal gravitational relationship of closer faster orbiting objects and more distant slower orbiting objects to be correct. Our own solar system has been scrutinized and mapped for millennia to prove this point and modern observations obviously back this up as well. Since we do know this to be correct then we can easily spot the flaw in dark matter particles as an explanation for the missing matter in the universe by looking at our solar system compared to our galaxy.

If dark matter clumps around matter and therefore affects the speed at which objects rotate about a given central mass then the planets in our solar system should be moving with respect to distance from the Sun at the same rate as any star anywhere in the Milky Way does with respect to its distance from our galactic center. In other words Neptune and Pluto should be moving at the same speed as Mercury and yet they are not. Remember the laws of physics and the laws that would govern the theoretical dark matter particles must apply equally everywhere in the universe with no exceptions to the rules if they are to be presented as an answer to the missing mass and flat galaxy rotation problems. The reality that stars around galaxies do not orbit the same as planets around stars despite the presence of dark matter theoretically being present and always interacting with the gravity of space as described above means that dark matter cannot be correct.

Dark matter being ever present in the universe and obeying the rule of clumping around areas of greater mass should mean that the Sun needs dark matter to hold the solar system together as the curvature of spacetime in our own solar system is scaled and relative to the spacetime curvature of the Milky Way galaxy. The Sun however does not need dark matter to hold the solar system together and the known masses, distances and orbital periodicity of all planets, asteroids and comets in the solar system fit neatly into known calculations of gravity without the need for an extra dark matter variable. Extra variables are a definite no-no when it comes to the universe as it does not ever create something that it does not require as this takes extra energy, matter or dimensions which never mesh with known and proven parameters. So dark matter is not real as the universe does not use it inside solar systems and would seem to only require it to be used on whole galaxies. The very dark matter that would permeate the entire galaxy and hold it together as a way of explaining the missing mass and flat rotation problem would and forgive the word usage but magically stop at solar system boundaries. This is again of course impossible and so dark matter cannot once again exist.

This seems to explain away the theory of dark matter with respect to detection events given the apparent abundance of dark matter particles and the inequality in orbital speeds of planets to stars and stars to the galaxy centers given that the same dark matter particles would permeate the galaxy in a smooth distribution following spacetime curvatures and hence apparent clumping behavior. What if we could also disprove the base properties of dark matter using so called dark matter objects in the universe? Well we can and this last proof against dark matter should be more than enough to dissuade anyone from looking for these theoretical particles. I will say here and before I present the final proof that I do not intend to attack dark matter camps in anyway that might be construed as malicious or otherwise nefarious as I only seek to prove the world round and orbiting the Sun such that all of science can move forward once again.

This last proof that I will give here is from galaxy cluster collisions and will focus on the Bullet Cluster collision which is a relatively old cluster interaction and one that has recently become the focus of much dark matter research. It could not be further from the truth that dark matter is at all responsible for any gravitational effects in the universe and therefore provide the subsequent missing matter. In this now famous collision event we see two galaxy clusters as they appear after their collision has taken place. One cluster has passed through the other or one has therefore passed right into the middle of the other such that it appears as though a bullet was fired through one cluster and hence the name Bullet Cluster collision. The dark matter theorists will present the image in three false color phases and each phase represents according to them different forms of matter. The most obvious one is the well defined spiral galaxies that cover the image and obviously represent large amounts of conventional stellar material and its constituent solar systems. The second is luminous hot gas and dust which envelopes the two clusters as one might expect due to the turbulent and almost chaotic dance of gravitational forces inside the collision event stripping away loose materials from each cluster and swirling it in gravitational eddies in space. The third phase is two roughly spherical looking blobs of nothing that extend beyond either of the galaxy clusters and are said to be made of dark matter.

So in the central areas you have normal spiral galaxies which are crisply well defined and easy to see and around that covering an area larger than the clusters is of course loose gas and dust glowing from emitted radiation. Then beyond this you see the two circles of dark matter or at least the two circles that are totally black in normal color images and theorized to be dark matter. The dark matter theorists argue that the dark matter has been ripped free from their parent galaxy clusters and flung through each other such that they have ended up on either side of the collision. In other words before the collision the galaxy clusters would have had in theory their respective clumps of dark matter surrounding and permeating them providing the gravitational super structure to hold them together as first theorized in the 1930s by Zwicky.

During the collision however the dark matter which possessed a moment of inertia similar to the galaxies involved in the collision did not experience a reduction in their velocities relative to the galaxies and normal matter in the collision and so travelled straight on forward. Since the normal matter was restricted by the gravitational tidal forces of the two clusters passing through each other we see that the galaxies and loose gas and dust have not progressed as far as the dark matter. This is of course due to the theory that dark matter so weakly interacts with normal matter that it is free to move unencumbered through the entire collision itself. The dark matter theorists have also proposed that dark matter interacts so weakly in the universe that it interacts weakly or not at all with itself either and this is why the two circles of dark matter were able to pass right through each other and out the other side of the collision also travelling outward beyond the entire Bullet Cluster area as well. This is the picture presented of the collision as believed by dark matter theorists and contains three distinct areas which are notably the galaxies and gas and dust as well as the dark matter circles further away.

Now that I have illustrated to you the basic construction and proposed dark matter rules of the collision I can begin to destroy dark matter once more using its own properties against it. The first major flaw with this collision using dark matter is the failure of dark matter to account for missing matter and gravitational forces in the universe. Remember galaxies will fly apart without dark matter and galaxy clusters will not possess enough gravity to hold themselves together without dark matter at least according to theory. The failure of dark matter particles here is that the collisions are many billions of years old as two objects covering as many millions of light-years as this would take billions of years to pass through one another. In this time span the dark matter circles have long been ripped away from their parent galaxy clusters and the constituent galaxies of each of the two clusters. The dark matter has had billions of years to exist in the circles outside of the collision as they did not experience a reduction in their relative velocities during the unfolding events. This means that for many billions of years the galaxies and galaxy clusters as well as the dark matter circles which were the clumps or halos have been separated from one another.

Here is where the failures of dark matter as a theory add up quickly. Firstly dark matter cannot exist without gravity and the mass from normal matter which creates this gravity. Just as dark matter clumps around a galaxy of normal matter to form a halo it is believed that dark matter clumps galaxy clusters in the same way following increased curvatures of spacetime. If you look outside a galaxy however you will not see a dark matter halo meaning that dark matter does not spontaneously clump to space or to other dark matter particles. Since this is believed to be the case then examining the Bullet Cluster collision provides evidence against the very theory of dark matter. This is due to the fact that over many billions of years the dark matter has had no galaxies or galaxy clusters to clump around and so just as space outside a galaxy is thought to not contain dark matter halos the dark matter once associated with the clusters would have long since dispersed.

The dark matter would not have survived the trip through the collision as the general shape of the two dark matter circles would have become fragmented during their period of dissociation with their local galaxy clusters spacetime curvature. The presence of two uniform areas believed to be caused by dark matter cannot be caused by any theoretical WIMP particle as it would not possess the ability to maintain its local coherence over billions of years. It would require normal matter of a typical MACHOS kind to maintain a roughly circular appearance in the image. Yet the normal gas and dust is seen as the two hot plumes and so cannot offer an explanation there meaning we see the normal tidal gravitational forces at work on the gas and dust keeping it close to the collision.

Further proof of this comes from the theoretical behavior of all dark matter particle candidates which state that they are both near massless and travel at neutrino or relativistic speeds. The dark matter particles over billions of years would have dissociated themselves from the local areas of spacetime curvature and once again started travelling at their near neutrino speeds in all directions from the epicenter of the collision. Again it must be remembered that no particle has ever stayed in some sort of orbit near matter due solely to gravity in the first place so why would anything ever clump up to begin with? Given the scale of the collision is far less than billions of light years which would also represent the time span of the collision the dark matter circles should have dissolved many hundreds of millions to billions of years ago leaving a void region of space rather than one supposedly lingering with WIMPs. The dark matter particles are believed using this very collision to be nearly incapable of interacting with even themselves and so cannot provide a mechanism for their maintaining a great enough density to curve spacetime such that they can retain their circular shapes beyond the areas of the normal matter in the collision image. In other words the relativistic speeds and non-existent interactions of dark matter with anything in the universe including other dark matter particles means that they can never survive on their own for billions of years without the presence of normal matter. So the theory that dark matter clumps were shed by the galaxy clusters passed through each other and then survived for billions of years beyond the two remnant galaxy clusters and loose gas and dust is false.

On a quick side note no other particle in the universe does not interact with itself at all and so the unlikely chance that dark matter is once again the exception to the rule of every other known particle in the entire Standard Model of particle physics is at best suspect.

To continue this line of matter and dark matter dissociation thinking we will now look at the galaxies and galaxy clusters proper. Just like dark matter needs normal matter around which to form a clump or halo it is believed by dark matter theorists that galaxies and galaxy clusters cannot exist without dark matter. This of course brings up the rather obvious fact that the galaxies involved in this collision have likewise been stripped of their dark matter for many billions of years. The galaxies were immediately stripped bare of their supposed protective dark matter halos as soon as the two galaxy clusters began to forcibly interact with one another and gravitational braking was occurring. This is again due to the fact that dark matter is believed to ignore the gravity effects of the collision and maintain their relative pre-collision velocities. So very early on in the collision the galaxies were denied their dark matter halos which is believed to keep them holding fast together. Over the ensuing billions of years the galaxy clusters and galaxies should have dissolved completely. Firstly the galaxy clusters cannot maintain their rough configurations as without the dark matter to hold them together they would have experienced erratic gravitational swells during the collision that would have driven each cluster apart. The individual galaxies in the collision would have faired far worse as these same gravitational swells would have intensified locally for each different galaxy depending on its position in the clusters during the collision. This means that some galaxies would have more deformation than others while others would have less deformation depending on the amount of local gravity they encountered during the event.

For example a galaxy with little deformation would look roughly spiral while a galaxy with high deformation would look completely destroyed. In fact all galaxies would fly apart due to the loss of mass believed to solve the flat rotation problem anyway so why is this area of space not simply a huge swath of random stars? When examining the galaxies within the image it is of note that all appear uniform regardless of their position to the epicenter of the collision as well as their position in the clusters. Galaxies at the leading edge of the clusters have had the most time to be stripped of their dark matter while galaxies still somewhat inside the remnants of the collision are younger in terms of dark matter dissociation. This does not matter however as no galaxies in the image are disturbed as they all should be to some degree if dark matter behaves as theorized and none are obliterated as many should long have been.

To reinforce this fact let us ignore gravitational tidal and sheering forces and simply look at the time that the galaxies as individuals have been stripped of their supposed protective dark matter halos. According to dark matter theory a single isolated galaxy needs a dark matter halo to hold it together and without this halo the galaxy will simply fly apart in all directions due to the flat rotation problem. In the Bullet Cluster collision all galaxies have been stripped of their dark matter halos and I need remind the reader that no less than 100 hundred percent of the pre-collision dark matter is needed to maintain galactic integrity. Anything less than this threshold value will result in mass loss due to decreased dark matter presence in the local spacetime curvature. So these galaxies which according to dark matter theory have been stripped of dark matter halos sufficient to register outside the collision billions of years after the event means they are grossly below the critical threshold value. As such the galaxies should have long over rotated themselves to death with the stars they possessed being flung outwards. A galaxy without dark matter would have become tiny and be populated only with its super massive black hole at its nucleus and a collection of stars that would now obey the normal rules for distance and orbital velocity seen with familiar objects like our own Sun and the planets in our solar system.

The interstitial space of the two clusters would not resemble neat orderly spiral galaxies but rather one roughly uniform distribution of wandering systems. All the space contained in the volume of the collision would be glowing with star systems and contain no distinctive black regions between galaxies and yet this is not what we see. If dark matter is needed to hold the universe together then how come the universe is holding itself together without dark matter as seen in the Bullet Cluster collision? The answer is simple it is because dark matter does not exist in WIMP form and no extra variables are needed to explain the events as seen in the collision image. So now what is needed is an explanation of both the apparent missing matter or more accurately the missing gravity problem and an answer to the flat rotation problem such that no extra variables are needed to make the universe work properly.

Galaxy Rotation

The observations that have led to the theory of dark matter are of course valid ones as no one in the scientific community simply said let us invent new matter just for fun. The theory of dark matter was born from a need to explain why easily visible objects such as whole galaxy clusters were sticking together when it appeared as if they did not possess enough mass to create the gravity necessary to do this. Likewise it appeared as though matter was missing as the rotation rates of stars around their galactic centers seemed too high for the given mass of the galaxy from luminous matter. This is commonly seen in the flat rotation problem which has been measured multiple times for multiple galaxies and as such the observations of this phenomenon are correct.

Also the rotation rates of objects like the planets in our own solar system around our Sun have of course been observed for millennia and are very correct, in fact perfect. This means that we know that given a certain mass and then at given distances from that mass orbital speeds can be calculated using normal matter and normal laws of gravity and nothing else. To put this another way our solar system does not need anything extra like dark matter to work. The Sun has enough mass to hold the planets in orbit and orbiting at velocities that drop off as distance from the Sun increases. This is perfectly normal and one hundred percent correct physics that can be argued by no one whereas theoretical objects like dark matter must be heavily scrutinized rather than simply accepted as something that simply must be real since the galaxies we observe do not behave like our solar system.

We have a model right here in our own backyard and in fact it is our backyard the very solar system we call home is working proof of the physics of the universe. Since the space and matter contained in our solar system is the same as anywhere else in the universe including the space and matter in galaxies then why would something extra be needed to make those galaxies work? The truth is nothing is needed to make the galaxies work any different than our solar system all you need is space and matter but once again they can do different things on large scales that they do not do on solar system scales. However the galaxies of the universe do not require ever any new sort of particle to accomplish this task and the number of variables in the universe does not go up. It was observations of the flat rotation problem compared to our own solar system that led to many believing that extra things like dark matter particles were necessary to make the universe work even though this is not the case.

The universe can once again be explained using only space and matter such that no new variables are needed and both the missing matter of galaxies and galaxy clusters can be accounted for. At the same time the proper explanation of space and matter interactions at large scales similar to whole galaxies can be used to solve the problem of why objects in the solar system rotate at one speed while objects in the galaxy rotate at another with respect to distance and orbital velocities. So we can solve both the missing matter and flat rotation problems in one go as it were. To begin with we need to address the problem of missing matter in the universe as this was very much the first thing that led to the theory of dark matter back in the 1930s.

The first form of missing matter is simply matter that is too dim to be seen with today's current level of detection capabilities. In other words, dust, rocks, planets, black holes and the like that we simply have not been able to see yet because the instruments with which we are attempting to locate them lack the sensitivity necessary to see them. This is quite normal in science throughout history and should not be ignored as proof of this concept has already happened with Hubble. Not the Hubble space telescope although that did prove this fact too but with Edwin Hubble the astronomer who was able to use a new and more powerful telescope to change scientific thinking. We have already learned how he discovered that the universe is made of multiple galaxies and that they are in motion rather than a single galaxy which was our Milky Way. Other galaxies used to be thought of as nebulae in the Milky Way but simply by improving our detection capabilities the whole universe grew by many orders of magnitude over night. A single year before the observations made by Hubble with any other state of the art telescope in the world failed to notice this as the detection limits of those instruments had been reached. All theories attempting to explain these little nebulae in the Milky Way were immediately discarded along with the static universe theory. This is simply occurring again today as newer instruments will undoubtedly lead to more accurate mapping of all the gas and dust all the wandering planets and black holes in our own galaxy.

All of these objects have mass and therefore contribute to the overall gravity of the galaxy and also to any gravity missing from galaxy clusters too. This being said dark matter in the form of what I correctly call normal matter is indeed a partial solution to the problem of the missing matter and subsequent missing gravity. It is not however the full solution as any amounts of missing normal matter will still fall short of explaining the lack of gravity in the universe as there is simply not enough of it missing that new instruments would later find.

Another reason why undetected normal matter will not solve the problem of things like the flat rotation mystery is because the matter would be distributed in such a uniform way that the whole galaxy should still behave according to the rules of normal gravity, distance and orbital velocity such as those observed in our own solar system. Simply putting more matter in the galaxy does not answer why the flat rotation problem occurs and this is true of either the MACHOS or WIMPS theories of dark matter as we have seen. Nevertheless additional matter in the form of normal matter that our instruments are not sensitive enough to detect will obviously yield more gravity and account for a percentage of the missing matter even if it is a small one.

The vast majority of the missing matter and flat rotation problem however are not solved through normal matter distributions in the universe but through variable gravity and ESS properties. The first application of variable gravity and ESS macht will be used to explain the simple bulk of the missing gravity from galaxy clusters and galaxies and the following ESS properties will be used in conjunction with variable gravity to explain the flat rotation problem. The first thing that need be addressed when discussing the problem of missing matter in the universe is that all calculations of missing matter are based off of the traditional three main views of the universe and how they treat both matter and space such that gravity is the result. What this means is all laws of gravitation from the three main views as I have explained previously are created to only account for space and matter interactions involving objects like our own solar system. Take general relativity for example and the decent job it does of calculating orbits within our solar system and then look at how it is utterly wrong as it cannot work for a flat rotation problem.

The behavior of matter is defined according to normal quantum mechanics that do not incorporate changes in space geometry and the behavior of space is defined according to general relativity which cannot understand changes in space and matter interactions such as those found in black holes. In other words the laws of gravity and the calculated values we expect to find from them always correlate perfectly with our own solar system but fail at galactic scales and this is where the real problem of missing matter and flat rotation lie. In order to address this we will for now leave our solar system where high levels of spatial compression do not occur and move outward to the rest of the galaxy where conditions can be met to incorporate variable gravity. The first example I have shown you of variable gravity is that of a black hole and recalling the rules for ESS under high levels of spatial compression and the resulting significance of those on quantum space permits for space physics which allows for gravitational effects outside of the norm found in our solar system. A black hole as it has been shown can have greater gravity than it should according to the three main views and this fact of the universe explains some of the missing matter problem.

The second half of the missing matter problem is not actually missing matter or missing gravity but a result of ESS properties and will be addressed later. The variable gravity created by black holes however can be addressed directly now as it provides for some of the missing matter directly inside a galaxy or between galaxy clusters. It has already been explained that the early dense and gas rich universe produced copious amounts of supergiant stars that burned through their supplies of fusible materials quickly and given their large masses exploded as supernova in only a few hundred million years. Such ancient explosions have long dispersed their supernova remnant gas and dust such that no such nebulae would remain to this day and therefore no evidence of any supernova event. The ancient black hole is all that remains and is quite beyond direct detection capabilities of current technology.

The numerous radio chirps, hisses and background noise that remain unexplained in the galaxy and universe are not a silent but a noisy reminder of these early black holes all around us. Differences in distance, relative velocities to us, and interactions with any number of materials or objects in the galaxy of the Milky Way or the universe in general will produce a symphony of complex background radio signals and this we do observe constantly. It only appears to be a mystery because there are so many sources all around us that no definite direction can be determined and the objects are black holes making direct detection difficult so these radio sources are simply said to be unsolved when they are quite normal. To illustrate this once again black holes suspected of being at the centers of old supernovas, binary systems like Cygnus X-1 and even the center of the galaxy avoid direct detection and yet we know precisely where to look with our most powerful and sophisticated instruments and tools. The ability to find a wandering or more correctly orbiting black hole that is simply somewhere in the blackness of space is even more beyond our current technology.

Leaving behind the problem of mysterious radio sources of the universe for a moment the reality of hundreds, thousands or millions of ancient or recent black holes orbiting at regular predicted intervals in the Milky Way or any other galaxy is a fact that will be borne out in time. This additional number of black holes all must obey the laws of ESS spatial compression and increased ESS macht and as such have increased strength to the power of their gravitational fields. The powering of variable gravity from both high levels of ESS spatial compression and the resulting combination of greatly increased ESS macht glow means that space behaves as though far more gravity exists in it than would be predicted by using normal gravity from the three main views alone. Space and specifically the ESS and macht in it always try to balance themselves out through volume changes of ESS and macht sharing from regions of ESS and matter interactions and the schales that surround them. These variable gravity increases lead to large scale balancing of the two space physics properties and extend great distances beyond predictions based off of gravity laws of something like our solar system. This will account for the missing matter along with currently undetectable normal matter in between galaxies inside galaxy clusters.

Any object the size and density of a whole galaxy will have a massive effect on the local area of spatial compression extending for millions of light years in all directions. The need for these massive objects to balance their respective ESS and macht glow with a neighboring galaxy or even an entire close by galaxy cluster will result in local areas balancing ESS and macht far beyond one galaxy. The result is a balancing between all galaxies in a galaxy cluster such that a common center for this balancing is made out of the combined local areas of spatial compression of each galaxy in the cluster with each other. The property of ESS and the macht it carries to be shared evenly amongst its schales extends for millions of light years and will easily distribute the additional variable gravity needed to create a single large meta object like a galaxy cluster. This explains the necessary gravity needed to hold galaxy clusters together along with undetected normal matter but cannot provide the full explanation of how an individual galaxy does the same.

The first part of variable gravity in galaxy clusters as well as undetected normal matter of course helps answer the question of where the apparent missing gravity or more accurately the matter has gone in galaxies but a second mechanism is needed to fully explain individual galaxies. Even this however cannot fully explain the missing matter in galaxies and the related missing gravity as it must also address the flat rotation problem. The flat rotation problem is still one that should be more widely cited in criticisms of all dark matter theories whether they are MACHOS or WIMPS. The reason for this goes beyond either of the two dark matter camps explaining where the extra matter might have gone. Neither of the two camps can offer any explanation for why relationships between orbital velocities and distances from a central mass like the Sun or the galactic center operate at two different speeds. To this end it is necessary to take another look at the nature of space and how space behaves. Indeed I must once again state the importance of a proper explanation of space as the missing component in solving the problem of how the universe works. The flat rotation problem is one of the greatest proofs that space is indeed made of ESS, contains macht and that interactions of space and matter due to varying levels of spatial compression is the only explanation for how small and large scale systems are incorporated with one another. To this end I will now explain how the flat rotation problem of the universe can be solved such that observations made of stellar orbital rates around galactic centers can be kept along with the known orbital rates of objects within our solar system. All will be explained using only ESS and normal matter and the addition of no new variables is needed either in our solar system or in the Milky Way or any other galaxy at larger scales.

The very first thing I must explain is the behavior of ESS to share macht evenly with one another and the geometrical way in which the ESS physically arrange themselves to do this. I touched on this when I spoke earlier of both ESS spatial compressions in our little room filled with spheres and when I explained the shape of the universe. First off let us recall the way ESS behave in the little room and remember that when spatial compression is very high around one of the mostly opaque orbs ESS from more distant areas of the room will move towards the area of higher spatial compression. In other words the ESS are not a rigid structure of the universe but can move inside of it as we have seen such that the ESS in the room form a continuous collection of little spheres from highest to lowest spatial compression. The ESS spheres must obviously be able to move in order to do this and this is likewise reflected in the shape of the universe at large scales. In my explanation of the shape of the universe I mentioned that the universe will always form into a spherical shape after the Big Bang and will not have great protrusions or angled sides as this is the least efficient way of sharing ESS macht; so no four dimensional cubes. Any non-spherical shape will create macht imbalances such that certain edges, surfaces or vertices have different levels of macht. The ESS as we now know share macht amongst themselves and each other in an effort to balance the total macht of the local area of space. This means that any non-spherical shapes will be naturally returned to a sphere.

If there were a protrusion of some sort and you can imagine a promontory like projection of space after the Big Bang due to small variations in ESS distribution magnified by the rapid expansion rebound phase of the universe this freely moving ESS fact becomes perfectly clear. The promontory of ESS is in an arrangement that does not permit the even sharing of macht and would require energy input into the system to keep the ESS in this shape. The universe as is well known tends towards expending the least amount of effort possible and in this case to alleviate the macht imbalance the promontory ESS will be pulled back down towards the surface of the universe. As the ESS are pulled down they do not simply experience increases or decreases in volume necessary to return the universe to a spherical shape they also move in any of three spatial directions to accomplish this balancing of macht.

So what you have seen is that ESS can move in space as necessary to balance macht and this means that ESS if under enough influence will in a sense follow macht in an effort to balance themselves. This is the second key piece in explaining the flat rotation problem of the galaxies that we see in the universe. In order to demonstrate how this mechanism works with both variable gravity and differences in solar system orbital speeds as compared to galactic orbital speeds I will first explain how a black hole affects changes in local spatial compression.

A black hole compresses the local ESS to such high degrees that massive increases in the amount of donated energy from matter to the ESS macht are made possible and variable gravity becomes a contributing factor to the laws of physics in the universe. So far I have only explained a stationary black hole in space and the way it affects local spatial compression. What I will now do is explain ESS properties in the presence of a spinning black hole which is the standard black hole in the universe. All black holes have high rates of spin about their axis which can result in relativistic speeds and this is certainly seen with materials falling into the event horizons of black holes which are accelerated to near light speed. The high rotation speed of black holes coupled with the variable gravity that they produce is needed to help explain the flat rotation problem. I mentioned earlier that the missing gravity problem is not entirely correct as it should be called the apparently missing gravity problem. Meaning that gravity is not really missing as this would denote an absence of mass from normal matter or dark matter.

The truth however is that in order to solve the missing mass of the universe with respect to galaxies and solve the flat rotation problem at the same time does not take just missing matter. Instead the behavior of space inside a galaxy can be used to account for some of the apparently missing matter. What if you could reduce the need for more matter to solve the problem? What if you could make space behave as though it had much more matter than it does such that you do not need to go looking for dark matter? The answer to both questions is yes you can simply by using ESS and variable gravity due to increased spatial compression and macht levels of black holes over millions and billions of years.

Galaxy Formation

To begin with let us look at how a galaxy is formed in the first place and in so doing I will also remove the need for dark matter in early galaxy formation models. In the early universe there was nothing but hot dense gas mainly hydrogen and gravity more or less. If we leave these two alone for a while they tend to make stars as any random quantum motions of the hydrogen atoms is sufficient to cause an imbalance in the uniform distribution of matter in the early universe to create quantum level differences in gravitational attraction. The minute differences in gravity will continue to grow as hydrogen becomes more and more compressed even at a scale of only quantum fluctuations over time. This leads to dense balls of hydrogen that will now rapidly alter the local gravity such that a typical stellar accretion disc is formed and we are well on our way to creating a new star. These stars as we know are all or mostly all going to become supergiant stars that will burn hot and fast and become black holes in only a few hundred millions years or less. These black holes also are quite close together due to the young age and decreased volume of the universe and so will easily encounter one another over the next little while. These black holes are naturally all sources of variable gravity which means they produce more gravity per unit of mass than the same mass would under normal spatial compression. The black holes which are formed close together will naturally seek to balance their increased levels of ESS macht and so move together under what is known as the force of gravity. As the black holes merge with one another two things happen and the first is simply making larger black holes. As two black holes merge they will of course grow in size such that a black hole of the maximum size capable of being produced through explosive means of a supernova will grow beyond what is normally the threshold value. In other words the new black hole will exceed the maximum size allowed under stellar formation limits. This is of course the beginnings of a future super massive black hole.

The second thing that will happen is a repeat of this process many times over such that the early dense black hole filled universe can easily have tens, hundreds or thousands of black holes merging early on in the history of the universe. These black holes all produce variable gravity and as one much larger baby super massive black hole will of course produce a huge amount of additional gravity from the mechanism of variable gravity. This provides the extra early gravity needed to seed galaxy formation such that you will end up with a galactic evolution capable of producing large fully developed spiral galaxies like the Milky Way or Andromeda that we see today. The need for dark matter to provide the extra gravity necessary to seed early galaxies as shown through computer modeling is incorrect. The apparent missing matter which corresponds to the percentage of missing matter believed to exist in the universe from dark matter is not needed to cause galaxy seeding.

The increased gravity from variable black hole gravity allows for the known amount of visible matter in the universe to supply the necessary force to cause galaxy seeding to begin as well as agree better with current theories predicting the amount of baryonic matter produced by the Big Bang even if those theories are not fully correct. No new matter need be theorized to have been created in the universe nor has any matter gone missing from the amounts predicted to be formed in the universe. Simply using variable gravity the missing matter is revealed to be nothing more than a misunderstanding of the nature and behavior of space and how it interacts with matter. So in the early universe super large but not yet super massive black holes and the variable gravity they generate provide the necessary force to seed early galaxies such that we end up with mature spiral galaxies exactly as we observe them in the universe today.

Now that we have removed dark matter from galaxy formation at the early life stage of the universe we can continue to grow on these powerful black holes as a way to explain the flat rotation problem and therefore remove the need for additional matter in galaxies as well. We will now revisit three concepts at once in order to make the jump necessary to solve the missing matter and flat rotation problem. The first concept is the increased gravity of black holes, the second is the free movement of ESS possible in the universe and the third is the rotation rates of black holes. Black holes have variable gravity which we by now understand and ESS can move in space which we also understand. A black holes rotation is now the last key to the solution of the flat rotation problem and the missing matter problem as increases in black hole size correspond to increases in rotation rates of the black hole at its surface.

This is a basic physical law of the universe as can be seen with a simple set of spinning discs of two different sizes. If one spins a disc of smaller size at speed X and a larger disc at speed X at the same time a simple property of physics is revealed. This is of course the well known fact that as diameter increases and rotation rate remains the same the velocity of the outer edge of the disc increases. So if both discs are turning at a speed of X the outer edge of the small disc will not be moving as quickly as the outer edge of the larger disc despite the fact that they are rotating at the same rate. A black hole behaves the exact same way and as black holes merge in order to conserve this law of the universe the outer surface of the newly merged black hole must spin faster than the surface of either of the two black holes that formed it. Repeat this process and the velocity of the surface of the black hole grows ever faster approaching the speed of light or at least a significant fraction of it.

What this does to the local ESS is profound as the local ESS inside the black hole is attempting to share macht with the schales around it in an effort to balance out total macht. The ESS as we have seen can move freely in the universe if given enough influence and this leads to the first schale around the ESS inside the black hole orb proper to begin to move slowly in the direction of the rotation of the black holes spin. This effect will not be significant around small black holes but becomes rapidly more applicable to space physics as black holes transition from normal to super large and finally to super massive black hole scales. The property of ESS to balance themselves with respect to local spatial compression and shared macht glow means that they will slowly start to feel enough force to move them inside space.

In other words if we look at a large enough or fast enough rotating black hole we will find that the ESS in the first schale are free to rotate inside the second schale as they are more influenced by attempts to balance macht glow with the ESS in direct contact with matter inside the black hole. The first schale is the schale of ESS directly in contact with the last ESS to be inside the matter orb of the black hole and the second and third schales are ever more distant from the surface of the black hole. This is due to the highest levels of spatial compression and ESS macht glow being found obviously inside the black hole and therefore the first schale obeying these ESS rather than the less influential ESS in the second schale which are less spatially compressed and have lower overall macht glow. This process is allowed as the ESS can move freely within space and since the first and second schales are still touching can still transfer macht between them such that no sharing disruption occurs. The schales are trying to balance the macht between them but they are trying to share it with schales that are now spinning and so they are slowly dragged in the direction of the spin in order to try and catch up to those faster spinning ESS so that macht sharing is made easier.

This process repeats as I mentioned slowly over time as the mass of the baby super massive black hole continues to grow, as the rotation rate of the black holes surface continues to rise and as increasing numbers of schales are allowed to attempt to balance out with inner schales. The end result is as a super massive black hole grows so does the size of the spiral galaxy with it and as time progresses increasing numbers of ESS schales rotate within the space of the black hole as well. Since all space physics properties take place at time scales much faster than energy and matter time scales and the growth of the galaxies due to accretion of additional materials takes place even slower than those of energy and matter interactions the ESS have essentially an infinite amount of time to build up spatial rotation velocity. In other words the ESS of space found within galaxies possessing super massive black holes will always begin to rotate as a natural part of the evolutionary process of a galaxy from the beginning of the universe. In addition it will always progress faster than any interactions of normal energy and matter such that the ESS have time to rotate at the same speed as each other due to the need to balance macht glow and the speed at which macht glow is shared. The end result is the space inside a galaxy will rotate as compared to the space in between galaxies. This is why it takes both variable gravity and ESS properties such as free motion to allow for galaxies to rotate where as it only requires variable gravity to hold galaxy clusters together.

This allows us to directly answer the flat rotation problem of galaxies as well as the apparent missing matter problems once and for all. The fact that space inside a galaxy rotates means that less matter is necessary to provide the necessary gravity to hold the galaxy together because matter is bound to ESS. Since matter is interacting with ESS it does not feel a rotational force that would seek to throw it out of the galaxy or as is often said before cause the galaxies to simply fly apart due to insufficient visible matter. The matter inside the galaxies such as stars, planets and everything else feels no such force as it is moving with the space in the galaxy. This means it is naturally following the space around the center of the galaxy and never trying to fly outward as would be expected similar to an object tied to a string and twirled over your head.

By using rotating ESS much of the matter theorized to be missing and thought to be needed to account for additional gravity simply vanishes from the universe. To put this another way you are no longer looking for five to six times the total amount of visible matter in the form of dark matter because it does not need to exist at all. This explains the rotational rates of solar systems as well because the solar systems are not rotating with respect to the galactic center as they are inside the moving ESS space. The planets inside our solar system for example orbit exactly as normal physics predict given the mass of the Sun and the orbital distances and velocities from it. This is because from our point of view the Sun does not spatially compress space like a black hole would and so does not cause our local area of space to rotate inside the ESS of the galaxy. Instead we obey normal laws of gravity as anyone would except because to us the space of our solar system is stationary and so objects made of matter move freely with that space.

The space of the whole solar system is inside the space of the galaxy which is rotating and so the galaxy can move our solar system as well as every other system and star in it at the same speed everywhere regardless of the distance from the galactic center. The ESS in motion around the galactic center are moving because a supermassive black hole with millions of solar masses worth of matter all with elevated levels of variable gravity are attempting to share macht. The Sun inside our solar system does not need to drag space in a circular fashion to share macht and so is not rotating the space with us in it. The space of the galaxy is in motion as compared to the local space of the super cluster and the space inside a star system is motionless with respect to the galaxy. These two motions of space easily explain observed phenomena concerning star systems, galaxy flat rotation problems and does so without adding any new variables only explaining how the existing ones actually work. The flat rotation problem of all galaxies is not a problem at all but a proof of the free motion of ESS in space due to balancing macht glow levels from the natural spin caused by large black holes over time.

The rotating ESS of a galaxies space means that all stars that we can observe and measure within it will rotate at nearly the same speed as they are part of the ESS that is in motion. The rotating ESS also explains away the need for more matter to account for gravity as the rotation of space takes the place of gravitational effects as predicted by the three main views. This rotating ESS of a galaxies space also allows for orbital velocities inside star systems to behave normally as observed from our own solar system for millennia. This use of rotational space inside galaxies also means that the universe does not need new or extra variables to be incorporated into its workings like dark matter. A single mechanism without additional variables and using the same laws as demonstrated for phenomena such as black holes, the Big Bang rapid expansion phase and the deceleration and acceleration of the universe can also be used to explain how solar systems orbit at one rate while galaxies orbit at another as well as removing the need for dark matter.

Facts

The following summation of facts will thusly continue to outline the properties of space and the interactions of space and matter with the universe. The first fact for this section is that additional gravity and matter is found in currently undetectable normal forms of matter in the universe. The second fact is that the number of early universe supergiant stars created copious amounts of black holes which all possess variable gravity. The combination of these two factors leads to the third fact which is the mechanism for galaxy cluster coherence. The fourth fact is that these ancient black holes are responsible in part for the symphony of previously unexplainable background noise of the universe due to their complex interactions with objects in the universe and even distribution all around us as we observe the cosmos.

The fifth fact is that ESS are free to move in space and this relationship is made evident through black hole spin rates and size. The sixth fact is that ESS movement will cause the space inside a galaxy to move at a constant speed that increases as the mass of the super massive black hole increases and occurs over time at scales much faster than energy or matter rates. The seventh fact is that the movement of space inside a galaxy as compared to the space outside the galaxy negates the need for additional mass to produce additional gravity as objects in orbit around the galactic center are not under any outward rotational force due to their being a part of the ESS that are rotating. The eighth fact is that the rotation of space around the galactic center means that all star systems in it will rotate as observed in the rotation problem.

The ninth fact is that the rotating ESS of the galaxies allows for star systems to orbit at flat rotation speeds while the orbits of objects inside the star systems can progress at normal orbital speeds with respect to distance. The tenth fact is that the same mechanism that drives the rest of the universe as we have seen up to this point permits two orbital speeds contained within a single galaxy with respect to star systems and inside systems without the need for extra variables like dark matter.

The one additional point worth mentioning here concerning missing matter in galaxies comes from the supposed dark matter halos that are theorized to exist around the visible mass of galaxies. Many claim that it exists as proof of dark matter which as we have seen cannot be the case but if gravitational effects are being noted around the visible matter of galaxies then one might ask where this phenomenon comes from. The answer is not a single one but rather a composite of numerous factors. The first factor must immediately be MACHOS that lie at the rim of all galaxies as matter existing in these locations should be readily obvious to all. Take for example our own solar system and realize that with the human eye we can make out various planets from here on Earth such as Venus, Mars, Jupiter, Saturn and Uranus which are the easiest historically to spot. Then realize that we see them due to them being made of visible light as we view them and that we cannot for example see Pluto with the human eye as this exceeds the optical light gathering abilities of Earth based life forms. This does not mean however that objects do not exist in the solar system simply because we cannot properly detect them using our given biological sensors. Pluto exists as well as the objects in the Kuiper belt and for that matter all the asteroids in the asteroid belt which are much closer to home and yet still elude our eyes.

The existence of normal matter objects at the edge of a galaxy is to be expected for two reasons the first of which is they simply lie outside the limits of even our best detectors. The second is that matter distribution will not suddenly drop off at the edge of a galaxy simply because visible matter appears to stop. A galaxy has a massive gravitational field second only to galaxy clusters and will draw material in from most likely millions of light years in all directions. This means the matter falling into the galaxy such as gas, dust, rocks and the like will not have been subjected to the internal radiation of the galaxy for long enough to emit electromagnetic radiation for us to detect. This is of course coupled with the fact that they are simply too far away from luminous enough sources to absorb sufficient radiation to be seen. Look at gas and dust in a typical galaxy in the universe and you will see a glow of the general haze of gas and dust closest to the galactic center. This glow is not made of millions of stars so tightly packed that they give the impression of being a continuous smooth haze but is made of gas and dust. Moving outward to the vary rim of the galaxy you see that this haze drops off partly due to decreased concentrations of gas and dust but also because the radiation present at these distances from the galactic center are too low to fully illuminate the materials found there. Again remember that a limit to our detectors exists such that finding any form of electromagnetic wavelength light at the very edge of the galaxy is near impossible.

To further reinforce this point I would like to discuss the recent size increase of our own Milky Way galaxy. It was believed for a century that the Milky Way was 100 000 light years in diameter and this fact was considered carved in stone by the most ardent astronomers. Then suddenly new and better methods of collecting and processing data allowed for a rescaling of the Milky Way such that we are now 50 percent larger and have a diameter of 150 000 light years. We were sitting right in the middle of our own home and measuring it for a century and we got it 50 percent wrong and this was due solely to limitations in detection capability. Do you still think that we are perfect at detecting all the normal matter that exists in a clump or halo form around galaxies? You had better not be as that would be at best foolhardy and instead a collection of perfectly normal objects surrounding and being drawn into a galaxy seems far more likely than exotic and impossible particles with impossible behaviors.

Think of this as a kind of galactic Kuiper belt if you need too. Rather objects such as black holes, rogue planets from long past star systems and other normal materials are far more likely to contribute even if only in part to the apparent halos that surround the galaxies in the cosmos. The fact that the outer edges of the Milky Way were also found to resemble a kind of wave pattern rather than a smooth flat disc also suggests that materials can exist in orbits around the galactic center such that the projected gravitational effects are higher than expected from the galactic plane. In other words galaxies can appear gravitationally thicker than they do when simply viewed in visible electromagnetic wavelengths. If you combine all of these factors with simple distance from us as observers it becomes clear that distant galaxies contain no mysteries of missing matter. Rather they are too many millions of light years away for us to make accurate maps of their true matter distributions as well as compositions. At these distances the extremely low energy photons needed to detect normal matter being emitted from them have ample time to be scattered or absorbed by any number of events before it reaches us further limiting our overall detection capabilities. Advances in technology will demonstrate the increased amounts of normal matter being found inside all cosmic structures with sizes ranging in the hundreds to thousands or more light years in diameter.

A quick summation of theoretically missing dark matter around galaxies is dark matter is not real as you are not seeing an invisible form of matter around galactic halos. The reason you cannot see it is because no additional dark matter is there; what you are seeing is the variable gravity effects of the supermassive black hole and associated galactic nuclei materials. It only appears that more gravity is needed as understood by general relativity and its curvature predictions because general relativity is wrong in what gravity actually is. The predicted curvature is almost correct but the gravity is made from elevated macht glows and spatially compressed ESS of the space around and inside the galaxy so it appears as though more matter is there in halo form and why people keep saying we cannot directly detect the dark matter particles. In reality this is direct proof of variable gravity and that is why no matter can be detected and all theoretical dark matter candidates are false. The variable gravity allows for predicted curvatures of space but in reality variable gravity allows the visible light matter that can be seen to be responsible for all gravitation fields. Newton and the standard model are more correct in saying that gravity is a real force and not a curvature and space physics using ESS and macht explain why galaxies appear to be missing matter, why they have the flat rotation problem in the first place and why objects like solar systems require no missing dark matter and orbital velocities are explained using only known masses and distances from the stellar center.

Chapter 19
ESS Macht and Matter Interactions, Certainty Principle, Quantum Entanglement

ESS Macht and Matter Interactions

Up until now I have been focusing mainly on how space behaves in the presence of matter and using ESS and macht to explain what space is and how it works. Now I will be doing a bit of the opposite and showing how the nature of ESS and macht also affects the nature and behaviors of matter. To begin with this should be a relatively obvious fact of how the universe works as we know that matter does alter the geometry of space and it is the interactions of matter with the ESS that causes this to happen on large scales. Therefore ESS and macht which is what space is made of can also affect changes in matter as is the case with any normal phenomena in the universe.

Think for a moment of the humble atom and how if you perturb any one part of it too much you will notice a change in another part. Imagine the interior of a star and how external forces such as temperature and pressure affect a change in two atoms such that they fuse to produce a new heavier element. At the same time that this happens the atoms will release energy to the external environment which will therefore affect the local temperature and pressure as well. So with a simple example of a star we see how all actions can work both ways to produce different behaviors or changes in each other.

The nature of space and matter interactions is of course the same and we can see this every day simply by looking at the Earth as matter altered space such that Earth's gravity increased greatly from the gas and dust that formed our planet. In turn the altered space in the form of gravity then changed the behavior of elements such that something like simple carbon can be crushed into a new crystal lattice shape of a diamond. We therefore see space affecting matter but to understand how this fully occurs one must look much, much closer and smaller than a simple diamond. Since all large scale objects like you or I, the Earth or the Sun are made of the smallest fundamental particles the nature of interactions between these fundamental particles and ESS must be properly understood. Understanding this will help to explain a few things which I will discuss in this section the first being basic interaction properties of matter due to ESS macht effects. This will be examined in an introductory manner here and more detailed explanations will follow in later sections. That being said we will be getting a very detailed explanation of at least two unsolved problems here the first of which is quantum motions of particles such that random probabilities are removed from the system and the second being the answer to what and how quantum entanglement works and why it appears as though information is resolved instantly at both locations at once. There are no spooky actions at any distances in the universe only normal space physics at work.

So to start off we will be returning to our room filled with hollow ESS spheres and the mostly opaque orb of matter. We have seen how in the presence of matter ESS is altered in two ways. The first is the volume changes in ESS size and the second is the macht strength that is shared between all ESS. Space in turn affects matter as much as matter affects space and the simplest way in which ESS affect matter is through degrees of interaction. When I say degrees of interaction I am once again faced with a limitation of language such that I cannot choose a simple and apt word for the description of ESS and matter relationships. So I will pick a few words that must not be taken at face value and not held accountable for a direct definition of what is happening between matter as it interacts with space. That being clarified I will now explain what I mean with respect to the degrees of interaction between space and matter. Matter can be made to spend more or less time so to speak inside ESS, or it can be said that matter will impart more or less of its potential to ESS, or it can also be said that matter interacts with ESS more strongly or to a higher degree with ESS depending on spatial compression. I will use the example of a black hole to help explain this concept as it is a necessary key concept that one must fully grasp before being able to learn more about how the universe works.

To begin with let us look at matter when it is under low spatial compression or normal spatial compression as one would measure here on Earth. In these local areas of space spatial compression is not very strong when compared to a black hole and you or I or anyone practically can easily overcome the space and matter interactions. To do this pick up any object around you and lift it or throw it into the air and see how you have been able to force matter to move away from space by a small amount. Most people would say that you have overcome the force of gravity and you have but gravity is not magical and does not simply arise as a force that affects particles in the universe just because. No, gravity is a force that must be governed by laws and calculable interactions with matter and this fact has been known for quite some time as people have attempted to quantize gravity. Since we now know how gravity works through the two factors of spatial compression of ESS and through macht sharing of ESS interacting with matter and the neighboring schales around them we can see what you really are doing when you are picking up or throwing something into the air. What you are in fact doing is not defying gravity which is merely a product of space and matter interacting you are in fact removing matter from space to a small degree.

The matter you have picked up is closer to the center of the Earth and therefore in a slightly higher area of spatial compression of the ESS it surrounds and it also feels a slightly higher macht effect. The matter closer to the ground is interacting with ESS and the ESS macht more and so can be said to be spending more time inside ESS, or more time interacting with the ESS or giving more of its energy to the ESS. When you raise the matter over your head or especially if you throw it into the air you are moving the matter into an area of lower spatial compression which means less density if you will to the ESS it encounters and less macht glow as well. This also means that you have removed matter from space a small amount and therefore have caused matter to spend less time inside the ESS, less time interacting with the ESS or imparting less energy to the ESS macht. You have actually caused matter to very slightly uncouple from space and we can see this because the entire local ESS macht of the Earth system has now shifted slightly towards the matter you have put into the air whether over your head or thrown upwards.

This is because the ESS dominate the universe and dictate more interactions than anything else and in fact far more than just matter related interactions as most of matters behaviors is a result of ESS as I will explain later. For now however the ESS of the Earth system has moved its most highly compressed ESS sphere slightly in the direction of the matter you have moved. This mechanism is the same one we have seen working before where ESS attempt to balance out total macht within themselves. You have also done something else which is ever so slightly move the Entire Earth as the Earth will now try and settle in the new ESS sphere at the center of the Earth system. Matter moving as a result of ESS macht balancing will be looked at in this section as well but for now understand that matter can be made to interact more or less with ESS.

To examine this point further now that we have seen that matter can interact more or less with ESS we will look back at the black holes of the universe as mentioned earlier. So if we apply the same rules to a black hole we see that under greater spatial compression matter is made to interact more strongly with ESS and we can say that in a black hole matter is spending most of its time inside space rather than outside space. When I say inside space I once again have to specify that this only describes what is happening and not a literal translation of physical events. The idea of matter being inside space means it is interacting more with the ESS of the universe than say other matter of the universe. Being outside of space refers to when matter is not interacting with ESS and again matter is always interacting with ESS under normal conditions this is merely a way of describing the degree of interaction of the system.

So in a black hole matter is interacting with ESS almost one hundred percent of the time which is very different than under Earth conditions. If you could stand on the surface of a black hole you would not be able to lift an object off of its surface and even though the mechanism is the exact same physics as on Earth you would not be able to remove matter from space as before. You would not be able to decrease the amount or degree interaction of matter with ESS in the universe. So what you should understand by now is that matter can be made to spend more or less time in space, more or less interacting with space or more or less energy given to space. Any of these three or a combination of all three should be sufficient to understand the fact that matter can associate with space in differing amounts and this is the key to space and matter interactions. From this we can explain perfectly all manner of different interactions in the universe from quantum motion to quantum entanglement.

To begin with we will start very small and only solve the problem of quantum motion. Quantum motion is the apparent random motion of particles due to unknown forces as it is understood by modern science. The fact that particles will move around in quantized jumps for no obvious reason and with no external energy input has created the unsolved problem in physics as what is and what causes quantum motion. Working out mathematically how often a particle is going to make a motion or trying to determine both its position and momentum through equations does nothing to explain what and why things are happening at this quantum scale. Much like general relativity only calculates the amount of gravity in an area of the cosmos but does not explain why matter curves spacetime so does quantum mechanics fail to offer an explanation why quantum events occur at all. This is seen in the unsolved problems of quantum motion, quantum superposition, the uncertainty principle and quantum entanglement to name a few. All of these can be solved using a basic understanding of quantum motion and building upwards from there. So in order to deal with the problem of quantum motion and solve it once and for all we will go back and look at the mostly opaque orb in our little room filled with ESS spheres.

The room is filled with ESS spheres and a single mostly opaque orb right now and we can also see the slight decrease in volume of the ESS inside the matter and in the surrounding schales as well as the soft macht glow of the ESS. We know why these things occur and we also now that this is in fact how matter and space interact as we have seen it answer gravity, black holes, and the acceleration rate changes in the universe as well as other problems. Now we will look at these rules and apply them to the effects on matter to learn about how space physics and ESS and macht create quantum effects in the world we see around us and even in ourselves.

To begin with matter comes in different shapes and sizes from fundamental particles to composite particles like a proton or neutron and these sizes of particles are what helps to create quantum motion. Since ESS and matter are of two different size scales we can see immediately that it will be near impossible for a particle of matter to ever have a perfect central ESS. Different sizes of particles definitely means that matter will never share the same central ESS as the center of an electron is different in size than that of a neutron. We also know that matter affects changes in the volume of the ESS which again creates a new variable in calculating a central ESS for any particle. These factors contribute to the reality that while some particles may coincidentally have a perfect central ESS not all can by default.

Additionally a matter orb as we see it in our room contains many ESS such that the outer edges of the matter orb will have some ESS running right through it while others might sit nestled perfectly inside or outside of the edge. What this means is again coupled with differences in matter size or particle size it is impossible to have a perfect collection of ESS inside of matter such that a central ESS exists with all edge ESS being perfectly balanced equally on the surface of the matter sphere. A combination of the two factors contributing to ESS dispersal inside matter orbs will result in one thing which is a non-central ESS for the matter orb. There are also other factors to be considered such as spin, charge and the like but for now I will stick to a simple explanation. The non-centrality of the ESS inside a matter orb which can be a single fundamental particle or composite particle is what causes quantum motion.

The actual motion results from the interaction of ESS with matter as matter as we know is naturally attracted to space. This is empirically obvious as one watches light bend around a star due to a high gravitational field being generated in the local area of space. We also now know that it is due to the ESS and the macht they carry in combination that dictates how much matter is attracted to space. The end result is matter in a sense tracks after space always following the highest concentration of ESS spheres and the highest macht glow in the local area of space. We have seen this explained when looking in our little room of ESS spheres and using two different sized orbs of matter. The smaller orb seeks to balance out the ESS and macht it creates with the ESS and macht of the larger orb through space. However if we combine this natural tendency of matter to track after ESS and macht in space and couple that with a non-central ESS inside the matter orb we get a very interesting result.

This result is of course quantum motion as in one instance the matter orb is situated over an area of space that at that moment contains a single most highly spatially compressed ESS that also possesses the greatest amount of macht glow. However since we have learned that no single ESS can be the center of the matter orb the orb will try and move over a second ESS and the result is a new more central ESS than before. The matter interacts with space and imparts some of its energy or potential with that space and so now it imparts the highest level of energy to the new ESS. Therefore the first ESS is now off center and will pull the matter back slightly towards itself repeating the process and once again causing the matter to move. This movement is not confined to just two ESS as space is three dimensional in its appearance to us and so the ESS can trade the matter back and forth and around from any number of ESS in a three dimensional shape around a rough ESS centered mass inside the matter.

Here we see motions occurring that cause the matter to forever track after different central ESS and their higher macht glows. The motion appears to us as randomly occurring in three dimensional space and will always occur in discrete jumps as predicted by quantum mechanics due to the fact that matter exists in space over discrete units of ESS. The ESS are not a perfectly smooth flow of space with an infinitely smaller number and size of gradients necessary to produce a continuous motion and instead must move in jumps equal to the size of the ESS in the given area of spatial compression. No smooth spacetime, no different geometries or shape of one ESS to the next and no fractal shape to space.

The jumps will appear to be of a significant size because as we have seen ESS are smaller than matter and so a central ESS is only a representation of what is really a central collection of ESS inside the matter orb at any one moment. In addition ESS transfer macht far faster than matter transfers energy or information and so a small grouping of ESS larger than a single ESS can rapidly adapt to the new macht flow and become the central region. In this way a slower working quantum physics ruled orb of matter tracks after the quicker moving space physics ESS macht glow. The net result is what can be observed readily and has been for 200 years or so as Brownian motion. Brownian motion is a result of much larger sized objects with much more complex interactions between local areas of space and other objects in that space but the base set of rules determining why this occurs is all due to ESS macht and matter interactions in space.

If you take for example a hypothetical piece of matter that is exactly one fundamental particle in size and make it absolutely stationary in space you can observe the motion in action once you let go of the object. In this probably impossible or at least exceedingly difficult to perform experiment the moment you release the particle it will try to track towards a newer central ESS and in so doing move a single ESS in a given direction. The macht travels faster than the matter and so the new ESS are now dominant to the old ESS and the particle is set in motion but since the matter is slower to move through space than transfer energy to the ESS even as it moves it is already trying to impart energy to a new and therefore third ESS different from the first and the second for which it is currently on a trajectory to meet. This satisfies classical mechanics as was dictated by Newton centuries ago that an object in motion will tend to stay in motion unless acted upon by an outside force. In this case classical mechanics is correct and the outside force comes from the new macht glow of an ESS which will act on the fundamental and classically moving particle to change its course.

The quantum mechanical laws are obeyed because changes in energy from the matter to ESS are always carried out at or below the speed of light and in quantized amounts. The single ESS that is the new ESS starts the motion of the particle in jumps of one ESS at a time but as the process is allowed to progress kinetic momentum of the particle allows it to move slightly farther than before. The result is a new central region of ESS being used each time a direction change occurs and will reach a maximum distance travelled per jump dictated by the strength of the macht glow in ESS regions.

This is because there is a maximum amount of energy being imparted to the ESS from the particle that will not change under static spatial compression and so the amount of macht that can be shared also has a maximum limit. So the macht travels faster than the matter and only in a discrete amount dictated by the amount of energy donated from the matter under the local area spatial compression. The final balancing of the systems results in finite jumps in random directions governed by quantum amounts from normal matter and then transformed into localized discrete amounts in the ESS macht. This results in a perfect and seamless fusing of both quantum mechanics, classical motion and space physics such that the problem of what causes and why random quantum motion occurs is solved.

The reason that the motion appears instantaneous to us with perfect uniform jumps being made immediately is due to the light speed transfer of energy to ESS and the faster than light speed transfer of macht between areas of ESS. Detecting these small but subtle accelerations for quantum motion from a perfect stop of any normal matter object can only be made using instruments powerful enough and capable of measuring changes in macht glow. Our current quantum mechanical instruments cannot detect movement faster than light speed and since the transfer of energy is at light speed we perceive incorrectly that quantum motion was instantaneous. That being said the acceleration is extremely fast and for all practical purposes happens almost immediately since the interactions are at a minimum light speed for matter and energy and faster than light speed for macht. All particles we observe are of course not perfectly isolated from everything else in the universe and so other forces will be interacting from all three dimensions of space around the particle. These might be charge, gravity, spin or any number of other factors that will cause the particle to change its central tracking ESS rapidly resulting in true three dimensional quantum movement.

Now we have seen where the apparently random quantum motions of particles physically comes from we can apply this to a number of other unsolved problems as each is built off of this general type of rule. Let us scale up from a single particle for a moment and look at something slightly more complex like a hydrogen atom. A hydrogen atom consists of a single proton at the nucleus and a single electron orbiting around it. The proton is a hadron meaning it is a composite particle of three quarks bound together by the nuclear force and is different from our previous example of a single fundamental particle in space. We will however be treating the proton as a single particle for this next example as we want instead to focus on the motion of the electron as it moves around the nucleus. To begin with what do we know about the motion of electrons around any nucleus?

Certainty Principle

Well first off we know that they travel not in a clean neat orbit but instead in seemingly random and chaotic ways inside an orbital. An orbital is a region of space surrounding an atom where at a given energy level a certain probability exists of finding the electron in it. The two factors most strongly associated with this are position and momentum which according to classical quantum mechanics may never both be known with 100 percent accuracy at the same time. The reason is that theoretically according to the Uncertainty Principle the more you know for example position of the electron the less you know about its momentum. The opposite exists as well within the principle stating that the more you know about momentum of the electron you must therefore lose that same percentage about the position of the electron. This basic rule is then coupled with the fact that electrons appear to also move in bizarre ways compared to normal objects in the universe as they travel in their respective orbitals.

What is meant by bizarre is that they can be moving in one direction and then change directions for what looks like no real reason. More accurately what is a reason that exists hidden to the eyes of classical quantum mechanics as the electron is moving how it chooses and that is part of the universe since long before humans tried to figure out why. This means the electrons motion is perfectly normal and does not violate any laws of the universe and does not exhibit any magical properties akin to something like spooky actions at a distance. The reality is that electrons always obey the laws that govern them and these laws do not reside in either quantum mechanics or relativity. To that end let us explain the nature of motions of electrons around nuclei such that they follow known rules of the universe that are already established using previous unsolved problems.

The first thing we must understand is that the electron is a particle no different than any other when it comes to interacting with ESS and will follow the same quantum motion rules as the proton around which it moves. An electron will impart a much smaller amount of energy to the ESS macht than an entire proton but will do so nonetheless. As a result the electron is never going to follow a perfectly smooth path as it orbits the nucleus and will always create a motion that must be described within an orbital. Even if no other factors were at work the electron would be slightly pulled in different directions as it moved around the nucleus such that even a simple circular orbit would be impossible. Instead a series of quantum motions as previously described using ESS macht and energy transfer between matter and space will result in the electron being moved in any number of orbits around the nucleus with some being closer or farther from it and some being oriented along the poles, equator or anywhere in between. This is due to the electrons attempt to track after the ESS macht in a three dimensional shape around the nucleus.

A second factor that will determine the motion of the electron around the nucleus is of course the quantum motions of the nucleus itself. The proton at the center of the hydrogen atom is itself anything but stationary and will move easily in space as it tracks a central ESS macht just as any normal single fundamental particle would. The seemingly random but really only complex motions of the proton around a central region of ESS will create a new central point from which the electron must now calculate its own orbital motions. Think for a moment of a spaceship in orbit around the Earth and realize that if one were to make the ship orbit the planet but not follow the planet the spaceship would soon be lost from the system. This is because the Earth is also moving like the proton and creating a new center from which the spaceship must orbit. If you look at the total motion of a spaceship in orbit around the Earth and compare that to the Milky Way galaxy you can see that it really creates a circular corkscrew type pattern in space which tracks after the Earth just the way the electron track after the proton and the way a fundamental particle tracks after the ESS macht. So now we can factor in the motion of a non-stationary central point being the nucleus to the already complex motion of the electron in orbit. These two factors will create an even more complex orbital around the nucleus where at any given moment the electron might be found.

A third factor is not just the motion of the proton in space but the motion of the quarks inside the proton as well. A proton is not a single rigid perfectly spherical point in space that possesses a totally smooth surface and from which all forces flow absolutely evenly at all times. Instead the proton itself occupies an existence in the universe as a globular probability field of three quarks and the strong nuclear force which binds them. This globular form will rotate freely in space around any axis it chooses and perhaps even experience inversion events of quarks inside of its probability orbital such that some quarks will swap positions with one another as needed to balance themselves with external forces. This adds another variable in the nature of motion of the proton such that the surface of the proton is no longer treatable as a smooth sphere. Since the electron is bound to the proton and is trying to orbit at a set distance it will attempt to maintain this set distance even over an irregular and continually changing surface topography of the proton. This irregular proton topography is what will add an additional layer of complexity to the total orbital path of the electron. As you can imagine the electron having to track up and down around the hills and valleys of the proton's quark probability field which results in the electron changing direction for what only appears as no reason or at least random.

The fourth and final factor that affects the motion of an electron around a proton in a hydrogen atom is changes to the external environment of the system. Remember a closed perfectly isolated system never exists anywhere at anytime in the universe as the universe is always connected to itself and cannot be isolated from itself meaning the universe will always exchange information, forces or macht regardless of any attempt to erect barriers to stop it. Put another way the universe is the system, is the barrier and is the laboratory so the universe itself and what makes it up will always flow freely and unencumbered through all at once forever. That being said our little hydrogen atom is never really alone in the universe and besides the cozy thoughts that creates in our minds it also creates a realization that external changes will have an impact upon it. External changes are spatial compression or rebound of the local ESS, increased or decreased macht glow of the ESS and influence from forces of other particles. A change in spatial compression or macht glow will as we know create differences in orbital patterns and quantum motion as we have learned previously but influence from other particles will of course also affect the outcome of the electrons orbital pathway around the proton.

Take for example two hydrogen atoms placed close to each other but not touching and we see our little electrons moving normally around the protons governed by only the laws of quantum motion using ESS macht but as we place them together such that the two orbitals are now touching a new factor must be considered. This is of course the electrons mutual repulsion force that they share between each other and will make them repel in directions away from the other such that energy is balanced with the least effort possible on the part of the universe. What this means practically in terms of calculating the orbital motion of the electron is that the effects of external forces must be weighed with those of ESS macht such that the strength of the two and all variables concerning them determine where the electron will move next. Calculating these factors is simple since we know both the ESS spatial compression and macht glow for the local area of space and from there can calculate the quantum motion of each particle in the system such as the electrons, protons and even the quarks making up the protons.

Once quantum motions have been calculated for each determining the orbital path for an electron at any given moment around the nucleus it is essentially reduced to simple classical mechanics where an object stays in motion unless acted upon by an outside force. The system is initially calculated at its absolute base level for the entire universe which is space physics and this will give us precise accounts of what is known as quantum motion. After this the overall motions of the system will obey a fairly ordinary set of normal classical mechanical calculations which are of course made using quantized amounts for each particle. So at any given time since we know the interactions of space and matter with each other through ESS macht it is possible to figure out where an electron is and where it will go next whether by itself around a single proton or in any number of increasingly complex calculation of atoms. The full mechanisms behind the exact nature of all matter and space interactions through ESS and macht balancing will be covered later as for now I am just introducing you to these facts.

If this sounds like knowing both the position and momentum of a particle at any one time it is exactly that. The uncertainty of quantum motion of particles is resolved using the same rules of ESS macht interaction with matter that governs simple forces like gravity. It is a well known fact that the universe always knows with 100 percent certainty where all of its particles are and what they are doing and what they will do next. If the universe did not then it would have dissolved itself long ago as nothing could follow rules such that matter maintained itself in coherent forms meaning you and I and any other intelligent species would not be here to question it. It is utterly ridiculous to propose that something in the universe does not follow a known law of that same universe such that the universe itself would not know what is going to happen next and why. The universe does not allow spooky actions or the like at a distance under any circumstances and the idea that not being able to know both position and momentum of a particle at the same time is one of them.

To this end all that was needed was a simple explanation of the rules governing particles at their most base levels or in other words with matter to space in order to account for the missing information necessary to fully understand a given system. The information available to quantum mechanics alone is insufficient to calculate both position and momentum but not to space physics as the need for the universe to balance macht of different types between ESS is what is pushing the electrons in their orbital paths. The motion of the electron around the proton is in a sense a rough map of both the three dimensional cartography of space at the smallest level which is the ESS and of the strength of the underlying macht glows for that level of spatial compression in the local area of space in question. Naturally all other unsolved quantum mechanical phenomena can be answered using space physics and the underlying ESS macht interactions with matter. As I said before we will build on the base idea of quantum motion through ESS macht in order to handle larger problems more directly and we will now do so by looking at and solving the problem of quantum entanglement.

Quantum Entanglement

Quantum entanglement is the phenomenon whereby two particles will become connected or entangled with one another such that upon separating them physically will allow you to still determine the state of the second by measuring the first. Here is a simple and generic example and explanation of how quantum entanglement may be perceived with a basic system consisting of two particles. As a measurement is made of the first particle the state of the second is instantaneously resolved and will always be linked to the first. Take for example a single particle of spin up that has been entangled with a particle of spin down and separate them a set distance. The spin of the first particle is up but the observer does not know this until the measurement is made and as such has an unknown spin which is also true of the second particle. If one desires to know the spin state of the first particle they will need to conduct the necessary observation to resolve what state it is in either up or down. The particles however are entangled and as such if one is spin up the other will always be spin down and this can never be broken as they are entangled. The observer does make a measurement of the first particle and the spin is found to be up but before any measurements can be made concerning the second particle the state of that particle is already known and it is spin down. The speed at which the state of the second particle is found to be spin down far exceeds the speed of light and the second particle under normal physical laws of quantum mechanics and relativity should not be able to do this according to modern science.

Once again the universe is doing exactly as it desires and nothing odd is going on and yet quantum entanglement has been called spooky actions at a distance. The moniker for quantum entanglement of spooky actions was given by Einstein who himself realized that the state of the two particles was being shared between them in the system faster than information could travel using the speed of light. It was Einstein after all who said that nothing in the universe could travel faster than light and yet even he admitted that something was. The unsolved problem in physics that has remained to this day since spooky actions at a distance were first realized is that of quantum entanglement and specifically how it works.

The overall problem can be sub-divided into three main areas which need to be addressed the first is of course what is quantum entanglement really, the second is why does it happen and the third is what are the basic rules governing these actions? The reality that quantum entanglement is real has confounded modern science for one simple reason and that is nothing according to the three main views should be able to travel faster than the speed of light. Information as it is commonly referred to by modern science only refers to matter and energy which are bound by relativity to travel no faster than the speed of light. The problem of quantum entanglement is that in order for a single particle to know what the other is doing instantly information of a kind must be being shared between the two and this means it cannot be anything that falls under the roof of the Standard Model of particle physics or relativity. So what is going on and what is quantum entanglement and how can it behave as it has for a century of numerous perfectly controlled and reproducible experiments?

The answer to what quantum entanglement is is a simple sharing of information by two particles or greater that have been made to become entwined with one another. This is of course fairly obvious as this is what has been thoroughly tested over and over for a century or more. It deserves to be restated here though as this central truth must remain in order for an explanation of quantum entanglement to be properly made and the answers revealed. Two particles are made to become entwined with each other such that one does in fact know what the other is doing and can do so at a speed that far exceeds the speed of light. This is all that is needed to examine the second part of the problem of quantum entanglement which is why does it happen?

This answer is a bit more involved but still very straight forward and will be making use of various rules governing the interaction of ESS macht and matter that I have already demonstrated. The first thing we must understand with quantum entanglement is the fact that matter can be made to spend more or less time inside of or interacting with empty space and in doing so imparts energy to the ESS macht glow found there. If we recall that matter does give some of its energy to ESS and that the energy is not given fully of course under normal spatial compression then we have the first piece of the answer which is a donation of energy that is not continuous but rather can be thought of as a percentage.

Take for example a simple system that would involve a single fundamental particle interacting with empty space and for this we can use our little room full of ESS spheres and the matter orb. The macht glow of the ESS is elevated by the donation of energy from the matter orb but if we look at the glow it fades in and out. Let us say that as the matter gives energy it increases macht glow for 25 percent of a single second and that for the remaining 75 percent of the time it is not donating energy. The macht glow remains in the ESS even when the matter is not in contact with it as some of it has been shared by the ESS with the surrounding schales. The matter is of course bound to space and so it has not lost its energy either as this would violate the law of conservation of energy. Instead the ESS macht and the matter orb are sharing a combined total macht/energy for the system which will always sum correctly such that no energy is lost from the system. This is of course already proven by the idea that matter can exist as a solid form that we can touch or interact with but also can exist as a form that alters the space geometry of the local area.

Think of the Earth as being very solid but that the Earth also deforms the space geometry it exists in. It is known that matter must be interacting on some quantum level with space as a simple explanation of matter curving spacetime does not explain what is happening on a particle, force and quantum level. General relativity relies on magic in order to curve spacetime and this is of course exactly the sort of thing that is not happening. What you have learned by now is the mechanism that accounts for the quantum level explanation of how matter does affect space and it is known that this is in fact the ESS and the ESS macht of the local area of space. So it is a behavior of space and matter to share a total amount of macht/energy and that this total is conserved for a given quantum object and region of ESS. The matter will be giving energy to the ESS while it is inside space and not doing so while it is outside of space. Once again I remind the reader that the concept of inside and outside space as well as time spent interacting with space and time spent not interacting with space are necessary visuals to describe the behavior of space physics in the system and not a direct translation of word usage so should not be taken literally. That clarification having been made matter will move in and out of space and repeat this process over and over as long as it is bound to space.

The interactions however take place at an incredibly accelerated time frame such that to us viewing these events from a human timescale it will appear to be continuous. It can however be subdivided on much finer timescales and for this I will use the idea of Planck times even though interaction need not be a single Planck time in length and will vary. To lay down one of the basic rules here spatial compression will affect the time matter spends interacting with space. This has already been demonstrated by looking at matter under normal spatial compression which is found arbitrarily on Earth and inside black holes where spatial compression is far greater. The massive power of a black hole means that far more ESS are interacting with any matter orbs inside it and so also has an increased macht glow. Matter is naturally attracted to ESS and so when the matter orb is filled with many more ESS it will naturally spend more time interacting with the ESS as the total ESS macht pull is so much greater. This as we have seen is what causes the increased energy input from matter to the ESS of a black hole and the resultant increase in gravitational strength through the variable gravity mechanism. Far more rules and answers of unsolved problems will follow in a later section and will cover matter interaction with ESS and ESS macht in more depth.

For now we can see that on a Planck length timescale matter can be made to spend more or less time inside space and therefore give more or less energy to space. What this also means is that sometimes matter is not interacting with the ESS and is said to be outside space and at this time the macht glow remains shared between ESS and the total energy is conserved between the matter and the ESS. This means that matter is bound to space and space is bound to matter and it is this key fact that fully explains why quantum entanglement works and is natural in the universe. In order to explain how matter is quantum entangled with other matter we need only combine the two particles through this ESS macht and matter set of binding interactions. Let us look at the first particle and see that it is for example inside space at exactly Planck time one and now we entangle it with a second particle. The first particle has donated an exact amount of energy that is quantized and unique to that particle given a number of factors. The mass of the particle, the type of particle, the particle charge, spin and all related properties will produce a fingerprint of that particle. The next step is to look at the exact moment of quantum entanglement and the degree of interactions with the local ESS and macht of space in that area.

If we think of the interaction event as an angular wave form where time spent inside space is below a horizontal line and time spent outside space is above the horizontal line we can find the exact moment of maximum degree of interaction of matter with ESS. The very bottom of the angular waveform represents this point and it is at this point that we will set Planck time to one. This exact amount of interaction also makes up part of the fingerprint of the particle but not just the particle of matter but the space and matter system possessed by this one particle in the entire universe. Each particle has a different set of properties and even identical particles will also have slightly different degree of interaction times inside space. In other words particles can have one of any number of Planck times of maximum degree of interaction with local ESS and macht. This length varies for local area spatial compression but is very exact to each particle no matter how much of the angular wave form for that particle is inside empty space there will still only be one maximum Planck time where it is at the bottom of the wave form. The exact Planck time is directly calculated using quantum mechanical rules for two reasons the first being all particles belong to the Standard Model and normal physics as opposed to space physics. The second reason is that known particles all behave under quantized amounts and one of these is the amount of energy donated to space and so will always occur in set jumps if you will.

This creates a huge number of completely unique signatures for each and every particle in existence. It is also the exact and final explanation of why two particles can never have the exact same properties and exist in the exact same space at once. It is the complete explanation involving space of the Pauli Exclusion Principle and details why at any given moment in space two particles cannot share complete information as it includes sharing ESS and macht perfectly. Now that we have seen that a single particle can have only one exact set of variables to describe it not only in the macro scale of quantum mechanics in terms of general properties like mass, spin and charge for example but also in maximum degree of interaction of the particle with the ESS and ESS macht to which it is bound. The fact that ESS and matter are bound also further explains the associated lag tracking phenomenon of quantum motion as the residual and conserved energy of the ESS macht and matter is summed for that particle and the space it inhabits.

Now we can finally tie these two particles together such that they are entwined with one another and will exhibit the property of quantum entanglement. If the first particle is exhibiting a specific set of exact properties and maximum degree of interaction with space then the second particle is also made to contain these exact properties for an instant during the entanglement process and the two particles will naturally separate. This is where modern science and the three views encounter the problem of spooky actions at a distance as the particles are no longer communicating with traditional information. Remember that the information of space physics is not the same as information as shared by the three main views as the three main views would treat information as a particle or energy being sent back and forth. The particles may now be separated to any desired distance possible before a measurement is made.

At the time of observation a measurement is made and the state of the first particle is now known and found to be spin up. The second particle is instantaneously solidified as spin down with no time delay between them and so a spooky action at a distance appears but has not occurred. The exact properties and maximum degree of interaction with the ESS and the ESS macht has produced a very specific signature of ESS macht glow held within the ESS which is invisible to normal quantum systems and mechanics. Remember again here that the three main views cannot see or deal with space physics and only deal with normal matter and energy as is described in the Standard Model of particle physics. To this end the particles are entangled because they are bound by the residual macht glow that they have become entwined with. Since matter is bound to space and ESS macht glow attracts matter the two particles have now become not entangled with each other directly but entangled with the macht glow created by and shared between the two particles.

This is a very specific glow that only pulses in synchronous fashion at the exact moment of maximum degree of interaction as seen on the bottom of the angular wave forms in the graph. This is the specific Planck time to which they are attracted to the macht glow and since they are both attracted to the mutual macht glow the glow affects both of them perfectly such that they are kept entwined right down to the specific Planck time they will reenter space again. The unique fingerprint of particles I talked about earlier means that only these two particles will be entangled with one another through the macht glow they share and will ignore other particles whose macht glows are not synchronized with them. Macht glow is also made of strength and type of energy donated by matter to ESS not just timing of interaction and so the fingerprint is unique and easily able to keep the two particles entangled. Since ESS are so small compared to matter orbs and since ESS extend and share macht in three dimensions any unique glow patterns can be essentially found to trace out a line between the two particles. These macht glow lines of entanglement are uniquely synchronized to the two particles and since ESS is the space through which all matter exists the ESS macht glow can be transferred through solid objects. The solid objects which can be thought of as laboratory equipment are made of particles that are mainly empty space and the line has a small chance of tracking directly through a solid nucleus of any of the matter making up the equipment. In addition even if the line were to pass through a nucleus of the equipment the particles making up the nucleus are not likely to be the same type and experience the exact degree of maximum interaction with the ESS macht. This means that the shared macht glow line for the two particles can pass straight through solid matter and maintain entanglement of the two particles.

The act of measurement of the first particle is an interference of that particle with its synchronized macht glow and so the coherence of the macht glow line is disrupted and the second particle feels the effects. Long have people thought that when they made an observation of one particle they were in fact measuring just that one particle and that it was a collapse of the wave function that led to this. In reality the wave function is of little importance in quantum entanglement as it is really the collapse of the macht glow line that entwines the two particles that is occurring. Dissolving the macht glow line instantly solves both particles spin problem immediately because the line is connecting both at the exact same instant. Information is being shared by the two particles and it does travel faster than light but it is not instantaneous. It appears that way because the ESS macht travels much, much faster than light ever could and over such short distances it appears to be instantaneous. Certainly to the slow timescales of quantum mechanics and special relativity it is immediate but in reality there is a small non-zero amount of time taken for information to travel along the macht glow line.

Nevertheless the dissolution of the macht glow line is the real measurement that is being perceived during quantum entanglement experiments and not an individual particle. The macht glow line however is sharing information as the disruption of one particle from the macht glow line means that the effects of the synchronized macht glow will travel down the line to the second particle. The macht glow line is pulsing with the energy input from the two entwined particles and will be forced to stop pulsing at the act of measurement. Since the macht glow line attracts matter the loss of the macht line means that pulsed and entwined energy from the first particle is lost and so the glow to which the second particle is attracted is also lost as the pulse is now gone for that entanglement.

This is the mechanism by which the second particle is forced to drop out of entanglement from and explains why the second particle appears to spontaneously decide to be measured.

The combination of all properties of the particle including maximum degree of interaction of the particle while still maintaining the Pauli Exclusion Principle such that it incorporates the rules of space physics is what creates quantum entanglement and also what determines the outcome of measuring the macht glow line of the two entwined particles. I use the term entwined to refer to an explanation of quantum entanglement in its true and complete form such that it incorporates the space, ESS and ESS macht that matter must interact with in the universe.

The speed at which macht travels greatly exceeds the speed of light and so information is not lost as it travels in the phenomenon of quantum entanglement and explains why the action seems to defy physics. As I stated earlier the universe is doing nothing bizarre as its inner workings are known well to it and does not constitute spooky actions at any distances. This is not quantum entanglement it is particles entwining and must include space as part of the interaction as matter and space are both needed to make the universe. In addition it is shown here that the unsolved problem of quantum entanglement permits information to travel faster than the speed of light as relativity is incorrect in stating that nothing in the universe may travel faster than the speed of light. Instead no matter or energy can travel faster than the speed of light but ESS sharing of macht only travels faster than the speed of light and is responsible for both the formation of apparent quantum entanglement events and the solution to why and how one particle knows the state of a second particle immediately.

Now that the unsolved problem of quantum entanglement has been answered a list of the rules and explanation for why they govern these phenomenon can now be given. The biggest concern here is of course what is known in quantum mechanics as decoherence of the entangled system. It has long been known that quantum entangled systems are inherently unstable and one of the greatest challenges in making entangled systems is getting them to last for long periods and theoretically indefinite lengths of time. What quantum mechanics does not know is the exact reason why the systems lose coherence or therefore undergo decoherence such that the two objects are no longer entangled. In truth there is only one reason why a system becomes decoherent but there are multiple rules for why this happens.

First the why of decoherence is from a loss of entwinement of the two particles with their common macht line. It literally is as simple as that and all decoherence occurrences are a result of that one fact. The disruption of the macht line means that the macht glow is no longer being felt by the two particles perfectly equal. This results in the energy input for each particles local area of space and the ESS to which they are bound is lost and so the macht line loses its pulse. Since both space affects matter and matter affects space a disruption of any of the components of the system will result in entwinement loss of the macht line and therefore subsequent loss of coherence. The net result is the particles losing entanglement and in a sense dropping back into the realm of normal non-quantum mechanics. The components of the system that can be disrupted are either of the two particles and of course the macht line itself. Now that we know why decoherence happens we can look at the rules determining why this does happen and happen frequently.

The first thing you must remember is something I mentioned just a little while ago about the macht line passing through solid objects due to the fact that they are mainly made of space not matter and also that the matter that makes them up are not necessarily in possession of all the complete properties of the two quantum entangled particles. This includes exact mass, charge, spin and degree of maximum interactions with the local ESS each inhabits. The remaining space in between is made of regular ESS that are not directly interacting with any matter orbs as we have seen in our little room examples. They are made of complex and overlapping schales extending outwards from all the ESS and matter interacting orbs of the local area of space that makes up the instrument being used to conduct the experiment. The shared macht glow they possess is very complex and does not possess any linear organization as does the macht line that entwines the two particles. Remember that entwining refers to the use of both properties from matter and space to fully describe what is happening during the event as opposed to entanglement which refers to just matter descriptions of the event and are incomplete. To this end the background glow of the system is not chaotic but is perfectly ordered as it obeys all the rules of space physics such that all ESS and matter interaction events, macht sharing between ESS events and overlap of each individual particle of matter and its local area of space are accounted for.

While all the interactions can be calculated for any given area of space you choose only the macht line of the entwined objects stands out as a significant linear source of macht glow that is separate and unique compared to the background glow of the entire laboratory. This being said the macht line possesses the seemingly impossible ability of creating these spooky actions at a distance as it maintains macht speed information flow between the two particles which maintain light speed or slower information exchange with the local ESS to which they are bound. This unique macht line is able to flow directly through solid matter with perfect ease but it is not by any means a form of indestructible information sharing. If something can cause the line to be disrupted even its remarkable ability to connect two particles over great distances will be lost and the two particles immediately do what one calls collapsing the wave function and revealing their entangled states. This means rules exist to prevent the macht line from being made independent of the background macht glow.

The first of these rules is of course another particle somewhere in the system or the laboratory itself that has the exact properties of the two entwined particles. While the fingerprints of the two particles are hard to reproduce statistically due to their unique properties and the extremely small chance that any other particle is at their own maximum degree of interactions with space given that it could occur at any number of Plank times is small it is a non-zero chance. This non-zero chance is statistically significant when coupled with the incredible number of particles that exist in the surrounding space of the test system, the instrument used to conduct the test and the laboratory enclosing all of this plus any materials around the laboratory like the Earth, solar system or entire universe. Given the incredible number of particles that exist at any given time around the entwined pair decoherence is not a chance but a certainty.

However all is not lost in terms of maintaining coherence for longer than time approaches zero. The strength of the particles interaction with space will determine whether or not the system becomes decoherent. For this you must look at distance, time and number of external particle probabilities of interacting with the system and the strength of the interaction such that they possess enough force to disrupt the macht line. First let us look at distance as a factor and realize that macht glow diminishes over distance as we have seen with macht shared between schales in the example of our little room. So the closer the two entwined particles are the stronger their macht line and the more difficult it would be to break entwinement. If we move the particles farther apart their shared macht line is weaker and it will take less effort to disrupt their entwinement. The amount of energy input to the local area of space around or between the particles needs to be high enough to increase the ESS macht there such that the newly imparting energy from a third piece of matter will disrupt the shared macht line of the two original particles.

Two particles entwined and separated by a great distance are far more easily disrupted owing to the weakness of the shared macht line. If only two particles existed in the entire universe and they were made to become entwined they would never lose synchrony with each other. This is not because the macht line between them can extend for tens of billions of light years in so much as with no other particles in existence to add macht glow to the universe there is no way for the macht line to become disrupted. The particles can be separated by distances so great that even macht speed information cannot reach each other any more yet they will have last been in a state of entwinement and will retain the last known properties they shared. The universe does not however only contain two particles and so the sheer number of possibilities that a different particle outside the entwined system can disrupt the macht line becomes significant at all distances.

The second factor that a decoherence event can occur is in strength of macht glow from external particles to the system. Not just distance but also strength of external particles macht glow can affect the entwinement event such that a disruption occurs again. This is also why not all particles in the universe can affect the system as a particle far enough away will not be able to have its macht glow reach the entwined pair of particles and even closer the glow it creates will simply be too weak to cause a disruption of the shared macht line. However a particle will impart a certain amount of energy to the ESS and the ESS macht it possesses and if that is strong enough then it can overcome distance to disrupt the line. Although a rare occurrence the total strength of the macht line can become lost in the background noise of the local area ESS macht if that macht glow is strong enough. This explains why a particle in a random object in the laboratory is far less likely to cause a disruption event than a particle actually inside the instrument being used to conduct the experiment itself. Different particles interact with space in different degrees depending on their make up and this could be in terms of how they are put together with other fundamental particles to create a hadron, hadrons to form nuclei of various sizes and the way atoms interact with each other such that they affect the timing and degree of interaction with the ESS they inhabit.

The fact that particles do interact with each other directly is also a way that decoherence can be lost. Remember that in a system as small and exactly precise as a quantum one any interaction will alter the state of the system unlike the way something can be used to observe a classical system which is far larger and based on averages. Any attempt to interact with a quantum system to perform an observation means that the object you are observing is now affected by the action of observing. This action is not mysterious and is determined purely by the rules of space physics as I will explain later but here it does mean that any interaction with another particle or energy will change the properties of one of the two particles entwined. In this way a disruption of the macht line comes not from the space side of the entwinement but from the matter side and as we have seen both sides feedback on each other to keep the macht line strong and this keeps the particles entwined. If a particle is introduced to the system such that it interacts with one of the two entwined particles it must alter the properties of that entwined particle. This will change the subtle way in which the particle interacts with space and therefore the ESS macht glow it creates and which is shared with the other particle. The end result is the shared macht glow is lost or at least weakened such that the chance of the background macht glow of the system or another particles overlapping macht glow can disrupt the now weakened macht line.

Even a single neutrino which is near massless and chooses to avoid frequent contact with regular matter is all it would take on the quantum level to disrupt the entwinement of the two particles. The decoherence of the systems is the final outcome of such an event and will once again result in the two particles dropping back into the realm of normal physics and revealing their states. The fact that the macht sharing is faster than light and therefore lost faster than information can travel between matter and energy means that the two particles will always drop out of entwinement at the same time. This is of course due to the fact that the exchange of information between the space and matter must occur at light speed or slower and so the disruption of the macht line always results in a slower than macht speed realization of the two particle states back into normal quantum mechanics.

The last main contributing factor to why coherence is lost from a system is simply the passage of time. The reason being that the longer a system is made to stay entwined the greater the probability that a disruption event can occur for any of the reasons stated above. Leaving two particles entwined for longer allows for more chances of their distance to separate weakening the macht line. The longer they stay entwined and their distance grows the greater the chance that a particle of any macht strength can interfere with the macht line and disrupt it even if the macht strength of that particle is weak due to the weakened overall strength of the macht line. The probability that another particle will simply enter the system and interact with one of the two particles also increases and since laboratory equipment is never truly free from outside influences this is an increased risk to the macht line as well.

Lastly a combination of any of these factors will greatly increase the probability that the macht line will be disrupted as near countless particles are in the immediate vicinity of the entwined pair of particles and have numerous chances to disrupt the macht line. The overall factors of distance between the two entwined particles and the time that they are entwined for will ultimately determine the length of coherence. The two factors of distance and time likewise will be made up of probabilities of interactions of other external particles either to the macht glow of the local area of space, the particles themselves or a combination but all result in a disruption of the macht line shared between the two entwined particles and will result in decoherence of the event.

Facts

This section has covered everything from ESS, ESS macht and matter interactions, to quantum motion, quantum certainty and quantum entanglement and as such requires a long list of facts to describe the rules for each. To begin with we will break each one down to a short set of rules for each section and the first is of course space and matter interactions and quantum motion. The first fact is that matter affects space and space affects matter meaning that not only does matter determine the properties of ESS and their macht but in turn ESS and ESS macht also determine the behavior of physical matter and energy. The second fact is that matter can be made to spend more or less time, interact more or less with or interact to a greater or lesser degree with space. This can be quickly described such that matter is made to spend time inside or outside of space although this is purely for descriptive purposes as matter is always held within the space of the universe it only refers to the interaction of space and matter. The total macht and energy for the system is conserved such that the law of conservation of energy is not violated and explains why particles are bound to local areas of space.

The third fact is that matter will attempt to track after the local ESS center to which it is bound. This tracking is delayed due to the sharing of macht at faster than light speeds and the donating of energy to ESS at light speed or slower. The acceleration is near instantaneous and continues in all three dimensions. The fourth fact is that this tracking delay is what causes quantum motion as can be observed in experiments such as those exhibiting Brownian motion. The fifth fact is that once set in motion the classical laws of kinetic motion are all that are needed to determine the motion of a particle tracking after the ESS macht glow. The sixth fact is that electrons and their motion around a nucleus can be explained using these rules of quantum motion and coupled with the complex behavior of a central hadron in the case of a hydrogen atom will create the associated orbital patterns.

The seventh fact is that all properties of a particle or system are now known such that no uncertainty principle can exist by using the additional information from ESS and macht of space. The ESS and ESS macht as well as the factors that govern them will explain what forces are acting upon particles such as electrons orbiting nuclei to the extent that all motions can now be calculated. This means that looking at the motion of something like an electron and knowing when and how the ESS macht will affect its course around the nucleus means it is possible to know both position and momentum with 100 percent accuracy. In fact if all information of the ESS and the ESS macht for that local area of space are known it is impossible to calculate the motion and properties of the electron to less than 100 percent.

Now we can list the rules governing quantum entanglement starting with fact eight which is two or more particles must exhibit the exact same properties in terms of their local area of space to pulse in synchrony with the ESS macht glow found there. The ninth fact refers to the importance of the maximum degree of interaction of a particle with space as can be calculated from the bottom of the angular waveform. The tenth fact is that the exact maximum degree of interaction with space at the bottom of the angular waveform is what fully explains the Pauli exclusion in combination with all other properties of the quantum objects like mass, charge and spin. The eleventh fact is that the sharing of a macht glow between the two particles forms a macht line which is what causes them to be entwined. Entwined particles is the true state of the event such that the effects of space physics are incorporated rather than the purely particle description of quantum mechanics. The twelfth fact is that the macht line is unique to those two entwined particles and separate from the background macht glow of the system allowing coherence to be maintained even when the line passes through solid objects.

The thirteenth fact is that the information for the entwinement is carried by the macht line which travels faster than light and since it feeds back into both particles disrupting the line allows for information to be shared at the speed of macht between the particles resulting in the observed phenomena of quantum entanglement. The fourteenth fact is that a disruption of the macht line is the sole reason for decoherence to occur in the event. The causes of decoherence of the macht line can be distance, strength of external macht glows, direct interaction with one or more of the particles involved in the entwining and time as a factor combining all three. This concludes the major facts that govern ESS and macht interactions with space as well as resulting phenomena like quantum motion and quantum entanglement. All of these arise from the fact that space affects matter as much as matter affects space and we can now continue to build on this fact and answer more of the unsolved problems of the universe. The first we will answer is what forces are and at a basic level what are the rules that describe their behavior as we observe them.

Chapter 20
Forces and Force Carriers, Fields, ESS Macht and Matter

Forces and Force Carriers

To begin with let us look at what a force really is as described by modern science both in terms of what it is and how it works to get a start on looking at what forces really are and how they really work. A force is first described as something that acts upon an object and changes its course of motion or properties in some way. To think of the simplest example think back to classical mechanics and the idea that an object at rest will stay at rest unless acted upon by an outside force. If you place a ball on a table and leave it alone it sits there for essentially ever. If you push the ball you start it rolling and the reason it rolls is because you have used a force on it that has changed its state from one of rest to one of kinetic motion. This basic idea of what a force is can explain any classical mechanical action for a given system.

It might be something like friction for example where the heat energy is generated by the force of motion input into the system as a rubbing motion of two hands together. It has long been known in cold weather that rubbing your hands together will generate heat and help to keep you warm but where does this heat come from? Is it from the chemical energy stored in your muscles that is then transformed into motion of your hands which generates the friction and therefore release of heat or infrared photons? Well yes and kind of yes. The yes part is true because it is indeed stored chemical energy in your adenosine triphosphate or ATP molecules that is used to make your muscles move when you break a phosphate bond and create adenosine diphosphate or ADP. However this is only kind of true because that is simply what powers your muscles and therefore hands to move back and forth against each other. The actual heat coming from the palms of your hands is not caused by friction force at a base level because friction is not actually a fundamental force of the universe. Here is where we need to look at forces as they are described by quantum mechanics which really is the study of the exact workings of classical mechanics.

The concept or idea of a force as described by quantum mechanics or the Standard Model of particle physics is that of the four fundamental forces. These are of course from weakest to strongest or farthest reach to shortest reach gravity, electromagnetism, the weak nuclear decay force or weak force and the strong nuclear or strong force. Gravity as we all know is the weakest force by many orders of magnitude and this is one of the unsolved problems of physics as it is a fundamental force but is incredibly weak compared to the other three. The electromagnetic force is as close as you can come to what one might call a medium strength force as it is much stronger than gravity but much weaker than the strong force; I include it second because of its long reach. Then we get to the weak and strong nuclear forces which reside solely inside the nuclei of atoms in the cosmos. The weak force is responsible for the decay phenomenon seen in things like Thorium, Uranium or any other radioactive atom and causes spontaneous decay measured in half lives. The strong force is what holds matter together in the nucleus of atoms and keeps quarks bound into hadrons and hadrons bound into nuclei.

The thing about the forces is they have different strengths and ranges which modern science cannot explain. Take gravity for example and see that it is extremely weak compared to the other three and yet gravity has by far the longest reach of any of the forces. The electromagnetic force is stronger than gravity but cannot extend nearly as far and you can think of the Sun's gravitational field as compared to its magnetic field to see that its gravitational influence extends far beyond its magnetic field. The weak and strong forces on the other hand are incredibly strong compared to gravity and yet have distances that can be measured in femtometers or the width of a nucleus of a single atom. What modern science does not understand is why the forces have different strengths and ranges to them only that they do have them. The facts that the ranges and strengths differ are well known and have been scrutinized intently by modern science for decades and no doubt remains that these differences exist. The exact mechanism at a universal fundamental level for why these things are happening and not just measuring them is unknown to modern science and the three main views of it.

To begin with let us look at the next part of forces as described by modern science which is the force carrier and we see that these are small particles that seem to interact with matter in specific ways. An electromagnetic force carrier is of course the photon and interacts with matter through the absorption and emission of photons by the nuclei of atoms. This is not the sole source of interaction but it is the best example to give as everyone reading this book is familiar with it whether they know this or not. In fact you could not read this book with your eyes if this were not so as the photons of light illuminating the page or screen you are reading this off of are interacting with the atoms of the page. The light spaces of the page are not absorbing many photons and reflect them back at you and this is of course in contrast with the dark ink that makes up the words which are absorbing numerous photons. This contrast between absorbed and emitted photons is what makes the words stand out from the page as the light parts of the page absorb the incoming light and then emit much of it back into the room. The dark ink holds most of its photons and you see these areas of the page as dark and all of this occurs because the atoms making up the page and the ink react differently to electromagnetic force carriers or photons. So the force of the electromagnetic part of the universe is carried by light or photons.

The weak force is mediated by bosons that have a much shorter range and exist primarily in the nuclei of atoms and will cause them to decay into things like, leptons or other hadrons. If we look at the strongest of the four fundamental forces we see that the strong nuclear force also has a force carrier and that is the gluon. It has a greater strength than any of the other forces and also one of the shortest ranges it is responsible for why everything in the universe in terms of matter is here as it binds quarks into hadrons which make atomic nuclei. The force of gravity is suspected of having a graviton particle which would be the force carrier responsible for keeping you firmly on the Earth's surface and the Earth orbiting the Sun. Without going into needless detail or giving countless examples of how these forces work we can simply note here that they appear to be utilizing force carriers to interact with matter and sometimes themselves. What this means is that the force carriers are not actually the forces themselves only the exchange particles for these forces and matter or energy. This leads to the other unknown in modern science concerning forces which is simply what is a force really?

Up until this point modern science only repeats itself by saying a force is a thing that does a thing to another thing but no where along the way does it ever sit right down and tell you what that thing is. What I mean by this might sound amusing but is completely serious as modern science will tell you what a particular force acts upon in terms of matter or energy and what the force carrier for that particular force is but it never says what the force actually is. The reason is simple, modern science does not know and this is illustrated by the lack of understanding of forces coming from the three main views. Classical mechanics we know deals more with collections of quantum events such that they can be calculated for macro scale systems using averages to explain everyday observations. To this end classical mechanics deals implicitly with the inner workings of the universe and we will not need to examine it in minute detail here.

Modern science however uses the two remaining views of the universe to try and understand that same universe and both contradict one another. Let us take for example the idea of quantum mechanics and look more closely at its idea of let us just say matter, force carriers and forces. Quantum mechanics and the Standard Model of particle physics does well in measuring things like matter and force carriers but does not understand forces or the fields they generate at all. General relativity does not understand particles but tries to describe one force or the field it generates which is gravity and again can only be used to measure but not explain what it really is. General relativity says that gravity is not in fact a force at all and is the result of a curvature of spacetime created by the presence of matter and that gravity is a perception based of frames of reference and motion over time. A dose of reality and humility here and that is since they are opposed to each other they cannot both be right so which is more right? Well quantum mechanics is of course more right by far although it still does not know what a force or the field it generates really is. Why is quantum mechanics the winner of the two views?

Simple quantum mechanics has explained the other three forces better than general relativity tries to explain just one. This is because quantum mechanics can accurately catalogue which forces have which forces carriers that work with which particles and the values for each. General relativity has created an illusion called spacetime that explains no interactions with any fundamental particles at all and this includes the newly discovered Higgs boson. The Higgs boson seems to give matter mass and mass is what Einstein said leads to gravity only the Higgs boson is a part of quantum mechanics and not general relativity. The other reason why general relativity is wrong is due to the fact that all the fundamental forces of the universe exist as real forces and not curvatures. For example if we look at the forces of gravity and electromagnetism we see that they extend outwards from objects with mass into space and can affect other objects. The difference is a curvature of spacetime cannot explain all the forces as it is not needed nor possible. Gravity is only theorized by Einstein to cause a curvature and only because his equations which are based off of classical mechanics from Newton which does use forces and not curvatures fit better gravitational observations. According to this then gravity is magic that does not fall into the domain of the Standard Model of particle physics and you must remember there is zero proof of any kind that spacetime is real or a true explanation of gravity it is simply the way Einstein imagined the universe to treat gravity.

According to Einstein and general relativity we must find and this is totally serious the electromagnetic-time curvature of space, the weak-time curvature of space and the strong-time curvature of space. The reason is simple the universe does not create different mechanics for different actions and the quantum mechanical view of the universe has proven this with measurements of the three other fundamental forces and their force carriers. Gravity is an invisible field that extends outwards into space around the Earth and so does the magnetic field generated by the Earth so why does the magnetic field which is carried by the photon not need an electromagnetic-time in order to exist? It behooves me to think that anyone seriously believes that spacetime is real and curves as a result of matter and that objects fall into and is the explanation of gravity such that no force is required. The reality is that quantum mechanics will find a graviton or other force carrier perhaps the Higgs boson is the force carrier needed perhaps not but it will find a force carrier for this force as it has for all the others. This is the reason we do not see a curvature of electromagnetic-time such that magnetically charged objects fall into it nor do we see a strong-time curvature of space such that quarks fall into it becoming hadrons.

The hadron is the proof of this concept in black and white as the attractive forces of the hadron are not a result of a curvature but of a real and measurable force carrier known as the gluon. The gluon is what holds hadrons together and this is a known fact meaning no curvature of a strong-time is needed. The same is true of the weak force and the electromagnetic force and can be demonstrated with the idea of electrostatic force. Objects holding charge have mass yet the strength of the charge is not dependent on the mass of the object or the extent of the reach of the field of electrostatic force. A proton and electron have wildly differing masses yet possess the same charge meaning their physical size does not alter how far their charge reaches. If we look at the proton as a collection of quarks, gluons and color charge mediated actions the way these combine to form a single proton with positive charge is far different than that of a fundamental electron. The two particles could not curve magnetic-time the same way and nothing can fall into anything else as a result because the two would simply be attracted together like opposite poles of a bar magnet and close the curvature becoming a neutron. This does not occur and so another mechanism is a work besides curvatures of space.

To put this another way if electromagnetic-time curvatures were real then things like bar magnets and the Earth's own magnetic field could not exist and would not flow as they would simply collapse into a central neutral region of charge. They do exist however and they do possess a well documented and measured force carrier which is described in the Standard Model of particle physics firmly ascribing them to a proper force. Likewise gravity will also be proven to be a force with a carrier particle that can be placed into the realm of quantum mechanics as the probability that this is not the case approaches zero at least until proven when it becomes zero. The list of proofs against forces being the result of curvatures is lengthy and we no longer need to flog this point here but what we now must do is look at what quantum mechanics says about forces.

Quantum mechanics is the clear winner between the two theories with the other being general relativity in terms of describing forces in the universe and accurately measuring three of the four fundamental ones. What quantum mechanics does not understand about forces is what they are once we stop measuring force carriers or the particles they act upon. To this end we return to the general non-existent description of what forces are as used by modern science which is simply invisible fields that extend wherever they need to get the job done. This is meant with no disrespect only an attempt to highlight the shortcomings of modern science when talking directly about the inner workings of fields and specifically what they are and how they work. The particles that these forces and the fields they generate act upon is well known and is not under dispute nor are the force carriers that mediate interactions between the fields and the particles. What is being questioned is this concept of an invisible field that breaks the known laws of physics such that it can reach where it needs to reach and interact with matter at great distance. To illustrate this we will look at the electromagnetic force, specifically magnetic fields and the force carrier photons for this force. The reason we will do so is that it provides a good balance between the strength and range of the field in question as well as a good understanding of the force carrier.

Fields, ESS Macht and Matter

To begin with let us look at the humble bar magnetic field as generated by any simple bar magnet or the Earth. What do we see? What we see is a flow of magnetism from the north pole to the south pole of any of these magnetic systems or objects. There is an invisible field that flows from one point in space or at one end of the magnet to the other end or point in space. Anything in this field is susceptible to magnetic forces as a result and you can see this with something like iron filings on a piece of paper as doubtless many of you have in school experiments. What quantum mechanics understands is the particles the field can act upon and in this case the atoms of iron in the metallic filings on the paper. Quantum mechanics also understands the force carrier for this action which is the photon as it carries both the electric and magnetic forces within it. If you used a magnet on a copper wire you could move the magnet and cause the electrons to flow along the wire and see much the same effect of a force carrier mediating an interaction between a field and a particle. What quantum mechanics does not understand is the invisible field that flows from the north to the south poles of the bar magnet or the Earth. It is after all the result of the photon is it not as it carries the electromagnetic force and so photons must be the field yes?

No not at all as photons are only the exchange particle between the force and the atoms that the force acts upon. The photons cannot be the magnetic field as this would require the magnet to emit photons from the north pole and somehow retrieve them at the south pole with no external help. This is a clear violation of the laws of physics as this would require an external force to act upon the photons to alter their course back towards the south pole. Remember no object spontaneously changes direction in the universe without an external force acting upon it. In order for the photons to change direction and move from north to south would require them to somehow thrust themselves as if they were thinking about where they were going such that they curved gently from the north to south ends of the magnet as traced out by the lines of iron filings. Photons do not have fins to steer, attitude controls or rockets to adjust their trajectory in space. In addition to this there is no actual physical stream of photons flowing from one end of the magnet to the other. So photons are not the actual invisible field which is the magnetic field in question and instead they are only the mediators of interaction with matter and this invisible field.

The Earth works the exact same way as it does not also produce a stream of photons that suddenly change direction from the North Pole out into space curve back down tens of thousands of kilometers out in space and then curve back up again and flow into the South Pole. This then is the heart of the problem concerning forces and the fields they generate in the universe as seen by modern science. Modern science does not know what these invisible fields are and how they work at a detailed level. What modern science has theorized is happening is that the fields exist simply as a part of space and that they extend everywhere at once and at every infinitely small point in space carrying all the necessary energy with them to be interacted with by matter at any moment and this is known as quantum field theory. According to quantum field theory the vacuum of space is made of these fields and these fields possess energy which can be exchanged with matter through force carriers to create a number of interactions such as iron filings on paper. This idea is actually a very sound one and certainly much better than general relativity in explaining the way the universe behaves but it is still incomplete and wrong in a number of areas.

The first area where it is lacking is simply the raw energy of the fields which leads to discrepancies between theorized vacuum energy and experimental measurements of the vacuum energy to create the infamous vacuum catastrophe which we will discuss later. The second problem with the theory is that it still does not explain what the fields are in space and what they are doing only that they get turned on or excited by interactions with matter. The third problem of the theory is that it does not explain things like directional flow of magnetic fields and the like as they would still only produce localized and balanced excitations according to quantum field dynamics. In other words no north or south magnetic poles and no reason to explain why field flow occurs at all. The fourth problem is that quantum field theory still does not explain why these fields come in different strengths. In other words if matter is matter and it interacts with space why does it not interact equally with space such that all the fields are the same strength or at least why is gravity so weak compared to the other three fundamental forces?

The overall problem that needs answering here in terms of what forces are in the universe and how they work is a proper explanation of how forces are transmitted through space and how matter interacts with space to create these forces. Think for a moment about how matter interacts with space to produce a force of gravity and not a curvature of spacetime and then see the need to apply this to the other three fundamental forces. Once again the universe is not trying to fool us and is simply telling us to look a little closer at the ever important interactions of space and matter.

So let us do just that and take a look at how space and matter interact at basic levels to produce forces and the fields they generate and along the way answer the basic questions of why forces are what they are. I will be using only a basic explanation here of space and matter interactions as what is needed is a build up of the facts such that observations can be explained as we see them in the universe today. A demonstration of the workings of these interactions such that they do explain observed events will allow for a later more in depth look at these observations. Therefore in order to start out we will look more closely at the properties of ESS and the macht they carry beyond just gravity.

First off space as we know is full of ESS and ESS carry macht which is the way they share energy between each other and between schales. We have seen that the ESS macht of gravity can be increased depending on how matter interacts with space and thus we know gravity to be one of the machts of ESS and therefore space. The ESS however contain other machts as well such as the electromagnetic macht which we will be using to explain why magnets have two poles, why field flows from north to south and why magnets exist at all for that matter. Gravity is a part of all matter and does not flow anywhere so why does only some matter have magnetism to a significant degree and why should it flow in a direction at all? For now however we know that ESS have different types of machts such as gravity or magnetism and machts are very different than forces. A force is only applicable to Standard Model non-space physics as it is incorrectly believed to be contained at every conceivable point everywhere in the universe all at once so it can be turned on at a moments notice by the excitations in it which are matter. This is of course what leads to the vacuum catastrophe and the ridiculously high zero-point vacuum energy and since we know this to be wrong so is the idea that fields behave according to the theories of quantum mechanics and the Standard Model of particle physics. What is correct however is that space does possess these properties often called forces or fields but the difference is they do not operate in any way described by the three main views.

The fields or forces or whatever you prefer to call them to familiarize yourself with the concept exist only in discrete packages inside ESS. The fields are not continuous and do not fill all of space but exist only at the precise placement of ESS. I will completely answer the problem of the vacuum catastrophe in a later section and demonstrate further why what I describe here about forces and space must be true. The use of ESS and the macht they carry are what make up what people call forces as the macht can be shared from ESS to ESS and follow mathematical rules concerning number of schales, strength of matter and ESS interaction and distance from the point of maximum macht for that force. This is why fields do not exist as theorized by modern science because they are only applicable to matter and energy and operate at light speed or slower whereas ESS and macht are not limited by those rules.

In addition traditional fields do not share energy between themselves nor do they create changes in the physical geometry of space and never do they incorporate actual energy transfers between space to matter and matter to space. Instead fields are simply fields according to modern science and matter is an excitation of the already existing field energy and strength. What this does is make the ways that fields can move and behave and more importantly the way these fields share with matter limited to zero possibilities. In order for the universe to work as we observe it there must be a free flow back and forth between matter and space in the form of ESS macht to explain why everything works as it does. A detailed explanation of matter and space interactions focusing on the particle nature of matter will follow and will prove these statements using examples of electron orbitals, black holes and more.

Here however it must be understood that force is not adequate to explain what is happening in the universe and that space and the ESS that make up space communicate and share between themselves using macht instead. Macht again is the correct terminology to use as it incorporates the idea of both space physics and matter into one model description just like entwinement explains what is called quantum entanglement rather than the matter only description of entanglement. So now we have space being made of ESS and ESS having macht and macht being the thing that carries what is known as the fundamental forces.

To look at why ESS and macht are needed to explain fields we will use the idea of a magnetic field similar to a bar magnet or the Earth's magnetic field for reference. First think of a bar magnet lying on a table with iron filings spread around it and the field lines visible. What you can see is a series of lines radiating out from the north end of the magnet towards the south end and this occurs anywhere on the magnet not just the very poles. So a field line can originate around the midpoint just above it and then make a very short arc back towards the midpoint and enter the magnet just below it. The reality is that magnetism is always and immediately flowing from all parts of the north end of the magnet to the south end and this happens in a three dimensional shape. The two dimensional shape illustrated on the paper is just for easy viewing using the iron filings. Now we must look at not the field but the magnet itself as obviously it is what is generating the field.

First of all we notice that the magnet is made of one solid piece of matter and does not suddenly change its composition at the midpoint. In other words a magnet is not two different materials such that one side makes a north pole and one makes a south pole. This is important because it removes any extraneous ideas concerning the creating of two poles and the flow of the field and places all the field generation on a single continuous piece of material. This raises the very obvious question of if a magnet is uniform throughout why should it produce non-uniform poles at all? The reason magnets can generate a magnetic field is due to how their electrons spin inside them or at least how some of them spin.

If an electron is unpaired it means its spin is up or down but obviously is only one or the other. So that electron is not balanced by another pairing electron which would have the opposite spin. The result is a single electron with a spin in one direction and in a magnet the material has its atoms arranged such that the multiple unpaired electrons inside of it are all aligned in a single direction. This alignment of electron spin is what creates the magnetic field to flow in one direction but does not explain why this occurs at all and why the field changes direction back towards the other end of the magnet. All this does is provide a mechanism for explaining why some materials produce a magnetic field and that is all. To explain why a magnetic field flows from north to south it is necessary to understand the ESS macht of the magnet and what it is doing to the surrounding schales, macht and local space.

The first instance of the magnetic field begins with the movement of the particles in the magnetic material and in this case we will say the unpaired electrons. Under normal conditions the materials found in the magnet do not possess magnetic properties as one might expect for a bar magnet but the spins of the unpaired electrons are aligned such that they produce a flow of magnetic force inside the material and this turns on the magnetic force inside the bar. The second thing that has happened is that these unpaired electrons have of course interacted in some way with the ESS of the local area of space. Just as matter gives ESS macht gravity more strength the electrons give the ESS macht magnetism more strength. We have seen that gravity is caused by two factors the first being macht glow amongst schales and the second being spatial compression of ESS.

In magnetism the latter does not need to occur although it can be amplified in the presence of more ESS such as in a neutron star, magnetar or black hole. What does occur in magnetism is an increase in the strength of the ESS macht that is responsible for what is seen as the invisible magnetic field. Just as gravity is an invisible field that acts instantaneously on matter meaning that if the field strength is increased for a local area of space the effects are felt throughout the field immediately from the point of the increased mass to any observer mass at distance from it and these changes are felt faster than light speed. So are changes in a magnetic field felt instantly such that the magnetic field reaches out invisibly to an observer object and attracts or repels it.

At any distance that an observer object is in the field it will interact with the field through the usual photon force carrier. Objects themselves cannot move faster than light in the field because they have mass and are bound to ESS macht for other reasons. Light speed and properties of light will be dealt with later as well as the explanation for why matter cannot travel at the speed of light using quantum level explanations of space. The magnetic field however is invisible and reaches out instantly to any object in it and is made of an influx of energy from the electrons. The electrons are particles that can easily interact with the ESS macht magnetism of space and do so inside the matter of the bar magnet increasing the ESS macht through their aligned spin property. Some particles will interact with various fields as will be described later in the matter section however what is important here is that different particles can interact with different ESS machts. This should be fairly obvious as all matter can interact with the gravity macht but not all matter interacts with the electromagnetic macht. Matter is not equal to all forms of other types of matter and matter is all a little different and when coupled with the arrangements of complex particles like hadrons or multiple hadrons and electrons will produce various types and degrees of interaction. This property of the electron is what occurs between matter and space and explains why a field is generated in the first place.

◄New Universe Order►

Now we must make a distinction between invisible gravitational fields and invisible magnetic fields as they both share many properties but clearly behave differently. To start off with gravitational fields do not flow as a magnetic field does and the idea that gravity flows or that space flows in a direction of a curvature and into a gravity well is flatly wrong. Space or gravity flowing from anywhere means that a limitless source of these two resources must exist somewhere else for them to be coming from and going to. On the other hand if space and gravity are finite then the universe is happily ripping itself to pieces all around us as gravity and space flowing into our solar system or a black hole or any other object must be taken from somewhere else. This is again of course impossible as space and matter are finite and therefore gravity in the universe is finite and destroys the concept that space itself is flowing into objects like black holes as theorized by general relativity. Remember that space holds the zero-point vacuum energy according to quantum theory and general relativity would therefore be eating this energy which according to special relativity is equivalent to mass and therefore any one black hole would have long grown to infinite mass and consumed the universe. Lucky for us the universe is still here and is clearly not being ripped to pieces everywhere we look and so yes it is proven that gravity and space do not flow anywhere.

Gravity is generated from a single central point and is a property of all matter so things like spin cannot affect it like a bar magnet. Magnetism however is different and can easily be made to flow and this fact has been understood for centuries. This brings up the interesting fact about magnetism that is the ESS macht that controls it can be made to flow and in a manner that requires no strange changes to flow but naturally moves from north to south poles of a magnet. In other words the idea that a magnetic field is under some sort of external force changing its motion according to the laws of physics is not required. Magnetism will naturally flow and move from north to south poles but only through a macht sharing explanation of space using ESS.

To begin with let us look at space with the idea that the local Area ESS have a certain level of macht naturally contained in them. Space as we know is full of ESS and these ESS have macht, if we remember back to our original little room filled with the hollow ESS spheres and the mostly opaque matter orb we can learn more about macht. To start with ESS always have some macht in them and it does not take the inclusion of a matter orb to the room for the macht glow to begin and be shared amongst the ESS. The ESS have macht already in them it is only the addition of energy donated from the matter that amplifies the strength of the macht that we see as an increased macht glow. So when matter of the right type is made to interact with ESS it can impart donated energy to different parts of the ESS macht in the local area of space. This explains and is proven why matter that makes up magnets and non-magnets can be made of the same particles and yet equally contribute to gravity macht but not magnetism macht. Only the behavior of certain particles in certain types of matter can increase the macht magnetism glow of that local area of space and we see this with the aligned spins of unpaired electrons in magnetic materials.

In the case of all ESS a certain level of macht remains no matter what you do to it and if we remove the matter orb from our room the macht glow of gravity decreases but does not go away. We can think of this as a macht pool in the ESS of the local area of space such that all the ESS in that local area of space seek to balance out the macht they carry and create a uniform macht glow. You have already learned how macht travels outwards in schales from the ESS that are in direct contact with matter and learned that attempts to balance out the macht travel a great distance for gravity but not in a way that makes all the macht glow uniform in the room. A uniform macht glow can only exist in the room when all matter is removed and all ESS rebound to their relaxed state. In the case of magnetism it is a direct consequence of this balancing of macht found in the reserve macht pool of the ESS of the local area of space that creates the magnetic field and allows it to flow from North Pole to South Pole of any magnet.

Now let us examine exactly how a bar magnet works using the two factors of electron spin in the magnets material and the ESS macht pool and its attempts to balance macht glow of magnetism. To begin with a bar magnet or the Earth's magnetic field for that matter will always generate a flow of electron spin in one direction as this is the basic mechanism by which magnetic effects are generated. The electrons in the bar magnet for example all align in one direction and cause a flow of energy from south pole to north pole. The electrons are capable of interacting with the ESS macht magnetism of the local area of space and will impart their energy to it. This in turn rather obviously increases the strength of the macht magnetism for the ESS in contact with the bar magnet material. The matter inside the bar magnet is a quantum physical action that will physically push magnetic energy up the bar magnet and move it towards the north pole. Since this is a physical property of the material of the bar magnet energy donated from the matter to the ESS macht will always be moving in this direction.

This is an example of the increased magnetic force from the electrons in a sense tracking to new ESS macht magnetic if you will. The result of this motion in the local area ESS is that the macht glow of magnetism in the bar magnet tries to be shared with the schales around it as usual. Since the flow of quantum energy from the electron is coming in a set path physically from the matter interacting with space the schales cannot share macht backwards inside the magnet and so you have macht sharing only moving towards the north end of the magnet. The highest concentrations of increased macht magnetism are at the end of the north pole of the magnet as this is the terminus point for the unpaired and aligned electron spins in the matter of the magnet that are physically contacting the local area ESS.

However this is only the matter side of the system and for the ESS sharing of macht begins immediately inside and in the schales around the ESS interacting directly with the material of the magnet. What this creates is a higher concentration of macht magnetism in the ESS around the north pole of the magnet and the ESS in the south pole are also attempting to balance the macht glow of magnetism so they are decreased in macht magnetism strength. If you think of the macht glow in the ESS around the bar magnet it will be brighter at the north end and dimmer at the south end. This is what causes the field to flow or move out of the north pole and also why the field then flows backwards to the south pole. The macht at the end of the matter making up the bar magnet at the north end is now free to balance macht through the schales of space that are not interacting with physical matter and only now can the free movement of macht occur in space. Space as we know is made of ESS which permeates the entire magnet and all matter that makes it up as well as the space around the magnet.

Therefore the ESS which have an over-represented macht of magnetism in the north end will begin sharing that macht with neighboring ESS and this means the depleted ESS in the south end. So we can again use the rule of sharing macht between schales to explain the flow of magnetism in a bar magnet the macht in the north end ESS are seeking to balance out with the closest depleted macht which is found in the direction of the south end. Since the ESS touch each other and since they share macht a direct contact mechanism of ESS can explain how the field curves through space. This is because it is not really curving in a perfectly smooth line but making controlled and discrete jumps along the spherical three dimensional ESS geometry that is space.

I talked earlier about the idea that a field cannot change its direction or that a photon as the electromagnetic force carrier does not have little attitude thrusters on it to alter its course from going north then turning 180 degrees and heading south again and by using ESS you do not need these improbable scenarios. The ESS are small and touch one another so the macht glow is shared amongst ESS schales as it would for any other force like gravity and can follow a geometric pattern backwards never actually requiring any external input to alter the direction of field flow. The two north ends of two bar magnets have increased macht glow of magnetism and so repel one another because the ESS inside and around their local areas of space are trying to balance out the over-represented macht of magnetism they feel. Two south poles are again trying to balance out their respective areas of local space and the ESS macht glow of magnetism they carry by repelling each other as they try to increase the macht in their depleted ESS. If the south poles were in contact they would block the incoming macht magnetism they desire and so repel each other to allow the macht to enter their local areas of space.

Magnet repulsion you feel in your hands is the result of ESS trying to balance out macht and magnetic attraction is the end of one magnet using the macht imbalance of another magnet to its advantage to balance the ESS macht increase or decrease of macht magnetism it is creating itself. Bar magnets allow you to feel space right in your own hands and feel the quantum interactions of matter with the ESS macht of space. You are actually feeling macht and not anything made by the three main views of modern science or anything catalogued in the Standard Model as a particle or force carrier.

This explanation of how magnetic fields work is only one use of ESS macht pools for explaining how the universe works and later sections will utilize them much more to answer more of the unsolved problems of modern science. For now this basic understanding of how ESS macht generates fields is sufficient to build on from here. A perfect example that we have already looked at is the force of gravity in its base form under normal spatial compression. In normal spatial compression as is found here on Earth matter can only interact with and therefore donate so much energy to ESS macht of gravity. The result is a very weak gravitational force and if one were to look for a stronger field then a black hole would give you one as well as an explanation for why gravity is the weakest force but not always so. The way matter chooses to interact with space in this section and generate fields has also been touched on in the example of an electron and bar magnet but later sections will deal with the most specific interactions of matter and space such that all the minute details of ESS macht might be explored further. In addition to this I will explain far more about what matter really is, how it behaves and how its actions are controlled by the ESS macht of its local area of space. To end this section and the basic introduction to fields we will list the pertinent facts concerning space and matter interactions here through ESS and ESS macht.

Facts

The first fact is that ESS possess more than just one type of macht and these can include things like gravity or magnetism. The second fact is that not all particles can contribute in significant ways to all types of ESS macht. The third fact is that ESS macht glow can be made to be directional if certain specific conditions are met such as aligned electron spin in magnetic materials. The natural tendency of ESS to balance their shared macht is fact four and leads to a flow of magnetism from north to south magnetic poles. The fifth fact is that macht will be shared by direct contact of adjacent ESS touching each other and can create a discrete geometric pattern of macht glow that explains the shape and directional changes of field flow for magnets. The sixth fact is that ESS possess and try to balance out their macht pools which is responsible for magnetic field generation as well as other space and matter interactions. The seventh fact listed last in this section is that ESS have pools of reserve macht to draw upon and are never completely drained of these pools. This fact is the last one listed for this section despite being mentioned earlier in the section because this fact will play in directly in the development of the answers of the future unsolved problems. While we are at it why not use this ESS macht pool fact to answer once and for all why matter is more abundant in the universe than anti-matter and why this was always going to be the case and is perfectly normal for the development of the universe from the first instances of the big bang.

Chapter 21
Charge, Electron Drop Down, Matter over Anti-Matter

Charge

In this section I will explain why matter is more abundant than anti-matter in the universe and solve the baryon asymmetry problem but first we need to examine a couple of things before that. The two things we will be looking at are the types of matter that interact with space and why they must do so in certain fashions. This will lead us to an understanding of charge in particles such as electrons, protons and of course positrons which we will deal with in the anti-matter discussion. The second thing we will look at is the solution to another unsolved problem in physics which is why electrons always drop down to their lowest energy orbital states if left alone in a system. By examining these two topics first we will be able to lay down the necessary rules for why normal matter is dominant over anti-matter in the universe.

So let us begin with looking at how particles interact with ESS and produce increases or decreases in certain types of macht. Firstly all particles are unique in how they interact with the ESS of the local area of space they inhabit and this is reflected in the differences in things like charge, spin or mass. An electron for example does not interact with the Higgs field in the same way as a quark or hadron and thus we see different masses for these different particles. Similarly we note that electrons and positrons do not interact the same way with space as one is negatively charged and one positively charged despite them being identical otherwise. These simple examples continue for every known particle and demonstrate how unique each of them is compared to the rest of their family. In this section I will be dealing mainly with electrons, protons and positrons and the property of charge as it relates to ESS macht and the workings of the universe. Other particles and their properties will be dealt with later. For now let us look at an electron and note that it has a negative charge which must be explained.

Charge for example is poorly understood by modern science beyond measuring it and noting its various forms of interaction with other objects. What charge is and how charge manifests itself in the universe is not understood by the three main views and must be addressed first before we can move on. Charge comes from space and not from particles and this is made clear by looking at a fundamental highly charged particle like the electron. The electron has a negatively charged field around it that extends for a great distance as compared to the size of the electron itself. The field is as we know made by the ESS macht glow of negative electrostatic charge. The field is not made of photons at all as this is massively impossible and utterly and completely wrong. Why is this so wrong?

Simple, fields of all kinds extend beyond the range of the object that generated them and the photon travels in a straight line through space meaning it cannot account for the actions of a field. This is easily proven by looking at an electron orbiting the central nucleus of its atom. The electron can be viewed as simply moving in a circular orbit for this example and if we take its initial position to be at 0 degrees and its motion is clockwise then we can prove that the field between it and the nucleus is not made of photons. Photons are not exchanged between these two particles to mediate the forces found their and keep the electron at a set quantum distance from the nucleus. If photons were responsible for keeping electrons away from the nucleus as solely explained by quantum mechanics then this means that as the electron emitted a photon at 0 degrees towards the nucleus and then continued towards its new position let us say at 45 degrees it should in fact fall back into the nucleus as no photons are there to hold it up and away from the center of the atom. This is of course not what happens or electrons would be colliding with nuclei all the time and destroy the atomic structures of any molecules in existence. Even multiple photons cannot solve this problem as a perfect stream of photons one touching the next cannot exist as this would block the return photons from electron to nucleus and nucleus to electron.

There is a second problem with this idea that photons mediate the force between the two objects and this is a problem of aiming. If we say that the nucleus is essentially stationary as compared to the fraction of light speed motion of the electron then it might be possible for the electron to perfectly fire a photon towards and hit the nucleus. However even this is difficult as a single atomic nucleus is roughly the size of a grain of sand compared to the volume of the atom itself as generated by the outer electron shell if the entire atom were the size of a stadium. Now begins the true failure of the photon as repulsive force between electron and nucleus and this comes from the return shot that must be made by the nucleus to the electron. The nucleus is thousands of times more massive than a single electron and if the electron hitting the nucleus was hard enough then the nucleus hitting the electron is close to zero probability. The grain of sand that is the nucleus needs to be cut into a piece several thousand times smaller than itself to represent a single electron. The analogy of hitting a bullet with a bullet is grossly inappropriate for an atomic system as even that difficult feat would be child's play compared to hitting a single electron with a photon fired from the nucleus. This believe it or not is not the biggest flaw in the photon theory and instead comes from a second problem of aim beyond simple accuracy and that is anticipating and leading the target.

The electron was last at 0 degrees and has now moved at a fraction of the speed of light to 45 degrees around the nucleus. What this means is that it is in motion and constantly changing its position in space from moment to moment so it is not a stationary target like the nucleus. In order for the nucleus to correctly fire a photon back to and hit the electron at its new 45 degree position in its circular orbit the nucleus must be intelligent and thinking. This is not meant as any kind of joke but is dryly serious as one must come face to face with the realization that no particle in the universe can think and therefore anticipate the electrons new position at 45 degrees. In addition to this the photon must be fired earlier than the electron arrives at the 45 degree position because photons have a limiting speed of the speed of light and do not travel instantaneously through space.

Like any normal bullet they are slow and must be made to lead the target so that the shot arrives in the future where the intended target will exist. As I just said in the future position which a particle cannot know, calculate or anticipate in any way. It is impossible for any particle to know or guess where another will be in the future and therefore compensate for travel time of the projectile by aiming ahead of the targets anticipated and not known trajectory. This is of unbelievable importance to the workings of the universe because it flatly destroys one of the central ideas of atomic mechanics as held by the three main views. This idea was already touched upon when we learned how magnetic fields can be made to flow without the actual field changing direction as if being controlled or acted upon by external forces.

Do you want to destroy this photon carrying nonsense one more time? No worries all we need to do is introduce two more atoms to the system one to the left and one to the right of the central and original atom and squeeze all three together. In our first example the only thing we had to examine is the electron and its motion around a central nucleus but now we can see what happens and how photons fail as we squeeze this first atom between to identical ones on either side of it. To begin with the electron is a single particle and does not exist in more than one location at a time despite certain quantum mechanical beliefs. This is rather obvious as two atoms demanding the action of the central atoms electron at the same time by means of resisting the squeeze force through electrostatic repulsion would mean that the electron must exist in two locations at once in order to push two atoms apart. This would create matter from nothingness as the electron and all of its properties would need to be duplicated in zero time in order to repulse both of the other atoms simultaneously as each electron is a quantized particle that exists and interacts with other particles in only a single set amount. Upon completion of this task which electron is destroyed and where is it destroyed to? Where does the extra energy held in the destroyed electron go? What mechanism can create a new electron without altering the total conserved mass or energy of the system at will and always precisely where it needs to be? How many more impossible conditions and extra variables must be created to make something like this work?

The answer to the last question is lots and by now this short list of obvious problems with that idea should be apparent now rendering the possibility of multiple simultaneous positions of the electron or any particle in their probability fields to be quite wrong and needlessly magical and complex. Remember the electron is only a single particle without the act of measurement it is still a single particle so all probabilities are not needed as the universe will allow the electron to exist without the act of observation solidifying one into existence for our convenience. The problem of also now firing two photons between electrons of the three atoms should also be made clear as there was only one exchange photon to begin with moving from the nucleus to the electron. In this new squeezing example we have three photons being required one to keep the original electron aloft of its parent nucleus and two more keeping the electrons from the other two atoms at a safe electrostatic distance. The problem of photons being the repulsive force for electrons orbiting atoms is made worse by the fact that they cannot generate spherical fields and exist in point form only.

Take for example the nature of an electrostatic field around a nucleus and we see that it is a rough sphere and this cannot be simply attributed to the fact that the electron moves quickly enough as viewed by a human that it seems to appear at every possible location and probability in space at once. The reason is that it only appears this way to us as humans or other intelligent creatures that exist at macroscopic timescales like us. To the creatures existing in a quantum system such as an atom the particles move at the same speeds as each other since they all move so quickly. In other words the electron is very slow moving around the nucleus and cannot fill all the probabilities of space at once. What this proves is that the exchange of photons from a central nucleus to an electron and back again even if this was at all possible would create not a large spherical structure like a stadium with a little grain of sand at its center but instead create a grain of sand with a long thin gossamer thread of electrostatic force reaching out to the several thousand times smaller electron grain. This would mean that the atom has essentially zero volume as compared to a sphere whose radius is equal to the orbital distance of the electron and therefore produce no useful spherical electrostatic field of repulsion. In other words you cannot use ever an exchange mechanism of photons to explain how electrostatic fields and repulsion works with particles in the universe.

The way that you do explain how a field can be spherical and exist in every possible location of space at once around that particle without the need to replicate the original particle at all and not need a future knowledge of where to send something like a photon is to use ESS macht glow. A particle as you have learned by now interacts with ESS to increase certain types of macht glow in that space and the macht is shared amongst other ESS through the use of schales. If you examine an electron then we see that it is interacting with the ESS of space to increase macht gravity and macht electric charge. The ESS macht will naturally spread out in all directions as it is shared in ever increasing schales and since these schales exist in a roughly spherical form then we can create the spherical repulsion field of the electrostatic force. If this seems simple then good it should be as the universe works only in the simplest ways and complicated photon exchanges between electrons and nuclei are not simple, they are impossible.

The addition to this is that a macht glow being responsible for the field generating solves all of the associated problems of creating a spherical field that exists everywhere without copying any particles from nothingness or expecting them to magically exist in ever possible quantum mechanical probability space at once. To start off look at how particles exchange energy with ESS as we learned earlier and remember that the total conserved energy for a macht/energy system of space and matter is conserved. This means that the force felt by anything attempting to encroach on the local area space of the particle will always feel an equal push away from that particle no matter where in space it is. Think of a central electron and a second electron trying to move into the space of ESS macht glow belonging to the first. No matter what point the second electron tries to enter the field the force it feels will be equal as the glow is being shared from a conserved amount of macht/energy of the first electrons space and matter system. This means the electron does not have to exist in multiple locations at once in order to fend off would be invaders to its personal space nor does it have to create extra copies of itself which would violate known laws of physics by creating matter from nothingness. Instead a simple ESS macht glow will create the necessary field strength and geometry to account for a single particles charge.

Likewise a single electron can now power the entire space around the nucleus without the need to do anything strange in order to prevent the two side atoms from squeezing into its space. If we move on with this fact we can also explain how electrons stay in orbit around nuclei without the use of photons. More importantly we can answer a problem for which modern science has no answer which is why do electrons not simply crash into nuclei? This is a puzzle which modern science does not understand as an electron is negatively charged and a proton is positively charged and so great they sound like a perfect match for each other let us introduce them to one another. They never tend to meet up though despite being attracted to one another and this makes no sense. If I have a positive and negative end so to speak of a magnet they will be very strongly attracted to one another. If I have a positive and negative object in a macro scale system they are certainly attracted to one another but why not electrons and protons? If we look at the simplest atom in the universe the hydrogen atom we see a massive problem occurring as one electron should be attracted to and collide with one proton yes? If we look at any atom for that matter this should be true as electrons are negative and nuclei are positive so why are they not physically colliding together and remaining stuck as one if they like each other so much?

To answer this question it is time to learn something else about atoms which is they take in photons to increase their total energy which pushes their electrons out to higher energy orbitals. Modern science has long understood that if a nucleus of an atom absorbs a photon the extra energy it absorbs will cause the electron or electrons orbiting it to make a quantum jump of a set amount to a higher level orbital. The reason this happens is not understood by modern science as it only knows what causes the jump up which is the absorption of a photon by the nucleus but not why it jumps up and modern science certainly does not understand why the electron jumps back down again. To solve this problem we will first look at the proton as a single positively charged particle and then tell you it is not really positively charged. Here is another fact of the universe and that is there is no such thing as positive charge just as there is no such thing as cold. Cold I have already explained does not exist in the universe as only heat exists and in various amounts. When a human says something is cold they just mean to say it has less heat than something else which is usually them. In reality however cold does not exist and instead you have systems with less heat. The same is true of charge in the universe as the universe only has negative charge in it which is good because two charges once again overcomplicates things. So how can we make two opposing apparent charges from only one real fundamental charge?

The answer lies in something we discussed in the previous section and is the macht pools that exist in all ESS. ESS everywhere in the universe contain pools of all the different macht and we have already learned that matter affects those pool levels. If we look at an electron it is fairly obvious that it is donating energy to the macht pool and therefore increasing the macht glow of charge in that local area of space. This is increasing the negative charge of space and we understand that perfectly. How a proton appears but does not actually have a positive charge is different. A proton utilizes the ability of space and matter to interact with one another and alter macht levels as well but in this case the proton does not donate energy to the macht of the local area ESS but does the opposite. A proton is a type of matter configuration that will instead of donating energy to be used as macht will draw off macht from its ESS to be used as energy. This is another clear example of space affecting matter as much as matter affects space.

Where these energies go I will not explain here as I will continue to explain charge and how it explains the baryon asymmetry problem of the universe but I will be explaining nuclear macht and energy interactions later. Here instead we see the proton taking energy from the local area ESS macht pool of charge and using it for itself. What this does is utilize the ever active property of ESS to balance out macht amongst themselves and the local area of space around the proton becomes macht charge depleted to a degree. This is made possible by and proves the existence of the ESS macht pools that are used in other interactions of space and matter. If we look at the area of space around the proton it is similar to a cold area of space which is not really cold only possesses less heat. So too does the ESS macht of charge in space around an atom have areas of more or less macht charge and examining the amounts of charge and the ESS macht pools available to the local area of space we can explain why electrons do not collide with nuclei.

An electron possess elevated levels of ESS macht charge and a proton possesses decreased levels of macht charge but it still does not have the ability to completely deplete the total ESS macht charge pool for that area of space and so some remains. What this means is that an electron and the local ESS it inhabits will try to balance the elevated macht around it with the local ESS around the proton and the decreased macht around it. ESS always seek to balance out their ESS and so the two local areas are naturally attracted to one another. The electron is prevented from colliding with the nuclei however because the local area of space around the nuclei which is denser and more massive than that around the electron still retains more than enough macht pool of charge to prevent the electron from fully merging with the nucleus. A nucleus is always thousands of times more massive than a single electron and so the number of ESS it inhabits and the strength of energy held in macht pools in those ESS over power the electrons own ESS macht imbalance and prevent it from coming all the way into the nucleus and remaining there. Upon reaching this area of negative charge around the nucleus the electron will then be repelled back into space and will not fly free due to the fact that it is still attracted to the ESS macht glows at the nucleus.

What this does is reveal why electrons have rest states whereby they cannot drop to lower energy levels as this would mean overcoming the ESS macht pools of the atomic local area of space. Overcoming this is what and why electrostatic repulsion occurs and fails given certain conditions of the universe or laboratory. What it does mean is that by adding a photon to the nucleus of the atom the total electromagnetic energy of the nucleus increases and therefore the ESS macht charge of the local area of space of the nucleus increases always and is reflected in a new quantum orbital shape. Since we know only negative charge to exist in the universe then it is obvious that the net quantum amount of energy brought into the nucleus will increase the ESS macht charge for the nucleus by a measureable quantum amount. Even though macht is not quantum and is governed by space physics the energy input in total to the macht/energy of the system is still quantized to whatever energy the photon possessed and so cause an increase in overall negative charge repulsion of the ESS macht of the nucleus of the atom. This set amount is therefore donated to the macht and then shared through schales to increase the volume and strength of the macht glow.

The electron and nucleus are now more repulsive to each other because the local ESS macht for the system are no longer as far out of balance as they were before the addition of the photon. If you keep inputting photons into the system you will move the electron farther and farther away from the nucleus until you reach a threshold point. That threshold point is the point where the nucleus has gained so many photons and increased its negative charge so much that the electrons and nucleus no longer are attracted to each other through the ESS macht balancing property and the electron can now fly away from the nucleus. This is the exact quantized point where the macht pools of charge from the electron and nucleus are now balanced and so the electron no longer tries to balance its macht glow electrostatic with the depleted macht glow electrostatic of the nucleus. The electron will then move off into the universe by utilizing the same quantum motion tracking rules as laid out before.

The relationship here is charge and distance of the two ESS macht imbalances and the total macht pool for that local area of spatial compression and matter interaction to macht glow. Spatial compression or how much matter and to what degree it is imparting energy to the system will determine what the total charge and distance is of the ESS macht imbalance and at what point the electron will move away from the nucleus. This explains why at high energy levels photons can force electrons away from their parent nuclei and move freely as was seen after the Big Bang before the era of recombination or first light or inside the energy rich and laden cores of stellar plasma. This mechanism explains why atoms do not collapse into themselves due to their possessing negative and positive behaving particles and why additions of photons to a nucleus will cause the electrons to move to higher orbital energies.

Electron Drop Down

It also leads us into explaining electron drop down and electron ground state orbitals. Another aspect of this fact of space and matter interactions is that it explains finally why electrons spontaneously drop down to lower energy orbitals with apparently no external inputs. If you look at a blade of grass you see green color to its leaves and this is because you are seeing a green photon emitted from it and reaching your eyes. This photon was emitted because a photon of light entered the atoms of the blade of grass and excited one of its electrons to a higher orbital which then dropped back down to its original orbital and emitted the green wavelength photon that you see. The extra electromagnetic energy from the incoming photon is not lost just converted for example to heat as an infrared photon or in the process of photosynthesis into making more grass. The reason the electron decides to drop back down is unknown to modern science and can be explained simply using the ESS macht balancing of all systems in the universe.

The photon that is absorbed by the blade of grass increases the ESS macht charge of the nucleus by a set amount such that the electron jumps to a higher orbital energy as we have seen previously. The reason for the drop down is determined by the factors I listed above that affect those same interactions. They include things like local area spatial compression, amount of matter and degree of matter interaction with the ESS in the local area and the natural property of ESS to balance out its macht and macht pools. To this end the local area of space under the spatial compression surrounding the system is such that it can only support a certain amount of ESS macht charge in pool form. ESS are always balancing their macht and so continue to do so after the addition of the photon to the system. The higher charge value of ESS macht charge for the nucleus is now imbalanced with the ESS macht charge of the electron and of the matter in the local area of space which after all contribute to the total ESS macht balances of the space of the grass blade and beyond. So a neighboring particle has not changed its local area ESS macht and is resisting the new imbalance from the particle that did absorb the photon. In order to balance out energy the ESS macht of the atom that absorbed the photon the local area macht imbalance of the electron and the parent atom must be resolved.

Since the local properties of space mentioned like spatial compression, and amount of matter and degree of matter interaction have not changed this means the total macht pool has not either and the local area of space has no room to hold onto the extra photons worth of energy. The extra macht must be removed from the system at any cost and the simplest way to do this is to balance out the ESS macht glow of the electron with the elevated ESS macht glow of the nucleus all based off of balancing negative charge. The result is the extra macht is donated back out of space into the form of energy in a photon and emitted to once gain travel through space. The atom that absorbed the photon has now balanced through natural ESS macht mechanisms the total macht pool of its local area of space and therefore the electron drops down to the original orbital energy it was at before the photon was absorbed. This explains the electron drop down problem seen in all atoms in the universe and explains why no external inputs are required rather it is a natural property of the space and matter interactions of the atom itself. It also demonstrates why more highly energetic systems can have electrons moving up orbitals until eventually they are lost from the atom.

Now that we have covered these basic interactions of space and matter the answer for why normal matter is dominant in the universe over anti-matter can be explained and accounted for as a natural property of the universe. Let us first start off by explaining what a positron is as an example of anti-matter. A positron is a fundamental particle that is exactly the same as an electron except for its charge which instead of being negative is positive. A positron has all the same other properties of an electron such as mass and spin in fact many consider anti-matter to simply be the mirror image of normal matter. This is partly correct as normal matter and anti-matter are very similar and an electron as we have seen will produce a negative charge by increasing the ESS macht charge of the local area of space whereas a positron will decrease the ESS macht pool by as much. The workings of an annihilation event are exactly the same as those of the attraction between an electron and a nucleus but with one important difference. Since the mass and composition of the matter particles involved are identical there is nothing to stop the two from colliding into each other which triggers the annihilation event.

Remember it is the macht pool of charge possessed by the much larger nucleus that maintains some macht charge pool to repel the electron before collision can occur. The motion of the electron is determined by the rules for space physics that explain quantum motion as seen in a previous section and coupled with the electrons ever attraction towards the nucleus. This causes it to continually orbit looking for a dip in the ESS macht balance through which it can sneak in but since macht speed is so much faster than light speed this is impossible as the ESS macht glow easily beats the electron before it can meet up with the proton.

Another part of the electrons motions is due to the fact that as the electron moves so does the macht glow that it creates as the center of the electrons macht glow is never truly centered over the exact center of the nucleus. The tracking delay I explained earlier between matter and space creates a changing motion for the electron in terms of its own macht charge glow that it must in turn attempt to balance with the nucleus macht charge glow. The combined effects add yet another factor to the calculations of the complex and not random motion of the electron and represents the final piece of the puzzle in terms of finally calculating a particles position and momentum fully which in this case is that of an orbiting electron.

Returning to the electron and the positron then no such barriers exist to the motion of the two particles and so the ESS they inhabit naturally balance out their macht between them and cause the two particles to meet at the exact same point in space. Think of the matter as having an up arc to a sine wave and the anti-matter having a down arc to the sine wave in terms of charge and you can see how when they meet the two arcs cancel to a flat line. In this case the point will always be the ESS with the highest mutual macht glow and of course there are no hypothetical points in space only precise and discrete ESS. What occurs during the annihilation event is well studied as both particles will attempt to in a sense cancel each other out during what can only be described as an explosion of sorts. The differences in how each particle affects changes in the local area of space and the macht pool there means the trouble of hitting a bullet with a bullet is gone and instead the two bullets are actively seeking one another out to have a perfect head on collision. This is the most accurate description of what is well documented to occur between the two particles according to modern science. In actuality what happens is space finally balances out the ESS macht imbalance it has been under since the two particles were brought together. As we have seen matter interacts with space such that it is bound to ESS and this is seen through the sharing of energy from matter to macht or ESS or from ESS macht to matter.

When two particles annihilate they do something normally not seen in ordinary life which is they completely balance out the ESS macht they possess due to them being exactly the same in all respects except for charge. The ESS macht in that local area of space is now fully balanced and this means it is balanced without input from matter which is something that has not occurred since the big bang and shortly afterwards. In fact an annihilation event is probably the closest science can reasonably come to create areas of space which would fit the hypothetical characteristics of the little room filled with ESS spheres but no matter orbs. Remember this is a hypothetical universe where space exists but not matter and as we know our universe has both so the conditions of perfectly balanced ESS macht during annihilation events is a remarkable one in need of much future study even though the working mechanism is now fully known. The working mechanism is due to the ESS being fully balanced as I said without energy input from matter which means that the quick and violent release of energy by the local area ESS will uncouple the two particles from empty space.

This is possible because for the first time in these particles lives since they were created they are completely removed from space as there is no longer any macht or energy exchange between them. The annihilation event will in turn reduce both particles to ones that do not interact with the local area ESS as before and result in two high energy gamma ray photons travelling apart from one another. I will explain more about particles, matter and energy and the rules that govern them in a later section here I will focus on purely matter and anti-matter answers. To this end the electron and positron have become fully unbound from the ESS as the macht is balanced and they are no longer interacting with space in the same manner as they were before the annihilation. This explains how matter and anti-matter such as electrons and positrons work at the most basic levels of space and matter as well as the reasons of why and how they annihilate with each other. Now the answer for why matter is more prevalent by far in the universe to anti-matter can be given and it starts with the Big Bang and the appearance of the first particles in the universe.

Matter over Anti-matter

At the moment of the Big Bang both space and matter were created and brought into the universe through a mechanism I will explain later in full detail. Yes I am aware you have entered another patch of me saying I will tell you later but I just need to tell you later. That being said the early universe in the very first instant after the Big Bang is nothing more than space and a dense hot collection of what will become matter as we perceive it today. As the universe entered the stage of its early life where the first particles of matter appeared the very first interactions of matter with ESS can begin to occur. These interactions take place following certain rules for each particle which are unique to them. Particles are obviously not all created the same as can easily be measured and have been using the Standard Model of particle physics where different fundamental particles have different properties. A quark does not have the same mass, spin or charge as an electron or neutrino and this is of course quite obvious. Each different type of particle also has its own unique ways of interacting with space based off of these intrinsic and natural properties they possess. Once again an electron will interact with space such that it imparts energy to both the gravity and electromagnetic machts of the ESS it inhabits. A quark will interact with the gravity macht and the strong nuclear macht of the ESS it inhabits and this is how it binds itself with other quarks to form hadrons. You can try all you like but multiple electrons will not bind with each other through the strong nuclear macht. In this way it is easy to understand that different particles of matter will create different interactions with the ESS machts in space and that space in turn has a preferred way of interacting with them such that it will not try to share strong nuclear macht with an electron.

Why this is so important for the baryon asymmetry problem goes beyond simply explaining why different particles have different interactions with the ESS they inhabit. It is important because it tells you that each particle has a specific way that space prefers to interact with it. This can be thought quite correctly of as the interaction that is most favored by space as it involves the least amount of effort to maintain. Proof of this has been found in the differing decay rates of matter and anti-matter in laboratories here on Earth where matter tends to decay a little slower than anti-matter in things like kaon decay events. This also is obviously one of the mechanisms by which matter has a better chance of surviving until today over anti-matter as the anti-matter created in the Big Bang would have decayed slightly faster meaning that it disappears slightly ahead of matter so a complete series of all possible annihilation events in the universe is avoided. If less than 100 percent of all possible annihilation events occur then some matter will survive and become the matter we see today. The tiny difference in decay rates however is only part of the answer as to why matter is more prevalent over anti-matter in our universe and it is also not the main reason.

The main reason is due to how easy it is for the universe to maintain matter in certain interactions with space. This is what I referred to earlier as effort on the part of the universe to maintain a particular exchange of macht and energy between space and matter respectively. It does not mean that the universe is physically exerting itself as though lifting a heavy weight only that the way matter can interact with space has certain preferred means. The particle of matter that will interact with the local area ESS such that it has the properties of mass, gravity and negative one charge will become an electron as this is the natural and easiest way for the matter and space to interact. It would take more effort to make the particle interact with space in some different way such as the same mass, gravity but positive one charge like a positron. After all space behaves in one set way throughout the universe where it contains ESS and those ESS share macht amongst themselves and this sharing is uniform and equal as all ESS are connected to one another.

The same is true of matter as the particles that make up matter likewise have one preferred way of existing that is simply one of their base properties. This fact is reflected in the array of particles we see today all around us as you cannot turn an electron into a quark because they are two different particles that by their very nature simply have different properties. No one questions that space has different fundamental forces that are simply belonging to space from the very beginning of everything so why would one reject the truth that particles also have different fundamental properties that make them unique? The different decay rates of matter over anti-matter prove this fact by demonstrating that certain particles do exist longer than others even though they are identical except for one property. This one property is telling us that it is a configuration of the matter particle that is now the normal one such that interaction with space is performed with the least effort possible. Matter comes in different forms and these forms will naturally interact with space in set ways which are always favored by normal matter over anti-matter.

To demonstrate this let us look at the first moments that matter comes into play with space after the Big Bang. In the first moments after the Big Bang matter is not yet interacting with space and this can be seen as the absence of matter as modern science understands it and that is to say matter as excitations of fields. In reality matter excites the fields as we have learned by interacting with space such that ESS macht is increased or decreased. We have also seen that matter is unique and will interact with the ESS macht in specific ways giving us specific particles. Some will interact more or less strongly with the Higgs field meaning they have more or less mass than others. Other particles will interact more or less with forces that create specific kinds of charge and as we have seen all of these are simply ways of saying increasing or decreasing macht types in ESS. If all the various macht types are viewed as one screen sitting on top of space meaning they sit on top of the ESS then you can picture how matter is made to interact with the ESS by passing through this screen. Like any screen objects fit through it easier if they are aligned with it and this is akin to the old saying of putting a square peg into a round hole. You can do it if you use enough force but it will do so much more easily if you use the right peg which is in this case round.

Matter as it came into contact with the ESS of the early universe began interacting with it by in a sense and this is just a visual aid mind you of passing through the screen in the way that takes the least effort. To this end matter is more easily able to pass through the screen if it interacts with space the way the macht wants it to. In other words it interacts with space in the natural way for that specific type of matter that requires the least effort to accomplish based on the natural macht properties of ESS. In this way the majority of matter must form into the universe as normal matter and only some matter that is part of a violent event during the forming process will be forced through the screen as anti-matter. It can mostly pass through the screen because it has most of the properties favored by the universe and the ESS that governs it. Forcing the square peg through the round hole takes violence.

Imagine a series of normal electrons being created and in the middle a single particle starts to become a positron but this would require interacting with space in a way that is less desirable. As a result the macht glow from all the electrons around it nudges the almost positron particle back into a normal lower energy arrangement that becomes another electron. This is a form of macht balancing during and just after the Big Bang such that the least effort is used to make particles which is directly linked to the favored balancing of macht glow by all ESS and demonstrates again how space interacts with matter.

If this balancing is overcome due to some violent event however we get an electron identical to any other but due to passing through the screen in an unusual way becoming positive. This is easy to understand because we do know that matter interacts with space through macht to create the universe we see around us. The way that matter interacts with space determines the properties of that matter as well. Take for instance the way that all electrons in a magnet are made to interact with matter in one direction and you see it changes the way the matter behaves by giving it the property of a magnetic field flow. So a particle of matter that is destined to become an electron interacts one way which is more natural with the ESS macht charge of space to have a negative charge. A particle of matter that becomes a positron is the exact same starting piece of matter but due to unusual violent events is made to interact with the ESS macht charge of space in a manner that is opposite to the electron. This is what makes an electron donate energy to the ESS macht charge and what makes a positron take ESS macht charge from space. The two particles are the same except for charge just as we observe today.

The early universe is a very dense and hot environment that has many particles in close proximity to each other and more than likely constantly hitting one another especially in the very early moments of the Big Bang. The ESS in this early universe is likewise very dense and since matter attempts to bind to it and it is still expanding rapidly there are ample conditions for matter to become unbound from space even if only temporarily when the ESS it inhabits are moved away from its center of attraction faster than it can track to its ESS new central position. All of this once again calls into play the light speed or slower movement of matter with the faster than light speed movement and transfer of ESS and macht. These early conditions are violent and will produce natural binding and unbinding events of space and matter many times over.

This provides the necessary violent forces needed to make some matter pass through the screen of ESS macht such that they become anti-matter. The natural tendency of this to happen however is low and so the total number of anti-matter particles created is always much, much lower than that of normal matter. It does not mean that the ratio is outrageous being quadrillions to one necessarily only that it is the most significant source of production of matter over anti-matter such that it far outpaces the differing decay rate contribution of normal matter to anti-matter in the universe. After all if we saw only one percent of the matter survive until today as a result of differing decay rates then the background of the universe would be awash in many orders of magnitude more energy than it is as this energy came from the ninety-nine percent of matter that annihilated.

This now is only part of the explanation as well as the early universe will no doubt have many violent collision of matter and matter, matter and anti-matter and anti-matter and anti-matter as well. These collisions each have a chance to unbind the early particles of matter from the ESS they inhabit and once again balance out the macht of the local area of space. This means the particles of matter are now once again free to try and pass through the screen again and since the natural tendency of the universe is to favor matter over anti-matter a recycling mechanism is now in place to allow particles that first appeared as anti-matter in the universe to be reformed as normal matter. Any annihilation events that take place would normally leave only two photons in their wake but due to the extreme density of ESS and macht at this time in the universe these particles again can be reformed as normal matter as they recycle through the same mechanism as before and again have another chance to become matter over anti-matter.

The conditions to recycle matter and anti-matter are lost under our normal spatial compression here on Earth and in most of the universe as well as the extreme densities of ESS and macht coupled with the high temperatures and multitudes of matter particles simply do not exist in just about every part of the universe. The universe does contain some but I will again discuss these at a later section and topic. The net result however is one that will always create conditions that are favorable to normal matter in the universe and the early universe is set up such that it wants to create normal matter. A universe of equal matter and equal anti-matter as is believed by modern science was never a reality nor would it ever be as the natural tendency of space to favor certain properties in matter are innate. In fact it would be impossible to create a universe such as the one we live in now where anti-matter was the dominant form of matter at all.

Proof of these natural tendencies to favor matter over anti-matter has been seen in the differing decay rates of particles in laboratories. Proof of the need for violent events contributing at least in part to the necessary formation mechanism of anti-matter are likewise observed here on Earth and in fact all around us. In particle colliders normal matter is subjected to incredibly violent collisions whereby new particles can be said to shower out of the explosions. Some of these particles are anti-matter which we know were created by the violent explosion and did not exist before as this would mean they existed inside the matter we collided. This is impossible for two reasons as the matter we know to be normal matter as that is what we started with and secondly the matter could not contain particles of anti-matter mixed with its normal matter as this would lead to the matter annihilating itself from within. As this obviously does not happen we know this is also incorrect meaning the only place the anti-matter came from was the collision itself and from the normal matter inside being forced through a violent event to interact differently with the macht of the ESS inside the space of the colliders vacuum tubes.

To this end the constituent particles of the matter we collide are for a brief moment subjected to conditions seen at the beginning of the universe or at least as close as we can make and so we have matter and anti-matter created as these particles form into being in the collider. What is missing however is the extremely dense ESS and macht levels found at the early stages of the universe as the only point of violent action in the collider is the single spot at which the collision occurred. If the entire collider and all the space in the collider tube around the explosions could be made to be under the same extreme densities and movements of the early universe ESS as well as the incredibly high temperatures and energies then the recycle mechanism would be seen as the particles travelling outward would pass through the necessary screen to make them form into normal matter after they are created or released in the collision.

However the external input from all the local ESS and the macht is not present in the collider due to the need to instead create as near a perfect vacuum as possible and so the violent collisions simply allow matter to become both matter and anti-matter out of proportion to what existed at the early moments of the universe. Without the screening mechanism of the early universe we can create the violent conditions necessary to force matter into interacting with the space around it as anti-matter would but we cannot create the dense recycling conditions to allow for the proper matter to anti-matter ratio. Nevertheless what the colliders do provide is an ample amount of proof that violent events can in a sense knock normal matter through space such that it does not behave as it should with respect to the various ESS macht.

Natural phenomena exist to demonstrate this fact as well such as cosmic ray events high in the atmosphere where anti-matter is again created as a shower of particles following a cosmic ray spallation event. Radioactive elements likewise demonstrate this fact by being unstable to the point where the ESS macht balancing inside of their atomic structures means that some particles knocked loose are done so with such force that they again fly off as anti-matter. This also demonstrates the power of ESS in balancing their macht as they can at the most microscopic scales create events that change the way a decayed particle interacts with space. The complex nature of calculating these events makes them seem impossible but again knowing the local spatial compression for an area of space and the macht contained within it plus the behaviors of all the matter involved in that system will allow you to determine exactly where and what will decay off from the atom.

The random quantum nature of radioactive decay is not random after all and Schrodinger's cat does not affect the outcome of the experiment through the act of observation. Instead the cat simply waits for you to open the lid as it is always alive inside the box because that is the known state you last saw it in and the decay event is not random. So unless you wait for a year to pass the decay event will not be both yes and no all at once and instead will simply be no waiting for the right conditions to occur which as I said will happen within a year or less but by that time the cat will have been freed from the box. To conclude this explanation of matter and anti-matter for now it is revealed that the natural ESS macht alignment of space which permeates everywhere and is uniform therefore only has one arrangement itself. Since matter as we know interacts with space it will always attempt to interact with space using this one same arrangement as this is favored by the universe as the easiest way to exchange macht and energy between space and matter. The net result is matter is the natural result of interactions with ESS and macht as opposed to anti-matter which takes more work to create and is not favored by the universe as seen in decay rate experiments.

Facts

The important facts for this section are firstly that matter can interact with space such that ESS macht can be increased or decreased. The second fact is that there is no such thing as positive charge in the universe and instead is only a decrease in the local ESS macht charge pool for that area of space. The third fact is that ESS have macht pools which explain how balancing events can occur. A local area increased in macht will naturally balance out with an area that has a decrease. The fourth fact is that the increased ESS found in the nucleus of atoms still contains macht charge in its pool and therefore will never fully allow the electron to collide with it despite their natural attraction. The fifth fact is that the electrons natural tendency to attempt to collide and therefore fully balance its macht with that of the nucleus also accounts for some of its quantum motion. The sixth fact is that photons do not travel back and forth between electrons and nuclei in an attempt to explain electrostatic field repulsion and instead the ESS macht glow associated with the electron completes this task by maintaining a conservation of macht/energy of its local system.

The seventh fact is that matter can be made to interact differently with space despite it being similar to other particles of the same type the result is anti-matter like positrons. The eighth fact is that the complete balancing of local area ESS macht imbalances is what leads to an annihilation event that creates two photons in an electron and positron collision event. The ninth fact is that balancing of ESS macht for local areas of spatial compression explains why electrons move up in orbitals as energy in the form of photons is input into the system and why the local area of space cannot hold this extra ESS macht charge in its pool and so releases the macht back as an energetic photon. The resultant need to balance out ESS macht is what naturally makes an electron drop down to its lowest energy orbital state. The tenth fact is that if you exceed the ESS macht balancing distance for a given electron around its nucleus by adding too many photons to the system such that the local area macht pool for the nucleus is too repulsive you will cause the electron to fly off into space away from the atom to which it was bound. This is what explains the inability of electrons to maintain orbitals around protons in the early universe before the era of recombination or first light as well as what and why ionization events occur.

The eleventh fact is that the early universe contained the necessary conditions to not only favor the formation of matter particles into normal matter but also the recycling mechanisms necessary to allow anti-matter to be reformed as normal matter over and over again leading to a universe that naturally started with more matter and would lead to the universe as we observe it today. The twelfth fact is that evidence for this can be found in differing decay rates of matter over anti-matter and of the need for violent events to force matter to form as anti-matter in laboratories or through observation of natural events like cosmic ray spallation particle showers or radioactive decay. The nature of matter over anti-matter and the answer to the baryon asymmetry problem in the universe is solved using the natural properties of both ESS macht and matter. In addition to this other unsolved problems such as what charge is and why electrons drop down orbitals spontaneously are now also answered. The next sections will deal with further developments of the exact workings of matter with ESS and ESS macht in the universe to create the cosmos as we observe them today but first let me explain a slightly different aspect of the universe. This is the answer to the number of dimensions in the universe and how the universe is actually made of fewer dimensions instead of more.

Chapter 22
Dimensions of the Universe, Time

Dimensions of the Universe

The nature of the world around us has been a question that far predates the usual 5 000 or so years the history books will tell you. Since the very first carvings or paintings were made in any ancient human settlement tens of thousands to possibly hundreds of thousands of years ago people have been trying to make sense of the world around them. The world at that time was the universe as no one could conceive of anything larger than what they could see from horizon to horizon and also by looking up. In reality the actual search for a meaning and order to the world is a quest to discover the workings of the universe from the largest objects imaginable right down to the smallest fundamental bits the cosmos has to offer. A human living one hundred thousand years ago and thinking the universe was just the little patch of land they lived and hunted on is no different than scientists a century ago that thought the universe was just the Milky Way. Both groups were not to be ridiculed as they were only trying to understand the universe it is just that the size of the universe has changed in our understanding over time. The kinds of objects we can detect billions of light years into space and therefore into the past were beyond the scope of the average person a hundred thousand years ago. Likewise the world of fundamental particles was also real but non-existent to ancient human thought as they simply lacked the tools necessary to study them not just in depth but at all. So what were people left with? Their senses of course and a human brain to begin solving problems they could experience around themselves.

The very first problem they ever tried to solve is the problem of how many dimensions there are in the universe. This is the very first problem that any creature has to solve as it involves locomotion and will be encountered in many species before they even open their eyes. Before a little kitten can open their big adorable eyes to peer out at the world around them they are already making sense of things through touch in terms of their parents, siblings and simply their own bodies. Humans are no different than kittens and as such must make sense of the world around them in a spatial way too and human babies explore in virtually the identical way to kittens. What an awesome triumph of all babies when they discover they have thumbs to suck. So naturally the first thing adult humans will do in trying to figure out the world around them is to make sense of dimensions.

The reason I have gone so far back in human existence and used analogies of kittens and human babies to begin a section on dimensions is because despite modern sciences understanding of the universe around us our knowledge of dimensions has not progressed factually for all these tens of millennia. In other words we can still make no more or less sense of the universe around us in terms of how many spatial or temporal dimensions exist than we could simply using our senses tens or more of millennia ago. I also said factually because despite a number theories espousing additional spatial and sometimes even temporal dimensions they remain nothing more than theory. This is true no matter how many variables are allowed, no matter how many given rules concerning the universe are permitted or how many different theories are put forth. None of them has ever been able to remotely prove the existence of more than three spatial dimensions. Maybe the reason we have not progressed beyond three dimensions in all that time is because there really are only three dimensions. So we must now look at the universe in a different light and see if we can answer how it all works using what is proven and what is not theory.

To begin with I am going to give you the answer right now and the answer is three. There are three dimensions to the universe and the universe needs no more or less to produce a cosmos as we see it today and that behaves as we see it today. The necessity here is to prove this while only using what we know to be fact around us from our empirical senses to what we have discovered through careful and repeatable experimentation. Why do I like three dimensions more than five, six or ten or more? The answer is simple I can create a universe as we see it today such that only three dimensions are required and use that same method of creation to answer multiple unsolved problems about the universe. What I mean by this is simple as simplicity is the key to the universe and this has been borne out time and time again. Take for instance matter and at face value it seems complex, a different atom for every element. Then look deeper and see that you can take just three particles to make every natural and man-made element in existence. All of a sudden the idea of simplicity reigns and complexity fades. Every time you can explain the universe using fewer rather than more variables you are on the right track.

The second reason this is so is by checking and seeing if the mechanism you have used to create a simple universe can be used to answer unsolved problems that are otherwise unrelated to it. If for example I can show you that ESS macht are responsible for just three dimensions this would seem like a good model but only useful in solving one problem which is the dimension problem. Using this central solution to answer other unsolved problems such as quantum entanglement, gravity or particle fields for example helps to prove the need for only three dimensions as the universe never chooses a path that uses extra variables and the other solved problems reinforces a system using ESS macht. For these reasons I feel that a central mechanism that explains the number of dimensions in the universe to be three and three alone is the true number of dimensions based on the information available as it is simple and can be checked against other unsolved problems. I mention this here and now simply because the number of dimensions is such a base level question and the theory of multiple dimensions proves impossible to investigate in our three dimensional world. Making a model that can use only three dimensions to solve multiple problems including those that would seek to use more than three dimensions in the universe tends to point to a true three dimensional form to the universe. I will therefore go on from here and look at why multiple dimensions are not needed and explain how you can create a universe out of three dimensions and where these dimensions are useful in terms of both space and matter. After this I will talk about time as time is often viewed as the fourth dimension and I will also discuss the topic of time head on and show once and for all whether time is real or not.

When considering how many dimensions exist in the universe it is necessary to examine the practicality and feasibility of multiple dimensions that exist beyond the normal 3+1 that is currently accepted by modern science. The reason for this is obvious which is leave no possibility unexplored and the theory of multiple extra spatial dimensions or even time dimensions should be examined before moving on. If we look at the idea of extra time dimensions briefly we see that this is quite a ridiculous undertaking as the nature and existence of a single time dimensions is still hotly debated in modern science. The understanding of a single forward moving dimension to time is still not concluded and so simply saying alternate time dimensions could or should exist simply to solve a problem is highly unethical. They cannot even be proven wrong. The only real practical application of an alternate time dimension is one in which time can be made to flow backwards in order to solve certain problems most notably the entropy problem.

◄New Universe Order►

The entropy problem is one in which the total entropy of the universe is thought to be conserved and no information lost and as such a reverse flow of time might solve this by explaining how you can start off with a well ordered universe from one of relative disorder such as the one we live in now. The idea of reverse time is nothing new as Merlin the famous and legendary figure in history was thought to live backwards in time. This made perfect sense as he was always aware of future events and therefore could help alter the course of what would be future events from our timeline. This idea is wonderful in tale, song and legend but wrought with complications that certainly do not resolve themselves in the three main views of modern science. The more scientific idea behind modern versions of backwards time seek to link all time in the universe such that a sort of account can be made of where entropy is going and where it came from in the hopes of one day making a universe that once again becomes more ordered. This idea is simply one that exists on paper as I have said a true understanding of what a single forward flowing dimension of time is still eludes modern science so attempts to formulate extra temporal dimensions such as reverse time are on very shaky ground indeed. The idea of reversible time also has another problem which is it solves no other unsolved problems in the universe and can only be applied to one or maybe two problems at best.

If we examine the idea of multiple dimensions of space rather than time we run into an equal amount of trouble albeit with far more understanding or basis to start off with. This is of course due to the fact that many of the properties of matter or the reactions that they can undergo is well understood by modern science and as such the very real three dimensional nature of them can be studied and extrapolated more easily. Take for example the theory of extra dimensions that might be used to explain matter such as in string theory. The basic idea is that extra dimensions hidden to us might exist on the surface, interior or anywhere else they can be made to work in parts of matter. A standard explanation is that if we look at a one dimensional object such as a string from a distance we see only a line. If we zoom in closer to that line we see that it can have a surface that wraps around the line like a cylinder and will be another extra dimension that is normally invisible to us. The extra dimension is really a fourth spatial dimension according to theory as it is hidden on a three dimensional object. If this sounds impossible to you it is as any object magnified sufficiently to show a surface feature such as this extra fourth spatial dimension is really only revealing a different feature of a normal three dimensional object. No matter how close you zoom into the object you can find new nooks and crannies but they still exist in three dimensional space due to the ability to reduce the system to its simplest components.

Much like in mathematics you reduce sides of equations to remove extraneous information such that the overall equation becomes simpler, cleaned up and easier to work with the same is physically true of objects in the universe. Not to mention the fact that the extra dimensions are only theorized to exist and would still result in particles creating three dimensional structures anyway. In other words not four, five or six dimensional effects just normal three dimensional effects that we see in the world around us meaning maybe there really are only three dimensions as that is all the world has. Any object can be constructed out of only three dimensions at the smallest level and so no matter how complex a shape it makes it is still able to be built from only three dimensions. This favoring simplicity is a constant held by the universe and is one of the true unbreakable laws of physics and far more importantly of the universe itself.

What about compactified dimensions you ask? Well these extra dimensions are believed to be folded up into complex hyper-dimensional shapes that can exist inside normal three dimensional spaces. Here is a thought, what makes up the extra dimensions, is it an infinitely larger number of dimensions? What this really means is nonsense as multiple flaws exist with this multiple-dimension theory. To start off with the dimensions are not at all compactified or folded in on themselves as many believe them to be because the theory of compactification was only created due to the need to simplify the number of dimensions into something that can even be remotely worked with in three dimensional mathematics and therefore real physical laboratories and space. In reality the dimensions would exist in their un-compactified form or un-folded form and therefore not be able to be neatly reduced to make life easier for those trying to work with them.

This is something I must stress perhaps more that anything I will state in this whole section and indeed is one of the most important points in all of modern science today and that is mathematics do not make reality. People need now and I mean right now to stop trusting and following the mathematics towards the dark side of science and must not be seduced by the easy path to answers for problems that require a better understanding of the universe rather than some trick of the equations. The universe does not change itself for us and instead we must change ourselves to fit the universe and this has been one of the core driving forces for my own work as I search to take what the universe has shown us and use only that to make a model of how it works. I will not simply hide behind impossible theories that are built upon our desire to change the universe to fit some mathematical trick we have pulled from a hat. I apologize to those of you reading this that are offended by this concept and I applaud those who stand by my side and know that the universe is very unimpressed by what we think of it and must learn its secrets rather than force ours upon it.

To that end I will return to the idea of mathematics having created seductive ideas concerning multiple dimensions and state flatly that these theories exist only in the mathematics and not in the universe. I do not doubt that those who have created these seductive theories have checked and double checked their calculations and have never once forgot to carry the one but that does not change the fact that the answers derived there from are nothing more than theories that represent at best physical impossibilities. I openly ask all to explain themselves and the creations they have made in their theories of multiple dimensions. It is not all about me trying to sway you to my side of thinking and it also directly involves those who would challenge me to explain in physical and non-mathematical terms how there ideas can be manifest as real properties of the universe. It is time for you to go on the defensive. You cannot use your mathematics and instead must climb straight into the real and physical laboratory to prove how multiple compactified dimensions can be used to build the laboratory you are in and how to test for extra dimensions. In other words forget the theoretical and grab something physical in the universe to prove yourself and no more hiding behind derived seductions.

This might sound harsh but I patiently and confidently await your inevitable failure not out of spite but out of the desire to move forward as one such that we can finally understand the universe. To those of you reading this who agree with me I am sure you will see that the sheer number of problems and complications of multiple dimensions makes the idea impractical to explain the universe as we see it today. The greatest hope of the theory of multiple dimensions is that a possible explanation for matter and gravity can be gained such that a theoretical particle can be used to represent the graviton. All of this remember is not needed with the other three forces and all of this is only to explain gravity and that is it just one little problem of all the unsolved ones.

In order to do this however mathematical inventions are used such as multiple extra dimensions or things like branes when string theory is mentioned. We have covered what a brane is previously and it is known that they are an extra entire dimension themselves outside of normal space and in fact are merely a scaffolding of sorts that strings can hold onto. Bringing even more multiple dimensions into the picture we have multiverses which stack every conceivable universe inside of every other one such that two are separated by mere fractions of a micrometer. All these universes and all these dimensions and all these branes exist inside the same cubic volume of space as each other according to the theory. The impossibility of this is staggering as the universe, dimensions and so on are not thought to touch but only ever share gravity in some way. Which means of course they are touching all the time as this is how they share gravity.

In addition the universes are depicted two dimensionally as sheets that stretch on and on only ever briefly touching for a fraction of a second every trillion, trillion, trillion years or so when a new Big Bang is due. In reality the universe is three dimensional so no convenient spaces exists between them to keep them from touching and creating more Big Bangs constantly as in literally every second. Here once again the necessity of simply saying the universes exist magically inside four or more dimensional space crops up as no logical explanation can be given outside of the mathematics that created them as to how such a feat is at all even remotely possible.

What about Pauli's Exclusion Principle? Surely the shared space of multiple dimensions, branes and universes all existing in the same area of space would be forever creating impossibilities in quantum logic. In order to even create many of these scenarios more are needed to explain that which is already quite impossible or at least very improbable. Take for instance strings existing as theoretical one dimensional objects that are either open or closed so in other words a line or a loop. Well right there the loop variety or closed kind is now two dimensional is it not? The zooming in of any object to sufficient levels would once again reveal the surface structure of the string which of course means the string is now three dimensional and therefore not one dimensional. Did I mention that the only reason these dimensions of the strings were reduced from three to one was to satisfy the mathematics and make them at all possible to theorize? In other words anything and everything that needed to be done to make the mathematics fit was in fact done.

That however is not even the most perplexing part of these theories as the strings require something to sit on in order to vibrate correctly to interact with the rest of the universe to create matter and gravity as we perceive it in three normal dimensions. The idea of a brane is needed as a literal fourth dimension and there can be more than one brane for the strings to sit on. Where do these branes exist in space and how were they created such that they expanded lock step with the volume of space from the Big Bang if they exist in a fourth dimension outside of space itself as they must in order for multiple universes to share gravity? Matter is made of matter or energy whichever you like, well what are branes made from as they are not part of regular space but exist as a fourth dimension scaffold and for that matter this begs the question what are dimensions made of? Are there dimension atoms? In other words the invention of improbable things like strings simply gives rise to a never ending chain of new inventions that are even more improbable like branes.

These theories look good on paper and can pull off some clever tricks in the mathematics but remember they are no more than tricks. The number of great minds whose work these theories were built on that also echoed the sentiment of do not trust the mathematics and do not try to make the universe fit your theories is long and apparently lost. Remember the universe does not like to create more or use more than it has to in order to get the job done as we have learned from things like electrons, protons and neutrons being only three particles that make up any atom you can imagine. How, why and what would be the reason or mechanisms of the universe to create multiple dimensions, branes and multiverses in the first place as the number of extra ingredients and methods of construction the universe would have to employ is absurd when faced with the fact that all of these would solve only a couple of the unsolved problems in the universe rather than all at once? On top of that the theoretical multi-dimensional theories only fit a tiny number of unknowns in something accepted and beloved by modern science which is the Standard Model. Why would the missing few pieces require all this extra effort on the part of the universe to make when none of the others that are long since known and studied do?

I will close this thought here by simply saying the theories and those who espouse their effectiveness at solving problems were created and are championed with good intent. It is not that any malicious acts were perpetrated in coming up with these theories I hope, but rather that they only were created as a way to try and solve the universe. However it is time that a far simpler and in many ways far less exciting answer be given for how you can build a universe as we see it today out of nothing more than three dimensions.

To start off with I will give empirical reasons why the universe only has three dimensions. Life tends to exploit everything the universe has even if we as humans cannot use a particular part of it either at all or as well as other animals. A few examples include bats and dolphins that use sonar which humans do not or whales that use infrasonic sound that we do not. Felines have built in night vision and can see in the dark far better than humans can and multiple animal species can see in both infrared and ultraviolet wavelengths that we cannot. Fish sense electrical fields in the water while many birds sense magnetic fields in the air. What I am getting at here is twofold with the first point being many things that seem impossible to us are utilized by other animals or plants. It may still be light but some wavelengths simply fall outside the range of what humans can see and yet life has found a way to see it. Even other forms of energy that we cannot use life still can such as sunlight to create food directly as found in plants. Other senses humans do not seem to possess or are so dormant and ignored in our modern society that we do not realize they even exist and this could be electrical fields or magnetic fields.

Life no matter what we as humans say always finds a way to exploit any kind of matter, energy or field found in the universe. Seeing in the dark is using quantum particles of electromagnetic radiation to stimulate quantum electric synaptic pulses in a chemical neural net called a brain and magnetic direction finding uses the quantum spin property of subatomic particles to induce field flow. So if multiple dimensions are indeed a part of our universe it would make sense that some life form no matter how simple or complex would have found a way to utilize them. This might sound a bit fanciful to some pure physicists out there and I understand that but you would be remiss in your duties as a pure scientist not to at least entertain any and all possibilities concerning the limits of the cosmos. To this end it can be at least inferred that biological proof exists to the fact that the universe possesses only three dimensions.

We as humans are certainly backing this idea fervently since our earliest days of questioning up until today as we have probed down to the fundamental levels of the universe and never found more than three. We might have postulated a greater number of dimensions but this is only theory that to date has never and likely will never produce results indicative of multiple dimensions. In addition we have been looking for them over and over again as we often exhaust a possibility before moving on yet have not been able to find them. What is worse for those who believe in multiple dimensions is that the universe is so three dimensional no matter what scale we examine that it is impossible to construct apparatus to test for multiple dimensions. The idea of multiple dimensions itself is only concocted as a way of creating theoretical and unfound particles that could possibly fit somewhere into the Standard Model of particle physics. A model by the way that has every other particle, force carrier and field existing happily in three dimensions. This would make it seem unlikely that the universe would need to create objects with multiple dimensions to fulfill only one task in all the list of unsolved problems and have that object fit no where into the rest of its family tree. Until everything in the universe is discovered however there is a small non-zero probability that something might exist with more than three dimensions. Realistically though if one can construct a universe as we see it with only three the need for exotic dimensions to make even more exotic particles to answer only one of the list of unsolved problems would become moot.

So let us start at the very beginning of the universe with the Big Bang and the explosion of both space and matter into our current universe. This is a point in time when we know two things to exist which are space and matter held within it and both are expanding rapidly with every passing moment. It can be agreed here that the universe must have three dimensions as it is spreading out in three directions which can be described as up and down, left and right plus forwards and backwards. This is the shape of space and the matter within it fills out this space uniformly and so the universe overall has three inarguable dimensions. The question of whether or not more than three dimensions does not really come into play until we try to build things out of the matter and fill the universe with more complex objects. So if we go back to our little room filled with the little hollow ESS spheres and mostly opaque matter orbs we see that the universe is still three dimensional.

We will look at space and matter separately and start with space. The space of the universe must be three dimensional for many reasons the first being empirical evidence. Even if you adopt the idea of the three main views of modern science space always has three dimensions. Classical mechanics obviously treats the universe as three dimensional with its macro scaled objects being free to move anywhere in three dimensional space. Quantum mechanics agrees with three dimensional space because it describes micro scale objects accounting for classical mechanical behavior which still exist in three dimensions. Even quantum scale fundamental objects are still capable of filling a volume and therefore must possess three dimensions as well. Lastly general relativity uses three dimensions constantly as it bends spacetime into curvatures which all describe motions of objects in three dimensional motions over time. So already modern science has three working theories that agree on the universe using three dimensions and has used the three dimensional ideas of these theories to better understand the universe empirically.

Now if we look at the little room we see that space extends in all three dimensions and that the little hollow ESS spheres are also three dimensional. They can perform their job with the use of three dimensions and they need no more to do so. The ability of each sphere to spatially compress or spatially rebound requires only three dimensions. In addition the ability of them to share macht amongst each other requires only one dimension as it is not a physical object that is being moved back and forth. The macht exists as a part of the ESS and so is not a separate structure in the universe. The arrangement of ESS and the macht they share can form ESS macht glows as we have seen and this is capable of creating very strong effects as they manifest themselves with other areas of space or with matter. In these manifestations however they are still only making use of two factors which are ESS volume as seen through spatial compression or spatial rebound and macht glow.

The ESS of the universe can be thought of as the ultimate micro scale objects from which everything else can be built and as such any large scale macro construction are not made of additional parts but merely repetitions of micro scale ESS. In other words with respect to space you do not need more dimensions than three to create large scale objects or effects as they will all at their core be made of the same small ESS and the macht they transfer which are of course only made of three dimensions. Since we know that ESS and ESS macht sharing are responsible for everything from gravity, the changes in rates of acceleration of the universe and quantum entanglement to name a few unsolved problems it is obvious that they do not and cannot arrange themselves into multiple dimensional configurations and also that they do not need too. The fact that they can explain numerous unsolved and seemingly unrelated problems without ever changing their shape, properties or behaviors means they do not need to make use of other dimensions for any other problem either. The ESS need only three dimensions and can scale themselves up to any size necessary to create larger space structures or effects while remaining the same at the most base fundamental level.

Now we will look at matter and see if matter needs more than three dimensions or if matter needs three dimensions at all. To begin with matter does not need three dimensions to behave as we see it in the universe today and instead can exist as a one dimensional object. To explore this idea let us look at how matter behaves with space a bit more closely in terms of space and matter interactions. We know by now that space and matter interact through ESS macht and through matter donating energy to that ESS macht structure. If we look at our little room again we see the mostly opaque orb is creating a macht glow around it and this macht glow is made from the matter contacting the ESS it inhabits directly. The matter particle is very small but still larger than the ESS and so we can say in space physics terms it exists in three dimensions but in how it behaves with other matter it can still be treated like a one dimensional object. This is because the way ESS share macht is dependant on both strength of macht and number of ESS in the local area of space. If we look at the mostly opaque orb and look at its ESS macht strong force for example we see that the macht glows brightly and that the reach of that macht extends beyond the matter orb itself through a number of schales. If we place two more matter orbs in the room and move them close to the first we see that they immediately are attracted to one another and form a hadron for this example we will say a proton. The three matter orbs share much ESS macht strong nuclear force between them and yet there is much space surrounding them. The proton itself is largely made of what is commonly thought of as empty space with three small particles inside it.

We know that all quantum particles experience quantum motion and we also understand what generates quantum motion from a previous section. This produces two unique properties of matter when viewed at the level of ESS the first being that the boundaries of each quark are not physical as we perceive them in modern science and the second is their motion adds more space to the system. The first fact comes from the fact that the shared macht between schales of a single quark are very strong close to the quark and then fall off geometrically with distance. Given a certain amount of spatial compression you cannot overcome this macht glow. In other words under normal spatial compression there exists a certain radius from the quark that the ESS macht strong force is too strong to overcome and acts as a physical wall to the interior of the quark and ESS macht glow. If spatial compression is decreased you will see that physical boundary grow in size and under high spatial compression the neighboring particles will be able to push that wall back and closer to the real quark inside. Each piece of matter as we see it exists inside a protective schale of ESS macht.

This can be seen perfectly clearly if we look at an atom and think of overcoming the electron degeneracy pressure or exciting the atom to increase orbital size. The ESS macht glow of the electromagnetic force as we know generates the orbital distance and energy of any electrons around a nucleus. If you want to overcome the electron degeneracy pressure what you are really doing is overcoming the ESS macht glow for that local area of space in terms of ESS macht electromagnetism. The electron orbital creates what we perceive of as a solid wall to the atom as a particle and you can demonstrate this right now by touching anything around you. The electron degeneracy pressure of the atoms in you and whatever object you decided to touch keep your atoms from merging or passing through the atoms of the object. The atom is not a real particle in the sense that it is a single object but it behaves as though it were one and the same is true of the quark. The quark may be a fundamental object unlike the atom but it still has a protective wall around it that like the electron degeneracy pressure must be overcome if contact with the matter particle is to be made proper. In other words the macht glow of the quark in ESS space gives it a larger three dimensional volume than the quark matter itself.

The second way matter appears to be physically larger than it is comes from the motion each particle constantly performs as it tries to balance with the local area ESS macht. We have seen that particles will not simply sit in one place in space for any length of time and prefer to move around which creates quantum motion. This motion is made from the ESS macht and matter attempting to track one another but cannot fully do so due to two different timescales each system uses and we have learned this in a previous section. The net result is a larger than rest ESS macht glow created by the particle and therefore a larger than rest physical boundary to the particle. When I say rest with respect to this topic I mean if the particle were perfectly at rest it would create a macht glow similar to the one we first examined in our little room. The particle is not at rest however in reality and so will move in all three directions from moment to moment which creates additional volume to the macht glow. If the particle were to move in all three directions it can increase its radius along each dimension by ten percent for example then it is obvious that this moving macht glow will be larger than the rest macht glow. This second fact coupled with the first mentioned above will create an ESS macht glow that acts as a protective wall to any fundamental particle in space similar to the way the electron degeneracy pressure creates a physical wall to an atom. Both of these reasons can account for a three dimensional shape to a one dimensional piece of matter because the overall three dimensional shape of the matter is made of ESS macht glow through schales which extends in all three dimensions.

The way we see matter both in our most sensitive laboratory experiments and empirically in our everyday lives is such that matter is three dimensional. Any large scale objects of matter are all built using this same scaling up principle of ESS macht boundaries surrounding matter particles. The way an electron makes us think something is solid when really it is mainly empty space all matter structures in the universe will appear solid to each other even though they are made of a significant amount of ESS macht glow instead. Similarly just the way you can construct any large scale space object or effect using small ESS which need only three dimensions to exist so to can you make any large scale matter object which are made of smaller three dimensional objects. In other words you still need only three dimensions to create any space or matter you interact with through instruments or your senses in the universe and since both of these can be constructed using simple ESS macht explanations which are three dimensional themselves no further extra dimensions need be used.

You might be able to describe the universe in terms of extra dimensions but you would not need them at the most basic levels of the universe itself. Think mathematically for a moment of the equation 2x=2y and you realize that it is fully correct 2x units equals 2y units but if you look closer there is extraneous information. In this case it is the two number twos in the equation and if we divide both sides by two we get x=y which is much simpler and does not need extra variables to arrive at the correct answer. The simple analogy here is that it might be possible to explain some problems in the universe using more than three dimensions but through ESS macht you do not need the extraneous information and so can explain the universe quite simply with only the true three dimensions needed. To extend this further no extraneous information is needed to then take the smallest level structures in the universe which are the ESS and scale them and their effects and interactions with matter such that they create all macro scale objects in the observable universe as we see it today. What this means is that you do not need to create anything new for the universe to appear exactly as we see it today beyond space and matter. Instead it all simply requires the correct explanation for how the two interact and then follow these interactions from the smallest possible scales to any large size you desire.

It is also possible to describe the universe as an object built with only one dimension using ESS and matter. Take for instance the idea that each ESS represents a single location in space that if we reduce its properties to simply its total power which is a combination of its volume due to spatial compression or spatial rebound and its total macht glow then we can treat this as a one dimensional object. It will have one dimension with a certain amount of power and if we then look at fundamental particles of matter as single one dimensional objects we can say that they have a certain level of power that they will give to the ESS. The ESS will use that power to create macht glows which are really just differing power levels of these one dimensional ESS and only appear to create a three dimensional structure. The ESS macht will now create a three dimensional shape to the one dimensional matter particles out of their own one dimensional ESS structures and macht sharing. The result is still a universe that appears to our instruments and senses as being three dimensional and yet is created out of two types of one dimensional objects interacting with each other. The use of one dimensional objects can then be scaled up to create anything of any size in the universe and even the entire universe itself.

This is of course only an example as we know ESS for example to be three dimensional and as I stated just a moment ago the real ESS are simply reduced to a one dimensional location in space for the sake of argument. What is interesting is that if treated as one dimensional objects both the ESS and matter can be used to correctly build the universe as we see it today. So rather than invent more than three dimensions to explain the universe we can actually take two away. This serves once again to reinforce the truth that the universe has and does not need more than three dimensions as it likes to keep things simplistic and use that simplicity as effectively as possible. This is also a physical explanation of the number of dimensions to the universe and I will be explaining the nature of matter particles more fully in a later section I only wanted to explain here how three dimensions are needed to build a universe. I will also demonstrate in the future matter section how particles of matter do not need more than three dimensions either to create the universe we see today. Now it is time we moved on to time.

Time

This next sentence will make me friends scientifically and also make me lose some as well as it is the answer to what time is and at the same moment whether or not time is the fourth dimension. So I am just going to say it here and get it over with so that I can begin explaining exactly why this is so. Time is not a fourth dimension and time is not a real thing in the universe only a construct of intelligent creatures' minds. Yes I said it and I will prove perfectly that this is correct. If I have just won you over as a scientific friend then thank you for your understanding and if I have just lost you then allow me to show you why this is true and hopefully I can win the open minded of you back again. To start off with let us start back at the moment of creation according to modern science and the three main views.

The Big Bang is the moment believed by modern science when the entire universe came into being from a singularity that is an infinitely small point that exploded outward and over time produced the universe as we see it today. In the Big Bang theory everything in the universe came from this moment and not before and this of course includes space, matter and time. Time was created in the Big Bang is a theory that should be setting off alarm bells in your head right now. According to many and Einstein as well things only occur because time passes and the events we see are relative to one another given a certain frame of reference. This is how Einstein arrived at his theory of relativity and from which he bases all his work including the singularity. Since time is supposed to have started or been created from the Big Bang this means according to relativity that the Big Bang never actually happened.

In order for any explosion to actually take place would require the passage of time for the singularity to go from being a single dimensionless point in space to an active and outward spreading explosion. If time did not exist before the Big Bang the singularity never exploded because it could not progress through relative frames of reference from a pre-explosion to a post–explosion state. In other words you cannot have a Big Bang because before the Big Bang time did not exist and so could not pass and so could not create an explosion. This is all of course false and so relativity is of course false as the universe is quite obviously here and therefore did explode. The nature of any theory is to explain the universe as we see it today and therefore from the very beginning relativity cannot explain a universe as we see it today and must be discarded as it is wrong.

To further prove that relativity is wrong we must again look at the idea of the Big Bang and where it came from which is an impossible object created from a singularity which because it cannot produce a universe as we see today is an incorrect theory. In a singularity many impossible things happen and these impossible things cannot describe the Big Bang. To begin with a singularity contains multiple properties such as gravity, time, mass and so on as described by Einstein's equations a black hole has a singularity at its heart and we see those properties there. If the universe which includes all space, matter and time and everything else came into existence from the Big Bang which was according to relativity a singularity then once again we see Einstein and his work cannot be correct. A singularity contains these properties and yet the universe was not born yet and so had no space, gravity, matter or energy or anything else so how could the singularity have those properties before they were even created? The answer is of course it did not as it did not exist as a singularity is already easily proven to be impossible. Remember a singularity is only a massively derived product of a mathematical equation which as we have seen cannot be trusted to produce physical objects in our universe. By the way even Einstein himself said science should not blindly trust mathematical equations and here that exact thing is being done.

A singularity cannot create the properties which it is claimed to have as the very creation of the universe brought these things into being according to relativity and the Big Bang. So time once again cannot exist as it is supposedly to have been born from a singularity. In addition to this a singularity is also defined as having infinite gravity and therefore zero time and if the universe started from a Big Bang singularity then this is true. All the finite matter in the universe would be confined to all the finite space in the universe in a perfect dimensionless space or a space where density is created by an object of zero volume. This will create a perfect one to one ratio if you will of mass being compressed into space where both quantities are 100 percent accounted for as they involve everything that constitutes the universe. In this impossible scenario you have true infinite gravity as it is impossible to have less since the volume of the singularity is zero and therefore according to the spacetime curvature of general relativity infinite curvature is perfectly acceptable and achievable to be infinite meaning time must therefore be perfectly zero. So we see from the very definition of a singularity as derived from general relativity and using the entire mass and space of the universe that time must be zero in a singularity and certainly in the Big Bang singularity. What this means once again is that time cannot flow or move or pass in any way as it is fully stopped and therefore no moment can progress from one to the next. Again this creates a situation where the universe cannot create itself.

Once again we are obviously living in a universe that exists quite happily and contentedly around us and so we know that time again cannot exist because if it did the universe would not be here based on singularities and general relativity. One last note is that too many people simply treat all of existence as being made in two parts which is the moment the Big Bang was contained in a singularity and the period afterwards that led to the universe today and has spanned an estimated 14 billions years or so. This has a fatal flaw which is fully ignored and foolish to do so which is how long did the universe exist in a singularity? One second, one day or a trillion, trillion, trillion years? How about an infinity of time before hand and since nothing existed by the known laws of physics the impossible infinity before the universe could somehow have happened.

The reason this is important is that you cannot simply look at the shortest time-span possible to get over the hurdle so to speak of going from just before the moment of the Big Bang to the time afterwards where physics behaves more as we would expect. If you are going to talk about time and by your rules then you must be ready to face an eternity of nothingness before the Big Bang. In this period of infinity before the Big Bang what happened? There was no time to pass so how did moment to moment actions occur that would eventually lead to the moment of the Big Bang or how, why and what exploded? The answer to all these questions are quite impossible to find using the three main views of modern science because they are fully unequipped to deal with them but they must be asked at least to show that they are in fact unequipped to deal with them.

I will now offer further proof that time cannot exist by looking at a moment in the life of both space and matter in the absence of each other. Spacetime is what I will be defeating here as I will prove that space and time are not in fact one and the same interwoven fabric as believed by general relativity. By the end of this section I will also have proven what accounts for the apparent time dilation that occurs in certain systems. I am not saying that a clock in orbit does not go slower than a clock on Earth I am only saying that it is not because of time and it is not because of spacetime. I will be explaining the true mechanism behind these effects such that time not be needed at all to explain them. For now however let us look at spacetime as a thing if you want to call it that such as it is believed to be a real solid tangible physical substance as believed by general relativity.

Yes I said solid to get across the point that it is not an abstract according to the theory of general relativity but an actual force that can influence physical matter. So if spacetime is a thing then it is rather simple to prove it wrong as we can dissociate systems from space and therefore time. Take for example space by itself in the absence of matter and you have no curvatures of spacetime because there is nothing there to curve it. In this example spacetime has no relative frame of reference from which to measure itself and so time becomes irrelevant. All of spacetime and general relativity is based off of looking at objects relative to another time frame. This could be two objects attracted to one another, an object under acceleration such as the famous and unfortunately scientifically blinding thought experiment using the elevator in space example or anything else but they all involve looking at something with respect to something else even if it was only that same objects previous position. If we remove that moment to moment playback of the history of the objects world line we get a moment where the object is perfectly at rest. In the example of space with no matter we have no way to see anything in motion because according to Einstein space is the backdrop for frames of reference and while it is true space can be considered the backdrop to the universe it does not need frames of reference to be described. Space by itself is still space and so now has no frame of reference meaning it has no time because nothing is being measured from moment to moment. Therefore time is not real as it is not needed as a dimension to describe space as it exists in a state without matter in the universe.

Furthermore general relativity believes incorrectly that there are no absolute conditions to the universe and this is in direct contrast with classical mechanics that saw space and time as both absolute and relative measureable time which is outside of absolute time. Einstein used the idea of only relative time to do away with absolute time in accordance to his theory and states that no such thing exists. According to general relativity only relative time exists and this is where the idea of frames of reference comes from and where absolute time is forbidden directly according to general relativity. General relativity allows for the forbidden to be made real and thus destroys itself as it contradicts itself by describing a system of space without matter. In such a system time now becomes absolute as moment to moment nothing changes in the so called time dimension as no matter curves the spacetime and so time is meaningless. Time takes on some undefined mathematical term that is an infinity which of course is physically impossible and so proves general relativity wrong.

To further prove general relativity wrong and also spacetime as well we can look at how we need to describe the system of space with no matter. In this description we can account for all information needed to define the system in terms of only three spatial dimensions and we do not need a fourth time dimension. The space will still have height, length and width but it does not have time and so can be described without time as a dimension. This means that spacetime cannot be real as space can be described without time. It further hints at the reality that any system can be described and explained without time and I will do so later in this section.

Now we will look at describing matter without space as another way to prove time does not exist. The universe is made of two things space and matter and as such it is believed by modern science and this belief is correct that you can describe each component separately. General relativity however cannot accept such an idea because if you remove space from the equation it can no longer describe matter as it has no frame of reference in which to move.

Matter existing outside of space will cause no curvatures in spacetime as there is no space for it to curve. The result is matter free from frames of reference as matter is made to move along spacetime curvatures according to Einstein and these curvatures are where the frames of reference come from that are needed to describe the system. Since the matter is removed from the only time component of the universe according to general relativity it obviously exists without time as a descriptive dimension and so once again only needs three dimensions to contain all the information for its system.

To prove this further let us keep matter in space and simply freeze the system for a moment as in a single quantum instant. We know quantum systems to be very much real as we can manipulate them repeatedly in the laboratory and demonstrate that quantum systems exist only in definite states and not continuous ones. In this instant of frozen matter we see that in order to describe the system you need only three dimensions as it is not moving or changing its state in any way so we do not need a time dimension to describe it. If the matter is frozen it is not travelling along the hypothetical spacetime curvatures and therefore is not moving in time. It will also not be alternating between excited and relaxed states or anything similar as it is absolute in its frozen state. Therefore it is outside of time as it is not using it and yet still exists in the universe. This means that time is not a part of the universe as it is not needed for all systems whether that is space by itself, matter by itself or space and matter together but motionless.

All particles in the standard model are described containing various properties such as mass, charge and spin but none are described with a time component as it is literally not needed for the well being of the physical particles existence in the universe. These conditions exist constantly in the universe as if a particle where to stop completely it does not simply vanish from the universe because it can no longer be described with a time dimension. This goes back directly and inexorably to the flawed thinking that the universe must conform to a particular theory or idea instead of the other way around. The universe does not employ a magical emergency force to eradicate matter from the universe just because it is motionless and then magically re-create it exactly where it was simply because it starts moving again and all just to keep a theory happy. The reality is the theory that pre-supposes this is simply wrong and will be discarded no different than Newton's ideas were discarded by Einstein centuries later. A small pun here but it is time to discard Einstein, spacetime and time altogether so that we can move forward as one to the truth of how the universe works.

To begin this process we will look at a much studied and well catalogued view of modern science which is quantum mechanics and the Standard Model of particle physics. The first point that must be made here is that at the lowest energy states in the universe theoretically possible general relativity fails. I am talking about the bottom of the potential well for an object such that the hypothetical absolute zero has been reached. In this extreme state of an object general relativity does not work anymore as it cannot explain certain events such as quantum tunneling. A particle still has a chance of making a small movement in space but does so through quantum mechanical rather than general relativity means. The particle will not move by frames of reference from point A to point B but will make a quantum jump which is quite different as it does not involve a world line for the object in question and occurs in zero time. In a quantum jump there is no motion smoothly through space and so no curving of spacetime similar to an object speeding up to the speed of light. With no curvature to spacetime there is once again no time in the system only three dimensions needed to describe the system. This kind of apparent quantum behavior again contradicts general relativity directly and shows again where the two main views are incompatible. The reason this incompatibility is so important is that it sets a sort of basis for demonstrating that quantum mechanics is far more accurate in describing the universe than is general relativity.

If we look at quantum mechanics we can easily see that all the forces in the universe are accounted for already from the weak force to the electromagnetic force to the strong nuclear force. Each of these three forces can be described fairly well in terms of quantum mechanics and the particles on which they act and the force carriers that mediate those actions are all well understood. General relativity is only an attempt to describe how objects move due to gravity and that is all. General relativity does not explain how this interaction takes place with the very, very solid nuts and bolts of the particles it is meant to interact with. In other words Einstein did not understand what was happening in the system only how to measure what would be the results of mass A and mass B or a certain mass under acceleration but he did not understand at all what was really going on at the fundamental level. Instead he simply invented spacetime to account for the kind of world line curvatures he was contemplating and said this must be what gravity is.

According to Einstein gravity is not real only a trick which causes things to magically fall from somewhere to somewhere in space without forces, force carriers and no way of explaining the magical spacetime that matter was somehow touching and moving along but not touching as a force? It is rather obvious by now that spacetime is not a thing as it must have a kind of influence of sorts on whatever fabric it is theoretically woven from and it must touch in some way and therefore interact with in some way the particles that make matter that are supposedly responsible for its curvature in the first place. In other words general relativity is wrong, spacetime is wrong and what is happening is the actions of a force and a force carrier no different than any other on the fundamental quantum particles of matter.

While I have already demonstrated how this occurs using space physics I am here to prove time wrong and the Standard Model can be used as a form of evidence in this regard. I have said this before and I will state it again here the Standard Model and quantum mechanics do not use weak force-time, electromagnetic-time or strong nuclear force-time to describe systems and to explain why these systems work. Therefore there is no such thing as spacetime as the universe would not create a separate substrate for gravity to be a part of simply to once again satisfy a theory rather than have a correct model explain the universe. This leaves the obvious question of if spacetime is not real and therefore time is not real how can effects such as time dilation be accounted for in experiments such that the apparent discrepancies in time are still accounted for but explained using more quantum like systems? I will now explain how these things do in fact take place without time by first explaining what we as humans mean as time.

To begin with no one said that time has to be real and therefore time does not have to be a fourth dimension at all. Humans are like any other intelligent creature and they all understand time as something that accounts for one moment to the next. Time has always been something that humans have used to measure things and the earliest examples are some of the ones I gave at the very beginning of this book. Stonehenge, the Goseck Circle and the Nebra Sky Disc are all examples of ancient humans using time to account for one moment to the next. The first attempts to account for the passing of moments is the use of large scale calendars that marked off whole seasons by telling people when certain solstices or equinoxes would occur. The use of this was to know when days would get longer and when to do important things like plant crops or slaughter your livestock. The point of these activities was simply to say we are counting time such that it serves us and we have chosen Earthbound arbitrary units to do so.

Take for instance the idea of a season and you only have to look at a couple of different planets to see that they differ widely. Orbital periods and rotation periods about a central axis are different for all planets in the solar system and this means the idea that a solstice will occur on Earth say March 21st will not coincide with a similar solstice on Venus or Mars for example. Likewise the three months we count as a season is different for Venus or Mars again and so the idea of time and time units becomes quite a planet-centric thing. So much so that humans have further divided time into weeks, days, hours, minutes and seconds that once again only apply to an Earth day. A Moon day lasts 28 Earth days and that is also how long a Moon year lasts which is quite different from Earth's. In other words time is simply something that humans have devised as a way to account for the passage of one moment to the next and in no way needs to be universal.

If one were to argue however that some sort of base time can underlie these individual planetary times then one is obviously talking about an absolute time that governs the clock of the universe. Absolute time is again forbidden in general relativity and so once again spacetime is incorrect and therefore time is again not real. No matter how you look at it time is simply something that humans or other intelligent creatures can use to keep track of the passing of moments in the universe. Even if time was to be real it must also be a construct of any intelligent creatures mind and the two could exist as one. However the two must always exist together you can never have a universe where time is absolute and not a construct of intelligent minds. You can however have a universe where time is not real and is solely the construct of any intelligent minds that decide to create it. What this means is that it is logically far more likely that time really is all in our minds as it were.

The very definition of time seems to give this fact away as one can describe easily a definition of anything else in the universe without making reference to itself in some fashion. This is not so with time and to help explain this try yourself to define time without using anything that explains a preconceived notion or understanding of what time is. Time is something you count with a clock or watch. That is no good it involves time already because a clock is understood to measure passage of time. Time is a second, month or decade and so on. Again no good now you are discussing units of time which involves understanding the passage of time. Time is the passage of the past, present and future in a straight line. Nope no good because all three of past, present and future are defined as parts of time and so require an understanding of the passage of time. We can stop here for now because you will not be able to fully explain time without using time itself in some form or other in your own explanation. This becomes a problem for time because everything else in the universe can be described independently of itself and these things are all physically real. We know three spatial dimensions to be real and can measure and describe them absolutely at all times anywhere in the universe and we also do not debate there existence. Time on the other hand we do not know is real, we cannot measure it and describe it absolutely at all times anywhere in the universe and we continually debate its existence scientifically and philosophically for millennia and more.

The reason it is so hard to account for time then as a dimension is because it is not really a true dimension and our own human difficulties in reconciling this fact are a huge glaring signpost that this is so. Our natural instincts are telling us that time is not real and yet our logical minds are being dragged down into the realm of wrong turns on the path to explaining the universe because we keep trying to treat time as real. We would know 100 percent that we are on the right path if we can account for things like time dilation without using time to explain them. In other words if it can be shown that things like clocks and reactions do seem to slow down under certain conditions in the universe and that these are linked not to a raw slowing of a thing called time but rather to a set of explanations that works seamlessly with other unsolved problems in modern science then we know it is the only correct answer. One of the reasons this is so is because the underlying mechanisms for the other unsolved problems in modern science have already been explained using this same method and this method serves to further reduce the complexity of the universe by removing a fourth unnecessary dimension called time from it. So I will now show you the way time dilation can be accounted for through normal interactions of space and matter such that no time is required to explain it.

The very first thing I will say here is that of course rooted in all science is a concept of human time which is not real, only in our minds and the minds of all intelligent creatures and is used to simply measure one moment to the next. Counting how long it takes for water to empty out of a bowl with a hole in it does not make time real only it is a useful tool for us to calculate rate of flow. To this end there will be use of this type of intelligent mind counting of one moment to the next in this and many examples in the universe simply because it is a tool we use. Time dilation is seen as the effect of time passing at different rates based on two things the first being the strength of the gravitational field that matter is in or the speed at which the matter is moving. The higher of either variables and the slower time seems to pass for a clock that is positioned there. This is normally explained as a product of time being a part of spacetime which is according to the theory of general relativity a curvature of the fabric of space meaning according to the theory that time is inside space in some bizarre way. This involves time as a thing and so simply time itself is slowing down and anything caught in it is also slowed according to the theory of general relativity but this is not true.

What is true is that matter is being made to interact more or less with ESS and the ESS macht that it possesses as a part of space. Some of the confusion surrounding time dilation and general relativity is that it seems to account for effects of clocks observed in space but what is wrong is why it is happening. No one will argue that a clock is running slower in orbit than on the ground but it is not due to spacetime, general relativity or anything of the sort and this is where the confusion lies. Time dilation of clocks does involve space which is true as do all interactions in the universe so this is hardly Earth shattering or cerebral splinteringly weird news and never should have been treated as so. How clocks slow down has everything to do with these interactions and nothing to do with time.

I gave an example earlier in my book about using a heat clock to measure time and create curvatures in heat-time just as curvatures in spacetime are seen as real. This is exactly what is happening again here as the heat clock was seen to go through certain physical changes notably the colder the clock gets the slower the gears move in it and the slower time appears to go and so one can argue that heat-time is real and therefore a separate part of the universe. Heat-time can be plotted as temperature against time on a graph and that line is now the world line of the system as looked at from one frame of heat reference to the next which gives the curvature of the heat-time itself. We all know this is not so and that a heat clock simply slows down in cold weather making the gears harder to turn and therefore makes time too apparently move more slowly when in reality there is a physical explanation for this based on quantum interactions of matter. The exact same thing is true of spacetime as physical interactions are responsible for the apparent slowing of time which can be explained without the use of spacetime the way apparent changes in time could be explained without the use of heat-time in the heat clock. The physical mechanism at work here is exactly the same as any other interaction in the universe which is how much does matter spend inside space and specifically the ESS and macht that make up space.

As matter exists in the universe as you have learned it must interact with ESS in space and in so doing donate part of its energy to the ESS and macht of space. This has been already demonstrated to explain a great number of unsolved problems in modern physics and the question of time as a dimension is ended here. Time is also related only as a measureable tool by us as how long something takes place and in space and matter interactions this is one of the key variables. In discussions of things like black holes and quantum entanglement it was shown that matter can spend more or less time, more or less degree of interaction with the ESS of space. In a black hole for example the matter spends most of its time interacting with space and less time interacting with other matter as the number of ESS per volume of matter has increased and therefore it is more attracted to space. It also donates more energy to space and so has less energy to donate to other particles of matter than it would under normal spatial compression. This was also shown in quantum entanglement to be true as the two entangled particles could share each others macht glow even though they are in far lower spatial compression than a black hole and therefore do not have the use of more ESS per volume of matter. They are still however able to spend more time interacting with each other rather than with other matter because the faster than light macht speed sharing of energy permits this. So in two different unsolved problems of modern science we see how ESS and ESS macht glow can account for varying degrees or times of interactions between space and matter. In time dilation events the spatial compression is higher than normal spatial compression which is the arbitrary spatial compression found here on Earth.

If we look at extreme environments like a black hole we have high spatial compression due to gravity and this is the first form of time dilation I will explain. In a black hole it is known that gravitational forces are very high and that time seems to slow down. To begin with we know that singularities do not power black holes and time does not stop inside a black hole and both of these are based off of relativity and so are incorrect as relativity itself created the impossible and self-contradicting singularity. After all a black hole is where according to general relativity time is supposed to stop meaning that inside a black hole spacetime is a forbidden object and so the whole singularity which needs spacetime to be created is again wrong and therefore time is wrong too. That being said time does slow down in high gravitational fields which are a sign of areas of space that have high spatial compression and thus high volumes of ESS per unit of matter and strong macht glows. The matter under normal spatial compression would be able to interact with itself at normal rates but near a black hole it cannot do so.

The reason this occurs is best explained by looking at a living organism which we will say for this example is a human. A human is near a black hole and appears to be waving very slowly as compared to when the human was waving on board the spaceship that placed it near the black hole. It is often thought that the raw thing which is time in the universe simply flows slower for the person near the black hole than the person onboard the spaceship and this is not true. First of all explain exactly what a flow of time is anyway? What is flowing, from where to where and where does it all pool up? These simple questions highlight once again the evasive nature of trying to describe something because in reality it is not there to describe. So how does the person near the black hole wave slower than before and why does the wristwatch they wear differ from the wristwatch of the other human onboard the spaceship?

The answer lies in how much the matter that makes that person up and all of the matter they are wearing at the time like the wristwatch spend interacting with the ESS of the local area of space where spatial compression is much higher. In this higher area of spatial compression the matter tends to spend more time and give more of its energy to the ESS that it inhabits. What this means is that all functions of the person such as muscle fibers contracting and relaxing to make their arm move or the motions of the gears in their watch is slowed. If we look at how their muscle fibers move at the atomic level we find that the necessary motions involve two molecules essentially gripping each other tightly and then quickly sliding along one another to then grip again. The use of actin, titin and myosin filaments allow your body to move as they interact with one another and these interactions are governed by simple quantum mechanics of atoms with one another. The person near the black hole will appear to age slower and this is also incorrectly attributed to physical time slowing down. The way people age is through tiny imperfections in the near countless replications of your DNA. A wrinkle for example is brought about by DNA not correctly replicating itself over time as it has been slightly damaged. In order for your DNA to work it must unpack itself, then unzip itself at a certain gene sequence, then replicate the protein it wants there at that specific location and then zip itself up again and pack itself away again. All of these steps, as is every process that involves DNA is a molecular one and molecules are made of atoms which are made of fundamental particles which are quantum objects.

So the aging of the person or the contracting and relaxing of their muscle fibers that make them wave are the same as the quantum particles in their wristwatch as they are all bound by the same set of space matter interactions. Any piece of matter that is brought into contact with a strong area of spatial compression will have its fundamental particles interact more with the ESS of the local area of space and with the ESS macht found there. This means that the atoms making the muscle fibers of the person or the DNA of the person spend more time interacting with space and less time interacting with each other. So it takes more effort and more time on the part of the atoms in the person to wave or age because they are not readily interacting with one another. The amount of time and also energy that they can spend interacting with one another is made easier if they are not interacting as much with space which is also what is found in areas of low spatial compression.

The biology that powers both people in this example is the same, the way they wave or age is the same yet they do so at different rates because the person near the black hole cannot at a quantum level interact as quickly with their own quantum particles due to higher spatial compression. Remember matter and space interact and so if you increase the interaction of matter with space you must therefore decrease the interaction of matter with other matter so to speak. It simply takes more time and effort for an atom in the person near the black hole to correctly interact as it normally would with another atom of that same person and so all the events occurring inside the person such as waving or ageing are slowed.

This explains how the exact same physical processes can occur in both people but at different rates without using time as part of the equation. Instead the amount of time flowing for both is the same but the rate at which they both process actions internally is different. You can explain time dilation of watches and people in areas of high spatial compression without the use of time. The mechanism at work is purely physical and not temporal at all as we only perceive it as a change in time and this is identical to the quantum mechanical reasons why the heat clock runs slower for a physical and not a temporal reason. In reality what is going on in terms of rates of interactions and degrees of interactions of both the space and matter for the system in question is in fact fully conserved at all levels. The matter orb and the ESS and macht within it are all conserving their total energy as I have already explained in a previous section where the total macht/energy of the system is constant and maintained. The exact same thing is occurring here only because it produces measurable changes in rates of reactions of fundamental particles it is incorrectly viewed as altering time by relativity.

In truth time is not part of the equation and the changes in rates of reaction can be fully calculated from the conserved macht/energy of the system in terms of the space and matter found in it. Think of a simple sliding scale if you will where the more spatially compressed the local area of space the less energy or time if you will that matter has to interact with other matter. The more you slide the scale towards matter interacting with matter the more spatial rebound you have. You can move the slider anywhere you want and you will always find the true amount of spatial compression of the ESS and the strength of the macht glow there as well as the degree of interaction of matter with space and the rates at which that matter can interact with other matter as the total macht/energy of the system is conserved at all times. If you want me to put this in a time context for arguments sake then I will and it runs as follows which is time slows for interactions between quantum objects in space but flows at the same rate for quantum objects in their own space.

Now I will use the same non-time dependant explanation to explain time dilation for fast moving objects. This explanation will be somewhat shorter as much has been covered in the previous high gravitational field example. To start off with any object moving through space is of course at its base level either a quantum object or made of quantum objects which are all matter orbs in our little ESS sphere room. If we look at the orb at rest it is contacting a normal number of ESS and can be said to be under normal spatial compression. As we move the orb through the room from let us say right to left we see that it is contacting more ESS in a shorter amount of time than usual. The amount of energy it gives to the ESS is usually the same for the system but as it encounter more ESS it will naturally give a little more energy to the ESS. Here I must remind you of the lag in the ability of matter to track to ESS macht that we spoke of earlier which as you all remember is the product of macht speed being much faster than the light or below light speed of normal matter reactions. As the orb moves faster and faster the matter it is made of encounters more and more ESS in the same space of time and so gives more and more energy to them.

The ESS that the matter orb leaves will still contain some of the macht glow which they must share through schales with the ESS that are newly contacting the matter orb. They lag between ESS macht and energy which is conserved for the local system begins to affect the orb because in order to remain rooted to the ESS it must not violate the conservation of macht/energy of the system. The answer to this is to give more energy to the ESS in order to obey the natural laws of the universe which state that matter is attracted to space. This begins a cycle where the faster the matter is made to move through space the more ESS it encounters in front of it and the more macht those ESS have. The ESS as we know spatially compress if given enough macht and will do so slowly at first but rapidly as the speed of light is approached. The speed of light is key here because it is the fastest speed available to matter and also the limiting speed at which macht/energy is shared in the system. The lag in the sharing of energy from the slow quantum speed of matter to the ultra-fast macht speed of space is not noticeable that much in normal conditions and explains why it jumps up suddenly at close to the speed of light as this is the point where the effects become a limiting factor in physical movement. Remember physical movement of quantum objects is what makes the person wave slower, age slower or their wristwatch move slower.

At speeds close to the speed of light the amount of macht glow from the donated energy of the matter is so high that it travels ahead of the matter due to natural macht sharing between schales and causes the ESS that the matter is approaching to spatially compress to a high degree. In this way the matter simply moving through space and obeying the natural law of the universe which is to conserve macht/energy for a given system will create a situation in which the space is highly compressed around it and therefore makes itself interact to a higher degree with empty space. If you try to make this particle interact with any other particle in its local area of space it will encounter the same difficulty as a particle close to the black hole and will find the abilities it used to possess to do so now diminished. The more energy you try to use to accelerate the particle only furthers its interactions with the ESS and the macht it encounters and so you are attempting to overcome the macht/energy conservation of the system. The speed of light is the fastest that matter can share energy and so if you try to move it faster than light through space you will be breaking the natural contact of matter with space and violate the conservation of macht/energy. The force carriers for the matter will no longer be able to keep up with the motion of the particles and the energy donated to space will be lost in the macht.

This of course cannot happen as the conservation of macht/energy forbids it and at the same time when you accelerate matter through space you only serve to strengthen the ESS macht glow of space meaning you only ensure that the matter will feel more contact with a higher volume of ESS per unit of matter and with higher macht glows. In essence the harder you push the more you guarantee you will not overcome the speed of light holding the matter to space. You are in a sense defeating yourself by fooling yourself into thinking you could win over the universe. For arguments sake if you did manage to move matter to the speed of light then you would create the same conditions as the Big Bang where matter is now fully dissociated from space and space can rebound at macht speed and you had better not be around that point in space if it happens. The use of spatial compression to explain time dilation in matter at high speeds also explains the slow waving, ageing or passage of time on a wristwatch in exactly the same way as the black hole example. Once again a physical explanation accounts for the time dilation without using time as a thing in the universe and once again the same slider rule applies here as well. So you can explain time dilation without using time for both areas of high gravitation and high speed in the universe and with the exact same physical mechanism of ESS, macht and matter interactions that govern every other unsolved problem in modern science.

The end result here in discussing the fact that time is not real means that time is not a true dimension in the universe and so we do not live in a 3+1 dimension universe after all. The idea of time can be kept for the purposes of measuring one moment to the next as done by any intelligent creature in the universe. So time can be kept as a tool of science, mathematics and general use in everyday life for things like timing the planting of crops as our ancient ancestors did or to make measurements of systems that do not in fact have time in them. Time as perceived by humans and all intelligent life will of course be used to describe all manner of events but at the most fundamental levels time is not a dimension and that question can finally be put to rest after untold thousands of years. The reason this is so is that beyond the time construct of a measuring tool you can describe and account for all the interactions and parts of the universe without it. Time is a wonderful invention no doubt but an invention as well and is not a dimension in the universe leaving only three spatial dimensions as the true number of dimensions.

Facts

The key facts for this section are firstly no matter how you describe a system physically you can always reduce it to three dimensions. The second fact is that extra dimensions can also be reduced to three dimensions making them a moot point in terms of absolute number of dimensions in the universe. Much like protons and neutrons are not fundamental particles because you can get a little smaller than them so too are extra or compactified dimensions not fundamental dimensions either as they can be described by the three true dimensions. The third fact is that the sheer number of flaws, inconsistencies and impossibilities ascribed to extra dimensions prohibits them from being a part of the universe as the universe always tends towards simplicity. The fourth fact is that both space and matter share in having three dimensions with the exception that much of matters apparent size derives from its point like motions in three dimensional space. The fifth fact is that the number of dimensions can be reduced to one dimension for the universe simply for the purpose of making information describing the universe reduced when ESS and matter are treated as point like particles each of differing size and motions.

The sixth fact is that time is not a real dimension and can be proven to be so by examining the flaws and contradictions of spacetime with respect to the Big Bang, singularities and when compared to other known forces of the Standard Model of particle physics and quantum mechanics. The seventh fact is that time is a construct of intelligent minds such as humans and is arbitrarily determined by each different reference point in the universe such as planet to planet. The eighth fact is time dilation is explained using physical means and not time itself. The ninth fact is that both areas of high spatial compression due to gravitational fields or speed give rise to the same interactions of space and matter that create time dilation. The tenth fact is that matter spends more time in contact with the ESS and macht of space than it would normally and so cannot interact as much with other matter. The eleventh fact is that this leads to decreased rates of reaction between fundamental quantum objects and so perceived time slowing in macro scale systems like the waving or ageing of people near a black hole. The twelfth fact is that the total degree of interaction, energy or macht for the actual matter orb and the ESS spheres is conserved. This conservation of the system can be viewed as a slider bar where increased spatial compression results in decreased interaction of matter with other matter. The thirteenth fact is that time dilation can be explained using physical systems and not time for both types of time dilation and this mechanism is the same one that is already used to answer other unsolved problems in modern science. The fourteenth fact is that time is therefore not a true dimension and exists only as a measuring tool leaving the universe as a 3 space system not a 3+1 system.

Chapter 23
Life History Data of the Universe

The Future of the Universe

The next unsolved problem of modern science I will answer is the fate of the universe or more accurately the life history data for the universe. Life history data is all the information necessary to describe the life cycle of an animal from its birth through to its death and what happens to it at what stages of its life along the way. To explore the life history data of the universe I will first talk about the death of the universe as we are approaching that point in the far distant future. Afterwards I can go back and explain the birth process of the universe and in so doing I will explain the what banged, why it banged and what was there before it banged for our universe as well. Those are also unsolved problems of modern science and will be addressed here but not until after the universes eventual death. To begin with we must look at the possible fates facing the universe as favored by the three main views and these are things like the Big Freeze, the Big Rip and so on that we have examined in the past. There are also two other fates called the Big Crunch and the Big Bounce but they are currently not favored by modern science as there appears to be no way to stop the universe from accelerating its rates of expansion and so the idea that it would stop and constrict inwards on itself again is not seen as plausible. I will explain the exact fate of the universe using simple space physics to show why the universe will indeed behave in a certain way no matter how far forward you look in its life.

The Big Freeze and Big Rip are similar in nature in that they have no mechanism to stop the expansion of the universe and as such do not have a method by which they can give birth to another future universe. In the Big Freeze nothing too drastic happens only all the energy in the universe tends towards a minimum as entropy tends towards a maximum and everything in the universe becomes so spread out that no new methods of heat generation can be found. The final result is all stars eventually dying out and their stellar remnants simply cooling to the point of absolute zero or evaporating as radiation if they were black holes in many trillions of trillions of years. This theory is the ultimate expression of physics when a system is left alone without any external influence and it simply cools off and all reactions in it stop, meaning it is at the very, very bottom of the potential well. It is not a flashy or spectacular theory in terms of the fate of the cosmos but it is a realistic one according to modern science.

The Big Rip is far more exotic and far-fetched than the Big Freeze as it involves the idea that the run away acceleration in the rate of expansion of the universe is never checked and only grows in strength. In this scenario the expansion of space never ends and eventually the space between atoms spreads out so much that even atoms and fundamental particles simply dissolve as they cannot hold themselves together. This idea is slightly less favored over the Big Freeze because although it matches current rates of acceleration of the universe it also assumes that the rate of expansion will never be checked and continue growing ever stronger forever. This of course involves the influx of huge amounts of additional extra dark energy that no one has any idea how it would be created from nothingness or where the force necessary to create it would come from. Also if it is seen as increasing in amount the dark energy should also be able to run out so to speak and therefore not have enough force to cause the Big Rip Afterall. In addition one must be wary of the human-centric view of the universe where it could be that we are simply observing the universe at a particular specific time where it appears as though expansion will continue forever and this is not so. Nevertheless as theories go it is being considered as a possible fate to the cosmos.

The next two theories involve the opposite fate of the universe where it does not expand forever but instead will halt its expansion and then collapse back onto itself. The Big Crunch is the simplest and most favored of these two as it simply involves everything in the universe rushing back towards the center of the universe probably from where the Big Bang itself started and simply crushing all matter, energy and space back down into a single infinitely small mass that destroys the universe in the process. This is followed by the idea of the Big Bounce where all the same matter and so on rushing back into the center of the universe simply explodes outwards again and creates another universe such that the universe lives in cycles. Both of these theories are less favored than the two expanding fate ones simply because no one in modern science has a workable mechanism for the way the universe would halt its expansion such that a collapse process could then become dominant. The Big Bounce is the least favored because besides the unknown collapse mechanism it also requires an unknown explosion mechanism necessary to trigger a new universe from the collapsed existing one. In short the four main theories of how the universe ends are so different from one another in terms of opposite ultimate fates that none are satisfactory in explaining the fate of the universe. None of these four theories can provide solid workable models by which they can use current information and observations from the universe to explain step by step how the universe will end and exactly why it will do so. In other words why is the mechanism driving the end of the universe correct?

So I will now explain how the universe will end and why it will end this way and I will do so in two parts. The first is to explain the properties of the entire universe such that it explains why it will keep expanding or stop and collapse. The second part will be an explanation of the mechanisms and exactly how they work that will allow this to happen. These two parts will lead naturally into an explanation of the Big Bang in terms of the what banged, why it banged and what was there before it banged. So let us start by looking at the first part which is the different properties of the universe and whether or not it will expand or contract on itself.

The first property of the universe that must be considered is the amount of space within it and what this means to the universe as a whole. Space is one of the two components of the universe besides the stuff within it like matter and energy and as such a crucial property of space helps determine the fate of the universe. This property is that space is finite in the universe as a set amount was created at the beginning of everything that does not increase or decrease as the universe ages. The finite property of space is perhaps the single most important feature in figuring out how the universe will age and as most people think of it eventually end. Matter is the other thing you need to make a universe or matter, energy and all the stuff that the universe uses to build things that we are familiar with to study, poke, prod and experiment on. Matter is likewise finite in the universe as you do not get the universe making more or taking any away as the universe ages. In other words everything in the universe is finite and as such negates one of the possible fates of the universe immediately and that is the Big Rip. The Big Rip works on the assumption that the acceleration phase of the universe will continue to run away levels eventually leading to a demise of all matter and even space itself will finally tear itself apart.

This is impossible as in order to accomplish such a feat two things must be ignored the first being the continual acceleration of the universe expansion phase. At one point in the past the universe slowed its rate of expansion and so can do so again; nothing says this slowing mechanism is gone like a whisper in the wind. The fact that we are observing the universe at the exact moment that it appears to be accelerating is hardly proof enough that it will continue until it rends itself asunder from its smallest particles of matter to even the space that binds everything together. The second reason that this is quite impossible is that in order for the Big Rip theory to work energy must be input into the system at cosmic scales continually and in ever increasing amounts forever until the universe rips itself to pieces and even afterwards more energy continues to pour in only the effects are now quite moot. Remember even if the universe is increasing its rate of acceleration without an eternal influx of energy the universe will not tear itself apart because if the energy input is not eternal you are simply left with a rapidly expanding universe that is no longer accelerating and so cannot produce the forces necessary to tear matter apart.

Where to start with this flaw? Well to begin with where does the energy come from to enter the system as the system is the universe. This means you cannot take energy from outside the universe which constitutes everything that is ever possible in existence and put it into the system. Second the energy must be poured into the system literally on a cosmic scale as in order for the universe to destroy itself from the inside would require equal energy distribution everywhere perfectly at once or imbalances of energy deficits would immediately manifest themselves such that areas of the universe could survive the expansion as unequal expansion would push space and matter into those energy depleted regions constantly. The depleted regions even if they had this magical energy being poured into it would have no chance to outpace the more energetic and faster accelerating regions of space and so would be compressed and not at all perish. So in order for the Big Rip to occur it would take energy from an impossible source pouring perfectly with not even a Planck unit of deviation anywhere constantly in unstoppably huge amounts for trillions of years to pull off this Big Rip. This is of course quite impossible and so the Big Rip fails once again.

The last way the Big Rip fails is that the energy is thought to be dark energy which is at best theoretical and requires matter to essentially die off in the rest of the universe so that it may take hold. Since matter is finite as well as space you cannot manage this feat especially because it has been shown that dark energy does not exist. Matter becoming energy as in the equivalency theory is not dark energy which is not electromagnetic radiation and exists only as a theoretical quantum construct of the vacuum potential of every infinitely small point of space. The deceleration and acceleration phases of the expansion of the universe have both been accounted for as a natural phenomenon of ESS macht and matter interactions following a Big Bang event class explosion. So now I can incorporate ESS macht and space physics such that the fate of the universe is simply and straight forwardly explained.

The Big Freeze will also not occur for reasons that will become clear and the mechanism by which the universe will halt its expansion and collapse inwards that is not at all understood by the Big Crunch and Big Bounce theories will be revealed as well. Even the idea of evaporating black holes leaving nothing but a sea of energy washing through the universe is impossible and can be easily dealt with using ESS macht. So let me jump right into it and tell you how the universe eventually ends up and it does not die for starters. The first thing that happens to the universe is that it slows its rate of expansion until the universe reaches a maximum size and stops expanding. The second thing that happens is obviously the universe begins contracting inward on itself heading backwards to roughly the point of the Big Bang. This point will no doubt have drifted slightly in absolute space coordinates from when it was born due to shifting masses and densities of spatially compressed ESS. The third things that happens is that the universe collects all of its space and mass back into one super region of everything and then something happens to that mass. The thing that happens I will explain later in this section once I have gone through the slow down, collapse and recycle mechanisms. What happens finally is not entirely what you might think but will make perfect sense once you hear it and the universe can be said to have more than one end although they all experience the same final outcome.

The universe as we observe it today is expanding still and accelerating as it does so. The acceleration mechanism I have explained in a previous section and will not repeat it here but I will say that it will not last forever. The finite amounts of space and matter that I mentioned earlier in the universe play a key role in stopping this acceleration. The finite size of space alone is enough to stop the expansion of the universe without the aid of matter due to the way space is constructed of ESS and macht. Remember that ESS is the true shape of the universe as it constitutes all the space within the universe and the universe does not extend beyond the boundaries of space. Space therefore is the limit of the universe and since that amount of space is finite all you need to do is show how space holds itself together. Space as you now know is made of ESS and ESS contain macht. The macht is how ESS share amongst themselves and each ESS has a macht pool that contains reserve macht if you will. Macht pools were dealt with as well in a previous section and are what actually hold the universe together. If we return to our little room filled with ESS spheres and no matter orbs as we first saw the universe many sections ago the ESS are quite relaxed. The ESS in the absence of matter are at their most spatially rebounded that they can ever be and this is the largest possible volume for all ESS in the universe. ESS will only ever decrease in size as they encounter matter in the universe and as that matter is made to interact more or less with them. Things like variable gravity as seen in black holes are an example of how finite matter can be made to increase ESS macht and therefore decrease ESS volume.

It is thought that matter will slow the expansion of the universe and that without matter the energy in the space will tear the universe to shreds. This is impossible for two reasons the first being the finite amount of space in the universe. A finite amount of space cannot ever spread out or expand enough to destroy the finite amount of matter in the universe. The second is that the space will eventually rebound to the maximum amount that it can given the amount of matter or energy in the universe and this will never be fully rebounded as in the room without matter in it. Even if the matter in the universe were somehow fully removed and did not interact with the space in the universe in any way the ESS would still only rebound to their maximum volume and sit there. Remember space is not carrying kinetic energy with itself at the edges of the universe as the universe only expands due to the finite ESS expanding. This means that the ESS can never be given an escape velocity sufficient to leave the universe. The ESS expand against each other and push one another farther out, the result is what we see as an expanding universe. So even if the universe had no matter and were fully expanded no momentum would remain to tear space apart. This is where the second factor of ESS macht comes into play.

ESS all contain a pool of macht that is never depleted from space and ESS all naturally share macht between each other. The result is a second line of absolute and unbreakable defense against the universe ever destroying itself. Due to the fact that space is the maximum volume the universe can ever achieve as matter is bound to it and given the fact that space is made of ESS all containing a macht pool that they must share with each other the ESS can never dissociate from one another because they are all attracted to one another in their need to share macht. This macht pool ever present even in the complete absence of all matter means that ESS will always need to balance macht and so will always remain firmly attached to one another forever. It is physically impossible for the universe to ever rip itself apart no matter what conditions are given or exploited. In addition since it is impossible to have the universe free of all matter and energy the ESS will always have an elevated macht glow which only further guarantees that it cannot ever destroy itself through run away acceleration of expansion rates.

This demonstrates how the universe cannot ever harm itself in any way and now we can focus on how the universe will halt its expansion and begin to collapse inwards on itself. The first method is simply by letting the universe expand as far as it can go and reach its ESS spatial rebound limit which as you have just seen will fully halt all expansion. This is the simplest method to stop the universe expanding and at this point the fact that the outward force of the universe is now incapable of getting stronger and all momentum is lost the universe will only feel one force in terms of expansion which is gravity. The fact that any matter is still creating gravity will mean that a very small non-zero force is being applied to all space that seeks to move inwards towards the source of the strongest gravitational field which in actuality is the source of strongest ESS macht glow. Over time the universe will simply pull itself into a dense space and mass once again which we will explore shortly. To explain that mechanism we will look at the second way that the universe can stop its expansion and do so before maximum universal volume due to ESS full rebound has been reached. This second method involves variable gravity and the evolution of objects in the universe.

The universe started out with nothing but space, energy and then fundamental particles not at all interacting with each other to form larger structures. Not even protons existed in the very beginning and this means that certainly no large objects like stars were present but as the universe grew up stars did begin to form. The early universe did not yet have black holes which have pronounced variable gravity however this would change as the total percentage of mass in the universe would begin to shift from 100 percent of matter creating what we think of as normal gravity to a small percentage creating variable gravity. The universe has been slowly shifting to more and more black holes over time and this means a shift away from the 100 percent normal gravity to ever higher percentages of variable gravity. Black holes as we know have near immortal life spans when compared to stars, gas and dust and so will always live long enough to consume more matter and energy which contributes to matter and therefore grow ever more massive. Black holes create variable gravity in copious amounts which when viewed over the life of the universe means that in the far future a huge percent of the gravity will be in the form of variable gravity. It is possible that 100 percent of the universes gravity will be locked in variable gravity at some point in the very distant future and if not then very close to 100 percent.

In the event of sufficient amounts of variable gravity the expansion of the universe will be easily stopped as the finite amount of space is now under the influence of the variable gravity. Space as I said before is not still being created by the universe as this would mean creating more ESS and macht from nothing which violates basic laws of physics stating you cannot get something from nothing. So it is a fact that space is finite and as such the ESS in it will over time only feel ever increased macht glows from gravity which they must share. As ESS of finite amount are forced to share macht they also spatially compress and decrease in volume. Since the expansion of space is caused by individual ESS pushing against their neighbors a spatial compression of ESS will halt the expansion and cause a collapse. The sharing mechanism of ESS macht also allows for deceleration, halting and collapsing of the universe to take place in non-uniform ways unlike the dark energy injection theories.

You can alter how much macht is in a given region of the universe and not create imbalances in expansion rates as seen in the dark energy theory because all ESS share macht. So any increase in macht gravity glow in one region of space will not remain localized as the ESS share the macht at macht speed and so will balance out for the entire universe. The net result is a large amount of variable gravity in any part of the universe of sufficient strength will always cause a collapse of the entire universe. Since we have observed that the universe is very uniform in nature the collapse events will occur roughly at the same times and in the same amounts across the universe given cosmic scales and lifespan. The collapse will also be seen as the reverse of this current acceleration phase and be non-passive. In other words the acceleration is apparently receiving more force to move faster and faster and is not doing so passively.

The collapse will be the same and not resemble free fall as proposed by the three main views. Instead the continual increase in variable gravity will actually cause objects not to fall towards the center of the universe but to be powered there by the continually increasing strength of the active quantum force of gravity. The sharing of macht glow will always lead to a collapse event and in fact it is utterly ridiculous to propose an eternal expansion mechanism once the nature of ESS and macht are known. The universe will halt its expansion either due to the maximum ESS volume being reached and macht pool sharing becoming the cessation factor or through increases in variable gravity objects throughout the universe sufficient to cause macht sharing to force ESS to decrease their volumes and therefore collapse the universe.

The very far future of the universe is no doubt filled with many super massive black holes that have billions or trillions of years ago consumed all the matter and energy they possibly can given their local region in the universe. The universe is still destined to collapse even if black hole evaporation were to occur to some of the black holes. The universe contains a simple recycling mechanism for gravity in the universe which is the ability of black holes to convert energy into matter far longer than regular matter normally can. As we know all matter can absorb a set amount of energy before it has reached its limit and must emit radiation as a dissipation product of new incoming radiation it absorbs. Black holes however use spatial compression to increase there capacity for absorbing and holding energy for near indefinite periods. As energy enters a black hole it is immediately absorbed into the black hole proper beneath the event horizon where it is forced to merge with the matter which comprises the black hole. A quick note here about black holes is that the way they work will be fully explained in a future section and will include the type of unique matter that black holes are made of as well as all properties of that matter and its formation.

For now that matter absorbs the energy which it then uses to fuel ever stronger variable gravitational forces inside the black hole. The black hole holds the matter and energy and so increases in mass which of course increases its gravity again and again. This ability means that free radiation in the universe is never free as soon as it encounters a black hole it is absorbed and added to the total gravity of the black hole. So in a universe filled with essentially nothing but black holes any evaporation taking place from one black hole which seeks to decrease that black holes mass and gravity will only be absorbed by another black hole which therefore will increase its mass and gravity in response. So here we see once again the total effects of finite space and finite mass in the universe as the universe will always contain the same amount of gravity no matter what you do to it. In a black hole free universe the space is able to expand by increasing the volume of its ESS as it has low macht glow of gravity. If you let that universe age the macht glow of gravity increases as black holes form and the finite matter is made to interact with the ESS more strongly or to a greater degree and so the gravity is shared such that the total volume of the universe is decreased. Again we see a sliding scale to volume of the universe and amount of gravity and this sliding scale is controlled completely by ESS macht and matter interactions with space by sharing or donating energy from matter to the ESS macht. The law of conservation of macht and energy guarantees that nothing is lost from our universe no matter how it appears only held in objects like black holes and the sliding scale demonstrates this apparent loss of information as well and will be dealt with later.

Therefore no matter what you do to the universe it has finite space and matter that will always mean that in the future black holes will conserve their total amount of variable gravity even if black hole evaporation events take place. Gravity due to variable gravity will remain constant in a universe filled with nothing but black holes as evaporation from one will lead to mass increase in another. The end result is a universe where the ESS must share large amounts of macht glow of gravity and will result in the universe collapsing. Remember that the universe in the beginning contained all of is parts in one meaning all space and matter and so must always be able to do so. This means that the universe always has and always will contain a balance of space and matter such that it can explode and collapse as the total amount of space and matter share gravity on this sliding scale which cannot be broken as it is possessing all space and matter in the universe.

Now the universe has stopped expanding and begun collapsing in on itself due to variable gravity and I can finally explain what will happen once the collapse reaches its conclusion. The universe at this point is made almost entirely out of black holes and any free matter or energy will be quickly consumed by the black holes present in the cosmos. The black holes will merge with one another to form super massive black holes just as we have at the center of our Milky Way galaxy. These super massive black holes will eventually be forced to merge with each other as the universe shrinks around them and the volume of space that is available for them to occupy decreases. Finally the universe will consist of little more than a single huge black hole which contains all the matter and energy in the universe. The final black hole will also pull any free matter devoid regions of ESS back into it at faster than the speed of light in a reverse look at how the space expanded faster than light without matter during the Big Bang. This black hole is sitting at the exact center of space and as such the space has spatially compressed the ESS to their limit and they do have a limit. Here it is necessary to explain what the space and matter of a black hole are in slightly more detail.

As we know matter is made to interact more strongly with ESS inside objects like black holes and therefore become locked there due to variable gravity. The matter is so infused with the space of the black hole that it cannot remove itself easily from the ESS it occupies. This also includes energy that is captured by the matter in a black hole and is the true reason why black holes are black. The energy coming into a black hole is absorbed by the matter of the black hole and under normal conditions would be able to be emitted from the matter again once it had reached its saturation or maximum carrying capacity for photons. The normal result is the matter glows in some wavelength of the electromagnetic spectrum but in a black hole this is not possible. The reason is that as soon as the matter of a black hole absorbs the energy it uses that energy to help fuel the ESS macht of the local area of space. Since ESS seek to balance macht amongst themselves the energy is locked into space and not able to be used again for emission purposes by matter under these spatial compression conditions. So matter cannot emit energy or light from a black hole because it does not have any extra energy to emit in the first place as the extreme spatial compression of the ESS the matter inhabits is locking away the absorbed energy. The need for an escape velocity equal to or faster than the speed of light is not necessary nor is it the reason black holes are black. This is a special set of conditions that only exists inside a black hole or similar object because of the need for extremely dense matter and ESS as well as bright ESS macht glow. The nature of the matter inside a black hole is unique and will be dealt with in a future section as it has properties normal matter does not. For now however the collection of matter is so infused with space and unable to free itself normally from this interaction that the space and matter become one.

This is where I get to have a little fun in naming things scientifically and use a very obvious and straight forward description of this unique interaction of space and matter as I call it for simplicities sake: space/matter. The space/matter is a new form of matter or space or both if you will in the universe because nowhere else do interactions of the type needed to make space/matter exist. Only inside a black hole or similar object do you have the necessary conditions of high spatial compression of the local area ESS and increased macht glows as well as high degree of matter donation to space occurring. Normally ESS try to balance macht and this is what we see in things like quantum motion or electron drop down but since a black hole has unique properties these balancing actions cannot occur. A black hole has for example essentially only gravity acting inside it with all other known forces turned off and this means the balancing of ESS is lost accept for balancing gravity. Since ESS balance gravity by sharing macht glow and since macht glow compresses ESS the space inside the matter never decreases and will create a circular system where by it holds the donated energy form the matter forever. The matter is attracted to ESS and so is forever bound to the ESS inside it and the macht glow they possess.

Again in the future section where I explain black hole matter I will explain why only gravity is the active force inside a black hole proper and why the other forces can still be active near what modern science calls the event horizon. The net result is space/matter which cannot be undone by any external forces in the universe but of course nothing in the universe cannot be undone. The key is in order to undo space/matter which is at the heart of a black hole whether it is a normal black hole or the single huge black hole left at the collapse of the universe is to undo the black hole from the inside. Only opening a black hole from the very center of its core can you hope to free the matter from the space to which it is so heavily interacting.

Space/matter is as I have said matter that is interacting to a very high degree with the ESS it inhabits but it is not fully interacting with these ESS. Another way of saying this is that the matter has not fully given 100 percent of itself over to space through energy donation, degree of interaction or time spent inside of space. So in space/matter the mass of the black hole is still located for a small part of its existence outside of space. This is the key to keeping matter and energy forever bound in a black hole and keeping the black hole stable. This arrangement however is not a permanent one and it can be altered for example taking matter that is normally 99 percent inside space and making it 99.9 percent inside space as more and more mass is added to the black hole. Space however will resist this as long as it can as a natural property of space is to rebound as we have seen from the rapid expansion phase of the universe examined earlier. Space compresses due to increased macht glow of gravity and naturally rebounds to a relaxed state when it can. In other words space of course has a natural tendency to expand if given a chance and this is commonly observed and understood in the cosmos around us.

◄New Universe Order►

The huge black hole mass at the center of the collapsing universe however has most likely the same dimensions as space itself or therefore the boundaries of the universe at this point meaning the maximum degree of interaction of matter and space can occur. The result is that a single point of space will no longer be able to resist the inward spatial compression of all the macht around it. At the very center of the entire collapsing universe will be a single ESS which as we have seen will always be the one single mostly highly macht glowing ESS for any matter orb. The entire mass of the universe represents the ultimate matter orb and so the single ESS at its center with the highest macht glow will no longer be able to resist spatial compression and for the smallest possible instant of macht timescales will become the rarest thing in the universe which is a single piece of True Space Matter. I will now switch to using SM for space matter and TSM for true space matter for ease of explanation. This single ESS will have for a single macht timescale become TSM as it is the only point possible that is sharing all the macht from all the surrounding schales of ESS in the universe that is overcome by spatial compression. It is the rarest object ever because the universe can only ever have one single ESS of it at a time where matter is no longer 99.9 percent inside space but is truly 100 percent inside space. In addition the TSM is the rarest object in existence because it lasts for only a single macht timescale and then destroys itself.

The detonation is not a new force, a new field or any new dimensions but comes about from the same physics that explains everything else in the universe from why electrons drop down, to why quantum entanglement occurs and why the moon and apples are attracted to the Earth. There is no need for a new mechanism of exotic physics because the same proven space physics work just fine. The detonation is a result of the ESS and matter becoming fully infused and creating space matter and at this point the little hollow ESS sphere does something it has probably never done before which is have its own spherical walls touch. The instant the walls touch the ESS has fully collapsed and the matter is 100 percent inside space which causes an outward explosion of all space and matter with it. The reason this occurs is because the macht limit for the central ESS has been reached. ESS carry macht and the more macht they carry causes them to reduce in volume more and more. Once however they have no more volume to reduce and the ESS walls are touching one another the increase in macht has no where to go and so can only leave the fully saturated ESS. A simple analogy for this is the atom whereby an atom can absorb more and more energy pushing its electrons to ever higher orbital states until enough energy has been absorbed and the effect too great such that the atom must break apart. The electrons will be too energetic for the nucleus to hold onto and so the energy and electrons are lost as the atom seeks to dump off this extra energy. The electrons are lost and the atom is technically destroyed as it is no longer an atom only a fully ionized nucleus.

The ESS can be thought of as doing the same thing where it absorbs more macht than it can handle at its limits and so must release this stored energy by dumping its macht as the atom dumps its photon into the universe. The net result is an expulsion of all macht from the ESS at macht speed and from this single ESS sphere we can start off a Big Bang using simple space physics. The macht released from the ESS can only go to one place which is outside of the ESS and that is where the single piece of matter that is sitting on top of the central ESS is located. The matter however cannot absorb the ESS macht released because of its specific properties which I will discuss in a later section. However the central ESS now is free from all matter interaction and so as we have seen before is free to rebound at maximum velocity in all directions to relax to its natural state. The ESS is surrounded by a primary schale of ESS which will now feel the pressure from the central ESS as well as the macht emitted from it. This outward transfer of ESS macht is what prevents the central TSM ESS from reabsorbing more macht as the schale ESS are now overloaded too and in addition to that because the Lion's share of the macht glow for the central ESS was coming from the matter that was surrounding it with that same matter lost the central TSM ESS will not have enough macht to contract on itself again.

These surrounding schale ESS will therefore also undergo an explosion similar to the first ESS as they too are now at their macht limits. The result is the entire first schale exploding just as the central ESS did and will start an obvious outwards chain reaction to the second schale which goes to the third and so on. All ESS contained in the SM will undergo Big Bang detonation events that cause them to lose all of their stored macht except of course for their macht pools and therefore will all begin rebounding at maximum velocity in all directions. The speed at which macht is transferred and ESS can adjust their volumes is far, far faster than light speed and so the universe and the space within it can explode outward schale by schale one detonating after the other from one macht timescale to another until all ESS are free from there excess macht. The law of conservation of macht and energy must be maintained so additional extra macht is simply absorbed by the matter as in returned to the original donated energy form meaning once again ESS have lost their macht and are free to rebound. This is the rapid expansion of the universe that can easily and quickly expand space to unimaginable sizes at speeds far faster than that of light speed.

The macht that has been freed from the ESS does not exist as free macht for long and of course reverts to normal energy and so the early universe has a sea of high energy permeating it as one would expect. In addition the matter has undergone a significant change as well during this explosion period. Remember that the matter orbs are larger than ESS spheres and so as the ESS spheres expand at macht speed they will literally explode matter orbs from the inside by spreading out within the orbs themselves. The orbs are of course bound to space and so as the ESS expand the matter orbs must be expanding with them to volumes larger than the original matter orb. This means the matter orbs must automatically breakdown into the smallest fundamental particles possible in the universe. This also produces the necessary sea of fundamental particles out of a sea of pure energy as one would expect in an early Big Bang universe. The matter particles cannot of course reform into normal SM matter and stop the explosion of the universe because of the limiting speed of quantum interactions which are all light speed or sub light speed in nature. The normal exchange speed of energy donated from matter to ESS is light speed or below but the transfer and motions of ESS are always many times faster than light speed and so the ESS move underneath or through the matter orbs at macht speeds which move too quickly for slow light speed quantum donations of energy to be made to the ESS. The net result is a failsafe mechanism built right into the properties of ESS and the ESS macht they share that prevent the Big Bang from stopping.

Another failsafe mechanism is the spatial rebound which is occurring throughout all the schales of the SM and therefore the universe. Since the only thing holding the matter orbs in their incredibly dense configuration was the high spatial compression of the ESS the matter orbs cannot again form SM as the spatial compression is lost in the local area ESS. The spatial rebound removes that critical factor of high spatial compression of ESS necessary to create SM as discussed in black hole physics previously. As I mentioned earlier the only way to undo a black hole is from the inside and a single TSM creation will provide the necessary detonation event to do just that. So in other words the normal interactions of matter and space that produce the SM in the first place will always create a Big Bang when a single piece of TSM is generated and once the detonation reaction chain starts nothing can shut off the explosion until all ESS have dumped their macht out again and into the universe.

Once the ESS have fully rebounded to their normal states the universe will have stopped moving faster than light can travel and so the slow limiting quantum light speed or slower reactions can finally take hold again. This is what I have talked about earlier with gravity being turned back on in the universe to stop the rapid expansion phase of the early universe's life cycle. At this point the four forces as modern science understands them can once again begin to take hold with the fundamental particles floating free within it as well as the sea of energy and begin to create larger objects like protons. I will discuss the turning on of the four forces and the interactions they produce later in a section along with the properties of matter that I spoke of earlier. What we are left with is a workable model for the end of the universe as such that the Big Rip is proven impossible. The Big Freeze is also proven impossible as a recycling mechanism is found for the universe that halts its expansion. The Big Crunch and Big Bounce are not complete theories as they do not have this crucial halting mechanism nor can they explain what happens to the matter and space as it collapses inwards at the end of the universe. The Big Crunch sees the universe as becoming another singularity which we know is an impossible object and involves ridiculous mathematical notions like all matter existing in a space of zero dimensions. The Big Bounce also does not have a complete theory as it has no explanation for the detonation mechanism that would restart the universe and of course it still lacks an expansion halting one as well.

The only model that can explain why the universe will not tear itself apart, why it will halt its expansion, why it will collapse, what the collapse result produces, if a new Big Bang event will occur and why that event will occur as well as how it is powered is through the use of ESS macht interactions with matter as described by my work which incorporates space physics and normal matter. In this way the fate of the universe as many people call it can finally be answered in a way that uses normal physics as observed in the universe as we see it today and controls everything from simple gravity to quantum entanglements without the need to create or utilize new forms of matter or energy, branes or dimensions and can be used to solve more problems than just one at a time. From here I can discuss additional unsolved problems concerning the universe specifically what banged, why it banged and what was there before it banged as well as is it possible to have more than one type of Big Bang event in our universe?

Before the Big Bang

The answers here have confounded modern science for a century but in explaining the end of our universe the beginnings of our same universe become perfectly clear. To begin with our universe is not the only universe in existence and it is every universe in existence. How this is so comes from the fact that the universe we live in is just one in a series of universes that have been exploding, expanding and collapsing onto themselves forever. This is why our universe which is the one we live in now is really every universe in existence as it is simply a part of this great life cycle of universes. Therefore the answers to those three questions of what banged, why did it bang and what was there before it banged really are all quite easy to explain.

The what banged was a huge mass of SM that was left over from the collapse of the previous universe. The universe that existed before ours really was not a separate universe just our own universe in the middle of its life cycle one cycle before us. This is also directly linked to people who say nonsensical things like when and how will the universe die? The universe is immortal and cannot die ever it is just that simple. Our universe will not die in trillions of year's time it will simply recycle itself in such a way that it creates a new universe. This has been explained through multiple mechanisms for halting the expansion of the universe and for the collapse events that will begin universal compression. The universe is made of near infinite ESS and these ESS can and do both undergo spatial compression and spatial rebound during each universe cycle.

The universe itself is basically one huge ESS that over the course of its life will cycle through phases of universal rebound and universal compression as well. So our particular cycle of the universe's life will one day undergo universal compression which will collapse the entire universe to a single large SM mass. This mass will be the beginnings of a new universe cycle marked by the period of billions to trillions of years or more of universal rebound. In other words the collapse of our universe into a large SM mass will be the what banged for the next cycle in the universe life. To this end it is rather simple to see that the what banged of our universe today is simply the SM mass of the last cycle of the universe's life. What banged was a large SM mass left over from the previous cycle of the universe's life which was also brought about by universal compression.

The why it banged is answered from the heart of this large SM mass as to be expected. The previous universe cycle collapsed to a point where all of its ESS and matter or most of it was locked in SM form until the conditions were right for a single piece of TSM to form at the SM mass heart or core. The single TSM is enough to trigger the events necessary to destroy SM from the inside as discussed before and as such unpack if you will the entire universe from its SM dormant state. The TSM exists only for one ESS and exists for one macht timescale only and then is immediately destroyed. Remember time does not exist and there is no smooth flow of time so events are ordered simply from one quantized moment to the next or in this case machtized moment to the next as the macht timescale, macht speed and ESS lengths are all far more refined than the much larger, slower, crude quantum and relativistic ones. The end result is a single macht instant of existence for the TSM before the next macht instant occurs where the TSM is now destroyed and has dumped out all of its macht into the universe again.

This is why as I have said TSM is the rarest of all objects in the universe because it exists one only for a single macht time and only once for an entire cycle of the universe in it's never ending life. The TSM is probably the only time one can truly say all four forces are united into a single force as they are for a single macht time carried perfectly and completely inside a single ESS. The unification of all four fundamental forces is made possible inside a single TSM and after that all forces will undergo breaking and dissociation from each other as needed. The ESS houses inside of its macht glow all energy of forces but only as I said for a single macht time. The natural breaking of the forces occurs after the macht is dumped back into the universe and requires no special mechanism as it uses the same space physics rules that govern all ESS, macht and matter interactions seen in anything from a black hole, to Brownian motion in a cup of tea here on Earth under normal spatial compression. So the natural properties of ESS and the rules of space physics are the reason for why the SM mass banged in the first place and will always create such future bangs as well. This is true of the universe cycle that preceded us and the universe cycle that will follow us as well as every cycle in the universe. In retrospect these questions concerning the Big Bang are not at all exotic and once ESS, macht and matter interactions are understood are in fact quite normal.

The what of the what was there before it banged question is really a simple one to answer as well as we know what was there before it banged. What was there was an older cycle of the universe that collapsed due to universal compression to form a single large SM mass. The true answer to this question is both scientific and philosophical as what the question is really asking is what preceded our universe. This was a question that was asked when the limited thinking of the day was that the universe simply did not exist before it exploded and so what was there? Sound a lot like the Milky Way is the whole universe and then a few years later we learn the Milky Way is but one galaxy in that universe?

The scientific community looked at the time before the Big Bang as a time before the singularity that was believed to have started it all. In this view physics breaks down and that is a clear sign that the current models and three main views of modern science were wrong. People were asking what the singularity was doing just sitting there inside a great nothingness for who knows how long perhaps trillions of trillions of centuries or maybe for a trillionth of a trillionth of a second. This kind of questioning is fully useless as singularities do not exist, the universe did not start from a singularity and the laws of physics do not break down simply because you are talking about a Big Bang. I could keep destroying these old models and singularity arguments but it really is quite an unrewarding endeavor as it has been proven that the universe did not and cannot ever start from a patch of nothingness populated by a single inhabitant that has been sitting there for an indeterminate length of time which cannot be counted because time does not exist at this point in the existence of everything and the object in question is infinitely massive in infinitely no space. Tell me when you hear it like that is does not sound like crazy talk?

My work offers a perfectly sensible explanation of the universe and how it came to be and this is based on the same ESS, macht and matter interactions of space physics that can be used to explain everyday occurrences as well as the more rare in the universe and do not require a breaking down of physics at certain scales nor does it fail to blend both the very small and the very large as one. To this end I ask the reader not to attack my work as an explanation for how the universe works or why it began the way I say it does but instead I ask the reader to examine that incomprehensible and contradictory claim of one paragraph ago that uses a singularity to explain the beginning of the universe and instead of attacking me prove in black and white how that mess can be made to even remotely make sense in such a way that it offers a simpler explanation than mine and does so with a single central mechanism like my space physics that can then be used to account for everything else in the list of unsolved problems of modern science. I mean no offense to any reading this and I am certainly not attacking anyone I am only proving a point here that the simplest explanation seems for more rational and logical than one that involves numerous hidden variables, extra dimension and the like. The goal here is not to tear down a theory without offering a new and better one but to defend your own viewpoint. I cannot see how the claims of a singularity mechanism that involves the breakdown of physics and has numerous, highly problematic and contradictory questions and variables can be defended as a serious explanation for the universe and its birth.

To this end the what was there before it banged question is far more easily answered with SM that can be rationally explained step by step, moment to moment from the existence of a universe like ours as we observe it today and simply reborn in the next cycle of the universe's life. The use of collapsing old cycles of the universe's life also involves normal observable and experimental everyday physics that are simply taken forward until the universe collapses and never involves outside forces, dimensions or causes the physics to breakdown at any point. This also answers the philosophical questioning of what was there before it banged as most would simply ponder what was there before the first cycle of the universe's life. The answer to this is simple as it is just another cycle of the universe's life.

There is no beginning of the universe simply because humans demand one or cannot think widely enough such that this is obvious to everyone. The universe is immortal and infinite and has no birth or death only life. I will prove this by asking another simple question which is why do forces exist? I could have easily asked why do fundamental particles exist or why does space exist? The reason all these things exist is simply because they do and that is that. If this answer is unsatisfactory to some then I offer a rather simple and glib piece of advice which is get used to it. The universe and everything in it is rather unimpressed with the human desire for things to make sense the way humans want them to and accepting an alternate model is the first and last step into thinking on universe sized scales. There are certain properties of the universe and the constituent parts it possesses simply because there are and this is massively obvious to all but the dullest of minds. I can also prove this perfectly and easily right here and now.

If everything in the universe did not possess certain unique properties then we would have no universe as nothing would vary in any way to create complex structures like atoms, animals and planets or stars. If everything did not have these innate properties on the most fundamental scales then the universe would consist of simply a homogenous mush of space, matter and energy all smeared into one indistinct and forever repetitious color. There is quite a lot of complexity in the universe and if you do not believe me go have an argument with anyone studying entropy. The point is that the universe has certain properties simply because it does and these are innate that require no explanation only an acceptance that they are real. The universe as an entire animal itself thus also has certain innate properties as it simply does with no explanation required. This should satisfy, confuse or annoy most people into an answer of the what was there before it banged question.

Now I would like to continue talking about the life cycle of the universe and put in a few points about other ways in which you can start a new universe off. The first thing I want to explain here is the nature of variability on ESS macht scales inside the SM mass and how these are responsible for creating all the large scale structures we see today in the universe. The universe as we know is one that is extremely uniform on the largest of scales with matter and energy being distributed fairly evenly. The matter and energy however is not perfectly even as we can easily see with stars and galaxies lighting up the universe with huge voids of empty space between them. So we know that matter had to have small imbalances in this uniform spread a long time ago so that the universe as we see it today can have clumps of stars forming galaxies. The variable gravity of the universe as we know is more than enough to create galaxies as small collections of black holes and stars and grow from there to massive spirals like today. However the necessary fluctuations in matter, energy and consequently gravity that would eventually lead to the sites where stars would form from huge accretion discs of gas came from the SM itself that detonated in the Big Bang. The SM mass and in fact all SM masses are of course highly uniform structures that have only a few variables dictating there overall composition but they do still possess some variability internally.

A mass of SM is as we know made of space and matter particles that have become almost fully merged into one. This follows the standard laws of space physics whereby matter spends more or less of its time or degree of interaction inside space. In SM we know that matter spends nearly all of its time inside space and is such is locked away from the rest of the universe and as already explained cannot be undone by outside processes. Any outside process would simply fall prey to the high spatial compression of the ESS and macht found there and have its matter quickly converted to black hole SM matter and repeat the cycle of being bound almost entirely inside space. What this means is that matter spends almost but not all of its time inside space and so the maximum degree of interaction angular waveform still holds true for SM matter as well as any normal matter. In this case the matter will spend a very small, let us say for example 0.01 percent of its time outside space and this is true for every single piece of black hole matter particles. Since a black hole has a plethora of particles bound into its SM and since the maximum degree of interaction is capable of manifesting itself over a huge number of Planck times each individual matter particle has a unique signature of interaction with the local area ESS. This signature is highly variable but not fully unique to that matter particle and will be shared as the number of variations is quite high for all particles it is not infinite and so a number of other particles will share this unique signature.

Sound familiar? Particles of uniform type with synchronized degrees of maximum interaction with space according to their angular waveforms? This is of course natural quantum entwinement occurring many times over on large scales in the universe. Black holes are simply one of the locations in the universe easily capable of creating natural quantum entwinement events and holding them for significant but not infinite amounts of time internally. It would in fact be possible for the black hole to hold the entwinements forever if the black hole could not gain or lose any mass whatsoever as no new variables would be introduced to the system that would cause the local ESS to try and rebalance their macht. With stable macht sharing between all schales in the black hole the macht lines of quantum entwined particles would not be disrupted and the two particles could remain together forever. This of course has an exceedingly low probability of occurring as black holes are almost constantly changing mass and it is usually higher. What it does mean though is that multiple particles inside a black hole can become entwined with each other as the phenomenon is not restricted to pairings alone. It is common for particles to form a three dimensional lattice of quantum entwinements inside the black hole and the SM will in fact have multiple lattices existing at any one time. These lattices are of course not permanent as the black hole has an ever changing mass and therefore ever changing balance of shared macht between schales of ESS.

The significance of this arrangement of entwined and free SM particles in the Big Bang is that these variations will play out in large scale later in the life of the universe. Think for a moment of the internal arrangement of entwined lattices inside the SM that will form the Big Bang and realize that these come from particles being at a massive number of degrees of maximum interaction with ESS. The variations are almost non-existent at any large scale because the composition of SM is itself so highly organized and uniform. There are however tiny significant differences in the structure of the overall SM mass and these can be amplified through the Big Bang. As the single piece of TSM forms at the heart or core of the SM mass the matter held there is stripped of all its energy capable of being donated to ESS. The ESS experiences ESS wall collapse and will overload dumping all stored macht back into the universe. This event happens at a single macht timescale far faster than any quantum timescale. The matter that was at that location of the TSM ESS is now awash in a sea of pure energy and the ESS inside it spatially rebounds outwards in all directions at macht speed. The ESS that was the TSM affects the schale around it to repeat the process as it dumps out its macht and they overload and so we have the simple and familiar Big Bang detonation cores firing outward in a predictable sequence of events. The ESS work all at macht speed and so will absorb macht, collapse their walls and spatially rebound outward essentially instantaneously as compared to the quantum controlled matter that surround them.

What this means is that the fuse is lit faster than matter can change its maximum degree of interaction with space and so matter is ejected from the ESS it inhabits in whatever degree of interaction they were in. This could be more or less inside space and along any number of degrees so that the matter and energy for that tiny but local area of the future universe is ejected at slightly different Planck times. This is all that is needed to create a very uniform universe from the highly organized SM composition but still produce small variations that will manifest themselves into things like star forming gas clouds later in the life of the universe. In addition to this once the matter and energy are free the number and complexity of interactions will also help to create the necessary imbalances in mass that will lead to star formation. The key point here is that because the ESS will spatially rebound inside the matter shredding it apart and spreading it throughout the galaxy the most important variations in matter and energy distribution occur first here on the macht scale with the quantum motion of matter and energy playing a much more secondary role later on in the diffuse gas of the early universe. You need these small variations inside the SM mass to create the uniform but still star forming universe we see today.

What this also does is finally explain in a factual way what is happening to create these slight variations beyond the tired old excuse of modern science which is simply that random quantum fluctuations must have occurred. You have no doubt heard far too often the simple glossing over of a factual mechanism for these variations as they are simply stated to be mathematical probabilities. Well instead of not understanding what is happening inside the mathematics these variations can now be fully explained using facts and not probabilities meaning the real reason is revealed and not hidden in mathematics, so called hidden variables or simply by saying we need fluctuations to make the models work so they must have been there.

Now it is time to add one more unique and incredibly important factor in the creation of our universe as we see it today with respect to matter distribution that can only be created from SM masses. This is the use of quantum entwinement that I explained earlier inside the SM mass. Quantum entwinement is made possible because the macht lines between particles glows in synchronous fashion with the particles involved in the event. This as you remember means that they share macht essentially along these lines as well as with the schales around them. The arrangement of matter particles inside a SM mass as you remember contains many lattices of quantum entwinements. What this means is that the glowing lattice arrangement of entwined particles inside the SM can actually feel each others macht at the same time. This of course becomes highly significant during the Big Bang explosion as the first TSM will experience a maximum load of macht for a brief instant. The macht increase and following macht dump back into the universe means that if that ESS were part of an entwinement the other matter particles would feel this macht shockwave through their local area ESS. The end result is that the lattice networks will have moments when the entwined particles attached to them will be forced to experience ESS spatial rebound or Big Bang detonation events slightly ahead of or behind others in the local mass of SM particles.

The differences areas gain are extremely small but the effects later in the universe are amplified. The familiar analogy of aiming a gun is perfectly apt here where any shooter or marksman will tell you that a small difference in firing angle at the gun will translate into a massive miss at the target as the round has a great distance to exaggerate this small difference. The same is true here and with all minute variations in the SM mass composition as they will be small in the first instant of the spatial rebound event but over the course of expanding to form the huge volume of the universe will become significant. Significant enough that a galaxy will form at regular intervals from another galaxy but overall uniformly throughout the universe. The Milky Way galaxy which is our home and the Andromeda galaxy which is one of our nearest neighbors are separated by millions of light years but are practically touching on cosmic scales. In addition to this the stars that formed them were far more uniform in the millions of light years between them after the formation of the universe following the Big Bang the small differences in SM composition saw to that. Over the billions of years since then these differences allowed some stars to be pulled towards one galaxy or another. Afterall the early universe would have had stars forming uniformly throughout the universe the apparent millions of light years between galaxies with no stars came afterwards. A combination of galaxies attracting stars and space expanding between galaxies is what leads to the universe in terms of galaxy distribution we see today. So in the end the necessary variations in distribution of matter and energy necessary to create a uniform cosmos still capable of forming stars that would lead to galaxies and a universe as we see it today all began inside the SM mass and was caused by ESS, macht and matter interactions once again.

BBB

The universe as most people think of it is a single mass of everything in existence and this is precisely how you should think of the entire universe but what if this universe we live in was only a part of that? This is the concept of the universe as one might think of it stretching on forever and involves what I call Big Bang Bubbles. This is how you can have an entire universe to us but this is only our view of what the real universe is and to every other intelligent species as well. So what is a Big Bang Bubble? A Big Bang Bubble which I will now refer to as simply BBB is an expanse of space that has undergone a TSM event within one part of the entire universe. In this universe which very well may be ours the universe can be a single BBB or be a larger infinite structure that is made of multiple BBB. The reason I am including this information is because what we see as the entire universe can still function through the same halting, collapse and TSM explosion events even if it is only a part of the entire universe as a whole. So the real universe is a continuous volume of space and matter with zero boundaries anywhere within it. In other words what we might call the universe really has no edge as it is not actually a universe but a BBB and at the edge of our BBB is simply more space extending outwards to more universe which in all directions will have more BBB belonging to other intelligent life looking at us as well.

In this larger universe ESS are contiguous from one BBB to another but just as a galaxy has dominance over the local area of space such that it has a central point which is its super massive black hole so does the local BBB have a central point to which it can once again collapse. The collapse mechanism is further realized by the ESS of neighboring BBB providing a braking mechanism whereby our BBB ESS simply cannot push against them anymore and halt leading to a collapse. In this way BBB are able to maintain a kind of local cosmic space for themselves but in no way deny travel of objects or information from one to another. Think if you will of a circular pond where the water is the universe space and soap bubbles floating on the surface are BBB. The water permeates the bubbles of soap and exists both within them as well as between them. Therefore the physics that operate in the pond everywhere are identical to everywhere else which includes the soap bubbles. If you were in a spaceship at the edge of our BBB nothing would stop you from crossing over into the next

BBB and existing in it. Information is likewise permitted to travel between BBB but at two different speeds. The standard slow speed of light which still obeys known laws of physics preventing us from looking beyond our observable horizon and the faster speed of macht which allows for information to travel from most likely the entire universe to us given its age of billions of years.

The Cosmic Microwave Background that we have so cleverly found is a kind of background noise that exists within our BBB and so will be louder than any CMB coming from left over BBB adjacent to us. Essentially the CMB of our BBB drowns out easily detectable signals from other BBB although extremely high energy events can penetrate this noise. What is not possible most likely is detecting messages of information encoded in macht lines between BBB or at least very distant BBB due to two factors. The first factor is simply macht strength which as you know decreases with schales as you move from the area of matter and space interaction. Therefore the chance of a signal being strong enough to reach billions of light years is near zero probability. If the signal was strong enough and a single macht line is given sufficient energy from numerous particles of matter all at the same exact maximum degrees of interaction of space due to their angular waveforms then yes you could send a signal.

The second factor is a clear line of sight such that nothing disrupts the macht line of communication. When we look into empty space we find more and more stuff such as stars, galaxies and everything else that fills the universe. So even if we peer into a truly empty from our perspective piece of the sky given enough distance it is highly likely something will block the macht line and disrupt the glow. Therefore the combined distance with probability of encountering a disruptive object between us and a neighboring BBB makes information travel of macht difficult to detect at best. What I have hinted at here is very real which is using macht lines to communicate much faster than light simply by listening in at different degrees of maximum interaction of space. In other words you can tune into the interactions of matter and space and if a message is being sent at a particular degree of interaction and the signal is strong enough you can automatically entwine matter here with whatever is sending the signal elsewhere in the universe. Without diverging too much here there is a reason other intelligent life does not communicate using electromagnetic waves as they are simply too slow and primitive a communication method. Back to information sharing however nothing dictates that information cannot be shared between BBB only that as we use electromagnetic waves to do most of our research we are limited to what we can see beyond 14 billion light years or so.

The importance of BBB in a larger universe is to allow for an infinite universe that can extend forever and not need boundaries. The larger universe might still have boundaries but they cannot be crossed. A quick and obvious explanation for that here is as follows. Space cannot as we have seen be separated from itself due to its natural properties of sharing macht glow and doing so from its macht pool which cannot be depleted. Since the pool is a part of fully spatially rebounded ESS macht always remains to hold space together. Matter as we have seen cannot exist in the universe without space and the two are inexorably linked forever. Matter has a natural tendency to interact with ESS and so will always be attracted to ESS despite being able to move freely within space. Matter only exists briefly outside space as it were during TSM explosion events or Big Bangs. However since the matter operates much slower than ESS the speed at which ESS spatially rebound and share macht far exceeds matter and so even if matter was moving in a direction outside of space into the nothingness void outside of the entire universe the ESS would easily spatially rebound through it far faster than the matter could ever travel. The ESS will also share macht glow at macht speed which again is far faster than that of light speed and so the matter will never be able to escape the spatial rebound of the ESS or the macht speed glow that it is attracted to. Again no matter how hard you try matter cannot leave space. Therefore since space cannot tear itself apart and matter and energy are bound to space nothing can ever cross over the edge of the universe and be lost forever this includes, ESS, macht, matter energy or information of any kind. The boundary of the universe is an absolute in existence and the edge may never be crossed. The boundaries I talked of being absolute pertain to the larger universe but not BBB as they may exist without boundaries being little soap bubbles held by a common bond of water in the circular pond.

A practical application of this is that it explains why everything in the universe is normal and involves no absurdities. Such nonsense would include whole universe where the laws of physics are completely or even slightly different than our own laws here in our BBB. So no more of this talk of anti-matter universes, universes where the four forces operate at different strengths or places where time runs backwards. The known laws of both space physics and quantum mechanics will continue to operate smoothly everywhere as they do here because the universe no matter where you go contains the same structures of space and matter; like the water inside and between soap bubbles is all the same water. The ESS in our BBB are the same as the ESS in every other BBB as well as the types of particles we have and how they interact with those ESS to affect macht glow. What can happen between BBB are an overlap of space and the effects of incoming shockwaves.

As a TSM event occurs in one part of the universe and a Big Bang is triggered the expanding space will spatially rebound which exerts a pressure through the ESS that can be thought of as a BBB shockwave. This shockwave is of course made of near unimaginable power and dwarfs anything known to date such as hypernovae or black hole mergers by several unknown orders of magnitude based on however big a BBB happens to be which can be larger than what we can measure for the observable universe of 14 billion light years or so. The result is that a new BBB can push a shockwave into an older BBB that might be nearing the end of its current cycle of life and will begin again. If the older universe is not set to collapse yet or is in some transitory phase of this particular cycle of its life then the shockwave from the new younger BBB can in effect cause the remnants of the old universe to start collapsing.

This is very similar to how clouds of gas and dust inside nebulae are churned up by shockwaves from a nearby exploding star. The shockwave will cause the gas and dust to go from a relatively stable and neutral state to one where gravity can once again take over and cause the matter to coalesce and begin collapsing into a new star forming region. The same space permeates through a nebula as does permeate through every part of the universe and so the wavelike properties of a simple nebula can be scaled up to represent the entire universe and any BBB interactions within it. Just as we can use ESS and macht to scale up from a single fundamental particle and never change the laws used to describe the system when explaining a much larger non-quantum scale macro system so to can the laws of ESS macht and matter interactions be used to scale up whilst changing nothing in order to explain the universe and BBB overlaps.

The remnant SM masses in the form of black holes in the old universe would simply start to collapse as the shockwave from the young new BBB pressed against the old BBB space and caused it to shrink allowing for the variable gravity there to act more effectively over a short distance due to the ease with which the ESS of the old BBB can know share macht. A possible use for this kind of BBB overlap existing in a universe much bigger than what we can observe from our own BBB at this time is that older materials such as the golden galaxies at the very edge of our own BBB can now be accounted for. The existence of galaxies populated by many golden and therefore older stars so far away from us and therefore so far back in time can be explained using BBB overlap within a larger universe. This goes hand in hand with any structure that currently defies explanation such as older galaxies, fully developed spirals and possibly the incredibly energetic and rare high energy bursts sometimes detected from the farthest regions of our BBB. Remember also that if a mass of SM is contained within the entire universe then the very outer schales of the SM will be in contact with space that is similar to space under normal spatial compression they way a normal black hole is in contact with normal space.

What this means is that the very outer schales of the huge SM mass that reached the critical TSM event point may not fully detonate themselves and remain in small SM fragments simply blown off the outer layer of the SM mass in all directions from the rapidly spinning SM mass. These are essentially black hole fragments which exist fully formed and ready to create stars and galaxies at the time the BBB is zero seconds old. Anything forming around these small SM fragments will be able to evolve as though the universe was much older at these points in space. If the universe were only one universe there would be no normal spatial compression ESS for these fragments to be blown off into and so this scenario only works as in the pond example where there is already space and therefore more universe for them to expand into. However since they are a part of our BBB they would not be outside the detectable information range of light speed.

Examining different types of SM fragments can reveal if we live in a single universe or just one BBB within a larger one. While this will not necessarily be the case for every anomaly many might be explained this way such as the older structures at the edge of our universe, what appears to be dark flow where something beyond our detection range seems to be pulling huge swaths of materials from our universe and high energy rare bursts that appear to come from the very limits of our BBB. Nevertheless fragments of SM from a BBB or small deviations in the ESS of the SM during the TSM detonation event can leave pieces of fully formed SM essentially alive and well in a universe only zero seconds old. These SM fragments take normally hundreds of millions of years to form but can exist immediately accelerating the rate at which everything in the early universe, but only at its farthest edges, can evolve. This even includes the overall shape of our BBB as other encroaching BBB would cause the surface of ours to deviate from a perfect spherical shape. Even in a single BBB universe any tiny imbalances in how the SM mass detonated would still lead to irregular shapes of a perfect sphere as the local area ESS and macht for areas of space near the edge would have more SM and therefore elevated levels of ESS macht gravity pulling them into non-spherical shapes. The lack of sensitive enough and far looking detectors is simply what is hampering our understanding of the edge of the known universe that is the known universe to us. Hopefully new instruments will solve these problems and allow us to look at least as close as we can to the edge of our own home BBB.

Our own home BBB and the way it has behaved in the past in terms of its rate of expansion can also be accounted for simply using a universe with multiple BBB in it. The apparent decrease and then increase in the overall rate of expansion of our universe can be explained if our universe which is simply one BBB is actually exploding into the rest of the real universe. To begin with you must realize that any substance under certain conditions can have different properties such as water and ice. Both water and ice are the same substance but under different conditions behave very differently so that I can float an ice cube of solid water in the liquid water of my beer on a summer's day. Given a certain set of conditions these two states of matter although the same molecule underneath it all will not interact with each other in a smooth and contiguous way. The exact same fact can be applied literally to space as believe it or not space will not interact with itself in a smooth and contiguous way in certain instances.

To this end let us blow up a BBB using the standard TSM detonation event from an ultra massive SM mass and follow all of the same unpacking rules for any normal BBB in the universe. The space is decoupled from matter and so devoid of all donated macht and of course spatially rebounds much, much faster than light. This SM mass is of course or rather was existing inside the whole entire true larger universe and therefore existing in what is normally thought of as normal space found at the surface and extending outward from the SM mass. Upon exploding, the BBB as you know is unpacking the space component of the SM mass too and therefore it is unpacking its own space along with it. This space however is not saturated with the same amount of macht glows as the universe around it and it is also not under the same level of spatial compression as the space of the universe around it. This means it is the same substance but in a different state of being just as water and ice are the same substance but in two different states of being.

Just as ice floats on beer so too the ice of the BBB space will not interact smoothly with the beer of the space of the universe and using ESS and macht we can explain the deceleration and acceleration phases of the universe. If treated as a fluid space can have any number of liquid properties in an event such as this and the first it will have is to produce two different densities to space. The best analogy to think of here and the simplest is of an explosion underwater whereby a small charge is detonated and the resulting gas bubble which moves as a fluid is formed and will expand outwards rapidly and violently in all directions for a time. After this initial blast however the pressure exerted from the surrounding water which is also a fluid will push backwards onto the explosion and slow it for a moment even collapsing the gas bubble for a second and then the bubble will expand again very rapidly once more. The process will be repeated in a series of decelerating and accelerating rates for the gas bubble until the total energy for the explosion has run out and the surrounding water wins and collapses the bubble back onto itself. In reality of course the air just floats up but in the series of explosions the water ultimately wins out because it is sapping the excess kinetic energy and heat energy from the gas and so the bubble will if held in place collapse again.

This all sounds very reminiscent of the universe and how we have observed it apparently exploding violently during the Big Bang then slowing down again then speeding up again and this is of course where we are observing it now. If we were observing the universe at a different moment in its life cycle say five or ten billion years in the past or the future we would see a different rate of expansion. Remember just because we are observing the universe accelerate its rate of expansion does not mean this is how it will behave forever as we ourselves know it did not behave this way in the past and so I must caution against human-centric thinking here when it comes to the universe. So the universe resembles an explosion with a series of expanding and collapsing gas bubbles inside a fluid. The ESS of space upon exploding as stated will expand rapidly many, many times the speed of light and this phase is obviously over for us right now. The ESS of space expanded into the rest of the universe where the spatial compression and macht levels found there slowed the rate of expansion. This occurred because the gravity being turned back on in our universe stopped the faster than light expansion phase and so the space of our BBB was suddenly forced to interact directly with the surrounding universe in a normal macht sharing through schales mechanism.

Once macht sharing through the universe was turned back on the differences in our BBB and the rest of the universe became significant and the exact same mechanisms for the BBB shockwaves took over. The energetic shockwave of our universe newly filled with macht glow from matter interacting with space again had a remarkably different macht glow level than the space around it. Just as cold water and warm water do not mix due to the different amounts of energy they possess so to does space not mix seamlessly if two regions of it have vastly different levels of macht glow and ESS spatial compression. Yet both are the exact same substance with literally no barriers between them only the differences in their internal properties of energy. So the outside universe pushed back against our BBB to decrease partly the rate of expansion of our observable universe.

The BBB of our observable universe did begin to accelerate again as the total kinetic force or residual momentum of our BBB is obviously not cancelled completely or in fact diminished that much at all. This is akin to the bubble expanding once again inside the water which you would normally think of as halting the bubble right away because water is so much denser than gas. Just as the bubble of gas expands again so does our universe which is not as dense with matter and space interactions as the rest of the universe and so our ESS continue to push against the rest of the universe ESS all around us. The macht sharing of glow between our BBB and the rest of the universe also helps facilitate this as it is known through the law of conservation of macht/energy that macht cannot simply be spread evenly away from the matter through space that donated it. To this end just because the macht created by our BBB at the edge of it and the rest of the universe can now share a glow the total macht pool for our universe cannot stabilize with the rest of the universe at macht speed. The macht of our universe is still bound to our ESS and so maintain the ESS macht glow for our expanding BBB of space against the rest of the universe. This essentially keeps one pool of water colder than the other and so despite touching will allow them to remain separate.

Since the matter in our universe is now bound to space and cannot travel faster than light speed the matter and therefore the donated macht it gives space cannot travel permanently beyond the boundaries of the ESS of our BBB and so once again keeps us separate from the rest of the universe. Remember that space is expanding first and the matter is being taken along for the ride so cannot travel faster than light even if space can. All of this of course is simply to keep two different sets of ESS spatial compression and macht glow levels to our BBB and the rest of the universe so that the shockwave push of our space with the rest of the universe can maintain this type of gas bubble explosion underwater. The evolution of the explosion is of course what is creating the physical motion of the deceleration and acceleration phases of our observable universe along with the previously explained mechanisms from another section on the same previously unsolved problem.

At the end of the explosion we see the energy from it depleted as over time the forces and temperatures of the gas bubble and the liquid have reached equilibrium and the water can win out and collapse the bubble. In the BBB in the true universe we find in the future the kinetic energy and macht glow differences of our BBB and the true universe are now balanced. The result is the outwards acceleration and expansion of our universe cannot ever possibly go on forever into a Big Freeze or a Big Rip and so another natural halting mechanism to the expansion is provided. It is provided by the universe itself and will simply be due to the energy of our BBB shockwave being dissipated into the rest of the universe and energy dissipation from any explosion in the universe whether chemical, nuclear or a TSM detonation event is all just because of balancing out ESS spatial compression and macht glows. So the universe using a BBB mechanism inside a true larger universe explains our rate of expansions and the future halting mechanism such that we do not simply vanish and from that point our BBB will become the backdrop for another BBB and our space is the stopping liquid to that second BBB expanding gas bubble shockwave.

This series of BBB gas bubble type explosion also helps to solve once again another of modern sciences unsolved problems which is the matter over anti-matter observation. The fact that our universe has far more matter over anti-matter is actually removed completely if you use a series of BBB inside a larger universe. The reason this is so comes from the fact that it is known that our universe whether a single large universe or simply a smaller BBB inside the true universe all came from an SM mass. If we look at the BBB model inside the true universe we simply see that all the matter of the universe collapsed into an SM mass which is made of one type of matter which will be explained fully later. For now however it is suffice to say that all the matter as we can observe today collapsing into a black hole which is really just an SM mass is normal matter and so the SM mass is made from only normal matter. This means that upon the universe unpacking itself from a TSM detonation event all the matter in the universe was already normal matter and so all the quantum predictions of matter and anti-matter being produced in equal amounts is not needed. The matter unpacked was normal matter with no ratio discrepancy from it and anti-matter meaning no need to avoid annihilation events as the normal matter simply cooled and formed things like atoms and molecules we see today. The same creation of anti-matter that we see today can normally occur such as violent collision of cosmic rays in the upper atmosphere and the same observed differences in decay rates can still occur it is only that anti-matter was never a part of the SM mass and so our universe and the true larger universe was ever only going to be normal matter as it has always ever been normal matter.

Facts

This is where I will conclude the fate of our universe for now and sum up the key facts explained in this section. The first fact is that the ESS of the universe itself which contains a macht pool will prevent the universe from ever tearing itself apart. The second fact is that variable gravity will also stop the universe from tearing itself apart. The third fact is that since space and matter are finite after the creation of the universe from the Big Bang the power necessary to halt expansion and start the collapse of the universe has always and will always exist as seen in the slider scale example. The fourth fact is that the universe will eventually collapse regardless of what halting mechanism becomes dominant at the point of maximum universe expansion based on matter composition through formation of SM masses. The fifth fact is that the universe will always recycle any free energy lost from one SM mass to another by absorbing the energy and trapping it in the SM internally. The sixth fact is that SM cannot be destroyed from the outside and must be destroyed from the inside only. The seventh fact is that all the SM and space will eventually collapse around a central ESS roughly at the center of the universe again based on matter drift and ESS local spatial compression and macht glow over the cycle of life of the universe.

The eighth fact is that a single TSM will undergo a 100 hundred percent binding of matter into space and hence create a single macht timescale moment of TSM and is the only time all four forces are unified perfectly as one and beyond that space and matter unified as well.

The ninth fact is that the TSM will dump all of its macht back into the universe and immediately begin spatial rebound at macht speed. The tenth fact is that the spatial rebound TSM event will be repeated through each consecutive schale until all the SM is unbound. The eleventh fact is that the universe is simply immortal and has no birth or death only life as a concept of limits is a product of primitive minds and not need be true. The twelfth fact is the uniform but imperfect spread of matter and energy needed to produce a universe as we see it is contained in the small variations of the maximum degree of interaction with space that matter has based on its angular waveform. The thirteenth fact comes from the fact that particles are uniform inside SM masses and so quantum entwinements events are relatively common and will also help to form the necessary subtle irregularities needed to create a universe as we see it today. In other words SM masses such as black holes are centers for naturally occurring quantum entwinement events.

The fourteenth fact is that the universe may be made of multiple BBB contained within a single larger universe. The fifteenth fact is that it is impossible to leave the true larger universe although information and objects can travel freely between BBB. The sixteenth fact is that the laws of space physics and quantum mechanics are uniform throughout the true larger universe and do not break down in nonsensical ways no matter the conditions of the BBB in question or the space between them. The seventeenth fact is that BBB shockwave induced overlap can be a mechanism responsible for collapsing other BBB such that they will become huge SM masses capable of triggering the necessary TSM detonation event needed to create a new BBB. The eighteenth fact is that the old and fully developed mystery structures at the edge of our known universe can be at least partially if not fully explained by using BBB in a true larger universe through the shockwave overlap mechanism.

Chapter 24
Matter, New Particle, Big Bang and Turning on of Forces

Matter

This next section answers once and for all what matter is as this still remains an unsolved problem for modern science. What matter is and some of its specific properties as well as new matter types will be explained here. To begin with what was our first impression of matter as humans? Matter is solid and matter is the stuff in the universe that makes up people, other stuff and the Earth. That is exactly how ancient humans saw matter as it was a solid thing which they could touch, see, feel, taste or smell. People were matter, to them as well as rocks, water or even the Earth. Something like air to them was not matter as they probably were not aware of it as a solid thing. We know now that air is made of very solid things only so loosely collected they act as though they were not solid at all. Ancient people would surely have known about air as the need to breathe is by far the most immediate of human concerns. The bottom line however is that matter to them was real and solid as in made up of stuff. As time went on people continued to think about this possible take on matter for millennia as matter went from various forms to various little bits to various particles.

Somewhere along the way however humans realized that the universe also contained forces and that these forces were connected to matter but not matter itself. An example of this is the strong nuclear force that binds quarks into hadrons as most of the hadron is not really solid but made of the strong nuclear force. Only three little quarks make up the solid bits of the much larger hadron and it is known that the remainder comes from the strong nuclear force that binds them. So in this way forces are related to matter but not physical matter yet they behave as though physical. An atom has an outer electron orbital shell that seems solid enough to us as we cannot put our hand through a rock. This is odd because both the rock and ourselves are made of mainly empty space and in fact we are actually near 100 hundred percent empty space by volume. So does this mean we missed out on being a vacuum by say 0.0001 percent or so? Well no of course not as the electromagnetic force that holds the electron away from the nucleus is the all important factor in making the rock and our hands seem solid. It is this invisible force that keeps the two from passing straight through one another. So matter and forces are linked as we have found out after some centuries and millennia of study which led to the next step in the theoretical understanding of matter.

This next step came when it was theorized that matter and energy are in fact equivalent. The familiar equation of energy is equal to the mass times the speed of light squared is what Einstein theorized made matter and energy equivalent. In this moment he believed that matter and energy was in fact one and the same thing and from there science theorized a new form of matter. This is the form of matter where matter is of course not matter anymore and not real or physically solid but instead is simply an excitation of the fields that make up the vacuum of space as theorized by quantum field theory. In this view of matter a simple excitation of a field will give rise to an area of space where all the properties of say an electron are met. Elsewhere a collection of three quarks or rather quark excitations will create a proton field excitation which will attract the electron field excitation. The whole lot can be made with excitations into a field excitation hydrogen atom as it is argued by quantum field theory that the necessary fields all exist in the vacuum and so by stating that matter is energy all you need is to excite the energy of the fields to produce any particle or atom you want.

From here of course it is theorized that whole collections of excitations can then be made into larger scale objects such as rocks, hands that try to pass through them or people owning said hands. The effect of these excitations is theorized to produce exactly what we sense as physical solid objects just as our ancient ancestors did. So this leaves us with two very different views of what matter is. On the one hand we have matter being treated like a solid object or particle and on the other we have matter being treated like an energy wave of a field excitation. To confuse this process even more experiments like the famous double slit experiment appear to show matter as being both a particle and a wave of energy. In other words modern science does not know which is correct in determining what matter is. However matter at the end of the day can fundamentally be only one or the other so which is it?

To answer this I will do two things the first is show what it cannot be and the second will then be to prove why the remaining option is the correct one. So essentially at the core of this question is the choice between matter being an excitation of a field in the vacuum as mainly described by a wave or if it exists as real solid matter outside of the vacuum as a particle. The evidence for both theories is actually fairly sound and a third option exists which is to describe matter as both a wave and a particle hence the rather scientific name of wave-particle duality. Wave particle duality is however incorrect because at a fundamental level matter may be only one thing or the other and the idea of describing matter as both a wave and a particle is very practical but can be thought of more as a macroscopic description akin to a classical mechanical description of a quantum mechanical object. The classical system is by far the most superior form of describing how macro-scale objects and systems behave and interact with one another but once broken down is a conglomeration of quantum mechanical systems. This is similar for wave-particle duality which can describe the apparent behavior of a particle usually while it is doing something or interacting with something else but at the most fundamental level when the particle is isolated from the rest of the universe it must choose one side or the other. To this end I will eliminate wave-particle duality as a true description of the fundamental nature of matter and keep it instead as an important experimental observation of a more macro-scale behavior of matter. This leaves two choices the wave option and the particle option for matter and from here I can begin to take apart the incorrect one.

The incorrect option is that matter is a wave which is nothing more than an excitation of the fields of the vacuum made manifest such that they behave as though a point particle. This theory of matter works mathematically but only partially and fails logically on all counts in the physical world. The reason this is so comes from how the matter is said to be generated which is by exciting field lines and in fact that matter is nothing more but excitations of field lines. By attempting to combine quantum mechanics with relativity quantum field theory explains matter as excitations of specific fields. Quantum electro-dynamics tries to account for electromagnetic forces and interactions of particles while quantum chromo-dynamics tries to do the same for nuclear forces such as those that make quarks. The theories are mathematically accurate and give good predictions for experimental values which lead many to believe they are in fact correct but at their core they still do not address what matter really is. To understand this further let us revisit general relativity which as we have proven is completely wrong as it cannot hope to account for quantum level objects, does not explain what a spacetime is and creates its own mistakes and contradictions like singularities.

What general relativity does do is a good job mathematically of predicting certain observed gravitational values but as I said it does not understand in the least what is really going on in the system. This renders it a useful little equation and nothing more. Quantum field theory is much the same as general relativity only far more sound to begin with. It also makes good mathematical predictions about experimental results which have been confirmed and repeated and it is using the fields of the vacuum to explain some of what matter is doing. The main fault in quantum field theory lies in that at its core it states that particles are nothing more than excitations of the vacuum fields and from there can be used to build a universe. The idea of using vacuum fields to account for some of the properties of matter is fine and well understood as obvious scientific fact. Saying that matter does not exist in particle form is where it fails to explain the universe.

The combination of certain forces into larger ones has led again many to believe in quantum field theory unwaveringly and this can be dangerous. A perfect example is of the electromagnetic force as it was shown centuries ago that both the electric force and magnetic force can be described as a wave and carried in light. While not confined to light this first step had many people theorizing that everything can be continually reduced until one super-force was found to rule them all. The electro-weak force is the next theorized force that can incorporate the electromagnetic force and the weak nuclear or decay force into one. The idea of combining all forces is appealing as it means that the universe can be described in far more understandable terms since the number of variables will reduce each time a force is united. This part of quantum field theory is perfectly logical but again the idea that matter which is where forces are found are caused by excitations in the vacuum fields is where the mathematics fails to meet reality.

Quantum field theory again falls prey to the seduction of using too much mathematics to describe the real world and in order to make predictions steps over the point where it addresses what matter really is. To begin with the predictions of quantum field theory are good ones and I am not suggesting that they be discarded in any way I am only saying that the fundamental reason these predictions can be made needs to be clarified. The fault lies in treating relativity as perfectly correct and needing to mix it with quantum mechanics. Energy and matter are not the same thing and this simple fact has led to a whole slew of misconceptions and scientific dead ends for over a century. To prove this rather simply show me a kinetic energy photon.

Do not fret I know you cannot and no matter how much you attempt to break down and compartmentalize each piece of a larger kinetic object falling and hitting the ground there is no such thing as kinetic energy. Kinetic work maybe but energy as in equivalent energy meaning the matter really is just photons well no of course not. Matter is matter and energy is energy and the equivalent formula is misread as meaning that they are one in the same thing when it really is only saying if you could convert one into the other it would be worth so much. According to Einstein himself nothing can travel faster than the speed of light and so mass cannot travel at the speed of light squared. Yes I realize that many reading this are beside themselves and muttering that this is only meant to show how much energy the mass has and that it would have that much energy if you could accelerate it to the speed of light squared.

All whilst saying that they miss out on exactly what was said that matter is only equivalent to the energy and not actually the energy itself. What is really being said here is that matter and energy are not the same thing because matter can never be accelerated to the speed of light squared according to relativity but what is being said is that the way matter interacts with the vacuum it acts as though it has the equivalent energy of its own mass times the speed of light squared. Another way of saying this is that the energy that matter has invested into space and the vacuum fields is the equivalent energy value. So a quark is still just a quark when it is bound into a hadron and the hadron is mainly made of energy acting as virtual mass meaning that the energy is not mass only treated as such. That may sound circular and obvious and it should for good reason as much of what is viewed as the rest energy of matter has nothing to do with the matter on a purely energy level and everything to do with space. I will return to this point later when discussing what matter is and why it is a pure particle at its most fundamental levels but for now I am simply stating that matter is not energy and therefore the concept of fusing quantum mechanics and general relativity such that in quantum field theory matter is treated as energy manifest as excitations of vacuum fields is incorrect.

The idea that matter is not real and that matter is only energy and that all must be equivalent under the restrictions of relativity means that the theory of matter being simple excitations of fields in the vacuum cannot be correct in assembling a universe as we see today. First off what is causing the field to become excited? If matter is energy and the fields are energy then what energy is causing the field to become excited at that point in space as everything is equivalent. The energy of the fields should never be excited as it cannot excite itself. That is like trying to make a cup out of water and of course make the water inside it be separate from the cup which you also made out of water. Your hand is water, the air you breathe is water and your stomach will also be water but never will these objects interfere with each other and simply flow together. Magic it would seem is keeping them apart even though through water relativity they are all exactly the same. Afterall it is a well known fact that waves will cancel each other out when a peak and trough meet so if matter is made of waves at all then cancellations of the same waves that made them should be happening all the time. Once again the waves cancel and you are left with a flat featureless expanse of fields that are not excited.

This might cause some physicists grief but it is the simple truth a field cannot be made of fields and excite itself to create something that is nothing but fields then interact with more fields and then not destroy itself as all are made of the exact same thing with the exact same properties at a combined force level which is energy. Does this mean all the predictions for experimental values made by quantum field theory are suddenly wrong? No of course not I am once again saying that at the core of quantum field theory there is no true explanation of what matter is. Mathematics cannot be trusted here to explain what matter is no matter how convincing the equations. If we listened to mathematics than as soon as you place one apple over an empty space you destroy the apple, or maybe you do not. In mathematics one apple divided by zero is a headache and causes weird things to happen to the apple as apparently it is no longer defined. Well in real life and not the magical land of mathematics the apple still sits there waiting for you to eat it and remember Newton and the moon. It should be obvious to supporters of quantum field theory that this is so and the need to continually restate the previous sentence can hopefully be lost as everyone grasps it.

To further confound the problem how would one create any objects larger than single fundamental particles from pure energy and fields? You would need a different mechanism for excitations of the vacuum fields for every fundamental particle and why would there be different excitations anyway? It is already impossible enough for pure energy of fields to for no reason manifest themselves as a particle excitation even if it were the only particle in the universe as the fields should always relax to a ground state that reduces the strain at the point of excitation on the field. Everyone knows including supporters of quantum field theory that the universe always without exception in any system tends to the lowest energy state and the simplest form of whatever object is being described. Field excitations not should but would always do the same meaning they would not instantly destroy their own excitations but that these excitations could never occur because that would mean an increase of energy and work into the system to make the fields create the excitation. The nature of reducing to lowest ground states would not permit any nearby fields to give the necessary energy to the part of the field that wants to create an excitation as they are all viewed in theory as contiguous. However if we do allow a fundamental particle to hypothetically exist as a manifestation of the vacuum fields through spontaneous and impossible field excitation then combining that with other excitations would destroy both.

Take for instance the idea that three quarks of differing type make up a hadron. Some are up and some are down quarks and are different excitations of the fields according to theory and so if they exist alone when they are combined they should destroy themselves into one single field excitation so as to reduce the strain on the vacuum fields in total. They would never create more strain by making virtual mass from gluons which of course according to quantum field theory are simply more and more and still further complex excitations of the same underlying super-force field that has undergone symmetry breaking. An electron and proton should naturally cancel out through their opposite electromagnetic charges so as to reduce the strain on the vacuum fields but as we know they do not. So why does a simple hydrogen atom even exist when the proton should reduce to a single excitation and then the electron should reduce with the proton remnant to from a new single excitation? This should make a stable neutron excitation made not of quarks and gluons but just quarks which we know to be false and a neutron is anything but stable as it decays in about eleven minutes. If the nature of matter is from field excitations and these fields contain a single unified force even if it is currently broken why would the stable electric charge cancelling neutron even decay as it should be the most stable form of excitation of a few respective fields? The answer is because field excitations are not what matter is made of at a fundamental level as the only way to explain these behaviors is not by using fields or energy or relativity but by using matter pure and simple.

So yes the answer to what matter really is at a fundamental level is a solid particle as it is the only way you can create a universe as we see it today such that small and large objects behave as they do and the predictions of quantum field theory may be kept to an extent. Again as I have said the idea that matter does create fields is not under debate but the true nature of matter cannot be described by quantum field theory. Just as general relativity does not describe space, gravity or the universe but we may still use it as a simple calculator for certain observations. To dive right into this, matter is a solid object that will be fundamental to the universe and can be seen as equivalent to energy under certain but not all conditions but in no way ever is it actually energy.

Specks/Motes

The universe is made of two things which are space and matter and this accepted fact seems to have been forgotten by many in modern science so it is well worth repeating now. Space is space and matter is matter and matter is not simply an excitation of the former. Matter is a particle that interacts with space to produce what we view as normal matter, anti-matter and fundamental particles of all types. To this end I will refer to what truly lies at the core of all particles as specks or motes of matter as these are the hearts of what modern science sees as say an electron, quark or any normal particle no matter how small. The most fundamental bits of the universe are what make up what people think of normally as particles. The smallest fundamental particle that appears to be completely indivisible may actually be made of smaller things yet as continued research into particle physics has shown for decades but at the very center of whatever the smallest particle is in the universe lies a speck or mote of matter.

It is these specks and motes that interact with the ESS macht of space to produce what modern science understands as a particle such as an electron, quark or composite particle like a proton. The tiny speck/mote at the core of a particle is what gives rise to the ESS macht glow that is mistaken for a real particle and remember the boundary glow of particles as described previously to understand this fully. Boundary glows as you should recall are determined by spatial compression of the local area of space, shared macht glow between schales of neighboring particles relevant to the system and degree of maximum interaction with the ESS of the matter orb creating the macht glow. These speck/motes are real and solid for two reasons the first of which returns to the fact that the universe does not care about what humans think and operates perfectly normal. Humans might sometimes want the universe to be made of fields and try to make the universe fit this theory but in reality nothing says this needs to be true as it is not true.

To put this another way why do you not prove to me and the world why the universe should be made of vacuum fields and why should they exist at all? You cannot answer this because in truth there is no need for the universe to do what you want as it simply does what it wants. The universe wants to have space and matter. Why? Simple because it just does. Afterall according to field theorists the universe has fields and why? No reason, just because. This may not be a satisfactory answer for some but that is a harsh reality they must deal with as the universe has created itself exactly how it wants because it is exactly how it is. Why is there a photon, gravity, any of the forces for that matter, why does anything exist as it does? Simply because that is the way things are. To this end modern science needs to stop trying to make the universe fit its theories and instead accept a model that explains the universe as it is through both observation and experimentation.

The second reason matter is a solid bit of the universe is due to the fact that this is how you can explain everything you see in the universe around you while keeping the predictions made by various theories of modern science. I have nothing against what modern science has recorded and catalogued in things like the Standard Model of particle physics only that the observations made are a what or how much happened event and not a why they happened explanation. Using matter as speck/motes that interact with ESS to produce a macht glow with unique macht glow boundaries for each speck/mote does fully explain the universe as we see it today. To delve into this further let us look at the idea of string theory as a way of confirming this notion. String theory uses strings of all things to explain how matter can manifest itself in the universe. The idea behind string theory is a good one only needlessly complex, fairly ad hoc and impractical. The reason it is complex is due to the fact that it seeks to use mathematics upon mathematics to finally derive a theoretical form of matter that can under certain but not all conditions in the universe possible explain the nature of fundamental particles. Different versions of string theory require different starting conditions and pre-conceptions of the universe which do not work or are not shared with other versions and so immediately become suspect.

There are literally millions of different plausible solutions to the string theory equations that all work mathematically only to explain matter. What this means is that string theory is incorrect because given enough assumptions about any system you can use any theory you desire to prove it correct. A rather quick proof of this is to assume that the entire universe runs on magic and once I have made that assumption it is easy to prove that magic wands can be used to explain anything observed as they are the conduits for this underlying magic. Since no one reading this can prove magic does not exist at the fundamental level of the universe and that these magic wands could be say open ended one dimensional objects through which the magic is channeled you cannot say this is wrong. It is however highly improbable as I could have easily said the working mechanism of the universe was based on any other number of assumptions and each would seem correct. To this end the complexity of string theory really is its own undoing.

The reason it is ad hoc is because instead of trying to solve the problem of how the universe works head on it instead derives ever more seductive mathematical objects that are quite impossible in reality but can exist in the world of mathematics. If you look at an optical illusion such as the three pipes that become two, or the stairs that go back onto themselves forever you can understand how something can exist in theory but not in reality. This is the same flaw that many have warned of in the past century as mathematics can create many seductive illusions but these cannot leave the magical world of the page and become a solid fact. The success of mathematics in creating objects that have later been at least partially proven to exist is not trivial as certain particles have been theorized and later found but they always exist within the normal and rational confines of things like the Standard Model which are excellent catalogues of the bits of the universe. It is when these mathematical seductions become multi dimensional so that even there own creators cannot hope to prove they exist in the laboratory as it is impossible to use three dimensional equipment to look at theoretical eleven dimensional objects. This is where the ad hoc comes into play as the string theories have grown ever more dimensionally deep.

The reason string theory is impractical is because it needs to continually increase the number of variables in the universe which is in direct opposition to the idea of simplifying things into a single unified theory of everything. Remember they were the ones who wanted to simplify the universe in the first place to a single super-force and now they are complicating it to almost no end. While the debate exists in modern science as to whether or not a unified theory of everything exists the fact remains that all in modern science do agree that the universe always tends towards a minimum number of components and lowest level of energy states. To this end the idea of creating more things in string theory is backwards thinking and the two biggest faults are with extra dimensions and branes. The extra dimensions are of course easy to understand as the complexity of having more dimensions is simply that they make the overall number of variables higher and therefore more complex and therefore not favored by the universe. In addition the extra dimensions are for no reason compactified in the theory other than to make them easier to work with and even if the dimensions could be compactified this would require something that string theorists have not considered which is a packing protein.

Think of DNA for a moment and you see a double helix shape in your minds eye that most people find unmistakable. What most people do not know is that DNA almost never exists in this form in your cells as the fully unpacked double helix molecule takes up more volume than the cell could hold and the cell would burst. So how does DNA stay in a tight form in your body? The answer is it uses a packing protein that it can wind itself onto to become denser and take up less volume in your cells. The dimensions of the DNA have now changed to new dimensions of the packed up DNA molecule. The DNA however cannot accomplish this feat alone and needs an extra mechanism, it needs the packing protein. Strings would require such an extra mechanism as well to fold themselves into a compactified arrangement as folding dimensions takes energy and this is something the universe forbids as it always relaxes to the lowest possible energy state. Therefore the string theory needs yet another variable in the universe simply to compactify its theoretical extra dimensions into a shape that is easier to work with. The strings would therefore also require signals to know when to make themselves folded and unfolded as they need to interact with other strings, branes or fields. The strings also are said to be open strings like a line or closed strings like a loop. Why? No reason.

Two shapes to strings are simply needed by string theory in order to work to explain theoretically things like matter and gravity, holding onto branes and possibly contact with universes in get this, still more dimensions that are completely untouchable. Again we have an example of things simply being stated as existing in the universe and not questioned as string theory simply says that is the way things have always been because that is the way the universe works according to it. The problem with assumptions like string theory over other logical assumptions comes from the fact that other assumptions do not require extra variables. For example simply assuming that the universe has space and matter which has been accepted in every theory from quantum mechanics, general relativity and even string theory itself for a century or more is fine and agreed upon. It is only when the number of variables needed to explain the universe as we see it today rises for no reason other than a mathematical one and especially by quite a few variables that the probability of the theory suggesting it being correct it falls to zero.

This brings us to the realization that the strings in string theory themselves cannot be one dimensional as the closed loop strings are immediately two dimensional. The reason for this difference in string types is to allow them to hold onto brane structures which are theorized to be extra dimensional structures that permeate the entire universe and are not part of space. So again string theory adds another variable to the universe in theory which is the brane structure and the brane structure was only created mathematically to accommodate the theoretical strings. Does anyone see the ad hoc here? The impossibilities of another dimension for branes as there may be more than one separate from the known and proven three spatial dimensions that lies outside of space and matter are simply ridiculous. This is precisely where one with an intelligent mind must sit back and simply say did anyone stop to think about whether this was at all even remotely possible or made any logical and rational sense?

This reminds me firmly of the idea of the singularity at the creation of the Big Bang. All of matter, energy, space and everything that will ever be in the universe spanning untold tens of billions of light years or many more of the entire universe was placed inside a theoretical object with zero dimensions and infinite density because the finite matter and what not is placed inside an infinitely small point actually made any stark ravingly mad sense in the first place? I mean this in all seriousness and as a cold wake up call but did no one once stop to say this is absolutely stupid and I mean dead stupid as well as impossible?

You should have as the same known laws of physics that we can experiment with and observe over and over again simply break inside a singularity. Do the laws of physics break inside a cup of water or inside a particle collider or anywhere else you look in the universe? No they do not so stop trying to make them break simply to fit a seductive mathematical creation from an incorrect theory and start to observe the universe and explain it for what it is as we see it today. This is the paradigm shift in thinking that is fully needed and has been screaming at the face of modern science for a century. String theory cannot hope to be fully correct as it simply requires the universe to fit it instead of the other way around. That being said string theory does have one promising feature and that it backs up the idea that the universe is made of space and matter. This is because even though strings are incorrect the fact that they represent a separate part of the universe from space is quite accurate. The universe is made of both space and matter at all times.

Matter is real and made of the specks and motes that interact with the ESS of space to affect changes in local area ESS macht glows. This is a perfectly normal truth and one that is not open for debate. I will pose this to you now to end such debate and ask you why should space be real? Matter in the universe is something that we have all been able to touch and physically interact with forever and space is the somewhat intangible second half to the universe that we cannot directly interact with at least in a classical sense. Matter however is treated by modern science to be highly suspect and made of millions of dimensions and other fabrications whereas space is taken at face value though no one in modern science has seen it or understands it. It is the matter that we know to be real no matter what causes it and yet when we probe into matter at the smallest levels and at the highest energies it is the space found there that gives us the biggest headaches in modern science and yet instead of saying that space is where the theories fall apart and so maybe space is not a real thing people for some reason attack the matter that they know to be real. It must be the matter that is not real, it must be the matter that is flawed or it must be that matter that simply is not a part of the universe and space for some reason is taken as a first principle.

It is far more logical empirically, experimentally and theoretically that matter is real and that modern science should be questioning space and not the other way around. So to solve the problem of what matter is in the universe take matter at face value as a real thing and equal part to the universe and build a universe as we see it today from that. Using matter as a real solid thing that has always been a part of the universe and always will be just as much as space solves the problem of what matter is and does explain why the universe is as it is and behaves as we observe it. To prove this let us finally delve straight into a black hole and forever unlock not some but all and every single one of its mysteries and to do that I am going to use specks and motes of matter and ESS macht. By the end of this how a black hole is made, what a black hole is made of and the properties of a black hole will all be proven.

Here is where I will explain at a fundamental level how the specks and motes of matter interact with the ESS of space and where some people will not accept the paradigm shift in thinking. For those who do accept it do not be deterred by those who do not as it has been long overdue that science move again into the future and the last future in the understanding of how the universe works. We are ready to move again away from the idea of a geocentric universe, away from the idea of a flat Earth and away from the Milky Way being alone in a static universe. If you want to move forward and be among those that understand how you can build a universe as we se it today using nothing but matter and space then accept the shift.

To start explaining the shift I am going to start with normal matter as we understand it today and use the simple hydrogen atom for reference. I will take normal everyday experimental and well understood matter and use it to create a black hole such that no laws of physics are broken and I can create and explain a black hole exactly as it appears in the universe and how this can then be used without changing laws or inventing new variables to end this cycle of our universe and begin another. All the interactions of the specks and motes of matter with the ESS and macht of space can be taken back and put into any of the unsolved problems that I have solved in previous sections and will provide the answers to any section that I said I would touch on in more depth in a later section.

The first thing I want to say is that the explanation is a simple one and this is good the universe likes simplicity. If you were looking for a cosmic onion of an answer I am happy to tell you that it is not needed. The universe works simply and this is the simple explanation of those workings. A hydrogen atom as we know is made of two things the first being an electron and the second being a proton. If we consider either the electron in the atom or the quarks in the proton we can learn the first thing about matter which is that matter is not a field excitation. Matter is not an excitation of a field but rather it is the physical object that excites the field itself. As I mentioned before matter cannot be made of field excitations because there is no reason or mechanism for a field to simply excite itself but if a catalyst of sorts is there then field excitation can occur. If you think of particles as modern science currently understands them then the analogy of a snowflake works perfectly.

Almost all snow flakes have at their centers a tiny piece of debris which is usually dust whipped up into the atmosphere by winds close to the surface of the planet. This piece of dust is what the ice crystals begin to form around and contributes partly to the unique shape of a snowflake. The entire snowflake can be viewed very easily as one particle but the piece of tiny dust at the center is too small to see with the human eye and so usually goes unnoticed. This means that while the snowflake appears to be made of only one thing which is frozen water in the form of ice it actually has a second part that is in fact the most important part as the entire snowflake would not exist without it. If the snowflake was frozen water people might say it is an excitation of the moisture in the air and the air resembles the field energy. In addition the type and shape of the dust in terms of what it is made of, its size and surface topography determine what the final snowflake will look like in part. There are other factors that determine what the snowflake will look like as it forms and of course in its final form as the number and severity of collisions it has with other snowflakes, the moisture content of the air and the length of time it stays aloft in the atmosphere all contribute to creating its final appearance.

Matter is identical to the snowflake in almost every respect from this analogy as matter too is seen as only the excitation of a field. Yet far smaller than the field excitation that is normally perceived as matter by modern science lays the speck or mote of real fundamental matter at its heart. This speck/mote is needed for the field to be excited around just like the dust is needed for the ice to form around in the snowflake. The field is large and represents the physical boundaries of the particle and its further reaching effects that make it difficult to see the interior. The interior is the speck/mote that excited the field in the first place and is ever present in all particles in the universe. Just as other factors affect the snowflake and its development over time until we observe it so do certain factors of the universe affect the speck/motes and how they appear to us when we observe them.

The first factor is of course the type of speck/mote as these come in different types after all which only makes sense as an electron has very different properties from a quark. The specks and motes must have different properties or shapes if you will because an electron speck/mote does not excite the strong nuclear macht of space the way a quark speck/mote does. In addition differences exist in how they create mass as some particles are heavier than others meaning that some interact more with the Higgs field. Combinations of how specks/motes interact with different ESS machts determine what properties they have, how strong these properties are and then in what way they can interact with other specks and motes.

The second factor that determines their appearance is their interactions with each other or with different particles altogether. Look at the common quark and it is always bound naturally to other quarks to form a larger object called a hadron. The field generated by a quark on its own is not the same as the field generated by it when in combination with other quarks. The field can be stronger or weaker and the field can also simply disappear as is most commonly observed. The third factor that affects the specks and motes as they generate particles is of course spatial compression of the local area ESS as this determines field size and strength. Using the example of our familiar matter orb we know that the field of gravity it generates is variable based on how many ESS it encounters. Combinations of all these factors will lead to what appears to be a particle to us as observed on Earth under normal spatial compression.

The result is a snowflake particle as observed by us which is made of the speck/mote acting as the dust and the excited field acting as the ice around the dust. The field excitations that we observe and measure are therefore correct only the generation mechanism is now revealed to be specks and motes interacting with ESS macht to generate the fields. The field excitations and force carrier particles that are commonly associated with the speck/mote of a particle are representations of the ESS macht glow caused by the matter itself. This is where the confusion of the virtual mass of a particle comes from as force carriers while they can be calculated through their energy to have mass do not have traditional specks or motes inside them; this will be explained in full later.

If a speck/mote is removed completely from interacting with space then the field disappears and since the law of conservation of energy states that the energy cannot be lost it simply returns to the field which is of course the ESS macht. This means the equivalency principle of the virtual mass holds true under normal matter and space interactions so that the particle appears heavier than it is which leads some to believe that the particle is made out of actual fields. The key is that the speck/mote once removed from interacting with space is still solid and real and has a set amount of mass to it as it is a set amount of fundamental matter. The speck/mote of the fundamental matter however is tiny and difficult to probe using normal means as it is most commonly seen as it manifests itself through a field.

Particle accelerators are only seeing the specks and motes as they interact with ESS of the accelerators local area of space. When high energy collisions occur and new particles are seen from existing ones that is simply the trails of various specks and motes that have been dissociated from their normal arrangement found in the ordinary matter accelerated in the collisions and then briefly interacting with the ESS macht as they trail away. Similarly certain collisions are the correct type, energy or chance to knock some of the specks and motes free such that the do not interact with the ESS before they are lost from the detectors range. This missing mass in the collision is not really missing it is simply not detected because it is not interacting with the ESS inside the accelerator again and our instruments detect only space and matter interactions as seen by ESS macht glows. If you want to detect a speck or mote you need to build a detector that has an area of no fields in it then an area of fields present and as the speck or mote moves through the void it will only reappear showing its properties upon reaching the filed space again.

The colliders we use also have another major flaw in their ability to probe matter at its deepest levels and that is their levels of spatial compression. A black hole for example is known to have high levels of spatial compression so any conditions existing inside it are maintained for long periods of time. A particle collider does not have a surrounding spatial compression similar to a black hole and so any high temperature and pressure conditions it hopes to create inside it will not be maintained meaning the matter never fully behaves as it does inside SM. The particles created in the collisions decay rapidly because there is no high spatial compression of local area ESS to allow them to keep their form inside the collider as they balance their macht. Remember spatial compression and degree of interaction of space are two key factors in determining how a matter orb will balance out its macht and so the collider does not provide a stable environment for these particles.

The particles will rapidly balance out their macht with the ESS found inside the collider and in so doing decay away to fully balance out. All decay events are of course based on a particles need to balance out its properties and ESS macht glow with the rest of the ESS in the local area of space as well as any macht glows from nearby particles. So different particles will decay at different rates based on these factors and in addition similar particles will decay at different rates if you change one or more of the factors affecting how the particle balances its ESS macht. In other words decay rates are not fully fixed for a particle everywhere in the universe and can be tested for easily by altering significantly some of the factors that determine decay rate. I say significantly because under normal spatial compression here on Earth where our laboratories would be the other factors affecting decay will need to be altered severely to see a measurable effect.

So the virtual mass of the field is of course caused by speck/mote interaction with ESS and the force carriers are simply what we observe as this process occurs. Take away all the mass from a particle as modern science understands it and you will not see any of the mass from field excitation and be left with the true mass of the particle in the universe which is the mass of the speck/mote which is the real fundamental matter. This is also what explains the double slit experiment as matter is made of specks and motes that we view as larger field excitations. In the double slit experiment matter seems to behave as both a wave as it moves and as a particle as it strikes the detector. This is caused by the particle following a seemingly random path over time instead of a set one as might be expected and creating the interference pattern observed.

The particles will move randomly because they are in a sense riding a wave if you will of the ESS macht glow inside the detector. The ESS macht of space remember exists before the slit, in the material of the wall that contains the slit and the space beyond. Macht flows freely through all of it but the particle cannot and so the particle will ride this macht glow that moves through the wall with the slits and explains the experiment. The fields generated by the particles are larger than the speck/mote that created them and so the motion becomes significant and observable. The field of the speck/mote particle attempts to balance its interactions with the ESS glow inside the experiment and so moves forwards from where it was projected until it hits the target detector.

The motion is as we have seen before through the explanation of quantum motion not at all random but instead highly complex but following normal rules for space physical systems. The random motion of the particle is not random and completely calculable from the moment the particle was fired into the experiment. The particle simply follows these rules as dictated by how it will attempt to balance its macht glow with the macht glow of the local area of space until it strikes the detector wall. You can always calculate exactly where a particle will hit the detector using space physics and all probabilities are fully removed. This explains the wave motion of the particle which is really not the particle being an actual wave but only balancing its macht with the ESS of the instrument. The particle will strike the detector as though it were a particle and not a wave in a point like pattern because it is in fact a particle. The detection is what finally stops the balancing of the particles macht glow in the local area ESS macht due to the fact that the target detector wall is solid and so it is detected as what it is, a particle that hits in one place with force. This is a very simple experiment and system to explain as it involves all the basic properties of space and matter interactions that can be used to explain any other property of the universe.

The act of observing that destroys the interference pattern is simply due to an input into the system that disrupts the internal ESS as they are no longer isolated as well from the outside world as they were before. This means that the natural balancing pattern they create when more isolated of balancing their macht glows is simply lost not due to strange effects or observers or anything like that but simply because the outside world is now contributing to the interior ESS macht balancing act. Remember you are trying to observe a single quantum object with another quantum object and all are created and controlled by how space and matter interact of course you would always get a disruption of the wave pattern as soon as you poked a finger into the experiment. Anything you do to observe a small particle inside the detector is going to feel very large ESS macht balancing changes and not behave as though isolated anymore.

Again the double slit experiment is not random, not magically destroyed for no reason by observation or interference but instead perfectly follows the normal rules of all systems using space physics. The lack of modern sciences understanding of the underlying nature of space and the ESS it is made of and the macht it shares between schales is what is missing from the equations attempting to calculate and explain what is happening. Add the missing space information and all random effects are gone and the entire system can be fully calculated from start to finish with no unknowns in existence. The double slit experiment will become a simple and basic grade school level explanation of space like iron filings on paper.

All of this was necessary to explain a simple hydrogen atom and how it constructs itself in the first place. From here we can take this simple hydrogen atom and build a black hole step by step from a star to a neutron star to a black hole made of SM. The first thing we must do is look at the normal hydrogen atom and list a few of its key properties. The first is of course that it is made of single positive and negative components which are the proton and electron respectively. Under normal conditions these two particles will balance out there ESS machts in the local area spatial compression and settle into a stable configuration. If the hydrogen atoms are however highly stressed under high temperature and pressure they will change shape from a dual system of electron and proton and become a single system made of a neutron. This happens when a large star finishes burning its supply of fusion elements and must collapse under the strength of gravity. As the star collapses material in the cores plasma is crushed until the electron capture mechanism can take over. Electrons and protons in this process are forced to overcome their electromagnetic limits and the electron is absorbed by the proton creating a neutron. The electrons and protons ESS macht glows as we have seen which were being balanced collapse under the high spatial compression and the two will finally merge which they could not due under normal spatial compression. The result is a new particle called the neutron which has roughly the same mass and now no charge.

The mass increase it has is due to the fact that the newly fused electron and proton specks and motes will of course interact different than they did before with the local area ESS. The most interesting fact is that the charge is lost from the system as the negative one charge cancels the positive one charge. We have seen before that there is no such thing as positive charge in the universe and so what we have is a final balancing of the specks and motes that create charge in the system with their local area ESS macht. The question then arises as to why the charge is lost? Why should it be that the speck/mote of each particle that beforehand was able to make the local area ESS macht glow not able to make it glow now? The answer is in how the specks and motes interact with each other in the neutron.

It takes energy for a neutron to be created and it takes energy for a neutron to be maintained as a free neutron will decay in eleven minutes or so. In other words the arrangement of the specks and motes inside a neutron under normal spatial compression is not a favored way that space seeks to balance out the local area ESS macht. The result is that the arrangement is not permanent as the particle slowly decays and we have learned that decay is the result of balancing ESS macht. A free electron and free proton are much more satisfying from an ESS macht balancing point of view to the universe and so are stable. In a neutron star however the specks and motes are kept together by higher levels of spatial compression which allows them to stay in their current arrangement without decaying rapidly.

This is where we can talk about another aspect of the speck/mote nature of fundamental matter and that is shape. Shape first of all is only a description I am going to use in order to get the base idea across and explain why matter behaves as it does in the universe as we see it today. The shape of each speck or mote is unique and if we look at a simple electron it has a shape that will have let us say two edges. Again all terminology I employ here is strictly analogous to get the idea across. The first edge is the edge that will interact or drag through the ESS macht of gravity in space and the second edge is the one that will drag through the ESS macht electromagnetic property of space. This gives the electron its properties and the strength of these properties can be seen by how much it drags its edges through the ESS. So an electron drags its gravity edge lightly and its electromagnetic edge heavily in space resulting in a light particle with strong charge. A proton is made of three quarks which approximately drag their three edges through, gravity, charge and the strong nuclear ESS machts of space. This is how the quarks differ from the electrons as they have mass from gravity, have charge in terms of color charge and how they bind themselves into a hadron with the ESS macht strong property of space. The final product is an arrangement of all three quarks such that they produce a positive particle known as the proton.

Now when we create a neutron from the electron capture method we cause the specks and motes of the electron and hadron to do something interesting which is arrange themselves such that the final object does not drag any edges through the electromagnetic ESS macht anymore. What this means is that the new particle the neutron has an internal arrangement of the speck/mote surfaces it is made of such that the electromagnetic edges are hidden from space in a sense. Think for a moment of a geometric object made of four parts one being the electron and three are quarks. The arrangement is now ordered such that each of the smaller geometric objects is put together such that the edges responsible for charge are now hidden inside the center of the object. This means that they can no longer drag into the ESS macht electromagnetic part of space and so the local area ESS macht glow for charge is lost. The object still has its gravity and strong nuclear dragging edges exposed however so the object can have mass as well as strong nuclear charge to hold it together. The specks and motes inside the larger object can still behave like normal particles only they no longer act to create a charge and this explains why something like a neutron or an entire neutron star has no charge.

The electromagnetic force is known to be two separate things that are combined under normal conditions but they can exist separately. A neutron star possesses the necessary conditions to dissociate these two forces again and it does so through the arrangement of the speck/mote structure of its neutron. Neutron stars have no electric charge but they can have a magnetic field proving this dissociation. If the collapse process of the neutron star is more ordered it can result in a much rarer object known as a magnetar. Magnetars are simply neutron stars that formed with their specks and motes arranged inside their neutrons in such a way that they ended up aligned with one another. This allows them to produce strong magnetic fields from neutrons that still maintain a neutral charge and therefore separation of the electromagnetic force into two separate components. The formation and properties of neutrons, neutron stars and magnetars can all be explained by using the shape of specks and motes of fundamental matter and there geometric orientation with one another as held in place by spatial compression. The analogy of speck/mote edges, shapes and degree to which they drag these surfaces through space can be used to calculate how matter will behave at all scales from a fundamental particle to an entire star, galaxy or universe.

The next natural evolution of matter from here is to go beyond a simple neutron star and create a black hole which is made of SM or matter that is near 100 percent interacting with space. Interactions with space and the analogy of edge dragging can be expanded upon using our familiar matter orb and ESS spheres from previous sections. Under normal spatial compression a matter orb will only have its speck/mote edges encounter so many ESS. As the spatial compression increases the number of ESS it encounters increases and so the number of ESS its edges will drag through also increases. This is exactly what is happening in a black hole as we have seen through facts like variable gravity previously. However a new arrangement of the specks and motes is necessary to create the correct conditions for a black hole to exist. The remaining particles inside a black hole would go beyond the standard quark/gluon plasma as they need to be arranged more densely in order to create their high gravitational fields.

It is known that quarks have a positive or negative charge associated with them and combinations of these charges will produce a proton or neutron for example. It is also known that the capture of an electron into a proton producing a neutron creates a different arrangement of quarks than what existed before in the proton alone. The neutron is not made of three quarks and an electron but three quarks only and so the charge carried by the quarks of the neutron reflects this absorption of the electron into the neutron. In quark gluon/plasma free floating quarks are held together by the strong nuclear force which means that the quarks are predominantly dragging only their gravity and strong nuclear edges on the surfaces of their specks and motes through space. When a black hole is created you once again have an increase in the necessary spatial compression required to hold matter in a form beyond a neutron star. If you did not then obviously you would only have a neutron star and not a black hole. Since the necessary spatial compression exists to create a black hole the matter speck/mote particles of the quarks must also now undergo another geometric arrangement that satisfies the local area ESS.

This was first made possible in the neutron star by cancelling out forces of charge to create a new arrangement of particles that was neutral. The same process will be repeated in the black hole where the quarks must be made to combine into a new particle so that they can become denser and create the necessary conditions needed to increase spatial compression and create the black holes incredibly high variable gravity. The quarks are forced together into a new arrangement of their specks and motes such that the strong nuclear edges are now also facing inwards and can no longer contribute to the local area ESS macht. This creates a new neutral charge particle exactly as the neutron was created as a new neutral charge particle in the neutron star. The new particle does not excite the strong nuclear force and so the quark/gluon plasma can be condensed into a much tighter object which is a black hole.

The amount of energy as I have stated before that can be donated from matter to space is never lost only transferred from form to form and so here we have only the gravity edges of the speck/mote geometric object dragging heavily through numerous ESS all at once. The matter is now capable of donating almost all of its energy solely to the ESS macht gravity of the local area of space and explains the very high gravitational forces of a black hole. In keeping with tradition of naming new particles after those who theorized them I cannot miss the opportunity to take advantage of the exact spelling of my own name. Electron, positron, neutron all have a similar sound and use an r in their spelling and so I would like to lightly offer adding an r to my own name to create this new particle.

The Clintron

The clintron will be the name I use from here on to describe this new arrangement of quark specks and motes into the geometric shape that constitutes the matter found inside black holes and SM. This clintron particle follows the same evolution of matter from a simple hydrogen atom consisting of an electron and proton to a neutron by way of cancelling charges. The speck/mote edges and how they drag through various ESS make the neutron possible and can be seen in both neutron stars and magnetars. The evolution of reducing charges and edge dragging is continued as the neutron star is surpassed and a black hole is formed with the quarks once again cancelling charge until they only contribute in their new geometric shape to gravity. So the energy of the three fundamental forces is now given essentially only to the gravity force and explains why a large gravitational field can come from a black hole without needing to curve theoretical spacetime into a singularity.

All specks and motes of fundamental matter that can be used to explain normal matter as we see it everyday are kept intact and all rules governing how they interact with space and each other are also maintained. In this way everything from a person, an apple or the Earth can be constructed from the same ESS and macht as well as the same specks and motes as a black hole. The universe therefore has no breakdown or uses any new laws of physics when dealing with exotic objects like black holes. The clintron is the last evolution of matter and space in the universe and shows how you can seamlessly move from the smallest quantum sized objects to the largest structure possible which is the entire universe while not changing the laws needed to describe any system along the way. The clintron will naturally be one of the fastest if not the fastest decaying particle in existence as it requires an arrangement of specks and motes that is only made possible under the incredibly high levels of spatial compression found inside SM. With the spatial compression removed the clintron will almost immediately decay into a more stable arrangement of specks and motes with respect to how fundamental matter balances itself with the ESS macht glow of space.

To prove this particle exists and that space and matter interact in these specific ways we can look at the evolution of the universe after the Big Bang and in fact inside the Big Bang. I have already shown how a single piece of TSM will cause an explosion powerful enough to start a BBB by collapsing the ESS walls and now we can add the clintron to the equation. In the instant that a piece of TSM forms the clintron is now dragging its specks and motes 100 hundred percent through the ESS it is inhabiting. The 100 hundred percent degree of interaction of the clintron with space means that at that one instant in the universe the fundamental matter is being made to exist or try to exist fully inside space in other words space and matter are trying to become one which creates an impossible situation where the macht balancing of the ESS cannot be handled anymore as the limits of both ESS macht glow and speck/mote interaction into space are both at their limits.

Space and matter as we have seen cannot fully exist as one and so the matter is ejected from the ESS and the macht is dumped back into the universe by the ESS. What you are left with is a standard Big Bang detonation event that will unfold as predicted. The ESS will using their faster than light macht speed spatial rebound expand inside the matter and shred it from the inside and what this means to the clintron is that it is dissociated in a spatial rebound decay event which is not the same as normal particle decay. Normal particle decay is created when ESS balancing occurs between matter and space which are continually interacting. In spatial rebound decay the matter is made to dissociate through purely physical means as the space inside of it expands and the matter can no longer maintain its geometric configuration of specks and motes.

In the TSM instance the clintron will naturally be forced to decay into the specks and motes that created it and in fact will decay into the absolute smallest possible specks and motes in the universe. I say this because it is possible that future experiments will discover small particles responsible for building things like electrons, quarks and the like. It would actually make sense that electrons for example are made of multiple specks and motes as depending on their interaction with matter and space may or may not need to excite more than one ESS macht type. Take for example a photon and neutrino which are very similar as they are both light speed or near light speed particles that appear to have similar masses. It might be possible that a photon is simply the same type of speck/mote as a neutrino except for the fact that the neutrino either has a different shape or is made of two specks and motes that allow it to weakly interact with the gravity ESS because it has a small non-zero mass. I will discuss this more later but for now it is highly probable that particles such as quarks and electrons or at least some of them are made of a combination of specks and motes in different geometric configurations. It also means that new binding forces would exist to hold things like electrons and quarks together as this would explain why you can have an up quark and a down quark which are similar except for color charge. So the strong nuclear force would be a compound force like the electromagnetic force or electroweak force if this is found to be true. These additional particles and binding forces in no way changes how matter behaves at the quantum level or how they interact with space and so would not disrupt the Standard model and could be smoothly incorporated to accepted physics.

Using only known fundamental particles for reference here however the shredding of the clintron by means of spatial rebound decay events will dissociate it into the smallest possible free specks and motes available to the universe. The dissociated fundamental matter speck/mote particles can now move freely through the universe which is under rapid expansion. Once the expansion has stopped due to the first specks and motes turning back on and dragging their gravity edges through the ESS again we see the emergence of the first force which is gravity.

The specks and motes will then be able to arrange themselves through normal gravitational attraction as well as the dense early universe into more complex shapes such as electrons, quarks and the like. These new geometric arrangements can begin to turn on other fundamental forces in the universe as we see for the first time the strong nuclear force become active now that speck/mote geometry has made quarks. The quarks have the correct edges to drag into now both gravity and strong nuclear ESS and increase the ESS macht in the local area of space there. Electrons are a different type of speck/mote and so do not turn on the strong nuclear ESS macht but only the gravity and charge machts of ESS. Here we see normal interactions of particles in high energy environments where the macht balancing of electrons and other particles like quarks cannot be balanced yet.

As the universe cools we see the balancing limits reached where quarks can finally arrange themselves into protons and so we have a universe filled with protons and electrons. The ESS macht balancing limit of the strong nuclear force is reached first because the speck/mote arrangements are very strong and overpower the influx of energy from the rest of the universe. This is one of the reasons that nuclei are so hard to split or fuse as the speck/mote arrangements of hadrons through the nuclear force is very stable with how it balances its local area ESS macht. The electromagnetic ESS macht is far weaker by comparison and so you can cause the balancing to easily alter as is seen in simple chemical reactions where the act of say burning something is all that is needed to rearrange the local ESS macht balances of different atoms or molecules. Trying to get atoms however to overcome their ESS macht nuclear balancing is far harder as it takes many times more energy to create fission or fusion events. So what we have now is a universe that needs to cool so that the electron ESS macht charge balancing mechanism described in a previous section with the proton ESS macht charge balancing can occur.

This is what is seen as first light as the new arrangement of specks and motes as well as ESS macht balancing energies for space and matter interaction systems is finally balanced. You can now have a universe filled with atoms as simple as the hydrogen atom where the arrangement of the specks and motes within it create a nucleus, orbiting electron and has all four forces turned on and able to interact normally as we would observe right here on Earth. Following the decay and full dissociation of a clintron from the moment of a TSM creation event and ending up with a universe as we see it today can be explained by following the arrangement of matter specks and motes and how they interact with each other to allow particles to have different properties based on their geometric configurations and by using the normal ESS and matter interaction mechanisms of balancing ESS macht levels.

All four forces can be turned on in sequence and by looking at how matter drags its speck/mote edges through ESS can explain the differences in the strength of the four forces. Gravity is the weakest by far under normal spatial compression because it not only has a weak dragging force through ESS but because it needs more ESS to drag through to create more strength. Gravity will become easily the strongest force if the correct levels of spatial compression are present as the speck/mote arrangement of the matter interacting with space automatically reconfigures itself until only the gravity edges it possesses are capable of dragging through space.

Since the amount of energy available to matter to be donated to ESS must be conserved all the remaining energy that used to be held in the other forces can be used for gravity dragging. In a sense the much stronger strengths of the other three forces are simply transferred to the gravity force by way of ESS macht glow of gravity by speck/mote edge dragging. This explains how gravity can physically overpower all other forces inside a stars core to overcome the normally stronger force of the strong nuclear force and why at the end of a stars fusion life gravity can fully overcome all other forces to create a neutron star or black hole. It also explains why even if spatial compression is elevated for matter inside a black hole the nuclear force does not simply become that much stronger than gravity again because it also has more ESS to edge drag through. The strong nuclear force edges of the matter involved inside the black hole are not available to drag through the extra ESS so normal matter can have an increase in gravity but not strong nuclear force allowing the gravity to from the perspective of modern science overpower the strong force easily. The arrangements of speck/mote fundamental matter inside of particles and how these interact with the various ESS macht of space is what explains the seemingly impossible unsolved problems faced by modern science every time a law of physics seems to break down or it appears that new exotic laws are needed to explain a phenomenon.

Facts

Here I will once again list the facts from this section with the first being that matter is a particle and not a wave or excitation of a field. The second is that matter is responsible for generating the fields associated with particles as understood by modern physics. The third is that matter is really a speck/mote of a fundamental matter that interacts with the ESS macht of local space to create a larger field. This is analogous to how a snowflake contains a piece of dust at its center around which the ice forms and grows. The fourth fact is that specks and motes have different shapes that allow them to interact with the ESS of space in different ways and to different degrees. This can be simply thought of as a shape to the speck/mote such that it has specific edges that can drag through the ESS and the more edge dragging that occurs the stronger the degree of interaction of matter with space. The fifth fact is that if completely dissociated from space matter still has mass in the form of the physical speck/mote and only obtains virtual mass as it interacts with ESS macht to produce what are known as fields and force carrier particles. A speck/mote for example has a fundamental mass outside the Higgs field and will use the Higgs boson as a force carrier to drag through space and have virtual mass that is felt as gravity.

The sixth fact is that the wave and particle natures as well as the confusion surrounding the double slit experiment is solved by using ESS macht balancing properties and speck/mote interactions with space. The matter is always a particle that moves as a wave as it seeks to balance its ESS macht with the local area of space and observer intrusion simply alters the ESS macht glow of the system destroying the balancing act seen before. The seventh fact is that matter can be rearranged at a speck/mote level to explain the formation of things like atoms, neutron stars, magnetars and black holes. The eighth fact is that black holes are made of SM which is an arrangement of quarks into the new clintron particle which permits the heavy edge dragging of matter through the ESS macht of gravity. The clintron is a particle without charge that possesses mass and has a very fast decay rate. The mass of the clintron is most likely equal to the combined masses of a neutrons quarks without the gluon force carriers but may differ as perceived here on Earth under normal spatial compression because its natural speck/mote arrangement allows it to use nearly all of its energy to edge drag through ESS macht gravity and would most likely result in a true mass only measurable under those high levels of spatial compression; or it might be a mass lower than that of a quark as quarks are broken down inside a black hole to their component specks and motes which are smaller still. Regardless it possesses the properties of mass but has no charge and is the final evolution of all matter forms in the universe as it is the last arrangement possible of specks and motes available to the universe before a TSM event occurs.

The ninth fact is that the existence of clintrons and how they interact with space can be proven by allowing them to be dissociated fully during a Big Bang event and the component specks and motes from which they were created are allowed to slowly turn the known forces of the universe back on in order based on their natural edge dragging properties. This can also account for why certain forces are seen as weaker or stronger than others given our local area spatial compression of ESS and macht glow here on Earth. Speck/mote properties such as edge type, degree of edge dragging through ESS and interaction with other particles will easily account for why each force possesses the strength observed here in laboratories. The amount of effort it takes to overcome the ESS macht balancing of matter and space for a given speck/mote type determines how strong each force is unless the speck/mote geometry is changed as in a neutron star, quark/gluon plasma or black hole to give a few examples. The tenth fact is that by using ESS and ESS macht as well as specks and motes of fundamental matter and how they interact with one another there is no longer a barrier between the smallest scale systems in the universe and the largest scale system which is the universe itself. No breakdown of physics or new exotic physics need be invented or employed to explain everything that is observable in the universe as we see it today.

Chapter 25
Entropy, Information Loss

Entropy Recycling

Some of the other unsolved problems in the world of modern science are more simple questions than massive dead ends or paradoxical situations. Some such examples of simple questions are that of entropy specifically why does it increase, will it ever stop increasing and can the universe naturally return to a state of lower entropy? The answers to these questions constitute one of the unsolved problems facing modern science and are fairly simple. The answer to the first question has been known for some time in macro scales from the classical definition and explanation of entropy and the reason entropy increases as a whole has been dealt with in a previous section. The more interesting remaining two questions have not but will be dealt with in this section.

First however I would like to look at the problem of increasing entropy as incorrect and state that entropy is decreasing as the universe moves forward. It is often argued that entropy and for these questions I will be discussing the quantum statistical form of entropy as most people understand it which is an increase in disorder and information of the universe. The general idea being that as the universe ages the complexity of the universe increases, the order that was once present at the beginning of the universe wanes and the amount of information needed to describe the system of the universe increases rapidly. To this I say that entropy is all about your point of view and to begin with let us look at the normal so to speak view of entropy.

This is the view that starts at the Big Bang where entropy is seen to be essentially zero as there is theorized to be no complexity or disorder in the universe only the perfectly impossible singularity. Since the singularity has no dimensions and really no true properties and is a single object there is no way for entropy to increase and since entropy is a measure of order from moment to moment this means that there is in fact no entropy at all as no changes occur in the singularity. So the universe it is argued started out very well ordered indeed and things only became pear shaped from there on so to speak. So now this no entropy singularity thingy explodes and we have a universe created from a huge and mighty explosion that is still very well ordered as it has only two parts which are of course space and matter in the form of a sea of energy.

It is a steady decline down the slope of order to disorder as the universe grows because every new occurrence of something in the universe brings more variables and therefore more entropy and it takes more and more information to describe it. The appearance of fundamental particles, the turning on of forces and the interaction of all of these really complicates things quickly. It is argued that the simple and neat and tidy order of the universe which used to be just a sea of energy and fundamental particles has been getting steadily messier as more and more new things like atoms for example are made. Atoms lead to stars and stars lead to supernovas which just tend to stir up the whole lot over and over adding even more disorder and yet still more information to explain it all. On the other hand the total number of particles roaming free in the universe at the very beginning of said universe has never been higher.

The fact that nothing was bound to anything can also be seen as a huge amount of entropy that only got smaller as particles found friends and mates and started to congregate in familial groups like three quarks to a hadron. Instead of describing three quarks separately I can now on a slightly larger scale describe only one hadron. So the number of variables in the universe for the system concerning this one hadron has decreased by two. As more and more of these events take place more and more free things and bits in the cosmos get bound up into larger more easily described objects. Eventually you get black holes which are in a sense order creators because they inhale everything that comes their way thereby reducing the number of free particles and objects in the universe allowing all of them to be described as a single object.

Now I realize that this is only true from a macro scale point of view and that internally you still have three quarks to describe and now you need to describe not just the quarks but their interactions with each other and so yes the total amount of information and entropy of the universe has increased at the smallest scales. I only wanted to pose this little thought reversal as a way of stating that depending on the scale you choose to use things can get much more ordered as the universe ages. A universe for example consisting of nothing but black holes at the very far end of the cycle of life for the universe is very well ordered. Space and black holes, that is all that needed to describe the entire cosmos and each black hole has an identical makeup to every other. So describing a universe with a small number of massive objects like black holes is far easier than describing a universe where every almost countless number of particles held in those same black holes are free to race across the galaxy anywhere they like.

If we keep things like this in mind while answering the second and third questions concerning entropy and information in the universe we can find a solution to the problem. The second question which is will it ever stop increasing is of course yes it will. No matter what fate for the universe you believe in if enough time is allowed to elapse then the universe will eventually reach the bottom of the potential well for every conceivable object ever created. At this point interactions with other objects will be impossible as the object cannot escape the well. The only motion remaining and energy in the universe therefore at least according to quantum mechanics that is comes in the form of random quantum probabilities. These probabilities theorize that even in the lowest part of the potential well for any object in existence a small but non-zero amount of quantum energy remains allowing the particle to jump from location to location in space around the well or more accurately from probability to probability.

Here it must be pointed out quickly that this is impossible as quantum motion due only to mathematical calculations means a form of perpetual quantum motion machine has been made which is of course wrong so in reality quantum probabilities at the bottom of potential wells is not true. Quantum tunneling events are due to inabilities to understand the macht balancing of the system. What this means is that apart from quantum motion everything in the universe will stop and therefore as long as you describe these probability jumps for each object in its particular potential well you can finally describe with a finite and stable amount of information all the disorder and therefore entropy in the universe. Once again because entropy is a measure of change of the state of the system in question if the universe does not change the concept of entropy is no longer relevant and once again entropy stops because it does not exist to change from moment to moment. So the answer to will entropy ever stop increasing in the universe is a resounding and proven yes.

Now we get to the meat of the matter and by far the most interesting and important question concerning entropy which is can it ever decrease? What this means is is there any way for a system such as the universe which is totally closed and isolated as it has no outside and therefore no where to decrease entropy internally at the price of increasing entropy externally of the system decrease its entropy? Can the universe decrease its entropy and therefore take less information to describe which can be seen as a sort of entropy recycling mechanism? To answer this unsolved problem it is necessary to look at two things which are the way the universe will meet its fate as many like to describe it and what happens to it while that fate is being met. The future of the universe is already known as I have answered this previously and then therefore we know the universe to eventually stop expanding and begin collapsing back onto itself until eventually a TSM event occurs and a new Big Bang begins.

The key to proving that entropy will in fact recycle itself and decrease comes from looking at what is happening to the universe during this expansion halt, collapse and eventual explosion. During the far future events of the universe a halt to the expansion of the universe will become real based on the mechanisms I have given previously. Using either mechanism once the universe is at a maximum it will have ceased increasing in volume and therefore one source of increasing entropy will be turned off. Entropy tends to spread out evenly in a system and the bigger the system the more spread that is possible so once the universe halts we can eliminate one source of entropy increase for this cycle of the universe's life.

The collapse portion of the universe happens for two reasons the first being any matter in the universe creates gravity and with the universe at its maximum size there is no more outward pressure and so gravity will slowly start to spatially compress the entire universe. The second mechanism is that an older universe will contain many black holes which are made of SM and therefore produces variable gravity. Any source of variable gravity exerts a significant spatial compression mechanism on the local area of space. So in either case the universe will begin to collapse slowly at first and then very quickly. During either collapse event the volume of space decreases faster than the volume of matter and so free particles and objects in space must exist closer together than during the expansion phase. This means that the amount of matter available to existing black holes rapidly increases and the black holes will not evaporate. Growing masses of SM will only add yet more variable gravity which again causes the universe to collapse faster reinforcing this positive feedback loop.

The SM in the form of black holes in the universe are made as we have seen of clintrons and those particles are made from a specific geometric arrangement of specks and motes that permit only certain forces to be active or at least for other forces to be massively dominant. The matter falling into the black holes is simply turned into more clintrons or will further impart energy to the already existing clintrons in the SM mass. Clintrons that have increased in energy can naturally donate more energy to the ESS macht of the SM matter which is of course where the variable gravity comes from. Aside from continually increasing the amount of SM in the universe which speeds up the collapse process entropy is also recycled in all SM and clintrons. This is because the arrangement of specks and motes is always the same and of a specific pattern.

This means two things the first of which is only a small number of forces are turned on inside clintrons which reduces the number of variables and therefore amount of information needed to describe the system. The second thing it means is that much of the information needed to describe free moving quantum objects is decreased because they are now bound up in ordered clintrons. The clintron particle itself can be described using a minimum of information as compared to all the constituent particles that went into making it. Think of three quarks bound into a hadron and you have the three free moving quarks, gluons holding them together and any motions or exchanges of color charge between them as well as all the fundamental forces they excite. A clintron removes the gluons, binds the quarks in a new geometric arrangement and turns off some of the excited fields. So SM matter is really a form of matter and space in the universe that naturally decreases entropy as opposed to increasing it. In this way the universe during its collapse phases of any cycle of its life will always take the maximum amount of entropy available and begin to reduce it.

The number of particles needed to describe the universe as well as the number of forces decreases without the use of silly notions such as universes mirroring our own with backwards flowing time or what not. This means that at the moment of the total collapse of the universe at the point where all space and matter are contained together in one huge SM mass entropy is essentially reduced to Big Bang levels. In truth it will reach below Big Bang levels and have less entropy than when the universe explodes as the SM locks all space and matter into neat repeating ordered clintrons that also incorporate space in a minimum volume. To prove this fact we will let the SM turn into a single piece of TSM which has the lowest possible entropy of all as it is the combination of space and matter, the fundamental forces as well as all matter and energy. At the point of TSM the universe has unified everything in it into the one sought after super force of sorts although in truth it is all fundamental forces, all forms of matter and energy and all space unified as one which is far more reaching than just unifying the four fundamental forces. This clintron TSM object is of course the lowest entropy possible in the universe and so even a single piece of it existing means that the universe must have less entropy than it does today no matter what.

The universe mass will undergo a TSM event that begins the Big Bang detonations all over again and creates a universe of space and matter once again. In other words the extremely low entropy universe experienced from our own Big Bang and to which the entropy of today and the entire concept of entropy in the universe and information comes from will once again be achieved. This means that the entire universe is at a lower entropy state again and so the answer is yes entropy can easily be decreased in the universe and the collapse and SM mechanisms are what will make this happen. Black holes and any SM structure are natural sources of entropy reduction as well as information reduction and the clintron particles they are made of explain why the amount of information needed to describe both the matter and number of working forces in them exist as they do. Entropy will decrease without unusual methods, objects, realities, dimensions or alternate universes and instead can be simply recycled with normal space and matter.

Information Loss

This is where I can start to explain about information and whether or not information can be lost from the universe as believed by some and rejected by others. To start off let us give an explanation of what is meant by information and from there I can explain different types of information and information loss. It is a long held belief in modern science that one of the fundamental laws of the universe is that information cannot ever be lost meaning no matter what you do you can always explain a system. Take for example a burning match and look at it before it burned and after it has burned itself out. The match is neat and ordered to a degree before it is struck on its striker pad and bursts into flame. The match burns for a while consuming the fuel source it has which is either wood or cardboard and then it goes out. During the burning process chemical reactions occur where energy is taken from the stored chemical bonds of the atoms and molecules that make up the match and its fuel source and released to the environment. Energy can be different wavelengths of photons like light or heat.

Also the fire of the burning match is a form of plasma that carries away atoms and molecules of fuel that will be in various stages of unburned to fully burnt. Smoke is also created in the process which is made of the oxidized fuel escaping the reaction and drifting off into the environment. The end result is a blackened and burnt matchstick that once cooled gives you no description of the match before it was lit. The idea of no information being lost from a system states that if you know the laws of physics and you know the information concerning all the escaped, heat light, smoke and so on from the match you should be able to work your way backwards to rebuild the match as it was before it was burnt. You do not have to physically rebuild the match only be able to determine what it looked like before hand. In other words all the information needed to describe the match before it was burned remains intact in the universe and can be retrieved from the post burning materials.

Science takes this idea to the farthest limits and says that everything in the universe in every system obeys this rule and you can take smoke and heat and light and using all the stored quantum information for each particle involved and this includes the smallest fundamental particles, forces and what not and you can recreate and explain any previous system. So no information is lost according to this rule and in theory this applies to everything in the universe from matches to black holes. Stephen Hawking somewhat upset the world of science when he proposed that information could be lost inside a black hole through pair production and capture of quantum objects near the event horizon of a black hole. The theory goes that quantum space is continually creating pairs of similar particles and anti-particles that pop into probability existence out of the vacuum energy and then annihilate with one another. A side note here is that the idea of vacuum energy making and annihilating particles is also incorrect as it would take free energy to make these particles and even the argument that a black holes gravitational energy is the supply source is still wrong because it would be free energy from the black hole if the particles annihilated neatly again and again.

Hawking suggested that any pair production close to the event horizon of a black hole had a very small chance which is slim to none of a pair of particles failing to annihilate close to the horizon. The reason this was said to be so comes from the fact that if one of the particles is produced near the horizon and crosses over it will be trapped in the black hole never to emerge again. What this means is that one particle travels off into the universe and the other is trapped inside the black hole. The reason it is thought this destroys information is that the event horizon is thought to not be in contact with the singularity of the black hole meaning the information of the absorbed particle is trapped inside the black hole that we as observers outside the event horizon cannot ever see. This is of course based on the idea that the black hole represents a part of the universe outside the universe which is not true. This is of course based on the idea that the event horizon cannot communicate with the singularity or rather the interior of the black hole and this is not true. Lastly this assumes that the black hole contains a singularity and this is not true as singularities simply do not exist. However Hawking is still very correct that things can be trapped inside a black hole and the event horizon represents the point of no return. It is in fact not Hawking that has been arguing over information loss all these decades yet he has been instrumental in starting the debate. So the main question exists in modern science today as an unsolved problem which is can information be lost in any mechanism from the universe? The answer is yes and to examine this I will explain the true nature of information as it relates to both matter and space in the universe.

◄New Universe Order►

The classic definition of information as described by modern science relates to information concerning matter and energy in the universe. Examples of this are things like atoms, photons and subatomic particles. Take for instance a hydrogen atom and you have three quarks, force carriers for the nuclear force and an electron moving around the nucleus as well as some photons according to modern science. The information in the system will allow you to describe all the known properties and behaviors of the constituent parts of the atom. So the hydrogen atom at one moment in time will be at position X and the electron will be moving let us say clockwise around the nucleus at heading 000 degrees. If we leave the system alone for a while we might find for example that the atom has moved its position slightly in space so that it is now at position Y and the electron is now at heading 270 degrees. Information theory states that we know the behavior of everything in the system so by knowing these behaviors and working from position Y and heading 270 degrees we can in a sense rewind the events of the system like you would rewind a tape in a VCR. Once the tape is rewound we find that the starting conditions of the hydrogen atom were position X and heading 000 degrees so the information involved in the system was not lost from the universe. This is the standard description and workings of information theory that assumes that no information is lost in the universe and applies to matter.

The reality of the universe and information however is far more complicated than just this simple idea of information and the description of matter systems only. The first thing I want to talk about is the idea of information theory in the very first place. Where is it written in the universe that information theory is even correct? What I mean by this is once again some human-centric view of the universe and how it must all work. There is nothing decreeing that information must be always preserved forever and can never be lost from the universe this is only an idea that humans have adopted as a way to make sense of things we see and experiment on. This idea also goes back to its roots many centuries before the twentieth century and quantum mechanics, general relativity and the Standard Model. Scientists of the day simply examined various reactions and said if you can go forward you can go backward and everything was maintained. From there on out science has become ever more obsessed with keeping this rule and has gone to great lengths to establish it in everything modern science touches such as quantum mechanics which deals with what is believed to be the most fundamental of objects possible.

The problem is that nothing says this must be so and as a result examining the possibilities of systems where this does not hold true has been considered heresy at best where the supporters are quickly and discriminately hunted down by angry pitchfork and torch wielding mobs of established physicists. Stephen Hawking was amongst those who dared challenge this rule and was hounded for decades until people felt they had concocted the necessary theories to disprove him. In truth however the idea that information need be preserved no matter what in the universe is simply a human invented idea and does not have to be held as perfect fact by the universe. Remember the universe does not care about our little humanesque concepts as they pertain to it and the universe should not be made to fit theory but rather a model constructed to explain the universe as it naturally is. So now we get right to it and ask whether or not information can be lost from the universe?

Yes quantum information can be lost from the universe in such a way that it cannot be retrieved back as quantum information. If you notice I said quantum information and not all information and there is a good reason for this. To start off I will say that Hawking was in fact right all those years ago as information can be lost from the universe but not as he described it. What this means is that first of all a black hole is part of the universe so even if the old incorrect version of black holes was true and the singularity could not communicate with the event horizon the information is still stored in the black hole and the black hole is still part of the universe so no information loss there. The idea of information loss however is that it is lost from observer view and so the events that led to the black hole gaining mass cannot be rewound on the universe's VCR. Hawking from that point of view is correct as information will be lost to the black hole and you cannot retrieve it from the outside but it is not lost so to speak because of the event horizon and singularities string being cut between their two Styrofoam cups. I will explain a simple quantum mechanical view of lost information here and then explain the different types of information and how they interact with one another.

Black holes as you now know are not paradoxical objects that contain impossible singularities where the event horizon does not talk to the rest of the black hole at all. Black holes are simple masses of SM that are built from normal quantum objects arranged in such a way that they tie up what modern sciences knows as particles, forces and fields into a simple structure that will consume anything they encounter based on perfectly normal laws of physics. As I have also explained black holes cannot be destroyed from the outside because any attempt to do so involves matter, energy or fields that will simply be smoothly incorporated into the SM mass and will therefore have no effect on disrupting the clintrons or ESS macht that hold it together and instead will only make it grow stronger and more massive. This is how you can have a smooth and contiguous structure to the black hole such that the very center of the object does in fact communicate with all other layers of itself and in fact the event horizon too. There is no broken line of communication which causes these paradoxes of physics in the first place only normal physical laws explained such that anything that crosses the event horizon cannot be retrieved until the black hole is destroyed. Black holes like all SM masses can only be destroyed internally and usually through TSM events so the information is technically lost from the universe that we can observe although of course it is not destroyed.

This now brings up the question that goes beyond the old if we cannot see it in a black hole it must be gone version of information loss and asks the real question of regardless of the object involved can information in fact truly be lost from the universe as in can it be lost from the observer such that past events cannot be rewound and calculated? To answer this I must explain the difference between two types of information the first is the only form of information known to modern science which is quantum information and the second is of course space information carried by ESS through macht. Quantum information is not lost from the universe if the universe only consisted of quantum objects and systems. So if the universe was made of matter only whether that be fundamental particles, force carriers or whatever then yes you of course would be able to always rewind any event on your cosmic VCR and see what came first. Quantum systems however are a single part of the universe and very much reminds me and hopefully you of early experiments using chemical reactions. It was understood that if you take two hydrogen atoms and one oxygen atom you get a molecule of $H2O$ and if you break that $H2O$ back down again you can reverse the reaction and get again two hydrogen atoms and one oxygen atom.

If the system contained only atoms as seen from a chemical point of view then yes you can rewind the events and get back to exactly where you started from in the system. In reality unseen factors exist in this system that are not described by atomic methods alone and with the hidden world of photons you cannot fully rewind the tape. This is exactly what is happening in quantum systems today as particles are only a part of the equation so yes if we use only quantum objects and rewind a part of the system we get what we started with. However the unseen interactions and properties of space which are a part of the reactions are not factored in and just like the invisible contributions of photons and energy to the hydrogen and oxygen example so too are space physics the invisible contributions of all quantum systems. So yes information is preserved from a quantum mechanics viewpoint of the universe and you can indeed back track all events to get to your initial starting conditions for any purely quantum system.

The other type of information in the universe is not quantum in nature but space physics in nature and can be called ESS information. ESS information accounts for all factors affecting the local area of space of the system and these can include ESS size, motion and macht. Since it is obviously proven that matter interacts with space and that these interactions are what produce the universe as we see around us then the necessity to include space physics in the concept of information for a system is a law and cannot be broken. ESS information therefore is shared between ESS through the use of macht and can be an increase or decrease in the macht pool possessed by all ESS in existence. ESS information does not travel in the same manner as does quantum information however and this difference is important for total information of a system in the universe.

Quantum information is a form of point like exchange between quantum objects such that a single particle will always pass information along to another particle in a point like linear manner. A single particle will have definite properties such as mass, charge and motion to name a few and when it hits another particle it will always transfer these properties in discrete and conserved amounts in a linear motion. If we ignore the mass and charge of the particle and focus only on the motion then according to the three main views of modern science a particles velocity and angle upon striking will send the second particle flying off on a vector calculated from those two factors. All quantum systems are similar in this respect to a simple Newton's cradle where the information of motion of the first ball in the cradle is conserved and transferred in a discrete amount to the last ball in the cradle. A particle will behave exactly the same as the balls in the cradle system and transfer information in a discrete manner that is conserved similarly for any quantum system regardless of the factors or information being discussed.

ESS macht however is not transferred in a single discrete amount throughout the system because the system is rarely if ever a single ESS with the exception of a TSM event. So a single particle will alter the macht of the local area ESS as these ESS naturally share the macht glow between schales of the local area of space. This is extremely simple space physics and something I have explained many sections ago but demonstrates another application of the macht sharing between schales as they can be once again used to correctly answer another of the unsolved problems of modern science. The ESS tend to share information in overall non-discrete ways as the glow is a sharing of information more like a nebel or mist in a local area of space. The rules for sharing information between a single ESS and another ESS adjacent to it will of course take place in a set amount and can be calculated for every such interaction but this is where traditional information loss occurs as I will explain by combining the quantum and space worlds of physics now.

From the point of view of the universe information is not lost as the quantum information for any particle or similar fundamental object can always be linked to the local area ESS it inhabits. That being said information from a quantum system can be and is routinely lost to ESS macht. The physical interactions of the system between the matter and space are not broken because it is obvious that they are still in contact and making objects that we can observe in the universe constantly. What is lost is quantum only information as the energy imparted by matter to the ESS is shared in the macht glow and so retrieval of this information cannot be done in a discrete quantum manner. You can put a quantum amount of information into a system but you will not always get a quantum amount of information back out of the system. What you do get out is a reduced and diffused form of information as it can be reincorporated into a quantum only system. Again I say quantum only because this is the form of information and type of system that classical information theory concerns itself with but is not the true and more importantly total information for a system in the universe. So from a purely quantum standpoint you cannot fully backtrack the sequence of events for a given system and get back to the starting point perfectly.

Imagine an electron and the seemingly random quantum probabilities to its motion as it moves around a nucleus. The electrons true motion cannot be fully backtracked using quantum only information because the apparently erratic jumps it makes as it moves is explained by information contained in the ESS macht glow it is creating and encountering with the nucleus and the surrounding system. This is why things exist as probabilities in quantum mechanics and not factual information. The reason this is so is because of three different but inexorably linked factors which are ESS macht glow, ESS size and macht speed. The simplest way to understand this is to look at what the quantum object is giving the ESS of the local area of space and what it is giving is a set amount of energy which exists in a normal discrete amount. This energy quanta is large and slow compared to all actions of space and this difference is why you cannot recover all quantum information perfectly from a system. The ESS are much smaller than matter and so cannot handle a single amount of energy directly and will naturally break it down to smaller amounts. The ESS share this energy with neighboring schales through macht glow and so the information is spread out from the quantum epicenter of where the matter donated it to space. Lastly the ESS share macht at macht speed which is far faster than anything the quantum world can achieve and so you cannot retrieve information back from the ESS to the matter at linear rates. The spread out nature of macht and the speed at which ESS share macht means quantum matter if it wanted to retrieve what it put into space which is of course a form of information will not be recovered at the same location or amount in a single frozen moment as when you put it into the system.

The fault lies in assuming that quantum speed or light speed interactions are the fastest in the universe as this means that a large slow light speed amount of time is the smallest time slice you have to describe the system. If that were the case then it might be possible for you to retrieve all information as the event is measured in quantum time only. However the universe is not limited by the sluggish pace of light speed reactions and so at a true single macht timescale one frozen moment for the system sees only some of the macht returned to the quantum object as energy. The macht speed transfer is incredibly fast compared to the slow light speed transfer of matter but it is still not fully instantaneous and so while information was given from a quantum system all at once in a quantized amount of time from the matter it will not be received perfectly back at once. It will be received over a series of macht times that will allow for the energy to be recovered in a more spread out and therefore non-quantum way which is only registered back in quantum mechanics when the full quanta is returned as a sort of particle action potential.

This is what prevents a scientist who examines the universe only through quantum mechanical ways to recreate past events of a system perfectly and is what will be seen as information loss. In addition if you throw in macht sharing from surrounding objects such as multiple pieces of matter sharing macht glow gravity then you are spreading the donated energy between particles. In the law of conservation of macht and energy the correct quanta of donated energy will be returned to any particle if it is removed from space but it will have shared its macht effects with the entire system which can cause subtle changes not seen on quantum scales and therefore is a form of information loss from a quantum only perspective.

The proof for this is seen in two ways and is fairly ordinary in the universe. To make things clear the above explanation is fully correct but does not pertain to all systems in the universe all of the time. If you look for example of conditions here on Earth where we have a normal spatial compression in our laboratories and hence where modern science has developed all of its laws then the information loss is not a problem. The required high levels of spatial compression needed to significantly decrease the retrieval of quantum information from a system are simply not present. This does not mean that the mechanisms for information loss are not present as they of course are only that all quantum actions available to us here on Earth are simply too docile to matter that much just the way relativistic effects do not manifest themselves readily on Earth. Therefore the two factors that are needed to cause what would be seen as a loss of quantum information from a system exist in more select environments in the universe.

The first can be a simple neutron star as long as it contains something like quark/gluon plasma. If you recall from the construction of a black hole from normal matter such as gas clouds through to stars and so on you remember that as you build the different levels of space and matter interactions you rearrange the speck/mote geometry of matter and therefore alter how it interacts with different machts of the ESS. Since the matter in a quark/gluon plasma is no longer following the same rules as seen here on Earth the effects of information loss become easily observed provided the quark/gluon plasma conditions can be maintained through continual spatial compression. I mention continual spatial compression because again the necessity of maintaining a local area spatial compression sufficient to create forms of matter similar to quark/gluon plasmas and SM masses is sadly not possible through standard particle colliders and so is highly unlikely that we can observe them any time soon here on Earth.

Nevertheless in a quark/gluon plasma the amount of energy being donated through the system from matter to space is immense compared to Earth conditions and the density of both matter and ESS means that the small size and ESS macht glow sharing of space will definitely begin to in a sense erase some quantum information from the standard definition of quantum information theory. The ESS are so close together and the macht glow is shared between them such that the donated energy from one piece of matter is transferred to another ESS and another until it reaches the local area of ESS for a different piece of matter. If these pieces of matter were taken out of the quark/gluon plasma the macht will not be able to nor does it have any desire to return to the exact piece of matter that donated it and so cannot be retrieved in any quantum manner to that original quantum system that donated it. The energy will be returned in total to conserve macht energy donation but the information specific to that particle at the exact quantum state it was in when it donated it is not. In fact the exact quantum state of most particles is never returned perfectly to a piece of matter and further explains the perceived nature of strange quantum probabilities.

This is a true loss of information from a quantum only point of view that is proven by ESS macht sharing. This is further reinforced by that fact that it is well known that quantum objects can exist in different energy states and need not be all similar. Two atoms for example can be possessing different amounts of energy such that one might have an electron in a higher orbital than another despite the fact that they are similar elements. What this proves is that a quantum object of a specific energy going into the quark/gluon system does not have to exit the system in the same way it went in and can have lost energy to other quantum objects through the general nebel of the macht glow of all ESS in the quark/gluon plasma mass as long as the returned energy is sufficient to maintain the atom at the lowest quantum degeneracy state of the atom for that local area of space under the specific ESS macht and spatial compression.

In other words the information needed to rewind the events is not from a quantum viewpoint possible because the retrieval of ESS macht from space is not possible and even knowing ESS macht events will not help you because the transfer of macht happens at a different timescale than quantum events. This also explains why on Earth the effects of these events are present but simply not significant due to the much lower spatial compression and therefore density of ESS per matter volume. In total however the macht contains the information of the system just not from a quantum mechanical viewpoint.

The second factor affecting information loss occurs in black holes or any true SM mass since these objects do not utilize all fundamental forces. Unlike quark/gluon plasma which uses more forces these are now absent from the system. Quantum mechanics is based off of the use of four fundamental forces and light speed limits in order to explain the physical world but these are not always present in black holes. One of the failings of modern science is in trying to keep all four forces working inside black holes. The properties of a black hole are such that the very high spatial compression and degree of interaction of matter with space as I have explained previously prevents matter from interacting with other matter as fast as it normally would. What this does is prevent information from being retrieved fully in a quantum sense at any moment in the system as once again macht speed and light speed do not match up. This coupled with the very, very high degree of macht sharing between ESS schales means that quantum information is once again lost in the nebel of ESS macht.

Now that I have explained these events as they pertain to both space and matter I will restate something I said earlier which is information is not fully lost from the universe it is only lost from a quantum mechanical viewpoint. The interactions between matter and space are of course real and ongoing and the energy taken from a quantum object is never destroyed only converted to macht glow. So at no point do you lose information from the universe only from a purely quantum system. Revisiting the classical information theory paradox of the black hole yes you do lose information as you cannot get it out of the black hole without using something like a TSM event. On the other hand no you have not destroyed information fully or lost it in any way from the universe rather it has been converted to ESS macht information which exists in perfectly normal space physics which modern science does not incorporate in the old quantum only view of information theory.

At the end of the information loss debate I will let you decide which viewpoint you like better either the information is lost from the observable universe or the information is preserved in total by the universe because from a human standpoint or the standpoint of any intelligent creature both are correct. The reason this is so is because both really are correct and so no paradox exists only a true explanation of how total information behaves in the universe and why it is not destroyed but also that under the correct conditions it is not possible to fully rewind the VCR tape of the cosmos and reconstruct a quantum system perfectly; so Hawking is correct concerning information being lost to a black hole until it is TSM detonated back into the universe. The exact sharing of macht is still known to the ESS of space however and so the information is not lost from the universe. If you rewind the

VCR tape of both space and matter interactions you can recreate any past event perfectly because you have incorporated both parts of the universe which are space and matter afterall.

Facts

The facts for this section are rather short as the mechanisms needed to facilitate entropy reduction and recycling as well as explain information loss are all built on previously used interactions of space and matter. These include SM masses and macht glow sharing between ESS and energy donation from matter to space. To that end the first fact is that entropy will stop increasing in the universe based on a halting of the expansion of the universe and the accumulation of matter in SM masses during the collapse phase. The second fact is that entropy is recycled through the arrangement of matter into clintrons as well as the reduction in the number of excited forces inside SM masses. The third fact is that using SM and TSM objects the universe can be recycled along with the entropy it contains to the same lowered states and ordered systems seen in the first moments after the Big Bang. This allows for the normal progression and increase in entropy understood by modern science today. The fourth fact is that from a standpoint of total objects in the universe versus total particles in the universe it can be said that entropy is in fact decreasing as more complex objects encapsulate numerous smaller systems in one. The fifth fact is that entropy can be easily and naturally recycled without the use of alternate universes or reverse time flows but instead by following the same unchanging laws of physics that produces the universe as we observe it today.

The sixth fact is that information loss does not need to be a real law of the universe and must be viewed outside the human constraint of simply wanting this to be so and attempting to get the universe to conform to it. The seventh fact is that two forms of information are used by the universe with one being that used by matter as seen in normal quantum systems which is what gives rise to the problem of information theory in the first place as well as unexplained quantum probabilities and the second belonging to ESS information as explained by space physics. The eighth fact is that information will be lost from a system if based on purely quantum mechanical explanations as it is lost in the macht glow shared by ESS schales. The ninth fact is that information is not truly destroyed as it is stored in a different form from quantum ones inside space in ESS macht information. This means that only if you incorporate both space and matter into examining information can you recreate a system entirely and explains why quantum only systems fail to do so.

Chapter 26
Light

Photons Explained

The subject of light has been one of considerable debate for many years and on the other hand has been something that has been altogether ignored. The ancient humans saw light simply as something that happened during the day and was brought by the Sun. Either this was simply a part of the world to them or it was the divine presence of some god or other. The fact is though that in general it was not the light that was of interest to them as opposed to the Sun which brought the light. The idea that rays of sunlight were travelling from the surface of the Sun's photosphere to Earth and to their eyes with which to see it was inconsequential to them. The Sun was the reason they built calendars, could see the world around them and every culture at one time or other believed it to be a god. The actual rays of sunlight comprised of various wavelengths of photons were not much studied in any serious and reputable manner for many millennia.

However interest in all of this changed around the 1700s and 1800s as new understandings of light exploded thanks to people like Newton, Herschel and Maxwell to name just a few. Everyone who has ever studied the history of science has no doubt seen a picture of Newton holding aloft a prism and observing a beam of white light coming into the prism only to leave split into its composite colors of the visible light spectrum. If you can imagine scientists of the day seeing this for the first time then you can understand the enthusiasm with which they attacked the questions surrounding light and its properties. One of the great breakthroughs came when light was discovered to have a speed rather than travelling instantly from the object to the observer as was thought for quite some time. This fact coupled with the spectrum of colors are probably the two most interesting sparks that ignited the study of light and has led to today's modern day search for answers of light and specifically the photon.

The nature of light is not as well understood as many believe as it raises some niggling points about the universe that do not seem to fit well with modern theory. A simple question that remains unanswered by modern science is what exactly is light? Not what is light but what exactly is light. There is a difference and this is easily illustrated by looking at something like friction. Friction has a simple explanation that is not fundamental which is it is the excess heat released when two surfaces are made to move against one another such as you rubbing your hands together. In more fundamental terms friction is not a magic property of hands or motion but comes from the resistance of atoms to one another as seen through the inability of one person rubbing their hands together to overcome the electron degeneracy pressure of the atoms their hands are made of. This is the same with light as the general understanding is that light is energy made of a photon that travels in waves. That is a very macroscopic rub your hands together kind of explanation as to what light is but it is not exactly what light is. A few examples might be what is a photon? What makes it have no mass? What makes it feel only the electromagnetic force? Why does it appear to be attracted to gravity but have no mass? What makes it red-shift or blue-shift? The list goes on and on and illustrates the lack of a fundamental description of what a photon is. Simply saying a photon is a wave and waves can be stretched or compressed is not nearly satisfactory enough as all such descriptions of light focus more on the what light is doing but not the why it is doing it.

◄New Universe Order►

To this end a better understanding of light is needed if it is to be removed from the very odd world of being energy which should seem very odd to everyone. The reason the idea of energy should seem odd is that there is only one kind of it in existence. Matter on the other hand has many different forms with many different sets of properties and can even have three generations of each in most cases. You can have an electron or neutrino or quark or whatever you like and each are very different because they all have very different properties. In addition to this each version has different generations such as a normal up or down quark and things like strange or charm quarks, top or bottom quarks. All of these forms of matter exist at different masses, spins and charges so there are a whole slew of various types of matter in the universe but energy seems to only come in one variety.

This is a strange set of circumstances if energy really is a separate entity from matter because it means that energy is not space or matter. Remember energy is not matter through the equivalency principle it is merely a calibration of both so that they can be calculated using the same units. Matter is not energy or it would violate the speed of light limit imposed by relativity because in order for it to be energy literally it would have to be moving at the speed of light squared. Instead this is simply a calculation of how much equivalent energy that amount of mass possesses. So this leaves us with the idea that the universe went to all the trouble of making space and matter and then an extra type of thing which has only one form. Matter has different forms and this is economical for the universe but making an entirely new creature called energy or photons and making them exist only in one type is not economical. Every time I see something in the universe that exists as a solitary object needed to fill only one job I get suspicious. So can light be explained away not as energy but as something else which would reduce the need for the universe to make extra stuff simply to fill one niche?

The answer is that yes it can and the idea that energy, photons, light or electromagnetic radiation or whatever one prefers to call it is a form of matter makes infinitely more sense. First this would reduce the number of variables in the universe which is always a sign you are on the right track and second it would answer exactly what light really is. Reducing the number of variables is always favorable because the universe always and without question tends to the lowest number of anything in order to construct a universe as we see it today. Lowest energy states for atoms, lowest density of matter and energy in space as space expands, lowest everything is how the universe by law exists. As for the second part it will finally answer what light exactly is beyond the simple it is a wave that has electromagnetic properties vagaries. Light is matter and to prove this we can explain a photon and all of its behaviors using a photon speck/mote and ESS interactions of macht with other objects in the universe.

First of all what do we know about light? One thing we know is that light travels at or below the speed of light meaning a photon can at maximum travel around 300 000 km/s and if it is traveling through a medium it can be slower. The maximum speed of a photon is therefore a quantum scale speed meaning that light cannot be a true part of space as macht is for ESS. Macht is the only natural thing to travel faster than the speed of light and therefore photons must be a part of the quantum realm which is made as we know from particles. All objects in the universe that travel in quantum mechanical ways are particles of some sort whether that is a boson, lepton, hadron or any combination of these and more. All of these particles have mass even if some of them possess virtual mass and all fit neatly into the Standard Model of particle physics. This alone stands to reason that light is a part of this model rather than simply being some strange visiting entity that interacts with tangible particles but is not itself a particle. Using this we can now construct light as a particle the same as any other.

To begin with all normal particles have mass because they interact with the Higgs field and the more they interact with this field the heavier they become. Further more all particles as we know are made of specks and motes of fundamental matter that interact with ESS in different ways to excite different fields that are commonly perceived as the particles themselves. Different speck/mote types and the arrangement of these objects in different geometric patterns determine what properties a particle has. One of the properties determined by the geometric arrangement of the specks and motes is that of interaction with the Higgs field. Certain particles in the Standard Model have heavier masses and this is because the speck/mote properties they possess allow them to interact to a higher degree with the Higgs field than other particles even if they share other similar properties. An electron for example has the entire single unit of charge that three quarks will have yet they do not share the same mass so one arrangement of specks and motes will give you a single electron and another will give you a proton. These two particles are incredibly different by mass which they get through speck/mote interactions with the Higgs field and yet the particular arrangements of both still allow them to have equal amounts of charge although opposite.

This should be obvious by now but it is worth repeating here with respect to light because it allows light to do something rather interesting. What that is comes from it not interacting with the Higgs field and so having no mass and yet still interacting with the electromagnetic forces of the ESS macht. In other words you can easily explain a photon as a simple speck/mote that will interact with the ESS macht glow electromagnetic but not with the ESS macht glow Higgs. This is a simple but effective explanation for how photons are in fact matter and therefore not a separate type of object called energy in the universe. This reduces the number of variables in the universe by one and keeps it firmly at space and matter.

Now I will explain why this matter construction of light is already proven in the universe and give examples of how other particles support this as well. To begin with let us look at a neutrino and see that it too is near massless, travels almost as fast as light and barely interacts with matter. A neutrino is quite well known to be a real particle of matter and not energy and so if we look at light and compare it to a neutrino the speck/mote reality of light becomes all the more clear. A photon is a speck/mote that simply does not interact with the Higgs whereas a near zero mass neutrino does so but hardly at all. There is a sliding scale correlation between the two as a photon moves at the speed of light and has no mass whereas a neutrino has barely any mass and therefore moves just slightly below the speed of light. Neutrinos strongly resemble photons that slightly interact with the Higgs field or the reverse is also true where photons look like neutrinos that simply lost their contact with the same field.

A simple speck/mote also explains how the same photon particle can not interact with one ESS macht but can interact with another as seen in the fact that photons easily interact with the ESS macht glow electromagnetic. As stated before different arrangements of specks and motes will allow one type of particle to interact with specific ESS macht glows while the same specks and motes in a different arrangement will interact with a completely different set of ESS macht glows to produce a particle of altogether different properties as seen in the Standard Model of particle physics. Now it is time to show how photons as matter can interact with space and other forms of matter such that it produces light in the universe as we see it today.

Firstly we must dispel a common falsehood concerning light which is that it is affected by gravity. Light does not ever feel the pull of gravity in the universe and gravity is not so infinite in a black hole that even light cannot escape it as we have all been told over and over and over again in that tired example of the power of black holes. Light does not react to gravity as it has no interaction with the ESS macht glow Higgs and so has no mass. Without mass it cannot be attracted to the force of gravity in the universe and is not attracted to ESS macht glow gravity created by matter orbs. Why light seems to follow gravity is because light is following the ESS macht glow electromagnetic of space and nothing more. Light has the speck/mote arrangement such that it is attracted to electromagnetic forces and will do so anytime a field is strong enough. This has been demonstrated countless times in laboratories using electricity, magnetism or both where by beams of light can be deflected depending on the field around them.

This fact is at work in space where all ESS have the macht pool electromagnetic contained in them regardless of the presence of nearby matter orbs which might cause strong gravitational forces. This explains a few key properties of light besides just not following gravity but for now we will look at ESS as the reason light bends. ESS contain all forms of macht in the universe in each ESS sphere at all times given the macht pools they possess. If matter is present then certain ESS macht glows will be increased or decreased such as the ESS macht glow gravity when normal matter is interacting with space. So if we have a large object like a black hole and light passes closer to it we see the light bend around the black hole but not because gravity has any effect on the light at all. Instead the same ESS that the light is naturally attracted to have undergone spatial compression due to elevated macht glows of gravity. The light is simply following the same ESS macht glow electromagnetic that it was before only now these macht glows exist in spatially compressed ESS and so the path the light takes reflects these compressed ESS.

To prove that light is further unaffected by gravity we can place a magnetic field in the same location as the black hole. The size of the ESS around the black hole are still spatially compressed and this is due to the increased macht glow gravity of the ESS. If we place an external magnetic source such that the light is in between the source and the black hole we see the light bend back towards the magnetic source. This is because the macht glow magnetic of the ESS that are still spatially compressed due to gravity is now increased and so we have two elevated macht glows in the same ESS. The photons will follow the normal electromagnetic force they always feel as they approach the black hole this time however they will be attracted to the magnetic source because the macht glow magnetic is elevated there.

So even though it appears as though light is following the gravity it really is following the electromagnetic force only and by introducing an external magnetic source we can show how two machts inside the ESS create a pattern where the light follows the shape of the ESS as spatially compressed by gravity but only follows this shape as it is attracted to the ESS macht glow electromagnetic instead. This proves that light is not part of space and not bound to gravity and what it also does is disprove the idea of pure spacetime curvatures as it demonstrates how forces can be transferred from place to place within space. In other words force which is really correctly and completely explained as macht glow exists inside space and can flow through ESS and therefore overcome the fallacy of a spacetime fabric.

As for the idea that light cannot move fast enough to escape a black hole or that a black hole has infinite gravity which causes light to be sucked in and never seen again all due to gravity this is simply false. A photon is a speck/mote similar to any other type of particle and as I have just explained will follow the shape of space into a black hole as it is attracted to the ESS macht glow electromagnetic. The photon as we know is a force carrier for the electromagnetic force and when absorbed by an atoms nucleus will cause the electrons around the nucleus to increase their orbital energies and move up an orbital level. This is because inside the nucleus the speck/mote of the force carrier must be directly interacting with the real particles of matter inside the nucleus. In order for this to happen the speck/mote of the photon is binding to the geometric arrangement of specks and motes that make up the matter of the nucleus allowing it to have more energy to give to space. This is seen in the fact that as the particles gain this virtual energy their relative masses increase meaning they are said to gain gravity and really what is happening is they are simply donating more energy to the ESS macht glow gravity for the local area of space by binding to the Higgs field interacting nucleus and the speck/motes it is comprised of.

In a black hole the same thing is happening as the photon follows the ESS into the black hole where it obviously meets the clintron which is in the tightest possible configuration for matter in the universe making any chance of avoiding a direct interaction with matter impossible for the photon. The end result is any and all photons that follow the ESS macht glow electromagnetic of highly spatially compressed ESS will always find their way into a black hole and once there be absorbed by any clintrons they encounter. The clintrons will absorb the speck/mote of the photon into their own speck/mote geometric arrangement and therefore have more energy to donate to ESS macht glow gravity. The fact that photons are made of specks and motes makes perfect sense as specks and motes of matter can naturally bind with them. Remember we are talking about the very fundamental objects of the universe and at these levels macroscopic thinking does not always apply such as in how would energy bind to a particle of matter if they are two different types of quantum objects? The easiest explanation and therefore the correct one is to make them out of the same types of quantum objects therefore eliminating any barriers to their ability to mate with one another. This also explains why all apparently different quantum objects can interact with various fields of space as two types of objects would require two interactions mechanisms with space as a result and this of course needlessly complicates the universe which we know the universe never allows.

We can further use the fact that light is made of a fundamental speck/mote of matter to explain some of its other properties such as its natural wave motion and red-shifting and blue-shifting in the universe. To start with we will look at the normal wave motion of light and explain how this is achieved through quantum motion no different than any other form of matter. As I have explained before in the section on quantum motion all quantum objects seek to track after the ESS macht glow they are capable of interacting with. This of course can also be a combination of macht glows and need not just be one single type. Light also tracks after the ESS macht glow electromagnetic that is a part of space and this tracking behavior allows for the fact that light travels in a wave and always at light speed.

A normal particle will track after the ESS macht it desires at less than light speed because although the exchange of energy between matter and space is at or below light speed any particle with mass is attracted to gravity and therefore has its motion slowed. So the quantum exchange of energy is light speed but due to the particle being attracted to the ESS macht glow gravity it cannot track to the next ESS at light speed and must do so slower than the speed of quantum exchange which is light speed. The more massive a particle is the more it is attracted to the ESS macht glow gravity and so the more effort is required to move it to the next area of ESS it desires to track to.

Light on the other hand does not interact with the ESS macht glow gravity and so is not slowed by any ESS it encounters as it moves. The light is also able to therefore exchange energy between it and the ESS macht glow electromagnetic at the speed of light which is the maximum exchange speed of energy between specks and motes and ESS. This ability of the photon to exchange energy at the maximum allowed quantum speed and the fact that it does not interact with the ESS macht glow gravity means that light is able to move instantly and at the speed of light. Only a particle with mass no matter how small would need to accelerate to its maximum velocity as it must balance out its tracking speed with the ESS to which it is attracted. Light on the other hand moves at the same speed as the energy exchange which is the speed of light and with no mass will move from zero to light speed in one instant because this represents one quantum exchange of donated energy to the ESS macht of space. This explains how a speck/mote of fundamental matter that makes up a photon allows it to move at the speed of light with zero acceleration from a quantum time perspective and why it moves at the natural top speed for matter which is light speed.

The same way that light tracks after one ESS area to the next and explains its light speed ability also explains its wavelike nature. This is a result of the light tracking through the ESS areas it is attracted to as it moves forward. ESS are arranged as you remember out of spheres and will always have paths through which the direct line of sight of any object is blocked by another sphere. This means that light must in a sense weave its way around ESS in its path as it really tracks the macht glow electromagnetic to which it is attracted. This makes perfect sense as light is no different than any other object in the universe that must obey normal physical laws such as an object stays at rest unless acted upon by an outside force. The wavelike nature of light is made real by the ESS and the way light follows them through space as no other mechanism can explain why light or anything else for that matter should oscillate up and down. A photon by itself cannot move up and down in a wave form as it moves forward in the universe as this means it is consciously changing its direction from up to down and then back again. Since light does not have little thrusters on it to slow its motion in one direction and push it in the opposite something else must be making it move like a wave.

The answer is the ESS of space that sits much like a backdrop to the motions of the photons passing through them. This ESS backdrop is not passive however and will cause the light to follow it instead. This mechanism requires no conscious thought or outside force acting on the object which of course is a photon causing it to move from one position to the next. In other words the laws of normal physics can be maintained without the need to add anything new to the system and explains why light should even move in a wave pattern anyway. If nothing was making the light move then the wave motion would be coming from the light itself which would require energy to do and cause the photon to lose energy which cannot be true as the energy of the photon is quantized. Since the macht glow is uniform given a certain level of spatial compression and macht strength the light will track in an overall straight line forward as it moves. The practical use of this is in explaining why red-shifting and blue-shifting occurs to light in the universe.

Since light is a speck/mote that interacts only with the ESS macht glow electromagnetic in the universe changes in the volume of ESS will cause shifting of the light's wavelength. If for example we start with a photon of let us say white visible light then any decreases in the ESS macht glow gravity will result in the ESS experiencing spatial rebound that increases the ESS volume. The light will therefore have to spend more of its energy in order to traverse the now larger ESS volumes it is encountering. Since energy cannot be destroyed from the universe and must be reserved in some recycling mechanism the energy possessed by the photon is now donated to the ESS it is travelling through and therefore the photon from our perspective loses energy and is red-shifted towards a longer length of the electromagnetic spectrum.

The ESS as you remember cannot hold a full donated amount of energy in a single ESS and so must spread it out which results in the macht glow being much larger in areas of low spatial compression. The reason is if for example 100 ESS spheres are needed to contain the total macht energy then smaller more spatially compressed spheres in normal Earth like conditions take up less total volume as compared to those in between filaments of the universe. So the much larger macht glow of the spatially rebounded ESS creates a much larger volume of space that the photon must track to and therefore creates a longer wavelength as the diameter of the macht sphere created is so much larger. Spatially compress the ESS and the macht sphere diameter is now much smaller meaning the wavelength of the photon has greatly decreased or blue shifted even though in each example only 100 ESS spheres are still used.

The ESS macht glow electromagnetic of space is increased but this property does not affect any changes in the total volume of ESS so does not cause a negative feedback which would end the red-shifting. The amount of ESS volume increase will correspond directly to the amount of red-shifting the photon experiences from the white part of the spectrum to the red. If we now allow the photon to encounter more highly spatially compressed ESS then the distance it encounters from ESS to ESS is decreased and so takes less energy for it to cover the same distance. The end result of this is the ESS macht glow electromagnetic of space can now be given back to the photon allowing it to regain its lost energy quicker due to the smaller distances and the differences in the macht speed to quantum speed of sharing. As the photon gains more energy it is now shifted back to the white light part of the spectrum and if given enough energy will then move beyond this to the blue end of the spectrum where its wavelength is further reduced. This accounts for how space and not gravity will affect the wavelength and therefore energy of a photon travelling through the universe. The gravity is only implicitly affecting the ESS to produce these observations but it is the macht glow electromagnetic, ESS sharing through schales and the differences in the speeds of macht and quantum worlds that causes this phenomenon to occur.

If a photon is generated here on Earth under normal spatial compression it will retain more of its energy as it travels through space as the ESS it encounters have already greatly spatially rebounded since the Big Bang. Photons from the period of first light may have started life as high energy gamma rays but were born into a still massively expanding universe and as such have had many billions of light years to not lose but share some of their energy with the ESS to which they track as they move through the cosmos. This also contributes to cosmic background noise. If we could watch a gamma ray photon that we create here on Earth we could observe its wavelength shifts as it moved through filaments of the universe as well as the vast regions between filaments. These two areas of the universe have massively different levels of spatial compression and therefore ESS volumes which would produce significant shifting of the wavelength of our gamma ray photons. Put one detector near the galactic center and another in the space between us and the Andromeda galaxy and you can easily see these proofs at work.

What this means also is that the speed of light is variable based on the amount of spatial compression of the local area ESS it encounters. Since light must still obey normal laws of physics where it takes X time to cover Y distance the actually velocity of light does not change for the speed of the speck/mote of light but will change for its ability to cover a straight line. If we take a meter of low volume ESS space we see the ESS are larger and this means the photon spends less time moving up and down as it moves forward so it can track faster to local area ESS. This means it will take Z time to cover one meter but this is not so for a photon in highly spatially compressed space. In an area of highly spatially compressed space you will find that much more effort is made by the photon to track up and down to the local area ESS than before. This means that it will be slower in moving forward as its overall velocity has not changed for it due to the fact that it still tracks to ESS at the speed of light because this is the quantum exchange speed for speck/mote geometries and ESS macht.

It will however take longer for it to cross this one meter than before because of the sheer number of ESS it must now navigate when compared to the spatial rebounded ESS and therefore its total transit time for the one meter will now be $Z+1$. The additional factor here is that the number of ESS the photon must exchange energy with or interact with increases in the high spatial compression and this means more time. The light still travels at the speed of light and the exchange of energy between light and the ESS still takes place at the fastest allowable speed for a photon but the number of ESS in the low spatial compression area is fewer than the high area and so only one exchange interaction need be completed between a hypothetical photon and a single ESS whereas the number of interactions of photon to ESS in the high area of spatial compression might be ten. This means no violation of the speed at which the actual speck/mote of the photon is moving nor a violation in the exchange speed of the speck/mote with the ESS but in order to cover the same distance more total interactions are needed to cover more ESS so the time it takes the photon to cover a set distance is slowed.

A photon can cover let us say one ESS per second as this is the time needed to exchange energy with a single ESS and if we have ten ESS in low spatial compression space and this is the distance of one meter then it will take ten seconds to track the ten ESS and cover one meter. If we have one hundred ESS per meter in an area of high spatial compression then it will take one hundred seconds to cover the same distance. These times are for large illustrative purposes only and the real time scales would be far shorter perhaps on the Planck timescale which is significant for small systems but virtually undetectable for human scales or experiments. This is because the speed of interactions is unchanging for the ESS and the speck/mote on what can be thought of as a fundamental quantum scale of a single interaction but when viewed on a macro scale of meters, kilometers or light years a more classical effect is seen.

This is what leads to modern science believing that the speed of light is constant when it can vary just as modern science sees gravity working on large macro scales but failing at quantum ones. Changes in spatial compression and ESS volume as well as ESS macht glow electromagnetic will affect how light moves but here on Earth or even in our own solar system where we conduct our experiments significant changes in spatial compression are simply not available. The differences in transit times for light over one meter are insignificant on a one meter to one meter basis but this is not so when talking about the entire universe. These small non-zero differences will become apparent when talking about a single light year which is approximately ten million billion meters. Over these distances the effects of light travel time in a straight line based on ESS spatial compression can be measured and will have important implications when talking about the entire universe.

Inaccuracies in measuring objects at the edge of the visible universe are now not a single light year away but ten billion or more making this effect one that cannot be ignored. This also explains why light is a particle rather than energy as this behavior breaks light free from the idea that it has a constant speed no matter how space is arranged which is obviously incorrect if it were a part of an impossibility called spacetime. Matter is regularly permitted to travel at different speeds and is only slowed by its relationship with space. Light also is limited by its relationship with space in how it moves and this is explained no differently for it than matter. Light is once again a simple piece of fundamental matter as a speck/mote geometry no different than any other particle and as such can have all of its behaviors and properties explained through the same mechanisms as normal matter which are simply specks and motes interacting with the ESS macht of space. The type of edge dragging and degree of edge dragging for each speck/mote and its natural property of tracking after ESS macht glows works for all particles, virtual particles and light meaning that everything in the universe that is not space is simply a form of matter.

An interesting side note here is that the hypothetical and seemingly paradoxical question of what would happen if you lit a light whilst travelling at light speed is easily answered. Modern science has long held that the question produces nonsensical answers that are themselves devoid of rational explanation for various reasons and this could not be further from the truth. The lack of understanding comes from general relativity itself which does not understand what is really going on in the universe in terms of matter, energy and space at high rates of speed. According to the theory of relativity absolute values do not exist and as such the ability to gauge what would happen at exactly the speed of light is impossible because you lose the frames of reference on which relativity is built. In truth relativity is dead wrong and this is fairly obvious to anyone as objects regularly travel at the speed of light in the universe and they are photons.

So once again the universe simply ignores humans and the theory of relativity in all forms as it does make things move at the speed of light despite the theory of relativity being unable to understand them. So in actuality objects can move at the speed of light and completely without two key components of relativity the first being frames of reference and the second being a relative observer. In reality the photon exists in space as we have seen not in a large scale system as viewed by the theory of relativity but in a small quantized system where a single photon speck/mote will exchange energy with an ESS as it moves through space obeying perfectly normal ESS macht tracking rules.

So turning on a light at light speed means that the light emitted from your light source stays in place with you as you both move at the speed of light through space and your specks and motes track their ESS in perfect step with the speck/mote of the photon. The light you lit up stays dark as it cannot travel faster than the speed of light which you are also moving at. In fact you do not even have to be present as an observer to prove this since we can see on a single quantized level that the photon must obey tracking rules before moving to the next ESS and so no light shines out to be absorbed and emitted by objects for you to see as with normal everyday light sources. The light source which normally emits a photon still does so as the photon is now dissociated from the matter that generated it such as an incandescent light source.

The difference is the photon simply sits in place in the exact space it was emitted and moves in lock step with the quantum particles that emitted it through space at the speed of light. Everything in front of you only appears as bright as it did before you turned on your light and the image is not distorted either only blue shifted. All the most fundamental workings of the universe and light are laid bare using ESS macht and tracking rules of specks and motes so turning on the light at light speed can be answered easily. So the paradox is shown to be non-existent as it should be because no paradoxes exist at all anywhere at any time and under any conditions in the universe and they never will either. The word paradox when applied to the universe simply means a lack of understanding of something that the universe ignores us for not understanding in the first place. Back to the drawing boards humans.

To further this we can look at why light always travels at the speed of light regardless of other velocities. If we use the example of a piece of matter moving through space at a velocity of 100 kilometers per hour we see that light emitted from it travels at the speed of light and not at the speed of light plus 100 kilometers per hour. The speck/mote and ESS macht explanation for this is quite simple as two factors of space and matter interactions are at work. The first is that light is a force carrier for the ESS macht glow electromagnetic and as such is generated from space as one normally thinks of a field being excited. What this means is that like all force carriers light comes technically from space and not from matter. Any normal field is dormant in the universe waiting to be excited by matter that moves over it and a simple example of quarks and gluons will suffice here. If you have three free quarks you will notice that they naturally are attracted to one another in an attempt to stabilize themselves into a hadron. You will also notice that three quarks by themselves are but a fraction of the total mass of a hadron as the rest of the mass comes from the virtual mass of the gluons that bind them together. The gluons carry the strong nuclear force and have appeared from apparently no where in the universe simply from the vacuum in order to bind the quarks. This is essentially true as the quarks need to interact with space and will excite the ESS macht glow of strong nuclear force as a result of them tripling up to balance their energies and respective ESS macht glows. The gluons are a part of the ESS macht glow that is holding them together and so are coming from space as the quarks donate energy to the ESS of the local area of space. In this way force carriers can be made to manifest themselves not from thin air but from the interactions of matter with space.

Light is no different and will be created as a force carrier by matter interacting with space as well and so no matter how fast the matter that created it is moving since it came from a space starting point it will have the velocity of the speed of light. Remember that only the matter is moving and that space is essentially stationary so the point of generation is seen as near motionless. This is like a person moving along in a car next to a series of poles with guns attached to them and each gun can fire a round at 1000 meters per second. If the person drives along it does not matter which pole they decide to fire the gun from while moving at say 100 kilometers per hour. The pole from which they fire the weapon is stationary compared to them and so the round will always be discharged at a muzzle velocity of 1000 meters per second. The car can travel at 1, 10 or 100 kilometers per hour but the gun always discharges its rounds at 1000 meters per second. The same is true of light which is always generated as an excitation of the ESS macht glow electromagnetic at an essentially stationary position in space.

The only time ESS velocities must be taken into account is around objects such as those made of rapidly rotating SM as these have the strength to drag ESS along with them giving them an orbital velocity of sorts. A large spiral galaxy will produce enough ESS rotation such that the velocity of these ESS when seen across tens to hundreds of thousands of light years will be made obvious. This still does not change the speed at which light travels meaning the tiny speed of the ESS does not impart any extra velocity to light only that this is an example of when the ESS that generate the photons are not truly stationary anymore. The ESS that make up the rest of the universe are the ones that are stationary and unmoving but both will still create photons travelling at light speed. If a galaxy was rotating sufficiently fast enough and for arguments sake towards us then as an observer we would notice the speed of the light from those galaxies as the speed of light plus the speed of rotation but this is simply a trick of the light so to speak and could be seen as an error in data processing.

The second factor that keeps light travelling at the speed of light is due to how light travels in the first place. This simple truth about light has been ignored by science forever as the idea that light simply moves has never been explained only taken as a property of any photon. As we know everything in the universe happens for a reason and must obey the laws of physics so there must be a fundamental explanation as to why light moves and does not sit still. The reason for lights innate ability to move and always move at its maximum velocity given the substance it is travelling through is explained using speck/mote interactions with the ESS macht of space. Light is a force carrier particle that travels following the ESS macht of the local areas of space by tracking the ESS macht glows in front of it. I have already explained how any speck/mote geometry will track to local ESS macht glows and light is no different meaning that despite the velocity of the object that caused it to be released from space it can never exceed its ESS tracking velocity. The instant the light is released from the matter as a free force carrier it must obey the tracking laws of all specks and motes in the universe which means that even if the matter that released it was travelling at just below the speed of light the photon is dropped back into normal space and must track its local ESS macht. The tracking of ESS macht by specks and motes never exceeds the speed of light as it is a quantum exchange process so the fastest light can exchange energy with ESS and track the next ESS glow region is at maximum the speed of light.

Taken together the stationary release point of light from matter as a force carrier in the universe and the limit to the photon tracking velocity of the speck/mote of light and its local area ESS insure that the speed of light is maintained instantly at 300 000 kilometers per second. This is the first rational explanation of why light behaves as it does from a truly fundamental standpoint that explains why light even moves to begin with and why it has this maximum velocity. The movement of light is exactly the same mechanism that drives all quantum motion only it is not happening in one localized area like an atom would but in a straight line that weaves through the ESS of space. The same movement mechanism explains why objects like free electrons refuse to stay in place either but move away in a line at the maximum velocity they can achieve based on their mass and the local area spatial compression. The similarities in motion and ESS macht interactions between fundamental particles possessing mass such as an electron and objects like light are not similar they are identical and offer further proof that light at its most fundamental level is simply a different geometric arrangement of specks and motes no different than any other particle or force carrier with mass.

Facts

The first fact concerning light for this section is that light is in fact not energy but matter and is constructed of a normal speck/mote geometry as are all particles. The second fact is that light is a geometric arrangement of specks and motes that does not interact with the ESS macht glow Higgs and therefore has no mass. The third fact is that light is not attracted to or affected by gravitational forces but instead tracks the ESS macht glow electromagnetic which is a part of all ESS and so spatially compressed ESS found in areas of high gravitational fields will appear to attract it. The fourth fact is that the motion of light as a wave is due to its tracking weave through ESS areas of space. The fifth fact is that light always travels at the speed of light due to two factors which are the stationary release point of photons from matter and space interactions and the maximum allowed tracking velocity of a speck/mote geometry that has no mass with ESS macht glows. The sixth fact is that while the velocity of a photon will never exceed the speed of light as this is the maximum tracking velocity for any speck/mote geometry it will take more or less time to cross an equal distance of differing areas of spatial compression being faster in spatially rebounded areas and slower in spatially compressed areas of ESS. The effects of this are not significant to any appreciable degree in normal areas of spatial compression like those found on Earth or at distances of only a few light years but will become very significant at distances of billions or so light years.

Chapter 27
Vacuum Catastrophe

Vacuum Solution

The world of modern physics is filled with triumphs as can be seen from such examples as Newton's genius in connecting the force of gravity with common objects such as falling apples to the same force that holds the moon in orbit around the Earth. The world of modern science is also filled with failures such as Einstein's impossible singularity theorized to exist in things like black holes or the singularity that is incorrectly believed to have started the whole universe off in the Big Bang. Each success or failure can be seen as having a certain quantitative amount to it in terms of good or bad. One particular failure of modern science has by far the largest quantitative value of the bad kind possible and it is known as the vacuum catastrophe.

The vacuum catastrophe is informally named after the discrepancy in the measured value of the vacuum energy of the universe and the predicted amount of vacuum energy as stated by quantum field theory. Quantum field theory is the belief that the universe is made essentially only of space and that the space is made of fields which can be excited and these excitations are what are perceived as particles. A field in empty space is in a non-excited state whereas a field around an electron is now excited and the electron is the resulting excitation based on the theory of relativity that matter and energy are equivalent. If matter and energy are equivalent in theory then the energy stored in fields can be excited and these excitations are the equivalent amount of matter. The idea behind quantum field theory seems to work at least theoretically fairly well for explaining the mathematics behind various phenomena.

There exists a flaw in the theory of quantum field theory however as it is this belief that empty space or the vacuum contains energy that has led to the vacuum catastrophe. It is assumed but not proven under normal conditions that the vacuum is in a relaxed or empty state for some of the fields and thus it can be said to be empty. One exception to this is the vacuum energy that is said to persist in the universe as a part of space and this energy is non-zero. A vacuum energy in the universe is assumed to be negative and be responsible for the apparent cosmological constant that seems to be driving the universe to expand. The catastrophe comes from comparing measured values for the vacuum energy conducted in experiments in laboratories here on Earth or with observed data on the rate of expansion of the universe with the predicted values taken from the mathematical understanding of the universe based on quantum field theory. The values differ by such large amounts as to easily place this failure of modern science at the number one spot of scientific mistakes. The value measured in the laboratory is very small and the large value predicted by the theory of quantum field theory is up to 125 orders of magnitude different. This is by far the single largest discrepancy between experiment or observation and theory ever in the history of science. In other words the theory is dead wrong as can be proven by physical results from experimentation and observation.

Firstly by experimentation using the Casimir effect which is an experiment whereby two very thin uncharged conducting plates are placed only a few nanometers apart and observed to move together for no apparent reason. The reason is not from any energy or work input into the system by experimenters and the whole apparatus is isolated in a vacuum such that the only energy is theorized to be coming from the vacuum of the universe. The idea is that the vacuum energy of the universe is pushing the plates together and thus proves vacuum energy is real. The setup can be thought of easiest in classical terms by thinking of two plates suspended in parallel in water where waves are introduced to the system. The space between the plates is isolated from the rest of the water and so no waves exist between the plates. Waves are free to move around outside of the plates and so will impart mechanical energy to the plates in the form of kinetic energy from the impact of water on the plates. The two plates will feel this pressure and be pushed together resulting in an external energy being the driving force for the plates' motion towards each other. In the Casimir experiment it is vacuum energy that is excluded from the space between the plates and so the only energy available to the system and the plates specifically comes from the vacuum energy outside the plates. Just as in the water example the two plates will feel the push of this vacuum energy and be drawn together. This value as measured for the vacuum energy is seen to be very, very small which does not predict a large cosmological constant by comparison.

The second way that the quantum field theory is proven incorrect in the vacuum catastrophe is from observation and in this case observation of the cosmological constant. The standard and accepted method of measuring red-shifts in light from distant objects as used by Hubble is still used today to measure the speed of the universe's overall expansion rate. The rate at which objects are seen to be moving away from each other is used to determine the cosmological constant of the universe. This is simply how fast the universe is flying apart from our Earth bound point of view and can be used to prove the quantum field theory predicted value wrong. The way to do this is simply to compare the two and see that the value predicted by the theory is many, many orders of magnitude higher than the observed. Since the observed value is correct as it shows clearly what is really happening in the universe the theoretical value is made immediately false. Any difference in the two values still makes the theory false as the theory must be correct the first time to be accepted.

The reason is simple as the observed value cannot be incorrect as it is simply a visual measurement made by instruments here in our solar system. The theoretical value is made only in the mathematics which as we know by now cannot always be trusted. Even if the theoretical value was close it would still be incorrect as the observations are not wrong and you can tweak a theory or fine tune a theory to make it fit the universe but under no circumstances ever can you tweak or fine tune the universe just so it fits a theory of human creation.

So why does quantum field theory predict such a high value when it obviously is not there and how does this affect the understanding modern science has on the universe? The first part is due to how quantum field theory guesses the universe works and as we have seen it assumes the universe is made of fields. The way the universe is divided according to this theory is through an infinite amount and therefore solid block of space. This is created because the theory sees every point in space and by this it is literally every conceivable point of infinitely small dimensions or at least a single Planck length in space. So every point touches every point and you are left with a single massive 100 percent solid block of quantum field space. This means that the universe has an infinite amount of space and therefore an infinite amount of energy as each point in space contains all the fields known in the universe and any amount of energy no matter how small when added to infinity is of course infinite.

The universe cannot exist in this state for two obvious reasons the first of which is that it is of course simply impossible and the second is that the universe would have infinite energy and mass and would render it unable to form and exist as we see it today. There would be too much energy and matter for anything to exist or operate as we see today such as atoms, molecules or even sub-atomic particles alone. Another reason this is impossible is that quantum field theory simply decides to explain away this belief by saying that in most cases the empty space simply balances out these fields for no reason other than it helps in correcting the mathematics.

Remember according to quantum field theory space consists of all fields and all properties such as the electromagnetic field for example in all points of space however for the sake of convenience the fields are simply cancelling out in empty space. These are the same fields through which an excitation of energy only gives rise to a particle according to the theory but whenever it is needed the fields simply do not carry that energy. This is a massive flaw in the entire theory of quantum field theory as it simply plays with the mathematics when it suites it and at other times staunchly holds firm that these fields exist and do what they claim. If we overlook this problem then quantum field theory has neatly left only the vacuum energy existing in each point of space and it is from here they derive their still too high value. Even the renormalization trick of quantum field theory used to try and negate the infinities created by their own theory cannot do away with the 100 or more orders of magnitude difference between measured and theorized vacuum energies; in fact the 100 orders of magnitude difference is after the renormalization because before it the value was infinity orders of magnitude difference.

The vacuum energy left in space according to quantum field theory still must contain huge amounts of energy as it is from this space that unproven pairs of particles are theorized to appear and disappear for literally no reason. The idea of particle pair production of a normal and anti-particle from empty space is firstly only theorized, secondly completely unproven by any physical means and thirdly only believed to exist as an artifact of the mathematics of quantum mechanics which by their own admission contains unknowns and numerous probabilities. The idea that a probability exists that an event might occur is the kind of thinking at work in vacuum pair particle production. This kind of event would create a particle that can be catalogued within the Standard Model and as such would contain all the necessary properties for a normal particle such as spin, mass, charge and the like. This means that the vacuum according to quantum field theory must contain at all points in space all the necessary energies stored at all times forever in order to cause a single particle pair to spontaneously materialize into existence. So the points of space in quantum field theory can never be empty, they can never cancel out as they are the very fields needed to first of all exist such that a particle can be created from energy and secondly they must exist to give that particle its properties.

In other words according to quantum field theory all points in space have all fields turned on at all times as it is fields that are energy and energy is equivalent in theory to matter and so gives rise to the physical nature and properties of these pair produced particles. The flaws with quantum field theory do not stop there however as the inclusion of only vacuum energy at all points in space still gives rise to an impossibly high value to the energy stored in space even if it were only coming from the remainder vacuum energy. On top of this the experimental results and the observational results tend to agree far more than either do with the theoretical predicted value of quantum field theory. In other words even if you do not accept the Casimir experiments or their kin and their results you are still going to lose against the universe which is never wrong and is expanding precisely at the rate it is and not a at much higher and therefore faster one predicted by quantum field theory.

To this end an explanation must be given as to what is occurring and something is happening to be sure. The Casimir experiments and observations of the rate of expansion of the universe are of course measureable and with respect to the observations are one hundred percent correct and impossible for them to be wrong. As far as quantum field theory is concerned it is the predicted value that is wrong and not the idea that the vacuum as modern science calls it does not contain energy. Quantum field theory correctly predicts that space has energy and indeed energy is there it is just a question of why the two values are so different. It is obvious of course that the value predicted by quantum field theory is wrong as anyone who studies quantum field theory will more than likely tell you is true with a smile as they know they simply made a mistake somewhere along the way.

However the question that arises from that mistake is important because it illustrates that a new model for how the universe works is needed as the quantum mechanical only model cannot explain fully observable events. The fact that quantum field theory has made this massive and erroneous prediction is actually quite a good thing as it tells us that the discrepancy is so large that no amount of fine tuning or trickery with the mathematics to produce a seductive derivative can fix the problem. In other words a new way of looking at the universe is needed to solve this problem so that the observed values can be kept and the known and correct laws of quantum mechanics can also be kept.

Like all of the unsolved problems facing modern science it is a lack of understanding of the properties and structure of space that has hindered progress. In this case the missing pieces of the puzzle are both the ESS and the macht they carry that make up space. Simple space physics of ESS and macht do away with the discrepancy of measured and theoretical values for the vacuum energy. First off stop calling it vacuum energy as there is no such thing as a magical energy that is a part of space apart from the known forces. There exists only macht of the known forces and it is shared between ESS and by using these the effects of things like the rate of expansion of the universe are easily explained without the use of an extraneous form of energy called vacuum energy. To start off with you must remember two properties of space the first of which it is made of ESS and the second is that space and matter interact to create the universe we see today. Space is a continuous connected work of ESS that extends from the edge of the universe through to where the Earth is and on to the next edge. Going way back to the original little room of ESS spheres a specific property of ESS can be seen immediately which is that they are hollow. Remember I said the fact that they are hollow would be of importance later?

The construction of space is thusly made of hollow spheres that touch one another but will also have spaces between their surfaces where they touch their neighbors. The reality of this is that space is not a solid block that must contain every conceivable point made of space over and over and over again. In truth most of space really is just nothingness with only a small percentage made of what can be called space. The long held belief that space simply gets bigger or smaller, more or less dense as time is rewound or fast forwarded and it only expands through itself is quite wrong. Space according the three main views as it expands simply gets larger meaning and this means the energy stored at every conceivable point of space stays high because space is still a solid block to them. Even views of the universe such as this have space expanding through a nothingness which exists beyond and outside the edge of the universe. This nothingness of existence is where the incorrect and theoretical singularity is thought to have existed and the idea that simply non-existence was beyond the universe is wrong.

The universe is made of space that is true but that does not mean that the entire universe needs to be solid space and even if you must keep your precious theory that nothing exists outside the universe before the Big Bang does not mean that inside the universe you cannot have a nothingness existing. That nothingness exists inside the hollows of ESS spheres and between the places where the spheres touch and is of extreme importance to the everyday normal functioning of the universe and in fact you cannot build a universe without it. Matter does not need to exist at every point in the universe so why does space? The answer is simple it does not and this missing piece of information is one of the facts of the universe that helps to answer all of the unsolved problems of modern science because these spaces or nothings are used by both ESS and their macht as well as matter to create the universe as we see it today.

If you look at a cubic centimeter of space and we say that a normal ESS under Earth spatial compression would exist such that the six sides of the cube are just touched by the ESS walls then you can see mainly empty space in the cube. The walls of the ESS are thin and at this scale would be far thinner that a human hair as the ability of the ESS volume to increase or decrease can vary by a large amount. The total volume contained by a sphere made of a thickness some fraction of a human hair compared to the total volume of one cubic centimeter is microscopic at best. Again one is not surprised that every part of the universe is not filled with matter and they are quite willing to accept that matter is mainly empty space. Well now it can be accepted that just as an atom is mainly empty space so too are ESS mainly made of nothingness.

To extend this fact let us construct a ten by ten by ten sided cube out of these little one centimeter cubes all stacked together. This now shows you the importance of the spaces between individual ESS spheres as they each occupy one cube only. Looking at the entire 1000 cubic centimeter block we see again that most of this block is not made of ESS but of nothingness inside and around each little hollow sphere. The total volume of actual space contained therefore in a 1000 cubic centimeter block of ESS space is almost nothing compared to a similar block of quantum field theory space. The simple realization that space is still made of space but also a considerable amount of nothingness immediately reduces the volume that must be made to carry energy. The amount of energy contained inside of real space which is now known to be made only of ESS is therefore automatically reduced many orders of magnitude. So the value of all energy no matter what forms are predicted to exist within a certain volume of space will be reduced to match those of experimental or observed amounts and still allow space to increase and decrease in size.

The second way that ESS and macht can be used to explain away the vacuum catastrophe is by looking at what happens when space and matter interact. As we know space is made of ESS and matter is made of specks and motes in various geometric arrangements. Collide two known particles in an accelerator and you can dissociate those particles such that the free specks and motes becomes various other particles or can briefly create new geometric patterns to create never before seen particles. These specks and motes will interact naturally with ESS in the universe and cause certain types of macht found in the macht pools of all ESS to either increase or decrease. In most cases ESS macht is seen to increase as has been explained in previous unsolved problems such as gravity and black holes. What this means is the amount of energy associated with a field is often times higher than it really is when the ESS are free from interactions with matter.

So for instance if you try and measure the strength of a field around a fundamental particle you will incorrectly guess that the field necessary to create that particle is always that strong. In other words if you measure the strength of the strong nuclear field in an atoms nuclei you will believe that the field that excited the particle if quantum field theory was correct needed to be that high or have the potential to be that high in order to create the particles as you observe them. Remember that particles are seen as pure excitations of fields and that they are made from energy taken from those same fields so all the rest mass energy must have been contained within those fields according to quantum field theory. Think of a hydrogen bomb and you see the incredible amount of energy stored in these bonds that is freed during a fusion event. According to quantum field theory this energy was all stored in the fields of the vacuum as it is from these fields that the particle was made real through the energy to matter equivalency principle. This is of course the incorrect view of space and matter and if you look at matter as a speck/mote that excites the macht pools of ESS then the amount of stored energy in space is far less.

Since matter imparts or donates energy to increase macht glow then this means the ESS macht is at baseline far less than seen in the excited field state. So when someone measures a field's strength from a particle they are in fact getting an artificially high reading all the time and this is not the amount of what is thought of as stored energy in space. Remember all that macht glow is what creates the seemingly physical boundary and volume of space occupied by the particle and since it is known that a small speck or mote lives at the center of the volume and has donated energy to increase macht to make that same volume it is the speck and mote donating energy and not the space that is responsible for all the macht glow. If you remove the speck/mote from space you no longer see an excited field as the fundamental matter has taken back the energy it was donating to the ESS and so the baseline ESS macht pool is always lower than expected. This means that overall ESS macht which is incorrectly seen as vacuum field energy is always lower than theoretically expected values and it is not until the interaction of fundamental matter in the form of specks and motes that you see the strength of a field as it can be measured in the laboratory on Earth under normal spatial compression.

Measuring the strength of the gravity in an area of highly compressed space like in an SM mass will give you very high readings for gravity but will give you low or no readings for other forces. This further illustrates the flaw in looking at strengths of excited fields and then applying that information to the baseline energy of the universe. Also a reminder here that in certain cases matter can take macht from ESS and turn it into energy for its own use and in these cases it will cause the value of the experimental vacuum energy to be lower than it should. These effects are tiny compared to the overall effects of ESS being mainly nothingness and speck/mote interactions being responsible for creating artificially high values for ESS macht pools but is still worth mentioning here. Every small adjustment in ESS macht values will close the gap of the vacuum catastrophe a little more.

Therefore if you combine the fact that space is mainly made of nothingness inside and around the very thin walled ESS and the fact that ESS macht is only elevated in the presence of and interaction with matter you can easily reduce the discrepancy between observed and theoretical values of the so called vacuum energy. This allows for the experimental results of laboratory studies as well as the observed rates of expansion for the cosmological constant to be kept very low. At the same time the quantum mechanical view of the universe that space contains energy and creates fields between matter may also be kept but the amounts have been proven to be far less by using the ESS and macht of space physics as well as the interactions between space and matter overall.

So nothing of fact needs to be discarded from the Standard Model of particle physics and by using ESS volumes and macht pools the obviously correct observations of the rate of the expansion of the universe may also be kept. The fact that vacuum energy does not exist is of no importance as it was never needed to cause the universe to expand in the first place. ESS are able to alter their volumes as they alter their rates of interaction with matter and so can naturally undergo spatial rebound which creates the steady expansion of space. Space is simply relaxing and increasing the volume of its individual ESS and so need not be made of infinite points of space contained in a solid block. The fact that the ESS are simply expanding as the universe ages and the spaces between them are really that, just spaces of nothingness allows for a model that not only erases the gap in the vacuum catastrophe but also continues to explain a little more about how the universe as a whole is built and works.

Facts

The facts for this section are few as the problem of the vacuum catastrophe is rather simple to solve as it is based off of already understood space physics. Therefore the first fact is that space is not actually all space and that much of space is the nothingness that exists within the ESS spheres and in the spaces between them. The second fact is that ESS spheres have very thin walls as this allows them to undergo large changes in volume depending upon the amount of spatial compression they are under. The third fact is that matter will donate energy to these ESS and increase their macht strength which is incorrectly measured as being the true strength of these forces. In reality the amount of macht pool in each ESS is not nearly as high as seen by measuring field strengths in the laboratory. The fourth fact is that simply by combining these previous facts of small ESS volume per volume of space as compared to the total amount of nothingness volume in that same volume of space and the elevation of macht strengths as measured in the laboratory the vacuum catastrophe disappears altogether.

Chapter 28
Absolute Values

Absolutes of Space

The universe has many laws and rules which we use everyday to construct our understanding of it but one rule has been neglected and outright dismissed and it is blinding modern science. This is the idea of absolute values in the universe which used to be taken for granted and for good reason as they form the lowest levels of the foundation upon which you build a universe and many exist not just a few. It used to be understood through classical mechanics that the universe had absolute and relative values such as absolute and relative versions of both space and time. Newton argues that the universe had absolute values for space and time and only humans and the systems we exist in or observe deviated from these values and produced the corresponding relative values. This was widely accepted and worked to solve a number of problems concerning the basic understanding of the universe. The idea of absolutes in the universe lasted until Einstein's theories were made public and accepted and the understanding of the universe appeared to take one step forward when in reality it took two steps back.

The reason is that Einstein's theories on the surface seem to fit the universe better than those of classical mechanics and to some extent they do. The equations of general relativity will produce more accurate predictions of things like light curving around a star and so blindly anything Einstein put forth concerning why this happened was accepted. This is the point where modern science took two steps back as the basic idea of frames of reference from which all of Einstein's work rests seems to work in Earth like conditions. The moon around the Earth, the Earth around the Sun or light curving to a greater degree around the Sun are all forms of motion of objects through the solar system that in our very medium level part of the universe appear to be explainable using relative and not absolute values. These fall apart once the medium level areas of the universe are left behind and you enter the small and weak or large and powerful.

Systems so small and energies so low as to be imperceptible to us on the one hand. On the other systems so large and energies so powerful that the laws of physics as known and accepted by modern science simply breakdown. This failure of modern science at both the small and large scales is based off of trying to accommodate relativity into the workings of either massive cosmological observations or repeatable tiny quantum worlds. In both cases relativity demonstrates its severely limited use at scales around the medium level of the universe which was the comfortable domain of classical mechanics before it. Remember relativity is only a fine tuning of classical mechanical systems such that a slight twist on how they progress through their respective world lines are mapped and equated. As a result the failure of relativity at both the small and large scales of the universe is a clear message that something about it is very wrong. This is obvious to most people immersed in physics these days as not only are the answers to the universe not given by relativity but also relativity creates its own share of impossible objects such as singularities. So what if one does away with the obviously incorrect frames of relative reference that relativity uses in order to explain itself? Can a system of absolutes exist in the universe despite the fact that the medium level of the universe that relativity appears to work on seems correct? The goal is to show how the equations of relativity can be kept such that gravitational effects for medium scale systems like a solar system can be remain but the limitations of relativity can be done away with allowing for the foundations of the universe to be uncovered; in other words let us get back to the classics.

The first step is to allow for the universe to do exactly what Newton said centuries ago which is be both absolute and relative at the same instant. This is not an impossibility and the fact that the universe contains absolutes is precisely how everything in it works in the first place. Without absolutes it would be impossible for something as simple as gravity to work at all. To understand this let us look at a single hydrogen atom and a single ESS as they exist at different instances in the universe. The behaviors of atoms are well understood and fully accepted as already having absolutes and so space and different systems in the universe also possessing absolutes should also be seen as quite normal. If you look at a single hydrogen atom in an excited state we see that this is a relative state of the atoms as the electron in its second orbital can exist at either the first or third as well and this is only a relative frame of reference in the atoms world line as the fact that the electron occupies the second orbital is not permanent. From this middle point of view one would say that the atom and therefore by extension all atoms only exist in relative states as it could also exist in a first, third or even fourth orbital theoretically. Hydrogen atoms exhibit relative tendencies but this is built around absolute values of the system as a whole.

If we look at quantum degeneracy states of electrons around nuclei we find that the electron will always tend to the lowest energy orbital around the nucleus and so contains at least one absolute value. This is the value of the atomic system at the point where the electron degeneracy pressure is resisting external influence through electromagnetic pressure. If the electron degeneracy pressure for the system is exceeded then the atom fails in maintaining its structural quantum integrity and collapses possibly through electron capture creating a single neutron. The atom stops being an atom as it is destroyed creating a single hadron and therefore has one absolute lowest energy value. Not to mention that an electron which is a form of matter is an absolute value in and of itself as it is a quantum particle.

The opposite is also true as the atom receives energy from the external environment and imparts it to the electron which jumps up to higher quantum level orbitals. The maximum amount of energy the atom can absorb before becoming ionized is seen as the second absolute value of the atom. There is no one and a half orbital to be found anywhere in a hydrogen atom or anywhere before ionization. The point at which the ionization energy for the hydrogen atom is reached destroys the atom again and so represents a second true absolute for a hydrogen atom. A hydrogen atom therefore has two absolute values in the universe and relative values in between thus all atoms exhibit the same properties. Single quantum scale systems such as atoms represent the matter building blocks of the universe and prove that you can have both relative and absolute values as part of a single system or object. If you move inside an atom and examine the nucleus or proton of the hydrogen system you see that these same rules apply for the quarks, gluons and interactions keeping them together as a hadron. Extend this one step further and you can breakdown anything in the universe until you have reached the last step in the chain and therefore the absolute limit of that system. Think about it for a minute, science has known that eventually you get to a smallest fundamental particle and by definition that is the absolute smallest value for a particle you can have

This is not just a conceptual idea it is a factual one as all objects in the universe are made of quantum systems which exist purely as absolute states. There is no such thing as a smooth transition between orbitals and they only occur instead at precise quantized energy levels.

- 565 -

◄New Universe Order►

This fact is somewhat apparent in most of modern science but largely ignored because relativity does not use any such system and relativity is held as being correct. Relativity is not correct and so doing away with a smooth fluid curving space and replacing it with the correct version of space which is one that can be divided into units doing away with infinities of relativity allows for the inclusion of absolutes into space that can now match those of the quantum world. Plus as mentioned a section earlier having absolute building blocks to space also does away with the smooth infinities responsible for curving a thing called spacetime into existing everywhere at once and having an absurdly high theoretical vacuum energy.

Using ESS space is divided into set structures that can now behave as one would expect a quantum system to. I have already shown that ESS are individual structures and that these ESS can change volume by absorbing or releasing macht. To understand ESS better I will use the same example of a hydrogen atom gaining or losing energy and the electron in orbit around it increasing or decreasing in orbital volume. Atoms such as hydrogen atoms when excited become more voluminous and when they lose energy they decrease in size until they are at their lowest quantum degeneracy state. Absorbing or emitting energy facilitates these changes and gives rise to two absolutes for the atom whereby exceeding these stops it from existing as an atom.

ESS work in a similar manner but can be best visualized as working opposite to matter. When an ESS gains macht it will decrease in volume and when it loses macht it will naturally undergo spatial rebound thereby increasing its volume again. So if we look at the macht degeneracy levels of an ESS we see that the lowest energy state for a single ESS is not when it is smallest like matter but when it is largest when it has dumped off as much macht as it can. If you remember the little room from many sections ago in which you saw the ESS of the universe for the first time you can understand this better. The room was initially filled only with the little ESS as hollow gently glowing spheres all of uniform size. In this system the ESS are devoid of all matter and do not have any extra energy donated from matter to them in the form of converted macht. This means that they have no extra macht and so can be said to have no extra energy meaning they do not decrease in volume at all. The state of ESS and therefore the true revealed state of space is one where with no matter present the ESS has spatially rebounded to their maximum possible volume. In other words the ESS or space has relaxed to its lowest energy ground state. This ESS level is held in place as you now know by the macht pool that each possess and is the only macht present as there is no matter elevating macht levels.

This is finally revealed as one of the absolute values of the universe as space itself which forms the entire background of existence cannot ever increase in volume. The ESS can still absorb more macht and therefore decrease in volume and we live in the medium part of the universe that always has elevated levels of macht so when we look at something we are always in the middle of the macht degeneracy levels. From an Earth point of view everything in the universe does appear to be made of relative components that have no absolutes. This is deceiving as it is only because anywhere we look or measure we see ESS volumes higher or lower than our own but we are never able to see the absolute limits and this is why one limit seems to destroy modern physics and it is the limit of the black hole. In reality the universe still has absolute values to space similar to but working in completely different ways from Newton's picture of absolute space.

It is not that space exists as an absolute for a given system so much as it has the ability to achieve these absolutes and this represents the ends from which things can be pushed so to speak. They are the foundations from which the universe can exert force either up or down in the overall scale of existence. From the fully spatially rebounded foundation the universe can push space smaller and smaller allowing for the kinds of interactions on matter that I have shown you. These include gravity and how it actually works, the limits for an atoms smallest and largest sizes before it is destroyed and limits of both space and matter in TSM events. The opposite foundation is held by the TSM form of space and matter which is the smallest absolute value for space in the universe as it is the point where ESS walls collapse. This represents the second absolute value to the universe a scale at which nothing can ever be smaller and therefore a second pushing point from which the universe can exert forces on objects.

When I say pushing points I also mean a point of reference from which all equations and calculations must be based to solve the mathematics of how the universe works and those mathematics must use, ESS, macht and specks and motes. Once you know the absolute smallest or largest values for space you can incorporate more easily all changes in spatial volume to things like quantum systems as well as place limits on forces in the universe as they cannot exceed these absolutes. In addition being able to use a top down and bottom up approach to space will help to narrow in on the mathematics and the final equations necessary to describe systems by attacking them in a pincer movement.

So the universe has at the very core of space two absolute values which cannot be exceeded for reasons previously explained and only appears to exist as a relative medium as theorized by Einstein. In normal everyday Earth conditions space and the things in it tend to exhibit frames of reference according to relativity but in actuality are only playing off of the absolutes of space in terms of its size and macht levels. This proves that the universe does contain absolute values and that some of these are related to space. Furthermore the objects that make up the universe in terms of matter and energy are bound to these ESS that at two different sizes possess absolute values and it is the properties of the ESS that shape matter into the universe we see today. One value is highest macht glow and smallest volume and the second is lowest macht glow or macht glow reserve pool and the highest volume. Therefore even if it might seem as though absolutes are not important to us in our medium level solar system they in fact are as the measureable limits of ESS and their macht set the boundaries from which we can measure and calculate matter interactions that make up us and the instruments through which we observe the entire universe.

Absolutes of Time

Another part of the instruments that we use to measure the universe is an artificial one that we have created called time. Time as has been shown is not a real force in the universe and as such not a physical thing of any kind that can be manipulated or by manipulation alters solid things like matter. The trick of the mind that believes time to be a real tangible force is simply a set of interactions between matter and ESS and no combined spacetime exists. Time does play a useful role as a creation of humans and all intelligent life as a way of simply measuring differences in reality. A perfect example of this is the illusion of time dilation as seen by Einstein who was confused as to what was really happening in the universe. Einstein simply thought that because he was trying to understand gravity which he equated to curvatures of space that when time got caught up in the problem that space and time must be one. This is of course utterly wrong as the idea of a single time is simply a human centric view and since Einstein could never understand what was creating gravity in the first place prevented him from understanding why time seemed to slow. This also forced the idea of Einstein's incorrect time onto systems and disciplines that did not want or need it such as quantum ones and created a whole slew of garbage for others to try and make sense of. The simple answer for general relativity was to ignore why the events were taking place and simply guess as to how they were by saying space and time were one so if you alter space you alter time.

The idea of time however as already explained is simply a human construct and so of course an arbitrary construct will have a very low probability of being correct when compared to the near infinite variances in ESS and matter interactions which give rise to what is seen as time dilation. To this end the arbitrary and Earth centric units of hours, minutes and seconds were unable to fit all degrees of spatial compression and therefore matter interactions with space that we can observe. To explain this a bit more bluntly think about Einstein's reasoning for slowing time relative to the observer using spacetime. Curvatures of space were seen as sources of time dilation because the fabric of spacetime was being curved. This makes no sense for a small particle of matter accelerated to near light speed as no curvatures of spacetime are occurring no matter what the mathematics trick you into believing. Time dilation appears to occur for both types of events yet only one is accompanied by a true curvature of spacetime as a single fundamental particle no matter how fast will not curve spacetime as it is not changing its density per volume of space no matter what confused people say. Ever heard that nonsense about the universe appearing flat before a photon? Time does appear to be slowing down however and this creates another Einstein paradox which is the observer sees time slowing for the object travelling near light speed and we will use the example of a spaceship here.

The spaceship is said to not even notice the time dilation however and so it travels normally and time passes normally for the people on it. To the observer though it travels normally but time passes slowly for the people on it. Most people reading this who have been brought up to believe Einstein will not see the paradox but those who know that Einstein and general relativity are wrong might see it. The paradox comes from the fact that the ship in both frames of reference is moving normally which according to Einstein himself is a no-no. The Moon is slowed in its orbit around the Earth according to general relativity because of time dilation meaning it physically moves slower through space than it should. The ship must be doing the same thing and yet according to the paradox of spaceship passenger and stationary observer it is not. The ship the faster it goes must slow down according to general relativity yet it is permitted to break relativities own rules and create a paradox where the people on board age slowly and observers do not. No this is not because you are putting ever more energy into the system in order to compensate for this so stop thinking that.

Think of it this way as well, a ship moves forward due to thrust which is the result of propellant being ejected from the rear of the ship. If you have a slow rate of reaction you get a slow speed and the faster the reactions occur the more power is available to push the ship forward. This is precisely where Einstein fails again as the ship itself is as he believes is in an area of slowed time and time that is getting slower and slower as the ship goes faster and faster. This is seen from the point of view of clocks moving more slowly and people aging slowly but the ship is breaking the laws of relativity as it moves faster. This is because in order for the ship to go faster at least one part of it which is the engine must act as though time were normal because the ship's engine must continual operate at a faster rate in order to produce more thrust. The ship must burn more fuel and at ever faster rates of reaction in order to continually speed up because if you apply a set amount of thrust you only get a set amount of speed; 1000 kilometers per hour of thrust cannot be added until you go the speed of light. In order to continually add speed you must create more thrust faster than you did in the moment before and this means the ships engine is not experiencing a slowing of time or it could go no faster.

If time is truly a real physical force which is a part of the very fabric of space as Einstein believed then it permeates everything and would trump all such actions creating a time dilation for the entire system and not select parts of it. I know what some of your are still thinking and to those who defend Einstein's work you must now look at an alternate view of the universe such that the limitations and failures of a spacetime curvature can be done away with. Yes you get to keep apparent time dilation only you must accept it for the reasons I have stated as this does get rid of things like singularities and prevent physics from breaking down in certain cases which is exactly what Einstein and his work created in the first place.

To understand this fact spacetime according to Einstein is a fabric with the two parts woven together and accelerating an object through space distorts that space creating a curvature. This curvature can be thought of as a gravitational field which is an invisible force that stretches well beyond the object in question and many hundreds, thousands or millions of kilometers into space. So if a ship going the speed of light or close to it were to pass by a stationary clock that clock should slow down from the spacetime shockwave riding in front of the ship. This means the clock on the ship should be going slower and the clock that is stationary should run regular then slow then regular as the spacetime shockfront moves past it and this is exactly what does not happen so Einstein is wrong. The stationary clock does not feel a slow down and will not feel any effects from the ship passing close to it at near light speed and what that means is that spacetime is not a woven fabric so once again Einstein is wrong because if it were it would produce time dilations for every part of the universe as the entire spacetime is supposedly connected.

So the pocket flaw of Einstein's spacetime fabric shows that space and time are not in fact one and that it is not due to any explanations or relativistic effects of Einstein that time seems to be slowing. Time dilation has been measured as a phenomenon but not due to the reasons given by Einstein or general relativity. Doing away with the idea of spacetime and all the nonsensical and flawed thinking of Einstein that creates impossible objects like singularities will allow one to still explain the phenomenon and not break the laws of physics along the way. Remember another problem with Einstein's work is the singularity which every physicist will acknowledge is a mistake and that mistake comes in many parts from the creation of an infinity in the equations. That same set of equations gives rise to the impossibility of accelerating an object to light speed which creates another infinity and again every physicist knows those to be a clear impossibility.

The way to explain these events correctly is to use time as a human construct only that simple measures one instant to the next or can be said to show the differences and changes in a system from one interaction to another. So time as we use it everyday is not a real thing in the universe but it is a very useful measuring tool for accounting for events that take place in the universe. Length in meters, temperature in degrees Celsius and time in seconds are all arbitrary Earth-centric units but still all useful in measuring things around us. To get to a time that is absolute is to get to a time that can be used as a base from which all larger measurements of passages of instants and events can now be measured. So an absolute value of time does exist and it will not only allow us to properly measure events in the universe it will do away with spacetime once again as absolute time is the only unit of time that is immune to time dilation. Again time is not a real thing only I am going to explain the smallest possible length of human measureable time and how long that would be from one event in the universe to the next.

Firstly it is not quantum in nature as quantum events are too large and slow and are themselves altered by the absolute time unit resulting in what is thought of as time dilation. The smallest measurable difference in events happens at the ESS macht scale as this is the smallest and most refined set of actions possible in the universe. No matter what physics might uncover at the very base of everything will be a single smallest set of space as made of ESS and this space must share macht. The ESS and the macht they carry are the smallest and fastest operating objects in existence and dictate how the much larger and slower moving quantum world behaves.

So understanding the space timescale will explain the quantum timescale and this can be measured at its most fundamental limits as a single ESS macht time unit. I say ESS macht time unit simply because macht is transferred between ESS and so can be seen as a way of measuring changes between ESS from instant to instant. The ESS themselves of course have a structure as well and this structure will change from instant to instant as we see their walls increase or decrease in diameter changing their volume. The components of ESS are also governed by actions which are dictated by changes in each of their macht levels so a single time unit of smallest size can explain both ESS and macht events from moment to moment.

Using a single ESS macht unit of time can allow for one to examine the step by step changes of a single ESS over time and from there scale up more and more time units until you can create a second. All actions of the ESS macht type and of space physics operate similar to quantum ones where things move about in step by step fashion or quantized amounts. This is because at the most fundamental levels everything in the universe has a base level building block which cannot be subdivided forever and so the exchange of information along it which can be thought of as time can also not be subdivided forever. Now here is where I can reveal a truth about the universe and time that has never been proven before which is time is not real and yet both absolute and relative all at once. Time remember is not real it is only a construct of humans and intelligent creatures as a way to make sense of the world around them and this is perfectly understood. Now I get to the fun stuff which is absolute and relative time and the proof for both of them.

Time does have an absolute value that is immutable and the unchanging unit of the entire universe and from it everything else about time is explained and cannot deviate. This absolute time is the smallest possible measurable interaction between ESS internally, with each other and through macht exchange and I will call it macht time for short although it is used to describe any parts of the system. The very base interaction of the universe is not based on the ESS being fully spatially compressed as in TSM or fully spatially rebounded as in space fully devoid of matter. It is only more or less macht that is shared between ESS in the universe and the rate at which ESS share macht between schales or the rate at which they alter their respective volumes does not change. These interactions are the base level interactions of the entire universe and must take place in step by step fashion meaning they never change the rate at which they occur. This makes perfect sense as an interaction is simply an interaction and cannot ever be subdivided as explained above.

What this produces is a unit of time, the macht time which cannot ever be decreased or increased and as a result is unchanging. Unchanging denotes absolute properties and no matter where you go or what object you observe in the universe you can never break free from this truth. Time on the scale of space which is made of ESS and macht is absolute and unchanging and will never alter as it works unseen in the background of the universe. Time from space is all powerful and creates the reality that we observe everyday as humans and all intelligent life understands and is explained using space physics not as theory, not as philosophy and ends the debate as to what time is once and for all. To do this I will build our reality from macht time and prove how this is fact and no longer a debate that needs to rage for another 5000 or so years.

We live in the world of matter and energy and we perceive the universe through these means and it is from this substrate that we have constructed the world and our understanding of it. Humans have forever seen the world through matter and energy eyes and when we learned to measure differences in time we did so with matter and energy eyes of machines. These machines learned from us and we taught them to think as we do and they have been good and measured everything we can point them at just as we would like to see the universe. From here we believed that time and space were one and that time is real and that time can be made to speed up or slow down. This should have been immediately obvious to us as incorrect as this implies that time is a lump of sorts that can be smoothly added to or taken away from at will in order to sculpt the universe. Energy does not behave this way and neither does matter as can be seen in the flawed belief that matter becomes more massive as it moves faster and faster through the cosmos.

As matter moves through the universe close to the speed of light it is theorized that it becomes more massive so that a single proton has the mass of ten protons and so on. In reality the proton does not actually gain more mass and is only said to do so through the mistaken literal translation and interpretation of the equivalency principle. The proton does not suddenly grow and create matter out of nothing to become ten protons made now of thirty quarks. Time similarly is seen to increase or decrease as certain conditions are met and yet this is equally as impossible as how would time simply be created from nothing in the universe to become more or less for a given instant in a particular frame of reference? There is no energy component to the matter component of time so you cannot create a time equivalence principle and so the answer is emphatically no, time cannot really increase or decrease. Time once again is a human construct that only appears variable in the universe. This is where macht time can be used to finally explain what we have understood as time for several thousand or more years. Macht time as has been shown is absolute and unchanging despite differences in conditions affecting space. The background time of the universe is therefore unequivocally absolute and unchanging.

However our reality is based on matter and energy which is a larger construct of the universe and has its actions dictated solely from exchanges of macht between ESS and the natural balancing properties of ESS in the universe. Rates of reaction for quantum systems are variable as I have explained in a previous section on time dilation which was of course the introduction to the answer of time. These changes in how quantum actions take place for a given system are perceived as changes in time when in reality they are based solely on physical laws between space and matter. There is no physical time in the universe despite what it looks like to us quantum creatures and yes we do see the time dilation effects as quantum matter and energy obey the laws of space physics underlying the entire universe. Our time is therefore perceived as variable as events that exist for us can be made to appear only to slow down or speed up but they are governed by the absolute time that does not slow down or speed up which is macht time.

So there it is finally answered time is not real and not a physical thing in the universe and it certainly is not a part of any spacetime fabric. Time is a construct of humans and all intelligent creatures as it is how we mark the passage of one instant to the next in the universe. In the realm of space ESS and macht time has an absolute form that is the only set of interactions in the universe that are not altered by any conditions. Time is variable in the realm of matter and energy of the quantum world as it is obeying the rules of space physics and so explains apparent time dilation. So time is not real, time is absolute and time is variable and can be used to construct the reality that we live in as life here or anywhere in the universe on any planet at any moment in the life cycle of the universe.

Losing some Black Hole Problems

I would like to provide one more answer to an unsolved problem of modern science here as it can be simply explained using space physics and do away with a series of mysteries faced by the three views concerning black holes. This is the series of problems that come about from using general relativity to explain black holes and create the observer problems of event horizons and the holographic principle as well as information loss. In the classic relativity paradox an observer on a spaceship sees a person fall into a black hole and as they approach the event horizon time appears to slow for the falling person and eventually stop on the surface frozen forever in place as time stops in a black hole due to the belief that a singularity drives the event horizon from the black hole interior. The observer sees the person falling and waving and as they approach the horizon they wave slower, fall slower and finally freeze on the event horizon literally forever. According to relativity the person falling feels none of this and simply passes through the event horizon and is consumed by the black hole lost forever in the singularity. The event horizon and singularity are believed to be incapable of communicating with each other because beyond the event horizon nothing can move faster than light and so no information can be transmitted back to the horizon to be seen in the universe again.

The number of failures with this view of a black hole and how it works is mind buggering and I will explain the truth about how all of this works here. A quick thank you to all still reading my work as I promised many times in many sections that I would develop your understanding of the universe and give you these secrets and answers in later sections and by now I hope you see this section building upon some of those. So to end your unease at thinking about black holes I will simply begin obliterating the flaws with the general relativity view of black holes and explain once and for all these perfectly normal objects in our universe.

To begin with a black hole cannot work as stated by relativity as everything would be frozen on the surface of the black hole meaning that black holes would be by far the brightest objects in the universe. This is because everything that they pull in would be frozen in holographic form on their surface and this means that light from them is escaping a black hole constantly so that the observer can see them. If everything never fully passes through the event horizon then images of everything would be frozen on the surface of the black hole meaning that it must also be radiating an infinite amount of energy. Infinite objects frozen in place because according to Einstein space and time are one will never pass the horizon and so will be seen forever. Even if the object does pass the event horizon the image which can be normal or holographic is still real and therefore the information about the object remains which is energy and so quantized. This will require an infinite amount of information to be emitted from the black hole in order to pass information along about what is frozen on the horizon. This creates another paradox as the point of freezing is where time stops and with no time no information can exist for us to see it and what is worse for relativity this violates its own law that nothing can escape a black hole and especially at the point where time stops.

In addition information takes energy to maintain and so the black hole is losing mass and increasing the rate of mass loss over time as more objects fall into it more information meaning more energy from inside the black hole must be spent illuminating it on the surface of the black hole.

The answer to this was the holographic principle which creates an infinite amount of new particles in order to spread the information of the falling object onto the surface of the black hole. The information created also takes infinite energy which again would be impossible for the black hole to maintain. All of this was simply because modern science could not abandon an incorrect theory which is relativity and an attempt to preserve the precious information that they simply concluded must never be destroyed. So the impossibility of the holographic principle was created which sought to account for the information thought to be frozen on the surface of the black hole. The holographic principle is of course impossible as the singularity according to the theory of relativity exists inside of the black hole and cannot communicate with the horizon. Since the object is thought to pass right through the event horizon it is impossible for any information of any kind which is limited by the speed of light to be transmitted from the singularity to the surface. So it is quite impossible using the rules dictated by both relativity and the holographic principle to have frozen images containing information frozen on the surface forever of a black holes event horizon.

Since black holes are in fact black with no images trapped on their surfaces then objects simply ignore Einstein's theory of spacetime and frames of reference and all relativity and obey the laws of absolutes held within ESS such as absolute space and time. This goes back to what I was explaining before about the correct explanation for time dilation and the actual causes of it. Black holes as we have seen are continuous masses of SM that have no boundaries between their cores, their surfaces and the space beyond them. A smooth transition of space and matter is constructed such that no laws of physics break down or are changed as an object transitions from normal spatial compression to high spatial compression inside the black hole. The object simply moves towards the black hole and as it approaches what is seen as the event horizon it loses the ability to emit light. The reason why people think of a black hole having an event horizon and this is the point at which it cannot transmit information anymore is simply due to the fact that at these levels of high spatial compression the balancing of macht between ESS disallows energy and matter to be in bound forms. Think of this as the point where electrons are stripped from their atoms and can no longer reform with nuclei. This is getting closer to what the beginning of the universe was like when electrons were unable to be bound into atomic form with protons.

The photons however are not causing photodisintegration as was the case in the early universe and are also being held in place by the ESS macht so they too cannot move freely. All energy and matter at this point are so heavily bound to the ESS macht of the space they inhabit that the concept of light escaping is not due to any sort of escape velocity it is simply irrelevant as the attraction and balancing of energy and matter with the ESS macht is too high. Objects will then be dissociated at their fundamental levels from normal matter like atoms and hadrons and such and have their speck/mote geometries changed into clintrons. This is the final locked form of matter in the universe as it is now apart of space and space is a part of it and this is the point at which SM is born. The information that appears to be lost is not lost from the universe but it is lost from the universe outside the black holes event horizon. The information I have explained previously is still in existence but is held solely in macht form between ESS schales in the macht glow and is now what can be called muted and dulled from a quantum point of view. So you do not see an object freeze on the surface of the black hole as the information is transmitted straight through it to the black hole proper. There is no slowing of time and freezing of the falling objects image on the surface because this is only a trick of general relativity.

In reality the perception of time in our quantum realm is slowed from an observer interaction perspective but not from a universe perspective. So yes the falling person will wave slower but they will never slow in their descent to the black hole as seen by the observer. The observer will see them pass smoothly to the point where their matter dissociates at the event horizon. Unlike general relativity which states something similar space physics never has the object freeze on the surface and the observer will easily see them disappear into the black hole if you remember the ESS and macht gravity explanation of black holes and why singularities do not exist.

There are also multiple event horizons for all SM masses as these are the different layers of spatial compression at which matter breaks down. First the electron event horizon, then hadron event horizon and then the quark event horizon at this point clintrons are formed and the final shape of matter and energy in the universe is made real. So the problems surrounding black holes are answered using space physics as absolute space and absolute time combined with normal ESS macht and matter interactions of specks and motes in various geometric forms are moved from areas of spatial rebound to areas of spatial compression. The key is the absolute time of the ESS macht which shapes the quantum realm we observe and can create time dilation but also allows for objects made of quantum matter and energy to pass through the event horizon of a black hole without causing any impossibilities or paradoxes. There is no freezing on the surface of the black hole, there is no holographic principle and information is locked in the SM mass and so is lost from the quantum universe but not lost from the universe as a whole. Everything is as it should be and no laws of physics breakdown at any point when traversing an ESS macht and matter interacting SM mass.

Facts

I think I will end this section here and so can now summarize the key facts that govern the universe and how it works as explained from these unsolved problems of modern science. The first fact is that absolute values exist in the universe and the entire universe is not based on frames of reference, relativity and observer information. The second fact is that space has two absolute values one for the TSM and one for the ESS under full spatial rebound. It is from these two bases that all effects of energy and matter can be pushed or pulled depending on your point of view and creates the universe we see around us today. The third fact is that time also has an absolute value which is the single interaction of ESS, macht and any combination of such interactions such that the smallest action is the smallest time unit called the macht time. The fourth fact is that time has two versions in the universe the first is the absolute version of space physics which never changes no matter what conditions are occurring in the universe between space and matter or energy. The second is the variable time that we humans and all intelligent life perceive and use to construct and make sense of the world and reality around us always. It is this time that is seen as causing what is known as time dilation and is created by ESS macht dictating how energy and matter must behave in the universe. The fifth fact is that black holes use both absolute space and absolute time to explain all effects on objects interacting with them and passing through their respective multiple event horizons. The falling objects move slower internally but not externally allowing them to pass quickly through the space around the SM mass and allows the observer to correctly see them move freely to the black hole and not freeze on the surface of the event horizon. This also obliterates all paradoxes, flaws and impossibilities concerning black holes as created by the theory of relativity and the theory of the holographic principle. Information is and is not destroyed from the universe depending again on your point of view of the universe.

Chapter 29
Stars, Fusion, Neutron star and Black Hole Formation

Star Generations and Stellar Fusion

The story of stars starts way back just after the cooling of the universe from the Big Bang and has continued from there until this very day. To begin with you have the Big Bang which in universe terms is like a giant supernova designed to not explode a star but explode a vast mass of SM. The precise nature and reason for this explosion is no longer a mystery as has been covered in a previous section. The long held belief that a central singularity existed from nothingness and with properties completely unknown to modern science which then exploded for no adequately explored reason is gone. One concept that is also gone with it is the randomness behind its explosion that for decades was simply accepted by modern science and yet fully unknown again. Think for a moment about the idea of a singularity exploding for no precise reason when the singularity had been allowed to exist for any desired length of time from a split second to trillions of ages of our current universe. The reason this was thought to be real is that time inside of and before the singularity exploded is said to not exist. Now look at the actual explosion and see that modern science has no reason for it actually exploding save for the idea of quantum uncertainty and probability. What this means is that the unknowns in the mathematical probabilities of quantum mechanics simply states that something can and cannot be whenever it feels like it.

This is of course untrue as the universe does in fact know what each part of it is doing whether or not humans are their to miss the big picture or in this case the very, very small one. So simply saying that a quantum system might undergo a mathematically random fluctuation which is then transposed onto the very real physicality of the universe in terms of space and matter is lacking at best. Simply saying that something happened for some random reason is akin to saying something happened because modern science does not understand what is actually at work. A precise and known factor not must but does exist to explain why such an occurrence will always hit a 1:1 probability. The usage of ESS and macht coupled with existing speck and mote information of matter will allow for this TSM detonation event to be calculated perfectly for any number of universe BBB. Stars also contain a certain element of unknown factors according to modern science and specifically things detailing how they form, live and end their fusion phases of life are just a few.

A precise mechanism for understanding why stars become neutron stars or black holes and the factors that lead to each is no longer a random quantum event but one that can be explained from start to finish. In order to get to that point I will start by explaining the evolution of stars after the Big Bang and after the era of recombination specifically. The very first factor that contributed to star growth as I have explained in a previous section is the unpacking of ESS through TSM detonation events that leads to spatial rebound and the variances in each ESS and overall ESS distribution in the universe. The maximum degree of interaction of clintrons with ESS in the SM mass prior to TSM detonation events will lead to a perfectly non-uniform micro-scale universe which is uniform at the macro-scale. These micro-scale differences are all that is needed to begin stellar materials becoming attracted to one another following the Big Bang and will begin even before the era of recombination. This is a perfectly easy to understand phase of star formation that will become exaggerated after the era of recombination where atoms can now form.

Atoms lead to a greatly accelerated attraction of matter to other matter as now the electromagnetic force can begin acting unhindered and in full. The early universe is of course filled with basically hydrogen and not much else save for some helium and a smattering of other light elements. So the first stars were made of basically only one thing which was hydrogen and as such the evolution of stars must be based off of this singular fact. Therefore the first stars were basically hydrogen and gravity and nothing else. The very core of the first stars is key in understanding how stars evolve as they were at their perfect centers nothing more that hydrogen slush being compressed to the point of fusion and there were no other elements present to any significant degree. Certainly there were no elements of any real weight when compared to modern stars which can have much more in them than just hydrogen. What this means is that the first stars would fuse various elements up until iron and then stop at which point the star would either slowly fade in luminosity as fusion stopped but the star did not go supernova or the star would go supernova. Most of the first generation of stars were of course huge as the universe was very dense and material was everywhere to feed the voracious accretion discs that formed the early stellar masses.

The end result is a small number of stars not going nova but most of them possessing sufficient size to trigger supernova events. The difference is that a hydrogen born star only possess iron at its core and nothing heavier and so undergoes a different kind of supernova which will produce almost always a neutron star and nothing more. This can be explained by showing the precise way that a star going supernova makes a black hole. By removing the uncertainty of this unsolved problem the reason why stars go supernova, what drives the explosion and how things like black holes can form is revealed. Currently modern science does not understand three things about supernovas with the first being why they happen, the second being how they work and produce things like black holes and the third is the lopsided nature of their final explosions. Each of these is explained by examining ESS macht and speck/mote interaction with the ESS to create what can be crudely though of as small Big Bangs. This is the reason for needing a precise explanation of the Big Bang mechanism at the very beginning of the universe and as I have explained only a TSM detonation event can destroy SM and it turns out that only a TSM event can create SM as well. To understand and appreciate this fully we will look at a star that comes after the first generation stars as it is a well known fact that supernova nucleosynthesis can and will create many elements heavier than iron as the star explodes.

The early universe and first generation stars contained essentially only iron and down in terms of raw elements. You needed the first stars to go supernova in order to make elements heavier than iron and in larger quantities. The first generation stars were of course large and as a result they burned through their nuclear fuel very rapidly and could spread elements into the universe. Once you have heavy elements in the universe something interesting happens which allows for black holes to be formed. The first generation stars left behind neutron stars which are of course not nearly as dense as black holes as they are made of large neutrons instead of clintrons and are still ruled primarily by the nuclear force. The change that happens when you have heavy elements available in the universe concerns how stars burn and therefore how they end their burning phases. Let me ask you this simple question: where inside a star do you think fusion occurs? If you said the center of the star deep in its core then you are following the accepted explanation of modern science that will lead to a lack of understanding on how stars live. Now I will tell you where fusion really occurs inside a star of second or higher generation in the universe.

◄New Universe Order►

Fusion on all second and higher generation stars does not take place in the center of the star but on the surface of the stars core. Fusion is not taking place in the center of our Sun but rather on the surface of the Sun's core as the Sun contains a massive amount of rocky material deep inside in most likely equal to hundreds of Jupiter masses. This is not a debatable fact but rather a fact that must be completely accepted as the Sun formed from the same stellar gas and dust that everything else in the solar system formed from. The Earth is of course made of many elements much heavier than iron and whether or not some of those can be considered as trace elements is not important as it does not make those elements disappear. To this end the Sun is also made of gas and dust that formed the Earth and most likely a higher percentage of these heavier than iron elements. The reason is that the Sun as all stars that are in the universe easily hold the Lion's share of the gravity and sheer power in any star system.

The result is that the Sun will have first pick and the greatest effect on everything in the accretion disc including heavy elements. The Sun does possess a huge solid core that is far too dense for fusion to occur in but not on. In fact that fact that the Sun's spectra shows low amounts of everything but hydrogen is simply a product of the fact that the hydrogen will float on the ocean of elements that exist deeper in the Sun. You cannot see what lies beneath the sea of hydrogen and so you cannot detect it as it is inert in terms of fusion and energy emission so spectral analysis of the Sun's light will reveal nothing of the Sun's deepest core. The truth is that the Sun has a core of non-fusion elements and this core is quite indifferent to the goings on occurring just at its surface. The fusion action if you will occurs just on this surface and takes place precisely as understood by modern science starting at hydrogen and ending at iron and iron peak elements. At the end of the Sun's burning phase of its life fusion simply stops and the Sun's mass is not large enough for anything more dramatic to happen. The neutron class star is different however.

In neutron class stars we see a larger mass that will obviously contain more gravity than our Sun and therefore produce a different end to its burning phase. Stars that are neutron class in our universe today are of smaller size than pure hydrogen containing neutron class stars of the first generation. This is because they contained nothing but hydrogen at their formation as the available accretion material was only hydrogen and trace other light elements. Stars of our time that become neutron class are made from many other heavy elements and since they are naturally many solar masses in size have naturally much larger non-fusion cores at their centers. With an initial dense and heavy element rich core today's neutron stars possess sufficient gravity earlier to have the hydrogen seas around their cores begin fusing. The reason for the size difference in neutron class stars can be explained by how they form neutron star cores after their burning phases are finished. A neutron star is formed due to the usual and fully expected mechanism which is that fusion and the outward push of heat stops and so gravity can as is commonly said win in the war between the two. Before I explain why some stars become neutron class and some become black hole class I will explain how gravity works in large objects such as stars which will explain why gravity can overpower the other forces of the universe even though it is said to be the weakest of the four by many orders of magnitude.

The concept of variable gravity will be brought back here and proven further by not only explaining why stars burn in the first place but also how things like black holes are created and literally everything stellar in between. A star begins as a huge nebula of gas or gas and dust that begins to collapse on itself under the force of gravity. As the mass of gas and dust coalesces into a central stellar pre-fusion mass the temperature of the cloud increases along with mass and gravity. Fusion begins once a critical density and temperature has been reached and the star starts to shine. The key feature of the star igniting is that no more matter falls into it as the matter is made of gas and dust which are very light and will be pushed away by the undeniable force of the solar winds. Our own star the Sun for example will not passively absorb gas and dust anymore as it has long blown off the cloud from which it was born.

This is evident by looking at not just the composition of space surrounding the Sun but also at the four inner rocky planets. These planets exist in a zone around the star where temperatures are too high to allow light particles or elements to exist as they are energetic enough to be blown off the rocky cores and into the outer solar system. Once in the outer solar system temperatures have plummeted and the fierceness of the solar wind has become diffuse enough to no longer push significantly on loose materials as it did in the inner solar system. This is why the outer planets are all gas giants and the inner ones are all rocky. What you are therefore left with is a star that ignites and shortly afterwards stops consuming fuel meaning nothing new is adding matter and therefore mass to it. This limitation of matter is key to understanding variable gravity from a stellar perspective as the rules of modern science state that the instant fusion begins in a star it should automatically turn itself off. This can be proven rather straightforward from the standpoint of matter, gravity, fusion and energy conversion.

Take a known mass of hydrogen in the form of a pre-stellar mass that is collapsing onto itself. This matter is not cold before fusion begins but actually fairly hot as the materials involved are under pressure and begin to heat up as gravity starts to force them ever closer together. The equivalency formula of relativity states that the cloud is losing mass quite rapidly as cold matter is emitting heat and since according to the theory of relativity energy is mass this means the cloud is losing mass. The addition of new material falling into the cloud allows for gravity to continue growing by adding more matter and so eventually we have ignition of the fusion core. This is where relativity and modern science fail to explain stars completely as no additional matter is being added to the system. With no matter added to the system due to the stars furnace pushing away the rest of the cloud the star should immediately turn itself off.

The reason is simple as fusion creates heat and converts a small percentage of the matter of the star into energy in the form of photons which is lost to space. Since matter is energy according to modern science the mass of the star has now decreased and therefore the gravity of the star along with it. The star has also shed excess heat from the fusion reaction process and so the star has lost both heat and gravity. Heat and gravity are the two driving forces behind stellar fusion and with those two forces decreased below the minimum threshold value for stellar ignition which the star just only recently reached due to accumulating gas and dust fusion should shut off and the star should never reignite. This is because the mass is now insufficient to start fusion up again and the gas and dust have been blown away from the star so no new material can be added to once again increase mass to pass the minimum ignition threshold. Remember the fusion process does not want to occur and must pass a threshold quantum value akin to an action potential to work and so a very cut and dry difference exists between fusing and non-fusing parameters. The funny thing is stars do not go out until they have exhausted their supply of nuclear materials and so an explanation for this must be given. The explanation is of course variable gravity which allows for a set amount of matter to produce variable amounts of gravity under the correct spatial compression conditions of the local area space ESS.

If you are still not convinced that variable gravity must be at work inside stars then let me explain the long term life of a star during its active burning phase. This phase is characterized by two things the first is the consumption of fuel and the second is the increase of the size of the star. Our own Sun can be used as a perfect example as it continually burns hydrogen fuel every second of every day and has been doing so for about five billion years. The Sun burns tons of hydrogen and so decreases its overall mass even if this mass is miniscule in comparison to its overall mass it obviously adds up over five billion years. The Sun has not stopped its fusion reaction though, even though according to modern science it should have long ago. The Sun will continue to burn hydrogen for about another five billion years at which time it will not be fully depleted of hydrogen but will switch to a helium fusing path. So the Sun's total mass has decreased by untold quadrillions of tons since it was first born and yet has not stopped fusing hydrogen.

At the helium burning phase something increasingly odd happens as the Sun will now burn a fuel that takes much more gravity and heat to fuse and yet this is what will happen as the Sun transitions to a red giant. So if the Sun has been losing mass which it needed to just reach the hydrogen fusion threshold ten billion years earlier by creating sufficient gravity to cause itself to ignite then how can it burn a heavier fuel with ten billion years less worth of matter producing gravity? Again variable gravity must be involved as normal gravity would have easily decreased to levels where the fusion reaction would have shut off. Plus the Sun will now become a red giant which means it increases greatly in size to the point where it will reach the orbit of Earth and envelop our planet. An increase in size means a decrease in density of the star as more matter and therefore more mass is pushed away from the very center of the star. According to relativity gravity is due to a curvature of spacetime with more matter in a smaller volume giving you more gravity.

Well the Sun is decreasing the amount of mass it collects near its core due to it not only converting matter to energy but also in spreading the remaining matter outwards as a red giant. This means that the star has lost the ability to create enough theory of relativity gravity in a small enough space to allow fusion of helium to occur. Again we see that stars should have long ago turned themselves off as the amount of mass and time involved is simply too large given ten billion years of fusion for modern science to account for. These are not insignificant values either but rather they are hugely profound as literally trillions upon trillions of tonnes of mass have been lost and the stars outer diameter has increased by hundreds of millions of kilometers and so no balancing or fine tuning of general relativity is going to work here.

Variable gravity however is explained by ESS and ESS macht inside matter such that matter can impart more of its energy to the force of gravity. If you remember gravity from the earlier sections is a force made of two factors the first being the number of ESS inside a matter sphere and the second is the amount of macht glow that is transferred between ESS schales. Using the two factors of gravity you can not only explain gravity but can allow stars to lose mass while still increasing their gravity over billions of years. The number of ESS and the strength of ESS macht are both increased as the star collapses such that even though stellar fusion has begun the matter inside it is now encountering more ESS than it was in the pre-stellar fusion mass that it formed from and this will allow it to impart more of its energy to ESS macht gravity. The increased levels of ESS macht glow gravity mean that the star will have an overrepresented force of gravity which exceeds the threshold limit for fusion. As material is consumed by the star it is converted to heavier elements which will occupy larger matter spheres and therefore will continue to encounter more ESS as they exist inside the star in ionized forms. A star will always gain gravity as it loses mass using ESS and macht as the natural tendency for matter is to encounter more ESS and therefore create variable gravity and this is something that cannot be done using relativity as the macht component is needed as well to make gravity work.

In addition to this the ESS inside the star will follow a positive feedback loop as the matter initially spatially compresses them from the pre-stellar cloud into a more dense fusion arrangement. As the star creates more heavy elements above hydrogen the matter spheres they possess encounter more ESS elevating gravity. The ESS share macht with all schales around them and this as we have seen causes the volume of ESS to decrease meaning the overall spatial compression of ESS inside the star increases. More volume decrease in ESS means more ESS per volume of matter sphere which again will repeat the positive feedback loop so that gravity necessary to create fusion continually increases. The gravity at the very center of the star will always go up and allow for the fusing of heavier elements with the resistance of the other three forces preventing matter from following the feedback loop such that gravity increases unchecked. So in the end despite the star losing overall mass and increasing in size which should turn fusion off variable gravity allows for the star to actually keep burning and burn heavier elements still beyond hydrogen. The combination of these two factors also explains how the final gravity of large stellar objects ten or more times the mass of our Sun will increase their gravities such that at the end of their fusion sequences the inward pull of gravity is almost ridiculously strong when it was never capable of that before.

Now that the mechanism by which gravity works inside stars is understood the exact way in which neutron stars and black holes are produced can be explained. I will start with neutron stars as they were the first to appear in our universe following the Big Bang and involve the first in a series of events that will eventually lead to black hole formation. To begin with I must restate the fact that gravity will increase inside a stars very core as time passes. This might seem strange as normally the total mass of a star decreasing means that the total gravity of the star does as well and from an Earth based standpoint this might seem so. The reason is that the increased gravity is only occurring at the exact center of a star because the star is a very loose arrangement of atoms when compared to something like a black hole.

What this means is that being able to perceive variable gravity inside a normal star from observations made around it at say an Earth based orbit will be difficult to impossible with today's technology. The amount of spatial compression inside a star is phenomenal when compared to the spatial compression of the local area of space around Earth but this stellar spatial compression is itself much weaker than that of neutron stars or black holes. To this end the center of the star contains the necessary spatially compressed ESS that can lead to variable gravity but these effects in a normal star will be felt roughly around the core. Remember that gravity is a macht like any other that is shared most strongly between ESS that are inside matter spheres so the actual amount of ESS sharing between schales inside the core is higher than outside the core.

This is because the core is so dense that the ESS inside it experience not only macht sharing but are also actively receiving energy donated from the matter directly which keeps the ESS macht glow strong. Outside the core and of course extending towards the photosphere the density of the star decreases rapidly such that the ESS there will not be as spatially compressed and will not be receiving as much donated energy either. The result is far greater macht glow gravity being felt inside and right near the core. This is the zone of the star that you will find the most pronounced variable gravity which can explain the fusion of heavier elements but also explains why detecting this increased gravity is so difficult from an orbital observer's standpoint. In a neutron star however the effects of variable gravity are much more evident and can be observed from an orbital standpoint. One reason for this is the outward push and masking effects from fusion, heat and energy is shut off so the inner workings can be seen properly.

Neutron Star and Black Hole Formation

A neutron star must first be formed and this is the result of variable gravity and after this process is complete the effects of variable gravity can for the first time be physically seen in the universe and without the aid of any instruments but instead with the naked eye. I will begin by explaining how neutron stars form in the first place from large stars many solar masses larger than our Sun. A neutron class star begins its life in the same manner as any other star and will burn its nuclear fuels until it reaches iron where fusion stops. At this point the star will collapse into a neutron star and the mechanism for this is fairly straight forward as its total mass is so high that the inward force of gravity can exceed the electron degeneracy pressure and creates neutrons through the electron capture process. The key factors here are the increased mass of the star which helps it to collapse and the arrangement of specks and motes inside the stars matter.

The first is obvious as modern science has known this for some time and that simply is a star whose mass is large enough will collapse inwards rapidly creating sufficient pressure to create neutrons. What modern science has never been able to explain is why this happens in the first place. Why is gravity so strong and how does it work on a quantum step by step process to collapse a star? The second thing modern science fails to explain here is what fundamental particles are and what they are doing such that an electron and proton can spontaneously become a neutron. Modern science understands that an electron and proton combine to form a neutron and have mapped these energies quite well but does not understand how a fundamental particle can join into a composite three hadron system to form a neutron. To answer these let us look at what gravity is inside a star once more.

Gravity is a combination of ESS and macht that affect matter such that it is attracted to space. Inside a neutron star or any star at the end of its fusion life the outward pressure from heat and escaping particles is halted. This allows the ESS inside the star to begin balancing their macht glows as they have wanted to do for many millions or billions of years since the star first ignited. The ESS share macht between schales far faster than light speed and between space and matter at light or below light speed. According to modern science neutron stars and black holes should never form as the force of gravity is by far the weakest of the four fundamental forces and so cannot overpower the other three. At the end of a stars fusion life cycle the star should simply dissolve with a small remnant core being left behind much like will happen to our Sun. Neutron stars and black holes do exist and the mechanism by which they are created is one of the unsolved problems of the universe that modern science cannot answer using any of the three main views. Space physics gives the explanation using nothing more than ESS and speck/mote fundamental particles.

In a collapsing star such as our Sun the ESS macht is not strong enough to exceed the electron degeneracy pressure and create a neutron star or black hole. The ESS are trying to balance their macht glows once fusion has stopped but do not need to rearrange the specks and motes of the matter that the star is made of to do so. A neutron class star is different in that the ESS macht is definitely strong enough to overpower the three remaining forces in an attempt to balance macht glows and so will easily crush the matter it is made of into a neutron core. This means that the way fundamental particles interact to allow an electron to fuse with a proton and become a new particle the neutron can be understood. The specks and motes of the electron and proton are as we have seen very similar and under sufficient levels of elevated spatial compression will allow them to rearranged their geometries such that once the specks and motes come into physical contact with each other certain edges are unable to interact with the ESS of the local area of space in the way they could before. If you remember the analogy of edge dragging you can visualize this fact easily as some of the edges are now facing inwards in the speck/mote geometry and are therefore unavailable to drag through space which would result in ESS macht glows.

So a neutron star will collapse simply because it has enough mass to over represent the ESS macht glows inside it sufficiently such that the ESS will have enough strength to over power the three other forces as they balance their macht glows. The ESS collapse quickly and simply pulls the matter which is bound to it along for the ride so to speak. The matter is in a sense stuck to the space which it envelops and so as the space crushes inwards the matter has no choice but to follow along for the ride and change its speck/mote geometry. Remember this is possible because the ESS are smaller than the matter that inhabit them and certainly many times smaller than the orbital shells of any atoms inside it and even of the spheres of matter made by three quarks into hadrons. What this means is that the natural repulsive force of the nucleus of an atom can no longer keep out the electrons which are trying to balance their ESS macht glow electromagnetic any longer. This electron and nucleus mechanism was explained previously and will now be overpowered as the much smaller total ESS and the vastly overrepresented force of gravity now seek to balance out. The ESS collapse inwards quickly to balance the macht glow gravity and this is now more powerful than the macht glow electromagnetic which has not changed.

Remember once again the variable gravity is increasing the strength of the gravitational force of the stars core but nothing is making the specks or motes increase the macht glow electromagnetic of the star and so this explains how a star can use the once weak force of gravity to overpower the normally much stronger power of electromagnetism. Neutron stars are easy to understand once you use ESS, macht and speck/mote geometries properly and can show how a normal star can transform into a neutron star without any missing steps or information. The mechanism by which neutron stars are formed is solved and does away with the unknowns of modern science that simply say the star was so heavy it crushed itself without explaining step by step how this happens at the smallest fundamental levels for both space and matter.

Now I can explain how this further proves that variable gravity is correct. The first additional piece of proof can be seen in the fact that the star collapses and is able to overpower the electromagnetic force which normally it should not be able to do. The second piece of additional proof comes from the fact that the neutron star is stable and does not dissolve. Modern science knows that a neutron will decay in approximately eleven minutes but does not know why this happens. Furthermore modern science cannot explain why a neutron on Earth will decay in eleven minutes but a star made of nothing but neutrons can persist for trillions of years. According to everything known by modern science eleven minutes after a neutron star is born it should go poof especially because gravity is not variable to modern science.

Think about this for a minute and you see that modern science says that a neutron star exists because of a rapid core collapse event and that the gravity of the star was strong enough to do this. This makes sense from the standpoint of a given mass in a set volume of space during the collapse event only as it allows for the electrons to move fast enough to be captured by the protons. After the collapse event however this should mean that the fixed gravity believed to exist by modern science no longer is pulling matter inwards and so the momentum responsible for the electron capture process is now lost. With the star now static in space the remaining three fundamental forces which modern science also does not believe can be turned on and off at will are still present and functioning. Since a neutron star does not according to relativity possess sufficient curvature of spacetime to allow for nothing to escape then gravity cannot offer an explanation for the continual existence of a neutron star.

I would like to state here that I have no problem with modern science saying that neutron stars or black holes exist only that modern science does not understand how they form, how they work and are making mistakes about the actual properties of such objects. To this end a neutron star should be able to decay in eleven minutes as the neutrons still posses the remaining three fundamental forces that allow for decay processes to take place and since the gravity of the star according to modern science has not increased given a set mass the decay particles are in possession of the required escape velocity necessary to leave the neutron star for open space.

The remaining fundamental forces are still in possession of enough strength to decay the neutron and so the surface of the neutron star should be decaying in continuous layers as the surface gravity is lower than that at the core. The star should decay rapidly with an increasing rate due to the fact that the loss of surface neutrons which is always possible due to a low enough escape velocity for the body of the object results in a total decrease of the mass of the star which cyclically will reduce the escape velocity further. According to modern science and the three main views neutron stars should dissolve very, very rapidly until the threshold is reached where the gravity falls exponentially and the pent up energy in the star is allowed to relax creating an explosion as the three remaining forces once again over power gravity.

Neutron stars however do not dissolve and we have proof of this in the form of pulsars which emit radio bursts at regular intervals. The fact that these pulsars exist is not just proof that neutron stars exist but solid proof that variable gravity is real and is constructed of ESS, macht and speck/mote geometries. These pulsars have been observed for decades now and in that time they due two things the first of which is persist and the second is they maintain their pulses at clockwork intervals worthy of any Swiss watchmaker. The fact that they persist means of course that they have not decayed and the fact that the periodicity of their radio signatures remains solid means they are not losing mass. Since they are not losing mass they are not decaying once again and so possess a mechanism that allows for the neutrons that comprise their cores to avoid the natural eleven minute decay process.

This mechanism is of course highly compressed ESS which are balancing their macht such that the speck/mote geometries for the neutrons is not allowed to revert to the previous collapse arrangements. The increased levels for spatial compression are simply not available on Earth and even inside particle colliders the pressures are only momentary and do not remain long enough to allow for effects due to increased ESS numbers inside matter spheres to be measured. This is the reason why mini black holes created theoretically in particle colliders are no danger to the Earth as the continual high levels of ESS spatial compression do not exist. That explains why the mini black holes created go poof so fast because the ESS quickly balances their macht ending the black hole.

The overrepresented force of gravity in the ESS macht glow gravity of a neutron stars core allows for persistent ESS spatial compression which means that the effects of variable gravity can be observed directly. In addition to this you must recall that a speck/mote has a set amount of energy available to it and that in the space and matter system in which you calculate a single fundamental exchange no energy is lost as this would violate the conservation of energy. So we see here that every single speck/mote of fundamental matter must be forced to conserve its energy within the local space and matter system. Therefore the energy wholly or partly in a neutron star that was held by the macht glow electromagnetic of the electron and proton can be calculated into the total gravity macht glow pool for the local area of space. I say wholly or partly because neutron stars can still emit pulsars so not all the energy is lost and certain stars are classed as magnetars which mean they have strong magnetic properties as well. Nevertheless it is known that the electromagnetic force is really made of two separate forces and neutron stars given their unique arrangement of speck/mote geometries can and do create regularly the necessary conditions to cause them to dissociate once again.

The degree to which the electromagnetic force is dissociated is unique like a fingerprint for each neutron star and allows for a wide variety of such objects to exist. In a magnetar for example the magnetic moments of the specks and motes inside that particular stars neutrons are arranged such that they highly power up the ESS macht glow magnetic. This is fully permissible since only the majority of the electric charge force need be turned off in the electron capture process. As for the reason behind a spectrum of neutron star types one only needs to factor in the composition of the star before collapse, its mass, its particular arrangement of ESS and any quantum entwinement events inside it at the time of collapse as well as stellar magnetic field flow which I will explain later. If you were able with today's technology to physically take a picture of a neutron star from a safe orbit you would be looking at the first physical visual proof of variable gravity as an object that should normally decay lasts and lasts right before your eyes. Thus the mechanisms by which neutron class stars create pure neutron star cores and how this process works step by step at the most fundamental levels is laid out along with the reason that the object does not decay. Further it once again demonstrates that variable gravity must exist and that the three remaining forces are overpowered by it despite the star having lost mass from its first ignition and certainly through its supernova phase.

To reinforce this proof of variable gravity and the inner workings of large stellar masses before and after fusion termination sequences I will now explain another unsolved problem facing modern science and this is the creation mechanism for black holes. At present modern science has no explanation for the creation of black holes as gravity alone is seen as insufficient to crush a stars core into a black hole. The simple explanation of gravity winning the war between the inward pull of it on matter and the outward push created by fusion products and the resultant rapid collapse of stellar materials which leads to a black hole does not offer enough power to crush the plasma into an area seen as infinite gravity in zero dimensions and with zero time. This of course brings up the realization that modern science does not know what a black hole really is in the first place save for the remnants of a supernova of a star many solar masses larger than our own Sun. The reason for this is the singularity which modern science does at least know to be impossible and a mistake created by the incorrect theories of relativity. Attempts to reconcile this idea of a singularity from a completely gravitationally collapsed object and the step by step workings of quantum mechanics has led no where for modern science as it simply leads to more infinities in the linear operators of relativity infused quantum equations.

Black holes however do exist as they are the only possible expected objects from large star supernova and are responsible for a number of incredibly large gravitational effects in the Milky Way. The giant living in the center of our own galaxy is proof enough from the orbital velocities of stars close to it and yet it itself is quite invisible to current technology. So the existence of black holes is not in question only what they are, how they are made and how they work is an unsolved problem for modern science. I have previously explained some aspects of black holes such as their general formation and the particles they are made from and exactly why these arrangements of speck/mote geometries and ESS spatial compression are needed to reconcile the very small world of fundamental systems with the very, very large world of well the rest of the entire universe. There is still more I need to explain about black holes but that will wait for a later section here I will deal with the unique and essential mechanism by which they form from normal stellar materials of an ordinary star.

To begin with we will first remember that a star of enormous size is needed to create a black hole as it must be much larger than a neutron class star in its pre-nova phase. The idea of supernovas themselves needs to be touched on here as yet another source of incredible mass loss by the star as it transitions from fusion to collapse phase of its life. Again it is necessary to state that a massive percentage and in some cases 90 percent of the star's raw mass is lost in a nova event. What this means is that once again the idea of normal gravity having the strength to hold a star in neutron or black hole phase is quite impossible unless variable gravity is used. For black holes however the supernova is not just key but is mandatory for its formation as you cannot create a black hole without one no matter how small it might be both the supernova and the black hole itself. I will now tell you the secret of creating black holes out of normal stellar materials not from a matter perspective so much as I have done so before with the creation mechanism for clintron particles but this time from a space perspective using ESS. I once told you a simple fact which was that you needed a TSM detonation event to destroy SM as you can only destroy SM from the inside out and now I will tell you another fact which is that you can only create SM with a TSM event as well. The required force that can crush matter into a configuration that will overpower the three fundamental forces other than gravity is achieving a velocity of spatial compression faster than light and reaching at least macht speed.

If you recall these events from a previous section then you will remember the creation of the universe in a BBB where matter was decoupled from ESS and as a result the ESS were able to spatially rebound faster than light speed. This rebound caused the ESS to push on one another as the universe spread out to its current state during a period of rapid expansion. A black hole's first moments of formation occur when the star coalesces into a pre-fusion mass as this is the moment when a star will accumulate sufficient matter to produce the future variable gravity necessary to create a black hole proper. This is also a star that must be formed from a second or later generation star as black holes will not be formed from pure hydrogen or light element stars alone; perhaps under extreme conditions of the early moments after the Big Bang matter was dense enough because the universe was small enough a kind of baby black hole could be born but not the type we see some 13.8 billions years later. The mention of heavier supernova nucleosynthesis elements that have settled to the core of the star is now important here. Despite the fact that the amount of stellar material at the core of the star is a trace amount when compared to the rest of the stars total mass it will be the region from which the black hole is created when the star collapses following its fusion life cycle. This amount of heavy material is what sets large pre-fusion stellar clouds apart and dictates whether or not you can make a neutron star or black hole when the fusion phase is completed. All first generation stars of pure hydrogen masses or other light elements will create neutron stars as they are missing this heavy core component. This is partly due to the very diffuse clouds of hydrogen which are energetic enough to remain spread out and dilute until fusion begins whereby the rest of the gas is pushed away from the star by the solar wind. A black hole class stellar cloud mass will be laden with heavier elements meaning they are less energetic as it obviously takes more energy to move them and so gravity can draw a greater amount of materials in before fusion ignites.

The black hole cloud therefore is larger to begin with making the star simply more massive which is still an essential factor in black hole formation. As the star burns during its fusion phase of its life the heavy and dense core becomes important as it is a site of non-fusion and is essentially static inside the star. The star can have a rotating core that is not at all stationary but it remains overall static in terms of the fusion events taking place above it. The star will only add more and more heavy elemental materials to this internal core as it moves up the fusion chain towards iron. Once fusion stops we can now proceed to the second step in the formation of a black hole which is the collapse phase. The speed at which ESS spatially compress is of course rapid as it was in the neutron star core as well. It will still reach greater velocities however as the star was far more massive in the first place and so the ESS are under a tremendous pressure to balance their macht glows inside the entire star and especially the core.

The ESS as you remember are moving faster than the matter can interact with them and so you have a moment when the ESS essentially uncouple from the matter enveloping them. The inward velocity of the ESS reaches a point where you will momentarily create the necessary ESS macht overload for a TSM event and in any TSM detonation event you find matter completely uncoupled from ESS leaving the ESS to once again spatially rebound at maximum speed. The amount of SM involved in a star is miniscule compared to a single ultra massive SM mass needed to trigger a proper BBB and so you do not have the effect become total to the point of destroying the star entirely. The uncoupling of ESS from matter allows for movement faster than light and represents the necessary TSM like conditions although as stated before a total TSM detonation is not necessarily needed. Some massive stars that go supernova do not leave a neutron star or black hole behind and it might be found that these stars were massive enough or had unusually high levels of heavy elements in their cores such that they did trigger very, very small TSM like events that blow the star to pieces and explains why some supernovae leave nothing behind. The trouble for observing such events is that if a small TSM event did occur the clintrons momentarily formed would during the TSM detonation event itself fully dissociate leaving not much of a spectral fingerprint to examine and perhaps supernovae devoid of large traces of heavy elements could be just as conspicuous as those that had too many.

Back to the star which does not destroy itself, in fact the opposite happens as the core which is super dense compared to the rest of the compressing stellar plasma which does move inward for a time before being uncoupled from the ESS is in fact turned into SM in the process. Remember that the core is not made of SM yet only densely packed matter that is enveloping highly spatially compressed ESS and contains greatly elevated macht glow gravity as well. The core which is made partly of elements heavier than iron and here static compared to the rest of the star is where the SM formation due to ESS spatial compression takes place. The heavy core is the key to creating a black hole as it represents a static and non-collapsing area inside the star where the spatially compressing ESS will create a TSM schale. Yes an entire very thin schale of TSM detonation will occur at the boundary layer of collapsing super heated stellar plasma and the static heavy element core. The TSM schale will detonate and follow normal but very short lived BBB detonation rules on the surface of the core. The spatially rebounding ESS will push equally in two directions one is outwards into space where the forces and ESS macht overloads can be balanced as the system wants them to be. The other pushing force is where? Inward of course against the super dense heavy element core where the internal ESS of the core and speck/mote geometries of the star have literally no where to go and so are made to feel the full brunt of the TSM schale detonation force. The inwards pressure which is nearly perfectly spherical in nature will uniformly crush the stars core into a very small and very real SM mass. The volume is small and the outward blast relieves the TSM strain from the inner core stopping the TSM chain from exploding the star completely.

It must be noted here that each star is unique given its total mass, the temperature at which its core burned and the heavy element composition of its core and so there is considerable overlap in what kind of supernovae and what remnant from that supernova you get. In other words a very small black hole has finally been created using step by step mechanics that account for the large scale effects of space and matter from an object as large as a massive star to the single fundamental quantum interactions of a single speck/mote or single ESS. The SM mass is of course constructed of clintrons and held in place following all of the already explained rules for black holes and ESS macht balancing. The inward ESS pressure wave created by rebounding space is what is needed to make a black hole from normal stellar material and gravity even though that star has been losing mass since it ignited. The heavy element dense core allows for a surface that the rapidly collapsing materials can accumulate on faster than they could normally as in a first generation star and therefore the TSM schale it itself creates is responsible for crushing the heavy element core into an SM mass.

As for those gamma ray bursts coming from supernovae the accumulated and hyper-compressed stellar plasma materials are consumed by the black hole core. The exact mechanism for all gamma ray bursts and relativistic jets will be explained later but for now this layer of collapsing stellar plasma underneath the inward detonating TSM is what the black hole eats to produce the very short gamma ray burst. The much longer lasting relativistic jets of things like active galactic nuclei will explain how these events occur and in more detail in a future section so please be patient. Here you can plainly see how creating a black hole is easy when using space physics and how it actually takes a TSM detonation event to create a SM mass where at the end of a universe life cycle it then takes a TSM detonation event to destroy a SM mass from the inside. The fact that the same mechanism can be used in two different ways to create or destroy SM masses is a proof that the use of ESS, macht and speck/mote geometries is once again correct in explaining the universe and answering the unsolved problem of how black holes form.

An interesting side note here is the use of heavy element cores to potentially explain magnetars and other rare objects in the cosmos. To begin with the core of a star destined to become a magnetar is made as we know now of heavy elements that are equal to or heavier than iron. The temperature of the stars core is obviously very hot and therefore has ionized partially or fully the materials inside that heavy core. What this means is that during the core collapse phase of the supernova event the electron capture process inside the heavy element core might occur only partially or not at all with any number of degrees possible in between. The end result will be a core inside a neutron star crust that is not actually made of neutrons and still exists as normal atomic nuclei of different types and in different layers from the heart of the core outwards to the neutron star crust. This allows for any number of interesting events to take place based on the properties of the core. The magnetic dipole moments of the core and the neutron star crust are obviously of importance but there are other factors as well. The number and type of layers in the core, whether they are rotating or not and the overall size of the core period as a potential power source of magnetic fields or high energy emissions from the star. The fact that the core remains intact with both protons and neutrons inside of it at high temperatures and under high rates of rotation means that a very large and powerful electromagnetic field can still be generated from this object as though it were a normal star only with a super powered magnetic field.

The type of star necessary for this outcome to occur might be always a small star that is just capable of going supernova but without excessive mass to force a complete crush and collapse of the core materials. It might be a star of specific elemental composition such that the free electrons are not able to easily permeate into the depths of the stars interior and donate sufficient free electrons from the surrounding stellar plasma to cause the formation of only neutrons or enough neutrons to turn the magnetar event off. Also it might be that the stars core after the collapse event regardless of composition is simply not rotating or rotating fast enough to cause the generation of an internal magnetic field from the star after the core collapse event is over. Of course it can also be a combination of these factors and the reality that this mechanism happens due to the incredible variability of stars in terms of composition and mass in the universe requires direct observation of supernova events and the resultant objects left afterwards to work out which details are most pertinent here. The fact that a stars interior can escape the electron capture process and remain intact after a core collapse event is fascinating and no doubt contributes to some of the seemingly exotic objects in the universe such as magnetars.

Supernova

The formation of both neutron stars and black holes comes from a supernova event and these events themselves are also not fully understood by modern science and represent another of the unsolved problems faced by modern science in understanding the universe. The problem can essentially be broken down into two general areas the first is the explosion mechanism and the second is the nature of lopsided explosions. The first part of the problem deals with not only the way in which explosions occur but also in generating enough energy to blow off up to 90 percent of a stars stellar mass in a single near instantaneous event. Neutrino masses were a partial solution to this problem in that they helped to explain some of the missing energy needed to keep stars from collapsing under the force of gravity after fusion was ceased inside the star. The neutrinos do not however provide enough push to actually blow off the outer 90 percent of the stars plasma mass.

The reason for this is that even when all three neutrino flavors are factored into these explosion models they simply do not interact sufficiently with the stellar plasma enough to push it outwards violently at near light speed explosions as seen in supernovae. So what can push matter at near light speed away from a massive center of gravity? More accurately how can anything by pass the very gravity that is causing the stellar plasma to collapse at near light speed towards the core? In other words modern science cannot explain how gravity can pull in at near light speed on the matter but the matter can also explode outwards at near light speed as they should cancel out and the star does not all that much really. In order to explain the supernova mechanism properly such that gravity can pull material inwards at near light speed but 90 percent of the stars material can be blown off at near light speed you can only use one thing and one thing alone which is ESS and ESS macht balancing with speck/mote geometries inside the matter of the stellar plasma.

Using the black hole formation explanation from a moment ago we see that 90 percent of the stars mass lie above the heavy element static core at the heart of the star. This explains how only a small fraction of the stars total mass remains after the nova to form a black hole and large supernova nebulae can be formed while not destroying the star completely. The single fact that proves that ESS are responsible for the formation of black holes and that they must be made to experience a TSM detonation event comes from the balancing of the gravity. You must recall here that the gravity of the star is pulling matter in at near light speed and yet the explosion seeks to blow the material off into the local area of space at near light speed which should normally result in the material going nowhere.

That is unless you use a TSM schale detonation which creates a very thin and short lived boundary layer of ESS decoupled from all matter speck/mote geometries. If your remember from the original BBB section I wrote earlier during a TSM event the ESS are allowed to spatially rebound at maximum velocity in all directions. The reason this happens is that in any ESS that are uncoupled from matter you no longer have macht glow gravity being turned on and what that essentially means is that gravity can be thought of as temporarily off. Not reversed or possessing any strange negative energy, forces, fields or other such needlessly extraneous workings of the universe just that it is not currently on. This concept makes perfect sense as I have already explained the fact that speck/mote geometries can be arranged such that certain fundamental forces are not used by them.

If you want solid proof of that then simply think of the photons that are entering your eyes as you read this as they are all speck/mote geometries that interact with every ESS in the universe but not all macht glows of those ESS. A photon possesses the forces of electromagnetism but not mass and so you have direct proof right there that not all ESS macht glows need be used always and all the time. Decoupling speck/mote geometries from the ESS macht glow gravity and exploiting ESS spatial rebound... a step by step explanation of hyperspace anyone? The universe regularly travels using hyperspace faster than light in parts all the time and they are all made possible by decoupling matter from the macht glow gravity of ESS. For that matter every part of your body exists partially at all times in hyperspace and travels faster than light too. This is because all matter that makes you up is made of specks and motes that interact by donating part of their energy through edge dragging with the local ESS that you are currently in. The donated energy is made by ESS into macht which has been proven to make up the macht glows that become particle volumes as thought of by modern science. Those same macht glows are a part of you and they all move always at macht speed which is almost infinitely faster than light and move at hyperspace speeds and exist only in hyperspace.

Supernova likewise utilize macht speed velocities regularly in order to create regions of TSM schales where the gravity is as I said best thought of as turned off temporarily. This boundary layer means that the ESS and macht glow gravity inside the core will not affect the ESS and macht glow gravity outside the boundary layer. This non core sided area of space is of course on the side of the boundary layer that is facing into open space and is a place of course where the ESS can balance its macht glows to reduce the strain on the system. The matter associated with the explosion will also be able to balance itself out in this open space facing side of the boundary layer as the ESS macht glows inside it can relax from the overrepresented state it is currently in and out into the low density space is where it can do this. The result is a mechanism for an explosion that can travel at near light speed away from the core which not just should be but actually is trying to pull everything in at near light speed. The explosion is further accelerated by the ESS rebound from the TSM schale detonation as well so more and I mean far more than enough strength is available to the supernova to blow off 90 percent of its mass at near light speed into space despite the black hole that is forming inside its core.

Nothing in the universe is more powerful than ESS macht balancing as it drives everything from why a star burns to why we stand on the surface of our planet and everything from why life can chemically replicate to why electrical impulses fire in our brains. In the form of a supernova the ESS macht stored strength will be the master of the system and partially push the stellar plasma from the stars TSM schale detonation boundary but also partially drag the matter that envelopes the ESS along for the ride. The speed of the explosion outside the boundary layer is light speed or below so we see how the much slower than macht speed quantum exchange of energy from matter to ESS is still of course perfectly conserved and the matter is allowed to exist perfectly normally as we observe it from outside the TSM schale boundary. The turning off of the TSM schale and the recoupling of matter takes some time and so explains why parts of the supernova happen first and multiple shockwaves can catch up to materials blown off first. Everything about the expanding supernova nebulae is normal and will produce both the well understood normal matter as spread through the interstellar medium as well as the now no longer mysterious objects that are born in the cores of the stars. So you have a simple space physics explanation step by step of how a supernova occurs from the point where fusion stops in the star right through to the necessary forces to blow off the stars outer layers all while explaining how the forces of gravity are balanced inside and the three remaining forces are overpowered.

This brings me to the second unsolved problem of supernovae facing modern science which is the lopsided nature of their explosions. Again we see that the current models of supernovae explosions by modern science cannot account for the irregular ways in which supernovae explosions play out. The main difficulty faced here is in explaining how the explosions occur near instantly and yet do not produce a neat spherical blast wave spreading outwards from the stars core. The reason a normal matter only approach to explaining this problem fails is that the traditional mechanism by which supernovae occur is through regular matter only explosions. In other words the force and shape of the explosion must come from using matter and only matter to produce the shapes we see in numerous supernova remnants through the release of stored nuclear energy. If matter were the only contributing factor here then all explosions should actually be seen as perfectly spherical.

The reason for this is that the matter inside a star is essentially perfectly uniform as the only construction of stars interior comes from single atoms either regular or ionized pressed up against each other in atomic layers. A star is basically a big ball of fluid in space and since gravity is pulling equally on a fluid it will pull it all into a nice neat and ordered sphere. The atoms are free to slide against one another and heavier ones to fall through the lighter ones so you have a nearly perfect sphere arranged in smooth layers of atoms from heaviest at the center to lightest at the top. The star also has far more gravity and pressure available to it than most other objects in the universe to force these uniform and smooth layers to form. The star also has millions or billions of years to make sure that every single atom is neatly tucked away where it should go. The result is that at the end of the fusion life cycle of the star the star is already super-uniform and the collapse process only makes this more so as the atoms being pulled into the center are now fully obeying gravity which seeks to only do one thing. What gravity is doing in the collapse phase is the strongest arranging of the atoms into a uniform set of layers as possible. The motion towards the center of the star means that the law of conservation of energy states that the atoms can spend as little time and effort moving side to side or in any almost random way they want and must therefore have most of their energy spent in moving towards the center of the star where gravity is organizing them into a theoretically perfect sphere. If this were of course the only way that stars existed and space was not a contributing factor then we should observe nothing but the reverse as the stars explode which is a perfect explosion as all the perfect atoms move outward from the core.

The observation of countless supernova remnants proves that this is quite untrue and so another explanation beyond a matter only one must be used. The reason that supernovae appear to always by lopsided or at the very least irregular is that space is also involved in the explosions. Here we must revisit the example of the marksman aiming at a target far in the distance. Any deviation from dead center no matter how absolutely small at the gun will result in a large deviation at the target as the effect is increased over distance. The same is occurring inside supernovae as the smallest parts of the system are the ESS and a tiny difference between highly spatially compressed ESS will be made large over time as they are allowed to spatially rebound to a much larger size. This is of course coupled with the fact that any area of space where matter is present is made of many ESS not just one so a small difference from area to area of highly compressed ESS will result in a great number of ESS showing this difference once they have been allowed to spatially rebound again.

Inside a stars core you have two things of course one being matter and the other being the ESS that the matter envelops. The ESS as you have learned from the SM masses responsible for the BBB of the universe are involved in the speck/mote geometries of the clintrons being at different maximum degrees of interaction with them at all times. In addition they are also under various conditions of quantum entwinement with other ESS and speck/motes as well. The TSM explosion that creates a BBB will produce from these highly spatially compressed ESS slight differences in how the universe forms due to miniscule imbalances in matter distribution over time once the ESS have spatially rebounded as well as having dragged the matter of the universe along with them. The result is a near uniform but not perfect distribution of matter through the universe which will result in places where stars and galaxies will eventually form. In a star however you must remember that the supernova explosion does not involve all the matter of the star at once like in a BBB event so you only have a small thin TSM schale detonation layer inside the star. This layer will provide the necessary SM rules for producing variability in matter distribution as the ESS spatially rebound. What it will also do is provide a push of overwhelming force to the remaining and non TSM stellar matter in the form of convecting plasma.

The micro-scale differences in the ESS as they rebound will not only create some of the lopsided outcome of supernovae but will also push on the macro-scale stellar plasma to immediately and quickly reveal any differences in its makeup as well. Running this combination of factors forward will provide the mechanism for a lopsided explosion right after the star goes nova. So any observers will naturally see an irregular or lopsided explosion of the supernovae and since the differences are formed in the ESS which are the smallest parts of the star you will see a near never ending array of supernovae each like a unique fingerprint of the stars heart. The stars' hearts do not stop beating after the supernovae are finished either as they continue now inside a neutron star or a black hole. Each of these hearts still has its own unique fingerprint that will reveal the inner workings of the core after the supernovae is complete and the nebulae it leaves behind has spread out.

Facts

This section has covered a few of the unsolved problems surrounding stars from how they work to how they reform into neutron stars and black holes but there is still much more to cover for all stellar masses no matter what phase of their life cycle they are currently in. The next section will deal with these additional topics in more detail and cover additional work concerning gravity, solar activities and black hole features but for now and for the sake of keeping each section somewhat compartmentalized I will review the key facts here. This sections main focus was on how stars like to transition from one life phase to another and the results that ensue which brings us to the first fact which is first generation stars in the universe contained mostly hydrogen and as such only produced neutron stars. The second fact is that first generation neutron stars were larger by volume than second and higher generation neutron stars as these latter stars contained heavier elements in their cores adding more gravity to reach the fusion ignition point sooner.

The third fact is that variable gravity explains how stars are able to achieve and maintain fusion no matter what size class they are in. The fourth fact is that variable gravity explains how all stars can transition to higher and heavier elements in the fusion chain and will lead to the only working mechanism for producing neutron stars or black holes. The fifth fact is that the increased ESS macht glow gravity of a collapsing neutron star allows for the three other forces to be overpowered such that the speck/mote geometries of all matter inside the star is locked in a neutron only phase. The sixth fact is that variable gravity and new speck/mote geometries must exist as the escape velocity for a neutron star is below light speed meaning the decay mechanism for neutrons should still be able to occur and dissolve the star from the surface inward at a rate of eleven minutes per layer. The neutron stars are in fact exceedingly stable and so no decay mechanism is occurring meaning there is an unknown to modern science mechanism keeping the neutron stars locked in place and this is of course space physics.

The seventh fact is that black holes are formed because they have sufficient mass surrounding a heavy element core that remains static for the life of the star and will provide the surface for the collapse event to form a black hole. The eighth fact is that it takes TSM to make SM which is a black hole and the heavy element core creates the necessary surface for the thin TSM schale detonation to occur on. The ninth fact is that the TSM schale event is needed to allow the core to pull matter in at near light speed and still allow the 90 percent stellar materials of the supernova to explode outward at near light speed as well as providing the necessary force to actually move that much star all at once. The tenth fact is the high levels of ESS spatial compression as well as maximum degree of interaction of the clintrons with space, quantum entwinements and macro-scale stellar plasma will produce the forces for irregular or lopsided supernovae. The answer to all unsolved problems concerning how neutron stars and black holes form as well as the nature of supernovae are explained in this section which allows me to proceed to the next where additional unsolved problems of modern science concerning stars and black holes will be addressed in full detail.

Chapter 30
Stars and Magnetic Fields

Magnetic Field Properties

This section will focus on once again stars not just because the universe is so completely full of them but because many of the unsolved problems of modern science of the universe are in fact answered inside of them. This is fairly obvious as one looks at the universe in its early life where everything in it is simply matter and energy in the form of space, gas and energy. The energy part is straightforward it is made of photons and happily spends its time playing around with any atoms it meets. The gas part is of course hydrogen, helium, lithium and beryllium mainly and these are just floating about playing with energy and talking to the space around them. The space that exists in the universe is fairly ordinary at this point to as it is just ESS that are only starting to interact in complex ways with matter and this friendly series of interactions will lead to all stars, planets and everything else we see today in the universe.

If you think of the early universe as a room filled only with gas you would be essentially correct as all of space constitutes the boundaries of the universe and gas and energy filled all space right to the edges of everything. From this point the ESS interactions between space and matter start to lead to gravity getting stronger and things like stars forming. In other words the once gas only room has coalesced into a room filled with stars. However the actual parts of what used to be everything in the universe from edge to edge are exactly the same they are simply now locked up for a bit in stars. This is why stars are holding so many of the answers to the unsolved problems of modern science and why stars in all of their life stages need to be further explained. This section will then cover stars in their active fusion phases of life and in their post fusion stages specifically inside black holes.

To start things off I will be explaining how magnetic fields work once again but this time from a stellar plasma standpoint as opposed to a simple bar magnet one. The explanation I will give for how stars use magnetic fields will also incorporate the more traditional bar magnet form as well and will help to prove once again how space and specifically ESS macht glow magnetic works in the universe. I will do this by explaining how stars hold onto their magnetic fields, how these fields generate a solar prominence and why exactly magnetic fields destabilize and do things like release coronal mass ejections as well as explain what variables go into a stars magnetic cycle. Afterwards I will use magnetic fields to explain some phenomena in the universe as well as how part of this overall magnetic explanation of matter and space interacting can explain the coronal heating problem which is partly answered by using ESS macht.

Stars contain magnets as all matter contains magnets so it does not really matter whether you are talking about a simple bar magnet, a rotating planet's core or a star. The difference is that while all matter has at its most fundamental quantum level a magnetic moment it does not always get to create a magnetic field. If you look at a star for example and take the various atoms that it is made of you will find very quickly in a laboratory that most of them are relatively non-magnetic. In truth they are all magnetic but in a lab and left alone with no special factors dictating how they behave they will not create a large, measureable and most importantly organized magnetic field. I have already explained how magnetic fields are generated in the universe as matter can be made to align its magnetic properties such that the ESS macht glow magnetic of space is made to flow in one direction. The ESS macht glow magnetic will be higher in some areas and depleted in others with the end result being that space tries to balance the ESS macht pool magnetic. This balancing is what you see in a simple bar magnet with the field flowing from north to south poles. The difference between a bar magnet and a star is that the bar magnet is solid and basically immovable on a physical level but the star on the other hand is exceedingly wobbly.

It is the wobbliness of stars that allow them to generate magnetic fields and allow these fields to move around something which a bar magnet cannot do. A bar magnet for example cannot reverse magnetic polarity nor can it make its magnetic field lines move in a twisting fashion. Both the ability to swap magnetic poles and the ability to have moveable field lines are something that stars do regularly. The most obvious of the two and the least well understood by modern science is how magnetic field lines in a star work to remain stable and to create things like solar prominences. Modern science does understand that the Sun like any star rotates at different speeds internally and so this can cause the field lines to technically twist around the equator of the Sun as time progresses. What modern science does not understand is how a magnetic field line can be made to twist without breaking and how these field lines break when they touch as it is often explained that once they touch they simply short out so to speak.

Yet no step by step mechanism is offered at a quantum level to explain how fields are generated and how they simply short out. After all a bar magnet can be made to interact with another bar magnet any number of ways you want and the field lines made to overlap so why do bar magnets within significant proximity to each other not simply short out destroying their magnetic properties? The reason stars can short out their magnetic fields and bar magnets cannot is not due to matter alone but instead to space and once again how space and matter interact. This recurring fact of space and matter interacting is again proving here as necessary to explain the universe as only using both of these parts together can magnetic fields inside stars be explained.

So the stellar plasma is wobbly compared to a bar magnet and this is due to the fact that it is comprised of a gas/fluid like substance. The reason it is both is that some parts of the sun are definitely a thicker fluid like layer while others are more like a very dense gas. The fact that neither is solid is the first step to explaining how magnetic fields work in a star as the ability to move atoms within atoms is key. To begin with we will allow the star to construct a magnetic flow the same way that a bar magnet would by having a number of atoms inside the star align their magnetic moments with one another such that they are arranged in a linear fashion. These linear arrangements will be for our example here constructed at the beginning of the solar magnetic cycle where the field lines run simply vertically from pole to pole. So far nothing is different between a star and a bar magnet as the star is behaving like a very large single bar magnet where a huge amount of magnetic moments contained in the solar atoms are flowing from pole to pole. The ESS macht glow magnetic is much larger than a normal magnet of course and yet obeys the same ESS macht balancing rules and so creates a normal field flow. The difference is that as the years pass the Sun's equator rotates at a different speed than the rest of the star so the field lines twist around and around the sun no longer resembling a single straight line from pole to pole. To explain how this happens it must be understood that the way a magnet works is by utilizing both space and matter not just one or the other.

So the solar plasma is a fluid much like on Earth how you can find rivers inside bodies of water like large currents running through the ocean. The plasma in the Sun that is part of the magnetic field is like a river of plasma inside the rest of the solar plasma and it is this river of plasma that carries the magnetic field lines with it. This idea of a river within the Sun is of course an illustrative one and should not be taken literally as the Sun and the ocean are still quite different despite both of them being fluid. The magnetic solar plasma however in the Sun is made of both space and matter and what you find here is that the atoms of the plasma that are a part of the magnetic field are of course aligned in a single direction to generate a field. These aligned atoms are also interacting strongly with the ESS macht glow magnetic of the Sun to increase the strength of the magnetic force there. The ESS of the Sun are of course in turn balancing their macht with the atoms that envelop them and will feedback positively to keep the matter aligned and a part of the magnetic field line. It is both the ESS and the matter of the solar plasma that move as one force throughout the Sun and create the magnetic field. It is the fact that both space and matter are needed to create a field line in the first place that keeps them bound to one another and keeps the field flowing strong despite being twisted.

Eventually however the field lines will twist after eleven years on the Sun on average and the lines will touch. This is the expected short circuit of the magnetic fields in the Sun and the same short circuit that is not understood by modern science. The step by step reason why this occurs is as follows and explains why liquid or fluid magnets short out but all others do not. In a bar magnet the atoms are held in rigid place by the structure of the magnet itself and so do not move, a stellar magnetic field is not however rigid. If you touch two bar magnets together end to end it does not matter whether two like poles or two opposite poles touch the fields persist and this is because the atoms making the bar magnet are held in place and so continue to imbalance ESS macht glow magnetic for the local area of space they occupy. In a star however the plasma is fluid and so when field lines touch it is not just field lines touching but the twisting rivers of magnetically interacting stellar plasma that physically touch. Unlike the touching of bar magnets though the plasma is a fluid which can be made to momentarily interact with each other. This touching means that at the single fundamental particle level the speck/mote geometries of the matter inside the plasma are able to balance the ESS macht glow magnetic that they possess with the other speck/mote geometry they touched. As you know magnets are just generators of ESS macht magnetic imbalances and space is always trying to balance out its total macht pools to even levels.

When the fluid magnetic plasma touches another fluid magnetic plasma flow the two are not held rigid like a bar magnet and so will not continue to keep their magnetic moments aligned with the ESS macht glow magnetic field flow they have created. In other words the plasma will allow for a balancing of macht glow magnetic between the atoms in the stellar plasma that will then balance out the local area ESS which diminishes the total field strength at that point. The stellar plasma is not normally aligned into a magnetic fashion and it takes both space and matter to create a magnetic field flow anywhere in the universe. So as a result you have the matter component of the stellar field line disrupted and no longer aligning with the total field line before contact was made. The exchange speed of energy from matter to ESS is light or below light speed and the speed of the atoms inside a star is exceedingly high due to high heat and pressure. This creates a situation where the atoms of the once stellar magnetic plasma to move out of alignment at near light speed on the quantum level which cascades down the area of the magnetic contact to cause all the physical matter present to become randomly moving and no longer aligned in a magnetic field.

This means that the space of the magnetic field line no longer has anything powering up its macht glow magnetic and certainly nothing keeping it flowing in a single direction. The result is a mass dump of all macht glow magnetic energies back into the stellar plasma such that the magnetic field is instantly destroyed at macht speed within the local area ESS and at near light speed in the stellar plasma. The dump creates an explosion in space but only an explosion in macht glow pools and not in physical ESS volumes. This is because the ESS do not alter their volume for macht glow magnetic only macht glow gravity and so the explosion takes place in space but does not move space at all as the macht glow magnetic travels near instantly at macht speed throughout all ESS to balance out the glow pool. The matter that was part of the field that envelops the ESS that was part of the magnetic field will now be out of quantum alignment with the ESS and so will move imperceptibly which can of course cause a change in the macht glow gravity which in turn can alter ESS volume for the local area of space but the amount is so small that little to no physical movement will occur.

This is in contrast to the movement of the macht glow magnetic which will explode inside space to a large extent and results in the total balancing of the ESS macht glow magnetic pools to pre-magnetic alignment levels. The bar magnet by comparison will do no such thing and still will have all matter interacting with the local area ESS to amplify the macht glow magnetic. This explains step by step how a bar magnet will hold its field intact, how the magnetic field lines in a star will maintain themselves as they twist and exactly how and why the short circuit mechanism works at a quantum level. The proof of this is seen regularly in solar prominences on the surface of the Sun and specifically the breaking of these loops of magnetic energy to create such events as coronal mass ejections.

The beginning of these events occur as the twisted field lines touch of course and follow the same rules for dumping ESS macht glow magnetic back into the local area of space. The difference is that the plasma is still partially aligned with the magnetic field and of course is still partially contributing to the magnetic field created by both space and matter. So as the total space and matter alignment is lost the stellar plasma which is held in the solar prominence is forced to move rapidly in an outward direction. The fact that the plasma is magnetically bound to space is further proved by the fact that an invisible field called magnetism can pull matter off the surface of the Sun against the invisible force of gravity and many kilometers into space. The total mass of the plasma contained in the solar prominence is certainly significant and should normally be held in place with gravity on the surface of the Sun and yet it is not. The magnetic field created from both space and matter is powerful enough to lift both the matter from the surface of the Sun and to extend the ESS macht glow magnetic outwards into space as well.

This total power however will short out in a very violent fashion just like it can inside the Sun but because it is now extending into space the result is quite different. The magnetic field is lost like before but the plasma is now lost outside the surface of the Sun and so as the field moves the plasma which as stated before is still semi-magnetic is drawn with it. The magnetic field is dumped away from the surface of the Sun and the temperature and pressure of the Sun itself is of course much higher at the surface and in the stars interior. This means that the explosive force of the now no longer magnetically aligned and bound stellar plasma must be expanded in a direction of lowest density and energy which is of course space sided.

So the ESS macht dump back into the local area of space leaves the stellar plasma exposed to space above the surface of the Sun and the last thing it can do is move away from the Sun outwards due to temperature and pressure differences. The result is the plasma being released having felt the last effects of its alignment and interaction with the ESS macht glow magnetic which is naturally polar in nature as macht glow magnetic always tries to balance itself out which of course creates a directional flow of magnetism in space that is always three dimensional. The last instances of ESS macht balancing occur as the ESS macht glow magnetic must obey the law of conservation of energy for the system and so dumps all unused macht glow magnetic back into the matter that was donating it in the first place. Since magnetic field flows are always from pole to pole in direction the stellar plasma last felt this pole to pole direction in three dimensional space and will result in the plasma moving to follow this macht dump back into space. The stellar plasma is not only ejected from the surface of the Sun but is spinning as it leaves due to the last motions it felt from the ESS macht glow magnetic trying to balance itself out with the local area of space. In addition this explains why some prominences simply collapse back to the Sun's surface without any spectacular consequences and why some become coronal mass ejections.

The solar prominences that become coronal mass ejections are caused because the total amount of magnetic field strength in the interactions of the space and matter that make the magnetic field is strong enough to allow the ESS to win out and balance the ESS macht glow pool as they want to relieve that strain on the local area of space. Since they are strong enough they pull the physical prominence together and we see the same short circuit mechanism where a fluid magnet cannot maintain the matter to ESS macht glow magnetic energy donation alignment. The identical mechanics to an internal short circuit occur and the field alignment is lost, the ESS macht glow magnetic explodes inside space and balances all local space macht pools. The matter is sent exploding off the surface of the Sun and spinning as a result and we see this as a coronal mass ejection. The size and mass of the solar prominence as well as the point at which it collapses in proximity to the Sun's surface, the degree of local area spatial compression coupled with all ESS macht glow magnetic and speck/mote geometries will determine the degree of rotation and size of the plasma as well as the speed at which it moves away from the surface of the Sun.

The solar maximum cycle of the Sun is eleven years on average and is another unique fingerprint of the Sun as it differs from all other stars. It is of course an average and might be close to but will not likely be the same as any other star. The reason is that you can calculate any stars solar maximum cycle by combining various factors concerning the star and the phase of its life that it is currently in. It has been shown that the ESS of a stars local area of space is necessary for a twisting magnetic field system and as such the star itself can affect the local area ESS directly. All stars are of course made of matter and all matter has a specific speck/mote geometry that it uses to edge drag with the various ESS macht glows of the universe. The amount of matter also affects changes in local area ESS and most of these are manifest in the form of changes to the gravity of the star and therefore to the volume of the ESS it contains. The gravity of a star is variable and will change over the course of the stars life for example the gravity at the core will increase despite the mass and density of the star decreasing as it expands as it ages.

The result is a chemical composition change to the star and so a change in both temperature and density of the stellar plasma. These factors will alter the speed at which the field lines rotate, twist and eventually cross inside the star. The combination of these factors will create a unique solar maximum cycle for all stars in the universe and these maximum cycles vary over the age of the star. While it is difficult at the distances we are from other stars besides our Sun and with our current level of technology it is possible to calculate these solar maximum cycles for any star we can observe. The solar maximum cycle in turn can be back calculated to learn about the ESS spatial compression inside the star as well as the gravity in it and at different depths inside the stars core. This is of course because all magnetic fields are not just created by matter alone but by space as well and the degree of interaction of all matter with the ESS macht glow magnetic is determined by the overall ESS spatial compression for the local area of space. To this end knowing about the ESS in a particular area of space will allow you to calculate the magnetic fingerprint of a star and likewise knowing the magnetic properties of a star will allow you to learn about the ESS inside of it.

Stellar Magnetic Problems

Another of the Sun's magnetic features can be found in the coronal heating problem which remains one of the unsolved problems facing modern science. The truth is the coronal heating problem is very simple but the reason it exists is somewhat unclear. The surface of the Sun is approximately 5800 degrees Kelvin and the corona is around 1 500 000 degrees Kelvin. The temperatures for both the Sun's surface and its corona can vary a decent amount but the averages remain fairly consistent over time. To this end the unsolved problem in physics faced by modern science is why is the corona hotter than the surface of the Sun when it is farther away from the source of heat generation? This seems backwards to conventional thinking as if you place your hand near a hot cast iron stove you feel a great amount of heat and if you move your hand away in 10 cm intervals you feel the amount of heat on your hand decrease. With the Sun however it is as if placing your hand near the hot cast iron stove you feel some heat and as you move your hand away from the surface of the stove you actually feel more and more heat as opposed to less and less.

Since the Sun and all stars in the universe do this on a regular basis clearly some normal physical process must be at work as the laws of physics are not being broken just to try and confuse sentient minds. To better understand this problem two facts about the corona need be stated here and the first is that it is much hotter than the surface of the Sun which by now you know. The second fact is that the corona is many orders of magnitude less dense than the surface of the Sun also known as the photosphere. The photosphere is so named because it is the layer where the visible light that we see here on Earth can be said to originate at least in terms of what parts of the Sun's makeup you can actually see. So the corona is hotter than the surface but also much less dense and this is important as it offers some help in providing a mechanism for explaining the coronal heating problem. The help is that the corona being so diffuse means that despite it taking energy to make it hotter than the surface of the Sun it also means that it does not take much energy as there is not much material in the corona to actually heat up. So the corona must be getting additional energy from the Sun in order to increase its temperature but it does not necessarily mean that it requires a huge amount to get the job accomplished. The corona is kind of like the high energy particles in the Earth's auroras only coming from internal influences from the Sun and not external ones as found on Earth.

The Sun's corona is hotter than the photosphere because of the magnetic fields generated by the Sun itself. This is evident for a few reasons one being that the corona is not a uniform feature of the Sun throughout its life and will change in size and shape over time. These changes are all linked to magnetic activity in and on the surface of the Sun. One example of this is the tendency of the corona to be most prominent around the equator during periods of less intense solar activity when the Sun has entered what can be thought of as a calm or resting phase to its magnetic cycle. The polar regions will see less magnetic activity and less coronal coverage so much so that coronal holes can be found over the polar regions of the Sun in these times of quiet. The Sun's equator during the same periods will still have a corona as the equator rotates faster than the poles of the Sun and is partly responsible for the twisting of the magnetic field lines. So in areas of highest magnetic activity like the equator you continually see coronal activity. More reasons why the corona is tied to magnetic activity is that in places where sunspots are found the corona is stronger as they can help form coronal loops which transport heated plasma into the corona along magnetic flux lines. So the corona overall in these areas is affected by solar prominences and associated magnetic activity. Lastly the corona is highest at the peak of the Sun's eleven year solar cycle. The solar cycle is the eleven year cycle of magnetic activity in the Sun which has already been covered but here again it is magnetic activity that is found to be linked to coronal characteristics.

To this end it is obvious that the coronal heating problem is solved partly or maybe even wholly by magnetic influence alone. This does not mean that a single mechanism is responsible for the coronal heating problem as it is far more likely that multiple factors help to heat the corona to its high temperature while allowing the surface to be much cooler. The difficulty in providing a heating mechanism for the corona from the photosphere is that the energy carried needs to be transported past the chromosphere which is a thin layer of the Sun's atmosphere that exists between the photosphere and the corona. Multiple mechanisms have been proposed to explain how energy can cross this gap but they all have some problems. Sometimes the energy is seen to dissipate before reaching the corona or be reflected back down to the surface as seen in certain wave theories. At other times energy can be delivered but not deposited evenly in the corona to account for even heating of the entire region. Nevertheless it is obvious that something is heating the corona and since the corona is so heavily linked to magnetic forces generated from the Sun it must be at least partly explained through magnetic means.

It is most probable that multiple theories used to explain the heating of the corona are all involved as the corona is so diffuse that not much energy need be given in order to heat it up and therefore rather than one single answer to the problem several smaller ones can add up to the amount of total energy required. In order to get heat to the corona you must transfer it in some other form as direct heat transfer is in violation of the second law of thermodynamics. In this case the law states that the Sun should be losing heat to the surrounding part of the system which is open space just as heat is lost to the room with the hot cast iron stove in it. So since direct heat exchange is ruled out another pathway for the energy must exist as it is impossible that the corona is heating itself for free so to speak.

The simplest way would be to inject heated matter directly into the corona in the form of solar flares but this is impossible as it would have been detected easily by now no matter what wavelength of light one looks at. A possible part of the solution to the coronal heating problem is made by using nanoflares which are seen as small solar flares that shoot upwards almost completely perpendicular to the surface of the Sun and outwards to the corona. These small solar flares however are not found in sufficient number nor are they long lived enough to impart the kind of energy necessary to provide a continual heating mechanism for the corona. This being said they have been found and are of course created through the regular magnetic flux pathways and so do impart some energy to the corona at least despite it being very small.

Waves are another possible part of the answer to the problem as it is known that the Sun's plasma can be made to move like and generate types of waves. These waves are not always purely physical as you might think of a typical wave on the surface of a lake or pond created by a stone dropped into its center. Instead some of these waves are magnetic in nature and can carry with them magnetic energy which can be used to heat the corona. The most important of these wave types are called Alfven waves as named after the Swedish scientist Hannes Olof Gosta Alfven. These waves are a form of magnetohydrodynamic that is carried by ions and the magnetic field which makes it capable of travelling across the chromosphere. The waves are generated from surface activity on the Sun associated with what are called granules. Solar granules are simply pockets of surface plasma that resemble small granules when viewed from above. These granules reveal turbulent solar activity beneath which can stir up the plasma and since plasma is associated with both electrical and magnetic properties disturbing the plasma too much will likewise create surface magnetic effects. Some of these effects are thought to generate the magnetic waves that are used by wave theory in explaining the coronal heating problem.

The magnetic wave theory states that any waves that travel into the corona can then be made to dissipate their magnetic energy into the materials found there. This dissipated magnetic energy will create a shockwave that will heat the corona and as long as the energy carried is high enough the temperature of the corona will also be high enough to solve the coronal heating problem. Magnetic waves and especially Alfven waves have been found to carry energy possible of crossing the chromosphere into the corona but are found to not dissipate energy sufficiently or fast enough to create a uniform heating required to explain high coronal temperatures. Still by offering some explanation to the coronal eating problem it is likely once again that magnetic waves are a part of the total solution to the coronal heating problem.

One last magnetic angle to the problem comes from the use of the Sun's magnetic field to deposit energy into the corona. This is a very attractive idea since the Sun's magnetic field easily travels straight through the corona and obviously well out into space and the entire solar system for that matter making it easily capable of crossing the chromosphere. The solar corona contains materials that can be affected by strong magnetic fields and the method of heating comes from the fields themselves creating electrical currents in these materials. The electrical currents get their energy from the magnetic fields from the Sun and so represent the source of energy with the Sun being easily strong enough to provide all the power needed for heating. The heat in turn comes from the collapse of the electrical fields which must transfer their energy to other forms in order to obey the law of conservation of energy. This new energy is made of both heat and wave motion which can then interact with the coronal materials to raise their temperature to millions of degrees Kelvin. The reason this is so attractive as an answer is that the Sun's magnetic field is everywhere above, below and side to side throughout the coronal layer meaning that uniform heating is not a problem nor is getting the energy through the chromosphere. The theory is known as magnetic reconnection theory as the energy comes from magnetic fields reconnecting with one another.

To explain this simply plasma creates small magnetic fields around matter and can be formed around something as small as a molecule or atom. The magnet behaves normally on its own but when brought into contact with another plasma magnetic field it will then join up in a normal north to south pole manner. So far so good but if you disturb the magnets making the field the field actually remains intact as it is now carried by the electrical currents associated with the field. This is because the plasma is a fluid substance and if more magnetic fields are introduced and the plasmas creating the two magnetic fields are disturbed enough the field will then technically dissociate from the matter creating it and reconnect with a new field. The fields will always connect to other poles in other magnetic fields such that normal magnetic rules are obeyed but the reconnection will then dump the heat energy and wave energy into the surrounding environment. It has been seen that everywhere on the Sun is plasma and that this plasma is always in motion so there always exists the correct conditions for magnetic reconnection events to occur. It is also known that the coronal heat levels and distribution around the Sun are linked to magnetic events so once again makes an excellent candidate for the solution to the coronal heating problem. Magnetic reconnection is not possible however as previously explained using coronal mass ejections when massive amounts of magnetic energy and stellar plasma is involved.

The main problem with magnetic reconnection theory is that it proceeds faster than all models used by modern science predict. Obviously the kinetic motion and number of particles involved in the event must be factored in as well as a literal physical impediment to the magnetic fields can exist if the particles that are generating the magnetic fields are restricted from moving as well. The restriction in movement means a restriction in how fast the magnetic fields can break connection with one plasma group and reconnect their field lines with another group. This being said the observations of things like solar flares shows that magnetic reconnection can take place many, many orders of magnitude faster than theory predicts. Since it is known that magnetic reconnection is occurring something must be allowing the process to speed up such that the normal results expected no longer be the norm.

The something is spatial compression of ESS in and around the Sun as well as all associated processes of ESS and macht. To begin with it is known that all fields are really just macht glows being shared from ESS to ESS and the magnetic fields of the Sun are no different. This is why as stated above the magnetic field lines can be maintained in space when the matter associated with them is removed. To start with the magnetic reconnection problem and the speed at which they take place should normally be governed by what can be thought of as matter only reactions because according to modern science and especially quantum field theory all things in the universe are simply excited states of fields and nothing more. This leads to the very obvious problem of hitting the speed of light limit to the universe which by now should be obvious to all as incorrect as it limits the way the universe moves and lives too much. The particles that are observed to be reconnecting are believed to show the path of the magnetic field lines that are also reconnecting and this is in general correct. However it is not the whole picture as the magnetic field lines are moving underneath the sea of particles as the particles are simply there to follow the magnetic force. Now to explain what is happening we can look at the particles as they follow the ESS they are enveloping.

The ESS contain the macht glow magnetic which is what is short circuiting out in reconnection events and is of course similar to the solar flare examples from just a couple of unsolved problems ago. The reconnection events play out in the same way and as I have explained perform a macht dump into the local area of space and some of this is seen as magnetic reconnection. In something like a large coronal mass ejection the short circuit is complete as the energies and machts involved are simply too large for a star the size of the Sun to reconnect them before the plasma is sent away from the surface of the star. Remember that all events in the universe take both space and matter together and so the plasma will physically leave the area of the short circuit as well making the old field lines impossible as they were a combination of the matter and ESS macht glow magnetic. This is a second way in which the magnetic reconnection will not be possible for a large single short circuit event. Overall the event is complete and no reconnection occurs on a large scale but very small reconnection events will take place where the main lines break from the surface of the Sun. However these events will be on a scale many, many orders of magnitude smaller than the entire single short circuit event of the coronal mass ejection.

For normal smaller magnetic reconnection events the standard ESS macht dump takes place as the matter and space making up the overall magnetic field are rearranged. Since space has not gone anywhere locally and neither has the plasma matter which will charge up the ESS macht glow magnetic new connections can be continually formed and broken for theoretically ever if the star never lost mass. The speed of the reconnections is many times faster than predicted because the speed at which ESS macht travels is many times faster than light. As the ESS balance out their macht glows the transfer of macht between schales is essentially instantaneous when compared to the slow speed of energy and matter. Magnetic fields are made of both macht and speck/mote geometries of the particles they are associated with and this means that the matter is tracking after space. The matter is tracking as fast as it possibly can in order to catch up to the macht speed shifting inside the ESS. The matter seen in the solar flare reconnection events for example is moving slower than the universe really wants it too and the ESS macht glow magnetic is moving faster than can be seen using particles as a guide for magnetic field lines. That is why observations outstrip the theories of modern science. The matter in the magnetic reconnection problems is simply being dragged as fast as the local area of spatial compression will allow it to be and therefore exceeds what is predicted by modern science. This is coupled with two other factors that help move the matter faster than expected and again both are related to ESS and spatial compression of the local area of space.

The second factor is the number of ESS in the local area of space for the star and for all magnetic reconnection events taking place there. The spatial compression of any star is much higher than anything on Earth and this must be remembered always as physics that exist in our cozy laboratories simply do not exist everywhere else in the universe. You cannot use Earth based spatial compression to experiment on matter and then interpret those results directly for areas of extremely high ESS density such as black holes or moments after the Big Bang. In addition to this the opposite is true and you cannot apply Earth based spatial compression and hope to understand the kind of space that exists between filaments of the universe or at the extreme edges of the universe many billions of years of expansion into the future. This means that the levels of ESS per volume of space are simply higher around and obviously inside a star than they are here on Earth so when events take place faster than expected it really is not to be expected. The higher number of ESS around the Sun means that macht glow magnetic can be much higher than normal since there are more ESS for the specks and motes that are increasing magnetic force there to be edge dragging through. More ESS means more magnetic energy and therefore a far higher pull on all visible matter that are following the invisible magnetic field lines. The amount of power that is predicted to cause the field lines to reconnect is simply not being estimated as high enough and so observations show events taking place faster than expected.

The third factor is also related to the number of ESS inside the Sun as the spatial compression goes up quickly as you travel closer to the Sun's core and so does the number of specks and motes that can interact with those ESS. The end result and one that is quite shielded from our instruments here on Earth is a massive reservoir of macht glows of all types from within the Sun's core. The influence exerted on all matter in the Sun is accelerated by higher levels of macht than we can experience on Earth so you find that increased levels of heat which is due to balancing of macht glow electromagnetic is accelerated. For those who can visualize this concept the answer is quite obvious as the Sun is slippery when compared to the same forms of matter on Earth and so movement inside the Sun is easily facilitated despite any increases in density. This is of course from ESS balancing their macht levels and will affect matter not just from a magnetic standpoint but from physical and heat standpoints as well. The matter is able to push past normal kinetic barriers because all of its properties are influencing the greater number of ESS inside the Sun and the ESS in turn are trying to balance their elevated levels of macht faster as well. The total end result is that the matter seen to be following the magnetic field lines will move faster and reveal the faster magnetic reconnection speeds underneath it all.

In short the magnetic reconnection problem is solved simply by looking at the elevated macht glow levels inside a star and factoring the increased number of ESS that are trying to balance those levels. Therefore the macht total reserve pool inside the Sun will influence what is happening on or near the surface as well. If you remember that macht sharing is highest between ESS that are in contact with one another and in contact with matter orbs you see that all balancing is highest inside the star and falling off as distance increases from the star's surface. To this end the macht reserve power inside the star will always do exactly what we see to magnetic events on the surface of the Sun. The first thing it will do is accelerate all events like magnetic reconnections and the second thing it will do is allow those events to collapse back to the surface of the Sun. A coronal mass ejection exemplifies this type of event as the plasma is ejected based on balancing its heat with the local area ESS but the magnetic field line which is dominant and embedded in the ESS of the local area of space will balance with itself and the highest macht pool for it to balance with is inside the Sun so we see magnetic fields flowing back into the Sun while matter is ejected away from the surface. Magnetic reconnection events are not problematic at all when using ESS, macht and speck/mote geometries to explain these events as why they occur, how they occur and the speed at which they unfold all obey perfectly normal and calculable rules for space physics.

So what this means overall for the coronal heating problem is that a combination of factors will allow the relatively diffuse coronal layer of the Sun to be heated to a higher temperature than the surface. Multiple factors such as nanoflares, magnetohydrodynamic waves and possibly others will provide some additional heat as well as magnetic reconnection events. The magnetic reconnection events themselves are now explained away and are no longer a problem so any heat donated to the corona by them can be factored in as well. The changes in ESS spatial compression away from the surface of the Sun also allows for decreases in macht glows as matter is not experiencing as many ESS as before. The distance of the Sun's corona will vary when compared with all other stars in the universe depending on their overall size and composition.

The importance of this is that at a set distance each star has a fingerprint that reveals the ESS spatial compression density unique to it where the corona of that star will exist. This area of space around the star has the right conditions for the speck/mote geometries of the matter there to experience a balancing of all of it's macht glows such that the macht glow magnetic associated with the matter has reached a threshold value where it will break allowing magnetic reconnection to occur. This means that each star has a region of space surrounding it that simply due to the speck/mote geometries of the matter it possesses will always cause that matter to balance its different macht glows at those distances based on ESS density. The type of matter whether it is an atom, an ion or a fundamental particle will all be reduced to the constituent speck/mote geometries they possess and these will be what reach this threshold point no matter what. This balancing is what provides the right requirements for the space and matter together to undergo magnetic reconnection and therefore heat the space and matter there to many times the temperature of the surface of the star.

Facts

The fact section for this topic is fairly short and begins with the first fact that bar magnets and plasma magnets are essentially the same at a base level. The second fact is that on a macro scale the magnets are different because the ESS and matter of plasma magnets can be dissociated from one another easily. The third fact is that the ESS macht glow magnetic will explode inside space to disrupt the previously stable magnetic configuration of both space and matter into a single recognizable field. The fourth fact is that the properties of all solar flares and coronal mass ejections is based off of how the matter was last interacting with the ESS macht to which it was bound. The fifth fact is that the use of ESS macht as the true magnetic force allows for magnetic reconnection to occur at speeds observed without causing problems.

Chapter 31
Black Holes, Relativistic Jets and Evaporation

Black Hole Atmospheres and Spectrums

This section will examine black holes once again and hopefully be the last time that I need to cover them. Simply because black holes are so useful in explaining the limits of physics as they exist everywhere in the universe and so there exists many angles at which to view them and I hope this section can contain the last few pertinent ones. That being said I am sure I will be explaining them many more times to come but for now let us look at their basic structures on a macro scale as opposed to the micro scales that I have been focusing on using ESS and speck/mote geometries. Black holes are first and foremost large objects in space despite many people erroneously saying how small they are. Yes, yes the star used to be tens of millions or more of kilometers in diameter and now it is the size of a small planet or so but that in no way makes it small. What is heavier children a kilogram of lead or a kilogram of feathers? Which glass is more full the tall skinny one or the short fat one? Everything is the same just distributed differently in three dimensions and this means that a black hole is still large as it contains a huge amount of matter in a smaller space. So despite a black hole occupying a much smaller volume than it did when it was a star it is in fact still a large celestial object.

Therefore it is necessary to stop thinking about it as a single fundamental object like a singularity. Anything even the size of an apple is monstrous when compared to the quantum world so a black hole must not be thought of as a purely extreme physical environment ruled by some phantom singularity which is by theoretical definition a single point object and so the smallest thing ever possible. Is the fact that the black holes mass having been compressed into a smaller space important? Yes of course it is as this is what gives the black hole its unique SM structure but it does not mean that it is a homogenous object when viewed at macroscopic scales.

So looking at a black hole at macroscopic scales the first thing you see is that the black hole is not made of just SM no more than the Earth is made of just its rocky surface and interior. The Earth has a thin atmosphere of gasses on its surface held in place by gravity and a black hole has a similar atmosphere held to its surface with dynamic forces all its own. Black hole atmospheres are of course very different than those you would find on Earth but in many ways are the same. A black hole atmosphere is made of loosely held together particles that are free to move around much like loosely held together atoms are free to move through the Earth's air. The particles in a black hole atmosphere are much more energetic however and are heated to much higher temperatures but obey the same physical laws as those on Earth. Need I say once again that the laws of physics do not break down and the universe does no strange things outside, inside or around black holes?

The accumulated gasses and dust in the atmosphere of a black hole are similar to objects in orbit as they are trying to fall into the black hole but are moving at high rates of speed which prevents them from falling inward directly. Objects however will pass into the black hole proper and become part of the SM inside but before they do there is always something interesting that can happen before the interesting things that happen inside the black hole of course. Some of the interesting things are the generation of a magnetic field from above the event horizon and the emission of various forms of electromagnetic radiation. Black holes are made of clintrons which are the final evolution of all speck/mote geometries for matter possible anywhere in the universe and under any conditions. Clintrons have long since held their electromagnetic forces inside themselves so that they are no longer edge dragging into the ESS of the universe. As such clintrons do not create magnetic fields directly but their presence always creates conditions that allow for magnetic fields to be generated around black holes. This is due to the unequalled amounts of variable gravity they create around SM masses and the effect this has on matter falling into those same masses.

The rotation of all matter falling into the SM mass will be forced to align itself with the ESS which are under high spatial compression just above the event horizon. Since matter is naturally attracted to ESS and the higher the number of ESS the greater the attraction the matter is quickly aligned with the ESS and held firmly in place as a result of this fact. The orientation of matter into a common plane around the black hole is what can create the necessary conditions for magnetic field generation by the black hole despite the black hole being made of clintrons. The main point of how black hole magnetic fields are generated is that they are created above the event horizon and not beneath it so that another interesting phenomenon can be explained. This is the generation of electromagnetic radiation from around black holes in the form of different frequencies of photons. These photons come in many forms but the most common to modern science are the low frequency radio waves emanating from around black holes. It is most likely that they started out as higher energy photons but lost their shorter wavelengths travelling away from the highly compressed ESS space of the black hole before reaching us.

This generation must again occur above the event horizon as anything below it will be incorporated into SM. The fairly violent environment that makes up black hole atmospheres is not only responsible for the creation of magnetic fields but also numerous collisions of particles in all regions and altitudes of the atmosphere. Any number of normal mechanisms can be employed here as normal physical laws still apply as the matter above the SM mass is not yet converted into clintrons. What is key about this feature of black holes is that photon emission will occur as different event horizons are passed on the way to the surface of the black hole proper and reconfiguration of speck/mote geometries turns all matter into clintrons. As matter passes down towards the black hole it will encounter areas where different aspects of the fundamental forces are stripped away as eventually they all are held within the clintron which is of course the one and only building block used in all SM structures. Each time a force is lost energy will be released as the ESS macht of the local area of space balances out in new ways to accommodate the way the new speck/mote geometries edge drag with space. So it can be seen that as the electromagnetic force between atoms is lost you will see an emission of photons, then as you lose coherence between the nuclear forces of atomic nuclei more photons are emitted due to even higher energies being required to overtake the nuclear force. Finally if you could detect the moment that quarks are rearranged into the next form of particle and it is possible that there are other speck/mote geometries before the clintron then you will see a different kind of photon released still.

Lastly no matter how many forms of matter and particles are created and destroyed the end point of everything in the universe is reached which is the clintron a combination particle and space object that is made of nearly perfectly fused ESS, macht and speck/mote geometries. The clintron is formed from all the necessary parts of both space and matter as both are needed to create it and as such it will not emit photons because of its speck/mote geometry so you cannot detect this form of matter directly ever using any range of electromagnetic radiation. Clintrons will also be thought of as moving very rapidly in place inside the SM mass proper as they are pressed essentially right up next to each other and will therefore move only slightly in place but very rapidly. From the standpoint of modern science they are seen as moving rapidly because all matter as it approaches a black hole speeds up immensely whether moving through space or simply moving through vibrations in static space and therefore having massive amounts of matter making them up because of the relativistic effects produced by their movement.

The actual reason they are more massive is of course because they are donating more of their energy to the ESS macht glow gravity of space as the clintrons are locked right next to one another with very little to no movement actually occurring as a black hole is really SM meaning that both the space and matter that make up a clintron exist as one and move in lock step with each other. This is also why a large object like a black hole can behave as though it were a single large quantum object because the clintrons are so tightly packed together that all movement of the black hole is shared around whatever single ESS is the true central ESS. The motion of the black hole like a point particle is a reflection of the movement of the entire SM mass around this one central and therefore can be thought of as a point particle which is the single ESS at its heart. An interesting note about clintrons is that they are immune to time dilation as they do not exchange the usual quantum information between them that normal matter does which is known to experience time dilation. There is nothing to be exchanged between them which can be bogged down by interacting with the ESS of space and macht more than other matter. This means the clintrons are the endpoint for matter again as they are a part of space and governed not by the variable quantum time perceived by us but by the absolute macht time of the ESS of the universe.

What you will get up to this point is a spectral fingerprint of a black hole that should always have distinct bands of photon emission at the various levels of event horizons as you approach the SM mass but before anything falling into it becomes a clintron. Therefore the recognizable signature of a black hole should be very easy to detect as long as you are sufficiently close to it to detect the faint radiation being sent away from its atmosphere which is above the last event horizon. Black holes are easily detectable despite the black hole proper being truly lost from normal instruments eyes. To this end black holes both produce magnetic fields and emit energy from their atmospheres but not from themselves directly.

This loss of different fields can also be used to explain the magnetic properties of things like magnetars. Magnetars are still possible with the electromagnetic force gone because that is only necessary to make atoms have electron orbitals. Once the orbitals are lost the electromagnetic pressure that was keeping atoms and such high volumes is gone and you can have much smaller neutrons as a result. Losing the electromagnetic force does not mean in any way that you have lost the magnetic force as the two are separate only viewed as the same under normal Earth conditions and is similar to how the electro-weak force can be combined but also are not the same single force but two joined together. The neutrons still have magnetic dipole moments however and even though their charge is lost they can exert a magnetic force without an electric component especially because it is thought that neutron stars have a quark/gluon plasma core meaning the quarks are free to align themselves along a common magnetic moment and be held there by the strong nuclear force. The motion of the magnetic dipole moments in the neutrons will transfer energy to the ESS magnetic power creating massive magnetic fields of pure magnetism and this is observed by matter or energy which still has both electro and magnetic properties. The observed event seems as though the magnetar has both electric and magnetic forces when it does not as it is only perceived to because electromagnetic particles are used in detectors and become excited in response to the magnetic only field. Remember here that ESS macht glow are fields and that the force carrier of the photon only is popping up when needed and not the other way around so you can still have a magnetic field without an electric component.

One interesting point about SM is that as you get closer and closer to the right conditions to make it you notice that temperature and motion increase rapidly as particles are forced to move ever faster. The reason for this increase in both temperature and motion is of course due to one fact and one fact only which is the ESS that the particles occupy are trying harder and harder to balance out their respective macht reserve pools. This is why all things happen in the universe in the first place such as something as simple as why a fire burns. The matter in the flame is interacting with other matter in the universe to trade off various forms of charge and bind into new molecules. This is a simple fact that can be seen as you combust anything like a match, the wood is the source of fuel and the oxygen in the air is the oxidizer.

In reality this is simply two forms of matter that are interacting to create new molecules and the way space responds is to try to balance out the ESS macht that was held in the pre-burning arrangements. What is normally thought of as chemical potential energy is really pent up ESS macht held in place by matter to space due to how the speck/mote geometries of its constituent particles are all edge dragging in space. The matter before it burned was made of atoms that are not large enough or unstable enough to be naturally radioactive which would give them the ability to balance out the macht imbalances from the specks and motes they are made of. As something decays it slowly balances the ESS macht glow created from all the edge dragging speck/mote geometries and in something like the wood in the match the atoms are simply too stable to decay despite them on a fundamental level trying to balance out their macht glow pools. So this is why anything burns in the first place as ESS macht imbalances between negative and positive charged particles attempt to balance out the ESS macht glow they are sharing.

The same sort of thing is happening as you approach the event horizon of a black hole where ESS macht is trying to balance out more and more before it becomes SM. Like in a particle collider where very high energies are needed to get particles to combine forces like the quark/gluon plasma high energies are needed to create SM. This is where SM stops being what many people would call normal though as SM has no temperature. Yes that is correct SM does not have temperature because it does not need temperature to exist. Does entropy exist for a quantum system at one instance in time? No of course it does not as entropy is a measure of change from moment to moment so with no changes in time there is no change in entropy. What about Einstein, relativity and spacetime? Without looking at the world line of an object you get no moment to moment happenings in the universe and so spacetime does not exist and Einstein and relativity are wrong. Moments in existence do happen and they happen all the time so many things can and do exist without some of the traditionally held beliefs in modern science.

To this end it must be understood that heat is merely the transfer of electromagnetic radiation from object to object such that it increases its motion in three dimensional space. In SM however the electromagnetic force is turned off and so it is impossible for electromagnetic energies of any sort to occur. With no photons you have no heat and as for motion the SM does not need to balance its macht with other SM as they are made of clintrons which are the endpoint particle of the universe imparting only gravity to the ESS of the universe. The maximum degree of interaction of each clintron does not have to be the same as its family members but it still has no need to balance macht with anything. This allows for SM quantum entwinements to exist as well as all related future unfoldings of the SM in for example a BBB.

So in no way are the physics of SM altered by not having a temperature only that it does not need one to exist. If you could get close to a clintron you would feel heat as you were rearranged on the fundamental speck/mote level of matter but this heat would only be coming from you and not from the clintron. The clintron would be providing the mechanism to heat you up in the form of high levels of variable gravity but it would not actually be heating you directly as you would be heating yourself from your own matter. So SM and certainly TSM do not have a temperature and instead exist perfectly normally within the universe breaking no laws of physics simply by using high levels of spatial compression of local area ESS, increased macht glow gravity creating variable gravity of enormous amounts and rearranged speck/mote geometries for all of its matter.

Think of the reverse of super forces like the electroweak force of the electromagnetic force and instead you have the properties of physics not combining but reversing. This is why you can have things like gravity turn off in a BBB and also how you can begin to account for entropy recycling in the universe on a step by step level. The speck/motes of the clintrons are turning off some forces and recycling entropy and can all be calculated technically from a quantum mechanical standpoint finally allowing for the first proper explanation of things like entropy recycling on a mathematical level beyond absurdities like mirror universes with backwards time.

Unlike absolute zero which can never be reached as matter is always in contact with ESS which will therefore always have a macht level above the ESS macht pool levels of the universe and so always be trying to balance itself out which is a form of heat SM and TSM can be thought of as reaching an absolute temperature. No it does not mean that SM and TSM are super cold and will freeze you instantly should you come into contact with it as this is a non-SM way of thinking. To understand the temperatures inside a black hole you simply must accept that physics do not break down but that temperature is no longer needed in these areas of the universe. What this also means is that there is yet one more absolute value in the universe as this is what can be viewed as the temperature of the clintron TSM as it is the last possible structure that can ever exist in the universe or the first possible structure depending on how you look at the universe and its life cycles. All properties of the clintron TSM are absolute values for the universe and once measured can be used as starting points off of which we can create a foundation and reference point for all other physics in existence.

Relativistic Jets

Now that you have learned a bit more about the macroscopic nature of black holes and some additional pieces of information concerning their microscopic make up as well I can explain some of the behaviors of SM masses known as black holes. The first thing I will explain here are the relativistic jets created by some black holes at the centers of active galactic nuclei. A relativistic jet is a jet of high energy materials in the form of particles and energy streaming away in basically narrow beams from the poles of super massive black holes. The particles can range from electrons to positrons and are commonly seen annihilating with each other as they travel upwards and downwards from the active galactic nuclei. These annihilations are detected by the pairs of high energy gamma rays produced as the two particles meet and annihilate each other. It is often that the hundreds and thousands of light year wide swaths of detectable radiation from these active galactic nuclei are simply spread out and diffuse due to interacting with the interstellar medium wash of the universe. The speck/mote geometries of the electrons and positrons are rearranged by their violent collision such that they lose many of their familiar properties and are no longer able to interact with the Higgs field rendering them as what modern science calls a massless particle. Thus they become photons and are able to begin ESS macht tracking at light speed which moves them at, surprise light speed. The streams of particles are known to come from black holes as they are the only objects in the universe with enough power to create such phenomena. The reason is all other known objects do not contain enough matter and energy to force particles at their fundamental levels to behave in a manner that will account for the unique and high levels of energy released.

How black holes produce these jets is not exactly known and how they work in some black holes but not in others is also unknown to modern science. To this end only two answers need be explained concerning these jets and black holes the first is how the jets are made and the second is the rule explaining why only some black holes have them. I must remind the reader here that relativistic jets are not at all the same thing as gamma ray bursts from supernovae as gamma ray bursts last only a few short moments as compared to relativistic jets which can persist for thousands of years or more. Certainly the mechanisms both generate high energy photons but obviously the output duration varies by a huge degree and so the two are not perfectly related. A relativistic jet that extends for tens of thousands of light years means that it has been working for quite some time as the particles in it can travel at a maximum velocity of light speed and so the size gives the age of the jet. In a previous section I said I would explain gamma ray bursts in a later section and this is the later section so if you have a modicum of patience more I will explain gamma ray bursts after relativistic jets.

The mechanism by which a relativistic jet works requires spatial compression of ESS and matter travelling with the ESS to eventually give you a single jet. The first thing required is an object with sufficient mass that it can highly spatially compress the ESS of the local area of space around it. The second thing you need is this object in motion such that it will cause the ESS to spin around it like the ESS that are spinning in spiral galaxies to account for the flat rotation problem. The amount of spin can be incredibly small in fact as it only needs to be moving space a little bit in order for a jet to have the necessary space conditions to exist. The third thing you need is matter to fall towards the object and be used to help form the jet and of course become the visible materials that we can see as the jet itself.

The start of the whole process is of course to have an object like a neutron star or black hole formed from a large star by way of a supernova. The object will then need a source of materials to help create a jet and so what you find is matter surrounding these objects sometime after they are formed as the process of forming them will almost always clear the local area of space of all remaining matter. This is of course due to the explosions that help create neutron stars or black holes and push stellar materials into supernova remnants. The object which I will now simply refer to as a black hole for simplicities sake is also in need of a certain amount of spin. The object is massive enough to spatially compress the ESS surrounding it and due to the ESS nature to share macht between schales will mean that as the object rotates the ESS surrounding it will eventually be dragged in the direction of the objects rotation as a way to follow macht glows.

An interesting note is that an object that is at rest as massive as a black hole can actually start spinning as matter coalesces around it. A perfectly still black hole in a cloud of gas and dust will over time begin spinning and the more material you let fall into the black hole the faster it will spin. The spin will increase until a relativistic jet is turned on. The mechanism for this is simple and can only work using ESS as the basic structure of space which will not break any known laws of physics or observations and will in turn explain the way relativistic jets are turned on. As matter begins to be attracted to the black hole an accretion disc will form which must obey the law of conservation of angular momentum. What this essentially means is the matter begins to spin as it falls into the black hole and the closer and tighter it gets to the surface of the black hole the more it spins or spirals around the black hole.

So now you have a perfectly still non-rotating black hole accreting rapidly spinning materials around it and this is where the ESS come into play. The ESS as you know share macht between themselves using the schale structure of space and the spatial compression of a black hole is high enough to firmly grip any ESS outside the event horizon in the nearby region of space. The ESS have a reserve macht pool that is increased as matter interacts with it in various ways and one of these is gravity. In order for ESS to accept donated energy from matter they share it amongst each other in a general field surrounding that matter and the closer to the matter the stronger the field and the more spatially compressed the ESS become.

What this means is the rapidly spinning matter is bound to the ESS it envelops in a sense and as the spinning matter spirals around the black hole the ESS are trying to share macht between all local area ESS. Since the black hole can pull objects downwards towards its surface at near the speed of light this means the spiraling matter is falling in a circular motion at near light speed. Therefore the spiraling matter creates an ESS macht glow and the ESS attempt to balance macht between themselves and this means between the ESS being enveloped by the rapidly rotating matter and the ESS that are non-rotating inside the black hole and on its surface. Over time the ESS attempting to share macht must also maintain the conservation of donated energy between the matter and themselves. These rules were described in a previous section and are of course at work here where the ESS cannot simply steal donated energy from the matter as the matter is not 100 percent interacting with the ESS yet as would be the case in a single piece of clintronTSM. The matter continues to rotate and the ESS obeying the conservation law attempt to follow the source of donated macht glow meaning of course they start to move slowly.

This movement is not going to happen in areas of lower spatial compression such as around the Earth because the Earth simply cannot spatially compress the local area ESS enough to make this phenomenon occur. In a black hole however this can and does occur even if the black hole is at rest in space because it must obey the conservation law of donated energy and so the ESS surrounding the black hole start to move around the stationary black hole. Since all ESS in the universe are connected to one another through reserve macht pools the ESS inside the black hole will eventually try to follow the macht glows of the ESS enveloped by the rapidly rotating matter outside the event horizon. This is further proof that gravity can only be explained using two factors which are ESS and macht as a single theoretical spacetime does not allow for a curvature to begin acceleration of a black hole. Only freely rotating and non-twisting space as described by space physics that can transmit a real Standard Model force of gravity through space allows for this to occur.

Once again I need to state that black holes are perfectly normal objects that do not require bizarre new physics to explain where time stops, dimensions drop to zero and gravity becomes infinite all due to false objects called singularities. The space made of ESS outside the black hole seamlessly continues right through the event horizons and down into the very heart of the black hole. There are no special cases, there are no exceptions there are no impossible derivatives or seductive non-existent mathematical descriptions to be found here or anywhere in the universe. Therefore the normal ESS macht glow sharing between schales which I have shown to underlie every process in the universe does not breakdown simply because you enter the event horizon and the black hole proper. The ESS in the black hole simply start to follow the ESS outside the event horizon and begin to rotate as long as sufficient pull is given from the matter outside the SM mass.

The rotating mass outside the black hole is pulling ESS with it and so those ESS pull on the internal ESS of the black hole and eventually the whole process will result in the static black hole becoming a rotating black hole. Over time the rotation of the SM mass will attempt to match that of the matter outside of the event horizon and therefore will start accelerating to however fast that matter might be moving. If that is close to light speed then the SM mass will attempt to match that velocity even if it never does achieve light speed itself. The mechanism necessary to start black holes spinning will however be turned off as the matter falling into the black hole becomes a relativistic jet and energy transfer between the SM mass and free matter changes slightly.

This process can best be explained by explaining how relativistic jets are created and so I will now begin constructing a relativistic jet from a rotating black hole. The rotation of the heavy object being the second thing needed to form a relativistic jet and the third being matter. A quick note here which is a neutron star or black hole can be easily formed already spinning and this would satisfy the second criteria straight away I only wanted to explain the mechanism by which a particular type of heavy object in the universe sufficient to spatially compress ESS to a high degree can begin spinning once matter starts to accrete around it and fall into it. It is highly probably that the object will be formed with a spin already part of its structure and as such will naturally begin to move the ESS around it in a circular motion. Since ESS are what space is made of then it is natural that the space around the black hole will be capable of affecting matter in exactly the same and different ways as normal space all at once. It will affect matter exactly the same as normal space due to the fact that no physics change near a black hole but it will produce a different observation for us here on Earth.

This is of course the formation of a relativistic jet which we do not see coming from space and matter anywhere else in the universe. No stars, planets or smaller objects produce these jets and yet all are made from the same building blocks as a black hole. The key to space and the creation of relativistic jets is in how matter interacts with it before being allowed to pass the event horizon of the object. In the case of a neutron star it is simply passing a threshold value where it accumulates onto the surface of the star rather than pass through an event horizon and therefore becomes a clintron. In addition there is nothing stopping neutron stars from reaching a carrying capacity for materials before it violently gets rid of some accumulated material once again.

As for a black hole which I will use as the main example here we find that matter falling into the black hole will meet up with rotating space on the way down. The normal path for anything falling into a black hole should be straight down as this is the simplest path caused by gravitational forces. However in a black hole the space around it is rotating and matter is bound to space so the matter must follow the space. To this end we remember that space is made of ESS which speck/mote geometries are naturally attracted to and as such the matter must follow these ESS wherever they go in order to satisfy macht glow tracking rules. The matter falling into the black hole will be forced to move in a circular fashion around the black holes surface as it follows the ESS which are also moving around the black hole. The accretion disc is primarily focused at the equator of the object and so we see the most matter accumulate there. So the equator of the black hole has a very large amount of materials in it and will therefore be the starting point of the relativistic jet formation process. Yes the jets start out at the equators in most cases at least in how they are formed despite the actual jet erupting from the poles.

Space is made of ESS and ESS can only absorb so much energy from matter given a set amount of spatial compression. In order to increase spatial compression the donated energy of matter must increase so that the ESS can once again decrease their volume and repeat the process of adding gravity to the local area of space. In order to add more matter to the black hole that matter must be allowed to enter the black hole and be converted into clintrons to complete the SM mass arrangement. Since the matter is held in a rapid circular orbit with the ESS around the black hole it is not really falling directly into the black hole and it certainly is not adding it's mass directly to the black hole. This does not stop matter from falling into the accretion disc however as the force of gravity from the black hole is of course held and shared through the ESS macht schales that reach out into the space all around the object. Therefore more matter than can be held in the orbiting ESS accumulates around the equator and there is only one place for it to go which is the space around the black hole that is not around the equator. What this means is the extra matter that the black hole is not consuming at the equator because of course some is slowly falling into the black hole proper cannot be held in the saturated ESS found there.

With the ESS saturated the matter must move away from the equator towards the poles as the space there is not yet filled with too much matter for the ESS to hold onto given the set amount of spatial compression for the local area of space around the SM mass. The matter slowly moves let us say upwards towards the top pole of the black hole through the not yet fully saturated ESS along the way until it reaches the exact top point of the black hole. This is where the relativistic jet is born as all the ESS surrounding the black hole are rotating with the fastest rotation at the equator but still some rotation occurring at the poles. Yet there is a single point for each pole that is not rotating as each black hole is encased inside and out with ESS as space permeates the SM mass fully and extends seamlessly outwards into space in all directions. This means that a single ESS at the event horizon of both poles of the SM mass must be in a non-rotating motion and is the only single ESS no matter how spatially compressed at this point to not be orbiting in a circular motion.

These two ESS, one at the top and one at the bottom of the black hole are motionless and stationary and all matter that has been slowly moving upwards to these points will eventually encounter a non-rotating ESS. The reason the matter has not been falling directly into the black hole up until this time is that it is like an object orbiting the Earth where it is falling almost as fast as it is moving meaning it stays fairly stable over time. This orbiting and falling trickery is over once the matter reaches the non-rotating ESS and there is no longer a mechanism to keep it from falling this time directly downwards towards the black hole. At these two points around the black hole it is as if the matter is falling straight through the space that it could not fall straight through anywhere else on the black hole as it gets caught up in the circular motion of the ESS there. So the matter that has been accumulating around the equator can finally fall directly down into the black hole and reach the event horizon at which point the relativistic jet is born.

Now then I will explain why this falling process creates a jet in the first place and then I will explain why it is a thin jet that is formed and not something larger. You have probably heard many times over the statement made by modern science that a black hole simply cannot eat matter fast enough and so it spits some of it back into space and this is what a relativistic jet is. What that statement is is another of modern sciences famous ways of saying it has no idea what is really going on at a fundamental level and so makes a sweeping statement that the black holes simply eats too fast yet once again offers no precise step by step explanation of why this is so. I will explain in this section exactly why black holes cannot simply eat all the matter at once and I will show you the precise point by point, quantum object to quantum object way of describing this. Why this explanation gives rise to a narrow jet and how these jets contribute to the life of the object creating them.

To begin with the ESS and total amount of spatial compression around the object must be considered as well as the matter interacting with it. The matter as you know is attracted to space and the space has a maximum amount of matter it can hold onto using ESS given the amount of spatial compression the ESS are under. This is what we saw earlier with the matter orbiting the equator of the black hole but being unable to fall into the black hole which results in a failure to increase SM mass and therefore a failure to increase the gravity of the black hole which would allow for further spatial compression of the ESS and therefore increase the carrying capacity of the ESS for matter. More matter is falling into the black holes orbiting space from the accretion disc than the black hole can absorb and so additional matter must move towards the areas of least saturated space which is found near the poles. The two ESS points at each pole are where the matter is no longer orbiting and so this is the point where the matter can fall into the black hole and pass the event horizons which allows the matter to become clintrons and add to the total mass of the object. Matter falling into the black hole will be added to the total SM mass but not all of it will fall into the black hole due to a process by which ESS absorb matter and increase spatial compression.

As matter interacts with ESS around the equator of the black hole it goes from a state of being cold and low in energy to one where it is heated to extreme temperatures and is high in energy. This is of course due to the fact that the matter as it existed in the universe was simply gas and dust floating about minding its own business. Out in space it would be very cold and not possess much overall energy and so the interaction with the ESS at the equator of the black hole is profound. As the matter falls into the black hole and meets up with the orbiting ESS it will experience an increase in temperature and energy from the conditions which exist close to the event horizon. Particles will be accelerated to near light speed, they will be compressed tightly against other particles and will receive energy in the form of kinetic energy from the motion of the ESS at the equator of the black hole. The matter is obviously in a very different state than it was before at what can be considered its rest state for the system. The matter travelling towards the poles is highly energized as a result but does not desire to remain so as it is trying to emit energy such that it can return to a more relaxed ground state. As long as the matter is held within the rotating ESS it will be forced to remain at higher energy levels than it would like to. However this of course changes as it reaches the area of space directly above the poles of the black hole where space is no longer rotating and is in a sense static once again.

The matter is now free to fall directly into the black hole following a straight line path and this is where it encounters other matter doing the exact same thing. As the matter enters this area of non-rotating ESS it is able to finally try to shed some of its excess energy and so will do so all at once along with all other matter in the local area above the poles. What happens is the ESS at these points above the black hole are forced to try and deal with an overload of donated energy from all the matter there and without the rotation of the black hole to lock some of that energy away it cannot. In addition this area of space is less spatially compressed as compared to the equators simply due to less material accumulating near the accretion disc and so cannot carry as much again. The differences in spatial compression from one area of the black hole's ESS space to another does not need to be much for it to have profound effects on any matter in the area.

As a result the excess energy released all at once from any matter reaching the non-rotating space above the poles is expanding and escaping in any direction it can and one of these is up and out of the black holes local area of space. The region of space where the matter is dumped over the poles and where it is allowed to shed some of its excess energy is above the event horizon and in a region of non-rotating space so nothing prevents it from travelling at or near light speed straight away from the black hole itself. No laws of physics need to be new or strange here as the matter and energy behave as though nothing odd were happening because nothing odd is happening in the region of space above the black hole poles.

Therefore one part of the relativistic jets is created simply from the normal desire of matter to return to its lowest energy state or ground state. Since the matter has been given this increase in temperature and energy from interacting with the black hole it is free to release it once again. The net result is matter simply exploding or expanding away from the black hole at near light or light speed. The area of space where this occurs is of course hot, violent and dense with fundamental particles and so collisions will be frequent and highly energetic. This means that just like in a particle collider matter will be partially or fully ionized often and forced to collide with other particles on a regular basis. As a result numerous collisions will produce copious amounts of fundamental particles streaming away from the black hole. These will include things like electrons as well as positrons as these regions of space are quite good at producing anti-matter due to energetic collisions. In addition high energy photons of all spectra will be created and sent away from the black hole along the paths of the jets as well. What are produced are the normal jets and plumes of x-ray, gamma rays and annihilation reactions seen around galaxies in the universe.

The strength of the black hole far exceeds any particle collider we can make today and probably for some time yet and so the number and type of particles created here is larger than any sample we have seen to date. If you recall that it takes violent interactions of speck/mote geometries with each other or with high levels of ESS macht to create anti-matter then you will see here the perfect place to find particles of all types in the universe. Only violent reaction on the most minute speck/mote geometries will rearrange them so that they edge drag in different ways with the ESS of the universe. It is known that different edge dragging geometries will cause specks and motes to interact with the ESS macht such that they produce anti-matter property particles. These heavier particles or less stable ones due to how the speck/mote geometries they are made of interact with the ESS and how in turn the ESS try to balance out their macht levels will not last long enough or exist in large enough quantities for us to observe them from Earth.

All that we can see are the frequently produced particles like electrons and positrons which have very small speck/mote geometries allowing them to be produced often and to not donate a large amount of energy to any local area ESS such that the ESS are rapidly trying to balance out their macht which would result in a short half life. Since they are produced so easily and in such large amounts they can remain stable for hundreds and thousands of light years out into space. In the case of the positrons it is likely that over time they will meet and interact with electrons and annihilate either from electrons in the jet itself or simply in the space around the galaxy or in the interstellar medium beyond it. This also explains why x-ray and gamma ray production is seen to be quite high around galactic poles well out into space as the amount of time and the volume of space allows for any number of interactions with the numerous particles found there.

This being said the energy released from the matter as it reaches the poles will be released in all directions as one would expect it to be. Therefore the expanding sphere of matter and energy released around the poles must be dealt with such that only the matter and energy travelling away from the black hole in jets can be accounted for. This is easily explained using the ESS in the local area of space around the black hole and inside the black hole proper.

In order to explain it I will begin by explaining that the black hole cannot simply absorb all the matter falling into it all at once and so again some energy will be in a sense thrown upwards in a spherical or hemi-spherical shape from the surface of the black hole proper. The mechanism for why this occurs I will explain later I only want to illustrate that even more energy is being sent away from the black hole in what should be a pattern much larger and more diffuse than the relativistic jet alone which is very narrow and concentrated by comparison. The first thing I must do here is explain the shape of the space above the poles of the black hole. Remember that the event horizon is where the single non-rotating ESS is located as below this is the SM mass of clintrons which are held together as one. It is only the region of space above the event horizon that is allowed to rotate freely and rapidly and therefore there can only ever be two exact ESS which are non-rotating one at each pole and sitting directly on top of even touching the event horizon.

If you think about how ESS are arranged in space you remember that they are like little spheres sitting snuggled up to one another so that they will share macht through schales. If you look at the relativistic jet with ESS in mind from a side view of the black hole you can easily see that the single non-rotating ESS will have other ESS sitting on top of it in the second schale of ESS. The non-rotating ESS is of course in the first schale as this is the first schale of ESS outside the SM clintron mass. Since the nature of space is to expand as you travel away from a single point and that point is a non-rotating one and the fact that the rotational forces are not coming from the pole but from the equator you can see that the schales expanding three dimensionally above the single non-rotating ESS will also be non-rotating.

This is again due to the fact that the rotational energy for the ESS comes from the equator of the black hole and this spreads out and dissipates towards the poles as well as dissipates in altitude from the surface of the event horizon of the black hole proper outwards through schales into space. If you need to take a moment to visualize all of that in your head with the little hollow ESS spheres in mind please do so as you cannot understand the formation of jets in physical space without it and this is the one and only way that space is arranged around these jets. There are two different directions in which the equatorial rotation force on the ESS around the black hole dissipate in three dimensional space through schales and only the poles are without these forces.

This results in a single conical region of space that is very concentrated and narrow but non-rotating extending upwards away from both poles of the black hole and can be thought of as a region of static space tunneling outward from an ocean of rotating space. The over lapping dissipative forces from the two rotational factors will attempt to close this space but cannot because of how they diminish from their point of origin around the equator. What they will be able to do however is further concentrate and narrow the conical shape from which the beam of matter and energy can emerge that is the relativistic jet. All in all this explains the way and the reason that the jets are so concentrated and narrow as they emerge from a black hole and travel out into space. It explains why it is so thin a beam but it is only the beginning of explaining why the jets are not hemispherical in nature. This is explained by the rotating region of space outside of the non-rotating cone of space as matter and energy are released in all directions at once yet are only seen to be travelling in one.

The reason that we only detect matter and energy in these thin jets is because any matter that is released in a direction that is not inside the angle of the conical structure of the non-rotating space will obviously enter the rotating region of ESS once again. This results in the matter and energy simply falling back into the same region of space where they were bound before as the rotating space was sufficient to hold them before and will do so again. Remember that the matter and energy falling into the static region of space were allowed to shed excess energy meaning they are now less energetic than they were before they entered the static space. Therefore if they did not possess the energy necessary to escape the black hole and the rotating region of ESS before they now have even less energy and ability to attempt the same feat.

As a result they are now in a less energetic state and will be forced to re-enter the rapidly rotating region of ESS where they will be once again held for a time just as they were when they first fell into the black hole. The only force that has not diminished in the rotating region of space is gravity and the less energetic materials falling back into it will have less ability to resist the force of gravity which will in turn pull it deeper into and hold it more firmly in the rapidly rotating ESS. This mechanism easily explains how matter and energy released in the static cones of non-rotating space are forced to re-enter the rotating ESS and be hidden from our view as we try and detect it. This matter will not need to stay here permanently though as it will eventually either travel once again to the pole where it has the chance to escape again or it will be pulled down into the black hole where it will become part of the SM clintron mass.

Remember once again here that all matter and energy in the universe are simply made of specks and motes with certain geometrical arrangements and the properties of a clintron are such that all geometries are rearranged so that only the edge dragging of gravity is dominant therefore anything falling into the black hole proper will be forced to rearrange its speck/mote geometries to become a clintron. This is regardless of what properties it had before no matter what they are such as mass, charge spin or whatever they will all be rearranged into clintrons in a single SM mass.

What needs explaining now is the characteristic of black holes to not absorb matter and energy instantly. This explains the long held statement by modern science that black holes simply cannot eat fast enough and so some matter and energy is burped or coughed out. Once again modern science offers no precise explanation on a fundamental level of why black holes cannot eat as they say fast enough. In reality according to modern science they should be able to eat everything without hesitation as according to modern science gravity is the result of a curvature of spacetime and there are no rules in general relativity to prevent spacetime from curving as fast as need be. The universe according to general relativity does not exist in quantized amounts at all scales as general relativity cannot be incorporated with quantum mechanics to produce a quantum theory of gravity. General relativity simply states that spacetime is a perfectly smooth and featureless curvature created from the world lines of objects from frames of reference. This means that so long as matter is added to a black hole the equations of general relativity simply produce a new curvature of space given the mass in that same area without a need for quantized steps.

So according to general relativity spacetime will curve as fast and as smoothly as needed to simply eat all matter falling into the black hole with no problem because the gravity necessary to do so exists instantaneously and spacetime can curve faster than the speed of light. Everyone reading this should know general relativity to be wrong about that fact simply because nothing no matter how small you look in the universe exists in a smooth and seamless fashion as proved by the Standard Model and quantum mechanics. So since modern science therefore does not have an explanation based on the three main views it employs to explain why a black hole cannot eat as fast as it needs I will show how this is possible using ESS, macht and speck/mote geometries.

As I have made clear simply saying something happens for a random reason is not at all an explanation for why that thing occurred in the first place. Too often in modern science unknowns are treated as real and then people build from there on out. Perfect examples of this are both quantum mechanics and general relativity which at different times say things just happen for a reason or just because or due to random probability which no one can know. Well all of that is nonsense as nothing happens just because and nothing happens due to a probability that no one knows. I have said this before and I will say it again everything in the universe works perfectly fine whether or not humans or other intelligent creatures have figured it out. To this end there is no just because effect in the universe and there is no random probability only a set of rules known to the universe and not modern science that makes things work the way we see them.

I would like very much through my work to explain these rules that are hidden from modern science but are not strange or odd at all as they are a normal part of the universe. When I think about problems I do not think about our understanding of them I think about the universe and how it works and to this end I think about how the universe puts itself together in a way that works and how much of modern science can mesh up to that is not important. New ideas about the universe did not mesh with modern science centuries ago when it was proved that the Earth was a sphere and the Sun was at the center of the solar system. So just because a new idea does not mesh up with the status quo of modern science is a red flag for modern science to catch up with the new idea and not the new idea being made ad hoc to fit the views of modern science.

A black hole and why it cannot absorb all materials instantaneously is exactly the same and has a precise mechanism by which this occurs using ESS, macht and speck/mote geometries just as they are used to explain every other unsolved problem in modern science. To begin with the ESS that exist inside and outside the black hole proper behave as any normal ESS does by interacting with matter such that the macht it carries is altered from the reserve pool level. If we remember way back to the first section of space physics we see the room filled with little hollow ESS spheres and the mostly opaque matter orb. The interacting of the orb with the spheres results in the spheres decreasing in volume as macht glow increases and this is very basic space physics. Now when we saw the orb move from right to left in the room we saw the ESS it approached on the left decreasing in size as the ESS on the right that it was moving away from increased in size to reflect the overall macht glow imparted by the matter orb with the hollow spheres. This process was gradual as it took time to complete and did not occur instantaneously because there is a light speed or slower exchange of energy donated from the matter with the space it envelops and there is a macht speed or slower sharing of glow between schales. The macht speed sharing is near instantaneous compared to the very slow light speed reactions but it still has a limit and therefore will not occur in zero time. This means that ESS take time to decrease in volume and then share that macht glow through schales with their neighbors. Applying this to the black hole we see that matter falling into the static space above the poles will encounter a small hurdle when trying to enter the black hole and this is the limiting rate of spatial compression of the ESS inside the SM mass.

The matter and energy available to the black hole is more than it can absorb because of a specific series of events in how matter makes space eventually spatially compress. To begin with let us look at normal ESS inside a black hole and see that they are very, very small as they have very, very high macht glows of gravity running through them. The inclusion of more matter would seek to elevate these macht levels which would in turn spatially compress more ESS but the first step in the series of events is to have a piece of matter interact with the ESS in the first place. So the first step is a slow light speed or less quantum interaction that donates energy to the ESS of the black hole and the ESS must receive this donated energy first in order for it to increase its macht glow. The second step is the macht speed sharing of the new donated energy between schales of ESS such that the macht glow gravity is balanced in the local area of space again. This is followed by the third step which is again a macht speed reaction and actually takes place beginning with the energy donation but spreads slowly as the macht glow is spread out amongst the ESS and it is the decrease in volume of the ESS. The fourth step is made possible finally now that the ESS have spatially compressed more as space can now hold more matter because it has more ESS to hold onto it with.

Remember that all ESS anywhere in the universe have a finite carrying capacity for matter based on the amount of spatial compression they are under and if that spatial compression is insufficient then you cannot put as the saying might go five kilograms of matter into a four kilogram ESS bag. Increase the spatial compression of a local area of space and you can put as much matter into the ESS bag as you like until you reach the point of forming a single clintron TSM and then the bag breaks open again.

This is why under normal spatial compression here on Earth it is impossible to truly recreate SM mass conditions or those of the clintron TSM detonation event as no particle collider no matter how large will ever be able to do one crucial thing and that is maintain a continuous spatial compression such that all particles collided and created or destroyed in the event are travelling through highly spatially compressed space. All events created here on Earth travel out in normally spatially compressed space that does not tell you at all how these particles truly behave under the high energy, temperature and spatial compression conditions that create them. This brings us finally to the fifth step which is the fact that space from outside the black hole in the form of less spatially compressed ESS now have enough shared macht glow from the use of schales that they can decrease their volume and move inside the SM mass. This allows the SM mass to finally increase its overall carrying capacity for matter as the number and size of the ESS is now sufficient to receive more matter. The sixth step will therefore be additional matter falling into the black hole from the static conical space above the poles being added to the SM mass.

There is a seventh step which involves the rearranging of speck/mote geometries such that they have reconfigured themselves to break apart from the forms they possessed outside the black hole and into clintrons inside the SM mass. This process is again mediated in a non-zero speed and will be a combination of quantum and macht speed events as both matter and space are needed to rearrange speck/mote geometries. This step happens while the others are occurring but takes some time as the matter upon passing through the last event horizon must be converted into a clintron geometry in terms of its specks and motes as this is the only particle that can form SM masses. The time it takes for this to occur also means that the black hole cannot simply absorb the matter in zero time as rearranging the speck/mote geometries will overall take some time and varies depending on what matter was breaking down upon entering the black hole proper. The combination of these series of events means that a black hole cannot absorb matter at an instantaneous rate as there are rules for how the universe incorporates matter into space. This also shows the precise mechanism by which all fundamental properties and parts of both space and matter interact to explain this phenomenon.

This entire process now that it is fully understood and explained is able to therefore contribute to the creation of relativistic jets from the poles of large objects like black holes. Part of the jet as we have see is simply created from the extra energy being released from the matter that was given to it by the black hole. That energy is under pressure to escape the system as much as any hot gas desires to escape a small space and so accounts for part of the materials leaving the black hole in the form of jets. The inability of the black hole to absorb materials in zero time creates another bottle neck effect for the matter that is trying to enter the SM mass and this point is the single ESS above each pole and the small region of space around it. The matter falling in simply cannot go through all the available steps in the series of space and matter interaction as described above. Since it is also only the static space that is readily accepting new matter as opposed to the rotating space there is a finite amount of ESS available to start the series of events. So more matter is trying to be absorbed by the SM mass than it has ESS available to do so and under these conditions normal physical laws can apply. Some of these laws are simply compressing too much material into too small a space too quickly as is seen in a fusion reaction or a particle collider. The matter either collides with violent results or tries to exceed the Pauli Exclusion Principle limits and must therefore move to areas of space where these violations do not occur.

What is commonly happening is akin to a series of micro-supernovae events that occur on small scales but are continuous and relativistic jets are in fact areas of continual supernovae explosions. The matter is accelerated to near light speed into a single point like in a collapsing massive star and at the center it behaves like a type 1a supernovae that must create a burst of energy to release the building pressure. In this case the point of formation is the single non-rotating ESS above the event horizon at the two poles of the black hole. The ESS macht degeneracy pressure can be said to be overloaded at this point if you would like to have a simple quantum mechanical analogy ascribed to it although it is more accurate to say the ESS macht levels are too overrepresented and cannot balance fast enough. This point is where the materials rushing inwards exceed all carrying capacities and will become a micro-supernova of sorts that ejects material into space and is immediately followed by more material falling into the black hole to repeat the process. Whether you look at this as a series of never ending small and controlled supernovae events like a type 1a supernovae or simply a never ending series of high energy collisions the result is the same, matter must be excluded from the SM mass until it can spatially compress its ESS enough to carry more.

The net result therefore is matter needing to escape the confines of the polar regions of the black hole which in many ways can be thought of as anti-vortexes as they are regions of non-rotating ESS as opposed to a normal vortex which rotates in an area of non-rotating materials. So as the matter explodes it can once again travel into the rotating walls of the ESS anti-vortex or upwards and out into space in what we see as a relativistic jet. The explosions are violent and largely nuclear in nature so occur at or near light speed depending on the end products of the reaction whether they are matter or energy respectively. This accounts for the high velocity and high concentrations of fundamental particles being created inside the jets such as electrons, positrons and very high energy photons. I must remind the reader here that I use the terms energy and matter to accommodate traditional thinking but in reality both are simply different forms of speck/mote geometries only appearing different as to how they choose to edge drag with the ESS of space. This is the last mechanism by which relativistic jets are created from objects like black holes and explains step by step the precise process through which space and matter interact to allow these events to occur.

Now here briefly as that is all that is needed is the explanation of gamma ray bursts from supernovae. As I explained before supernovae are created from TSM schale detonation layers that push materials into a heavy element core and the core becomes the black hole. The compressed stellar plasma I mentioned before is the fuel for the gamma ray burst and comes about simply because as the stars core collapses into a black hole the law of conservation of angular momentum holds and the black hole spins rapidly. This spin creates the necessary rotating ESS and will form two anti-vortexes at the poles of the new SM mass with the stellar materials on top of it. Using the rules for how black holes absorb matter into their ESS and can then increase the carry capacity of the ESS for more matter we see the delay mechanism for how SM masses consume materials played out. There is not much stellar plasma left for the black hole to eat so it is consumed quickly which explains the short gamma ray burst as a short supply of fuel is all that is available. So here finally is the very simple explanation of how gamma ray bursts are created from supernovae, why they are so short and due to the massive amount of energy stored in the ESS by the heavily interacting speck/mote geometries with the local area ESS why they are so energetic. Space and matter shape the ESS anti-vortexes, space and matter fuel the jet itself from the materials of the accretion disc and the motion of ESS around the SM mass and finally the way space and matter need to interact to increase spatial compression create the conditions to release concentrated and narrow beams of matter and energy out into space at relativistic speeds.

This brings us to explanations of the relativistic jets and active galactic nuclei as we observe them from Earth in the universe today. To think of the jets is to think of some of the speck/mote geometries that have changed in the universe as they are needed to construct things like neutron stars and black holes. A neutron star may have a weak magnetic field or no field at all while a black hole always has no field. That is a field that is coming from the black hole proper as opposed to materials accreting around it. However without accretion materials the black hole has no magnetic field and with the use of ESS in space these astrophysical jets are explained without the use of magnetic models. It will always be the influence of the black hole on its surroundings that cause the generation of a magnetic field such as a black hole whipping the local area of space around it into motion and energizing it but the magnetic field will not be generated from inside the black hole proper.

Magnetic fields will obviously have some influence over materials around the black hole and I am not saying that magnetic fields contribute nothing to the overall behavior of materials and black holes. What I am saying is that magnetic fields by themselves cannot hope to explain the jets as they simply do not offer an explanation as to how these jets are formed and operate alone. Where magnetic fields will become more influential is in how materials are pulled into the black hole specifically before they pass through the last event horizon. So if you look at a typical black hole as it consumes materials there are many chances that the path the materials will take on their way down into the black hole will be anything but perfectly smooth. The model of a smooth and totally uniform accretion disc is highly useful in understanding how parts of the universe work from anything as simple as a solar system forming to a black hole creating an astrophysical jet.

These models of uniform accretion discs however are rarely played out in the practical universe as we have seen by examining things like planet formation around stars. Take our own solar system for example and you see that there is no neat tidy order to the planets such as smallest to biggest, or vice versa. This simple observation is repeated regularly in astronomy as we look at planets in other star systems. Despite not being able to see these other worlds directly we can see their effect on the parent star to which they are gravitationally linked. What we see is that each solar system in the universe is unique producing its own version of what a star system should look like. Some systems move their stars a great deal from side to side while others do not and this alone tells us that accretion discs are non-uniform because if they were then we would always see the same side to side motion in each system. Without observing this we know that each system formed uniquely and therefore there is no simple rule for size of accretion disc and distribution of planets by mass in the star system.

A black holes accretion disc is undoubtedly both more uniform and more disorganized than a star systems all at once. This is due to the fact that the black hole has a very long reach to its gravitational field that can extend outwards and begin influencing materials to fall towards it for quite a ways. The materials therefore have ample time to interact with each other on the way down to the black hole. Closer to the black hole, as in very close to the event horizons the strong forces of the black hole will in a sense pulverize matter sufficiently to create a smooth distribution throughout the ESS close to its surface. The reason this is important is that it allows for observational differences in outputs from objects like black holes that produce jets. The jet may move slightly over time, the light emitted from the accretion disc can vary and overall changes to the system will persist. In the case of the first we see that the black hole itself can be made to rotate about its absolute center if given enough external force.

What are the chances that a black hole even a super massive one at the center of a galaxy will ever encounter another black hole? The odds are pretty much one to one that over the very long lifetime of a black hole it will merge with another of its kind. Anytime you have two SM masses in close proximity to each other it is a foregone conclusion that they will move each other in the process. A small black hole is still an awesome object that will definitely affect a much larger one even if it is of course absorbed by the latter over time. Such mergers are likely to be along slightly different planes and therefore will cause a spinning or wobbling effect between the two.

If the larger SM mass is absorbing the Lion's share of the materials in the accretion disc it will probably have a relativistic jet being produced from it. The smaller SM mass will tug on both the materials in the accretion disc and the larger SM mass directly before and especially during the merger. What this means is that irregularities in the emission of energy as seen in high energy photons or any wavelength electromagnetic radiation will be visible to us here on Earth. It has been noted that the x-ray spectrum of certain heavy objects in the universe exhibit a quasi periodic oscillation over time as certain peaks in the x-ray spectrum brighten and dim. The simple motion of a merger between two large objects such as two SM masses or even an SM mass and a star will cause fluctuations in emitted spectra to be observed. The example of a star orbiting close to a black hole perhaps in an elliptical orbit provides a mechanism for some of these oscillations as when the star is at perihelion to the black hole or perhaps periholeion (admit it that is at least half clever) the increased force of gravity felt by the stars outer atmosphere will allow for materials to be consumed at a greater rate for a brief moment in the system causing fluctuations in observed energy outputs.

If a star is in a circular orbit with a common center of gravity with the black hole then the quite regular and normal solar prominences and coronal mass ejections of the star will provide short regular bursts of extra material to perturb the accretion disc and once again create quasi periodic oscillations in high energy spectra as observed. The stars normal magnetic cycle could be found to be hidden in the x-ray emissions of these black holes and using our Sun for example would yield a series of oscillations that revolve around a regular eleven year cycle. The underlying point is that the inclusion of any number of external factors to the SM mass and its surrounding area of local space can cause momentary increases in the background noise of the black holes emitted spectra. Since the number of possible interactions available to an SM mass in the universe is too great to list it is suffice to say that over the entire observable universe it should be perfectly normal to see changes in oscillation of SM spectra.

The materials that are entering the accretion disc of the black hole will be one source of observed oscillations but the materials that are leaving the accretion disc will also be another. Materials leaving the disc are those entering the black hole proper as once inside the event horizon they become a part of the SM mass of clintrons and can no longer emit anything into the universe that we can observe as visible photons. What these masses can do is alter the local area spatial compression and the ESS and macht found there to a degree that will in turn influence the accretion disc and any materials in it that we therefore can still observe using electromagnetic detection. Think of it this way as matter is spiraling around the SM mass it is also travelling with the rotating ESS found in the local area of space.

This matter is akin to anything in orbit around the Earth as it is moving at almost the speed it is falling but this is merely a trick as it will eventually fall all the way to the surface of the Earth or the SM mass. So despite the fact that materials may have been in orbit of the SM mass for quite some time they will land at the event horizon where they will be converted to clintrons. At this point the clintrons will force the ESS of the SM mass to follow the series of steps outlined above to increase spatial compression and draw more ESS into the SM mass. This will in turn change the rotational velocity of the black hole slightly, change its mass, changes its diameter and also and most importantly change the ESS macht in the schales around the black hole. Remember the ESS inside and outside the SM mass still share macht all the time as long as the ESS holding the matter do not violate the law of conservation of macht/energy for that area of space.

What this means is that the ESS will feel the increased macht and since they are bound by the rotational speed of the ESS and the transfer of quantum speed donated energy from matter to space they must conserve the total macht/energy for the system. The ability of ESS to share macht is far faster than quantum speed so what happens inside the SM mass will travel outwards quickly and will affect the materials in orbit around the black hole proper. This new tug on the materials as felt by the increase macht in the rotating ESS will interact instantly with the matter in orbit moving it slightly and since it must maintain the energy of the space and matter system it will have to shift this energy slightly as it moves to its new location closer to the black hole. The normal amount of matter being absorbed by the black hole is not very high and so the shifting of matter produces a low back ground noise of detectable energy for example x-rays. The energy and motions of everything around the SM mass is of course great and so energies are naturally expected in the x-ray ranges. If a large amount of matter that has been in orbit for some time which might be all the mass from a solar flare for example enters the SM mass all at once it will of course produce a shifting effect many times larger than just a steady trickle of gas and dust from the accretion disc. This will cause a fluctuation in the emitted light as we observe it that can be made to look like the quasi periodic oscillations observed today. The energy released might even have been as gamma rays and over great distance and travelling from high to low areas of spatial compression have lost energy to become x-rays.

The shifting effect produced will pull on the matter in the accretion disc which since it is at high temperatures will have some plasma like properties and will assuredly cause events similar to the magnetic reconnection found in the Sun's atmosphere. The same kinds of energies released in these magnetic reconnection events will be found here only on a scale far greater and therefore containing far more energy than in the Sun's atmosphere. The magnetic reconnection events will stir up the plasma again causing particles to collide with each other in highly energetic collisions that occur above the event horizon therefore allowing normal electromagnetic radiation to escape the black holes gravity and be useful for us to detect. Also the dump of energy from the magnetic reconnection event back into the plasma will easily be enough to push particles and atoms inside the accretion disc to higher energy states that will be capable of releasing quantum amounts of energy when they return to a more relaxed energy state like a black hole aurora. Again we see normal physics being used to create these necessary explanations for observation as long as they are combined with a proper explanation of the space physics at work as well.

In addition since things tend to stay in orbit around SM masses for so long an entire star could have been consumed and held in the accretion disc of a black hole of sufficient size. What this means is that a black hole that appears to be devoid of any nearby stars such as a binary or trinary system could years after it accreted the stars matter absorbed it into the SM mass which we are now just seeing. Any number of shifting effects from the absorption of matter into the black hole will cause the materials in the accretion disc to react in a very energetic way that we will then observe here on Earth. Even the possibility of a binary black hole system will cause the materials in a shared accretion disc to react as they are whipped up by and absorbed at different rates into the two black holes. This will cause a great number of changes to the local area ESS and macht between the two SM masses that will provide additional shifting forces in ESS macht as well as simply moving the hot particles, gasses and dust around in the accretion disc that will release x-rays of oscillating nature that then fade over time as the accretion disc calms down again.

Events of this nature are likely to be common at different times in the active galactic nuclei of many galaxies as over billions of years the number of stars, wandering planets and rogue black holes that will be absorbed by the super massive black hole will no doubt have an impact on how it looks to us as we observe it. Our own Milky Way galaxy is no different as we have already mapped the stars orbiting close to it but we have not mapped anything else as we simply cannot see that level of detail yet. So this means no regular black holes, no planets or anything similar and therefore just because we do not notice them does not mean they are not there. Over the entire observable universe many different galaxies are in many different stages of having their super massive black holes absorb things and so those that are we can detect as having oscillations for example while those that are just finished or about to consume something we see as stable. Overall however a great number of factors with fairly normal explanations can play out to explain quasi periodic oscillations as we observe them here on Earth.

Black Hole Evaporation

The question of stable is one that remains as one of modern sciences unsolved problems when it comes to black holes as they are believed to evaporate over time. The idea of black hole evaporation deals specifically with the gradual erosion of a black hole over many billions or trillions of years in which the black hole essentially turns into radiation. In one of the incorrect theories of the future of the universe the Big Freeze, it is thought that the future of the universe in the very distant future that is, will see nothing but black holes remaining. Over time it is believed that these black holes will slowly fade away and as they do so they dissolve into radiation that will eventually spread out in the universe and cool which gives the name the Big Freeze its meaning. How a black hole is thought to dissolve according to modern science begins with a black hole that is no longer consuming materials anymore as this would lead to it growing and not shrinking.

The black hole would then be vulnerable to the effects of quantum space where pairs of particles pop into existence from the vacuum and annihilate with each other completely removing any trace of their existence forever from the universe. The odd time this happens close to the event horizon of a black hole one of the pairs will enter the horizon leaving the other to not annihilate and therefore fly off into the cosmos. The black hole will lose energy in this process because it is thought that the black holes gravity is what excited the vacuum to make the particles appear in the first place. Since it took two particles worth of energy to create them and the black hole is only getting one particle back it has therefore lost the energy of one of the particles. What this means is that according to the energy and mass equivalency formula the black hole losing energy means it has lost matter and therefore mass as well and so has shrunk a tiny bit. Over time the black hole will repeat this process until it has dissolved into nothing and finally disappears with a violent explosion as it is thought that this is what happens at the end of the evaporation event. All of that evaporation is of course theoretical and only believed by modern science.

There are of course a large number of impossibilities with this theory that render it incorrect fairly quickly. To start with information is lost from the universe during pair production and annihilation events and information loss is thought to be an unbreakable rule of modern science. The reason for this is seen in the fact that no trace of the pair production particles can be detected in the universe ever before the event occurs as it is only the flat-line vacuum energy from whence it comes. This is impossible because information theory states clearly that all events in the universe must be preserved in information form and therefore this means past, present and future ones. So you must be able to know that the particles are going to be produced and yet you cannot so information theory has been violated thus proving itself wrong. Furthermore the particles if they do not fall into the black hole are forced to annihilate perfectly which do not produce photons like in normal annihilation events instead they produce nothingness again.

This is again a direct violation of information theory as you cannot back track and find out that particles were ever produced, what they were, what their properties were and how they affect the local area of space because according to quantum mechanics the particles just perfectly vanish into the vacuum. What this also means is that for no reason the universe has a single perfect entropy recycling mechanism which is free because it requires no energy for it to work. The entropy of the system is perfectly recycled as the particles are sent back into the vacuum and this means that it took free energy to make them and then free energy to recycle them hiding all the evidence and the entropy changes all of which is again a complete violation of all held dear by modern science.

Quantum mechanics also is the real culprit behind the idea of pair production and all this quantum space and quantum energy nonsense in the first place. Yes nonsense because once again the entire escapade rests solely on the failure of the mathematics to predict the correct one to one probability of the system and so a huge error that has plagued science since its inception remains which is that quantum objects can be two things at once which I am stating here they fully cannot be. A quantum object doing two things at once, being in two places at once or simple choosing to exist one moment due to probability and then choosing to not at a different probability is just more of the human flaw in understanding the universe as at no point does the universe not know what is going to happen. The universe cannot surprise itself and never sits there and all of a sudden says wow as something it did not expect suddenly happens. The universe is everything and therefore the universe is the cause and effect of every action possible meaning that all driving actions behind any human misunderstood quantum systems are fully known to the universe as they have to be in order for the universe to make them happen. The long and the short of it is that pair production spontaneously from a vacuum simply because the mathematics says it must occur is false. So once again quantum mechanics itself proves information theory wrong and so proves pair production wrong.

If you need far more physical and logical reasons before you believe that pair production from the quantum vacuum is wrong then so be it as here they are. First of all the particles annihilate yet produce no photons and this proves quantum field theory wrong right away as according to it the entire universe is simply made of excited fields and this includes energy such as photons and matter such as electrons and positrons that can annihilate too. So if the pair production particles do not produce photons as they annihilate this means they are not made from energy or vacuum as this would mean they are different than a field excitation. So vacuum energy cannot be quantum field energy as this would mean that as these particles annihilated they would give rise to a normal photon that we can detect.

In the event of a pair production close to the event horizon we see according to theory a single particle which is not trapped by the black hole flying off into the universe. This particle is now magically made of regular energy and therefore the kind that can be converted into a photon through the equivalency formula whereas if it had annihilated it would have been made of non-photon producing magical vacuum energy? Since black hole evaporation pair production gives nothing away it means that the information is again lost and that the particles were not made from vacuum energy in the first place as they would be made of the same stuff as all field excitations. Not to mention the fact that even if you let the magical pair particles annihilate you cannot get free energy from the universe and so even if no trace of them remains the black hole had to expend real energy to make them and so lost mass once again.

Another flaw is that the pair production would always lead to a decrease in black hole mass as the event was caused by gravity from the black hole itself. This is due to the fact that even if the particles were somehow made from normal vacuum energy then annihilating them would make two photons or two vacuum photons or something but they would not neatly go back into the vacuum anymore than electrons and positrons go back into the vacuum disappearing and leaving no trace and no information behind. Before any of that, what caused them to be created anyway? No, saying gravity is the reason is not enough as gravity is supposed to be according to modern science a curvature of spacetime and not a tangible force and therefore not able to interact with quantum field energies to make things pop into existence. We do not see this effect taking place here on Earth and we certainly have been looking at the most fundamental particles for ages now. Even if gravity could somehow provide the force to make these pair production events take place what is the particle that is the trigger mechanism and why are no other particles anywhere in the Standard Model capable of the same feat despite all particles being theorized to be part of the same excitation mechanism?

The answer is that no possible mechanism can be given for why gravity should simply make something pop into existence and what I am talking about here is a solid piece of evidence and not a phantom lurking in the mathematics as a real tangible object of quantum size and nature must be produced to explain it. Since we are talking about popping into existence what does that even mean? Again the answer is simple as it means a lack of understanding on the part of modern science as to what is happening and a falling back onto the incorrectness of quantum mechanics where it is simply believed that things happen simply to satisfy mathematical probabilities.

The pair production of particles also means another huge flaw and that is the ability of the particles to find one another perfectly right after they are produced and annihilate cleanly again. The particles would have to be somehow aware of each other and be able to do what nothing can do on its own which is change direction in violation of the basic laws of motion without an outside force. This is necessary for pair production as the particles cannot at any time no matter how small be acted upon by an outside force to get them to change direction and head towards each other to annihilate again as this would alter the universe around them and leave a trace of information behind which they never do. So the particles which would be produced at near light speed and probably travelling perfectly away from each other along a 180 degree line as remember they are opposite particles so splitting their properties save for one means keeping their spatial vectors in accordance with the Pauli Exclusion Principle and the conservation of energy. In addition the particles must be produced travelling away from each other because in the hypothetical black hole absorbing one particle the other travels into the universe meaning it had sufficient motion away from the event horizon to leave.

I should also mention here the razor thin margin for error that is needed for two incredibly small quantum particles to be produced with a tiny amount of quantum space between them and yet this space is somehow sufficient on a relativistic scale of the event horizon that the space between them is large enough one can leave the system. The particle altitude if you will above the event horizon is mind clenchingly short and perhaps only a few Planck lengths which makes for a massively precise set of conditions all to satisfy some quantum mechanical inaccuracies.

This means that they cannot even self attract due to their opposite charges and we know this to be factually correct as never in a tremendous number of particle collisions here on Earth in our laboratories have we ever seen two anti-particles simply stop, change direction and go back without interacting with anything around them so they can perfectly annihilate. The environment in which these pair production particles are believed to be produced is one of the most dynamic places in the universe which is a single femtometer or so above the surface of a black holes event horizon where space is thought to be incredible fast, dense and energized. What this means is that there is no way two particles could be allowed to freely move after production and cleanly annihilate with each other again to leave no trace behind. Something in the environment in which they apparently popped would interfere with them long before they could annihilate and therefore make this theory once again incorrect.

One other reason why pair production is quite impossible is that it requires energy to make it happen and you do not get this for free. Saying gravitational energy is what supplies this is again false as gravity does not excite the Higgs fields which you would need at the point of pair production in order to make two particles. Gravity just is the simple result of the curvature of spacetime according to modern science and cannot hope to excite a field as if this were the case then we would regularly see pair production events in the laboratory and we do not. Furthermore the black holes gravity varies as distance from the black hole increases and the theory of pair production is linked to energy being used to make particles which are of a set energy or mass and so this means that at only one exact distance from the black hole would the energy be correct to exactly give the mass of these theoretical pair production particles.

This means the very idea that you have an exact mass down to the quantum fundamental level states you have an exact distance which might be only a few femtometers or so wide in space. This immediately limits the concept of pair production at different distances from the black hole where some neatly annihilate with each other as they were formed far away and some enter the event horizon as they were produced much closer. So again the theory is proven wrong and there are even more reasons why this pair production from the vacuum is completely false but I will leave them for a future discussion and for now let us look to see if the basic idea that black holes can evaporate is even remotely feasible.

The particles are produced and one is absorbed and the other flies away lowering the overall mass of the black hole and we detect the radiation known as Hawking radiation in the process. Well half of this is true and the Hawking radiation does exist in the sense that material can be radiated away from a black hole and in this the theory is correct. The theory is incorrect in how this process takes place however as pair production has been proven to be quite wrong and with it I must also bury the theory that space is filled with this probability driven quantum energy that makes things pop into and out of existence all the time. So I will now explain the process using no hidden quantum probabilities, no mysterious pair produced particles and still allow for materials to be radiated away from the black hole such that it loses mass.

First off if you recall from an earlier section I said that TSM events are needed to destroy SM masses and this is still true but there is a small condition to what I have said. This condition is that I said you cannot destroy a SM mass from the outside and that a TSM event can easily do so from the inside which means that you cannot do anything externally to the SM mass in order to destroy it. This is still correct as the only way to undo a SM mass is from the inside and a slow dissolving mechanism for black holes can exist but again only from the inside and not from anything you can do to it on the outside. Try to do anything to a SM mass from the outside and all you have done is give it more mass in the form of anything whether matter or energy being converted into clintrons and so the black hole will grow and not dissolve.

Furthermore the time scale needed for black holes to fully dissolve is so long that this should have a close to zero if not completely zero chance of occurring due to two different factors. The first is that the universe is always filled with stuff moving about it here and there and so the conditions necessary to clear an area of space around a black hole one hundred percent for trillions or more of years simply will not occur. At some time after the area of local space around the black hole is cleared materials will come into contact with the black hole again and be added to the SM mass. The second reason this will probably never occur is that in either of the two universe models where we have a single BBB or exist in a series of BBB in one large universe the black holes will be recycled into a huge SM mass that will undergo a TSM detonation event long before they can dissolve completely. In either case a typical black hole is quite stable and will easily last longer in terms of lifespan than the recycling period for the universe and whichever life phase it is currently in. Instead I will focus here on describing the mechanism as it can happen and will pretend that it can exist for a sufficiently long enough period of time that no new materials are absorbed by the SM mass and thus it can dissolve fully and for this I will use the example of a hypothetical mini-black hole which will require less time to fully dissolve.

So here you have a system consisting of a local area of space around the SM mass and of course the SM mass itself. Nothing will enter the system and only materials dissolved from the SM mass will leave and are in fact allowed to leave the system completely for this example as it will not alter the final outcome. The word dissolve is used instead of evaporated because the actual process that is occurring is far more similar to dissolving something than causing it to evaporate. In this case I will show how a black hole can essentially dissolve itself into the solution of empty space around it. If you would like to think of this another way think of diffusion where materials spontaneously move from an area of high concentration in all directions to an area of low concentration in a volume.

It is interesting to note here that diffusion is of course just another example of quantum motions in action just like Brownian motion which again is caused by the need for speck/mote geometries and ESS to overall balance the donated energy and macht of the local area of space. This is similar to what is happening with the SM mass only on a much more impressive scale as an SM mass can be said to be balancing itself with the local area space ESS macht. The first thing you must realize is that nothing can ever destroy a SM mass other than a TSM detonation event as SM masses are made of clintrons which are the most stable speck/mote geometries ever possible in the universe and so once again I am correct in my previous statement from the previous section where I said you need a TSM event to destroy a SM mass. However SM masses and clintrons are only stable when they are SM masses and I will explain what I mean by this now.

The clintrons held within the SM mass are perfectly stable as they create a feedback loop whereby their particular speck/mote geometry allows them to donate energy through edge dragging to just the ESS macht glow gravity of space. The ESS macht glow gravity in turn provides the attractive force to hold the clintrons in this unique speck/mote geometry and so we see how the inclusion of both space and matter are needed to create and maintain clintrons. Now we can get to the part where the SM mass might want to dissolve into the space around it and this comes when we have an area where the two necessary conditions for clintrons are not fully met.

This area is the last event horizon or the actual physical surface of the SM mass itself which is in other words the solid surface of clintrons themselves. SM masses are actually quite interesting because they are the first things in the universe you can ever physically touch as opposed to simply feeling the electron orbitals pushing against you with regular matter. Since clintrons have no orbitals of that kind and are the densest possible arrangement of speck/mote geometries possible touching them is actually touching them and in some ways this is the first and only time you can ever touch space itself too. This is of course due to the fact that the ESS are a physical part of the clintron as well and so touching a clintron means you can actually touch space and the ESS which exist throughout it everywhere at once. This is an interesting fact but one that obviously you will never remember as soon as you touch it you will be completely rearranged into the SM mass as well and render you quite dead in the process.

That being said the SM mass requires both space and matter in order to maintain itself and if one of these two resources is in less than one hundred percent supply the conditions can exist to dissolve the SM mass. The last event horizon or exact surface of the SM mass is an area where these conditions are met as you have the matter present in the form of clintrons and you have space in the form of ESS. What you do not have is a continuous SM mass anymore as the space surrounding the black holes is devoid of SM mass. The space around the black holes has no matter in it of any kind as is needed for black holes to dissolve and so the ESS outside the black hole cannot be SM in nature. These outside ESS are not as highly spatially compressed as the ESS inside the SM mass and have no clintron speck/mote geometries to hold them together through donated energy which means they have decreased macht glows as well. So here we have conditions where the absolute top surface layer clintrons will have nice warm safe SM mass on one side of them and ESS devoid of all matter on the outside. What this means is that I must remind you of what SM mass is with respect to one property which is the maximum degree of interaction of matter with space.

As you recall everything in the universe is made of matter interacting with space to different degrees and can be described as maximum degrees of interaction, time spent in space or edge dragging through space or energy donated to space. What this means is that the matter spends a certain percentage of its time if you will inside space in the form of interacting with it and a certain percentage of its time outside space or not interacting with it. Again you must remember here these are not literal definitions only descriptive ones to help different readers visualize this concept in a way they find most easy. Nevertheless matter spends a percentage of its time interacting with space and the other percentage it is not and so we can use this fact to explain SM masses and how they dissolve. SM masses have their clintrons spending near one hundred percent of their time interacting with space and so not much time outside of space so to speak. The clintrons however do not spend a true one hundred percent of their time in space as this would make them for one instant a clintron TSM and lead to a clintron TSM detonation event.

The black hole is real and quite stable so we have direct proof that the clintrons are not one hundred percent a part of space yet. If we use the example of the clintrons being 99.99% percent interacting with space this means they are 0.01% not interacting with space. These numbers are not literal and again only for illustrative purposes as in reality the clintrons would be spending far more of their time interacting with space I am simply not going to write out all those little zeros here. So if we look at an individual clintron we see that 99.99% of the time it is inside space with of course 0.01% percent of its time spent at the maximum degree of interaction and 0.01% of its time at the minimum degree of interaction which is when it is outside space. Here you should recall that during quantum entwinement events two speck/mote geometries need to be in perfect maximum degrees of interactions with the ESS of the local area of space as each other. This fact will play out again here shortly but first we can focus again on the 0.01% of the time the clintron is not interacting with the ESS.

During this time we see that the conditions for the SM mass are not being met 0.01% of the time and so a chance exists for it to dissolve into the universe again and radiate materials out into the system and even beyond. SM masses however are the most stable things in existence and so if left alone will not dissolve by themselves leaving the clintron to avoid dissolving the SM mass at that time when it is 0.01% out of space interaction. Here we see the synchronous behavior of the quantum entwined particles example coming back into view. If you have two clintrons become perfectly synchronized in terms of their maximum degrees of interaction with space then this means they also share a minimum degree of interaction with space as well or they share they exact same 0.01% time outside of space. These two clintrons must also be on the surface of the SM mass where the two necessary conditions for creating and maintaining SM mass is not being met in the form of the outside ESS not having matter associated with them. At this point all the conditions necessary except for one are met to have the clintrons dissolve from the surface of the SM mass and this last condition is proximity to each other. The clintrons must essentially be so close to each other as to be neighbors such that they share overlapping ESS between them and this allows for the maximum amount of ESS macht glow sharing through schales as well. Why this is significant goes back to the same macht balancing rules followed by ESS for all events in the universe and once again underscores the fact that the rules of space physics are not being altered nor are any new variables being added and yet these same rules continue to explain another unsolved problem in modern physics.

Standard space physics macht balancing rules are applied here when the two clintrons are synchronized at their minimum degrees of interaction with space and their close proximity causing them to share ESS strongly. At this very short lived 0.01% time and with all other conditions met the possibility that the two clintrons will interact with each other exists such that they balance the shared macht glows between them through the local area ESS. What happens to the two clintrons is now based on the other clintrons around them as they also possess ESS macht glows that are shared throughout the SM mass. The two synchronized clintrons will need to move in space in order to balance their macht glows and this will move them closer to or farther away from the clintrons in the area next to them. Both events can produce a dissolving event as if a clintron is pushed to close to the SM mass it will mean that the other synchronized clintron is now left in a slightly less dense area of the SM mass or to look at it another way it will have less of the necessary spatial compression through shared ESS macht glow to keep it in the clintron speck/mote geometry which requires normal SM mass conditions to achieve. The clintron that is now technically farther away from the SM mass will be beyond the event horizon of the black hole again and so can decay.

Here once again I will remind the reader of the properties of clintrons as they are simultaneously the most and least stable objects possible in existence anywhere at anytime in the universe. This is due to the fact that as long as the two necessary SM mass conditions are met nothing can undo them but as soon as those conditions are not met the clintron will decay. The decay rate of a clintron is the fastest possible again for any object anywhere at anytime in the universe except of course for a single TSM clintron object. The decay of a TSM clintron is not really a decay process but instead a detonation process for a BBB event whereas the decay event for a clintron is exactly the same as the decay for any particle in the universe that is unstable. Remember all particles in the universe are in fact unstable no matter how stable we can measure them today as they are all made of different speck/mote geometries that are constantly trying to balance their ESS macht glows and so are always under pressure no matter how small to relieve built up imbalances on the local area ESS macht in which they exist.

A completely stable electron, neutrino or photon is actually slightly unstable and wants to decay but cannot because the local area ESS is not spatially compressed enough to force the macht balancing rules to break these objects down to their component specks and motes. If these objects are brought into an area of space where these high levels of spatial compression do exist such as in a SM mass then they will be overwhelmed by the local area ESS macht glows. The result is the particles decaying in order to balance macht glows and being rearranged into a clintron which is how something like a photon as seen by modern science to be massless can in fact add mass to a black hole. A single fully decayed speck/mote is the only object that can exist in a state of not needing to balance its ESS as it is creating an ESS macht glow but it cannot be decayed any further and so will be impossible for it to balance any lower its ESS macht glows.

A quick fact here that helps to solve another unsolved problem in modern science is that decayed particles will remain in the universe in various speck/mote geometries that our instruments cannot see and therefore contribute to the mass of the universe as well. This is due to the fact that unless these decayed speck/mote geometries which can be smaller than what we view as fundamental particles interact with sufficient energy with other speck/mote geometries or fully formed as modern science views them particles, then they will escape detection.

These speck/mote geometries will exist however and of course must edge drag with the ESS of the universe nonetheless and so can have many unseen gravitational effects as they are not bound in detectable forms but will interact with the ESS macht glow gravity of the universe. This ESS macht glow is of course shared through all ESS around it using the normal schale sharing mechanism and therefore detectable objects that we can see are in fact feeling this gravity. All other ESS macht glows such as electromagnetic, weak decay and the strong nuclear are also elevated by their edge dragging and contribute to the overall ESS macht glow of the universe. Once again there is no dark matter people only regular matter decayed into its component speck/mote geometries and contributing to the total ESS macht pool of the entire universe and that is the largest macht reserve pool possible. In addition these unseen decayed speck/mote geometries also constitute the apparently invisible effects that can sometimes happen to normal Standard Model particles as they are observed travelling through the universe.

Back to the SM masses as they dissolve clintrons and we see that the decay rate of the clintrons is incredibly fast as the ESS macht glows shared by the clintrons with space and the donated energy from the clintrons to space are so high that the desire of space to balance its ESS macht glows is the highest possible in existence. The result is that the ESS will balance out the macht glows quickly which will make the speck/mote geometry of the clintron decay all due to the fact that the sufficient levels of spatial compression of the local area ESS are no longer met and as we have seen the two necessary conditions for SM masses are lost. As soon as the SM mass conditions are lost nothing holds the speck/motes of the clintrons together anymore and not only the ESS of space but also the speck/motes themselves seek to rearranged through the decay process to reduce the overall need to balance energies and machts. The clintron then decays rapidly into all of its component specks and motes which will obey normal decay rules such that conservation of energy, velocity, direction and the like are maintained. This results in some of the specks and motes falling back into the SM mass and some decaying away from the surface out into space. These decayed specks and motes that leave the system of course are what is seen as lost materials radiating away from the black holes and represent the lost mass of the SM mass itself.

Any number of interactions by the clintrons at the start of the process whereby they balance macht glow levels will result in decay processes of one or more clintrons. Any of these decay processes in turn will result in different numbers of specks and motes being sent outwards into space as remember each decay event will be slightly different from one another. What this means is that in one decay event a single speck or mote might be lost to space whereas in another event two specks or motes might be lost. This means in the first event the black hole loses one unit of mass and in the second it loses two units of mass so to speak. The decay events however are always occurring in three dimensional directions and so the black hole will never lose a single total clintron mass in the process but only a fraction of one. Since the clintron is made of the smallest possible particles in the universe which are fundamental specks and motes the total mass lost for any single event is very small and is as close to zero when compared to the total mass of the black hole as is possible. This is one reason why the black holes dissolve so very, very slowly.

Another reason is that it is difficult for the necessary conditions to be met such that the clintrons are synchronized to their minimum degrees of interaction with space, on the very top surface of the SM mass and located right next to each other such that they share ESS macht glows strongly enough to force them to move in space and allow for the decay process to occur. The window for the decay process is ultra short as well and if it is missed the clintrons will simply reorder themselves with the total ESS macht glow of the SM mass and remain stable meaning the black hole loses no mass during that potential event.

One more fact here is that any free speck/motes from a decay event that are not incorporated back into clintrons will be bound to the surface of the black hole and further cause irregularities in the total macht glow sharing of the SM mass. A single speck or mote however is smaller and therefore contributes even less macht glow to space than a clintron and so will have little effect by itself to cause significant imbalances in the total ESS macht sharing of the SM mass that could contribute to a clintron decay event. These differences are of the smallest scale imaginable but do represent a very, very small chance that a pair of synchronized clintrons will either undergo a decay event or will cause them to miss the window as the decay chance is as said above very short. These free specks and motes unless they come into contact with more of their kind will not form clintrons but if they do will again recycle materials into the total SM mass which once again makes it overall stable.

The speed of the SM mass dissolving is something that will occur faster as time goes by and I will explain these mechanisms here but after I talk quickly about the ESS spatial rebound of a dissolving clintron event. A single dissolving clintron event will see the matter become more unbound with the local area of space it is in which in turn decreases the ESS macht glow there. As the particle decays as fast as possible for the universe the ESS there will also rebound quickly which means its volume increases. Since it cannot expand into the SM mass which is already at maximum spatial compression possible except for a single clintron TSM it must expand outwards away from the surface of the SM mass out into space. What this does is further help to propel the specks and motes that are the decayed remains of the clintron into space as these specks and motes are travelling with the spatially rebounding ESS. This is similar to what happened after the BBB as space was moving away from the single TSM detonation event the matter associated with it had to travel along with the ESS as well. Here in the case of the black hole the ESS are rebounding quickly as they were under high levels of spatial compression inside the SM mass proper.

The decayed clintrons have emitted specks and motes at high speeds and these speeds are made a little bit faster by the spatial rebound speed of the ESS that they envelop. This serves not only to increase the overall velocity of the specks and motes but also to move them further away from the event horizon of the SM mass which would once again incorporate them back into the SM mass. With the ESS spatially rebounding and the specks and motes moving away from the black hole the SM mass has once gain dissolved a very small amount.

This need to balance macht glows as the driving force behind quantum motion of particles also explains why black holes are black and not necessarily need an escape velocity of the speed of light or greater. The reason being that inside the black hole there are fewer macht glows to balance and so a particle as viewed from outside the black hole in the normal space modern science understands cannot use macht glow balancing for propulsion. With no ESS macht for it to balance with it simply has no propulsion to leave the black holes event horizon. This also explains why particles above the horizon can move away from a black hole as the necessary macht glows exist to facilitate quantum motion.

This need of the ESS to spatially rebound is what helps to drive the speed at which the SM mass will dissolve completely and again the chances of this happening for a normal black hole in the universe is close to if not actually zero. The surface area of the SM mass decreases meaning that the number of clintrons decreases there and so the distance between them decreases as well. Since the clintrons are now closer together and the number of variable maximum degrees of interaction is reduced in total for the SM mass the necessary conditions for a clintron decay event will slowly increase. As the SM mass shrinks more and more it also has less total mass available to it to increase the macht glow gravity in the shared ESS schales around it. With more decay events of clintrons occurring the total mass of the black hole will decrease faster over time until the ESS spatial rebound effect takes over.

This ESS spatial rebound effect works in two ways the first being as a transport assist mechanism for the specks and motes decayed from the clintrons and sent outwards into space. As the black hole dissolves more and more the spatial rebound of the ESS can also help disrupt the surface ESS macht glow of the SM mass a little more making decay events slightly more frequent. Any imbalance in the total ESS macht glow of the surface of the black hole will cause the neighboring ESS to try and balance out. Since the matter there is already slightly out of balance with open space on one side of it it will feel the need to balance the total ESS macht which as shown above can lead to a chance for materials to be dissolved from the event horizon of the black hole proper. This is another way that the rate of black hole dissolving increases.

The second way this ESS spatial rebound mechanism affects black hole dissolving events is by causing the SM mass to explode from the inside. This again illustrates but in a different way how only SM masses can be destroyed from the inside and not by anything done to them from the outside. The way this works is simple as SM masses have a threshold value at which they can no longer be SM masses made from highly spatially compressed ESS and clintron speck/mote geometries. Once gain you must remember that Einstein and his theories of relativity are incorrect and we know this to be true as it is Einstein's own theories that create paradoxes with his work and impossible objects like singularities. Another impossible object created by relativity is the micro-black hole which is simply based off of the incorrect theory of Einstein's that if you compress something sufficiently into a small enough space you can create a high enough gravity.

This is impossible because matter has a smallest fundamental size and so cannot be compressed to the necessary levels in the necessary small volumes of space to create the micro-black hole curvatures of space. What this means is that the minimum size of a black hole is far larger than a single point in space as suggested by Einstein and modern science. To put this another way it would be impossible for any particle known or un-known to be compressed to a size small enough that the curvature of spacetime created by it could be capable of producing the infinite gravity of a black hole but exist in a size smaller than an atom. Even a single clintron which incorporates matter and space cannot by itself be a black hole as the necessary conditions are not maintained for it to exist and if you remember the single clintron black hole would be surrounded by a schale of non-SM mass ESS meaning every side of the clintron is exposed to space.

One other point never answered by Einstein or modern science is exactly how does one crush a fundamental particle to infinite smallness anyway? All known particles have a set size and mass or did people suddenly forget what the word quantum actually means and just decided to throw it around for fun? The point is that particles are points meaning they occupy a certain volume of space that is quantized for them and cannot be changed for any reason or they stop being that particle. Think of an atom and you have a nucleus and an orbiting electron and for simplicities sake think of a hydrogen atom. If you crush it into a smaller volume than it wants to occupy when the electron is at the lowest quantum degeneracy state you do not simply make a smaller hydrogen atom. Instead you force the atom to be something else and in this case it undergoes electron capture by the proton and becomes a single neutron. Crushing the neutron will force it to be come quark-gluon plasma as the increased pressure raises the temperature of the neutron until it dissociates from its comfy hadron shape to one of free roaming quarks and gluons.

Again you have forced the neutron to stop being itself and become something else and the key here is that eventually you run into the smallest possible particles possible in the universe and since they are quantized cannot be changed. So the act of crushing those same fundamental particles into a smaller volume of space simply to make the impossible creature existing in general relativity mathematics come to life is quite wrong. Therefore you cannot ever create a micro-black hole as the properties of matter themselves forbid its existence and therefore there is no need for a dissolving black hole to shrink down to a single last particle in an infinitely small space known as the singularity. This proof by the way shows a practical reason why singularities also cannot exist as they cannot crush matter whether it is many particles or a single one into an infinitely small space as this would mean destroying those same particles which is of course impossible.

Therefore knowing this and looking at black holes as rational objects made of perfectly normal matter that is simply interacting with space in a way that it did not before allows you to see why black holes that are evaporating will end their existence with an explosion. The SM mass that is the black hole proper is made of clintrons interacting with highly spatially compressed ESS such that the two create a positive feedback loop that maintains the structural integrity of the object. A certain amount of matter and space are required to make this happen as can be easily understood from looking at the very dense cores of objects like super massive stars and neutron stars. A super massive star for example is more massive than a black hole as it has not lost up to 90% of its mass to a supernova yet. This huge mass is pushing inwards into a single point that exists in a non-fusing core because the core comes from a second or higher generation star that contains elements beyond iron.

So in this stars core where fusion is not taking place there is no outward pressure from the fusion reactions in the form of heat, neutrino emission or anything at all. Yet this core does not collapse and so upholds itself nicely despite being more massive than the black hole and this is due to the fact that it has not experienced a TSM shockwave that has rearranged the speck/mote geometries of its core atoms into clintrons yet nor has it forced enough space into that clintron mass either. The net result is that without the right conditions matter and space will try to avoid being a SM mass pretty much all of the time and the same can be seen in neutron stars which do not contain clintrons.

A black hole however does have the right conditions for SM masses to exist but they must have sufficient ESS spatial compression and clintron geometries for their specks and motes. The ESS mass is held together by donated energy in the form of elevated macht glows between schales from clintrons and that in turn keeps the ESS highly spatially compressed which can keep the matter in clintron form. If you decrease the total mass of the SM mass far enough you will find a threshold value where the ESS are no longer spatially compressed enough to hold the clintrons in their current speck/mote geometries because there is not enough clintrons to donate enough energy to keep the macht glow at SM levels of ESS spatial compression. As this factor is lost from the system the ESS must spatially rebound which means the clintrons lose the ability to maintain their speck/mote geometries and they will obey the normal ESS macht glow balancing rules for any particle with space as explained before. The clintrons are as stated the fastest decaying particle possible in the universe and so they quickly dissociate themselves into their component specks and motes.

This of course draws macht glow from the ESS quickly which allows the ESS to rebound further and if the withdrawal of macht from the system is fast enough it will allow the ESS to rebound faster than the slow quantum interaction speed of matter with space. Since this quantum interaction speed is at best light speed this means the ESS in the local area of space that used to be the SM mass will be free to spatially rebound faster than light speed just like during the Big Bang. With the ESS spatially rebounding at macht speed the conditions for a small BBB will exist but will not under any circumstances possess enough space, matter or overall power to actually create another universe. Do not think this is a place for mini-universes, multiverses or other bizarre forms of universes within other dimensions to exist as it is simply an explosion. Why would someone want to invent any of those strange other universe conditions anyway? They all only solve one problem of modern science and in order to do so create nothing but impossible objects, extra variables to complicate the universe further and are based off of theories that can never be tested and proven wrong. It always struck me as odd that so many people were attracted to such nonsensical ideas and never stepped back and simply said hey now enough is enough.

That explosion however will be powerful and will push the space around the black hole dissolving event outwards just as the rapid expansion phase of the universe did. It will also contain on a small scale the same post BBB conditions whereby you have matter solidifying back into the universe again and technically all the associated events along with it right through to mini first lights. The explosion is of course the result of the fact that SM masses all have a threshold value whereby exceeding this will not provide enough space or matter to create and maintain a SM mass. Therefore a black hole dissolving event will proceed slowly at first and remain slow for a long time. As more time passes it will be possible for this dissolving to speed up until this threshold value is reached. At that time the dissolving of the black hole will proceed briefly at near light speed and once the ESS is free of interactions from matter will briefly proceed at macht speed. The last phase of the black hole dissolving event is the small scale production of BBB events such as matter solidifying once again in the universe so to speak and then behaving in a normal everyday way afterwards.

So in this manner the black hole will dissolve all at once from the inside out which again shows that a SM mass can only be destroyed from the inside and since the final dissolve mechanism is an ESS spatial rebound event similar too but not identical too a BBB it can be said to be a form of clintron TSM detonation. The interesting thing about clintron TSM events is that they are the ultimate unification of all forces at once in the universe and this is all four forces, space and matter. In addition to this they are at the exact same moment both the creation of a clintron and its destruction and so are both giving and taking away life at the same time. As far as the universe goes they are also perfectly existing at one moment as the destruction of one universe and the creation of another. They literally combine not only all forces, space and matter but also creation and destruction in a single object for a single macht time which is the absolute limit to the universe under any conditions.

Facts

The key facts concerning black holes in this section are all building upon previous explanations of matter and space interactions from earlier sections and so I will keep them short here. The first fact is that black holes do not create their own magnetic fields from materials inside themselves but only from materials in orbit in their atmospheres. The second fact is that as matter falls to the surface of the SM mass the matter from which it is made will slowly undergo changes in its speck/mote geometries such that photons can be emitted due to processes ranging from loss of electrons to loss of neutron stability or decay events and all photon emission stops at the surface of the black hole proper. This results in unique black hole spectra which will be recognizable as long as accurate detectors can be placed close enough to the actual black hole and will vary based on black hole size, rate of rotation and what types of matter are infalling to its surface. The third fact is that SM masses have no temperature as the need for thermal energies does not exist inside a mass of ESS held clintrons.

The fourth fact is that SM masses or other large objects can create relativistic jets because they have enough strength to highly spatially compress the local area ESS around them. The fifth fact is that the spinning motion of matter around the equatorial ESS spatial compression imbalances will move materials towards the poles. The sixth fact is that energy imparted to the matter orbiting a black hole is what is needed as one half of the jet mechanism. The seventh fact is the anti-vortex of ESS above each pole is the second half of the mechanism needed to create a relativistic jet. The eight fact is that the simply release of energy over the static ESS from the orbiting materials gives some power to the jet. The ninth fact is that the rate of matter absorption by ESS is what provides the rest of the power for relativistic jets. The tenth fact is that the spinning ESS around the anti-vortex is what consumes the energy and matter from these reactions leaving only the non-rotating ESS as the escape route for the jet materials. The eleventh fact is that the gamma ray burst mechanism of stars is created the same way as the materials on the inside of the supernova TSM detonation schale and are consumed quickly in the same way by the newly formed black hole.

The twelfth fact is that all quasi-periodic oscillations of the black holes accretion disc are caused by changes in both ESS inside the SM mass and materials falling onto the event horizon. The thirteenth fact is that black hole dissolving events are possible but unlikely given the vast amounts of material in the universe and the long lifespan of black holes meaning they will rarely have the chance to completely evaporate before gaining more mass. The fourteenth fact is that dissolving events of SM masses are possible and all are mediated by the need of SM to be made of both highly spatially compressed ESS and clintrons. The fact that the ESS and clintrons are both needed to maintain SM mass is what allows for slow loss of matter at the surface of the SM mass. The fifteenth fact is that it is the imbalance of macht from the interior SM mass and the space above the event horizon that allows for rare events where clintrons are moved into areas of ESS slightly less dense than needed for SM matter to exist and the result is the rapid decay of the clintrons which further decreases the SM mass. The sixteenth fact is that this process speeds up as the surface area shrinks more imbalances between surface clintrons are possible and so more dissolving events occur. The seventeenth fact is that the final moments of the black hole upon dissolving fully are reminiscent of a TSM event as space reaches a threshold value below which the overall ESS spatial compression and macht glow pools are not high enough to maintain clintron speck/mote geometries and the system must decay all the ESS and clintrons. The eighteenth fact is that the free speck mote geometries will be released into a surrounding area of space far less dense and energy filled than the early universe meaning that some of the specks and motes will not fully rearrange themselves into recognizable particles to our instruments available to modern science today. This means that a fraction of real matter is in the universe in a form that simply does not appear as real to us meaning it also provides some extra mass as well to the overall universe.

Chapter 32
Uberfluss und Uberlauf or Abundance and Overflow

Gravity again

I would like to start off by saying how much fun it is to get to use German words like that simply because they sound so cool and science rarely gets to do Uber anything. With that in mind this section should be much fun as I will be focusing on not a single large problem or a series of smaller ones that are linked but instead I will handle the abundance and overflow here. To be fair many of these problems can be incorporated into previous sections except that in order to do so would be to reveal too many of the inner workings of the universe before I have explained all the others. This is where I remind the reader of my need to begin by slowly building an understanding of the universe section by section and reveal more of the universes beautiful secrets as the pages turn. This uberfluss and uberlauf is not just contained in this chapter but will continue till the end of the book as previous problems are expanded upon here.

Could I have explained the minute details of speck/mote geometries before I have even explained what ESS macht tracking or balancing rules were? Well technically yes I could have hit you with everything at once but that would have been overwhelming which would doubtlessly mire the reader in too many concepts all at once. So I would like to say thank you here if you are still reading as this means you have stayed with me over these past sections and built upon your understanding of the universe and now I feel comfortable in simply hunting problems down one by one in any fashion I choose. Anything I explain now will hopefully seem simple and obvious to you as I will be applying any of my space physics rules at any time.

This section will also hopefully be welcomed as I hope the previous ones were by anyone working in the sciences with open arms and smiles as I will be answering more of the unsolved problems in modern science. This also means that some might dislike the true answers as they might not fit with a preconceived view of the universe. To this I must remind the reader and thinker of the need for humanity to once again take another step forward and away from our past notions of a flat world in the center of the cosmos. Here it is necessary to let go of any of the held beliefs of the three main views of modern science as they have not nor are they ever capable of answering the unsolved problems of modern science beyond an ad hoc approach. This is all well known to you or you would not be here reading this after I have already discarded so many traditional beliefs of the three main views. So either you are bored or curious or both but no matter what thank you again and let us get started with some properties of gravity such that we can disprove an old belief and accept a new one.

Gravity over the centuries and millennia has been poorly understood at best and the greatest first step taken was that of Newton and his falling apple. Without doubt the most important of the concepts involving our early understanding of gravity goes beyond the general concept of stuff falls down and not up. Here I will not resist telling you my favorite gravity joke and yes this is how science minded people have fun and it goes like this: There is no such thing as gravity…the Earth just sucks. Either you are grinning or your daft but no matter what this is actually a fantastic and valid account of how humans thought about gravity for millennia. It was not until Sir Isaac Newton sat beneath his apple tree and experienced a falling apple bump him on the head that he looked up at the sky and had a sudden realization. What Newton pieced together at that moment was that the same force that caused the apple to fall is what held the moon in orbit with the Earth.

This was a great surge in human thinking as the two seem completely different and from this he worked out his laws of gravitation and a great number of other truths about the cosmos. This first step in our understanding of gravity shaped the modern world of physics for centuries as the theory of general relativity is just a tweaking of Newton's genius. That being said the one thing that general relativity did was provide a good equation for calculating the motions of objects in space under fairly moderate conditions. Here is where general relativity stops being useful however as it is nothing more than a tool for simple measurements as all of why gravity works in the universe is unknown to general relativity.

Here as I have said before is a moment in human history where we must step away from the general relativity past and into the future away from the geocentric model and towards the heliocentric. The reason for this is simple general relativity is dead wrong and nothing can be done to salvage it so stop trying. First of all general relativity fails not a little bit but completely at quantum scales and quantum systems are quite real and measurable. The second reason is that general relativity does not work on any scale beyond the moderate as it fails at not only small quantum scales but also at very large cosmic ones as well. This is seen in observations made not only at the edge of the universe but in the spaces in between and some of these spaces are filled with black holes. This is the third reason why general relativity is false as it itself creates objects which are impossible and the most heinous of these is the singularity.

The singularity is by far one of the most damaging of ideas ever conceived by modern science and should be discarded but not forgotten as it must not be repeated. I have spent enough time in sections before here and there proving why singularities are impossible and so I will not repeat those here but I will say that they simply serve to show how general relativity does not understand what gravity really is. Plus I have already destroyed spacetime by proving time is not real either so there is yet more proof and now I will give one more reason why the basic idea of general relativity is incorrect in explaining gravity.

This comes from a simple exploration of what we know about the universe and how general relativity attempts to fit gravity into that. To begin with let us look at the Earth as what it really is which is a large collection of matter floating through space and for the purposes of this example it can be thought of as a single object within a system as we do not need to picture the Sun or moon or anything else around it for now. The Earth is made of many different elements and has a solid core of iron and various other elements like nickel however the inner core is pure iron. The Earth is in fact about 32 percent iron and the core is layered such that the heaviest elements are at the center and despite the Earth having very heavy elements on it like thorium, uranium and the like they are not at the inner core most likely and certainly in trace amounts at best.

The reason is simple as the Earth is over 4.5 billion years old and these radioactive elements have half lives much shorter than that meaning they are no longer as abundant as they used to be. In addition to this the way the Earth formed was under near constant bombardment from asteroids, comets and collisions with other planetoids. This means that the inner workings of the Earth were heated to very high temperatures sufficient to liquefy all elements and as such the heavy elements like iron sank straight to the center of the core. All of this has been said simply to establish that the Earth's core is made of one element alone and that is iron and that core is a solid single sphere or very close to a sphere structure.

◄New Universe Order►

This is very important to proving why gravity cannot work as described by general relativity and why gravity must work using both ESS and macht. According to the theory of general relativity all gravity is caused by a curvature of spacetime where space is curved and any other objects close to it will fall into the gravity well created by the larger object. Spacetime is said to be curved into a single gravity well and if a second object is close to it the bottom of the well moves to a central gravitational point between the two masses and that way they seek to curve spacetime. Spacetime curvatures will move as the matter moves so you will technically have an effect on the Earth just as the Earth pulls on you although the Earth is clearly winning. If you move away from the surface of the Earth the gravitational force you feel diminishes so again we have proof that according to general relativity mass is the source of gravity. This means that according to the theory of general relativity the gravity at the very center of the planet is in fact zero as there is no way for gravity to exist at a central point in space according to it. This is of course wrong as the closer you go to the Earth's core the stronger gravity gets but according to the theory of general relativity the Earth at its very core is a place where you can float perfectly weightless. Why is this seemingly absurd situation possible you ask?

Simple, according to Einstein and his theory of general relativity mass curves space time and as I demonstrated in the two object system the center of gravity moves to follow mass. So the Earth's core at its heart must have no gravity as the mass above it is pulling spacetime away from the core's center and creating a spacetime curvature inside the Earth itself. If the Earth were made of only one element throughout and that it was a sphere 10 kilometers in diameter then the center of gravity and the strongest force of gravity you could feel would be roughly 2.5 kilometers below the surface. Yes I am aware of this simple analogy and the fact that a spherical object would contain a distribution of mass such that its total matter by volume would not produce a neat 2.5 kilometer mark of delineation I am simply using it as an illustrative tool. That being said the amount of matter found at the surface and to a distance 2.5 kilometers below the surface would curve spacetime up from the core and outwards into space. The 5 kilometers inside that line would seek to curve spacetime down towards the core and away from space but since the amount of mass on both sides of the line is equal this means that the center of gravity is at the line and at the core a perfect curving of space exists in all directions away from the core towards that 2.5 kilometer deep line. So it is easy to show that general relativity is wrong about spacetime being the source of gravity and the fact that the Earth has a nice neat solid iron core only further proves this.

The point of mentioning that the Earth's core is solid iron is that it negates the effects of increasing density towards the cores center. What this means is that since it is impossible to compress a solid due to the electron degeneracy pressure of the iron atoms the core cannot be compressed so the idea of increasing density at the center being used to save general relativity from failing is not possible. Iron cannot be compressed by an object as massive as the Earth and so spacetime cannot continue to curve further at the center of the planet as you cannot compress the solid iron core and so at some point in the Earth's interior exists according to general relativity this line at which gravity is strongest and the center of the planets core must still have zero gravity.

So the Earth obviously does not have zero gravity at its absolute core and objects do fall faster as you approach the surface so how can this be possible and the answer is by using ESS and macht to explain gravity. Spacetime has no mechanism to share curvature of space and will as shown above curve away from the core which is of course impossible. So in order to get a central gravitational point out of objects in the universe you must share gravity freely inside space allowing for the total gravity contribution of the mass involved in the system to be concentrated onto a central point. This is where you can return to the little room we saw so many sections ago with the faintly glowing hollow ESS spheres and the mostly opaque matter orb. At the very center of the orb is a central ESS which is the most spatially compressed of all the ESS but not to a massive degree more than the rest either. That is the key to an object like the Earth as it is not so spatially compressed that the central ESS is orders of magnitude more spatially compressed than all the other ESS inside it.

Even a black hole will not have a massive degree more spatial compression to its central ESS as compared to the rest as the ESS inside are all bound into SM matter. The SM matter is made of clintrons and ESS and due to the way the two interact as explained above they will be comparable in size no matter where they are in the SM mass. The Earth of course cannot hope to ever compress an ESS to the same degree as an SM mass but it does of course behave in exactly the same way gravitationally as I have stated before there are no breakdowns of the laws of physics in the universe. To this end the Earth or a black hole will always have a certain degree of spatial compression throughout them and certainly the central ESS in each will be the most highly spatially compressed but this alone cannot explain gravity.

A second mechanism is needed to explain gravity such that the errors made by the theory of general relativity do not appear again and of these errors is of course the gravity problem of general relativity using the spacetime curvature from before. In order to explain how gravity will not create this zero gravity effect at the heart of an object and have a true center of gravity existing somewhere within its surface like a sphere a second mechanism is needed to account for gravitational forces. This second mechanism is of course macht glow as this is how space itself talks to all the ESS within it. The ESS will not change in volume sufficiently to create a linear center of gravity as general relativity wants at the very center of an object but if you can transmit some of the gravitational macht through space you do not need to deform it immensely. This is of course the solution to the problem of singularities as described in a previous section whereby singularities are proved not to exist as you do not need an infinitely small volume of space where density and gravity explode to infinity and time stops completely.

So here again is another explanation of why gravity is in two parts as it can be shared amongst all ESS through the use of schales and transmitted to the heart of the objects core. Since ESS always seek to share macht and since I have shown that macht glows are how space creates everything from gravity to the perceived boundaries of a fundamental particle it can be shown again here that macht is the true second mechanism to gravity. The ESS receiving donated energy from the matter they are in contact with will seek to share that macht glow with their neighbors and this results in all ESS surrounding the central ESS giving it some of their macht as well. The result is the central ESS has the most spatial compression and the most macht glow of all the ESS in the local area of space concerning the object in the system. This means that a central point to the total gravitational field of the object can finally be created such that the core of the object does not have a zero gravity effect as is stated by the theory of general relativity.

In addition the ability of ESS to share macht means that you do not see these incorrect gravity wells popping up and therefore do not have a layer inside the Earth below the surface that is not the core where gravity is bending spacetime the most. Macht sharing allows for all points of the Earth's mass to have gravity and therefore does away with these gravity wells which also serves to solve more unsolved problems in physics namely the black holes immense gravity and the singularity once more. This is because according to Einstein all gravity is from one factor only which is the curving of spacetime and this leads to the flaws with excessive curvatures producing areas of the universe with near zero dimensions through volume and in some cases zero dimensions like a singularity. The ability of ESS and macht to create gravity will mean that the continual curving of space never occurs as all ESS share macht glow and far more evenly than a curved spacetime creates a gravity well. To the outside observer it looks like gravity is falling into a well only because we cannot go below the surface of the planet or to its very core and so this trick of the light appears correct.

However using ESS and macht the curvature of spacetime and gravity wells are done away with and so disallows curvatures like black holes and singularities while still allowing SM masses to exist. The SM mass is made of ESS and clintrons as explained before and since they always share macht you have a normal and stable object like a black hole without a need for an impossibility like a singularity and without having the laws of physics breakdown. Using ESS, macht and clintrons you can seamlessly scale up or down through the universe from a single fundamental particle to a planet and straight through the different event horizons and into the very heart ESS of a black hole using one set of rules.

In addition to this the explanation of gravity is more satisfactory to observed events and experimental results than if the theory of general relativity was correct. Remember back a few hundred years when Newton said a force existed like a tension between two objects? Well he was right all along as a string like tension does exist between objects and this better explains the observables in the universe when coupled with quantum mechanics and the Standard Model of particle physics. Since macht glow is real and since macht glow extends beyond the boundaries of the mostly opaque matter orbs of the universe it can be a force that reaches out from orb to orb. This should also keep quantum physicists and the Standard Model happy as once again gravity is a force and so can have a force carrier that interacts with other known fields like the Higgs.

This means that instead of the two curvatures of spacetime that are centralized on two objects and should cause space to curve away from each other you have a normal distribution of space between the two where by the transmitted macht glow is the shared force. The macht glow is being shared by the two objects and each objects internal ESS are trying to balance with not only themselves but with the ESS of the other object as well. So a force of gravity which is really the macht glow component of gravity is being shared like a string between the two objects just as Newton theorized. Not only is gravity explained perfectly this way such that it satisfies not only observations and classical mechanics but it also serves to fit the universe better as a whole.

This occurs in two ways the first being that classical mechanics are on the whole one hundred percent right as they do describe a system in macro scale. No, classical mechanics alone do not describe gravity correctly and I want all readers to understand that I am in no way saying classical mechanics is correct only that a part of it is when thinking of larger systems. That part is the behavior of objects on a macro scale to always obey classical laws of motion even if these laws break down at the quantum level. Think for a minute about how a ball will roll across a table and hit another ball transferring energy to it. This does not follow laws of quantum mechanics or general relativity but classical only and we can certainly measure and repeat these observations perfectly as many times as we want to. The laws of quantum mechanics and general relativity will never be able to do this and so some truth remains in classical systems.

Not to mention the fact that if classical mechanics has been usurped by general relativity as describing larger systems and how they behave and yet does not explain the quantum world then so to does general relativity fail and become just another theory for describing some motions in the universe. At the end of the day both classical mechanics and the long revered and incorrect general relativity as well should both be relegated to simply making a few macro scale observations and the real workings of the universe be left to the realm of the tiny. To that end explaining gravity in a way that suits normal observable and repeatable results that can be performed with no ambiguity or guess work as is needed in both quantum mechanics and general relativity is more satisfactory than not. Everything else we see follows classical laws so why not gravity as well?

This brings us to the second reason that an explanation of gravity of this sort is correct over the theory of general relativity and that is it uses no bizarre and impossible variables like spacetime in the first place. I have already proved spacetime to be quite non-existent and will reiterate here a crucial point in the explanation of gravity such that it fits observables rather than theory. We can observe and repeat experiments all day long on quantum objects and know that quantum objects are real as modern science has measured them thoroughly and with a high degree of accuracy. No one has ever been able to measure or detect spacetime at all no matter what outlandish claims they make. The best they can do is to say that they observe indirectly the effects of spacetime but without seeing spacetime or testing for it in a real tangible way the apparent effects of spacetime can be accounted for in superior and correct ways. I have already explained all observable effects of spacetime such as curvatures of space and time dilation without the use of spacetime itself and have been able to use the same rules for my mechanisms and models in other problems such as quantum entanglement something which the theory of general relativity simply can never do.

This is the point I am making here which is that quantum mechanics are quite real and observable and I have shown how an explanation of gravity that follows a classical explanation so to speak better fits with the Standard Model of particle physics. This is accomplished by the fact that all known forces in the universe are accountable through particles, forces carriers and forces but according to general relativity gravity is some kind of magical exception to the rule. I ask you in all seriousness what is more probable that quantum mechanics is wrong and general relativity is correct that a magical spacetime with no forces, force carriers or anything usual is the true explanation of gravity or that gravity is just one more normal force that can be incorporated into the Standard Model? Please tell me you chose quantum mechanics and the Standard Model. While I know that quantum mechanics and the Standard Model are imperfect they are far more credible and factual than the theory of general relativity ever will be. To sum up gravity cannot be explained using the theory of general relativity as it simply creates too many problems from normal explanations of gravitational forces on Earth to immense objects like black holes and the impossible and damaging concept of the singularity.

Chapter 33
Specks and Motes

Bauen Matter

This next section will deal with some of the fundamental properties of specks and motes to explain a bit more about how they work in the universe to create matter and how they interact with space as well. To start off with specks and motes are not matter as modern science thinks of it such as electrons, quarks, photons, neutrinos and so on. Specks and motes are smaller than any of these particles as they are arranged such that multiple specks and motes will create a larger particle known to modern science as a fundamental particle. This fact about matter in the universe is an essential one to construct a universe as we see it today, to explain why all objects in the universe we see today behave as they do and to decrease the number of variables needed to make a universe in the first place. Take space for example which has received next to no serious thought over the ages as a perfect example of how people simply think of space as one thing while they think of matter as many things. Why cannot space have a zoo of properties like matter has a zoo of particles and why could matter not be reduced to a single object the way space has? Well the answer is simple space is simple and matter is likewise simple and everything is simply nice and simple to put it simply.

Space has exactly one necessary object needed for it to do everything we see around us and that is the ESS with its macht glow. Matter also needs only one object to explain the universe as we see it today and that is a speck/mote. The ESS of space can carry all the known fundamental forces and depending on how matter interacts with them will respond in kind. If matter is capable of interacting with gravity you get gravity of course and if matter cannot you get light. If matter can interact with the electromagnetic force you get the generation of a magnetic field and if it cannot then of course you get no field. Gravity and magnetic fields are two examples of invisible non-particle objects in the universe that matter can and cannot interact with based on its properties.

Well matter is made of specks and motes in certain geometrical arrangements that will be capable of interacting with space in various ways and you can construct every form of matter in the universe using a single speck or mote. The specks and motes are easily thought of as a small three dimensional object say for example in the shape of a triangular pyramid which has four sides. Every speck and mote in the universe is identical just as every ESS in the universe is identical and no variation needs exist anywhere at anytime between these objects. You do not need multi-dimensional strings that exist in extra hyper-folded and compactified dimensions, you do not need two kinds of strings which are open or closed, you do not need an extra dimension through space called a brane for them to sit on and you do not need them to vibrate with unknown string energy in different ways.

What you simply need is a single piece of normal three dimensional bauen matter or translated from German construct type matter as in to construct or build. Bauen matter is the smallest building block for everything in the universe and is what specks and motes really are. I have referred to specks and motes for some time as they are easiest for people to quickly picture as simply being very small. However it is now time to introduce the bauen matter as the proper form of specks and motes from which all properties of matter are derived and from which the entire universe is made along with ESS. Bauen matter is the true fundamental matter in the universe but require their own name as fundamental has been used ubiquitously with Standard Model particles so long it would now be indistinguishable just as power would be confusing to use and so macht is used instead. Think for now of bauen matter as small three dimensional triangular objects with four sides although this shape is just for analogy and can be changed later. Ooh boy now comes the fun stuff as I will explain to you all of the properties of bauen matter and show how I can construct an entire universe with just these little pyramid triangles.

If you have been looking for a simple and easy to understand model of the universe then sit back and get ready to smile as I will do away with the extraneous extra dimensions, branes and ad hoc that has mired modern science. Bauen matter to begin with has four sides and each side is capable of increasing the macht reserve pools for a specific macht in ESS. The very first side of the bauen matter contains a new force unknown to modern science which is the bauen force and exists on one side of the triangle but also inside of it with the one side example for ease of illustration. The bauen force is responsible for one job and one job alone and that is attracting multiple bauens together and holding them in place. The bauen force is the most powerful of all forces in the universe far exceeding the strong nuclear force and also having the shortest range of effect by far. The bauen force will interact with the ESS macht bauen of space to cause two or more bauen to join together to create new speck mote geometries or bauen geometries. The effective necessary range for the bauen force is likely only as far as the next bauen and so is only one or two bauen units in length or reach.

The bauen force is capable of moving multiple bauen into different configurations such that various numbers of the three other sides are exposed. Exposure in bauen matter terms means faces of the triangular pyramids that are not facing each other and are exposed to space on the other sides. Again I remind you that the faces of the four sided pyramid are a good analogy of the bauen matter but do not need to be a literal one suffice that they get the explanation across to the reader. To continue then bauen forces exist at far higher energies than are probably possible in any of today's particle colliders even if they were fully redlined with no care for the consequences. The bauen force therefore is the hardest to currently observe if at all possible with current technology of the early twenty first century and so the end products of bauen forces on bauen matter is all that can be observed. Just as the decay products of man-made elements like ununoctium are observed without the ununoctium being observed directly so to are bauen matter and bauen forces illusive to see with only the remainder products visible.

These remainder products are what modern science calls fundamental particles as they are detectable with current instruments and by activation of currently detectable forces. In reality all fundamental particles are made of multiple bauen being held together by the bauen force the resultant bauen geometry is capable of activating a particular set of forces which again are really just ESS macht glows. Remember that all matter as perceived by modern science is really just the creation of a spherical schale of shared macht between ESS as these ESS interact with matter as can be remembered from the original section on ESS spheres and matter orbs. The perception of modern science of point like particles is only due to the large crude size of the quantum world and the bauen geometry inside is far smaller and also because treating objects like a point is easier for mathematical equations of the system but the particles are not really points. The perceived fundamental particle represents the boundary volume of the ESS macht glow and exceeding that macht glow would be for example to break the electron degeneracy pressure of an atom, or neutron degeneracy pressure of a neutron. These fundamental particle volumes are large in comparison to the actual bauen geometries which exist at their hearts and the individual pieces of bauen matter are of course smaller still.

So in a particle collider for example or other form of decay process in the universe you will observe objects like electrons, photons and neutrinos being emitted for example from the matter being observed. In a violent particle collider event you see a shower of small particles emanating from the impact point and these particles are all seemingly fundamental. They are of course not fundamental and the reason they appear so numerous in their variety is simply the way they are decayed from larger particles. The overall bauen geometry of the large particle is fragmented into many smaller bauen geometries. I said earlier that it is impossible with today's colliders to separate particles by breaking bauen forces and this remains true because despite the fact that bauen are being disrupted during the collision this is due not fully because the bauen forces are being broken. Remember that all larger particles are made from ever increasing numbers of bauen matter and so the complexity with which they must balance their overall ESS macht goes up as well. Highly complex and larger objects like the atoms of radioactive elements are so desperate to balance their ESS macht glows that under normal spatial compression we observe half life decay rates as measured here on Earth.

A quick fact about the universe is that under extreme amounts of spatial rebound or spatial compression decay rates and consequently half lives of all elements and particles changes. The decay rates also change based on the amount of macht glow in the local area ESS as long as these changes of macht glow are significant enough as compared to the internal glow of the object in question. I do not have time to explain the properties of various elements at very high or low temperatures here but their behaviors are all linked at the fundamental levels to ESS macht glow balancing. As for the bauen forces and particle colliders the extra strength needed to overcome the bauen force is not coming from the collider itself or any energies imparted from it but from the matter in question and its need to balance out local area ESS macht glows. So the force needed to disrupt the bauen force comes from the matter itself as well as the collider but to truly overcome the pure bauen force alone with no help from internal ESS macht balancing mechanisms requires more energy than humans can currently generate or harness.

One important note about the bauen force is that it is so short ranged it effectively closes certain sides of the bauen matter from interacting with ESS as they are simply too close to other bauen particles to allow edge dragging of these non-exposed sides into space and thus they do not increase macht glows either. So exactly what does all this exposing and non-exposing of bauen sides mean? Well it allows for the creation of all of the zoo of particles and an explanation for why they all have the properties they have and why you can create all of them no matter how different they seem with just a single type of matter which is the bauen.

Before explaining exactly how all matter and all particles as understood by modern science can be created with just one type of object which is bauen matter I will explain the nature of the bauen matter quickly. Bauen matter is simply put matter at its most basic level and is made of a single three dimensional structure that requires no extra dimensions. This is important to restate here as matter does not need to operate on multiple dimensions nor does it need to utilize extra branes or move in different ways such that different particles are created as a result. This is where the universe is once again simplified into having fewer parts rather than more as you can explain the universe without multiple and un-testable dimensions. You cannot test for let alone prove the existence of or explain physically how extra dimensions can exist beyond three and this is because the universe is whispering to us that there are no dimensions beyond three. Likewise you do not need a new brane upon which these objects will sit that exists as an extra dimension within space but does not react smoothly to changes in the overall shape of space. You do not need to have two version of strings which are open or closed nor do you need for them to vibrate in different ways either. Lastly you certainly do not need any and I apologize for shattering some people's illusions of the universe here but the existence of multiple universes that exist everywhere around us, inside us and beside us separated by the tiniest of impassible barriers. No you need none of these things which I for one am glad as you can explain the universe with normal three dimensional non-vibrating, non-brane sitting objects that are made of simple three dimensions that can sit perfectly still if you let them. Welcome home normal matter.

What this means and what I hope is utterly clear to all reading is that you can reduce the number of variables in the entire universe again and this means you can simplify the entire universe once more. What this means is that you can finally create a universe out of exactly two things which are space and matter, ESS and bauen. This means that suddenly all uncertainty, probability, and un-testability of the universe disappears forever and you can finally explain everything you observe around you in one simple model from the smallest to the largest of experimental or observable phenomenon. To prove this I will now begin to build a few types of matter from single bauen matter and show how they create the known fundamental particles observed and measured by modern science in the Standard Model of particle physics.

Let me start by taking a single bauen and allowing it to exist free in space and observing what it does to the universe. The first thing you will notice is that the bauen will increase all macht pools for the local area ESS it is in contact with. This is due to the fact that the bauen like the ESS must possess all necessary forces to build a universe the reason we do not see all matter in the universe exhibiting all forces at once is due to the first force which is the bauen force. This also explains how breaking matter of much larger size for example a neutron can be made to decay into particles that now excite some of these properties. In the neutron configuration some of the forces are not available to the particle such as charge but if you split the neutron you will create two separate objects which are the electron and the proton. The neutron splits along the easiest to balance ESS macht glows which results in a bauen geometry for an electron and a bauen geometry for a proton. The energy needed to balance out through splitting the proton along ESS macht glows takes more energy and so does not occur unless more energy is given. So if we build upwards towards an object like a neutron for example we can see how a single matter type which is the bauen can build all particles and energy.

I will now take two bauen and allow them to come close enough such that they are attracted to each other by the bauen force and held into a new bauen geometry. For this example I will create a photon of light and it does not matter what wavelength. In order to create light simply take two bauen and have the side of each bauen responsible for gravity to face inwards. The bauen have a very short range to the bauen force and so the attraction is halted once two sides of the bauen are facing each other and thus close these faces off from interacting with space and increasing any ESS macht glow gravity found there. The new bauen geometry can now maintain the exposed sides of the electromagnetic force and fully obey all ESS macht tracking rules allowing it to move instantly at light speed but will have no interaction with the Higgs field and so no mass and no gravity. By the way in case you were wondering the Higgs field can partially be thought of as the force carrier for gravity like a graviton as some of you might have clued into that already.

If we have one bauen face up and one face down then the charge properties of the two bauen cancel out allowing photons to not exhibit a charge. Photons do still have charge faces on their bauen geometries however and so can respond to outside charge influences. A difference in elevated macht glow electromagnetic or a decreased one will likewise push or pull the photon as needed to respond to a particular field and this explains on a step by step model why photons do move as they do in response to electromagnetic fields. Similarly the charge faces of a photon can be subtly altered in certain bauen geometries allowing for changes in photon speed. So it is not at all difficult to make light and you can do so by simply placing two bauen close together with the right sides facing inwards blocking them from being exposed to the space around them. Once we deconstruct matter to its bauen components it might be found that light is in fact more than two bauen objects and again this is simply to illustrate bauen geometries for the initiate.

This simple constructing of light cannot be overstated as I will now explain what happens during a matter and anti-matter annihilation event. First off the old term of annihilation is fully incorrect as it implies that matter is destroyed creating energy when in reality the two are the same thing only utilizing different bauen geometries and therefore altering how they edge drag with the ESS and alter macht glows which in turn is detected by the observer. Matter has mass and so an exposed gravity side to some or all of its bauen geometries and electrons and positrons also have charge so their bauen charge sides are exposed as well. If you remember that matter and anti-matter are attracted by their mutual effects on the ESS macht glow charge for the local area of space and the local area ESS are free to balance out the macht pool imbalances so direct collisions are allowed for matter and anti-matter particles with no obstacles like degeneracy pressures to overcome.

The result is a direct collision of the two anti particles at maximum allowed ESS macht tracking speeds and so a naked bauen collision event at closest possible range where the bauen themselves are forced to impact fully which can allow for a moment where the short range of the bauen force is challenged so to speak. Since the impact is so severe and the particles are perfectly evenly matched for each other you see on a vey small scale Newtonian mechanics take over as energy is transferred directly between the two particles much like in a Newton cradle. The bauen geometries will not fully break as the bauen force is too strong to be fully disrupted here but it will allow the bauen to move around in their current geometries. Think of the impact being strong enough to not break the bauen apart but to spin them around a central bauen force point. The final outcome of the collision is not an annihilation where the two particles are destroyed and magically changed into energy which modern science views as a separate entity in the universe. Instead it is simply a forced rearrangement of the bauen geometries along the conflicting charge sides which results also in the gravity sides of the bauen being positioned inwards against each other in the non-exposed and therefore non-edge dragging configuration of light.

Since the amount of energy measured from the light emitted in the collision is equal to the amount of bauen in the original particles it is seen that the equivalent energy is conserved for the system but this does not mean that relativity is correct in saying that energy and matter are the same thing. Matter and energy are not the same thing for a few simple reasons the first being that they possess different bauen geometries and so are not in fact the same thing. This also reflects that relativity only said that energy possessed by the two were equivalent as in the rest mass of matter is equal to a certain amount of energy. The reality is that a second factor separates the two which is that the bauen in each different type of matter or energy is capable of interacting with space in different ways which can alter the overall macht glow of the local area ESS to which they are in contact. So on large crude macroscopic scales of fundamental particles it can be measured during annihilation events or nuclear reactions that energy and mass seem equivalent and therefore this must hold true everywhere in the universe. This is not at all true as on the smallest scales of the universe the ESS and bauen will seek to balance macht glows at the smallest levels. Here it is seen that depending on how bauen sides are exposed to the local area ESS certain forces will be elevated or disappear altogether meaning a direct conversion of energy to matter is not possible.

If you cannot picture this think about individual bauen and the different sides they possess and then picture how they are smaller than light. They cannot be equivalent to energy as they are not energy which is a bauen geometry photon. If you extend this further to events after the Big Bang when matter was momentarily dissociated from the ESS of the universe then you see that the matter still possessed mass without activating the Higgs field. This is because the matter is still a bauen but it is simply not interacting with space and so cannot be said to have an equivalent energy as it is not ESS macht tracking. It is easy to rearrange bauen such that you have bauen geometries that will not ESS macht track as normal fundamental particles do and so will not produce an equivalency that fits into the theory of relativity. Again this cannot be seen in normal large scale sizes of fundamental particles and is reserved for true breaking of the bauen force but suffice to say matter and energy are not the same thing and a physical explanation of bauen geometries proves this. Physically explaining how matter and energy are constructed from different bauen geometries proves they are not the same thing despite the ability to measure an equivalent energy. The ability to measure an equivalent energy should not be ignored as it is still a useful tool for calculating reactions in the universe even if all theories of relativity do not understand what is happening at the smallest levels.

To continue the annihilation events further it is shown that larger objects like the anti-proton will follow a similar system of events like a collision between an electron and a positron. I use the word system because unlike a much smaller fundamental particle such as an electron or positron which have much smaller bauen geometries the proton and anti-proton are far more complex. Each of these larger particles are made of three quarks and held together by numerous gluons all occupying a far larger volume of space meaning they also have a more concentrated macht glow involved in their existence. What this means is that the annihilation event between a proton and an anti-proton is going to take more time and more steps to complete. This is due to the fact that since they are hadrons and therefore composite particles it is impossible for every part of them to perfectly touch the counter part of the anti-particle at exactly the same time.

Without a single instant in existence where all composite pieces of the hadrons can meet it is not possible to have a perfect instantaneous conversion of all particles into photons. One quark will touch first, then a second, while some gluons are lost in between and then the last quark with the rest of the gluons will finally meet and annihilate. The shower of photons that emerge from such an event is evidence of such a breakdown and the breakdown is the perfect example of how hadrons are made of numerous smaller particles. The annihilation must be a non-zero time event because even at light speed the macht tracking rules still must be obeyed. All of these smaller particles are of course constructed from fundamental particles which are made of one thing and one thing alone which are bauen geometries. Bauen geometries are made of one thing and one thing alone which are of course bauen matter and this presents the opportunity of building some larger particles from single bauen matter.

Bauen Matter, Bauen Neutrinos and Bauen Charge Electromagnetic

The bauen matter have of course multiple sides as described before which all hold a particular property associated with matter in the universe and coincide with all known forces of the Standard Model of particle physics. The key to the power of the bauen is the near limitless number of ways they can be arranged inside unique geometries all that will affect changes in the ESS macht glow pools of space in different ways. These differences are what make things that appear as fundamental particles due to how they create the macht glow spheres which represent the so called physical boundaries of the known fundamental particles. Two bauen can come together to create light and multiple bauen come together to create a hadron such as a proton putting the two together gives the rather obvious answer of what happens when a photon is absorbed by a nucleus inside an atom. Simply saying the relativity explanation of energy is matter and therefore adding a photon to a nucleus gives you more equivalent mass is wrong and vague at best. I believe in the truth of the universe which is everything happens for a precise reason without vagaries or random quantum fluctuations and probabilities in the mathematics as the universe is quite aware of what it is doing at all times. Unknowns in modern science to me are nothing more than modern science saying that it simply is at a loss with no precise answer for why something happens and therefore resorts to random probabilities as predicted by mathematics.

When a child or youngling asks why this and why that over and over the trail of answers will eventually fade away until modern science is forced to say either we do not know or just because. I would like to offer a complete picture of the universe that can follow the why this and why that questions until they run out of energy and not the answers. To this I again remind the reader of the power of using a single form of true matter, the bauen matter as a building block for everything in the universe that is not ESS or macht.

Bauen matter makes normal objects of mass and bauen matter makes light as well with no mass but they are of course the same bauen matter underneath it all only changing the geometries will produce different objects in the universe. One bauen geometry will make light and one a hadron so adding a photon to a nucleus only really adds more bauen matter to the nucleus. The geometry of bauen matter photons arranges on hadron bauen geometries to expose the gravity side of the bauen matter in the light. The result is an increase in the mass of the nucleus by exactly the amount of the bauen photon geometry and the overall mass of the nucleus is now reflecting that by really being more massive and not just being the equivalent more massive as dictated by relativity. This also proves the precise mechanism by which photons can be held inside an atomic nucleus in a step by step, bauen matter by bauen matter way that explains what is really happening beyond the vague explanation given by relativity. If matter and energy are different creatures according to modern science then how can matter hold the light as needed? I am not looking for a simple mathematical calculation of times and maximums which is traditionally given I am looking for a precise step by step method which all of modern science has none.

You have probably heard these phrases a too few many times before and they are the energy simply goes up, the atom can only absorb so much energy before it must give some back to space or since the energy goes up the equivalent mass goes up too. The question for all of these phrases is simply why? Do not make calculations of the rate or times at which these things happen as that tells you nothing of the why. Using bauen matter tells you the answer to all of these exactly with no guess work as seen in the photon and hadron example. To answer why an atom can only hold so much energy before it must give some up is beyond simple as the bauen matter are all interacting with the local area ESS and affecting changes in macht levels. This is where the ESS macht balancing rules from many sections before are brought back into play as the atoms nucleus has now gotten more bauen matter which means more interaction with the ESS and more macht glow for the local area of space. The local area of space has not changed its total amount of spatial compression to any significant degree and the matter has not undergone any significant changes such as becoming clintrons.

This means the total carrying capacity for the local area ESS has remained essentially the same and so extra bauen matter being added to the nucleus will eventually pass the threshold value for this carrying capacity. Once passed the ESS will force the balancing of the macht glows they possess and the extra bauen matter will be forced to leave the nucleus. Since it entered and bound itself to the nucleus as a photon which is a quantized amount of energy it is a quantized amount of extra energy that is in turn ESS macht balanced. So the ESS macht balancing forces out the same quantized amount that just entered the system in order to balance the ESS macht glow reserve pools. This means the exact same bauen geometry photon will be emitted in order to accomplish balancing and also explains precisely why each element, molecule or object has a unique spectrum.

The surface of each nucleus is a complex system of bauen matter in specific arrangements and these unique arrangements will also create unique patterns of ESS macht glow in the local area of space. Therefore each of these unique surfaces will naturally have areas where the macht glow favors more bauen matter than does not and since the total amount of extra energy tolerated by the local area ESS macht reserve pools is under constant pressure to ESS macht balance the accepted bauen geometries will be few. So few in fact that the spectral patterns are of course very thin when viewed in comparison to the total spectra being observed. Take this atom and that atom and then a few more to make a new molecule? It really does not matter as each new molecule will be forced to obey ESS macht balancing rules such that they create a new set of favored bauen matter geometries and therefore a net set of spectral lines. The bond angles of all molecules are here also explained step by step by the bauen matter and the arrangement of all exposed bauen sides with space and how they interact to increase the macht glows there. The spectral lines of elements and molecules and such can also be used to therefore back calculate the properties of the ESS in the local area of space as well as macht glows. The size of the ESS spatial compression, the strength of the macht glow and what macht glows are being elevated can be calculated for a quantized amount of macht or a single ESS.

Here the universe is explained at its absolute smallest limits finally as things like gravitational fields are explained for quantum systems by not only altering ESS volume but macht glow strength as well. The size of space does not appear to change much in a quantum system and yet gravity is clearly present and this is because ESS volume changes are not seen under normal spatial compression here on Earth but the strength of the macht glow is. The macht glow is what is being missed by modern science when looking at gravity in space of quantum systems. To prove this further the gravitational effects of things like relativity or the magnetic fields of an object like the Earth are really just the crude macroscopic versions of what is happening on the step by step precise scale of space physics using bauen matter and ESS macht.

Think of the macroscopic description classical mechanics affords to a system and then the quantized version of the microscopic system from quantum mechanics. ESS, macht and bauen matter give you the smallest possible explanation of the universe for any system and by understanding these variables all of quantum mechanics, relativity and classical mechanics are scaled up. Since space physics explains these smallest of interactions and all of the universe is built upon those interactions then the very idea of something like the uncertainty principle simply disappear as they are impossible to be made real. This proves here what I mentioned in an earlier section about the position and momentum of a particle being known at the same time. Since all of the universe is now known from ESS to macht and bauen matter and the way they interact are explained then the nature of quantum motion and the position and momentum of everything in the universe is laid bare.

If you find an unknown in modern science or anywhere simply solve the problem using ESS, macht and bauen matter as there is nothing smaller or more basic available to the universe. Continuing along the series of explanations of matter in the universe this also shows the precise mechanism by which electrons will drop down an orbital energy level and cause the nucleus to emit a photon. Explained in a previous section and using the same ESS macht glow balancing rules the bauen matter geometries can now be added to explain how the photon is absorbed into the atom. From there the electron now has to contend with the extra ESS macht glow imparted to the nucleus from the bauen matter and moves to a higher orbital but the local area spatial compression has not changed significantly and so the electron will drop back down to balance the total ESS macht glow reserve pools. This leads to the emission of the bauen geometry from the nucleus and the precise mechanism by which the electron drop down phenomenon occurs.

This kind of mechanism is also what properly explains the photoelectric effect from decades ago beyond the simple explanation of a threshold value must be reached as modern science can tell you what will be released when but not why. For decades the vague reasoning of energy must be absorbed sufficiently in order for the effect to occur but without a step by step explanation of the system has persisted. This is a point that I have touched on before which is much of what is known in modern science is not really known only measured and so the why is it happening of the system is ignored. Yes modern science has nicely measured over and over this effect and yes it has generated the idea of a quanta of energy but never has it ever described why on a fundamental particle level this actually occurs. Now that I have explained bauen matter to you this is really a simple application of both macht balancing rules for the local area of space and the bauen geometry of the object in question.

The whole process of the photoelectric effect is in fact a demonstration of the threshold macht value for the binding of an electron to the system. In other words the reason why a material will take a certain threshold value of photons coming into it before an electron is emitted is due to how much the photons can increase the macht glow of the system. The photons are imparting energy to the material which is trying to hold itself together by balancing with the local area ESS macht and once this threshold value is reached the ESS macht will force the bauen geometry to rearrange itself to balance macht for the system. In this way the easiest thing for the material to do is shed an electron which brings the entire system back into balance. Therefore the exact nature of what is happening at the fundamental particle and quantum system level are finally explained using space physics which are smaller than quantum mechanics. The hidden reasons why the quantum effect is noticed are due to the way the smaller world of ESS, macht and bauen matter operate inside the system of known fundamental quantum particles. Quickly this also explains why generators can make infinite electrons as the bauen matter being input into the system is being converted physically to new bauen electron geometries by the local area ESS macht levels. So a step by step explanation versus the old energy is matter vagaries.

I mentioned a moment ago that whenever you come across an unsolved problem in modern science or anywhere to simply use space physics to find the answer and I sincerely mean that. If you are fan of science and will never use my work so be it but please enjoy it and learn how the universe works. If you do work in science then please use my work whenever you hit the wall as you will find the answers you are looking for and they will be explained using ESS, macht and bauen matter. The random motion of electrons around the nucleus of an atom is also explained this way as is the random seeming disappearances and sudden course corrections they make as they journey about in their orbitals. The ESS macht balancing act of the system and the bauen matter it is made of explains why these motions appear random or why they might seem to disappear but in reality they simply follow simple rules. This is a basic example of where my work can be applied to an observed but not understood problem facing modern science and the number of cases of small problems I could explain here are simply a bit too long for now so I will move on to building up matter again and also build up the clintron in the process which is at the core of what is in the SM mass in terms of matter.

I have shown how two bauen can be put together to create a photon by keeping their gravity sides effectively hidden from space and so unable to elevate the ESS macht glow gravity of the local area of space. If more bauen are added say a third bauen to the photon but this time with its gravity side slightly exposed to space and this can be accomplished in your mind by having the gravity side not facing straight out from the bauen geometry. Again I remind the reader this is an illustrative model for your minds eye and need not be taken literally but it is close to what happens as the gravity side of the third bauen must be not fully exposed to space. The result is that the total shared bauen force between the three bauen matter objects is such that the total amount of gravity elevated in the ESS macht from the exposed side is reduced.

The total shared bauen force of the three bauen geometry and the short distance over which it acts keeps some of the gravity pent up inside the three bauen geometry. Placing the third bauen gravity side as far as possible from the center of the shared bauen force will place it at the limit of the range of the bauen force and allow that side to be able to interact more strongly with the ESS around it. The purpose of all this three bauen geometry and of being very specific of how the gravity side is arranged with respect to space is to explain how to make a neutrino. Yes three bauen geometries can be used to construct a neutrino which has virtually no mass like a photon and a top speed close to that of a photon as well. Even if the third gravity side was fully exposed to space it is only the gravity contribution of a single bauen gravity side so will not create a large effect with space in the form of elevated ESS macht glow gravity.

I must once again point out here that the real matter in the universe is the bauen and that all known particles are simply virtual particles as they contain a composite of what is thought of as real and virtual mass. The total spherical shape of the excited field that modern science views as the real particle is of course an illusion hiding the real bauen matter inside. This is necessary for me to explain here because once again the concepts of modern science serve to confuse the truth as all the conventional words for things like mass, matter and particle have been used and engrained in most reader's minds. This is why I have chosen to use German words to explain the true properties of the universe and to distinguish between what is my work and what is modern science. Thank you to my German readers for understanding and putting up with my use of words that might still be similar to you, hopefully English words have been exchanged where necessary to achieve the same effect for you. The point of explaining this once again here is because I have been recently and will continue to use the term matter only when referring to bauen objects or bauen geometries and I will not use the term mass. A photon of light still is made of matter despite not having what modern science calls mass or gravity. This is because photons are still made of bauen matter which as I have shown is the only true matter in the universe so you see the need for me to be careful with wording as much as possible in this and all sections. If I do make a slip up here or there please forgive me as the words mass, matter and particle get used probably way too much when talking about physics and the universe.

This being said a neutrino will have a very small and almost no amount of mass as viewed by modern science despite having three bauen objects worth of matter inside it. A photon is capable of interacting with the electromagnetic force and objects that also interact with the electromagnetic force. A neutrino has a third bauen object arranged onto it and so one side of the previous bauen geometry photon is now covered and thus it loses its ability to interact with the electromagnetic force to any degree similar to a normal photon. The arrangement of three bauen in this manner will give you an object that is near massless, travels at almost the speed of light and tends not to interact with much as it travels through the universe.

In a supernova the light is highly electromagnetic and so will be attracted to numerous atomic nuclei or ions on its journey out from the exploding star. A neutrino is just as small almost as a photon but does not tend to deviate from a straight line on its journey out from the star and so can pass easily through the mainly emptiness that makes up an atom as atoms are mainly empty space and not solid lumps of matter. So a three bauen geometry can be used to not only build a neutrino but explain observations such as neutrinos escaping earlier from a star going supernova and being detected here on Earth earlier than the faster light that is generated at the same time but gets bogged down on its way out of the star.

The three bauen geometry also allows for neutrino flavor oscillation to occur as the third bauen while strongly bonded to the surface of the two bauen geometry is not of course affixed to it like a solid crystal structure. The bauen force holds the third bauen to the two bauen geometry but the third bauen is still free to move slowly in place as it is not solidly bound to the original bauen geometry. This slow rotation explains the cycling of the neutrino flavor oscillation and since the bauen matter objects are thought of as triangular pyramids this naturally gives them four sides. One side is hidden from space and that leaves three exposed sides and they turn slowly as the neutrino moves through space and feels the local area ESS macht glow all around it. This gentle tugging of the ESS macht glow and the nature of all matter in the universe to need to balance ESS macht glows explains why the third bauen slowly moves in place.

The last piece of this unsolved problem comes from the fact that the total bauen force of the bauen geometry object is able to extend slightly around all the bauen in the object. Just as the gravity side is slightly decreased in its ability to fully interact with space and this is due to the total bauen force glow around the object so to are the effects of the other sides of the third bauen as well. Not to mention that the bauen forces from the two bauen geometry object are also now being slightly altered as to how they interact with space. This means that overall the total ability and way that the neutrino interacts with ESS in the local area of space and therefore increase, decrease or feel the tug of nearby macht glows changes over time but must cycle in a specific order as this is the way the third bauen object turns on the larger two bauen object. Depending on what side is hidden from space when the third bauen object is added to the two bauen object you will get a different neutrino flavor that will again have its own unique cycling order. One bauen side hidden of the third bauen object will give you an electron neutrino, another the tau and yet a third the muon all have the gravity side slightly exposed however. This use of bauen geometries shows exactly how you can build a neutrino and the precise workings of neutrinos oscillations which will vary based on local area ESS spatial compression and the strength of any macht glows in that same local area of space as well.

If more bauen objects are added to smaller bauen geometries you will of course make larger and more complex fundamental particles. Working up from a photon or neutrino and you can make things like quarks which are of course more massive as viewed by modern science. Other particles include things like electrons which are most famous for their negative one charge just as the proton is famous for its positive one charge. Quarks make up protons and electrons make up electrons but the end result is the same a particle that has a full one charge either positive or negative. All charges are of course derived from the charge side of a bauen object and only the way the bauen geometries come together alter whether it is plus one or minus one so to speak. An electron has all of its bauen charge sides exposed such that it produces a strong negative charge while a proton has the total charge values from all three of its quarks such that they produce a strong positive charge. The explanation for the electron is quite simple as it involves only exposing bauen charge sides to greatly increase the ESS macht glow charge for the local area of space and this produces the familiar to modern science electron that has a strong negative field surrounding it at all times.

The way a positive charge is produced is more complex but again created solely from bauen geometries and how the charge side of the bauen objects in those geometries are arranged. If you remember the explanation of charge from a previous section you will recall that there is no such thing as positive charge in the universe just as there is no such thing as cold. I explained how electrons are attracted to but kept from merging with protons using this negative charge only explanation of the universe. The creation of a second charge does not make sense from a practical standpoint for the universe as it involves making extra stuff which is a no-no as it creates extra unneeded variables. The addition of a positive charge to the universe also creates a problem as if electrons are negative and protons are positive and they are attracted to one another naturally why therefore do they not fully collide? What keeps them apart and why would adding a photon to an atomic nucleus push the electron farther away?

There is no force that keeps positive and negative things away from each other and you need look no further than two magnets for that answer as opposite poles attract one another and do not suddenly stop just before touching due to this magical unseen force. Likewise if an electron is negative adding a photon to a nucleus which then pushes it into a higher orbital means the nucleus has become more negative as like charges repel each other. This means the photon has a slightly negative effect on atoms and therefore proves that protons are likewise negative or at least using the same charge mechanism used by electrons to affect changes in the local area ESS macht glow responsible for charge when binding their smaller bauen geometries to large nuclei ones. Without reiterating everything from that previous section here I can build up a proton from bauen objects to satisfy this condition and simple quarks as understood by modern science cannot do this.

To start off with you will need to understand that there really is only one kind of charge in the universe and this is extended beyond just normal electric charge to also include color charge of quarks and gluons and what not. There is no such thing as color charge in the universe as this adds once again more and more extraneous variables that only serve one purpose and complicate the universe more in the process. The idea of color charge is a perfect example of an answer in modern science that produces loads of new particles, variables and complications but only answers one problem and poorly at that. Remember color charge is called charge but described as not being a charge as one would think of a normal charge and instead is you guessed it, some new magical extra type of charge in the universe just like string theory involves extra dimensions that magically are compactified and separate from the three normal ones.

So doing away with all this extra and unneeded color charge complication we can build a quark from simple bauen objects and go on to build all types of quarks, hadrons and even generations of matter. Think of it as photons are electromagnetic radiation which responds to electric or negative influences and while these photons can be manipulated using various forms of energy the idea that all charge, electromagnetism and magnetic forces are based off of one root object makes more sense in the universe anyway. First of all the basic shape of the bauen geometry will be discussed only concerning charge sides in detail as this is what we are really after and like the electron they will produce the necessary final charges to create a universe as we see it today.

Let us start with a down quark as this most closely resembles an electron due to its negative charge of one third. The reason a quark can have a charge of less that one which according to modern science and the three views it favors should really be an impossibility is because of bauen objects building that quark. All of modern science and especially the Standard Model of particle physics uses the idea of quantum mechanics which states that things come in discrete non-divisible packages hence the tern quanta. In other words like in biology everything occurs similar to an action potential which is either all or none and there is no middle ground as you do not get a 1/3 or 2/3 third action potential. The neuron either fires through its normal ion discharge mechanism or it does not and that is simply how these systems work. The world of quantum mechanics works the same way as things are discrete packages like a photon and there are no half photons, or neutrons and it is impossible to make a half neutron and charge should always be one as a quantized amount. Quarks are considered by modern science to be a fundamental object that cannot de subdivided any further and yet they according to modern science itself are allowed to produce a non-quantized amount of charge.

Obviously the explanation offered by modern science is wrong and something else is happening at a level smaller than a quark to explain this result. What is happening is the quark is not at all a fundamental particle as none of the particles known to modern science are actually fundamental particles at all. There is only one true object in the universe and it is the bauen matter from which everything else is made and from which particles like quarks can violate quantum rules and have charges of 1/3 and 2/3 which are not a set discrete single unit amount.

The down quark is created in this way by being a larger more complex object than an electron first of all as it has a higher mass and therefore is made of more bauen objects as they are capable of increasing the ESS macht glow gravity. Since each bauen object has the ability to increase the ESS macht glow charge of the space it is in, it is easy for the quark to produce the same negative type charge as the electron. The fact that the quark is much larger than the electron allows for it to have a greater number of and therefore more subtle system of charge interactions with space and can result in the production of a charge that is less than one. In order for this to happen bauen sides must be arranged such that they increase the negative charge of the local area ESS but not too much and a large bauen geometry for the quark can let this happen by the same mechanism that allowed the neutrino to have a small gravitational force only. Just as the neutrino has a very small mass and this mass is the result of only one gravity side of the bauen object being slightly exposed to space so to can charge sides be only slightly exposed to space. If only some charge sides are exposed and not fully to space then you will get a muted charge produced from the total bauen geometry of the quark. A bauen object is the only object that has a true single charge that cannot be altered but a quark is a composite particle of numerous bauen objects and so by using more than one bauen object the larger composite can produce a charge of 1/3 which is what is observed.

This use of bauen objects to create what modern science believes incorrectly to be a fundamental particle which is the quark allows for the quark to behave as quantum mechanics wants it to but still have a non-quantized charge value. This explains how something like a supposed fundamental particle can produce a charge that is not equal to one. It also explains how similar particles like up and down quarks can have different masses as they are made of composites of bauen matter and so will interact with the ESS macht glow gravity in different ways. Now to explain up quarks or things with what is seen as a positive charge and again bauen objects and how the sides of them are exposed to space is what creates these charges as viewed by modern science.

A positive charge as explained from a previous section is not a real charge but rather it is the change in the ESS macht reserve pool charge of the local area of space. I will remind the reader that all ESS have a reserve macht pool that allows them to function and certain particles will not increase this macht pool but instead will decrease it effectively drawing macht from space and around the matter that is present there. The total system will look like a hydrogen atom for example where the charge of the local area of ESS space can be seen as a wave form with the positive peak above the baseline representing the total increase of negative charge of the local area ESS macht glow charge as elevated by the presence of the electron. The negative trough below the baseline is seen as the decrease in negative ESS macht glow charge created by the presence of the proton and shows where negative charge has been depleted from space and drawn into the matter.

The baseline is of course the value for the normal ESS macht glow reserve pool for charge in the system if no matter was present. The up quark has been said to have a positive charge and also the proton has a total positive charge but this is really a backward kind of thinking as the universe deals only with what is seen as the negative charge. Therefore from the total baseline graph of charge in the universe it is said that more negative charge produces a positive value on the graph where as more positive charge produces a more negative value and in reality all this could have been avoided if modern science had used opposite charges for electrons and protons. Modern science has not however and so I will endeavor to explain through this confusion by showing how bauen objects can be used to draw out charge from the ESS macht glow reserve pools.

To begin with quarks are of course large objects as compared to electrons and so have much more complex bauen geometries and this allows for a series of different arrangements of the bauen sides on these objects. If we have a simple bauen geometry like in an electron it is easy to create a negative charge as the sides are all facing in such a way that they increase the ESS macht glow of space and create a negative field. The quark is large enough however to have some of these sides facing in ways that seek to cancel out these increasing effects and this is what creates the positive charge as viewed by modern science. A quark will have multiple bauen objects arranged on its surface and some of these will position their charge side away from space which will leave the opposite side to the charge side exposed to space.

Here I will explain one of the secrets of the universe and using this secret all forms of charge and magnetism can be explained using a single bauen object. You do not need multiple types of objects or particles, you do not need multiple types of charges or fields you need only one single type of matter which is the bauen object to explain how all electromagnetic forces work. Charge and magnetism are known to be related through a single force according to modern science and that is the electromagnetic force. The way that these both work is through the flow of donated energy from the bauen object to the ESS around them.

The bauen object has multiple sides and one of these sides is the well known charge side which allows the bauen object to donate energy to space and therefore increase the ESS macht glow charge in the local area of space. The opposite side of the bauen object has a reverse side to the charge side and this in no way means that it produces some opposite charge as this again would complicate the universe it only means that there is a side opposite to the charge side that does not donate energy to the ESS. So to picture this simply one side of the bauen object which is the charge side seeks to flow energy out into space to the local area ESS and therefore the opposite side must flow energy from space back into the bauen object to conserve the energy of the total space and matter system. In order to obey the law of conservation of space/matter for a given system the total balanced macht or energy of either space or matter respectively must be maintained. This means that the flow of charge moves from space through the non-charge side of the bauen object then back out into space from the charge side of the bauen object. Bauen objects have these two sides and the universe must possess them as it is the single simplest mechanism by which all charge and magnetic effects in the universe can be explained.

Take for example the idea of negative and positive charged fields and the fact that they are really just increases or decreases in the ESS macht glow reserve charge of the local area of space. Now look at magnetic fields as they exist in the universe always flowing from the north pole to the south pole of any magnet whether a bar magnet or a fluid magnet. Since both the electric force and the magnetic force are known to be one and the same the mechanism by which both work are one and the same and so if magnetic fields flow so must electric ones too. The bauen object allows for the precise flow of macht and energy in both space and matter such that both electric and magnetic forces will naturally move in the observed north pole to south pole manners. Yes bauen objects are the smallest forms of matter in the universe and are not made of particles which are simple fields as viewed by modern science. These bauen objects are free from fields as they are what create the fields viewed by modern science as fundamental particles and they do this by sharing energy with space and space sharing macht with them. So all magnetic fields are simply created by the normal flow of macht and energy through bauen objects and all scaling up of magnetic fields are created in the same way such that all matter as viewed by modern science has some magnetic dipole moment. A single piece of bauen matter is like a small magnet with one end flowing energy out into the macht and the other end drawing it back in again as space seeks to balance all macht glow imbalances.

In fact the numerous single pieces of bauen matter in the universe that exist literally everywhere in space but are so small they do not excite fields locally enough for us to detect them yet help to contribute directly to things like large magnetic fields. In the creation of things like large magnetic fields that extend far beyond the objects that created them such as planets and stars the use of free bauen matter in the universe helps to strengthen and carry the field charge as well. Remember that all magnetic fields are simply flows of the macht glow magnetic through the ESS of space from areas of high concentration to areas of low concentration by macht sharing through schales. A single piece of bauen matter near an existing magnetic field will naturally and always align itself with the direction of the field flow. This is because it has two sides to charge as explained before and so a natural flow for the macht glow magnetic. Each single piece of bauen matter is really a tiny and in fact the tiniest possible bar magnet. These free floating as in unbound from larger bauen geometries pieces of bauen matter will align themselves and help strengthen field flow similar to how magnets work in stellar plasmas. These pieces of bauen matter will naturally work themselves into the total field flow for the system just as magnetic field lines flow through matter as seen in things like magnetic reconnection events. This can be seen to help stabilize the field, help explain flow reversal and also to allow for immediate interaction between the field and other particles such as in the coronal heating problem as these particles contribute to but are free to move easily through space.

Single pieces of bauen matter in the universe can also contribute to the overall macht glow of the ESS of space as they still have their bauen sides fully exposed to space and yet space is sufficiently expanded enough that the necessary temperatures and pressures are not present to force them into large bauen geometries. In this way the normal elevated macht glows of these free bauen objects can help strengthen what is seen as the fields of space and at the same time account for some missing gravity as well as further decrease the vacuum catastrophe due to the fact that they are once again raising the strength of the energies in space beyond what is expected. The fact that they are present is to be expected as the universe moves forward in its current life cycle as things like decay processes or violent collision will always result in some single pieces of bauen matter being knocked free. This means that a percentage of space always has matter in it helping to further elevate macht glows beyond the normal ESS macht glow reserve pools and therefore ensuring things like the Big Rip or Big Freeze can never happen and therefore these free bauen matter pieces will help contribute directly to the halting mechanism of the expansion of space.

As for the creation of a positive charge in the universe this is made simply by having the bauen objects on the surfaces of the quarks face inwards so that the charge sides are all hidden from space and the opposite sides are exposed to space. This will have the effect of the natural flow of macht glow charge moving from space and into the bauen geometry of the quark which is a large enough object to effectively store the macht glow charge. Since the macht is flowing into the quark and the corresponding bauen object responsible for charge has no charge side facing out it is impossible for the macht that has flowed into the quark to exit again back into space as there are no exposed bauen charge sides able to allow the energy to pass through again. So the quark which is matter can accept macht from the local area ESS in the form of charge which will of course decrease the ESS macht glow reserve charge around it. This means the total reserve pool of macht glow charge for that area of space has decreased and will do so in exactly the same fashion as observed by modern science to produce what is seen as a positive field around a fundamental particle. The bauen geometry of the quark has a set size however and so can only decrease the local area macht glow charge so much and then it stops drawing in macht and this is why the quark has a finite charge and not one that keeps inhaling macht forever from the universe. In fact if it helps you think of all bauen geometries better they are like little ESS macht sponges either absorbing or wringing out macht into the local area ESS and as a sponge they have a set size which defines their exact and quantized total values in all of modern science for everything described in the Standard Model of particle physics.

This satisfies the need for a single object to explain both negative and positive charges as well as the corresponding field excitation as understood by modern science that creates a known Standard Model particle. The field excitations are of course much larger than the actual bauen geometry that created it and thus gives rise to the macro scale electrons, quarks and what not of the particle zoo. Simply arranging the number of bauen around the quark will allow it to have a charge of less than one which explains how it can be a value that is not quantized as is desired by modern science. The charges of 1/3 and 2/3 are now obtained as the much smaller bauen objects can be arranged such that in total on the surface of the quark they create either a total 1/3 charge effect with space or a total 2/3 charge effect. So a down quark would have one bauen object facing with its charge side exposed to space and this is similar to an electron that would have three bauen objects charge side all facing towards space. An up quark would therefore have two inward facing bauen arranged so that the charge sides are hidden inside the bauen object and held in place by the bauen force. The actual arrangements of bauen objects on a single quark or electron might not be just two or three but instead some other total number so long as the number of bauen objects used produce the necessary 1/3 and 2/3 charge for quarks and the 3/3 charge for electrons. The total number of bauen objects in each of these structures is likely to be higher based on the masses found for each fundamental particle in the Standard Model of particle physics.

This is due to the fact that the total mass of these objects is larger than that of a single bauen and even though the virtual mass of these particles does not reflect the true matter only portion of the mass for the particle which would be the bauen matter itself without field excitation it does at least indicate how many bauen objects will be present. This must be tempered by the fact that different bauen geometries will excite the fields in differing amounts and can make a smaller bauen object look more massive to modern science by creating a larger virtual particle field as seen in the known fundamental particles. Therefore the mass of particles as measured by their field excitations alone is not a true indicator of their actual mass which comes from the bauen matter only as this is the matter that does not incorporate equivalent energy mass. Think of it this way as a single photon if it were made of two bauen objects appears to modern science to have zero mass which as we know is incorrect as it has two units of bauen matter for its real absolute mass. So just looking at what modern science views as mass is not literal for the absolute matter it is made of and the different arrangements of bauen sides exposed to space will fool modern science.

The inward facing side of the bauen objects responsible for charge is what allows them to create both negative and positive charges as well as magnetic fields. It also explains the precise mechanism touched on in an earlier section relating to magnetic fields and how they flow. The normal flow of magnetic energy through the bar magnet is explained at a sub-quark and sub electron level by using the sides of bauen objects in order to move the macht through the local area ESS. This also explains what modern science views simply as invisible fields such as magnetism and gravity again by proving how space and matter interact to create these observable effects. In reality these fields are not mysterious or invisible and are instead perfectly understood using ESS macht glows and precisely explained by using those same macht glows and bauen objects and how their sides are exposed to or hidden from the ESS with which they interact.

The flow of ESS macht through the bauen matter is what causes the ESS macht glow magnetic to be higher at the north end of a magnet. This is what causes a decreased amount of ESS macht glow magnetic at the south end of the magnet and will create the second observed magnetic pole. The way matter and space share energy and macht inside the magnet creates a flow of magnetic energy through the solid bar magnet and this in turn creates a flow of macht in the space surrounding the bar. Since the field energy cannot flow backwards through the one way macht/energy flow of the bauen objects sides it must flow back the only way it can which is through the ESS.

Here is where the standard rules for ESS macht sharing through schales once again appears and the distribution of macht glow flowing back to the decreased region is seen as the overall magnetic field. This is the precise mechanism by which magnetic fields work in the universe as this accounts for both the matter and the space factors in magnetic materials. The desire of ESS macht to balance out due to magnetic flow through bauen objects also is explained in stellar plasmas as the magnetic fields created are three dimensional in shape and free to join up with other magnetic fields they encounter. The desire of the ESS to balance its macht glow magnetic means that fields can flow any direction they need to in order to balance the total macht glow magnetic. The macht glow magnetic is still tied to the magnet that created it as this is in conservation of the law of space/matter macht/energy for a system and so the field does not flow away from the source of creation but rather moves the magnet through space towards another magnet in an attempt to balance the total ESS macht glow magnetic of both of them. In stellar plasma this is what results in the pockets of magnetic fields that align themselves with each other yet are still fluid enough to move around. It also demonstrates how the magnetic fields will reconnect during magnetic reconnection events and shows how the matter behaves as well as the space in the system as the two are needed for creation of the magnetic field. Understanding the motions of magnetic reconnection in stars shows the fluid way that both space and matter move when controlled by magnetic fields which are a product of magnetic flow through bauen objects and ESS macht.

Chapter 34
Sightseeing at the Zoo and Elsewhere with all Three Generations

Particle Zoo

Now I will solve two more unsolved problems facing modern science by looking at bauen geometries and ESS macht balancing rules. This is a reminder here that once again the same laws that govern the universe everywhere once again apply here and with no change in forces, particles or any other variables such that new physics need to be created or that random probabilities in mathematics must come into play. The two problems I will solve are the three generations of matter and the zoo of particles which specifically comes down to why are they there? In the case of the former why does matter appear to have three generations and in the case of the latter why is there a huge slew of different particles especially those produced during high energy particle collider events. To start off with we will remember that all atoms are made of various particles such as electrons, protons and neutrons. Next modern science views these particles as either fundamental or composite and an example of a composite particle would be a hadron.

Hadrons are things like protons and neutrons which themselves are not fundamental although at one point in history they were believed to be. Today modern science understands them as collections of three quarks held together by gluons and this is why the protons and neutrons themselves are not fundamental as they can be broken down into smaller units. The quarks are the particles considered to be real matter as they contain actual mass and interact with the Higgs field to create a real and non-virtual mass as a result. The gluons are the force carriers for the strong nuclear force which binds the three quarks into the hadron composite particle and are said to be responsible for virtual mass as they do not have real mass themselves but contribute to the total mass of the hadron through the use of the energy equivalency formula of relativity.

Therefore the total mass of the hadron is made of technically real matter particles in the form of quarks and virtual matter particles in the form of gluons although all are considered to be particles in the Standard Model of physics. The virtual mass of course really comes from the gluons still influencing the ESS macht glow of space and so trying to move them even if they are said to have no mass is still trying to overcome the ESS macht balancing for the total hadron and we perceive this as virtual mass when really only the bauen matter in the hadron is real mass. These quarks and gluons however are considered by modern science to be like the electron as fundamental particles that cannot be broken down any further. From here I can explain why matter has three generations and why a zoo of particles exists as seen from events like particle colliders collisions.

I will answer both problems by simply explaining both at once as the two are inexorably linked to one another in their inner workings. If you look at that simple atom or rather ion that is being made to collide with another particle inside of a particle collider you will find that the fairly stable structure of the hadron is rather easily smashed to bits. I say rather easily because it is a difficult engineering endeavor for humans to actually delve into the secrets of the atom at such small levels yet we do manage this feat on a fairly regular basis and so it has become easy to look inside what makes up the stuff of the universe. This contrast of easy and difficult is important as we can see inside atoms and look at what we find there yet it is difficult to get down to the most base levels as what we do in the laboratory pales in comparison to what the universe can achieve on a daily and for it mundane basis. So if you look into a particle collider collision you will see that the atoms are easily destroyed and that a shower of particles is all you are left with.

There are two interesting facts about the particles that come out of the collision the first being they are of an extremely wide variety and certainly more varieties than just up quarks, down quarks and gluons. The second interesting thing you will note is that all of these particles tend to exist for about the length of time as a blink of a blink of a blink of your eye. This is to say a very short time indeed and a very short time for a very great number of different particles all of which are part of the vast zoo of particles and many are not fully understood or catalogued. This shower of seemingly strange and short lived particles is one of the unknown problems facing modern science in that the nature of these particles is not fully understood. They are of course particles as understood by modern science in that they have mass, certain other properties and excite various fields by which they are detected. The unsolved part of the problem comes from the fact that science only gets to see them during very violent and very short lived events like those found in particle colliders. The particles themselves are different from the known fundamental ones and thus the question arises as to how fundamental and quantized particles can be broken down when of course this goes against modern science and its belief that fundamental particles are of course fundamental. It is not that modern science is repulsed by the prospect of something being made of smaller bits as this has already been accepted during the use of quarks and gluons for hadrons. It is simply that in the modern Standard Model of particle physics and the world of quantum mechanics breaking down fundamental particles is again a no-no.

This shower of zoo particles is explained using bauen geometries as it has already been shown that quarks are not at all fundamental particles any more than an electron or a photon. The size of a quark compared to a simple electron for example is quite great and as a result the complexity of the bauen geometry reveals that the total number of bauen objects in the quark geometry is also fairly great. As you smash atoms together in a particle collider you effectively overcome the various degeneracy pressures found in an atomic nucleus. In doing that the fundamental particles involved cannot move out of the way of one another fast enough and so collide head on which is done with enough momentary force to effectively break them.

Some of the particles resulting from these collisions are of course known and well catalogued ones such as quarks and gluons but many others are not. The well known particles are those that escaped direct hits with other particles and managed to simply be unbound from the strong nuclear force which previously held them into a hadron configuration. The other particles that come out of the shower are those that came from direct hits such that the quarks and other fundamental particles inside the atoms themselves were broken down into smaller objects. Even gluons are broken down as they themselves are simply bauen geometries that have the gravity side of the bauen objects hidden from view and the strong nuclear side exposed. This means that gluons can still have the effect of virtual mass by interacting with the quarks but not need to be real matter as viewed by modern science. In reality however all particles including gluons are made of true matter which is bauen matter no different than a quark or light.

These smaller objects are made entirely of the fragments of the larger bauen geometries from which they came such as the total bauen geometry of a quark. So since quarks are made of a collection of bauen objects all held together and producing the properties of a known quark the fragments will total the number of objects of the quark but have different properties. Think of it this way a single quark can give rise to multiple zoo particles that have suddenly come into the view of the detectors and hence the observer all being smaller than the quark itself and with a whole plethora of unique properties. Since the bauen objects that made up the quark have now been dissociated from their original geometry these new fragments will have various numbers and types of bauen sides exposed to space. One fragment might have four gravity sides, four strong nuclear sides and no charges sides exposed. Another might have one gravity, one strong nuclear and two charge while a third might have no gravity two strong nuclear, and three charge but one charge for negative and two for positive based on bauen sides and ESS macht flow with the fragment. Still more might have one of each side, two of each side or three of each side exposed and of course the list of possible combinations goes on.

The combinations are further complicated by whether or not the sides are directly exposed or partially exposed such as a side exposure for a single bauen object on a larger bauen geometry fragment. The total bauen force and overall effect of exposed sides is a factor as well as the strength to which a bauen object is held to the fragment geometry. Is the bauen object being held loosely enough that it can move as in neutrino oscillation meaning that you would be looking at the same particle fragment only seeing it in a number of different flavors? This also means that the particle could actually oscillate while the detector is recording the event if the oscillation is fast enough and therefore would give rise to apparently different particles even though it is simply the same one oscillating between flavors. The number of variables is large and the number of fragments larger still plus the fragments will edge drag with the local area ESS of the collider in different ways which will not always give an obvious description of the particles properties.

Think of the photons which appear massless as they do not edge drag with the ESS macht glow gravity of the local area of space in which they move, yet photons are made of the true matter which is bauen matter rather than simply being some magic unknown thing called light which to be blunt just does whatever it wants as viewed by modern science. So some of the fragments will appear to be one thing in the brief time the detector can see them and in reality be something slightly different because only the field excitations are seen by the detectors.

This is something I must talk about for a moment here which is a failing of modern science to see things beyond field excitations as that is the only way humans currently can observe the universe. Particle colliders and their detectors rely on seeing the excited fields of bauen objects and cannot see the objects themselves and this leads to a series of problems in modern science and its understanding or rather lack of understanding of the universe. This is not at all a fault of modern science as they are simply and only currently limited by the ability to see fields and in the future modern science will of course adopt new detection techniques.

However today the reality is that field excitations are all that can be seen by modern science and this hides the truth about both ESS macht and bauen forces from view but make no mistake about it both ESS macht and bauen forces are quite real. They are both in fact necessary to explain the universe exactly as it is observed without the use of extra variables and in fact reduce the variables in the universe to exactly two things the first being ESS and the second being bauen matter, the communication between the two being made possible through macht and bauen forces. During observed decay events of higher generations of matter and things like the shower of particles created in particle collider collisions the truth is when individual bauen objects or small bauen geometries are dissociated from a main bauen geometry such as a quark they can be missed. What this means is that through no fault of modern science at all just its limitations in only detecting the universe through excited fields parts of the matter that was collided or decayed appears to disappear.

During collisions it is seen often that some things are simply gone and this is not true the matter is not gone only in a bauen form that cannot be detected by the colliders instruments as the bauen object is either not interacting with ESS macht at all for a brief instant or it is in a geometry that masks it during the detection event. Think of how light masks its true mass in the form of bauen matter by simply not exciting the gravitational field around it. A single bauen object takes with it the macht glow it used to help create during its time in a larger bauen geometry but if it is free from geometries after the collision it will create a macht glow through edge dragging so faint that no human detector can currently see it and so it is said that something weird has happened and mass has disappeared. This includes missing matter in the universe as well as any bauen objects free from known geometries will simply float through space interacting with the local area ESS in incredibly small amounts and with properties unknown to modern science but they will be there.

Also they are not the missing dark matter for which people have been searching as the sheer number of bauen objects would not exist in free form sufficient to account for the missing gravity. Free bauen objects will also have all sides exposed to space and thus be capable of interacting with other particles readily as long as the local area ESS can balance their increase in macht to whatever existing bauen geometry they attached themselves too. Even high energy decay events will likely shed some small number of bauen objects that will not be detectable to modern science and coupled with the quick violence of particle collider events only makes detection of these single bauen objects that much harder.

In addition I have mentioned this before in many sections but the local area of space needs to be under sufficiently high spatial compression and for sufficiently long periods of time such that the necessary conditions to create these fragments or keep them in stable forms long enough to detect be met. Currently human technology simply cannot create any areas of spatial compression nearly high enough or for long enough to allow for stable bauen fragments to be created and detected. In fact in order for our colliders to work we do the opposite and make them as close to a vacuum as possible which only means the local area space the particles collide in has even less spatial compression available to them to persist in. The fragments however that are created during particle collider collisions are the true insides so to speak of fundamental particles such as quarks and are explained using bauen objects and excited fields which are of course macht glow shared through schales in ESS. These fragments will be very different and quite numerous and so constitute the zoo of particles as seen by modern science yet will appear to excite familiar fields and therefore contain familiar properties.

The overall geometries of something like a quark are stable and repeated an incredible number of times as there are almost countless quarks in the universe. Therefore upon breaking a quark you will almost be guaranteed of getting particles that are similar if the numbers of quarks are broken often enough. A single quark break will result in a small chance of seeing duplicate particles or particles that can be grouped in similar ways as the number of fragments is not high and the number of ways the quark can fragment is. Repeat breaking quarks over and over and you will start to see patterns or at least groupings of similar particles as you get certain fragment geometries from the same starting quark geometry. The similar particles are coming from the fact that a quark geometry is always the same for a particular type of quark and so the fracture lines if you will inside it are similar too. What this means is the total bauen geometry necessary to make a quark is held together through the bauen force along certain specific paths as this is what makes a larger bauen geometry quark. The entire quark like anything else in the universe is under continual stress to balance its macht glows with space and certain parts of it will be more able too and certain parts of it less able too. What this means is the parts of the bauen geometry for the quark will when collided with another particle produce certain fracture patterns as some bauen matter or smaller bauen geometries that are less stable for the entire quark will be sheared off first.

This is how we can start to see inside a single quark and reflects on a fundamental particle scale the exact same mechanism that is happening for all decay events of heavy unstable nuclei in the universe. These nuclei follow the same rules for ESS macht balancing and shedding particles or bauen objects to balance out and this is where half lives come from and why eventually you are left with a stable object as that is the point where the ESS macht is balanced for all bauen geometries in the atom. Furthermore particle collider events and decay events are not set as if you alter the amount of spatial compression or ESS macht glow for a local area of space this changes the limits at which the bauen geometries can be stable through macht balancing rules. So you will get different results for collisions and decay events under different conditions and although the fragment bauen geometries released from these events might be the same the rate at which they occur and the threshold values will be different. To truly map something like the quark more fully it would be exceedingly helpful to collide only one type of quark over and over to clean up the data so you do not have for example up quark collisions and down quark collisions releasing all their fragments together making the sorting out process for bauen geometries released almost impossible. This is the first step in assembling a quark geometry from its fragment objects during multiple collision events and can be used to look for similarities in quark construction between differing quark types. This leads us now directly to the unsolved problem of generations of matter and so I will get right to it.

Generations of Matter

All matter or at least a great deal of it seems to have three generations and yet contain all the same properties as the previous one except for mass and the unsolved problem is why and why do these exist at all? The answer is simply starting with a bauen geometry for a known particle like a quark or an electron and then adding fragments to it. These fragments can either be single bauen objects or groups of bauen objects similar to the fragments from the previous particle collider proofs. Since the starting geometry which was for example a quark is always the same no matter what as an up quark is an up quark it was shown over repeated breakings that similar fragments will be created as the known quark breaks along consistent lines as it is made of a single bauen geometry. So these fragments are essentially quark like in nature and not so much for how they interact with the ESS to excited fields as observed by modern science but for how they can bind to existing quark geometries. In other words a quark fragment has a chance of binding to another quark which results in a quark of similar properties yet higher mass as the number of bauen objects has now increased. The chances of a quark gaining fragments or even individual bauen objects is of course rare and takes not only special conditions as compared to normal conditions on Earth but also high energies.

Yet these conditions do exist and can happen as three generations of matter have been observed. Creating quark particles in the laboratory can be thought of as taking pieces of the above mentioned quark fragments and briefly conditions are right to allow them to join to an existing quark to make a higher generation of matter. What this means is that an explanation of how matter can have three generations is quite simple as the ability of a quark geometry to exist is easily known and so adding more bauen objects to that geometry will increase its mass but not its overall properties such as charge or spin. The fact that there are three is simply the ability of the necessary conditions to add enough bauen objects to create two higher mass versions beyond the standard and there is nothing saying you cannot find a fourth generation either.

Now I will explain why the generations exist in three discrete tiers and not as a continuous transition of masses. This will also serve to explain why matter can exist in three generations in the first place as the Standard Model, modern science and quantum mechanics in reality says they should not as these particles are described by them as fundamental and therefore set. After all a quark should be a quark with its known properties why would the universe create and how would it create higher generations of matter with exactly the same properties and make them all fundamental? The answer is of course that it would not as once again this is too complicated and for no reason and we all know the universe simply does not make itself complicated. So instead making higher generations of matter from smaller ones by simply adding more bauen objects is the answer as it involves no new variables and only uses the same ESS, macht and bauen matter to accomplish along with unchanging and known rules of space physics.

So the reason that matter exists in three discrete steps of generations is linked to the fact that bauen fragments from something like a quark break are so short lived. All fragments from quark breaking events for example will decay very rapidly into smaller objects until they escape detection. The reason this happens is of course due to the ESS macht balancing rules that all matter in the universe must obey and bauen objects are no different. Since all matter is simply made of bauen objects arranged in different geometries it is easy to see that if the type of increased macht glow imparted to the local area ESS by the bauen geometry is not stable given the physical arrangement of bauen objects into a larger bauen geometry then the bauen geometry will decay in order to balance the ESS macht for the local area of space.

What this means is that you simply cannot make a stable bauen geometry any way you want given the local area ESS spatial compression and macht glow. The universe however can arrange bauen matter into stable bauen geometries and ones that are so stable they endure since the BBB such as photons and the like. You had better believe that other stable configurations of matter were possible inside the BBB and can be if the universe is diffuse enough as well but only because in both of these examples the spatial compression remains consistently high or low enough for those particles to exist for extended durations. These bauen geometries are of a set type however and others are not stable and so decay rapidly. The higher generations of matter are stable but not for our level of spatial compression or for the total amount of macht glow in the ESS as found on Earth. Alter the ESS spatial compression and the macht glow for the local area of space and all matter decays at different rates or not at all and this is another fact of the universe missed here on Earth. So higher generations of matter that we see must obey these universally stable configurations of bauen objects into larger bauen geometries but due to the amount of spatial compression and macht glow available to us here on Earth and inside our instruments they cannot maintain the larger geometry based on the need of the local area ESS and its balancing of macht.

The result is that discrete generations of matter are possible and they will always have the same properties as the previous but also that they will decay rapidly simply because the ESS balances the total macht of the system which is able to overcome the bauen binding forces of the bauen matter in the higher generation bauen geometry object. So here the very large and very massive third generation forms of matter are prone to the same ESS macht balancing rules that cause them to decay as fast as the much smaller and fewer in number bauen geometries of the fragments. These simple facts are what explains why you have generations of matter that follow discrete jumps and maintain similar properties, why you get smaller and numerous zoo particles from particle collider events as well as all violent and high energy events in the universe not listed here such as anti-matter creation and cascade particles from cosmic ray events high in the atmosphere and where some of the small fragments and individual bauen objects go in the universe.

Strong Nuclear Force and the Interior of Hadrons

This use of bauen objects and what sides they expose to space as well as the total ESS balancing of macht for the local area of space also explains precisely how quarks and the strong nuclear force works. It is known that if you separate two quarks the nuclear force between them gets stronger and eventually a new quark pops into existence to create a quark pair again. This means that according to modern science a fundamental particle can be created from nothing or at least from the quantum fields that modern science believes to exist in space and from which all particles are simply excitations and of course according to modern science all of that is quite impossible. On the other hand the explanation for both why the field gets stronger and why the new quark appears to pop into existence is perfectly, precisely and above all easily explained using ESS, macht and bauen matter.

To begin with remember how modern science finds quarks and this is of course in a technical sea of gluons which are the force carriers for the strong nuclear force. The quarks are known to space physics as bauen geometries that are simply made from bauen matter arranged such that certain bauen sides are exposed to space and can thus increase the ESS macht glow found there. The gluons are also made from bauen geometries and do act as modern science states heavily employing the strong nuclear force of its bauen sides with space. The stable arrangement of the bauen objects and how they edge drag with space and the total ESS spatial compression and macht glow for the local area of space is balanced under Earth conditions into a hadron. This is because the bauen objects also have as explained earlier their own bauen forces and these allow them to balance bauen forces across their geometries. These bauen forces also coupled with the ESS macht glow are what keep particles separate as the necessary energy to create larger bauen geometries does not exist as the particles simply move about under normal conditions and so quarks stay quarks and gluons stay gluons.

If you move two quarks apart however the ESS macht balance is being strained and this feeds back on the bauen geometries in the system and so they must react as well as space and matter always interact in the universe. So the gluons respond by slowly exposing the strong nuclear sides of their bauen matter from a slightly exposed to more fully and finally fully exposed position to space. This increases the ability of them to increase the ESS macht glow strong nuclear for the local area of space. Since all ESS macht is interacting with matter in the form of bauen objects and bauen objects have specific arrangements of bauen sides to space a change will occur. Also since the total macht and energy for the ESS and bauen matter system must be conserved due to the macht speed sharing of glow between ESS the greater distance between the two quarks will cause the macht and energy to be locked to a certain bauen geometry. This will in turn cause the gluons in the space between the two separating quarks to feel the full force of this separation and of the ESS balancing macht glow and will result in the gluons having one or more of their exposed bauen matter rotate on the larger gluon object just as a single bauen object rotates on the neutrino geometry and changes flavors through the oscillation mechanism.

The rotation of one or more bauen objects on the gluon geometry will result in it balancing the ESS macht that was being maintained in the system before separation and the only way to do this is to create that balance again. To create that balance the gluon rotates one or more of its bauen objects to literally become another quark which in turn will now have all the same edge dragging properties of the separated quark which was supplying the ESS macht balance to the system. With the system now containing the newly created quark geometry the donated energy will return to normal and so will the total ESS macht glow of the system and so the system will self balance itself in order to maintain the total ESS macht glow. Any apparent mass difference between the gluon and the quark geometry simply reflects the way the bauen matter are arranged on the larger bauen geometry and means not all the edge dragging sides are equal between the two. Some sides that were hidden are now exposed and some exposed are now hidden with slight exposures produced along the way. Remember bauen geometries will produce different field excitations as viewed by modern science and interpreted through the incorrect energy to mass equivalency formula of Einstein and so the total energy for a gluon and quark need not balance through relativity as it is the differences in which sides of the bauen object are exposed that fools the Standard Model.

The overall effect is that the bauen geometries can arrange themselves under the influence of ESS macht in such a way that a new quark is created from an existing gluon and the surrounding system. This also demonstrates the similar properties and abilities to regularly interact between quarks and gluons which makes perfect sense as they are both found in hadrons, they both deal with the nuclear force and both are not unified until very high temperatures and pressures are met which keeps them similar to each other. Likewise things like light, neutrinos and electrons share more forces and properties and so are more similar and also unite under similar energies such as the electric force and magnetic force into the electromagnetic force and the force carrier the photon. Similarly W-, W+ and Z bosons are explained as well with their charges coming from normal macht generating mechanisms of the exposed sides of bauen matter as described before and with no charge as in the case of the Z variety. The ability of them to take part as absorption mediators helping conserve things like momentum, charge or by sometimes making higher generations of matter all are explained using the normal macht balancing and bauen geometry rules seen before. All of these interactions and any others seen in the universe or laboratory are illustrations of bauen matter interacting with ESS and the two coming together to balance forces. In this way the truth about numerous forces and particles in the universe is explained and the precise mechanism by which these interactions take place is shown step by step.

This roiling interior of the hadron is also where the well known phenomenon of CP violation occurs as certain weak decay events do not obey the standard rules as do the electromagnetic and strong nuclear interactions. The reason for this is already explained in the above proof of quarks and gluons during quark separation and so a small outlining of what is happening is all that is needed here. In truth CP violation is simply modern science seeing but not understanding what is going on in the complex hadron system and the use of bauen geometries, bauen geometry fragments and bauen matter. Since it is now known that actions inside a hadron occur as a need to balance various macht glows for a set level of ESS spatial compression of the local area of space as seen through bauen matter objects the CP violation is actually an expected result of the weak force.

The reason this is so comes from the need of the ESS macht to balance glows by causing some bauen objects to be shed from larger bauen geometries in order to relieve built up macht stress. Just as this need was seen in quark and gluon interactions it is simply a result of the fact that the weak force deals with the bauen objects that are least stably bound to the larger bauen geometries and so will result in them experiencing whatever changes are necessary to balance the system. Kaon decay resulting in CP symmetry violations is simply because these events involve common bauen geometry fragments subject to the macht balancing rules and reveal one common type of bauen geometry released as well as showing modern science the reality of the world of space physics beneath it. The world of space physics and the need to balance macht and thereby altering bauen geometries should in fact be well accepted as the violent events needed to do so are seen all the time in anti-matter production which is simply a rearrangement of bauen matter and bauen sides in larger bauen geometries.

These changes might be the result of spatial compression and macht influences from other materials, the laboratory or a set of quantum interactions that are simply unknown to modern science because they are governed by space physics. The lack of understanding that space physics are underneath all quantum mechanics means that certain violations appear as if from nowhere when in reality they are simply following macht balancing rules. The fact that these bauen objects appear to fail in their reversal test is not due to some strange quantum working of the universe but simply because the complete list of space physics interactions between all ESS, macht and bauen matter of the system is not mapped yet by modern science. For these reasons it is no wonder that the CP violations all involve systems with much shuttling of bauen geometries to balance macht glows such as those involving quarks of any generation which are shown to be unstable and things like neutrinos which are already known to have an oscillating bauen matter side as well as neutrinos being a common object emitted by systems all in an attempt to balance macht.

The end result is that upon knowing all ESS, macht and bauen matter details for a system it will be shown that CP violation is simply explained and can be perfectly predicted using space physics for any given system. In addition under sufficient stress from increased ESS spatial compression and elevated macht glows all the known forces will experience this CP violation. CP violations of this kind will be found to be quite common in objects such as stellar interiors and certainly in neutron stars, quark/gluon plasmas and the like. The only place you will not find these violations is inside an SM mass as the matter there is more or less locked in place and will be explained further such that it is shown no bauen objects are being balanced internally. The particle that makes up SM masses is balanced.

To look at a few more of these examples quickly I can explain that certain bauen sides are obviously more influential over the local area ESS macht than others. Looking at the strong nuclear force it is dominant to the other forces and so the bauen geometries of the universe are more attracted to one another inside the nucleus than they are to influences from outside. This means that it is most difficult to break the strong nuclear force and therefore helps keep particles and atoms stable. If the hadron is examined and you look at the three quarks present you can see that they are large bauen objects that are very stable in their triplet configuration. The gluons are also stable but less able to maintain a balanced macht glow with the ESS of space and so they are the easiest to move around inside the nucleus helping to balance all macht glows.

The reason they simply do not fly away from the hadron in a type of balancing maneuver is due to the fact that this would then in turn create an imbalance in the total macht glow for the hadron. This is not permitted under the systems levels of ESS spatial compression and so the gluon is never actually lost from the hadron. The strong nuclear force is also still very powerful and so in order to balance the macht will move the gluon but not permit it to leave the hadron and so this demonstrates not only macht balancing but also how certain forces and their respective strengths can be used to build matter. I can now use these rules to also explain some of the nucleosynthesis of the universe no matter where or when it occurs in the life cycle we are currently observing.

The real nucleosynthesis of the universe must always occur in places of violence and persistence in order to make new particles. Going back to the limitations of current particle collider technology available to us as humans and only one of the conditions is barely met. This condition is the violence in the collisions which does have the strength to break matter apart beyond simply separating atoms from nuclei or separating nuclei into protons and neutrons. The protons and neutrons are split and the quarks inside of these are split again and so on. This means that the violence necessary to see inside atoms is present but not the persistence of the conditions needed to create new elements. The local area spatial compression of the particle collider is far too weak to allow for these violent events to last long enough or for the collisions to occur in spaces where matter can stick together long enough for the bauen objects inside them to create new arrangements and geometries that are then held firm by the bauen force inside them. The only place in the universe known to us where these events are both violent enough and persist for long enough are in places like Big Bang Bubbles, supernovae or in things like neutron star or black hole mergers.

These events certainly have the violence required and since they do not end in a fraction of a second like particle collider events the length of duration is present such that the bauen forces can hold bauen objects in new geometries. The particle colliders we have are working in quite the opposite way as nucleosynthesis for despite containing the bare minimum of violence needed for nucleosynthesis to occur they lack the duration which leads to the complete undoing of nucleons. Think of it this way as whenever you have a violent event you get bauen geometries being dissociated from the larger object from which they came. The bauen objects are still interacting with the local area ESS just as they would normally but now they are in geometries that are not normally found freely in nature which means they are not normally stable under the local area spatial compression. The total bauen force shared by the object is now forced to balance itself with the local area ESS macht and at the same time try to maintain the geometry it has which is held by the bauen force. The bauen force may be the strongest force in the universe but it is still only supplied in small amounts per bauen geometry when compared to the numerous ESS surrounding it and all the macht influences it feels.

Remember that a single small bauen geometry that is in a form not normally found to be stable in the universe is small in terms of true matter which are its own bauen objects and how much effect they have on the total ESS macht of the area. Now compare that small amount of force the geometry has with the total ESS macht coming from everywhere around it including other bauen objects, other particles and the particle collider and the Earth surrounding it. This surrounding macht glow is massive and can easily overpower the bauen forces of the small geometry that is newly created. You can easily overpower gravity by standing up or the electromagnetic force by burning a match to release stored chemical potential too.

This results in one thing which is the ESS macht balancing of the local area of space breaking the bauen geometry into fragments or individual bauen such that the total ESS macht balance is achieved. Single bauen objects can be made in this way as a larger bauen geometry that is not balanced can have a single bauen object balanced from it by the local area ESS macht. This results in a stable bauen geometry which might be for example a photon, neutrino or an electron and a single bauen object which escapes detection due to limitations in our current detectors. So particle colliders provide the violence and the local area ESS macht balancing provides the decay mechanism for the resulting fragments of bauen geometries. If you were to add the persistence of these conditions to the experiment as well as higher levels of spatial compression you would actually see nucleosynthesis occur in the laboratory. This is because the same conditions exist in naturally occurring areas of nucleosynthesis in the universe such as BBB, novae and heavy object mergers. These naturally occurring events allow the bauen geometries time to meet up with others of their kind and arrange themselves into new geometries larger than before. Think back to the different nucleosynthesis processes in stars and how time is needed for the process to be fast or slow for example.

The persistence also allows for the bauen forces to have time to then hold these new geometries in shapes that will become stable which will be one of the normal forms of bauen geometries that we see as stable today and modern science calls fundamental particles. These particles also need time to form into larger objects like hadrons and these hadrons to in turn form into nuclei. A neutron alone will decay as it is an unstable object and at roughly eleven minutes it has a very short half life and the quarks and gluons inside it will decay even faster if they are alone. In this way it is shown that persistence of violence and high spatial compression are needed to create new elements through the use of nucleosynthesis. This also helps to prove the bauen fragments are what make up normal fundamental particles as you can reverse the decay process seen in particle colliders to actually make elements. The entire zoo of particles that can be extracted as it were from normal matter all have different properties that are quite different from one another and certainly different from the matter they came from. Yet if you combine all these zoo particles no matter how different or strange they seem from one another they will always give you once again normal atomic matter. In addition to this the matter that was made came from things like electrons, photons, neutrinos and a whole slew of other decay chain particles and energy according to modern science and yet they will always combine into only two things in the nucleus which are quarks and gluons. Think of this as putting a complex block puzzle together.

This proves the existence of multiple bauen geometries and the fact that these geometries are made of single bauen matter with multiple sides that interact with the ESS and increase or decrease the macht for the local area of space. The way this is proved is by looking at the supposed fundamental particles of the nuclei of atoms according to modern science. According to modern science fundamental particles are fundamental and cannot be broken and in addition to this matter and energy are seen as two distinct things which are only equivalent to each other but not the same. Since the nuclei of atoms are made of hadrons and hadrons are made of quarks and gluons then how according to modern science can something like a photon or neutrino or the like from the decay chain be forced together once again to create a quark if the process was reversed and from only field excitations to boot? In other words inside a hadron is nothing but quarks and gluons and you simply according to modern science cannot stick a photon or neutrino in there and have it stick yet these nuclei must be constructed of the reverse decay process products. In order to do this bauen matter and bauen geometries are used to take something like a photon and actually incorporate it into the nucleus or a neutrino into the nucleus as the decay process is reversed.

In this way known fundamental particles can actually be incorporated into a nucleus because in the nucleus they are not fundamental particles anymore as they have had their geometries rearranged so that they can become part of the existing particles needed to make up a hadron. To put this another way simply reversing the chain of events from a particle collision and reconstructing the nuclei that were involved can be accomplished simply with bauen matter as each bauen geometry is not fixed the way a fundamental particle is believed to be fixed according to modern science. So the events in a particle collider and its collisions allows for a look at how matter is created in nucleosynthesis such as that found after the BBB and can be explained using bauen matter and ESS macht with the end result being new elements. Understanding these chains of events will allow you to better understand both ESS macht and bauen matter as well as how bauen forces work to hold bauen geometries together.

This is how everything in the universe is built as you can build our universe with only two things which are ESS and bauen matter. Think about how incredible this is as you start off with a big bang bubble and end up with us. Literally. One BBB and then you have an entire universe filled with everything we can observe and all of it is made of just two things depending on how they are put together. If humanity has been searching for a simple explanation of the universe and one that always seeks to reduce variables then this is it. There is no need of multiple elements either, the kind from thousands of years ago or just the ones we put into tables today. One type of space the ESS and one type of matter the bauen and the universe evolves into the most beautiful creature anyone could hope for.

Chapter 35
SM Masses and BBB, Inertia and Some of Matters Workings

SM Masses and BBB

After the BBB these bauen and the ESS through which they moved began to interact such that nothing more than simple edge dragging of exposed sides of bauen matter and ESS macht balancing rules arranged the bauen into geometries that created larger composite particles and structures. These structures in turn fed back into space to create various macht glows of differing size and strength such as magnetic and gravitational fields extending for millions of light years and more. On the smaller scale and using the explanation of particle collider fragments along with ESS macht balancing from before and we can construct things like electrons, photons and quarks just after the BBB. Other geometries like gluons, neutrinos and all forms of anti-matter filled in the universe as well allowing for not just more complex interactions of all the stuff in the universe but also in making simply bigger stuff. The now strongly increased ESS machts of the universe are holding the quarks in hadrons through the ESS macht glow strong nuclear and the electrons to them using the ESS macht glow electromagnetic. Atoms in turn will follow internally all ESS macht balancing rules regarding the machts they utilize and will balance these when joining and interacting with other particles and atoms. These atoms will now follow the mainly ESS macht glow electromagnetic balancing rules to create molecules, crystals and the like.

All stars and planets are made from these and on some of these planets warmed by those stars life evolves. This life of course comes from the combination of particles and atoms that follow the simple ESS macht balancing rules underlying all interactions in the universe. Eventually that life turns into intelligent life that can look at the world and universe around it as well as up towards the sky and out into space. This life will make sense of everything it sees the best way it can and eventually ask how does it all work? No matter how complex we or others like us have become we ourselves are all based off of simple bauen geometries and ESS macht balancing and nothing more. That fact, not idea but fact should be the most incredible thing you can ever comprehend as it links us and every thought we have ever had to the universe itself. There is no barrier between the bauen matter and the ESS they inhabit and the macht glows shared here and the farthest reaches of the galaxy and the most distant parts of the universe. There is literally no boundary layer anywhere to separate you from the very central clintron inside an SM mass or the thoughts of every living creature in the universe. The entire universe is a single living life form that we are all a part of and I mean that in the most scientific way possible I have shown on the most basic levels of the universe that it and we are nothing more than two things space and matter, ESS and bauen objects.

This universe is in the middle of or roughly somewhere between the beginning and end of one of its life cycles and eventually it will recycle itself. All the objects we see around us will eventually become part of an enormous SM mass that is quite comfortable being much larger than singularity. This SM mass will eventually create a single TSM clintron at its very center and another BBB will be made starting all over again the process of building a new and equally beautiful universe full of things as complex and curious as us. Before this happens though the universe must make this SM mass and the key to an SM mass is the particle it is made of which is of course a clintron. A clintron is the end point of all matter in the universe as it represents the last bauen geometry possible in existence and the geometry that is also the key to the next life cycle of the universe starting again which quite literally makes it the TSM clintron seed from which all life grows both organic and inorganic. Organic life is anything like us and inorganic life is anything like the Earth we live on or the Sun that gives us heat in order to drive all the reactions we need to live here. Making a bauen geometry for a clintron is exceedingly simple as all you need to do is obey the bauen force binding rule for bauen matter.

For bauen matter the rules are simple and for a bauen clintron geometry the only rule to obey is facing all gravity sides of the bauen matter fully exposed to space. This will produce the necessary properties of the clintron which are all forces turned off except gravity. The exception to this is that the bauen object might have some sides other than gravity exposed to space but they will be in the slightly or partially exposed positions and will produce essentially no effect on the local area ESS macht as compared to the total increase in ESS macht glow gravity. The total bauen force of the clintron geometry will also follow the balancing rule for bauen objects where the total energy donated through edge dragging in space will be calculated and once again produce a perfect donation to the macht glow of gravity or as close as possible. The mass of the clintron is the only property of significance in the SM mass as it is what is holding the ESS in the highly spatially compressed form and this high ESS spatial compression is in turn holding the bauen objects in the clintron geometry.

This is the precise mechanism explained for what I touched on in a previous section about both space and matter being needed equally to make a clintron and therefore an SM mass. The fact that the bauen objects inside the clintron are not able to expose any of their sides to space other than gravity means that the macht glow gravity is not only stronger and therefore able to hold the bauen geometry in the clintron shape but also that there is no macht glow available to push the bauen apart. If the other ESS macht glows were available then the normal pattern of electromagnetic and strong nuclear forces would be present and this of course would allow the matter to follow ESS macht balancing rules that can create stable atoms which are very high in volume as opposed to matter. Normal atoms cannot hope to create the spatial compression necessary or the bauen geometries needed to make an SM mass.

So the clintron is held in place so to speak from the outside by the highly spatially compressed ESS and macht glow gravity as well as internally by the lack of other increased ESS macht glows. The necessary properties of the clintron are what allow ever more bauen objects to enter and become a part of the SM mass as various event horizons are passed. As I said once before the interior of an SM mass is essentially all clintrons with zero space between them all touching at once in a perfect clintron crystal of sorts meaning it is possible to interpret the black hole as a single massive particle. Any new matter added to the SM mass will simply become free pieces of bauen matter that leave only the gravity side exposed to space and so it can be thought of as one single SM particle. The reason this is possible is because just as in the generations of matter unsolved problem where adding more bauen geometries similar to the original particle makes it heavier so too is the addition of any new material to the SM mass converted to the clintrons. In this way it is almost but not quite as if the entire SM mass was made from a single central clintron seed to which all new matter simply create ever larger generations of the clintron bauen geometry.

This also explains perfectly what happens to matter as a degeneracy pressure is exceeded as the ESS macht balancing of the object is overridden by the local area of space ESS macht and the way the bauen matter expose sides to space. If you exceed the electron degeneracy pressure all you have done is forced the electrons and nuclei to balance total ESS macht of the system in a way that accommodates the bauen geometries that they are made of. If you exceed the bauen geometries you will get the electron and the proton interacting to again balance the ESS macht glows and this happens by forming a new neutron geometry from the bauen matter of both the electron and proton. Exceed the neutron degeneracy pressure and you repeat the process all over again and if you repeat all these processes enough you get the only object in the universe strong enough and capable of handling the incredible ESS macht balancing and high spatial compression of ESS as well as bauen geometry binding forces in the system. This object is the clintron and can be fully built using the same balancing rules and rearranging of bauen geometries that precisely explain something as normal as exceeding a degeneracy pressure.

Each time a pressure is exceeded or a geometry rearranged you will pass through a new event horizon on your way into the SM mass proper. The journey from normal space and matter that we can understand and explain in our laboratories on Earth is undisturbed as you move into an SM mass. There are no breakdowns in the laws of physics and no new physics needed as often touted by modern science all you need to do is simply balance ESS macht and rearrange bauen objects. It might sound silly but we are the same thing as a SM mass and a SM mass is the same as us. What I am reinforcing here is the truth that everything in the universe is simply made of ESS and bauen matter and that the same laws that work here on Earth work for everything in the universe no matter how small a quantum scale or how large a cosmic object. The clintron is proof that this is so and is the single most important object in the universe when it is in its TSM form as this is the only mechanism that can literally breathe life into the universe itself and the beautiful thing about all of this is that the universe breathes life back into itself forever. Be impressed, the universe is cool.

As magnificent as the universe is in its large scale life cycles there are still many more little details that need some explaining and the ones I will show you here are about light. Light as you remember is an arrangement of bauen objects with the gravity sides of their surfaces facing inwards to block them from donating energy to the ESS macht glow gravity of the local area of space. The bauen geometry photon still has its electromagnetic sides exposed to space and will of course respond to any ESS macht glow electromagnetic it encounters. This is why light can be bent by a simple magnetic field for example as the ESS balancing of the bauen photon object is responding to the local area macht glow electromagnetic from the applied field source. Since a photon has a neutral charge but still uses the electromagnetic force this explains not only how they are attracted to an atoms nucleus for example but also why they can do what an electron cannot and plunge straight into the heart of the nucleus to add its bauen geometry to the geometry of the nucleus. Of course inside the nucleus it will be attracted in a variety of complex ways to whatever particle that makes up a hadron finally adopts it.

This is how particle accelerators work as they create strong ESS macht glows within their vacuum chambers that particles are then made to move through. The particles all have a charge of sorts whether it is from a single particle like an electron or an ion such as a nuclei but they all have strong and one sided ESS macht glows. The accelerator is actually increasing the macht glows of space inside of itself and these glows are what the particles feel. They respond by balancing their macht levels with those of the accelerator but since the accelerator is far more massive and powerful obviously it wins. The result is the particles being made to continually balance ESS macht glows with each new field the accelerator creates or manipulates. So a simple exploitation of the ESS macht glow electromagnetic is all that is needed to make an accelerator work and the use of radio frequencies in accelerators is really just a quick way to make particles follow something that already moves at the speed of light. The behavior of light however is not bound to simple machines like particle accelerators and can be observed throughout the universe. Some of lights properties and how it interacts with space can be seen by looking at different wavelengths of electromagnetic radiation in space and through the ages.

The photon is bound to space and proof of this relationship and of why it exists comes from looking at how bauen photon objects react to differing amounts of spatial compression. If we look at the early universe just after the BBB we see that many high energy photons were created in the form of early gamma rays and these were spread fairly uniform throughout the universe. These photons all lost a lot of their energy over the billions of years since the BBB and now many are lower energy microwaves. The only way this can happen is through the bauen object explanation of light using ESS macht glows of space through which they move. If light was simply pure energy which was quantized and not a physical object made of bauen matter then it would be impossible for the light to lose energy at all. The reason is the light is seen by modern science as quantized into discrete packets meaning they have a set value which if altered would mean they are no longer a discrete quantized amount and therefore violate its own definition. Light being made of energy and quantized cannot red or blue shift as this is impossible for it despite the rather crude analogy of saying it behaves like a wave. Behaves like a wave but not actually a wave are two very different realities.

The concept of wave particle duality or whether light is simply a wave or a particle is answered by looking at how light behaves and using space physics. Light cannot be a wave which is known but light does behave like a wave sometimes and so cannot be a particle or can it? What if you could have a discrete quantized particle that was always a discrete quantized particle yet behaved exactly like one would expect a wave to when needed? Well actually this is possible and it is not wave particle duality as that is also wrong but it is instead something different. All objects in the universe whether they are what modern science sees as matter or energy are really wave motion particles meaning they are true particles that move as waves but not one or the other or somehow both at once. This is because all things in the universe that are not what you would think of as space are made of bauen matter which are the only solid matter in the universe as fundamental particles as catalogued and understood by modern science are not real particles just the ESS macht glows from bauen object geometries. These bauen geometries are real and solid and can be thought of as discrete quantized objects they will however always move in accordance with the ESS macht balancing rules of the universe as shown in a previous section.

The motion that results from this ESS macht balancing is what makes things move as they sit perfectly still if that makes sense to you and it should by now. Think of the fallacy of superposition which of course is not real only the macht speed transfer of glows through ESS and the quantum speed tracking of the matter with the macht glow. The particle appears to be in two states at once because the transfer of energy and macht as the particle spends time or degrees of maximum interaction inside and outside of space is occurring at the maximum allowed speed for the system. All instruments used to observe such interaction which includes intelligent beings themselves operate at the same speed as we and our instruments are made of collections of bauen objects that also move in and out of space at maximum speed. So of course when you make what is called an observation you can only ever sense the target object being either in or out of space as your equipment made of slow quantum speed matter will always only be able to see one or the other position. Your mathematics might make things look like it should be both at once and you can try and interpret results to favor this belief but in the end it is folly as superposition simply does not exist.

Quantum computers can use a single electron to be both a yes and no at the same time? No they cannot and this is why they encounter such high degrees of decoherence in the quantum states induced by these machines. The electron vacillates rapidly between the two states as it is in space it is seen as if in the higher energy state because it has now donated its energy to the ESS macht glow and you see this increased field as a more energetic particle. When it is out of space the matter has donated less energy to space and so the overall field excitation is seen as smaller and it is said to be in a lower energy state. The idea I must point out here that a discrete and quantized particle can naturally switch between higher and lower energy states is absurd as well as this means it is again not discrete. They only way to account for the changing energy levels of light or the alternating higher and lower energy states of a particles is if it is in a sense putting its energy somewhere for a moment. This place is of course space in the form of ESS and this allows the particles to always be discrete and quantized at its heart which is the bauen matter geometry but still allows for the changing levels of field excitation viewed by modern science which explains things like superposition.

As for light the early BBB gamma rays were in areas of high ESS macht glow and highly spatially compressed ESS allowing them to have quick back and forth ESS macht tracking motions. These shorter motions in a high macht glow means the photons were possessing what is viewed as a short wavelength and high energy but as space expanded this meant that spatial rebound occurred and ESS macht glow electromagnetic was spread out through larger schales. The bauen photon geometries were no longer able to move in short waves and interacting with as much ESS macht glow electromagnetic so their wavelengths increased and their apparent energy levels dropped. Looking at the wavelengths of photons in a given area will tell you about the ESS spatial compression there and the amount of macht glow as well. Both factors will alter the total photon wavelength but yes you can study space directly for a given area by looking at the wavelength of a known energy photon created in it as it will reflect some of the size of the ESS volumes and the macht glow strength for that area of space. In spatially rebounded areas less ESS mean less macht glow reserve pool for photons and so less energy. The bauen geometries at their hearts never changed only the amount of interaction with the ESS in the local area of space and so the light remains as I stated both a discrete bauen matter object and yet as a changing wave when observed using the methods of modern science.

So light can be discrete without losing energy which would mean it would lose mass through the theory of relativity and its equivalency formula and therefore no longer be a discrete quanta of energy. In addition to this it also explains exactly why light behaves as it does and how it interacts with space beyond the bafflegab of relativity which simply states it changes and does nothing to explain precisely why these things are happening on a quantum level with quantum objects and the universe. Again the inaccuracies and mistakes of relativity in all its forms and that of spacetime are laid bare as neither can account for point by point what is happening to things in the universe. This is of course akin to the failure on all counts of all theories of relativity to explain what gravity is doing to real quantum objects in the universe and instead it says something wrong and vague like they fall into a curved spacetime. The gamma rays at the beginning of the universe had to lose their energy somewhere and so they lost it to space but by showing how this happens and why space physics explains once and for all precisely how these events occur between real and measurable objects such as those well studied in the Standard Model of particles physics and the entire universe in which we live and observe.

Since I am on the subject of bauen objects creating fields which are viewed by modern science as fundamental particles let me show you another way the discrete and unchanging bauen matter can interact with space precisely. The importance of how bauen objects interact with space has been shown again above using the behavior of things like photons but now I can show you how these interactions also play with the concept of fundamental particles. The arrangement of bauen matter in a particular geometry has been shown to account for all particles including light and this is due to how the bauen objects expose their sides to space or not. This means that in different configurations bauen objects can alter certain properties of fundamental particles all while allowing them to remain quantized.

Remember it is the bauen matter that is discrete and unchanging inside the field excitation that is viewed as a fundamental particle but the amount of macht glow which the bauen increase or decrease can change. This macht glow is the volume viewed as the field excitation of the fundamental particle so it is proven here quite simply that a particle can at its heart be discrete by having a set bauen geometry but the field excitation does not need to be discrete for that same object. In other words you can have something like a quark for example with the same amount of mass and spin but differing volumes to the space it occupies. If you look at a normal quark under normal spatial compression and normal macht glow strength which is defined for simplicities sake of humanity as that found in the local area of space surrounding the Earth. Like standard temperature, pressure and the like standard spatial compression and standard macht glow are useful to us as we have made all of our particle measurements using standard spatial compression and standard macht glow which modern science never knew they were doing.

These two factors determine all of what modern science understands as the size of a particle in three dimensional space as calculated from the total macht glow volume of the bauen geometries edge dragging their exposed sides with space and increasing or decreasing respective macht glow reserve pools found there. So from here I can explain what happens to a standard quark of let us say volume one here on Earth and compare this to a quark residing in quark/gluon plasma as could be found in the heart of a neutron star. The conditions inside a neutron star are quite similar and different from those of Earth as the star is much denser and made entirely of neutrons. It is essentially the same because the laws of physics are identical there as here on Earth but they are different because the spatial compression and macht glows are so much higher. Inside a neutron star the total number of ESS per volume of matter is very, very high as compared to Earth and therefore the total amount of donated energy in the form of elevated macht glow is also much higher. Any quarks found here are living in a space that is much more cramped so to speak as compared to Earth and each quark is very close to its neighbors. What this means is that the empty space found inside a normal atom which allows for a very relaxed local macht glow is now absent. The increased number of ESS per matter volume means that the quarks are essentially shoulder to shoulder with a higher degree of maximum interaction of their exposed bauen object sides with space.

The end result is a quark that is now smaller in volume than one found here on Earth as the nature of all fundamental particles is to reflect how the bauen matter inside them contributes to the local area of space and its macht reserve pools. Inside quark/gluon plasma the quarks are now compressed as they are still donating only a set amount of energy through bauen matter edge dragging of exposed strong nuclear sides with space. This means that they have not changed how much strong nuclear energy they are donating to the ESS macht glow strong nuclear and since the ESS are so much more highly spatially compressed they can still only share this macht through a set number of schales surrounding the matter orb proper. So if for example a quark on Earth uses 100 ESS to create its macht glow so to a quark in a neutron star uses only 100 ESS as well only this time because the ESS have decreased so much in volume compared to Earth the 100 ESS of the neutron star quark occupy a total smaller volume in three dimensional space of the universe.

Therefore they will occupy a smaller volume of space due to higher spatial compression and likewise the total area of space will as always obey the macht balancing rules of ESS and so each neighboring quark will still try to resist the quark degeneracy pressure placed on them by balancing their own local ESS macht schales with their neighbors. This places a macht glow pressure on each bauen objects own ESS macht glow strong nuclear field which is what is seen as the quark volume. The end result again is a pressure from all sides of a single quark that pushes against its own macht glow strong nuclear from within a static ESS space that does not move.

If you remember the way that space can explode from inside itself using macht shockwaves of ESS macht balancing as explained in the rules for magnetic fields in stellar plasmas in a previous section then you will easily understand what is happening here. Space is not moving yet space in the form of ESS macht glow strong nuclear is moving inside of itself. This is what causes the macht pressure to be exerted on the quarks and further compresses their total volume to one that is smaller. These two factors of ESS and macht glow will now result in a quark that has a volume less than one or less than the volume of one of a similar quark under standard spatial compression and standard macht glow. This example of changing particle sizes is important because once again it reveals that while a particle can retain all of its original properties one such as volume can be changed depending on the conditions of the universe around it. This can also be translated to other particle properties besides volume I have simply used volume as an example here as it is probably the easiest to visualize for the reader and utilizes the movement within static space example again which is key to these particle changes. Space can compress why not matter?

The concept of particles that are not always as we perceive them here on Earth is a key fact of the universe and one that must be accepted in order to move beyond the current stagnation in understanding of the universe by modern science. The fact that things such as particles can change should also be immediately accepted as this type of phenomenon is observed constantly throughout the universe and the laboratory here on Earth. Think of two hydrogen atoms bonded to a single oxygen atom for a moment and then think of phase changes of matter. Water is always water which is made of the three atoms mentioned yet if you cool it enough it will freeze into a solid. Heat that solid and a phase change of matter occurs such that it is now liquid and if you heat it further now it will phase change to a gas. This kind of reality of the universe is no different just because someone is looking at what they believe in modern science to be a fundamental particle.

I will now explain another unsolved problem of science that utilizes this idea of a changing particle which is thought to be fundamental. The idea that gluons are fundamental in nature and are the color charge carriers of the strong nuclear force as well as how they change color charge and how they impart this to a quark is easily explained using bauen matter and ESS macht glows. The idea that a gluon has a color charge and an anti-color charge is simply a way of complicating things and even more complicating is the idea that quarks absorb differing charges, balance color charges to clear or neutral and the like. Want to explain all of this using only one variable and make the universe simple as it should be? Of course you do that is why you are still reading this either you agree with me and I thank you for that or you disagree with me and want to see me fail and I welcome that as well. However I am sure you will agree that simplifying the universe is always the way to go and this idea of color charge can be simplified again to one variable. In order to simplify all color charge variables all that is needed is to examine the exposed or hidden sides of the bauen matter strong nuclear as they edge drag through the ESS of the local area of space.

The concept of color charge is no different than regular charge which is positive or negative and made from the same single side of bauen matter charge. Only the change in exposed to hidden side of the bauen matter side charge is needed in the overall bauen object geometry to make the charge field either negative or positive as explained in a previous section. The exact same mechanism is used here as a single bauen object gluon can have exposed, slightly exposed or hidden bauen sides strong nuclear in its overall bauen geometry. One geometry will produce a blue color charge, another will produce red and one more a green charge for example. The way this works is simple and exactly like the normal charge where one geometry will increase or decrease a certain amount of the local area ESS macht. In the case of hadrons the local area ESS macht being affected is the macht glow strong nuclear and each gluon type will alter that balance. A blue color charge for example might work like an electron by increasing the strong nuclear macht glow of the system while a red could decrease it like a proton and a green is a mix in between. The blue color charge is accomplished by exposing the bauen side strong nuclear to space and the red is achieved by turning them inwards while the green is in the slightly exposed position. The total bauen force of the object will be factored to determine the precise amount and type of effect each geometry of exposed, slightly exposed or hidden sides has on the macht glow strong nuclear for the local area of space.

So in this way the gluon leaving one quark will take with it a certain amount of macht glow strong nuclear and this will create an exact deficit of amount and type of macht glow strong nuclear in the quark. The quark will then be perfectly receptive to the exact type of color charge gluon that will balance that change in macht glow strong nuclear left by the change in the quark from the previous gluon leaving. The second gluon that has a different color charge will then move from a second quark to the first so that the total macht glow strong nuclear for the first quark is balanced as the ESS want it to be following the normal macht balancing rules. Remember how a particle can decay if left alone? Well the quark if isolated is known to decay and so is unstable and therefore wants to balance its macht and bauen geometries nicely. The stable way for a quark to exist is in a triplet form of a hadron and so it is always decaying over and over again and it is decaying in a sense gluons. The way the gluons were made to become a new quark through the strong nuclear force of two separating quarks is again used here. The nuclear force and the desire of space to balance macht was so strong that it could rearrange the bauen geometries and the same is happening here as the gluons are emitted just as they need to be and so on through what is wrongly seen as color charge.

These are the same rules that balance out everything else in the universe already and are no different here and do not need to contain a separate type of complicated color and anti-color charge system. So this system of using gluons and quarks will always be able to balance out their local area ESS macht glow strong nuclear simply by different bauen geometries for gluons. One quark will always lose or gain exactly what it needs to through emitting or absorbing a gluon of the corresponding bauen geometry type that will have the effect of balancing the total macht glow strong nuclear for the system. This also explains simply by using a total balancing of ESS macht glow for the hadron how quarks and gluons know exactly what kinds need to be sent to which other particle and never make mistakes. The problem of why a quark would spontaneously emit a certain kind of gluon and how it knows to go to which corresponding quark is answered through simply macht glow balancing as well as what provides the necessary force to move these gluons around as they are simply following the macht glow of the ESS they inhabit. The reason that this balancing is even happening in the first place is explained using space physics as it simply shows that quarks as they exist in the universe need to balance their macht glows in the local areas ESS they exist in because of their unique stable configurations of bauen matter. The easiest way for them to do this or the way in which they will remain stable with the least energy or effort on the part of the universe is to group into hadrons.

These hadrons now have a total ESS macht glow created by the three quarks and this in turn creates a larger area of space that now seeks to balance the total ESS macht glow found there. In order to do this smaller bauen matter geometries known as gluons will move from quark to quark in an effort to balance this total ESS macht. Remember that some quarks have a 1/3 and some a 2/3 charge associated with them which is already in need of total macht balancing for the ESS they inhabit. This means that any three combinations of quarks whether they make a proton or neutron will have a need for dissimilar types of quarks to bind together into a hadron but also to balance the dissimilar effects on the total macht glow strong nuclear in the ESS they inhabit. The result is that while a stable configuration of three quarks is made there will be micro-imbalances in the ESS macht balancing rules around each quark that are translated into the total hadron ESS macht glow.

Part of this hadron macht glow you must remember is in a non-spherical form as the three hadrons will not produce a total ESS macht glow equal to a sphere. It will be as if three large diffuse spheres are joined together and all moving around a central point of total macht glow charge as created by the combinations of 1/3 and 2/3 charge quarks. This need to balance the dissimilar charges will result in a slightly off center central point of charge glow balance. Coupled with the normal quantum motion for all particles in the universe as they continually ESS macht track as described in a previous section and the quarks cannot remain perfectly still in space. As a result they will be moving and rotating around this central charge point of ESS following the ESS quantum tracking rules. The motion of the hadron in space is enough to change from moment to moment the central point of ESS macht glow charge as shown for any composite particle in space there will be a central ESS that holds the highest strength macht glows. The central ESS will not last as the particles motion will create a new central ESS and then the quantum tracking rules alter the system to match this. This creates a changing balance throughout the local area ESS as to how the ESS are balancing the macht glows for each quark in terms of the macht glow charge. This imbalance in ESS macht glow charge is what causes the quarks to need to balance overall macht glows for the local area of space again due to their 1/3 and 2/3 charge differences.

This allows the strong nuclear force which is binding them to also need to balance as the central ESS for the system keeps changing. The strong nuclear force is present in the hadrons as the mechanism for how they are bound but it is the charge force that drives the exchange of gluons. The macht glow strong nuclear force must balance out its local area ESS macht within the ESS of the local area of space and so must send or receive gluons carrying the macht glow imbalance with them. The quarks keep moving in space and so the need to send or receive specific gluons to and from specific quarks will also change. All these changes are directly in accordance to the position of the quarks over the ESS of space that they inhabit and the need for these ESS to balance their respective macht glows. All gluon exchange mechanisms, why gluon change occurs, how the quarks decide which gluons to send or receive and the total motion of the hadron is all determined by the ESS and macht glows that are being continually balanced for the local area of space. So all color charge, all strong nuclear interactions and why they even exist at all and are needed in the universe are perfectly explained using space physics such that only one variable need be used. That variable is simply the position of bauen matter sides on the gluons bauen geometries and the natural resulting way in which the quarks making up the hadron interact with space and space in turn balances its ESS macht glows. So there you have it hadrons, quarks and color charge all solved neatly in one go using nothing more than ESS macht and bauen geometries and with no different rules than anything else in the universe.

Inertia and more Behaviors of Matter

While I am on the subject of how matter interacts with space I will also explain another unsolved problem in physics and this time it is the precise explanation of what and why inertia exists. The first thing you must realize is that once again a concept, law or observation about the universe is taken for granted without being fully understood by modern science. This time the concept is that of inertia which is originally coined by Newton and used in his laws of motion to help explain the behaviors of objects. It is the familiar law that states that an object in motion will remain in motion unless acted upon by an outside force. This also applies to objects at rest which again tend to sit put unless you impart a force to them to move them from their sitting position. This realization of Newton's is incredibly important in our understanding of the universe as it helps explain how things happen. Much like another of Newton's strokes of genius which is the realization that gravity holds us to the Earth the same as the moon and lets apples fall on our heads so to is inertia a concept that helps free us from old and traditionally Earthly ideas. The Greeks for example believed quite wrongly due to the ideas of Aristotle that objects would always come to rest on there own as rest was to Aristotle the natural state of matter or anything with mass. Aristotle believed that objects only moved forward because something was pushing on them and that in the absence of that pushing force they would therefore return to their rest state which to him was of course rest. Taking the example of an object thrown like a projectile Aristotle believed that the medium through which the object moved was responsible for the pushing action on the object and that the pushing action simply faded away over time and distance. This is of course dead wrong and as dead as Greek thought in terms of science and the universe but it does serve to illustrate the far superior and infinitely more useful ideas of Newtonian thought.

This is of course reflected in how Newton was able to see the universe beyond the Earth such that he could not only see that gravity was an invisible force that attracted objects to one another based on mass and distance but also that inertia explained that outside the Earth an object would travel in a straight line conceivably forever. So it can be said that inertia is a measure of how much force is needed to overcome a masses desire to move in a straight line. On the Earth things like gravity and the resistance of the air through which a projectile moves slows and stops an object in motion but this is not of course the reality of how objects want to move according to inertia. Later Einstein tried but failed to explain inertia using his irrelevant inertial frames, world lines and spacetime to mention a few. What Einstein sought to do was to explain motion using spacetime and the curvatures through which objects' world lines were made to deviate whether accelerating, at rest or from the perspective of the object or the observer. The theory of general relativity seemed to provide the answers by accounting for accurate motions of objects in large areas of space but fails for two reasons.

The first is that the universe does not require world lines or spacetime to exist as both of these are simply inventions of Einstein to help him make sense of the universe in a way that he could understand. Einstein however does not understand the true workings of the universe as world lines, spacetime, observers or anything else held to be pivotal to Einstein's work is not needed or used by the universe in reality. The proofs for these have been given before and anyone still reading this by now should realize that these proofs are correct as they not only disprove Einstein and the theories of special and general relativity but they also explain the universe without them. The universe for example exists independent of observers or the need for the passage of time rendering spacetime and frames of reference dead wrong.

The second reason Einstein is wrong in his understandings of the universe and concepts like motion and inertia is that he does nothing to explain why it exists at all. In fairness to Einstein this is actually the central problem when talking about inertia and motions of objects in the universe and of course motion of the universe itself as they are not at all understood using any of the three main views of modern science. Inertia and motion are simply taken as fact that they are in some way a fundamental property of matter in the universe but modern science never stops to explore why this is so and once again simply makes observations, laws and equations to catalogue the phenomenon so that it can move onwards from there. Therefore in order to explain what it means to have an object in motion stay in motion unless acted upon by an outside force it is necessary to explain why something like inertia even exists.

To start off with inertia is not an inherent property of true matter but rather a property of virtual matter and for this section true matter refers to bauen objects and virtual matter or just matter refers to the results of bauen interactions with ESS producing familiar fundamental particles which are of course in no way actually fundamental. If you have a bauen object not interacting with space then the idea of inertia simply vanishes as it is impossible for it to have inertia. If you could somehow interact with a single piece of bauen matter and that bauen matter was not at all interacting with any ESS you could use zero energy to move it. This is easiest to understand if I explain how inertia is born and to do this I will explain why virtual matter or matter has inertia and where it comes from. To begin with we will allow the bauen object to interact with the ESS around it and inside of it which of course turns on or rather elevates various macht reserve pools.

I said turns on because despite space always having macht stored in its ESS macht reserve pools the object has not yet started to feel that macht and the macht has not yet started to feel the bauen either. When matter interacts with space you get the elevation of macht glows and you also get the matter being involved in the workings so to speak of the space around it. This allows for the matter to feel the macht glow of space which is something it could not do before interacting with space properly. The bauen objects are therefore now feeling space and the ESS are feeling the matter and this alone is what makes inertia. In fact everything about inertia and real and virtual mass can be figured out using this simple interaction for bauen matter with ESS macht. There is nothing in the universe to make an object resist being moved and I will use the example of a stationary object for all explanations of inertia and motion simply because it is easy for people to picture.

◄New Universe Order►

If you think about it why should anything resist being moved as in why should it actually take any effort to move something if you push on it? A single person pushing the Earth should realistically make it move away from the person as no magical force called inertia actually exists to simply say that the object cannot be moved. Here of course is where modern science steps in and thinking it is rather clever explains all about total mass of the object, gravity and the very weak mass and force of the person pushing so that using lots of nice and repeatable experiments it can be shown that the person will not in fact move the Earth. That is all very nice but you have never once even begun to explain the precise why of this occurrence. Explain point blank on a fundamental scale using fundamental particles and forces why inertia exists. Modern science obviously cannot or it would not make these sweeping statements that sound absurd such as inertia is simply a property of matter. Modern science might try to explain it saying something about frames of reference and things magically sticking to curved spacetimes but this again is all smoke and mirrors as all of Einstein's work fails utterly at explaining what is going on to known fundamental particles with known fundamental forces and this incorrect thing called spacetime. The trouble here is that inertia does exist as something we can repeatedly measure and so an explanation must be given of why it is so.

This explanation using precise mechanisms comes from the simple interaction of ESS and bauen matter to create elevated macht levels felt by both space and matter. With the macht levels being felt by both space and matter you can see that the matter is of course investing energy in the ESS it is interacting with. As explained and used to solve every unsolved problem of physics before this the fact that the matter is now bound to space in the form of bauen objects being bound to the ESS with which they share energy and macht is now in effect. This means that in order to move an object such as a fundamental particle which represents virtual matter or matter you must make the true matter or the bauen objects at their core stop interacting with some ESS and interact with others. In other words you must get the central ESS which possesses the highest macht glow to give back that same macht glow to the matter so that the matter can then move to another central ESS and make that the highest glowing ESS. This is using the quantum tracking rules for quantum motion, superposition and the certainty principle once again as it is shown that due to how these bauen objects are interacting with a central ESS they resist change in position.

There is a certain amount of effort required to physically make the macht shift away from one ESS and towards another and this is of course the macht that is bound to the matter and so the matter is resisting as well. The total number of ESS due to spatial compression and the amount of macht glow will determine how hard or easy it is to move any object from rest to motion. The net result is that the bauen object central to the fundamental particle as perceived as virtual matter by the ESS macht glow field is naturally attracted to a central ESS and of course the surrounding ESS of the central one. The reason it is difficult to push an object and overcome its inertia is because you are trying to overcome the attraction that space and matter have for each other using the ESS macht and bauen objects. The more bauen objects you have or the greater the spatial compression and macht glows the harder it is to move the object. So moving a black hole is near impossible besides the fact that it will have fatal results if you lay a hand on it. This is again reflected using the same precise explanation of gravity on a step by step scale. The speed at which the macht is shared between schales and the speed at which the energy is donated to space is limited and therefore will result in a maximum speed at which you can push the object due to these two macht and quantum speeds.

If you were to have a perfectly solid and perfectly stationary object floating in space with nothing else around it or you it would take effort to move it when you push on it. The elasticity of the object is gone so it cannot be said that the initial attempts and forces you apply to the object to move it are simply being absorbed as the object deforms slightly in response to your touch. Let us say the object is large at 1,000 kilometers in diameter and made of diamond which will not be elastic in any way so again the delay in the object moving is not simply you deforming the object before it moves. Let us also say that you are held perfectly still in space by an invisible force with no mass whatsoever and therefore no gravity associated with it. It is as if you are simply standing with your back to a wall that is utterly immovable and massless so that the only gravity that exists is that of the object and yourself. Now if you push on this 1,000 kilometer diameter orb of perfect diamond you will notice that it does not move right away. The reason this is so is that the object wants to remain at rest, but rather than this innate desire to remain motionless being some sort of magical base property of anything with mass it is solely due to bauen matter being bound to ESS macht. If this were not so then since there is no gravity involved in the system except for that of the diamond then you should be able to push it with essentially a light touch only. According to modern science as long as you apply any non-zero force to the object in free floating space it will begin to move.

This is not however what happens and you will notice that it takes a great deal of effort to move it and in fact more effort than an average human can manage. That is because a normal human can only impart so much force to the object and this force of motion is not enough to cause all of the bauen objects inside the diamond to move from their central ESS to another central ESS farther away from you in the direction you apply force. Every single bauen object inside the diamond is being held in place to the ESS it inhabits from the macht glow it feels and so there are far more bauen objects in the diamond simply interacting with space than a normal human can make move. This simple example is an easy way to see that all inertia and motion in the universe is caused by matter and space interacting. Similarly if the diamond were in motion and a normal human tried to deflect it from a straight line and of course assuming they were moving along side of it at the same speed and again with an immoveable wall behind them they would once again find it very difficult or impossible to deflect the diamond object.

The reason again is exactly the same as that of the stationary diamond as the bauen matter in it were exchanging energy and macht with the ESS around it so that the quantum tracking rules were maintained. In the moving diamond the bauen objects have kinetic energy to aid them as they are already moving forward and so the total number of bauen objects inside the diamond are again following the quantum tracking rules. These rules if you remember will due to the differences in macht speed and quantum speed of macht sharing and energy donation keep the object in motion as the matter tracks to the next central ESS macht glow.

This is just another example of the quantum motion of matter explained by space physics only this time it is scaled up from the tiny micro quantum world to the large macro classical world. The objects in question here need not be 1,000 kilometer in diameter diamond spheres but stars, planets, whole galaxies, black holes and whatever else you might care to examine on a cosmic scale. The classical laws of motion and the relativistic equations will still account for observations made of large scale objects but only the ESS macht and bauen matter of space physics explains for the first time what is actually happening and why on the sub-fundamental level. Trying to move the moving diamond still means overcoming the quantum tracking of each bauen object with the ESS of space and the macht glow they feel.

Instead of starting the object moving this example has the diamond in motion and therefore carrying the total quantum tracking motion energy of the diamond with it which resists change in direction from the normal human.

Another proof of this explanation of inertia and motion comes from particle accelerators or anything that can move matter to a fraction or higher of the speed of light. If matter is made to move near the speed of light it is said that its mass increases due to the equivalency equation stating that since its energy has increased so has its mass. For example inside a particle accelerator it is believed that a particle such as a proton which has a known mass will increase in mass as it is accelerated and at near light speed its mass has increased significantly. If the proton could be accelerated to the speed of light it is said that the protons mass will become infinite and that all the energy in the universe cannot actually supply enough fuel to move a single proton to light speed.

The idea comes from the theory of relativity which according to it states that energy and mass are equivalent and so giving a fixed mass such as a proton more energy must also give it more mass. This is of course wrong as simply accelerating a proton to near light speed will not actually make the proton copy itself in some strange way so that more protons are made to account for the increase in mass. Protons or anything else in the universe for that matter cannot inflate like a balloon simply because you move it faster and faster through space. This is understood by modern science and it is only said that the virtual mass increases while the actual mass of the proton stays the same so even at near light speed you still only have one proton. This also means that the gravity of the proton is the same despite the speed of the proton through space and so a single proton has the same gravitational contribution to the universe whether at rest or in motion. Since we already know all theories of relativity to be wrong in why the universe works but not in calculating values found within it we can explain why this is happening.

The values for accelerated particles in terms of virtual mass found using theories of relativity are accurate and can be repeated the only thing wrong with theories of relativity is that they cannot explain what is actually going on and simply say energy and mass are equivalent. Energy and mass are not actually equivalent at all and so a different explanation of these observations is needed and the same inertia and motion laws for classical objects applies here as well and do so without using spacetime. In the case of accelerated particles you are once again looking at small objects within the universe at the most fundamental levels possible. In this case the bauen matter at the heart of all fundamental particles and the ESS macht glows they create in space. When a particle is at rest it creates a set amount of gravity because it is interacting with space under one amount of spatial compression for the local area ESS. In turn it will produce a set amount of macht glow gravity and simply moving this object will maintain its macht glow gravity as its bauen object geometry has not changed. All of the other macht glows are also set at a fixed rate for this object in terms of the physical bauen geometry as simply accelerating a particle does not actually make it change the exposed, slightly exposed or hidden sides of the individual bauen matter that makes up the overall bauen geometry.

In a previous section I showed that accelerating an object will result in things like time dilation and explained part of why objects seem to get more massive and therefore overall are affected by space the same way they would be in a high gravitational field. This does not mean of course that they actually get more massive and therefore create more gravity as if they had grown copies of themselves. What it does show is that by being made to interact more strongly with space the matter has less time to interact with other pieces of matter. I can now explain this fully as I have introduced you to the bauen matter that makes up all fundamental particles.

The start of all of this is again the same as before with particles simply interacting with more and more ESS as it is accelerated through the machine. The number of new ESS the particle meets per second is insignificant compared to the overall quantum speed tracking of the matter through its donated energy to the ESS and its macht glow amongst the schales of space for the local area. As the speeds get closer to that of the maximum tracking speed of matter which is near light or light speed the effects become more and more noticeable. This is the real explanation that these effects are noticed as speeds become close to the speed of light not simply that things are curvy in spacetime or some such other vagary. The precise explanation comes from the fact that the slow speeds of Earth and anything we can build other than particle colliders operate so far below the maximum light speed quantum tracking speed that it is simply not noticed other than a few trillionths of a second are being lost from an atomic clock every now and then. The quantum tracking speed is so fast that anything we do to move objects in normal time is easily compensated for by the fast quantum tracking speed. Only once an object such as a particle is moved close to the speed of light that the limitations of the quantum tracking speed are seen.

Nothing in the universe does not allow bauen matter to move infinitely faster than light and in fact moving matter faster than light is rather simple as all you need to do is simply decouple it from the ESS glows to which it is bound. Specifically the ESS macht glow gravity which is really the only one you need to decouple from as this is the macht glow that is responsible for increases in virtual mass. In a typical atom for example it is known that the force of gravity is by far the weakest but this is only due to the bauen geometry inside the fundamental particles in the atom and the low levels of spatial compression present in the region of space in and around the atom. In a particle accelerator the other forces are so much stronger than gravity that they are creating stable macht glows that are more interested in balancing ESS macht with other bauen geometries in the local area of space. This is akin to what happens right before the actual collision takes place as the matter in question is still talking if you will to itself and other particles but this is overridden as soon as the particles collide and the ESS macht glow becomes highly overburdened and then the particles are forced to deal with the space around it and the non-gravity forces must now talk to space and each other in ways not seen in the moments before the collision. Again this is where the zoo of particles and little bits of missing matter which are really just small undetectable bauen geometries come from as the collisions are quite complex despite always following set rules.

The strong nuclear force is called strong because it takes so much effort to disrupt it, far more than gravity. This reflects what is occurring in a particle accelerator at near light speed as the bauen geometries inside the particles are still more interested in talking to each other and do not concern themselves so to speak with the increasing number of ESS as the forces such as electrons binding to nuclei or nuclei holding together are far stronger and shorter ranged than gravity. The force of gravity is not interacting much with other bauen objects due to how weak it is using non-clintron bauen geometries and low levels of spatial compression and so will be free to feel large increases in speed and number of ESS encountered far more easily. So easily in fact that this is where the falsehood of increasing virtual mass comes from and why particles inside accelerators appear to get more massive as they go faster and faster.

Therefore in particle accelerators matter seems to take more and more energy to move to the speed of light but matter is not really bound by the speed of light in actuality. Bauen matter after the BBB was decoupled from the ESS around it and was free to flow as it wanted through space so this is nothing new only explained here precisely for the first time. In particle accelerators and particle colliders however the matter stays interacting with space and cannot be freed using current human technology and so will experience an increase in virtual mass. When this happens the total energy that it takes to move an object must increase as what you are doing is trying to overcome and hurry up the natural quantum tracking speed of the matter.

The ESS that are bunching up in front of the moving particle is only part of the reason, the other part comes from the fact that the matter must maintain its donated energy and so will be forced to donate from ESS to ESS the energy it can given its particular bauen geometry. So the matter is in a sense forced to donate its energy then reabsorb it then donate it ever faster as it is made to interact with all of the ESS in front of it and this is where the limit of the quantum tracking speed is reached at light speed. It is not that light speed is as fast as matter can move through space but rather the maximum speed that matter can donate and share energy with space in a perfectly stationary position. The fact that the matter is moving is simply a reflection not of a maximum speed limit of matter but a maximum speed at which energy and macht can be shared using matter. ESS macht on the other hand has a limit to its macht speed sharing between schales far faster than light so the limiting factor here is not the ESS macht glows. In fact the need for macht speed to far exceed quantum speed is necessary in order for macht to share its glow through schales creating fields and spatial effects such as particle volumes. The macht speed must be this fast so it can take the donated energy from the matter share it then allow it to be returned to the matter if the matter is in motion such that the matter can follow tracking rules to the next central ESS of the matter orb. The macht speed glows are what allow matter to exist as we know it for example a quark field to make a quark, a strong nuclear field to make a hadron and an electromagnetic field to make atoms.

Everything that we as humans have been looking at in terms of what modern science calls fundamental particles is nothing more than the simple result of how bauen geometries increase macht glows of the local area of space ESS. Matter as understood by modern science cannot move as fast as or faster than light speed as this means the quantum tracking speed would be exceeded. Exceeding the quantum tracking speed means the bauen objects would not be able to maintain the normal force fields as understood by modern science and so not make atoms, hadrons or even quarks. The result is matter simply disintegrating into bauen geometries and single bauen matter objects as they would not be able to increase the ESS macht glow reserve pools and make what are thought of as fundamental particles. This is where matter and that is to say the only true matter in the universe which are bauen objects can actually move faster than light.

This is what was partially occurring after the BBB and when the universe slowed down enough for bauen objects to once again interact with space the familiar to modern science fields turned back on and fundamental particles could start to form with of course the bauen force being the first to turn on. In particle accelerators however the quantum tracking speed cannot be exceeded using electromagnetic radiation which itself has a maximum tracking speed of light and so just as you cannot cool something to absolute zero because of diminishing returns with the substance you are using to cool your test sample neither can you here accelerate anything to light speed as the substance you are using for the acceleration is a quantum speed object called light which is bound by the same maximum light speed quantum tracking speed. So in order to move even a single proton or the like to near light speed means you must use a tremendous amount of energy to push on that tiny object and force it to try and exceed its own quantum tracking speed.

The other major factor in why these effects are observed is that the bauen geometry for the matter is not changing as the particle is accelerated. Any bauen geometry that creates a known fundamental particle will only ever have certain sides exposed, partially exposed or hidden and these are fixed and set amounts that is why all particles of one type have identical properties. This set geometry means the bauen object is only donating so much energy to the local area ESS macht glow reserve pools. A piece of matter is exchanging and sharing energy donation with ESS most easily during complete rest. All of this is because the matter is under a set amount of spatial compression and if the spatial compression is changed the matter will be forced to behave differently as a result. Particle accelerators force matter to experience a type of very short lived spatial compression as they are made to do two things the first being encounter more ESS as they move through space and the second is during the collision which forces the matter into a small area of space that is compressed because the macht glows of the matter are shared over normal amounts in a small area. The actual space in the collision of the particle accelerator is not really compressed it is still at normal Earth spatial compression only a momentary region of space is created that is over burdened with donated energy from matter in the form of the two colliding particles ESS macht glows.

In both of these cases however the bauen geometry that makes up the field glows of the perceived fundamental particle does not change and so neither does the amount of energy it wants to donate to space. What does happen during particle collider collisions as a result of this over burdened ESS macht glow is that the bauen geometries are forced to deal with unusually high ESS macht glows after impact. The way in which these bauen geometries reabsorb their donated macht and then settle into new patterns of ESS macht glow balancing always obeying all quantum tracking and ESS macht balancing rules results in the zoo of particles that is observed. The high number of variables for a single collision means that often results are dissimilar from one collision to the next in so far as they create their own unique fingerprints of particle trails. Any unusual particles or absence of particles during these collision aftermaths are simply the result of larger original bauen geometries being fractured and the component geometries being forced to exist briefly in a part of space with high levels of macht glow but no increases in spatial compression.

The particle as stated is most easily maintaining its macht glow in space during absolute rest but in motion and when that motion is near the quantum tracking speed the number of ESS that it must now share its energy with increases. The geometry of the bauen object has not changed and so it has not been able to expose anymore sides to space and therefore feed all the ESS it is now encountering due to higher spatial compression. Since the geometry is not changing the macht glow is not changing either even though space has more and more ESS around to absorb more and more energy. The matter cannot donate more energy as it cannot change its bauen geometry such as in a clintron which exists only at the highest levels of spatial compression possible which are all technically at light speed if viewed in traditional ways by modern science.

Normal matter cannot increase its macht glow gravity and so it appears as if it is getting more massive because of the energy required to move it but in reality you can only increase macht glow gravity by creating other bauen geometries up to clintrons or increasing spatial compression significantly which does not happen truly in particle colliders. Clintrons are of course able to donate far more energy to the ESS macht glow gravity exclusively and so can exist at higher spatial compressions which in turn provides the ESS macht glow necessary to maintain the clintron bauen geometry and these two factors together create a positive feedback loop that is what allows SM masses to exist and be immune to external forces.

This limit of bauen geometry interaction with local area ESS macht glows is why everything matter related happens as observed and can be explained using a simple match. A match when unlit has certain arrangements of bauen geometries that create fundamental particles which balance their respective ESS macht glows by being chemically bonded. As soon as the match is lit and these bonds are lost due to an increase in bauen geometries of photons the resulting fundamental particles release what is called chemical energy which is really just more bauen geometries following ESS macht balancing rules. The newly altered arrangements of fundamental particles freed from their previous bonds have different ways the bauen objects in their geometries are able to interact with space through ESS macht glows. The ESS for the local area of space in and around the match is now experiencing a rapid increase in electromagnetic macht glows and small amounts of others either increasing or decreasing. These macht glows of electromagnetic force inside the ESS of space and how they need to balance themselves is why when you apply an electric or magnetic field to them they move in response. The ESS macht glow as I have shown before will balance itself out and so the match burns and burns hotly. The result is a plethora of fundamental particles increasing macht glows to unacceptable levels for the amount of spatial compression available to them and so the macht glows must be decreased as the spatial compression of ESS is not high enough to maintain them. This results in the matter of the hot burning flame to move apart from each other as this increases distance between their bauen geometries and their neighbors' bauen geometries. This increase in distance is accompanied by an increase in number of all available ESS to these particles and so they are able to balance their macht glows with space. The more the macht glows are balanced the more the particles cool off by decreasing the macht glow electromagnetic in small spaces and the motion of the particles goes down as well as the macht balancing becomes less and less.

This explains why hot things burn from stored chemical energy, why hot things expand from these burning results and why they both cool and increase the entropy of the system. All of these changes are occurring only because bauen matter sides are exposed to space in certain geometries and therefore certain macht glows are permitted to be balanced for the local are spatial compression of all ESS in the system which is both in the matter and around it as well. So understanding why things experience changes as observed in something like a particle accelerator is not only useful for explaining high speed phenomena on matter but also in explaining why matter does everything it does as the two are inexorably linked.

The use of bauen geometries and macht glow is also useful in explaining why atoms and molecules bond and transitions in phases such as between solid to liquid or liquid to gas occur. If we begin by looking at something like a single atom we can see easily that the electromagnetic force is really just the macht glow electromagnetic for the space of the atom and as the need for balance of all macht pools by the atom is continued photons are used as the exchange particle. The photon is of course just a bauen geometry photon that is shed by the nucleus in order to balance macht glow for the system and likewise when the electron momentarily possess the photon it too will need to shed it to balance the system again. This is because the electrostatic forces of the atom between the electron and nucleus are trying to balance through their mutual attraction for one another and this results in a slight imbalance at each location. If we start at the nucleus the bauen geometry there must shed a photon in order to balance macht glows and then the electron which is in balance for the local area of space will be out of balance upon receiving the photon. This now means the bauen geometry electron needs to shed the briefly attached bauen geometry photon to balance macht once again and sends the photon back to the nucleus.

The mechanism of balancing macht glows does something that was previously unexplained which is how does a photon know where to go? In other words how can a particle that travels in a straight line and obeying normal laws of motion know to be emitted at a precise trajectory to intercept either the electron or the nucleus? The answer is by following the macht glows of the ESS that are smaller than the particle and exist in space. The straight line of the photon is moved as needed anywhere in three dimensions to allow it to follow the macht glow that needs balancing. In this way the photon can still obey all normal laws of motion and forces and yet move as needed at the slow light speed of the quantum world to intercept a tiny electron or tiny nucleus in the huge vastness of empty space that makes up an atom. This explains how a bauen geometry photon can be used to help be the force carrier as understood by modern science in the electromagnetic force of an atom or molecule.

If we continue looking at molecules now we can see that when two molecules meet they can bind or not. The reason molecules bind can be explained simply first as imbalances in each molecules overall macht glow electromagnetic. If you have a positive or negative atom or molecule it will be attracted to the opposite charge atom or molecule and bond into a more stable configuration. A stable configuration is denoted by one that minimizes all macht glow imbalances for the two objects involved. Since a positive and negative charge set of objects are not neutral they need to balance macht glows and so will join up and begin sharing bauen geometry photons to mediate this desire. The rules for macht balancing are again simply employed here to explain which photons will go to which atoms or molecules in order to best share macht glows in space. This also explains all bond angles in every substance in the universe.

The reason why they do not bond to each other sometimes is a bit more complicated but still uses all the familiar rules learned so far. To start with let us look at two molecules or atoms and realize that they are once again simply products of both space and matter interacting to create macht glows that are perceived by modern science as fundamental particles, atoms or molecules. These macht glows are made of energy donated from matter and the macht glows of the ESS of space and must obey the law of conservation of macht/energy for the system. As such if you have an atom or molecule that is not bonding to another atom or molecule this is because the easiest way for the system to share macht and energy is inside of itself rather than reach out to another object to share macht glows with it in order to balance the system. Remember that each object and the space it occupies are sharing macht and energy and so it is easier for the object to share those machts and energies with itself rather than donate it to another.

The reason this is so comes from the fact that macht can freely be shared amongst bauen geometries in the universe so long as when a bauen geometry is fully dissociated from space it can receive the equivalent donated energy it gave to space. So for the act of sharing of macht which is what is really occurring anytime anything in the universe bonds to something else you must have the two objects strongly attracted enough to each other that they can share macht between them. What this means is a single particle in one part of space can donate energy to the macht glow of the ESS and have a second particle absorb that macht glow back again. This is like exchanging blood between bauen geometries through space as one particle can be thought of as transfusing some of its energy to a second particle through an ESS macht artery and receiving it back again through an ESS macht vein. The same blood need not come back to the first particle though as long as it has enough blood from the second when they break their bond or dissociate from space fully. So when anything bonds to anything else in the universe it is sharing macht because the macht glow balancing for those two objects favors sharing a common macht glow pool between them and if they do not bond together this means it is easier to maintain the donated energy through the law of conservation of macht/energy with the bauen geometry object that donated it as opposed to sharing it with a second object.

These general examples also explain why things like solids melt to liquids and liquids evaporate to gases. In a gas phase for example the macht glows of each object are so high that they have too much energy for the local area spatial compression for the ESS they inhabit and so they generally refuse to bond with other objects due to high temperatures. They of course will bond through a process such as burning and if the electrostatic differences between them are great enough but a single gas of one element or compound will not become liquid due to these high macht glows from high temperatures. Once the system has lost these extra bauen geometry infrared photons it will have lowered the macht glows around the atoms or molecules within the gas and they can now share macht more easily through a liquid bond and hence we have a phase transition from gas to liquid. The atoms or molecules are of course still filled with a decent amount of energy and so follow the exact same rules for when a substance does not bond as they are still more easily balancing macht glows by not sharing with another object. The release of still more heat means the objects are now moving slowly enough which reflects the macht balancing induced quantum motion from a previous section that they can come into contact close enough and long enough to share macht glows more efficiently between themselves in a solid form. Looking at phase transitions for various compounds in the universe under standard spatial compression will help to show how much spatial compression really is occurring there and what macht is present all of which will help to narrow in on the ESS found there. This is the simple truth behind all phase transitions between states of matter in the universe and can be further examined to help explain certain properties of materials in the universe as well like electrical conductivity.

Electrical conductivity can be seen in another phase of matter which is plasma and also in metals and the mechanism by which they work overall is the same as a need to balance macht glows in ESS space. I will explain metallic electrical conductivity because it is easiest for people to picture mentally and also because it is more applicable to everyday life in the twentieth and twenty first centuries. Let us start out by looking at a metal that is a conductor of electricity such as in a wire. All metals are said to have a loose sea of electrons inside of them that take very little effort to move if an electrical current is applied. The exact mechanism by which this works is through the input of bauen matter geometry electrons into the system which the metal can easily absorb due to its own need to balance macht glows.

The introduced electrons however will obviously carry with them their own macht glows that will now imbalance the overall system. The flow of current from one end of the wire will simply cause the electrons that are loosely bound to the metal atoms to move out of the way of the macht imbalance and so an electron flow is physically created as each electron moves to balance macht glows. This occurs from a single electron in the chain starting out feeling an increase in macht glow electromagnetic from the current applied and will seek to balance macht glows for the entire system by moving to the space where the glow is lower which is away from the applied current. This will create a stress on the next electron in line and so it will move as well and so on until a flow of bauen geometry electrons or electric current is realized.

From here it is easy to finally explain a couple of the properties of metals as current is applied to them beyond the simple explanation of it is a property of all metals or something equally as vague. The first property is resistance as is seen in how easily a material can be made to move electrons. This is explained by the affinity each bauen geometry electron has for balancing the macht glow pool of its parent atom. In other words how easy it is for that electron to balance macht glow with the atom to which it is a system before the above explained bonding rules are used to share macht between a second object. The easier it is for an atom to maintain its balanced macht glows with the local area ESS internally the less it will want to balance with another object and therefore the lower its electrical conductance or higher its resistivity. A metal atom is fairly easy to conduct with meaning it has low resistance as can be seen from the fact that it is often easier to balance macht glows from one atom to another rather than strictly internally. This leads to metals being used as easy conductors and the less balanced the macht glows internally for an atom the better it will conduct as it has less resistance.

This leads to the second property of conductors from resistance which is the generation of heat as a result of the applied current. This heat comes directly from the lag time in the increase in macht glow of the system in question and the ease at which it balances with the local area ESS macht glows. The more it balances internally with the macht glows the longer the lag time and therefore the greater the build up of excess energy seen as macht glow in the system before it is released through the macht balancing systems. So a poor conductor stores more energy for longer in its atoms and therefore tries to balance macht glows any way it can and not just by transmitting the electrical energy down the line so to speak. The fact that these poor conductor materials have a high lag time means that they can shed the excess energy in the form of bauen geometry infrared photons before electrical conductance can carry away the excess macht glow for the system. So a better conductor has less lag time and will more easily shed excess macht glow through the use of an active electrical current as opposed to a need to release a bauen geometry infrared photon or bauen geometry visible light photon if the heat is hot enough and a glow is noticed.

The very best conductors which are currently known are superconductors where the ability to balance excess macht glows is most easily facilitated with transfer of electrical current as opposed to shedding bauen geometry infrared photons. This is also the precise quantum reason why these conductors operate as they do and must be fully understood in order to make higher temperature superconductors. The solid arrangement of atoms in the superconductor means that all donated energy from the matter to space is more easily balanced through sharing the macht amongst each other using most of the bauen geometries in the object. A tiny number of bauen geometries on the surface of these atomic or molecular materials are in a state where the parent atoms are sharing the Lion's share of the macht glow internally with only a small amount being held in the loose electrons. Therefore when a current is needed the loose electrons are barely needed to balance the macht glow forces with the local area of space and so can move freely.

In the case of zero resistance to superconducting materials this is because the macht glow balancing for the superconductor is done one hundred quantum percent but not one hundred macht percent internally and the electrons in this solid state are no longer needed to balance macht glows between objects as they would be in a liquid or gas phase. The fact that the electrons are held to the parent atoms means that some macht energy is still present but due to the fact that all space physics involving ESS and macht are so much smaller and more refined than the large, slow and crude quantum world the overall resistance is perceived incorrectly as zero. A simple step would be to alter bauen geometries of the material in question to allow for electrons to be bound as absolutely loosely as possible to the parent atoms so all macht balancing occurs through electron transfer as opposed to emission of bauen geometry photons and the resulting resistance and heat.

Some Quantum Fun

One other point of interest in quantum mechanics and modern science is the accepted theory that particles simply appear, disappear and then reappear again at will due to quantum probabilities. This is quite impossible for a number of reasons and I will explain exactly what is happening here now that ESS and bauen interactions are better understood. This all goes back to familiar ideas such as the uncertainty principle which is based off of quantum mathematics that simply reduce to probabilities and unknowns on quantum systems due to a lack of understanding of space and matter. Initially these ideas were sound and very valid and have served science well for decades but due to their inability to give a complete model of the universe are therefore ultimately wrong or at least inaccurate.

The first problem with the idea that a particle can disappear or reappear at will in the universe is that this takes energy. Specifically free energy as the driving force behind these events must be forced to happen as they cannot simply exist on their own as this would constitute a perpetual motion machine of the quantum level. Remember something in the system would have to spend energy to makes these things happen and yet each particle in the system retains all energy as they are quantized and so no energy loss occurs meaning an impossible free energy system is in use at least theoretically. Free energy is illusive as its nearest kin the free lunch and you will find neither despite looking high and low.

The same is in effect in the idea of appearing quantum objects as this means something behind the scenes is pulling the wheels, cogs and levers necessary to drive the machine of the universe forward. Even the false idea of pair production near black holes cannot answer this as two more problems crop up here the first and most obvious being that black holes do not permeate all of the universe and so cannot be said to account for all observations of quantum disappearance here in laboratories for example. The second reason is that the black hole is not a free lunch as it loses mass in the process so a source of limited and very depleting energy is used to make this false pair production happen at least in theory. So without pursuing this farther yes the idea of free particle production is impossible as the recycling of the anti-particles to account for the energy is still impossible as it takes energy to create it in the first place. Nothing changes what it is doing in the universe without being acted upon by an outside force and this is not just applicable to classical objects, but all quantum objects and space itself. So you will not be getting free energy along with your free lunch anytime soon.

Another problem with the idea of popping into and out of existence at will violates the laws of entropy in the universe as well as conservation of information. If particles are allowed to simply appear and disappear again perfectly such that all energy is conserved for the system this means it happens in isolation of the universe and not only does not produce an increase in entropy but also destroys all information along with it. The same modern science loves both the idea of increasing entropy along with the arrow of time and the conservation of information in the universe and so it is slightly behooving as to why that same said modern science would elect to believe in quantum appearance and disappearance of particles. It does appear that particles appear and disappear from the universe however and nothing says that they cannot according to the known laws of modern science so what is needed are my laws which are not known to modern science but are perfectly well known to the universe itself. If you want particles to still behave as you observe them in experiments and yet not violate any laws of the universe then you need simply to look at the maximum degree and minimum degree of interaction of matter with space.

As explained in a previous section matter can be thought of as spending time inside and outside of space or to have a maximum degree of interaction and minimum degree of interaction with space respectively. When matter is at its maximum degree of interaction with space it is inside space so to speak and will create the strongest macht glows within the local area ESS and all schales surrounding it. ESS macht glows are what create the observable fields associated with all particles according to modern science and so maximum degrees of interaction are what are detected in laboratories here on Earth as a particle. When the matter is outside space it is at its minimum degree of interaction and this is when the macht glows of the local area ESS and schales are at their weakest. If this sounds like the explanation for superposition from a previous section it should as it is directly linked to it.

When modern science makes an observation of a particle it uses instruments which are of a very large crude scale as they are constructed of what are thought wrongly of as fundamental particles. These particles are themselves governed by a series of interactions inside and outside of space as well and so can never be more accurate than the particle they are attempting to observe. Since the instruments and probes are all made of and operate on the idea of recording quantized events they act as neural impulses in a human body and this means they act like action potentials. Action potentials as explained before cannot partially fire or fire at any percentage of their minimum threshold value and will always exist only at 100 percent. Quantum objects are no different as the very nature or them is to be made of quanta or discrete and set amounts so by definition they are action potentials.

This explains why superposition is so strongly believed to exist as the instruments recording these events as well as all mathematics involved in creating models for them are based off of real or virtual objects that exist only in set amounts. To this end the act of observation which in reality and once and for all is explained simply as a quantum level action potential interaction between two quantized objects will repeatedly see the object as being in both states at once. In reality it is never both states at once as the bauen matter at the heart of the fundamental particle is simply interacting more or less with space over time and this subtle below quantum level flow of macht and energy is simply not capable of being detected using the instruments and mathematics of modern science on Earth today. Also you must remember that the instrument being used to make the observation has at its heart a quantized object in the form of whatever probe is used to observe the sample. That probe can make observations while it is itself in maximum or minimum degrees of interaction with space also complicating the reality of the bauen object in the particle and therefore continuing the incorrect conclusion that superposition is all states at once.

So simply expanding upon this further it is rather simple and basic to see that the particle never disappeared in the first place and always existed in the universe. What has happened is that the particle is at some point below the threshold detection level of the instrument being used in the quantum experiment and so the particle cannot cause the action potential observation or falsely thought of act of observation to collapse the wave function and make the particle materialize into the universe from every possible probability. The shared macht glow of matter and the fact that matter as modern science understands it as fundamental particles are made from both ESS macht glows and bauen geometries means nothing can ever simply appear and disappear in the universe. Not only has this been shown for a particle but for a larger object like a human the total number of particles making up a single person create a shared macht glow to space and since all matter shares macht through donation to ESS and that ESS share it between themselves with macht a general macht glow like a glue is created that keeps us very much real. So real in fact that we cannot simply disappear one quantum instant as all our donated energy is being shared in the ESS macht and so ensures we cannot simply pop out of being one moment and then come back another.

There is no every possible existence of a particle or world or universe spreading out infinitely in every direction and dimension everywhere forwards and backwards in time so just stop it already. All these many possibility and many worlds theories and the staggering amount of information, matter and energy it would take for them to exist in a beyond infinite amount of space and universe is simply absurd as if no one simply sat back and thought about whether or not this at all sounded like something other than flying mouse feces mentally unstable; you can translate that one on your own. I mean to be abrupt for good reason as this is how best to clear the befouled air surrounding these theories and settle on the true ESS macht and bauen matter interaction explanation that does explain what is being observed without violating the universes laws. When matter is in space modern science can see it and when it is out of space modern science cannot and no other explanation is needed as it all comes down to the threshold ability of quantum detectors working on quantum objects. Underneath all of this is the far smaller and faster realm of the universe comprised of ESS and the macht glows they share amongst them. The differences in macht glows are only sometimes above or below the quantum thresholds of modern science and so all variation in macht for the system overall is simply missed. This is why everything is binary to modern science as particles are either off or on for example a zero or a one. All unseen changes in electron orbits or anything of that nature is explained the same way along with the so called act of observation and observer bias.

Matters of the Matterless

Here I would like to discuss the unseen and unsolved problem facing modern science known as the missing gravity. This is going to be a look at why dark matter and dark energy cannot explain the missing gravity of the universe whether that be the missing mass of the universe or the seemingly strange anti-gravity expansion of the universe over time. To begin with I will look at why dark energy is not a thing so to speak as it simply is an absurdity of modern science that must be done away with. The idea of dark energy is even less well theorized or understood than the imaginary dark matter which no one has ever found. The idea behind dark energy is that matter causes a positive increase in gravity through an increase in mass that interacts with the Higgs field to create a curvature of spacetime. Ignoring the fact that there is no such thing as a spacetime in the first place as proven in previous sections again and again dark energy therefore must have the opposite effect on said spacetime. It must seek to create areas where it flattens out the curvature of spacetime and even bends it backwards such that the universe behaves as though negative mass exists there and so causes the space to expand rather than contract. That is the idea of dark energy along with the caveat that it must be increasing over time as dark matter decreases since the Big Bang and from here it is easy to prove dark energy does not exist looking at the life history data of the universe as well as all experimental observables.

◄New Universe Order►

The very first thing that need be addressed here is that the word energy is used in contrast to the word matter for the various dark theories. Energy means photons as electromagnetic radiation is the only real form of physical energy in the universe known to modern science and the Standard Model of particle physics. Kinetic energy, potential energy and the like do not exist but are macro scale classical descriptions of concepts that can be broken down at the fundamental level by quantum mechanics and so do not constitute real energy. That being said only electromagnetic energy is known to modern science and so dark energy must be a form of electromagnetic energy according to dark energy theory. Electromagnetic energy cannot be made to affect any sort of change in spacetime as it is rather the opposite that is true which is to say space affects changes in electromagnetic photons. As space expands and contracts it is the photon that responds to this by lengthening or contracting its wavelength and hence the red shift and blue shift observations of photons throughout the cosmos. As such dark energy cannot cause space to expand as this is by the definition of electromagnetic radiation impossible.

This leaves the theory that dark energy is something else and here I must call foul upon all supporters of dark energy. Simply saying that some unknown force of any type needed to account for the expansion of the universe suffices to become dark energy and hence prove the theory correct is poor science and an outright dodge at best. If one day in the future aliens came to Earth and said that magical faerie dust was the driving force behind the expansion of the cosmos it would not be a vindication for all those in modern science today who simply say that dark energy is therefore correct.

The truth is dark energy theory is wrong and that is all as it simply cannot account for the expansion of the universe let alone be accounted for itself on a level that satisfies all other known forces and particles which are fully accepted by modern science. The theory of dark energy runs into another problem which is that according to modern science it is a field of the vacuum that is responsible for the pushing force. This brings up a host of problems the most heinous being that according to modern science the strength of fields never changes and so the strong nuclear force is always the same no matter what for example.

Well according to dark energy theory it is a property of the vacuum and therefore a field but it is getting stronger as time goes by and therefore means the dark energy field of the universe is not set which is a massive no-no according to quantum field theory. Fields are not allowed to change strength according to modern science and yet modern science is saying that the dark energy field is doing just that and so immediately falsifies itself. If you add to this the fact that the dark energy is increasing in the universe this means that the energy it possess must be taking it from somewhere else in order for the law of conservation of energy of the universe to hold and yet we do not observe this fact. We do not see the other four fundamental forces getting weaker in the universe or in the laboratory. On top of that any strong field according to quantum mechanics contains sufficient force to create particles in the form of field excitations of that kind.

So why are we not literally tripping over these dark energy field excitations if they account for some 70 percent of the apparent gravitational forces in the universe? Where are the dark energy particles and the dark energy force carriers? You cannot simply create an unknown force and ascribe it to the vacuum and so it must be real because it is doing what you think you want it to. The idea that an unknown force field for an unknown force carrier and the like are simply evading modern science yet can fit into the Standard Model and be the necessary dark energy field is simply lacking in sufficient proof.

As far as the caveat that dark energy must be increasing over time as dark matter decreases is also preposterous as it relies on a mechanism that breaks down unknown dark matter particles into dark energy through a kind of matter and anti-matter annihilation reaction. Once again matter annihilating or decaying naturally will only produce more photons or small particles like neutrinos and so once again dark energy fails for the reasons proved above. In addition to this the amount of dark energy theorized to exist in the universe today is about 70 percent of the universe mass and dark matter is about 25 percent and so going back in time the dark matter must have been very dominant to the energy. What that means is the early universe would have such a high critical density since dark matter is thought to outnumber regular matter today five to one, it must have been something like 30, 40 or 50 to one in the past.

This means the universe would have shortly after it was born collapse into a massive black hole and be quite dead and again we are here so that is not possible. So it is fairly obvious to all and many scientists out there as well that dark energy is of course non-existent. It should be noted here that many scientists in modern science do not believe in dark energy already and I do not mean to include them generally in the group of scientists who are dark energy supporters. In addition I am not attacking dark energy scientists for wanting an explanation for the expansion of the universe only they should not be championing the dark energy theory as it of course is very, very impossible at best.

Here I will give a quick explanation of something I touched on before which is the increasing rate of expansion of the universe and apply simple bauen matter to the problem. To begin with the amount of bauen matter in the universe is finite just as the amount of space in the universe is also finite and from here we can solve the problem of the acceleration phase of the universe and the halting mechanism for the overall expansion such that the following collapse is permitted. To begin with normal matter as modern science sees it interacts with the Higgs field to make gravity through mass and this can be thought of as a normal particle like an electron for example. If the electron meets a positron it will annihilate along with the positron and become two gamma ray photons.

This is an example as explained before of bauen geometries of fundamental particles changing such that the gravity side of the bauen matter is no longer edge dragging through the ESS of space. Modern science sees this as a decrease in matter and an increase in energy which is not quite right as underneath it all the bauen matter in all associated geometries is still quite unchanged. It is however interacting with the local macht glows of space different than before and so as these reactions for example occur we see the number of edge dragging sides for gravity of the bauen matter in the universe decrease and therefore the overall macht glow gravity of the universe as well.

The universe is currently in an active star forming phase of its life cycle and will remain here for some hundreds or thousands of billions of years or more but it will eventually stop and become far more SM mass dominated. Right now however a free photon will simply encounter a star or other object that can only hold so many photons before emitting them again and so the total gravity of the universe never really increases and can be seen from this mechanism as contributing partially to the overall decrease in macht glow gravity and therefore acceleration of the expansion of the universe. Nothing exists in significant amounts right now to bind photons back into space such that they increase gravity other than black holes. The future of the universe however will have numerous black holes all far more massive than those we see or rather do not see directly today. The future will allow for ample opportunity for photons to be held by black holes. This of course means a part of the finite true matter of the universe which are the bauen matter objects to be incorporated into a particle that can allow them to edge drag through the ESS macht glow of gravity again and these particles are of course clintrons.

The clintrons will alter the bauen geometry photons such that they become clintrons and will allow for the force of gravity to once again increase which will lead eventually to the halting mechanism for the expansion of the universe. For now however the decrease in bauen geometries that edge drag with the macht glow gravity means that more space is allowed to spatially rebound than before. This is due to the fact that as the macht glow gravity decreases, as we remember from our little room from the first section on gravity we see the macht glow decrease and the hollow ESS spheres increase in volume through spatial rebound as the matter orb moves away from them.

The same thing is happening here as bauen objects have their geometries changed to not edge drag through the macht glow gravity and so the ESS in the universe will spatially rebound accordingly and this helps to explain on a step by step basis how part of the acceleration of the universe works. This mechanism will tend to play out more in the spaces between the filaments of the universe as this is where the least amount of black holes exist and will work opposite too but in concert with the SM masses in the filaments that are helping also to pull matter away from the space between filaments.

Now that dark energy has been squarely dealt with it is once again time to disprove dark matter as it also is an incorrect theory. Here I will explain some of the intricacies of why dark matter cannot account for missing gravity and does not work as a mechanism for anything in the universe at all. I will be explaining using both observations of the universe as we see it today and as it appeared billions of years ago in observations and not in theory. All that is needed to explain away dark matter is an examination of celestial objects today and those seen through telescopes over ten billion years away and ago. Dark matter is theorized to provide extra gravity by simply being around in copious amounts and yet interacting weakly with normal matter making them nigh on impossible to detect. Well after 80 years, tens or hundreds of thousands of scientists working on the problem and tens of billions of dollars dark matter has never been found. This has been dealt with in a previous section and for good reason as dark matter has not been found because there is simply no dark matter to find. So without reviewing those sorts of proofs against dark matter let us once again pretend that dark matter is correct and use its own properties to destroy itself.

To start off with dark matter is said to account for about 5/6 of the physical matter in existence which our own solar system can easily disprove using pure observations and zero theory for once. The Earth and every body in the solar system is calculated to have a certain mass given its size, density and elemental composition from which we can determine how much gravity it has. Now first of all I would like to say that despite classical mechanics and the theory of general relativity not being able to account for why the universe works as it does they are very good at calculating how much gravity is in existence between two objects. Newton and Einstein after him based their calculations of gravity on observable objects only and did not include theoretical mass derived from dark matter. Since this is the fact then how come both Newton and Einstein were able to nail almost perfectly the motions and gravitation forces of bodies in the solar system without dark matter? Should they not have both been off by at least 5/6? Well yes according to the theory of dark matter they should have been as they would have gotten inaccurate calculations for the forces and motions of objects based on only 1/6 the data. Remember that whether or not dark matter exists it also obeys the arrow of time and so even if dark matter existed it would be part of the calculations used in governing the motions of the observed bodies of the solar system. Since Newton and Einstein new nothing of dark matter they would have had there 1/6 masses forced to also move through the arrow of time and thus give false results.

Now to knock the feet out from under the dark matter theorists one more time as they are all giddily awaiting a chance to tell me that the dark matter known or unknown is naturally incorporated into the classical and general relativity calculations of motion whether known to Newton and Einstein or not. Sounds like I am in trouble right up until the point where the motions of the planets around the Sun follow a dark matter free pattern of motion. Yes it is observed that motions of bodies around the Sun decrease their orbital velocities in relation to increases in distance from the Sun itself. The amount of increase in the length of orbital periods and increase in distances correspond to values that are calculated without dark matter and yet dark matter is theorized to be everywhere in the universe. Dark matter is especially theorized to be attracted to objects with mass such as the incorrect dark matter halos around galaxies. Dark matter is so integral according to dark matter theory only of course that the motions of stars around galactic centers hinges on dark matter being used to solve the flat galaxy rotation problem.

So the dark matter according to theory is everywhere in the galaxy and so prevalent that it can account for tens of thousands of light years and trillions upon trillions of solar masses worth of visible and MACHO matter moving in a flat rotation curve. Yet it is no where to be found or worse needed through observation and not theory within our own solar system or any alien solar system where we have observed the motions of the planets and their respective distances from the parent star. The solar system and every other star system we have measured and not theorized does not need dark matter and has no flat rotation problem and this is the biggest proof possible that dark matter is completely false. So Newton and Einstein are certainly on to something more than dark matter theory is.

The proof against dark matter theory extends not just through space but through time and observations of the early universe are used here. If you are as tired as I am of hearing the mundane science fact that looking across space billions of years in the universe is also looking billions of years backwards in time well at least we can use that little bit of trivia in disproving dark matter. Yes looking across the universe is like looking back in time and so let us use that shall we for a contemporary proof of why the universe is as we see it today without the existence of dark matter. So once again let us pretend that dark matter is real and that dark matter was present at the beginning of the universe and that it is needed to provide the right amount of extra mass to let galaxies form. The early universe must have had a precise value to the amount of dark matter present for a simple reason and this is the well accepted idea of the critical density of the universe. Too high a density of matter to space and the universe collapses back into itself or at least the attractive forces are so high that a universe of black holes is born and we are not here to observe it. Too low and the universe cannot form any complex structure beyond fundamental particles and possibly atoms meaning that the universe is a huge ocean of gas and energy that never forms star systems and once again we are not here to observe it. Well good news for us as we are very much here to observe it and so the critical density is not too high or too low and we are in the comfortable creamy middle bit where life is possible.

Here I must follow a small tangent which simply says that the universe is two things the first of which is not magically perfectly balanced such that everything can exist without such and such a value being a trillionth higher or lower and disaster striking or that we are simply at the perfect time to observe it. The idea that everything is magically balanced such that the critical density is just right is beyond nonsense for multiple reasons one of which is of course it is because we are here and so it does not matter as the end product is a stable universe. The second reason it is not magically balanced is that space is changing over time and the particles in it are not and so a very simple proof exists such that the density of the universe is fluid over time and therefore the magical values of things are obviously not a factor. This is explained since the values of these objects such as the strength of forces, energy and fundamental particles does not change over time and so a smaller early and dense universe would live or die in opposition to our universe today being supposedly just right or a future universe where it is far less dense. You cannot have magical values as the universe only allows magical values if it is utterly static and unchanging and since it is obviously not say goodbye to the theory that we are in some perfectly balanced system.

The second reason is that we are not simply observing at the right time for everything to be balanced because once again the universe is billions of years old and so any observation made one, five or ten billion years ago would be one in which the universe again destroyed itself in black holes or empty space and gas. Looking back in time by looking at distant galaxies shows that all the necessary forces to create normal stars and galaxies existed just fine without our perfect timing and human observer bias so once again say goodbye to the theory of perfect timing. Remember I have already shown how the universe can have a changing size to its space and interactions with matter without the need for dark matter or energy and one that will always allow the universe to persist and not destroy itself by crossing some dangerous critical density threshold using just two things which are ESS and bauen matter.

This proof of a healthy universe despite human bias also allows us to use looking back in time to destroy the theory of dark matter on its own terms. In the early universe tens of billions of years ago we make observations and not theory that stars formed and galaxies formed after them. The theory of dark matter says that the only way this happened was that dark matter provided the extra gravity needed to in effect seed star and subsequent galaxy formation. Well that destroys dark matter theory right there. Did you see it? Dark matter according to theory is attracted to large masses of normal matter such as galaxies and galaxy clusters and so dark matter cannot be attracted to matter before that meaning it cannot provide the seed needed to make these objects. Dark matter theorists believe they see dark matter halos around objects as massive as a galaxy but no where else and certainly not in some random space in between and so according to them it needs large amounts of matter to form but that matter cannot form according to them without the dark matter.

Also dark matter is thought to only be attracted to matter in large amounts due to the fact that it interacts with matter barely or with new experiments in detecting dark matter daily failing possibly not at all. This means that the only way dark matter is still weakly theorized to work is by being attracted to the gravity of space created by normal matter and once again curving spacetime to account for the missing apparent 5/6 of mass in the universe. This fails again as gravitational attraction on systems using the theory of general relativity which dark matter relies upon does not exist in modern science on quantum systems. This is once again the failure of modern science in finding a theory of quantum gravity and so once again dark matter cannot interact with matter on the quantum level in the early universe to create additional curvatures of spacetime. Any quantum system we observe would appear to have 5/6 more mass than it does to effect changes in the behavior of the system and yet we do not see or calculate them. Bottom line here is that there is no possible way dark matter can provide the seed for star or galaxy formation at any time in the universes life whether it is around large preexisting structures which are thought not to be able to hold themselves together without dark matter or at the most fundamental quantum scale systems either.

In addition computer simulations simply inserting the necessary values for the 5/6 missing gravity and finding neat galaxy formation over time proves nothing as it simply cheats its way to an answer. These simulations do not prove the existence of dark matter they only cheat and fill in the holes of the missing gravity by shoving in 5/6 more mass. You might as well say this missing gravity comes from the magical faerie dust as you, I or anyone can simply call the inserted 5/6 mass whatever we want. Remember that previous simulations without the theory of dark matter showed a failure in producing galaxies. Additionally all simulations using theorized values for dark matter failed as well and these are based off of numerous projected values for dark matter particles. So only by throwing away theorized values for dark matter particles could the data be fudged enough to produce stable galaxies and then they were suddenly and rather quickly accepted with some bias.

Why should we not have used the simulation before showing failures of dark matter to be the long sought after proof against dark matter? Why did the entire scientific community not accept and run with that? Instead dark matter theorists simply ran with whatever fudged data they needed and prematurely claimed success in proving dark matter as real. This is of course utter nonsense and must be included in any discussion for dark matter so that dark matter supporters simply do not begin discussions from the biased first principle that dark matter is some magical given in the universe. So looking at early star and galaxy formation we can prove that dark matter is not real and not needed in the universe and we can even use more direct observational proof and not theory to once again accomplish this rather pedestrian of tasks.

The early universe is again said to be in balance so as not to upset the critical density and cause a cosmos in which we do not exist and cannot question why things are as they physically are. This means that early galaxies should be all of the form of young proto galaxies filled with young stars. The typical candidate is small, barely spiral and filled with hot young blue stars and yes we do find these billions of years away and ago. The trouble for dark matter theory is that we also find two other types of objects in the early universe as well the first being stars much older than expected in other words much sooner after the Big Bang and also fully formed huge spiral galaxies filled with older gold and red stars.

The reason this is a full defeat of dark matter theory is that dark matter is thought to have been needed in precise amounts and strengths to account for early star and galaxy formation so as not to upset the critical density of the universe. If we look at accepted dark matter theory and use the values for the strength of the particles and the abundance of them as a baseline for the production of a universe as we see it today we can move above and below these values as needed to prove dark matter wrong. Mainly because dark matter would need to exist in multiple forms that are not needed to produce a universe as we see it which is of course the impossible extinction of 5/6 the supposed particles in the universe. One amount and strength of dark matter to make the typical young galaxies and another to make the older golden star filled ones.

The second reason is that the dark matter would need to exist in amounts that according to theory and the precious computer simulations would provide for critical density imbalances that again in theory would destroy the universe before we could be here to observe it. The existence of older stars and galaxies by hundreds of thousands and millions of years closer to the Big Bang means that the baseline value of dark matter is wrong. Dark matter is theorized to be a set particle with set properties like every other fundamental particle and so cannot change over time and this means it will only provide the correct amount of extra gravity to allow for star formation from the early universe which is very uniform, low in entropy and devoid of large scale objects once or at one time. So this value is wrong as soon as earlier stars and galaxies are found as this means the dark matter would have had to have a much higher baseline value in terms of its strength, size or ease of interaction with normal matter to cause early gas and dust to create stars and galaxies much earlier than predicted by theory.

This means that the density of the universe would have had to be much higher and the overall effect would have easily been a total collapse of the universe into nothing but black holes and crushing back into itself and once again preventing us from being here to observe it. This is a simple fact about the critical density of the universe that cannot be avoided as dark matter is supposedly existing in amounts that account for 5/6 of the mass in the universe and so an increase in its baseline value capable of creating stars and galaxies so much closer to the time of the Big Bang means that the total dark matter spread evenly throughout the gas and dust of the early universe would have had zero difficulty in upsetting the critical density as it exists evenly throughout space and create black holes and halt the expansion of the universe immediately in terms of cosmic timescales. Dark matter is wrong.

To further add insult to extinction of dark matter theory galaxies that are large old spirals of fully formed size, quantity of stars and age of stars is simply the final word on the end of dark matter theory. Nothing in the baseline values and age of the universe can account for two types of discrepancies in dark matter theory. The first being stars and galaxies that exist too early for dark matter theory and the second are large fully formed mature spirals that exist too late for dark matter theory in terms of time needed for development. The change in dark matter baseline values necessary to account for these mature galaxies in the early universe means that the critical densities of the universe needed to account for normal early galaxy formation and also the even earlier seemingly anomalous stars and galaxies must both be wrong and a third value is needed.

There is no way the universe can have a single type of matter account for all of these effects without destroying itself due to imbalances in the critical density and is this a good time to mention that all three of these impossible values need to be worked into the fact that they must be decreasing over time so that dark energy can increase meaning there is no way a large old galaxy could have the mass necessary to hold itself together as dark energy increases over time and the rest of the galaxies in the universe are still somehow in their infancy depending utterly on the existence of ever heavier and more prevalent numbers of dark matter particles? Yes that would be rather dirty of me to mention that would it not? Still I need to mention this as it is once again not a direct attack on those who believe in dark matter or dark energy despite it probably sounding like one.

As the universe ages it creates larger heavier structures which according to a theory of dark matter and dark energy is impossible as dark matter needed to hold these large structures together is being lost from the universe and the destructive expansion force is increasing. Instead it is simply a wake up call to those who do or do not believe in the dark theories as in the end all scientists are concerned with simply figuring out how these observations can be explained. There is no harm in trying to explain the universe using dark matter or dark energy but there is a time to admit that the dark theories are incorrect as they simply contradict themselves, create impossibilities and cannot account for both observables and experimental data in both micro quantum systems and macro cosmic systems at the same time. This is once again why I ask all reading this to consider my model of the universe using a fluid and nearly seamless structure of space that can naturally account for all observations of the universe whether they are ten billion years or more old or made just today in solving the problem of how the universe works. If you are sitting in your laboratory right now and cannot solve the problem in front of you please give it a try through the use of my models I promise you will be well rewarded for doing so as you have nothing to lose.

Bauen Matter and Macht Meetings

Now I will use the explanation for why hot things move faster and emit infrared photons as a way to explain a few other properties and points of interest about bauen matter. So people know that if things move faster they get hot and if the get hotter they move faster and if you rub two objects together or compress them they get hotter. The trouble is the exact reason about why these things occur is unknown to modern science. Like many things modern science is great at measuring these events and saying the how and what happens as in if you rub your hands together they get hotter. Ask why and modern science simply says because you rub them together and anytime you create friction you get waste heat. Yes but why asks the curious mind and then modern science says well at the fundamental level of space and matter we simply do not know as we have no way to incorporate quantum objects with space.

The reason why things get hotter is of course known and it is known well to the universe and I will share with you now the reason this happens on a space and matter level right down to the smallest parts of each which are ESS and bauen matter respectively. You have learned much about the universe by now and some of the things you have learned will play out here as in any previous problem you care to apply them too. Just remember that the universe can not only alter its space, but its matter as well through changing bauen geometries and therefore can alter the strengths of its macht glows or forces such as in gravity. So in reality the universe has changing space, no particles that are fundamental except for a single bauen matter object and the four forces are not always set either in certain special conditions.

Things get hot and move faster for one simple reason only and that is the system in question is trying to overload the threshold macht levels for the local area of space ESS spatial compression limits. In other words every part of the universe has reached some balance or equilibrium between the matter and space found there. The matter is determined by what type of bauen geometry you have and how much total bauen matter is in that area. A black hole has a different amount of total bauen matter per cubic meter than the Earth does and so will affect the system differently. Also black holes are made of SM masses which consist of clintrons and so are quite a different bauen geometry than the normal atoms making up the Earth. The Earth is not made of any clintrons whatsoever and instead is comprised of various atomic elements that have a lot of empty space inside them. In addition to this the atoms have very different bauen geometries making up their protons and neutrons which are in turn made of quarks and gluons. In addition they have electrons and photons as well and these too have very different bauen geometry when compared to clintrons. You must remember that the number, type and degree of exposure of all bauen matter sides to space also determine what properties a particular bauen geometry will have.

So the type of matter and the amount of it present in a system will affect the total behavior of the system along with the space in it. The space within the system determines how much the macht glow varies from the macht reserve pools by the number of ESS in the local area of space. The more ESS present the more ESS are able to interact with the matter found there and will cause the macht reserve pools to typically increase. The more ESS in the system also means that the ESS have a smaller total spherical volume and therefore tighter schales through which they can share macht glows. Denser schales means that the macht glow between particles might be limited as explained in the quark volume example from earlier. So the density, volume and number of ESS in the system also affect the total behavior of the system. The system is therefore determined by both space and matter and space and matter will interact with each other and produce changes in the other until this balance or equilibrium is met.

Think of the SM mass containing highly spatially compressed ESS which are needed to maintain the bauen geometries for the clintrons and then think of the clintrons geometry which is necessary to make space compress sufficiently in turn. This is why black holes are a perfect balancing of both space and matter into a single space/matter particle which is the clintron. That is why black holes are properly described and explained as SM masses as they are large continuous collections of clintrons which are made of both space and matter.

What this does for the universe is create a place where the local space and matter within it will eventually reach a balancing or equilibrium point from space and matter interacting with each other. This point for any place in the universe creates a certain amount of spatial compression and ESS macht glow which as explained is dependant on both the properties of the space and matter there. The system therefore has a maximum allowed amount of macht in it that is best maintained when the system is left alone and not acted upon by external forces or influences. It is at this point when the system is not trying to balance its total macht beyond the degree of space and matter interaction. Any external influence or force will cause this perfect balance to become skewed in some way and the system will react accordingly if the force is enough to pass the threshold value for it. This is why things get hot as the external force will act upon the bauen geometries within it and force them to try and pass the specific threshold value for the system.

This is the single reason why all things in physics vary such as the boiling point of water not being a fixed 100 degrees Celsius, the temperature at which electrical conductivity is most efficient or the rate of perceived quantum interactions taking place will all vary in different amounts of spatial compression, macht glows or if the bauen geometries of the matter involved are affected. The example of something getting hot and moving faster is due simply to the fact that as you rub your hands together for example you put an external force on the electrons in your hands as they push against the nuclei of your hands atoms. This can now be viewed as a simple single atom as the electron is trying to be compressed beyond the electron degeneracy pressure for the atom in question and we will say a hydrogen atom. As external forces are applied to the atom the electron is being forced towards the nucleus and the nucleus is pushing back. The motion of the electron will increase and the motion of the nucleus will also increase as they are made to interact with each other more strongly.

The electron will rapidly shift to a part of its orbital that relieves excess strain on the total ESS macht for the atom which as stated is trying to maintain its optimal macht levels in a relaxed state. The electron moves to the new position of least macht pressure only to find it is now up against an adjacent atom within your hand. This forces the electron to move to a new and also lower macht level area of its orbital around the nucleus. This will continue in a shifting pattern as your hand is made of numerous atoms and so as one orbital position becomes undesirable for the macht glows of space another becomes desirable. This will cause the desirable one to become undesirable as a new orbital location is favored and the process repeats. The force applied is not sufficiently focused to cause the atoms in your hand to stop bonding into molecules and so they simple vibrate in their current positions. All of the vibration naturally occurs in accordance with the normal quantum tracking rules that are explained in the section on quantum motion.

The center of the atomic system will therefore move rapidly around as the entire atomic system tries to balance macht glows with its neighboring atoms. The result for all atoms in your hand is a sharing of the external force that results in the local area ESS macht levels attempting to evenly balance out the total macht reserve pools and maintain them at the desired level for the total spatial compression of the system and the bauen geometries present. When you rub your hands together you do not effectively alter the total amount of spatial compression nor do you effectively alter the bauen geometries of the particles in you hands and so nothing from your perspective changes except you notice your hands warming up. The local area of space does notice on the ESS macht level the external forces and so are responding by trying to balance the total macht despite not being able to reach a new balance or equilibrium point as the total ESS spatial compression and bauen geometries available to the system are not changing. This is now where infrared radiation is born and I will explain its role in your toasty hands now.

The increase in heat comes from emission of infrared photons from atomic nuclei that we feel as heat and is why a warm cast iron stove radiates heat through the room to us. If we rub our hands together we create heat and if a stellar cloud collapses it will eventual fuse and both create heat in a similar way. Stellar masses will release heat through the fusion burning process but that heat is also captured by neighboring plasma where it is absorbed and then radiated once again. This absorbed radiation causes the plasma to speed up exactly as described above and so this part of the stars heating is the same as rubbing your hands together. When you cause matter to create imbalances in the total system ESS macht reserve pool levels you will eventually reach a threshold point where heat is released. Before this threshold point your actions will not release heat although they may increase motion of the matter involved. On a truly fundamental level motion comes first then the release of heat as the threshold value for the system has not yet been reached. This threshold value is the point at which the particular bauen matter geometry in question along with the specific amount of spatial compression and macht glow reserve pools are balanced.

If you move a small particle it will only increase in speed and will not release heat despite what you might normally think. All quantum systems work on quanta or set values that change only in step by step action potential like fashion and so it is obvious that you will need to exceed this quanta value of motion before you get a release of heat. This is made possible by small amounts of ESS macht which are more subdivided than anything in the quantum world to build up to an action potential and once again we have the real explanation behind the photoelectric effect. In other words the old rule of increasing motion always means increasing heat is not true and only holds true if the ESS macht balancing threshold is exceeded. The way this works is simply the external force on the system which we will say is a single hydrogen atom causes the atom to increase its quantum motion in space. This motion is the result of the atom trying to balance its macht with the system and will be permitted up to a point where the bauen geometry of the hydrogen atom is no longer sufficient to maintain the ESS macht balance for the system. At this point the threshold value has been exceeded and heat can finally be released in the form of infrared radiation. The hydrogen atom does this as a necessary step to maintain its current arrangement of one electron orbiting one proton without destroying the electron degeneracy pressure. The electron is a fundamental particle which will not be diminished without considerable force but the proton is a composite particle made of fundamental particles and so takes less force to diminish.

The bauen force is still the strongest force in the universe but a single bauen piece of bauen matter can only apply one bauen matters worth of bauen force to holding itself onto the larger bauen geometry. The ESS macht that is being balanced in total with the system is drawing on all the donated macht from all the bauen objects and bauen matter present and so even though the four known fundamental forces are weaker than the bauen force the total applied force from them to the bauen geometry that will become the photon is enough to overcome this bauen force and cause the geometry to be shed from the larger geometry as a whole. The choice of which bauen object will be shed is simple and follows the exact same set of rules that the electron followed in its desire to move to a lower area of ESS macht as described in the example above.

Just as a certain position around the nucleus will be lowest in ESS macht glow and so the electron moves to it the bauen geometry of the proton will always have one small bauen object that is the most difficult to maintain as a part of the total geometry of the proton or the most easy to shed in terms of balancing the total ESS macht. This is how the bauen object is determined and how it is shed as the sides it has exposed to space as well as the sides exposed to space of all bauen objects in the proton will balance against the total ESS macht reserve pool level for the local area of space and its spatial compression. So simply looking at which bauen objects have which sides exposed and to what degree with space and comparing that to the total bauen geometry for the system and the ESS macht present will shed one bauen geometry photon.

The driving force is simply the total ESS macht of the four weaker forces overcoming the least tightly bound bauen geometries' bauen force. This is no different than gravity which is the weakest of the four fundamental forces being strong enough under the correct conditions to overcome the strongest of the standard four forces which is the strong nuclear force and cause fusion to occur inside a stars core. To this end the proton can shed a small amount of bauen matter in the form of a bauen geometry photon from its composite pieces and these can be thought of as the gluons or virtual mass. Remember a single proton is near infinitely massive in size compared to a single photon and therefore can happily shed numerous small bauen geometry photon objects for quite some time all the while losing the apparent virtual mass of a single photon at a time. The amount shed is exactly the amount needed by the system to balance out the total ESS macht levels and in doing so the atom will now be at a new stable state for the system based on the external pressure placed upon it. This is why if left alone a hot object will cool down and finally explains precisely what is happening as described by thermodynamic systems as the local area space relieves the extra pressure on the macht glows associated with it.

If no new external forces where applied to it the system would emit no more photons as the balance of ESS macht glows would be at a new threshold level. This is very difficult to observe in reality as two problems occur the first being interference from matter near the experimental atom. Nearby matter will possibly be heating up as well and therefore emitting its own photons which can be captured by the experimental atom and once again impart the necessary force to exceed the threshold value.

The second reason is that the ability to isolate a single atom and then apply force to it so that its quantum motion increases by only a small amount sufficient to exceed a single ESS macht threshold without exceeding a second is incredibly delicate. The likelihood of a macro-scale effect being observed where numerous atoms such as those in your hands are present is far more probable and so in everyday practice it will appear as though things only get hotter the longer the force is applied. This is really a macro-scale observation however as a human rubbing their hands together lacks the precision to notice a micro scale increase in heat and then stop applying additional external forces. Nevertheless the fact that a single threshold value has been surpassed and that the result is the release of a single infrared photon does not change.

The bauen geometry for the photon is released and will interact with the local area ESS such that it decreases the virtual mass of the atom reflecting what is seen as a decrease in overall atomic mass. The atom can shed many of these bauen geometry photons which also demonstrates another fact about matter in the universe which is that all fundamental particles are massive collections of bauen objects overall. Another way of looking at this fact is that bauen matter is microscopic when compared to large macro-scale fundamental particles as understood by modern science using the Standard Model of particle physics. Bauen matter is incredibly small and so many, many bauen objects exist within a single fundamental particle and many more within a composite particle such as a hadron.

This is why a proton inside a hydrogen atom for example can be made to emit photon after photon after photon over time as the element heats up more and more and yet the atom does not stop being a hydrogen atom. The released photons can be absorbed into the geometries of nearby atoms in the system which for example can be the interior of a stars core. These photons will be temporarily absorbed by the larger bauen geometries of the nuclei there and will be maintained until the threshold of that new nucleus exceeds the allowed values for the ESS macht reserve pool levels of the system. The nuclei will then obey the similar rules for motion described above so that the new location or new bauen geometry is no longer favorable. Just as the electron around the atom was forced to continually move to continually balance out ESS macht levels of the system so to the photon will move from nuclei to nuclei to relieve the strain for a local part of space within the system. Overall however the photon is still moving to new geometries and this is how heating through absorption of infrared radiation works as the atoms themselves are heated on the ESS macht and bauen matter level.

Even heating the atom to the point of ionization only demonstrates the threshold value at which the electron to proton bond of the hydrogen atom is now forced to be broken once again as the current configuration of matter cannot be supported by the local area ESS macht through the normal balancing rules for a system in the universe. So for example if you heat a system enough the ability of the electron to move from one area of macht glow to another in order to balance the total macht glow of the system might not be fast enough and then the kinetic motion of a bauen geometry can overcome certain normal forces. A perfect example of this is how speed increases for an object as it attempts to balance macht glows and this speed can become fast enough that other forces like the strong nuclear are overcome. This can be seen in how a stars core fuses materials and overcomes the nuclear force to release heat which is a result of increasing macht glows that balance so fast the nuclear force is overcome and the only way the system can balance the total macht glow reserve pool for that system under the current level of ESS spatial compression is to release various bauen geometries like photons and neutrinos. This heating process can be extended beyond a simple atom and the electromagnetic and nuclear forces and be applied directly to particles whether they are composite or fundamental.

Heating the proton further allows for it to shed many, many more bauen geometry photons until eventually this is dissolved into quark/gluon plasma. This quark/gluon plasma is still capable of shedding more photons and yes if you heat it further you will actually melt the quarks and gluons and even photons into smaller bauen geometry arrangements. The last level of heating sees all bauen objects dissociated into single pieces of bauen matter which will now have all sides equally exposed to space as none can be hidden. At any point along the way the same process is happening over and over again as the threshold value for the system is being exceeded given the current bauen geometry of whatever matter is being observed. The only way the matter can continue to balance with the system is to shed a single bauen geometry infrared photon at a time. This will allow for the matter to reach a new balance point with the ESS and macht of the local area of space based on the total amount of spatial compression present.

This is a fact of the universe and is never broken no matter what matter is being discussed as all matter must balance with the local area ESS macht and obey any threshold limits for that same system. This fact should be seen as perfectly acceptable for all as it is observed in various forms daily and reflected in the classical laws that describe macro-scale systems around us. Think of putting pressure on the surface of a balloon and the balloon deforming as it must balance its own forces and properties with the external force you are now applying to it. If you exceed the threshold limit for the balloons system it will break as the threshold limit is the maximum elasticity available for the material making up the surface of the balloon. In other words an object in a system when faced with sufficient external force will reach a threshold limit and something must give in response. For a balloon it is the balloon popping to relieve the external force you are applying to it and so the system is now balanced as the air inside refuses to be compressed against the atmosphere outside the balloon.

In an atomic system such as a single hydrogen atom the balance between the electron and proton is like the balloons walls. As pressure is applied it will begin to deform as seen through an increase in quantum motion as the atom tries to balance itself with the universe around it. The threshold point similar to the balloon breaking where something had to give in the system is the point where the bauen geometry photon is emitted. This is the something that must give inside the atomic system in order for the hydrogen atom to balance out its quantum motions with the total ESS macht of the system. So both a micro-scale quantum object and a macro-scale classical object obey the exact same rules for the universe at all times. Classical mechanics can be said to be a collection of the space physics that are the true workings of the universe and are explained using ESS macht and bauen geometries. This is another demonstration of how space physics, ESS macht and bauen matter seamlessly transition from the smallest possible objects in the universe to the largest and of course the entire universe itself.

What this demonstration also does is provide the step by step explanation for Brownian motion as observed centuries ago. In Brownian motion the bauen geometries are seen moving around one another the same way a fundamental particle like an electron is moving to relieve macht glow stress around a nucleus of an atom. In this case the bauen geometries are in a fluid and so are able to move freely within the system and this results in each particle, atom or molecule of the fluid trying to balance the macht glow it has with any neighboring particles, atoms or molecules it encounters. The result is the total macht glow for one object trying to balance with the macht glow of another object and since all objects in the system are held in place by the walls of the container the motion will never stop as they cannot escape the system to fully balance the macht glows they all share. This also explains why a cold liquid moves more slowly than a warm liquid as the system has shed infrared photons as heat and so lowered the total number of bauen matter objects in that area of space.

The decrease in bauen matter means less ESS macht glow present in the system and so less need for the liquid to balance macht glows and so a slower Brownian motion.

Behaviors of Matter continued, Anti-matter and Light

The fact that the local area ESS macht which is from the total amount of matter and therefore all bauen objects in the area overcoming much smaller bauen geometries such as single photons is responsible for a whole host of things in the universe. The decay of all matter in the universe is directly tied to this fact as all decay events are nothing more than the total ESS macht overpowering bauen objects and the bauen force holding them to the larger bauen geometries. The weakest point on the larger bauen object will always be located around a smaller bauen geometry that is the most weakly bonded to the larger object and therefore least able to maintain its position on the larger object. Since the local area ESS macht is not changing from its total reserve pool levels and the spatial compression of the ESS found there is not changing any bauen geometries that are not fully stable for those conditions will decay. The more severe the conditions the faster the decay rate and the conditions can be thought of as the local area of space and the ESS spatial compression and macht glows found there. This is why something like a neutron wants to decay as it has a large geometry that is too complex to be maintained under normal spatial compression and macht glow.

Likewise anything somewhat exotic created inside a particle accelerator is also existing momentarily in an environment that is too hostile for it to exist for long. The rapid need to balance ESS macht tears the bauen object apart into many smaller bauen geometries until base level geometries are produced that are stable for the local area of space. This is different on different planets, stars or any location in space and the results of experiments we see here on Earth will not be the same elsewhere provided the changes in ESS spatial compression are sufficiently different from Earth along with macht glow levels. A neutron will decay but not always at the 11 minute mark more or less depending on where you are in the universe and this demonstrates a changing value for what was seen as a constant.

Much like light can have its speed varied throughout the universe this is also explained by bauen geometries and ESS macht balancing rules. Light is nothing more than a bauen object that has its gravity sides hidden from space and so produces no mass but nothing says the other sides of its bauen matter must remain in a set geometry. As long as the gravity sides are hidden the electromagnetic sides can move from hidden, partially exposed or fully exposed. The total macht glow from any bauen object is a combination of all the sides exposed from all the single pieces of bauen matter used to create it and therefore is totaled for the entire bauen object. What this means is that the bauen geometry photon need not always have the same amount of electromagnetic interaction with space.

If we look quickly at something like a radio wave and a gamma ray we can explain this rather easily. A radio wave and a gamma ray are both photons but have different energies and wavelengths yet travel at light speed. As mentioned in an earlier section the quantum tracking speed for matter to space is light speed and so all particles will if they are not exposing bauen sides of macht gravity to space will move at light speed immediately. A radio wave has lower energy as it spends less time interacting its bauen sides electromagnetic with space and so can be thought of as in a partially exposed and almost hidden geometry. The result is the forward kinetic motion of the particle will have it travel long distances forward while it only moves side to side slowly as it is not strongly interacting with space and therefore elevating the ESS macht glow electromagnetic. So you get a long wavelength with low energy as we perceive it because as you know all our instruments are interacting with space using their own matter. What that means is that we feel in ourselves using our senses and in our instruments the amount of ESS macht glow for anything we interact with and so a radio wave does not produce a strong electromagnetic macht glow as it has low energy and we therefore register it as having low energy using laboratory equipment.

If the bauen sides electromagnetic rotate a bit more and this obviously can occur on a remarkably fine gradient that far exceeds any quantum gradients as we are dealing with a single bauen matter piece moving about the central geometry and so the scale is that of space physics matter and not large macro scale quantum fundamental particle matter we see a near fluid spectrum of wavelengths and energies to all photons. This explains exactly why using precise bauen matter objects all electromagnetic radiation can change its interaction with space, why space does affect these changes in the first place as they are as you recall simply obeying all standard macht glow balancing rules for a bauen geometry through edge dragging and why these changes result in a near continuous as viewed from quantum eyes spectrum of wavelengths and energies.

The result is our gamma ray still moving at light speed but interacting heavily with the local area ESS it is in and so tracking rapidly side to side with all ESS macht glows found there and creating a strong ESS macht glow electromagnetic. So when we detect a gamma ray its total ESS macht glow electromagnetic is highly energized and carries a lot of force that we see as having a great and often destructive impact on our instruments. The destructiveness comes from a large amount of electromagnetic energy in the true form of macht glow electromagnetic for the bauen object gamma ray photon being dumped all at once into whatever it hits. What it hits now must deal with all normal macht balancing rules and threshold values for the local area of space ESS spatial compression and as described before produces macht threshold exceeding limits that harm the matter we use to detect it. All of this comes from changes in the arrangement of the surfaces of the bauen object photon and these changes can also be less fluid as explained next.

If the bauen geometry of a normal photon is altered through something in the universe or in the laboratory it can have the bauen objects on its surface subtly changed from that total bauen exposure mentioned before. This means that a permanent change in how the photon talks to space is made possible and this can alter the top speed of that same photon through a vacuum. The photon maintains its properties of no mass and no gravity but will no longer be able to move through space at what is thought of the speed of light but will be altered even if only a small amount. This is different than a normal photon simply changing its wavelength as it tracks after ESS centers as described in a previous section. Wavelength changes are simply a reflection of how a set bauen geometry photon will move through different levels of ESS spatial compression in the universe. A photon that has been altered to have a different bauen geometry from a normal photon and possesses a different top speed will also undergo its own wavelength shifting as it encounters different levels of spatial compression. So the idea that light can change how fast it moves through space is perfectly normal and should not be unexpected at all. A very simple change in the physical geometry of the bauen object making up a photon is all that is needed to utilize a different degree of edge dragging through space and this will alter the new photons speed. The reason that speed is affected is also perfectly straight forward as it simply means the affinity the new photon has for space is more or less than before and so it feels less interaction with the corresponding macht glows of space except for the gravity one of course.

The way that bauen matter sides interact with space is also what causes things like anti-matter to exist. The way that anti-matter edge drags has been already covered in much the same way a proton is positive and an electron is negative and now the way in which anti-matter is created can be simply dealt with. As shown earlier anti-matter is only ever created from relatively violent events in the universe and as such the reason for this is explained the same way as any change in bauen geometries takes place. The total force exerted on a small part of the bauen object must be higher than the bauen force that is holding the whole geometry together as in the decay example or in exposing sides as in the anti-matter example. Changing the exposed, partially exposed or hidden sides of a piece of matter can result in bauen objects not being decayed from the total bauen geometry but move or rotated on the surface of the larger bauen geometry. This will result in the piece of matter having some of its properties changed and this is exactly what happens in the case of anti-matter. So a violent event will produce enough external force such that the bauen force is no longer able to maintain the original geometry anymore and the anti-matter bauen geometry is created.

In the early universe after the Big Bang the high density of the universe allowed for multiple violent collisions of these geometries and therefore the production of the anti-matter partners to the normal baryonic matter. The early universe did also have its ESS spatially rebounding quickly however and so matter and anti-matter collisions were not a guarantee simply because theory predicts matter and anti-matter were produced in equal amounts. The fact that space was pulling matter and anti-matter apart means that the distance between potential annihilation pairs is increasing which gives the universe time to recycle the bauen geometries of anti-matter. The recycle mechanism comes from the faster decay rate of anti-matter as it takes more effort for the universe to balance out these edge dragging bauen geometries as opposed to those of normal matter with whatever local area ESS macht glows are present. Recall here the production of positive charge to understand which bauen sides are exposed as a need to balance macht glows for a set ESS spatial compression.

The second recycle mechanism is the breakdown of anti-matter into free bauen matter which can then be reformed into the universe through the right handed DNA so to speak of the universe. The right handed DNA is the tendency of the universe to favor matter arrangements of bauen geometries for particles over the anti-matter arrangements. This is a bauen object by bauen object explanation of why the universe favors normal matter over anti-matter as is simply explained by the fact that certain matter geometries have the sides of bauen objects exposed to space in a way that requires less effort to balance macht with. The universe afterall always tends to lowest energy states and the least amount of effort to make everything in it and this includes bauen geometries and how many and what sides of the bauen matter is exposed to space and to what degree for the total of all pieces of bauen matter in the total bauen geometry.

The tendency to favor the least effort in balancing macht levels is also seen in the three generations of matter. The same properties for matter can be seen in all three generations except for mass and this simply reflects the higher generations of matter having acquired more bauen objects that add to the gravity but not the other macht glows of space. This should not be weird or scary to any at all as a normal bauen geometry photon can have all of its properties remain the same but changes in the bauen matter making it up by exposure of bauen sides electromagnetic create changes in wavelength and energy and this is perfectly understood and accepted already. If more bauen objects are added to a larger geometry the total mass can be increased but this will come at the cost of the new object being forced to try and maintain its geometry against the same levels of ESS spatial compression and macht glow reserve pool levels. The original object was stable under those ESS spatial compression and macht glow levels so that denotes what generation of matter and therefore bauen geometry is stable there. A bauen fragment for a particular quark will have for example charge sides exposed to create a positive or negative charge. Since you know charge flows through bauen matter and space it is obvious that fragments will attach themselves to the original quark like opposite ends of a bar magnet. This arrangement is unique to that quark and so only similar fragments can be added like a key which explains similar charge but increased mass of the new generation of quark. The new geometry will not be stable and so will decay rapidly as the number of extra bauen objects is shed from the larger whole at the speed of quantum interactions with macht speed glows between schales. The total decay speed is calculated from the number and type of bauen objects being shed and the total ESS spatial compression and macht glows for the local area of space. This is why particles of varying masses have decay rates that are unique as the type and number of extra bauen objects that must be shed in order to balance the local area ESS macht glow varies.

Similarly constructing particles using bauen matter also accounts for the properties of all known fundamental particles as the number of particles with inwards facing sides for example as seen in the explanation of charges and matter can be amplified or decreased for a larger geometry. More exposed sides on the surface of the larger geometry or more sides stacked on top of each other as in more bauen matter stacked on more bauen matter but with the exact same degree of all sides edge dragging will amplify certain effects. This is how you can alter large geometries to create things like charge either negative or positive or gravity. For example it has been shown that exposing certain sides to space will create a negative or positive charge in either the electron or proton respectively. If you add more bauen matter to the first geometry for example on top of the original you can in theory create particles with negative two or three or whatever charge you want as long as you can make the bauen geometry stable for the local area ESS spatial compression and macht glows. The only rule is depleting the macht glow reserve pool for the ESS available which is not allowed for a single particle. This is not impossible as humans routinely create things not seen normally in the universe and for this example I will have you look no further than a humble and delicious cheeseburger.

Yes it may seem silly but all the elements inside a cheeseburger exist naturally in the universe but will never be arranged that way without the help of intelligent beings. You can wait for trillions of cycles of the universes life to pass and look as far and wide as you want but these elements will never naturally occur in the form of a cheeseburger. This may seem silly but it is solid fact and the same is true with the various elements of the universe as well as we create things like man made elements for study although they will not be stable. You can also go deeper and create custom bauen geometries although they will most likely not be stable this does not stop you from taking normal bauen matter and producing geometries that are not normally found in the universe. In other words we can create whatever we want in the particle zoo. So the ability to create particles with negative or positive four charges for example is possible and simply illustrates how normal particles are constructed by altering the number of bauen objects on the surface of the larger geometry whether in a spread out or stacked configuration.

This is just another explanation of the fundamental objects in the universe as to why they occur beyond simply saying the universe contains things like electrons and three generations of them. Instead of taking these particles at face value as if they are simply a given in the universe ESS, macht glows and bauen matter can be used to show why they exist and how they are constructed as well as why we see them in the ratios we do in the universe. In other words to produce a universe as we observe it today that is constructed from the smallest number of variables possible. The number of variables is exactly two which are ESS and bauen matter and simply looking at how they talk to each other and interact with each other is all that is needed to make this universe no matter how complex it seems.

Another useful aspect of the balancing limits of ESS macht with matter in the universe is that it tells us something of the local area of space. The simplest way to think of this is using a familiar equation such as F=ma and if you know two of the variables you can solve for the third. The same is true for any local area of space as you can learn something of the space and matter in it by looking at how that space or matter behaves. If the particular absorption spectrum for a particle, element or molecule is observed you will see certain lines present in that spectrum. The same is true if you look at the emission spectrum for the same sample and the two different types of spectrum will always show you the same lines just with different backgrounds. An absorption spectrum will be multicolored with a few dark lines in it and an emission spectrum will be the opposite of a dark background and a few colored lines. In the absorption spectrum you will see the wavelengths of light that are absorbed by the sample and as the sample returns from an excited state as in the emission spectrum you see the wavelengths that are emitted as the sample returns to the ground state. These are the wavelengths of bauen geometry photons that are accepted by the local area of space so that they do not upset the total ESS macht of the system.

If an element has two or three dark lines associated with its absorption spectrum these are the two or three geometries of bauen objects that the sample can bond with even if only temporarily. All other wavelengths of radiation being sent to the sample are of wavelengths that cannot be bonded to the sample as they are immediately causing unacceptable imbalances in the total ESS macht for the local area of space. The accepted wavelengths will be held by the sample for a short time as they are able to not only be accepted but they also show wavelengths of bauen geometry photons that can interact with the ESS in the local area as well. Remember from a previous section that the ground and excited states of everything in the universe are reflections of when that particular object is interacting with space, when it is inside or outside space and when quantum energy is being sent to the ESS macht of space. So any absorption lines for a particular sample show where certain bauen geometries of photons can not only bond to the larger bauen object but also exchange energy with space from matter to ESS macht glows and the schales found there. These are temporary bonding arrangements however as the local area ESS macht is still exceeded by the addition of the bauen geometry photon that has entered the larger bauen object.

The normal ESS macht balancing rules apply here once again and the weakest point is the bauen force holding the new bauen geometry photon to the larger object. Despite being much stronger than other forces the bauen force at the point of bonding is being overpowered by the collective and total ESS macht from all the other forces in the local area of space just the same way that large amounts of weak gravity can overpower the strong nuclear force of a single nucleus. So the ESS macht balance is being exceeded by the added energy donation from the bauen geometry photon and the local area ESS macht will balance out once again transmitting all new macht glows back from the schales around the particle to the particle itself. The law of conservation of macht/energy states the bauen photon geometry will be shed from the larger bauen object but retain all of its total energy and once again travel as a photon back through the universe. The delay between the quantum speed energy donation of the bauen geometry photon and when it was bonded and shed is the quick quantum jump that is seen incorrectly as superposition and also is what explains once again the electron drop down phenomenon seen in atoms. This also explains exactly on a bauen matter by bauen matter basis what is happening in these systems rather than simply measuring them and saying energy in must equal energy out as is the limit to the understanding of these events by modern science.

The practical application of this besides finally explaining what is actually happening at the sub fundamental level is also to illustrate how parts of the universe can be solved for much like the F=ma equation. By knowing what matter you are testing and what bauen geometry photons are used in the interaction it is possible to then learn about the ESS and macht glows found there. A first step will be to calculate ESS density using this principle as the local area spatial compression is one set value and as explained before a normal spatial compression can be calculated for Earth based laboratories. Absorption spectrums and emission spectrums will allow us to see even if only part of the answer for how space is behaving all around and in fact right through us. Most likely this will not be able to tell us the whole picture of how dense the ESS are around us but it will at least be a tool for beginning our understanding of the building blocks of space which are the ESS.

From there we can alter what macht glows are present in each area of ESS to see how these affect changes in matter and they will as space and matter talk to each other constantly. The changes in ESS spatial compression and macht glows will result in changes in the spectrums or in the time between absorption of radiation and emission of radiation all due to how the new ESS macht levels are met and exceeded. The most promising will come from measurements of time between the absorption and emission of radiation as increase in ESS density or macht glows should cause the time between the events to decrease. Then differences between ESS spatial compression alone, macht glows and then both ESS spatial compression and macht glows can be conducted together to see how the time differences are mapped.

Gravity from a very early section is a combination of ESS spatial compression and macht glow gravity so altering other macht glows while leaving the spatial compression alone will allow us to zero in on the exact properties of gravity for the system as well by manipulating macht glows that are non-gravity. Allowing us to change ESS spatial compression but not macht glows will let the gravity change while the other properties of the sample stay the same but they will be forced to produce changes we can measure in their behavior as they must also interact with space to create their macht glows. Remember that what we perceive as matter today is not truly matter alone but actually real matter which are the bauen objects interacting with space to produce a combination of space and matter which we have for millennia confused for real matter and matter alone which it never was. The changes in the combination of space and matter that we see as just regular matter by modern sciences view is how we will tease the secrets of space in the form of ESS and matter or really bauen matter out from the universe.

In truth we are seeing space all around us constantly in our experiments here on Earth but they are missed because it is assumed that they come from a matter only system. The missing holes in the knowledge of the universe that are encountered by modern science are caused by a lack of realization that space is an equal part in what is commonly thought of as matter. Remember that space is ESS and matter is bauen matter that come together to form things called fundamental particles which are according to modern science matter but in reality are the perfect blending of both space and matter. The only difficulty in conducting these experiments will be in altering background spatial compression and ESS macht glows from normal Earth ESS spatial compression and macht glows. It might simply be beyond our current human technology to sufficiently alter these two factors of the universe enough to see any real changes reflected in any test samples we choose. This means that the changes are for those who are patient but those who are patient will be rewarded with proof that the universe is made from space in the form of ESS and macht glows and matter in the form of bauen matter and geometries.

Taking the Universes Temperature and Multiple BBB Persistent Universe

If we continue looking at electromagnetic radiation for clues to how the universe is put together we can look once again at the cosmic microwave background radiation. The microwave background as mapped across the sky shows what is thought of as a fairly homogenous amount of microwaves. This is due to the fact that the temperature variations on these maps reflect temperature differences of only a thousandth of a degree Kelvin or so and so the entire map is seen as uniform. The map overall is fairly uniform but upon closer inspection it is actually not that uniform anymore and more importantly the cause of the differences are not the typical candidates that modern science thinks of. To begin with you must realize that the microwave background is near uniform but actually fairly non-uniform in the temperature differences. The reason this is so is because each small patch of color which highlights a particular temperature zone in the sky reflects what is actually quite a large region of space in the physical universe. What is seen as a small spot on the map is really a region of space that is on the order of hundreds to thousands of light years across. A single ten thousand light year patch of space when cubed to find its volume becomes a trillion cubic light years of space and now begins to show why the temperature differences are not so small anymore.

So if you calculate how much heat is in each one trillion cubic light years of space region of the map you realize that the differences are now massive. On top of this the heat that you see is a reflection of energy as this map is also nothing more than a remnant energy map of the big bang. So despite the individual microwaves being close to each other in energy if you sum all of their energies over the entire one trillion light year cube of space the very small energy difference now becomes a very large one. The energy is also old energy and must be backtracked to its birth close to the big bang itself. The energy back then was in the form of gamma rays that are far more powerful than microwaves today and so the energy stored and the differences in each one trillion cube of space today are magnified as they are condensed backwards in time to a point where each region of space had not expanded as much. Now you are left with many dense high energy cubes that are very different from each other as the fading of time has not diminished these differences to a gentle background glow of microwaves. Each cube is very distinct with a large difference between its neighbors in terms of total energy and this difference is real and so must be accounted for.

Before I explain where these energy differences come from and why they do I would like to offer an alternate explanation for the idea of the cosmic microwave background coming from a big bang of any sort. The idea that the universe is not coming from a single large explosion and rather is simply an eternal and roiling mass of stuff is something that has persisted in different forms for millennia. The truth is that with no one from Earth today having proof of how the universe started it is fun to contemplate the universe simply not exploding and being around for eternity. So how can the universe if it did not come from an explosion have a background temperature? To begin with think of the idea of a room in summer and that room having a small chemical explosion detonated within it. If we wait the heat from the explosion will expand and following the normal rules of thermodynamics and entropy will fill the room evenly with heat to say 25 degrees Celsius. It can be concluded that the explosion was of such a magnitude and force and occurred at such a time in the past for the waste heat to have fallen from many hundreds of degrees in the very violent and small explosion to the temperature we see now. It must have been that the room was heated from an explosion because we now see the waste heat in residual form and therefore that is the only proof needed to say that the room was heated by an explosion and nothing else. Well that is how it seems at least to big bang theorists who use the cosmic microwave background as proof of the big bang.

I would like to remind you of a small and key piece of information from my little room example which is I said it was a summer day. In summer the temperature is usually 25 degrees Celsius and besides being a great temperature to barbeque steaks and drink cold beer it also provides an explanation for the big bang not being needed to heat the universe. It is entirely possible that the room has had no explosion in it or that it occurred millions of years ago far long enough to rid the system of waste heat. The simple fact of the matter is the ambient heat in the room is nothing more than a natural product of the radiation from the sun making that nice warm summer day. In other words the natural background temperature and energy of the system was always 25 degrees Celsius and never needed an explosion to be created in the first place. The universe might be exactly the same as an eternal universe that simple ebbs and flows here and there on scales that are almost unfathomable. The natural exchange of matter to energy and back again will always produce waste such as microwaves or higher energy electromagnetic wavelengths that diminish over time and distance. The waste energy we simply see as it travels through the universe and because it is uniform and coming from all sides evenly we say there must have been a big explosion therefore a big bang therefore case closed. Balderdash.

The universe might look homogenous in every direction with no obvious center to an explosion and no obvious direction that all material is moving in because there was no explosion. The energy is simply the near absolute zero energy that is prevalent throughout the cosmos and coming from all sides because it is being generated from all sides and not from a center. This does away with the big bang, the singularity and inflation theory in one simple sweep. The universe seems to be expanding you say? Yes that is permitted in an infinitely large universe with no explosions as I said it will be forced to ebb and flow inside itself as the universe on a macro scale is homogenous but the tiny and mean tiny 100 billion light year or so bubbles that modern science can comprehend are practically fundamental discrete quantum objects compared to the whole. As far as the differences in the microwave background they no longer need to be summed and condensed to moments after the big bang as they are naturally slightly different simply from where certain large scale objects are emitting energy in the universe. Maybe black holes make the CMB radiation?

All radiation and matter is recycled because the universe is homogenous and infinite everywhere so energy we lose from our part of the universe is replenished by energy from outside anywhere our instruments can look. If there is an imbalance space will ebb and flow and can create large scale disturbances as space compresses or rebounds and you see things like the dark flow. The dark flow could be nothing more than an imbalance in a neighboring part of the universe or even a part beyond that part of the universe. The resulting movement of ESS through spatial compression and rebound will cause a massive swath of what we can see to flow for no reason. Dark flow is probably the best case for a non-big bang universe model.

As for recycling of matter and energy so that the universe never faces a big freeze that is precisely why we have SM masses made of clintrons. A clintron explanation for SM masses is the only way to create a stable unchanging universe that requires no big bang and will defeat the theory of the big freeze. All energy and matter in the universe are made from bauen matter alone and simply interact with the ESS macht glows in different ways to affect changes in the macht glow reserve pool. This has been proven in previous sections and so if we build clintrons from bauen matter we get a region of the universe that perfectly incorporates both space and matter together. The resulting clintron bauen geometry is a mix of ESS and bauen matter that will take all forms of matter and energy and always convert them into a single creation of both space and matter which is after all all that there really is to the universe. Upon the SM mass becoming too large to support the ESS any longer a single TSM detonation event will occur that will spew forth ESS space and free bauen matter. The bauen matter will undergo all the normal interactions with the right handed DNA of ESS and ESS macht and settle into a new region of the universe that need not be 100 billion light years across as theorized by the big bang approximately. It can be a very small region of space perhaps a hundred thousand or so light years across and if the universe is infinite in all directions this also explains exactly where these mysterious high energy gamma ray burst are coming from always so far away. The high energy burst can come from small TSM events and we do not see them because they are at the edge of our BBB. The change in spatial compression cannot be easily detected here as the energy is obvious while distortions in space billions of light years away is less apparent. The number of gamma ray bursts are too high you say?

Well it is a TSM event and they are almost mind shatteringly strong and the universe is infinite so they can come from an infinite amount of directions. Also it makes sense that the universe is uniform on a large scale and eternal and infinite based on gamma ray bursts alone. Gamma ray bursts created in space that is not spatially compressing or spatially rebounding will not have the wavelengths lose energy and so can travel vast distances and still be full strength gamma rays. The cosmic microwave background radiation was never stretched it is simply weak background radiation like the example of the room on a summer's day. So an eternal universe leaves the gamma ray bursts free to travel long distances without altering their energies and still explains why the cosmic microwave background is weak as it was always weak in this eternal universe. In other words all this space stretches and light stretches with it is simply not needed and has been looked at wrong for all these decades. Hey look at all these trees, you know what I see, I see a forest.

Any such large scale spatial distortion will easily bend light around themselves or through themselves and at the same time alter the predicted red shifts of things like distant galaxies. This is because various protrusions from such explosions will move pockets of space that are higher or lower in overall ESS spatial compression into our field of view compared to objects behind it thus altering the nice straight path of the light we see. In other words these can also be used to do away with the idea of dark matter playing with light from distant regions of the universe before they arrive here as well as anomalous readings in galactic red shifts at cosmic distances. Events like the Bullet Cluster collision also demonstrate a multiple BBB universe. The two galaxy clusters involved are massive and moving towards each other not accelerating their respective rates of expansion away from each other. Multiple BBB can account for this along with the fact that these clusters are massive and take untold billions of years to be created and collide over billions more meaning was there enough time from only 13.8 billion years ago of our universe alone for this to occur? So all in all a universe that is eternal and simply roiling around can be easily explained using ESS macht and bauen matter because the SM masses consume all and convert it into clintrons that can undergo TSM detonation events and create small BBB that easily flow into ours and restore entropy. The edge of our BBB will be populated mostly with new galaxies but have some old gold ones as well, not be a perfect shape due to ESS shockwaves overlapping from neighboring parts of the universe and create the necessary force to collapse materials beyond our BBB and recycle another part of the universe. The universe may not have started in a big bang but be an eternal structure flowing through everything like the eternal ash tree Yggdrasil.

Now that I have explained a universe of eternal ebb and flow and why it can make perfect physical sense let me explain where variations in the cosmic microwave background can come from in a big bang model of the universe. The cubes of space as mentioned before are large and the amount of energy contained in them after the big bang enormous. The view of modern science is that the universe came from a singularity and quantum fluctuations in that singularity are what we see today in the expanded universe as the temperature differences. This is impossible as two things are wrong here the first being there is no such thing as singularities and the second being the nature of the singularity forbids quantum fluctuations. A singularity is thought to be a dimensionless object that exists as a single point in space and inside it there is no time at all along with infinite density as everything has become this perfect singularity according to modern science. This means that the interior of the singularity cannot contain quantum objects as they do not exist and as such there is no room for quantum fluctuation so no mechanism for the observed differences however small of the universes temperature. So this means the fluctuations must have come from the moments after the big bang and this is also impossible as the moment after the big bang is seen as the inflationary period.

The inflation period cannot produce quantum fluctuations either as space is expanding and matter is not able to exist and therefore interact with itself or space in any meaningful way. So during the theoretical inflation period you have space expanding and as it gets larger the space between matter grows meaning the matter cannot hope to interact as the space between it and any of its neighbors is expanding faster than light. Matter cannot move within space faster than light according to modern science and so there is no chance for it to interact and create fluctuations either. Since it takes violent events like a Big Bang to make a fluctuation it is impossible for violent events to occur during the inflation phase as the space is growing between matter preventing the matter from interacting in any violent way with other matter. The limiting speed of matter as compared to the faster than light speed expansion of the universe is also what is responsible for allowing the anti-matter to decay faster than matter and keep from everything annihilating theoretically in the early universe resulting in fluctuations. The space between particles does not allow matter and anti-matter to interact until the anti-matter has decayed and been recycled through the ESS macht filter of the right handed DNA of the universe again. More proof that the universe was never going to become a simple sea of energy all along. This means that the fluctuations cannot exist according to any of modern sciences main views as the simplistic reasoning of mathematically convenient quantum fluctuations is never possible.

The only way to have a universe be created with temperature fluctuations as we see them and still use some sort of inflation mechanism is to use a large SM mass that detonates all

ESS and clintrons inside it. The fluctuations must be born in this way as the universe is not like our little room with the small chemical explosive in it. The chemical explosion and resulting fireball is non-uniform due to quantum fluctuations on the small scale and random particles, winds, impurities in the chemical explosive and oxidizer on the large scale. The universe as it explodes in the big bang and afterwards is thought to have come from a perfectly pure and uniform singularity devoid of impurities and there is no oxidizer for the explosion to react with. The universe is the explosion and the space is created along with the matter so there are no imbalances in mixing or burning of the cosmic propellant. Again the singularity cannot exist so all fluctuations must come from a point before the actual explosion and still allow the rapid expansion of the universe to unfold.

In order to do this you need a large SM mass as described in a previous section that contains clintrons but all at slightly different degrees of maximum interaction with the ESS inside of the SM mass proper. Along with quantum entwinement events inside the SM mass all the materials inside will be very uniform and very similar but not quite perfectly synchronized with each other. This will be all that is needed to allow for the variations observed today to occur as the SM mass explodes in what is thought of as a big bang. The differences for the cubes of space are made inside the SM mass and before it explodes and this allows them to unfold with the rapid expansion phase still intact. This is a necessary step as these differences will produce different localized TSM detonation events that unpack a region of the SM mass quickly and will evolve over time into one trillion light year cubic structures for example. This is another reason why single small quantum fluctuations using a singularity or the inflation theory are wrong in accounting for the differences in temperature of the universe.

Because the cubes are now one trillion light years in size it is impossible for a single small quantum object fluctuation of a singularity or post big bang nature to account for what covers all of space and all energy and matter for a trillion light years cubed. Quantum by nature is a single event and would not affect all energy and matter in that cube evenly and at once using the singularity and inflation theory model. In order to create this localized but uniform change a mechanism is needed that can have a quantum variance but still reach out to many particles around it and certainly enough to fill one of these cubes. The quantum entwinement and maximum degrees of interaction of clintrons in local areas of the large SM mass account for this by creating different times and rates of TSM detonation occurrence to the overall explosion of the SM mass. A single quantum fluctuation in a single clintron can cause it to TSM detonate before another region and based on how those localized clintrons around it are interacting with space will create the different cubes of the universe as the TSM detonation event there unpacks the clintrons at macht speed. This macht speed dump of all matter outside of the ESS allows for space there to expand rapidly to fill the cube and create numerous different cube powers throughout the universe. The matter was made of clintrons all in slightly different degrees of maximum interaction and so will create natural imbalances in the distribution of matter with space. This seeds the formation of all stars and galaxies as the universe for that cube cools again meaning dark matter is not real as it is not needed to seed star formation. So using a single large SM mass made from ESS, macht and clintron bauen geometries the mechanism by which fluctuations in the cosmic microwave background are created as we see them today and on the scale of cubes is explained before the big bang as is needed to allow for the rapid expansion phase of the universe to be possible.

Chapter 36
The Universe Rules

Why New Rules are needed

So with the conclusion of the previous section and all the uberfluss and uberlauf it contained it is now time for me to summarize not just a simple section but the universe itself. This is exactly what many in modern science have been asking for a century or more which is a simple explanation to solve the problem of how the universe works. People have been clamoring for a model of the universe that reduces variables, that does not require the creation of new ones and can be an all encompassing simple answer that is applicable to all unsolved problems as well as observed phenomena in the universe. This is what I have given you here; a simple explanation of the universe that reduces variables and remains constant no matter what problem is thrown at it. To those who have been seeking an answer like this or to those who have been hesitant or simply serious about embracing one then I invite you to fully embrace my work here. In this last section of my book I will try to show you why my work is the answer to solving the problem of how the universe works beyond the proofs I have been providing in all previous sections. To do this I will be looking at the state of the theories of modern science broadly and than at my own work as they are thrown at the universe.

To begin with I must remind the reader that all three main views of the universe as accepted by modern science all fail at explaining the universe flat out whether this is classical mechanics, general relativity or quantum mechanics. The reason this is so comes from the simple fact that modern science has within it unsolved problems and these should not exist if an existing theory was able to account for the entire universe. Classical mechanics is a good view of the macro scale world that we encounter everyday and is the best at bridging what happens on the small to large scales. General relativity to be fair was never going to stand a chance at solving the universe as it only ever claimed to understand gravity and never attempted to take a whack at things like quantum entanglement, the Big Bang or fundamental particles so to speak. General relativity however even in its understanding or rather lack of understanding of gravity fails miserably on all counts within the scope of the universe.

Flaws in the theory, impossible objects like singularities and a complete inability to even begin coping with the quantum realm make it by far the most incorrect of all the three main views despite many in modern science desperately clinging to it. This unhealthy fascination with the special and general theories of relativity in the face of their obvious short comings reminds me of how people must have clung to the notion that the Earth was the center of the universe and that everything moved in perfect circles rather than ellipses. The evidence was ever present that the geocentric circle view of the universe was dead wrong yet people tried to make that stick no matter what. No one today has a problem with ellipses and the Sun being the center of our solar system so in years to come I know for a fact that no one will have a problem with accepting something other than theories of special and general relativity and will look back at the time when they were so heavily defended to the point of stupidity as just that, stupid.

This leaves quantum mechanics which has by far enjoyed the most solid base of any of the three main views as it can be nicely tested and tested again and measurements made based off of its teachings. The whole of the Standard Model of particle physics is pretty much gleaned from it and practical applications of technology have been made from quantum understandings of particles such as photovoltaics or solar panels where as nothing has been built from a relativistic understanding of the universe. So quantum mechanics shines bright in the realm of modern science but still cannot account for everything and this is its downfall. While good at certain small scale things it cannot account for all and is guilty of creating some strange events similar to how general relativity created the singularity to name one. The misunderstood act of observation, superposition, quantum entanglement and the many worlds view of mathematical probabilities are some of the limitations created by quantum mechanics hindering itself. Coupling these facts with the reality that none of the three main views can be integrated with each other is an obvious signpost in the road to finding a correct answer to how the universe works.

What this series of limitations and failures does to finding an answer to the universe is it breeds ad hoc theory after ad hoc theory which do not answer anything without doing two things. The first is creating more variables to the universe and the second is creating more impossibilities that do not mesh with other theories or solve more than one specific problem. The addition of variables can be seen in any of the newest theories on the universe from string theory to the many worlds theory. If you look at the string theory for example you must ask yourself what is it actually trying to solve? The answer of course is the nature of matter and it seeks to explain matter through mathematics rather than through a practical approach. In other words it attempts to find any proof existing inside the equations to explain matter instead of just looking and observing the universe around us to see if a model can be found that explains matter as it already appears.

To this end string theory is famous for giving us multiple extra dimensions and the various versions of string theory use different numbers but they all use more than the usual 3+1 space currently accepted by modern science. The 3+1 is of course three spatial dimensions and one temporal one although some theories seek to add extra time dimensions as well. Nevertheless the mathematical theory of extra dimensions can exist on paper but do not manifest themselves in the real world as in you and I or anyone and everyone has never physically seen a spatial dimension numbered four or higher. In addition to this we have never been able to watch the passage of time flow backwards, or sideways or inside out or any other such mathematical fantasy of time.

The very idea of extra dimensions is even acknowledged by string theory proponents as being something of an unfortunate problem that they would rather discard if they could find new equations to explain the universe without extra dimensions. The argument goes something like we as humans cannot perceive extra dimensions due to our large macro-scale size. If you were to shrink down to the size of a speck of dust however something like a single line which to us is one dimensional would suddenly be large enough for you to see that it could have more dimensions. A single line to us can be thought of as a very thin cylinder which if you were the size of a speck of dust you could see the cylinder is three dimensional as you can now move around the surface of the cylinder in extra dimensions.

This is somewhat silly thinking however because you can simply play the one up game forever here where once a string theorist says that is all you get no more dimensions someone else simply comes along and says well let us shrink down a little smaller and find extra dimensions on the strings and so on and so forth forever. This leads to an infinite number of dimensions and therefore an impossibility like any infinite value in science and therefore is wrong. To further prove this wrong just as infinites in quantum gravity which uses relativity and relativity as well do exist and so show us that they are incorrect we can look at the fundamental particles for proof of only needing three dimensions.

Humans are large macro-scale objects compared to a single atom and our senses do not readily allow us to be aware of single atoms and the way they work. So from a human viewpoint the universe has three dimensions but that does not mean that more do not exist because it could be a bias of our perception of the universe. Now that being said animals like humans are made of the universe itself meaning atoms, energy and fields and so we actually are aware of however many dimensions exist in the universe as in order for us to exist we must interact with the universe. In addition to this there is always a life form somewhere that can utilize whatever the universe has created even if we cannot and a perfect example of this is the acute sense of the Earth's magnetic field used by many animals while humans do not use it. Nevertheless even if we disregard these proven facts of biology we can still entertain the idea that there might be more dimensions beyond three if we look at the building blocks of the universe beyond our human scope. Well we have already done that and done it in a scientific and proven way using repeatable experiments and measurable results. What I am talking about is the examination and exploitation of the atom and its various components such as protons, neutrons and electrons.

These particles and smaller ones still all exist physically in the universe and make us up and yet we cannot perceive them directly and yet we can perceive them directly using our technology. What we have found is that this micro-scale unseen world which could contain extra dimensions or might need extra dimensions simply to work do not. That is the fact of the matter as we know exactly how a proton behaves with an electron and then how that atom will behave with another atom and so on and so forth. So here we have proof that even the most fundamental particles as understood by the Standard Model of particle physics do not require more than three dimensions to operate normally and so simply because they are too small for us as humans to perceive even this quantum world does not require extra dimensions. What this means is that if we now play the one up game to infinity we do not actually get an infinity meaning if we continue to go smaller and smaller than say electrons and quarks we will only ever find matter, energy and fields using three dimensions and never ever find extra dimensions. So finally we have the correct answer of a finite value of dimensions obtained even if we theoretically continue shrinking things down an infinite number of times or ways.

This is not the sort of clean and problem free answer you get when exploring something like string theory and the impossibilities go further with string theorists admitting it might be impossible to ever test for extra dimensions as they cannot be detected using laboratory equipment that we can perceive only in three spatial dimensions. This of course allows for us to invent any number of extra mathematical tricks or seductions that never need be proved and so fall into the familiar realm of theories that cannot even be proven wrong. The inability to falsify a theory is a big red flag that it is probably wrong and full of needless ad hoc. So string theory is a very good example of a well known theory that illustrates how the current accepted theories of modern science contain vital flaws to their existence that create more problems than they solve. Making extra dimensions only makes for extra problems in a universe that is thought to run simply and beautifully on its own terms.

Not to mention the fact that these extra goodies can only exist in the mathematics as can be illustrated from the idea of the many worlds hypothesis. The many worlds theory states that all possibilities described in the wave function probabilities of quantum equations are made manifest upon the act of observation of the system in question. What this does is create the largest possible set of extra variables to the universe and ones that like extra dimensions can never be tested for. As the wave function collapses in our universe all other infinite possibilities are made real as the wave functions of every other universe collapse at once and solidify into reality according to the solutions to their respective quantum equations. To make a bold statement that will likely earn me no friends in the many worlds camp did anyone ever stop to listen to that reasoning and simply say this is crazy?

I mean this in all seriousness and knowing that those people backing the many worlds theory are simply trying their best to find the answer to how the universe works. That being said many of modern sciences theories are fairly crazy and no one seems to have said we can explore them a bit but not too much. Many worlds, the singularity either in black holes or the entire collection of space, matter and energy in the universe being in something smaller than an atom simply because the mathematics said it is possible on paper or extra dimensions of matter and branes everywhere you cannot look. It is by far time that modern science broke free from these ad hoc theories as each and everyone of them are leading us nowhere fast and worse they are divergent from one another which leads me to the second flaw with all of them.

This second flaw comes from the fact that there is no overlap in the current theories adopted by modern science and they cannot hope to solve more than one problem. What I mean by overlap is a failure of general consensus between competing types of theories and even between similar groups of theories. Take string theory as a perfect example and you see that there are not one but multiple versions of string theory and they are all different. There is no point in describing all of them as there are simply too many and the important fact here is that if string theory were correct then it should be immediately apparent how many extra dimensions are in fact real to use just one example.

I have discussed differing string theories earlier and that is where you will find the pertinent bits about each and also the crux of the problem with many of the modern theories currently favored by modern science. There should not be different versions to a theory as this means that the basis for the theory is wholly unsound to begin with. Think of it this way there is no difference of opinion between mathematicians when it comes to adding two plus two and getting four. There is one immediately obvious answer and so it is known that two plus two is four and we can all shake hands, agree on this and then move on with the rest of our day content in the knowledge that wherever we go in the universe adding two plus two gives us four. Any theory attempting to explain the universe should be held to the same rules and the fact that multiple and vigorously defended string theories exist with no one side being able to nail down anything to their sides in favor over the other theories is disconcerting to say the least.

If this simple fact is extended to other theories unrelated to string theory then the problem becomes even more exacerbated and to explain this let us look at string theory and dark matter or dark energy. String theory is seen as a way to explain the nature of matter and perhaps attempt to help with explaining why gravity is so weak and the like which has nothing to do with dark matter and dark energy. Dark matter and dark energy are a completely different type of answer to observed phenomena in the universe which require completely new and different types of variables in order to work. You cannot take what you know which is essentially nothing from string theory and transpose this onto dark matter and dark energy.

Dark matter is a mysterious theoretical particle that simply does whatever magic its proponents need it to do in order to fudge the numbers in favor of apparent observations lacking gravity in large scale structures of the universe. Dark energy is not even remotely theorized other than it is the mysterious and unknown force that just pushes everywhere in the universe. Even from the mouths of dark matter backers you will get nothing more than it is strange and we think there is a lot of it but we have no idea what it might actually be. Golly I can feel the confidence welling up inside of me already. Terse my tone perhaps or perhaps a bit acerbic? I apologize again to those who mean nothing more than to try and solve how the universe works but this section must be slightly abrupt and harsh at times as not only is it a review of the universe but a wake up call and a shocking one if necessary to call out the ad hoc theories and dispose of them before they do more harm than they already have to the minds of science.

To that end if you try to combine string theory and dark theories you get nothing, in fact the mathematics of one cannot be made to fuse with the other and the concept of one cannot be made to explain the other. This is the foci of the second point I am making here which is any string theory does not offer an explanation for parameters outside of its model. To extend this neither do dark matter nor dark energy attempt to explain what string theory is explaining and this is repeated throughout the currently accepted or at least entertained theories of modern science. Many worlds will not ever explain quantum entanglement, loop quantum gravity will never explain matter, string theory will never explain the Big Bang and dark theories will never explain the workings of a black hole to name just a few and so show the inability of any of these theories to cross over and explain something else other than what they were designed to do.

Designed to do with the inclusion of multiple new variables that seek only to further complicate the workings of the universe something that history has never borne out. As history progressed and as science advanced the number of variables needed to explain the universe decreases over time and not the other way around. So making ad hoc theories that only solve one problem and only with multiple new variables is obviously incorrect. In other words the theories accepted currently by modern science need twenty different theories to answer twenty different problems. This is outrageous as the problems they are trying to solve are huge and would present immediately substantial proof in other areas of unsolved problems and this is precisely what they do not do.

This is why I have presented my single answer theory to the world in a non-mathematical form and instead used a model that can be explained practically problem by problem. In other words I want to show how my work can use one answer to solve twenty problems and at the same time reduce the number of variables in the universe to the smallest number possible which is two and they are space and matter. You must let go of the limitations of thinking you have been taught since you started to study the world and the universe around you. You must ignore the preconceived notions of mathematics for now and focus instead on simply understanding how the universe works using my ESS macht and bauen matter interactions. Afterall Newton first had the idea of gravity and how it worked through the model he created in his mind and the mathematics came second.

If you can open your mind wide enough you will be able to see the universe even if only partly as I see it and be able to fully manipulate the ESS, macht glows and bauen matter geometries freely until you can understand how they easily work together to create any structure on any size scale from the smallest up to the entire universe at once. I have to be bold here and I have to make assertions here because I am not afraid as I know this is the single explanation to the universe that can account for all the accepted observations, experiments and phenomena as well as account for all the misunderstood observations and unsolved problems of the universe all without altering the laws of physics and without adding new variables. You know as well as I do that space is real and matter is real so if you can understand how the ESS macht of space and the bauen geometries of matter interact you can finally see how the universe works.

Rules

So here are a few but not all of the simple rules and facts about the universe that will be needed to construct the sought after mathematical understandings that modern science requires for the universe. The first rule is size and it must be understood that space has the smallest size of all coming from a single ESS. This fact is extremely important because all interactions of matter with space involve matter being larger than space and space changing size while matter does not. Remember from a previous section here that I explained that known fundamental particles like a quark for example will have different volumes based on different conditions in which they exist in the universe. This is because what modern science views as a fundamental particle is not fundamental but just the macht glow created by the bauen geometry inside of it and interacting with space. So what this means is that fundamental matter as understood by modern science is always bigger than space as it is always bigger than the bauen matter which is the true matter of the universe.

This fact explains why all quantum systems to date cannot incorporate gravity and changes in spatial compression in their calculations. It is partly due to the fact that gravity is very weak compared to the other three fundamental forces but it is also because modern science treats fundamental particles as point particles. This error puts the particles on the same size scale as space and therefore of course does not allow room for the space to compress or rebound as needed within the system no matter how slight that may be for the local area of space and its amount of spatial compression. The reality is that the bauen matter is creating a much larger volume of macht glow that represents what modern science sees as a fundamental particle. This large macht glow boundary can encompass many, many ESS and so the system finally contains a significant number of ESS so changes in ESS volume due to increase in macht glow gravity will now be seen and yes they will still be small but no longer invisible. Therefore by using ESS macht glows and bauen matter geometries on quantum systems containing fundamental particles the solution to gravity being incorporated fluidly into all size scales of the universe is now complete. This also allows for all four fundamental forces to be incorporated into a single equation that will scale up or down as needed from a single ESS to the entire universe.

The next step now is in measuring ESS volume under Earth levels of spatial compression and macht glows in order to determine a starting point for ESS volumes throughout the universe. Once ESS volumes have been accurately measured for Earth they can be compared to the total volume of a known fundamental particle to create a scale of space to matter for Earth conditions. The matter can then be weighed against the number of ESS inside it and the total macht glow through the surrounding schales of space in order to learn about the bauen geometry inside of it. This knowledge now of ESS size, macht glow and bauen geometries can be realized because the base properties of the fundamental particle are well known. These properties have already been fairly well measured and catalogued in the Standard Model of particle physics and are thought of as things like mass and charge.

The use of bauen matter to create what are thought of as fundamental particles makes for a clean answer to how the universe works and this is something that string theory cannot do. String theory is said to have a huge number of solutions perhaps ten to the 500th or even

1000th power which is of course absurd as what it really means is the people doing the calculations simply got tired and said okay we have proved our point. The reason this is a flaw of string theory is that the strings require extra dimensions to vibrate in and depending on how they vibrate they create different particles. Altering how they vibrate in any number of ways will result in a mathematical only answer for say a simple electron but that can be calculated in near countless ways which is another obvious sign that string theory is incorrect. In my work an electron is only ever created one way using the bauen side electromagnetic and therefore cannot have multiple types as seen in multiple string theory electrons. Besides this failure of string theory there is another that still complicates the universe further and that is the idea of strings themselves and their tendency to vibrate.

According to string theory different vibrations of strings will produce different particles so a string vibrating one way will make an electron and a different string vibrating another way will make a photon. What this means is that string theory has only further complicated the universe as the total number of known particles in the universe must now all possess their own unique string type in order to exist. No open and closed strings are incorrect as it is not important that there are two types of strings what matters is how they vibrate for eternity which is what makes them far more than two types. A string photon remains a string photon forever since the big bang until the end of the universe so it is a whole unique type of string as it requires a photon type vibration to exist which is different than all other vibration types for all other particles. So it is meaningless to say there are only two string types as this is wrong as there are as many string types through their unique vibrations as there are particles in the zoo of particles. This is once again simply complicating the universe and adding variables which is a no-no remember.

Using bauen matter instead and allowing those bauen matter pieces to arrange themselves into stable configurations based on their exposed sides and balanced with the ESS macht allows for a single variable to explain all matter. There is no need for a whole zoo of particles created from a whole zoo of different string types and vibrations. Think of the electron and the heavier version of it which are the tau and muon and realize that string theory states that it is not simply an electron string that is vibrating to make a tau or muon but it must be another type of string that is making that tau or muon. Couple this fact with the necessity of string theory to require multiple extra dimensions which break all known laws of physics and are completely untestable and therefore unfalsifiable and strings are not the correct answer.

You need a zoo of particles to create something simple like a drop of water using strings or something more complicated like an animal for example perhaps a lion or a human which is living, moving and thinking as each part of the atoms in that structure require multiple strings to make each particle type. Using bauen matter in specific stable bauen geometries you can take a single photon and if you had enough photons make a lion or a human from it even though it has a very simple bauen geometry inside it. The reason is the photon is made of bauen matter so even if a lion or a human is made of more than just photons you only need bauen matter from the photons and with enough bauen matter pieces you could simply build the geometries in normal three dimensional space to make a lion or a human or a star or universe for that matter.

What this means is that you can make anything in the universe without strings, extra dimensions and 10 to the 1000th versions of a theory as you only need the normal three dimensional space we know and a single type of matter which is bauen matter. One matter type, the bauen matter can be used to easily construct all fundamental particles and so does away with the many types of string vibrations and the need for open or closed strings for that matter. Bauen objects are held together with simple bauen forces and these require no extra dimensions just as the electrons, protons and neutrons of atoms are held together with normal forces that do not require extra dimensions either. Remember we looked at a macro-scale human and then went micro-scale to see atoms well now we can go even smaller to ESS macht and bauen scale and build the universe up from there. This is how you can uncomplicate the string nonsense, uncomplicate the mathematics and create every single particle in the zoo of particles from a single type of matter and therefore a single variable. All known and unknown particles are built this way along with every seemingly but not really strange result from all particle collisions and there is another plus to this bauen matter model as well.

Once bauen matter is understood and the first bauen geometries are constructed and assembled for known particles like electrons, photons and quarks and the like we can construct a periodic table of particles so to speak and predict what undiscovered particles will fill in the holes present in the table. This will allow us to easily predict exactly what we will find when looking at ever higher energy collisions in particle colliders giving the proof of these missing periodic particles in the flesh as it were. The use of bauen matter will therefore remove all the undesirable effects of string theory as a single bauen matter is needed to explain the universe not two strings and therefore the number of variables is already cut in half. Bauen matter does not vibrate nor does it need to and so the 10 to the 1000th number of solutions for strings is gone there is only one solution to bauen matter. The zoo of particles is now perfectly cleaned up with each animal in its happy habitat and any new animals will be predicted by looking at there empty habitats.

The last thing about bauen matter and understanding how their geometries are put together is that we will be able to do what we can do today with atoms which is make new molecules, crystals and compounds at will. In the case of bauen matter we can make new particles that are stable with the ESS macht for the local area of space they are in. We regularly make things from atoms that never occur anywhere in the universe and this goes back to my example of a cheeseburger and cold beer on a summer's day. There is nothing unusual about the atoms in our summer meal only that the universe cannot ever naturally make them arranged in that fashion. There is a one hundred percent chance that we will be able to create new stable bauen geometries that would normally never occur in the universe and therefore create our own new particles. Hints of this have been stumbled upon rather accidentally and without an understanding of what the universe is up to at the smallest possible levels with the possible creation of things like stable four neutron clusters known as tetra neutrons. We can then take these new particles and build our own custom atoms, molecules and structures such that the true future of humanity is made real.

Further proof of this comes from trees, rocks and humans and starts with the strongest natural thing humans could build from naturally without perturbing the universe and this was wood from tree trunks. Then humans got to understand the universe a little better and we started to heat rocks for really who knows why but we did and out came metal. Now we have done something to the universe that normally would not happen and this is called mining and smelting ores and metal is stronger than wood but we did not stop there. By pure accident and without understanding atoms, electrons and nuclei and the like we began to learn that mixing certain different elemental metals will give us much stronger alloys. Further along if we heat treat them, temper them and anneal them and so on we can make them stronger still and this involves atomic manipulation of the crystalline structure of the metal. The bottom line is that we can take natural elements from the universe and make things much stronger than before, shape them and build with them as we choose and allow them to conduct other particles like electrons to harness electricity.

So ever wonder how the aliens do all the cool stuff they do? Easy, they manipulate matter on the smallest levels possible which are ESS macht and bauen matter. Humans will be able to manipulate bauen geometries into new stable particles that we can use to make particle alloys if you will that will unlock hidden forces modern science simply cannot understand yet. This goes beyond simple building materials but to explanations of superconductors and making room temperature superconductors is one example. The manipulation and complete understanding of bauen matter and how they interact with space will also allow us to exceed current accepted limitations of physics without actually breaking the laws of physics. Hyperspace and hyperdrive are probably the most sought after of these applications and yes they will be real as bauen geometries can be made to maintain their normal bonds creating things like humans and spaceships but to uncouple them from the ESS macht glow gravity meaning no more relativistic nonsense. If space can move faster than light using macht speed physics then so can we and it will only occur once bauen matter is fully understood such that we can alter bauen geometries at will. The rules are fairly simple decouple matter from gravity which still leaves the remaining forces holding the object intact as if nothing were different and this also solves the zero to million light speed problem of being crushed in your ship.

I will refrain from expounding on this point here as the possibilities are now literally as big as the universe because unlocking ESS macht and bauen matter will give you the power of the entire universe in a single force. To use this universal force the binding rules for bauen matter into bauen geometries will need to be incorporated into the interaction of bauen sides with ESS macht. The understanding of ESS and ESS macht under Earth spatial compression levels will be the first step in making this a reality and the bauen matter rules can then be added to the overall equation of the universe. A simple new bauen geometry created by humans that is stable even if only for a few seconds will be enough for us to better understand these rules and so open the door for all future bauen geometries to be created as well as altering how these bauen geometries will interact with space either more strongly with ESS macht glows or less strongly and this will lead to the total harnessing of the universe force as a whole.

The last piece of the puzzle is the rate of interaction of ESS macht and bauen matter and so this raises the question of time once again and how it will be handled in a new universe equation. Time as proven before is not real and does not exist as it is in no way a kind of physical construct or force and it certainly is not a part of any theoretical spacetime. Time is only seen as variable from the viewpoint of quantum objects and its observers but the time underneath all energy and matter which is macht time is absolute and unchanging. What this means is that in order to construct a single universe equation all interactions will have to be calculated in a two step fashion. The first step being all calculations involving macht time as macht is shared between ESS and throughout the surrounding schales. The second time will be a variable time that is used to calculate changes in large quantum systems and will be secondary to the macht time as the absolute value of macht time will be used to construct the rates of quantum actions between objects or singly within a system.

So just as in mathematics certain operations are carried out first like the familiar BEDMAS where things like brackets and exponents come first and the division, multiplication, addition and subtraction follow afterwards you will do the same thing in the universe equation. The ESS macht calculation will be first and the quantum mechanical calculations will be second so a type of EMQM system can be used. ESS and Macht first with Quantum Mechanical second and this type of universal formula can be used to determine the effects on quantum systems as we learn to study the ESS and macht components of the universe better. The conservation of ESS macht and quantum energy laws state that the rates of perceived quantum actions from one object to the next especially when two objects that are not interacting are made to now share macht or the reverse when two objects are sharing macht are made to be separate again will reflect how fast macht is being shared between the ESS in the system.

A simple example would be to decouple matter from space as in the BBB where bauen geometries were 100 percent outside of or not interacting with ESS and so macht glow reserve pools were at base levels. The act of decoupling will cause the ESS macht and energy conservation law to transfer macht back through the ESS and the ESS must share macht back to the bauen objects that donated it and so this speed can be measured. Macht speed is not instantaneous and ESS spatial compression changes macht glow volume making one macht glow larger than the other and so the larger glow is slower overall to return the donated energy.

By measuring the speed at which macht flows back to a bauen object we can determine the local area spatial compression for that object as it exists in the universe. This same object can then be taken to another part of the universe with significantly different spatial compression levels and the speed at which the macht flows back to the object can be measured again. In this way macht flow which is of course based off of macht time which is absolute can be used to recover information about space and the bauen object it was donated from. This catalogue of objects in different areas of spatial compression will finally allow us to compare all known particles against each other and therefore all known bauen geometries against each other. The speed of macht flow coupled with spatial compression information will allow us to construct detailed geometries of each bauen object and specifically what sides of the bauen matter making up the object must be exposed to space. The end result is a complete mapping of all space and matter through both ESS macht and bauen geometries which will give us the definitive guide to how every object whether it is currently understood as matter like a quark, virtual matter like a gluon or energy like a photon behaves along with space that sits around it all and the perceived fields that exist within the universe.

The completion of this information set about both space and matter in the universe will finally allow us to construct the simplified equation for the universe that will do away with the perceived time of humans also known as quantum time or regular time and instead become a mathematical formula that deals with only macht time. The reason this is so stems from the fact that quantum time or regular time is just a construct of events governed by macht time and so once all aspects of the system have been measured the precise nature of all components of the universe can be equated through a single new macht time equation. In this equation ESS, macht and therefore space are still calculated using the same macht time as before but now the old notions of perceived, quantum or regular time can be expressed as simply interactions measured in macht time. This means that both parts of the equation which now include macht time governed bauen matter can be simplified into this final universe equation.

The balancing of ESS macht glow as I have discussed many times before is now the easiest to calculate as increases in macht glow reserve pools can be effortlessly shifted to and from space or to and from matter. When matter donates more energy to space the donation amounts as well as the rate of donation can be factored using macht time making everything simple. So in the end simplified equation there will be macht time, ESS and macht and bauen matter with all the respective flows of energy/macht going to and from, increasing or decreasing as necessary to balance the total law of conservations of macht/energy of the system. As one part goes up another part goes down, as matter moves more into space the macht glow goes up as the macht glow goes down matter therefore must be going more out of space. As ESS balance macht bauen geometries must change shape and therefore macht glows will be affected as a result again either up or down. Since all ESS, macht and bauen matter interact with no boundaries throughout the universe all systems can now be incorporated with each other and the entire universe as a whole. No more black holes are the edges of known physics or the universe has multiple worlds within it but separated by unseen dimensions in other words no boundaries to anything the universe is finally free to be a single living breathing creature no matter where or how you look at it.

So the basic rules in case you forgot them are here real quick: ESS increase and decrease in volume, macht glows increase or decrease from the macht glow reserve pools, bauen matter in different bauen geometries interact to alter these properties of space and space alters the allowed geometries of the bauen matter and all are governed by macht time ultimately or by macht time and variable quantum time in the first version of the universe equation. Measuring ESS for the levels of Earth spatial compression as well as macht glows and learning the first simple bauen geometries will be the first of some of the total steps needed to get at the final simplified equation of the universe. Modern science has not finished with the universe now that I have explained how it works but rather modern science has just reached the point where a whole new set of experiments and mathematics will be created in order to find the values needed to get the final simplified equation. This is the equivalent of heavy scientific job security for years, decades and maybe centuries to come and finally those in science know they are working towards the correct answer.

The significance of the simplified universe equation cannot be understated as any aspect of the universe can be immediately calculated once and for all. So for instance there will never be guess work, random variables or seemingly random events and all multiple probabilities will be impossible and resolve into the correct event or observation every time. So the outcome of every single photon in the double slit experiment will always be known before the photon is sent on its journey. The superposition of all particles will be known such that superposition is shown to be not a real phenomenon. The position, momentum and all other aspects of every particle in any system will be known with one hundred percent certainty. In other words anything known to modern science today as a random quantum event or probability will simply disappear for any system experimented on or observed in the universe. This might be kind of a sad day for some and maybe for others it will make the universe somewhat boring as there will be no new surprises as there can never be again once the simplified universe equation is calculated.

The practical side of this is that anything about any system can be known in perfect detail and so for example you can know the position and momentum in any particle even if you do not care about that aspect of your experiment. Perhaps you are looking at the entropy of the system, or rate of change or simply mixing two unknowns together and seeing what you get like so many scientists have done for centuries it does not matter any aspect of the system or experiment in question can be known. So it will be impossible to ever have unexplainable results as even if you appear to get one for that particular system all you need to do is go and examine the other factors, properties and information of the system and there will be something in there somewhere that accounts for whatever seems unexplainable. To solve one of the unsolved problems of modern science you will be able to finally look at any system no matter how small and simply scale it up and up and up until you are dealing with objects the size of the entire universe so the micro and macro have been merged and the factoring of something like gravity from ESS, macht and bauen matter systems to quantum systems is now complete and calculable.

This leads to the obvious questions of where do the unexplainable variables come from and where do the unsolved problems come from? The answer to both is from the currently accepted views and theories of modern science which are both quantum mechanics and general relativity. If you know anything about the state of science in the twentieth and early twenty-first centuries then you know that it is filled not with answers but unsolved problems. In fact whole lists of problems have been compiled that come from multiple sets of observations and in many cases appear to be where the laws of physics breakdown and the universe does the impossible. Human nonsense. Nothing breaks the laws of physics only the incorrect and currently accepted laws of physics as understood by modern science. It is equally ridiculous that the universe could ever do something impossible as the universe is everything and therefore does everything and so nothing is actually impossible. What this all means is that the theories of special relativity, general relativity and quantum mechanics are incorrect and incapable of ever giving humans the answer to how the universe works.

To begin with let us look at the theories of special relativity and general relativity and see where they are dead wrong. Special relativity makes a set of assumptions about the universe that are not proven and the most obvious is that light is the speed limit to the universe. No one can ever argue competently that this is true as no one in modern science knows what the speed limit of the universe is. The idea that light in a vacuum was the fastest possible speed available to the universe is only an invention of Einstein and in no way proven yet it underlies the rest of his theories. The types of assumptions made by Einstein with the universe have unfortunately permeated his later theories and the thought processes of many other scientists who for some blinded reason think everything he predicted is perfect when it is obviously wrong. Those that have rejected Einstein and his full body of work are largely ignored which is horrible as much of Einstein's own work creates impossibilities or limits the options available to the universe for how it can operate.

If you need proof of this look at the singularity created by Einstein and the obvious fact that it is wrong. To make matters worse people over the last century or so have for some reason or other worshipped the equations of Einstein despite the fact that they do not attempt to explain the universe but are only a fine tuning of Newton's original work which attempts to explain gravity and nothing more. From this it has been difficult to shake people from their fascination with the work of Einstein despite it being quite obviously wrong and no matter what the mathematics of general relativity say no one stopped to say enough is enough. The extreme curvatures of an unknown substance called spacetime producing an obviously incorrect object called a singularity have resulted in a century of ad hoc garbage as many well meaning and intelligent people have been trying to work around the flaws of the very theory they themselves have succumbed to. You simply cannot fix a broken theory and Einstein's theories are just that, broken.

Quantum mechanics does not fair much better as it has also produced impossible predictions in the form of things like quantum entanglement, superposition and the uncertainty principle. None of these events are actually impossible as the universe cannot contain the impossible in the first place and this is directly reminiscent of the problems of

Einstein and singularities. Singularities might seem real to Einstein and his mathematics but that is just a fault of Einstein's and not the universe, just because something appears in the mathematics does not mean it is in any way real in the universe.

It must be starkly stated once again here that you should never try and make the universe fit your mathematics but rather make your mathematics fit the universe. If that means literally throwing out decades or even centuries of cherished and adored equations then do it, just throw them out as they are garbage. This might be hard for some to do and a bitter pill to swallow in the face of having spent so many years following these equations down the wrong path but in the end this is no different than people that had to accept a hard fact that the Earth orbits the Sun and the bitter pill that they had just spent their lives fighting the opposite and imminently doomed battle.

This is exactly where we are today in modern science as people are desperately clinging to what they know is wrong and we can all see this from the plethora of ad hoc theories that may have been vomited forth with the best of intentions but remain waste regardless of that fact. The minds behind them should feel no shame in working on a different path if it means answering the question of how the universe works at least you tried it out. To this end quantum mechanics falls along side both the theories of special and general relativity as being full of impossibilities and producing ad hoc theories.

To sum a few of the quantum mechanical flaws up quickly think of the mathematical inaccuracies of all quantum systems that rely on probabilities and do not know definite outcomes. The basic idea behind these quantum mathematical equations basically goes that you can never know something for certain until an observation is made and before that all possibilities exist. Wrong on every level as the universe does not allow things to exist in every possible configuration despite how it looks to us and despite how convinced people might be that this is so. The universe cannot fool itself and knows precisely where it put everything and what those things are doing all the time.

So simply because the theory of quantum mechanics cannot predict accurately the outcome of a system does not mean in any way that these unknowns cannot be known. The universe knows them all along and so this should be a huge warning sign to modern science that quantum mechanics will never be able to give us an answer to how the universe works because built into it is a system that can never give a definite answer. This mysterious act of observation that seems to lie at the center of all quantum uncertainty is absurd with people exclaiming the system materialized into reality upon us interacting with it and only then we find the true nature of the system. Well obviously you know the answer once the universe has given it to you. The inability to know before hand has become the focal failure of all quantum theories for the last century and no matter how much the ad hoc continues you will never be able to get a model of how the universe works from quantum mechanics.

The problem only intensifies when the unthinkable is done and the theories of special and general relativity are crossed with quantum mechanics in an attempt to solve the universe. The multiplying of two failed theories with multiple problems and impossibilities inherent in each will of course never work and always fail again no matter how much ad hoc is added to the combined equations to solve the universe. The simple message here is to stop it, just completely stop using both of these theories as literal explanations of the universe that can in some way be salvaged or patched to clean up the mess. You have a boat with a hole in the hull that lets in one liter of water per second and you have a bucket that can bail one liter of water per second out. You can try for eternity but you will never save your boat from sinking the best you can do is maintain your sinking status quo as it were and get nowhere fast. Modern science has had over a century to play with both theories and have got nothing for good reason which is there is nothing left to get form these theories. They served there purpose and will no doubt be partially retained in the immediate future and especially quantum mechanics will be useful but in the end neither will be able to last indefinitely and neither will ever produce the answer for how the universe works.

This also means that everyone can stop saying things like the laws of physics break down here or there because quite honestly they cannot. So watch out for the space police as they will not allow the laws of the universe to be broken whatsoever and this only reinforces the fact that the known laws of physics to modern science are not the real laws of physics of the universe. The universe never does anything it cannot and therefore the laws are never broken if you know the right ones which I will state that I do.

So where is the answer then? Is humanity at the end of its road for understanding the universe after tens of thousands of years? Well no of course not as all that is needed is a way to start putting our thoughts into those that move forward despite having to lose some of what many have cherished and held dear for so long. The planets move in ellipses not in circles and this had to be accepted centuries ago but that does not mean that all the data collected or all the calculations used to predict circular orbits were obsolete or needed to be retired. The data was kept and the calculations still had useful applications but in order to move forward people had to let go of long held beliefs as they were nothing more than beliefs and never fact no matter how fervently they thought of them. So we need to do this today as well we need to look at what we can keep and what we need to do with it so we can let go of that which does not work and accept what does. How we find out what we need to keep begins by looking at what we need to understand about the universe which is quite simply the whole thing at once. It is finally time to stop looking at things in only small detail or only large detail and look at the whole of everything all at once at the same time because this is the only way we can solve the problem.

So the biggest question possible in the universe is simply how does the universe work? To find that answer you have to look at the two components of the universe and at the two scales in which they exist. The first component is space and this is the least well understood part of the universe by modern science and the second is matter which is well understood but not completely. Space has been guessed at over the millennia but not with much success and matter has been smashed apart until we cannot smash it anymore with our current technology and still matter contains many gaps or holes in the complete knowledge base of modern science. The effects of space have gone undetected inside matter and yet all that has been done is try to understand everything from a matter only perspective which is rife with holes. So looking at both is a must and this brings us to the two scales they must be looked at.

The obvious scale is what we have always seen which is the large scale and this is anything from objects on Earth or anything we can see in deep space to the entire universe itself. The large scale still contains observations which seem to go against our current theories and what we know from smashing matter to little pieces. Large scale objects in the universe also seem to disobey the laws that govern Earth sized objects like you or I or anything we have experimented with classically for centuries. What this means is that these laws are not in fact laws as if they were they could not be broken and so we must look at the small scale to see how they must be altered if they can be kept at all.

The small scale is therefore the other extreme at which we must study matter and space. Here on the smallest of scales is where both space and matter stop so to speak as they cannot get any smaller and this is where space in particular can be understood best. The reason for this understanding of space on a small scale is because unlike matter space rarely forms what can be thought of as large structures. On the other hand space and matter always form large structures because it takes both space and matter to create anything in the universe as what is thought of as matter is actually the interaction of bauen matter with ESS to produce changes in the macht glow reserve pools of the universe. This key fact shows how the universe always joins space and matter together so that you get the universe itself exactly as we see it today not just out in space as is commonly thought when one thinks of observing the universe but also in the very atoms and bauen matter and ESS and macht of our own bodies. So looking at both space and matter at both small and large scales at the same moment is not only needed to understand the universe and how it works but is actually the only way to ever look at the universe. ESS, macht and bauen matter.

Something From Inside My Head and the Future

I cannot see anything anymore except different bauen matter geometries interacting with the ESS and macht glows all around and through me and everywhere in the universe anymore and I know this is how the universe works I promise. I hope that anyone reading this can reach that point as well as once you have you will know that this is the only way the universe can work and instantly all problems you might have encountered in modern science are solved without effort at all. To some my explanations will not be sufficient as they are blinded by and bound in chains to the mathematics which they need like air. You must remember however and this is a fact that all discoveries and understandings in science have always come from an idea first and the mathematics follow. In some cases the mathematics follow years or decades later so do not be afraid of this as if you are you will remain blind to the true understanding of how the universe works using ESS, macht and bauen matter.

I have chosen a model only approach to explain the universe for three utterly simple and powerful reasons. In no order the first is to allow anyone reading this to follow along as a deep understanding of mathematics in cosmology and theoretical physics is hampering. The second is to do away with the preconceived beliefs in flawed theories and the mathematics that have been derived from them directly or in any ad hoc theories. The third reason is the most powerful and it is to show anyone reading this whether they are a lay person or someone heavily ensconced in science and physics whatever branch or discipline you ascribe to, to see how the universe works logically, practically and rationally. This is why I have slowly built up my explanation of the universe from simple space and matter to ESS and specks and motes and finally to the fully formed series of interactions between ESS macht and bauen matter. I wanted to show anyone and especially those trying to solve the universe how you can take space and matter at its most absolute level and account for not only the unsolved problems in modern science but also keep those exact same rules and explain all normal or what can be thought of as mostly understood phenomena all around us.

At no point change what I have given you as the working mechanism for interaction between space and matter despite the evolution of terms over the course of my book. Also I do not add any new variables at any point to my work or any of the unsolved problems of the universe and this results in what some might feel is a repeated explanation for each different problem. Good. I am glad it does because that should be the biggest proof you could ever ask for as simply using ESS, macht and bauen matter and the space physics interactions I have given you everything can be explained. Each section outlines a problem and shows how in each problem you can use space physics to answer the problem and how these problems are ridiculously simple when viewed using ESS, macht and bauen matter.

◄New Universe Order►

You wanted a simple answer, you wanted a model that uses as few variables as possible and you wanted a work that does away with problems of the old and I have given this to you. I cannot ask anything from you other than you look at my work with the most open of minds and with a thirst for moving past this final challenge to human understanding of the universe. If you want to hate me or think I am insane then you do that but whenever you get stuck trying to solve one of the unsolved problems of the universe or encounter a seemingly impossible result in your experiments then all I ask is that you look at it through the filter of my work. For those of you who do accept my work and see that this is how our universe works you have my thanks as I cannot in my current life prove this to the world alone. I do not work in one of the world's elite institutions and there is no reason for anyone to listen to me. I do know that the universe works exactly as I have explained and I ask for your help in helping me to save science and push humans to the point in our history where we will all look back in the decades, centuries and millennia to come as the point when the Earth was no longer flat, the center of the universe and frozen in scientific Fimbulvetr or the Norse winter that is to precede the end of the world and in this case the winter that embodies the eternal frozen state of modern science.

This is the point where the human race becomes a galactic civilization even if we cannot freely travel amongst the stars yet because at least we are now among those species that truly understands the universe at any place or scale. If you need to put my work in a historical context then I am very much like one of the first descriptions of an atom with electrons orbiting the nucleus of protons and neutrons. I have given you the model of how the universe works and shown you all the rules for how space and matter interact but just as in the early models of the atom the mathematics and empirical evidence came later and for my work some of that is here and some will come later. Much of what I have explained has empirical evidence in the form of observed events in the universe whether they are in the laboratory here on Earth or out somewhere in the vastness of the universe itself as I have finally explained why these things are as they appear or why they happen at all.

Yet some evidence and mathematics are still to come concerning my work and these are the specifics of ESS, macht and bauen matter and for that we must build on the empirical evidence until the final universe equation is realized. If you look for it you will find it because I promise you it is there. Thank you to all who stuck with me this long and whether you love my work or hate it at least I hope you are thinking about it and how it solves the universe. The long talked about but scarcely cared about paradigm shift in thinking is here and it is about time, which I showed you time does not exist anyway but nevertheless it is time for us all to move into the future and I am quite simply asking if you are going to be one of the many that are unable to leave the past or one of those who walks with me unafraid into the future? Because I am already there and it is simply beautiful.

Glossary

Absolute Zero. The theoretical lowest possible temperature in the universe at which no more usable work can be done.

Antiparticle. All normal matter particles have an opposite called anti-matter that will annihilate upon contact with normal matter usually leaving only high energy photons as the result of the event.

Atom. A unit of matter consisting of a nucleus made of a single proton or multiple protons and neutrons orbited by one or more electrons and comprise the known naturally occurring 92 elements as well as synthetic elements.

Baryon. Particles that constitute what is considered normal matter such as protons and neutrons which are hadrons found in nuclei. Baryons are governed by the strong nuclear force and have a spin value of one half.

Bauen Force. The binding force between bauen matter and the strongest of all fundamental forces in the universe as well as the shortest ranged of all forces. It is responsible for holding bauen matter in specific bauen geometries that create what are thought of as fundamental particles and the zoo of particles.

Bauen Matter. The true form of matter in the universe which cannot be subdivided any further and is the construct matter from which everything else is built. It is the second piece of the universe along with space that is needed to make the entire universe. Bauen matter has all the necessary exposed sides to increase or decrease ESS macht glow reserve pools of space and will therefore form larger particles.

BBB. Big Bang Bubble or a single traditional Big Bang expanse of space that exists within the true larger universe. The universe can be constructed from a single BBB or multiple BBB existing throughout it of which our BBB is one only.

Big Bang. The theory of where the universe came from which involved a single large explosion from a singularity that brought both space and energy into the universe.

Black Hole. A completely gravitationally collapsed object created from normal matter that has theoretically undergone sufficient decreases in density to curve spacetime in a singularity. Nothing is thought to be able to escape a black hole including light.

Boson. Class of sub-atomic particles which carry certain fundamental forces and have a whole spin integer commonly known as gauge bosons. The Higgs boson is a scalar boson of spin zero and is theorized to give particles mass.

◀Glossary▶

Brane. An extra dimension of space that exists usually outside the normal 3+1 space and onto which strings can be attached. Closed strings will not attach and open ended strings will which leads to the shape of what is sometimes referred to as a D brane.

Bremsstrahlung. Braking radiation that results from one particle striking another particle and the first particle transferring kinetic energy to the second as well as being transformed into a photon. One of the most important of two factors that are responsible for shaping the universe from its creation until today and the future with the second being gravity.

Clintron. The bauen geometry which consists almost entirely of exposed gravity sides from the bauen matter it is made of. It is the last arrangement of true matter possible in the universe consisting of the equal parts of both space and matter to create it.

Conservation of Energy. A law in modern science that states that energy cannot be created or destroyed only changed from form to form.

Conservation of Macht/Energy. The law that states a system of both space and matter must always conserve the total donated or received energy from the matter with the increase or decrease in macht glow reserve pools of ESS. Responsible for the overall behavior of matter in terms of composition, interaction with other matter and limiting factors such as light speed movement.

Eigenstate. German for inherent it is a property of a quantum system or object that is known and therefore not subject to probability. Quantum objects that exist in mathematical probabilities of the wave function can have their eigenstate revealed upon observation.

Electromagnetic Force. The force that involves the interaction of particles that carry electric charge either positive or negative and is the second strongest of the four fundamental forces known to modern science.

Electron. A fundamental particle that carries a negative 1 electric charge and orbits the nucleus of an atom with a mass 1/1836 of a proton.

Electronvolt. A unit of energy commonly used to describe how much mass a particle has through the equivalency formula. Typically eV which stands for electron volt has a prefix to denote the magnitude of the energy such as MeV or mega electron volt equaling one million normal electron volts.

Electroweak Force. The unifying of two forces, the electromagnetic force and the weak force and occurs at energies around 100 GeV or higher.

Entropy. The measure of disorder as it is commonly thought of in a system where entropy increases over time. As time progresses the system becomes more disordered requiring more information to describe.

◄Glossary►

Escape Velocity. The necessary speed needed by an object to escape the gravitational forces of a larger object. If this speed is not met the object will remain attracted to the larger object.

ESS. Empty Space Structure or the smallest possible structure of space in the universe and the true form of space. It is the first piece of the universe along with matter that is needed to make the entire universe. ESS always possesses a macht glow reserve pool of the fundamental forces that is increased or decreased by interaction with bauen matter. ESS can also increase and decrease in volume as well as move freely around each other to allow space to behave as observed.

Event Horizon. The point around a black hole where it is believed by modern science that the spacetime curvature becomes so severe nothing can escape from the black hole.

Exclusion Principle. Also known as Pauli's Exclusion Principle it states that no two particles may occupy the exact same quantum states at the same time. For example two similar ½ spin particles cannot occupy the same coordinates in space being their position and momentum.

Fermion. Particles such as quarks and leptons which obey the exclusion principle and have a half spin integer. They can have three generations of matter such as electron, muon and tau for example.

Frequency. The number of cycles per second that a wave will repeat itself such as physical waves, sound waves or light waves.

Gamma Ray. The shortest wavelength and highest energy waves known for the electromagnetic spectrum which can be a result of natural production, radioactive decay processes or annihilation events.

Grand Unified Theory or GUT. Any theory that can unify three of the four known fundamental forces and include the weak force, electromagnetic force and the strong nuclear force.

Gravity. The weakest of the four fundamental forces and involves all particles with mass.

Information Loss. The idea that information necessary to describe a system can be lost from the universe such that knowing the current conditions of a system will not allow you to back track to a previous state.

Ion. An atom that has gained or lost electrons from the normal amount present. The result is the addition or loss of multiple electrons from orbitals around the atom such that the atom becomes negatively or positively charged.

Isotope. An atom containing the same number of protons as the normal arrangement but a different number of neutrons.

Light. Energy of the electromagnetic spectrum in the form of waves ranging from radio waves through visible light to gamma rays and carried in quanta.

Light Year. The distance light travels in one year or approximately 9.5 trillion kilometers. The Milky Way Galaxy is 150 000 light years in diameter. The visible universe is approximately 28 billion light years in diameter.

Locality. Idea that a physical link such as a particle or information must be maintained between two objects to mediate a cause and effect response for a given system. Non-locality is the property exhibited by quantum systems undergoing entanglement events where information is being transmitted between objects faster than light.

Macht. The force, energy or life blood of space that is shared freely between ESS and is increased or decreased by interactions with exposed bauen matter sides.

Macht Glow. The shared macht between schales of ESS that results from the increase or decrease of the ESS macht glow reserve pool of space through interactions of bauen matter and ESS. Macht glows are what is typically thought of as the fundamental particles as understood by modern science and the Standard Model.

Macht Speed. The speed at which ESS share macht between themselves through schales. The fastest possible speed attainable in the universe for any system regardless of all other considerations. The speed at which ESS increase or decrease in volume as well and is part of the expansion phase of the universe.

Magnetic Field. A field involving the interactions of particles through magnetic forces and has been incorporated with the electric field into the commonly known electromagnetic force.

Mass. The total amount of matter contained in any object that is not energy and can be described as inertia or its resistance to motion.

Microwave Background Radiation. Also known as the CMB or cosmic microwave background it is the theoretical left over energy from the era of recombination or first light from the early universe at about 380 000 years old. The original gamma rays have been greatly red-shifted until they are low energy microwaves as we see them today.

Milky Way. Our home galaxy in which the Sun and Earth are located approximately 25 000 light years from the galactic center. The Milky Way is 150 000 lights years in diameter and has a super massive black hole at its heart.

Nebel. The mist or fog of misunderstanding surrounding concepts of the universe as understood by modern science.

◄Glossary►

Neutrino. A near massless fundamental particle which travels at near light speed and has three flavor states of electron, muon and tau. It is only involved with the forces of gravity and the weak force as it has no charge.

Neutron. A neutral atomic particle found along with protons in the nucleus of atoms and has the approximate weight of a proton and electron combined although not exact. It will decay in roughly 11 minutes and is not a fundamental particle as it is a hadron made of three quarks.

Neutron Star. A star that has been made from neutrons and is the result of a core collapse supernova from a stellar mass approximately 10 times the mass of our sun. It is made by the electron capture process of a proton and an electron during the collapse phase of the supernova and resists becoming a black hole due to the neutron degeneracy pressure not being exceeded.

Nucleus. The center of an atom consisting either of a proton or a combination of protons and neutrons. It is the region of positive charge in the atom and around which the electron or electrons orbit.

Particle Accelerator. A machine used to move particles or atoms to high velocities in a near vacuum using electromagnets usually for experiments involving collisions of particles or ions inside detection chambers.

Photon. Light as it is commonly known and consists of a single quanta of light energy as part of the electromagnetic spectrum from radio waves to gamma rays.

Planck Length. The theoretically smallest size possible after which anything smaller would destroy predictions made by both quantum mechanics and general relativity. It is approximately 10 to the negative 33rd power centimeters.

Planck's Quantum Principle. The theory that waves especially light waves will only be absorbed or emitted in a set amount or quanta where the energy is proportional to the frequency.

Planck Time. The theoretically smallest unit of time possible after which anything smaller is nonsensical to general relativity and quantum mechanics. Specifically it is the length of time light needs to travel one Planck length in a perfect vacuum which is approximately 10 to the negative 43rd power seconds. Also the time when the universe was one Planck length in diameter according to theory.

Proton. A positive sub-atomic particle found alone or with neutrons in the nucleus of atoms and is not a fundamental particle as it is a hadron made of three quarks.

Quantum Entwinement. The proper description of quantum entanglement events such that space is included in the description and calculation of the system and allows for the explanation of the phenomenon such that no laws of physics are broken.

Quantum Mechanics. The theory of energy and matter developed from Planck's quantum principle and Heisenberg's uncertainty principle and is one of the three main views of modern science.

Quantum Speed. The slow light speed or lower velocities possible by matter and energy in the universe and the fastest speed attainable by modern science giving rise to numerous unsolved problems. This is the maximum speed at which bauen matter can donate energy to ESS and track after ESS macht of space and therefore when unbound from the force of gravity as in light will achieve the maximum tracking rate and speed for matter which is C.

Quark. An elementary particle that is used to make hadrons such as protons or neutrons. The particle itself carries the theoretical color charge and is governed by the strong nuclear force.

Schales. The shells of ESS surrounding those ESS that are in direct contact with bauen matter geometries and are carrying shared macht to create a larger macht glow volume of space. Macht glow strength decreases with distance from the point of interaction between bauen matter geometries and ESS.

Singularity. A theoretical point in space where the curvature of spacetime becomes so exaggerated it is infinite. A singularity is theorized to have zero volume, infinite gravity and a complete stoppage of time.

SM. Space matter or the near perfect interaction of bauen objects with ESS such that the two are needed equally to exist. SM needs both space and matter to exist and a positive feedback loop ensures this with the loss of one component immediately resulting in the dissociation of the two SM components from one another. Space matter is the true form of a black hole and is made entirely of clintrons.

Spatial Compression. The amount of decrease in volume of ESS for a local area of space and is governed by the interaction of matter with space as well as the type of bauen geometry inside the matter. The highest level of spatial compression possible is that of a TSM clintron whereby the volume inside the ESS decreases to zero but the total volume of the walls of the ESS are non-zero which prevents impossibilities and infinities from occurring in the universe.

Spatial rebound. The amount of increase in volume of ESS for a local area of space and is governed by the interaction of matter with space as well as the type of bauen geometry inside the matter. The highest level of spatial rebound possible is that of an ESS fully dissociated from all bauen matter and not receiving any shared macht from neighboring ESS through schales. Typically only possible during TSM detonation events it is also the type of space that can undergo faster than light speeds and is responsible for the expansion phase of the universe.

Spin. One of the characteristics of all quantum particles inherent to them and is the angular momentum of that particle. The values range from 0, ½ to 1.

String. A theoretical particle belonging to string and M theories that is used to account for what creates matter as well as explain the weak force of gravity by giving rise to a massless plus 2 spin graviton. Strings exist with multiple dimensions compactified inside of them and vibrate in numerous ways to theoretically create particles. Strings are one dimensional objects that can be open or closed as in a loop.

Strong Force. Governs interactions involving particles that feel the strong force such as quarks and gluons in the nucleus of atoms. It is responsible for holding hadrons together and is the strongest force of the four known fundamental forces and also has the shortest range.

Symmetry. A system, force or particle that possesses a quality or property that will remain unchanged after a specific transformation. Symmetry breaking refers to the loss of this property as is seen in the breaking of the electroweak force into the electromagnetic force and the weak force in the early universe.

Theory of Everything. Also known as a unified theory is a theory that links all forces in the universe and removes all discrepancies of experimentation and observation between quantum mechanics and general relativity.

TSM. True space matter or the perfect interaction of bauen matter with ESS such that the matter is made to exist 100% within space and the walls of the ESS spheres contain all donated energy from the matter and collapse onto themselves. True space matter are made of a single TSM clintron which is the absolute last form of matter and space in the universe and exists for exactly one macht time. TSM clintrons are also the single most ordered objects possible in existence in the universe at any part of its life cycle and contain for a single macht time all fundamental forces, space and matter bound into one.

Uncertainty Principle. The creation of Heisenberg that states you cannot know the exact position and momentum of a particle at the same time. The more you know of one of the two properties the more information you lose of the second.

Universe. The entire collection of all possible space and matter in existence which encompasses all. There is nothing outside the universe as anything there would simply become a possession of the universe.

Virtual Particle. A particle that is not directly observed in quantum mechanics but due to its presence can produce measurable effects on the system such as a gluon being a particle with virtual mass inside a hadron where the quarks have physical mass.

Void. The true nothingness that exists in the universe in the spaces inside ESS and between their spherical walls as they touch one another. The void is the nothingness which contains absolutely no space, matter or forces of any kind whatsoever and is similar to the area outside the universe. It is necessary to allow ESS to spatially compress and spatially rebound as well as in helping to solve the vacuum catastrophe.

Wave Particle Duality. The idea that matter can be both a wave and a particle at the same time or more accurately exhibit both wave and particle properties simultaneously.

Wavelength. The distance between two similar points in any wave such as two troughs or two peaks. For light as the wavelength decreases it is blue-shifted and as the wavelength increases it is red-shifted.

Weak Force. The force responsible for decay events in particles and especially in nuclei. It is the second weakest of the four known fundamental forces of modern science and involves boson force carriers.

Wave Function. Mathematical linear equations that assume all known possibilities for a system in quantum mechanical calculations and which must always sum to 1. Collapsing the wave function reveals the one true possibility of the quantum object or system with a 1:1 probability and is said to be influenced by observer interaction.

White Dwarf Star. A star that is the remnant from a star that has left its main sequence phase and is supported from further collapse by the electron degeneracy pressure. It is a part of the Type 1a Supernova and used as a standard candle in cosmology.

Index

◀Index▶

◀Index▶

◄Index►